Springer Series in Optical Sciences

Founding Editor

H. K. V. Lotsch

Volume 236

Springer Series in Optical Sciences is led by Editor-in-Chief William T. Rhodes, Florida Atlantic University, USA, and provides an expanding selection of research monographs in all major areas of optics:

- lasers and quantum optics
- ultrafast phenomena
- optical spectroscopy techniques
- optoelectronics
- information optics
- applied laser technology
- industrial applications and
- other topics of contemporary interest.

With this broad coverage of topics the series is useful to research scientists and engineers who need up-to-date reference books.

More information about this series at https://link.springer.com/bookseries/624

Masataka Nakazawa · Masatoshi Suzuki ·
Yoshinari Awaji · Toshio Morioka
Editors

Space-Division Multiplexing in Optical Communication Systems

Extremely Advanced Optical Transmission with 3M Technologies

Springer

Editors
Masataka Nakazawa
Research Institute of Electrical
Communication (RIEC)
Tohoku University
Sendai, Miyagi, Japan

Yoshinari Awaji
Photonic Network Laboratory, Photonic
ICT Research Center, Network Research
Institute
National Institute of Information
and Communications Technology
Koganei, Tokyo, Japan

Masatoshi Suzuki
KDDI Research, Inc.
Fujimino, Saitama, Japan

Graduate School of Waseda University
Shinjuku, Tokyo, Japan

Toshio Morioka
Technical University of Denmark
Kgs. Lyngby, Denmark

ISSN 0342-4111 ISSN 1556-1534 (electronic)
Springer Series in Optical Sciences
ISBN 978-3-030-87617-3 ISBN 978-3-030-87619-7 (eBook)
https://doi.org/10.1007/978-3-030-87619-7

This Springer imprint is published by the registered company Springer Nature Switzerland AG
The registered company address is: Gewerbestrasse 11, 6330 Cham, Switzerland

Preface

On January 25, 2008, a new collaborative study group called "EXAT Initiative" focusing on space-division multiplexing (SDM) technologies in optical communications, the first of its kind in the world, started in Japan after a half year's preparation. The group was named after an acronym of "**EX**tremely Advanced **T**ransmission" also implying exa=10^{18} bit/s and was organized by the National Institute of Information and Communications Technology (NICT), Japan. The initiation was driven by grave concerns over capacity limits of present optical communication systems based on single-mode, single-core optical fibers in coping with relentlessly ever-increasing data traffic. Top researchers from industries (network operators, equipment vendors, optical fiber manufacturers), academia, and national institutes convened and discussed intensively breakthrough, scalable and sustainable technologies, which would enable a giant leap in optical communications. What was special with this initiative was that it also included researchers who were experts or rapporteurs in international standardization activities in the areas of optical fiber cables at ITU-T, as well as in the areas of optical connectors, optical amplifiers and laser safety issues at IEC. This was because it was thought that the new technologies need to align with and evolve from already standardized and being standardized technologies in order for the new technologies to be deployed in real systems in the foreseeable future. In the initiative, it was focused to identify ultimate physical limitations of the present optical communication systems, i.e., the amount of optical power (capacity) that can be transmitted safely in optical fibers, the bandwidth for optical amplification, and the capacity of optical submarine cables systems which is limited by the electrical power consumed by optical amplifier repeaters. More specifically, the use of the last degree of freedom, "space" for multiplexing and the need to develop new optical fibers (multi-core fibers (MCFs) and few-mode fibers (FMFs)) and new multiplexing schemes, namely SDM and mode-division multiplexing (MDM), were proposed, and three major fundamental technologies, namely "Multi-core Fiber," "Multi-mode Control" and "Multi-level Modulation," named "3M technologies" were identified. In 2010, the initiative developed into a technical committee on Extremely Advanced Optical Transmission Technologies (EXAT), the Institute of Electronics, Information and Communication Engineers (IEICE), Japan, and has been a central workforce to accelerate the research and development of ultra-high capacity optical communication and optical transport network based on the 3M technologies. The initiative has led to the creation of a series of Japanese national projects since 2010 to further develop the ideas proposed by EXAT, such as Innovative Optical Fiber Technologies: 2010–2012 (i-FREE), Innovative Optical Communication Infrastructure: 2011–2015 (i-ACTION) and Innovative Optical Fiber and Communication Technology for Exa-bit Era with SDM: 2013–2018 (i-FREE^2). In 2016, another new project "R&D of Space-Division Multiplexing Photonic Node" (2016–2020) started. Furthermore, an EU-Japan coordinated R&D project on "Scalable And Flexible optical Architecture for Reconfigurable Infrastructure (SAFARI)" (2013–2017) was created linking the Japanese EXAT community and the European relevant partners, commissioned by the Ministry of Internal Affairs and Communications (MIC) of Japan and EC Horizon 2020. These projects have been leading the most advanced research field of SDM technologies in the world, especially MCFs, related SDM devices and multiplexing technologies, achieving major world records in transmission capacity such as 1-Pbit/s transmission (2012), 1-Ebit/s·km transmission (2013), 2-Pbit/s (2015) and 10-Pbit/s transmission (2017), fully utilizing the SDM technologies developed from the original ideas discussed since 2008.

This book reviews such innovative research efforts conducted by EXAT over the last decades. There are nine chapters with concluding remarks, including chapters on optical connection technologies (Chap. 3) and on high-power issues (Chap. 7) which are important aspects in deploying SDM technologies in the future systems. In Chap. 1 "Introduction," physical limits and prospects of the present optical communication systems, EXAT Initiative and 3M technologies are briefly described, after which future demands towars optical communication systems are given from two major applications aspects, namely, ultra-realistic communication and 5G technologies. State-of-the-art terrestrial optical transmission systems and submarine optical transmission systems are also described. In Chap. 2 "Optical Fibers for Space-Division Multiplexing," design principles, transmission characteristics and measurements technologies of MCFs, FMFs and few-mode (FM)-MCFs as well as cabling technologies are described with future perspectives. In Chap. 3 "Optical Connection Technologies," recent progress on splicing technologies, optical connectors, fan-in/fan-out devices for MCFs and mode multi/demultiplexing technologies for FMFs are described. In Chap. 4 "Optical Amplification Technologies," a comprehensive review of optical amplification technologies for SDM transmission is given where history, targets, categories, evaluation methods and ongoing issues of SDM amplifiers are given. The chapter not only describes erbium-doped fiber amplifiers (EDFAs) and Raman-based SDM amplifiers, but also covers amplification techniques based on bismuth and rare earths that are expected to expand the transmission window from the conventional WDM transmission bands. In Chap. 5 "Optical Transmission Technologies," overview of transmission technologies, advanced modulation formants for higher spectral efficiencies, transmission technologies for MCFs, FMFs, FM-MCFs and signal processing technologies are given providing a variety of functions in high-speed optical transmission systems. In Chap. 6 "Network Technologies for SDM," basic technologies for network nodes and recent research efforts across the world which have extended the SDM concept to optical switching and utilized SDM-specific features to facilitate advanced network control are given. The second part of the chapter focuses on short-reach systems and discusses an application example of SDM for data-center signal protection. In Chap. 7 "High-Power Issues," firstly, the ultimate limiting phenomenon, fiber fuse is described in detail ranging from its basic properties, to fiber fuse of optical communication fibers, its detection methods, its halting methods, fiber fuse-based incidence as well as fiber fuse-tolerant fibers. Secondly, safety of optical communication systems is described, which is essential in designing and operating future high-power SDM systems. In Chap. 8 "Japanese National Projects on SDM Technologies," major national projects arising from the EXAT initiative mentioned above are described in detail. Lastly, In Chap. 9 "Concluding Remarks", challenges of SDM deployment and its major application fields for the future are identified.

We believe that this book offers researchers working at the forefront, as well as advanced Ph.D. students in the area of optical fiber communications systems and related fields, an essential guide to state-of-the-art optical transmission technologies. Finally, we would like to thank all the chapter editors for their great dedications in organizing and managing each chapter so nicely, and all the contributors for their excellent manuscripts. We also thank Drs. Fumito Kubota, Masaki Hirabaru, Naoya Wada, Kenichiro Miwa for their great support in establishing the EXAT initiative at NICT, Japan, the Ministry of Internal Affairs and Communications (MIC), Japan and the National Institute of Information and Communications Technology (NICT), Japan for their constant support, and all the committee members of IEICE Technical committee on EXAT for their continuous efforts and enthusiasm in advancing this most exciting field.

Sendai, Miyagi, Japan — Masataka Nakazawa
Fujimino, Saitama, Japan — Masatoshi Suzuki
Koganei, Tokyo, Japan — Yoshinari Awaji
Kgs. Lyngby, Denmark — Toshio Morioka

Acknowledgement

The following contributors have prepared the following sections.

Chapter Editors

Chapter 1: **Toshio Morioka** Technical University of Denmark, Kgs. Lyngby, Denmark

Chapter 2: **Masaharu Ohashi** Osaka Metropolitan University, Sakai, Osaka, Japan
Shoichiro Matsuo Fujikura Ltd., Sakura, Chiba, Japan

Chapter 3: **Ryo Nagase** Chiba Institute of Technology, Narashino, Chiba, Japan
Kiichi Hamamoto Kyushu University, Kasuga, Fukuoka, Japan

Chapter 4: **Makoto Yamada** Osaka Metropolitan University, Sakai, Osaka, Japan

Chapter 5: **Itsuro Morita** KDDI Research Inc., Fujimino, Saitama, Japan (presently with Waseda University, Shinjuku, Tokyo, Japan)
Toshihiko Hirooka Tohoku University, Sendai, Miyagi, Japan
Hidehiko Takara National Institute of Technology, Okinawa College, Nago, Okinawa, Japan

Chapter 6: **Werner Klaus** National Institute of Information and Communications Technology, Koganei, Tokyo, Japan

Chapter 7: **Toshio Morioka** Technical University of Denmark, Kgs. Lyngby, Denmark

Chapter 8: **Yoshinari Awaji** National Institute of Information and Communications Technology, Koganei, Tokyo, Japan

Chapter 9: **Takeshi Kamiya** The University of Tokyo, Bunkyo, Tokyo, Japan
Yuichi Matsushima Waseda University, Shinjuku, Tokyo, Japan

Contributors

Yoshiteru Abe NTT Access Network Service Systems Laboratories, NTT Corporation, Tsukuba, Ibaraki, Japan

3.4 Fan-In/Fan-Out Device for Multi-core Fiber

Kazi S. Abedin LGS Innovation—CACI, Reston, VA, USA

7.1.3.1 Fuse Detection Through Monitoring the Electrical Spectrum of Back-Reflected Light
7.1.3.2 Real-Time Monitoring of Fiber Fuse Using an Optical Time Domain Reflectometer (OTDR)
Yoshinari Awaji National Institute of Information and Communications Technology, Koganei, Tokyo, Japan
Preface
1.1 Physical Limits and Prospects of Optical Communication Systems
5.5.2 SDM Processing Technologies
5.5.3 Orbital Angular Momentum
Chapter 8 Abstract
Kazumasa Enami Tokyo Institute of Technology, Meguro, Tokyo, Japan
1.3.1 Ultra-Realistic Communication
Kiyoshi Fukuchi NEC Corporation, Minato, Tokyo, Japan
5.2 Overview of Transmission Technologies
Kiichi Hamamoto Kyushu University, Kasuga, Fukuoka, Japan
3.5 Mode multi/demultiplexing technologies for few-mode fibers
Nobutomo Hanzawa NTT Access Network Service Systems Laboratories, NTT Corporation, Tsukuba, Ibaraki, Japan
7.1.6 Fiber Fuse Tolerant Fibers
Tetsuya Hayashi Sumitomo Electric Industries, Ltd, Yokohama, Kanagawa Japan
2.3.1 Theory of Crosstalk
2.3.3.1 Preferable Characteristics for Spatial Capacity Enhancement
Toshihiko Hirooka Tohoku University, Sendai, Miyagi, Japan
Chapter 5 Abstract
5.1 Introduction
Koji Igarashi Osaka University, Suita, Osaka, Japan
5.4.1.2 MCF Transmission Technologies for Submarine Systems
5.4.3.2 Multi-core Multi-mode Transmission Experiments
Katsunori Imamura Furukawa Electric Co., Ltd, Chiyoda, Tokyo, Japan
2.3.5 Fabrication technology of MCF
Takeshi Kamiya The University of Tokyo, Bunkyo, Tokyo, Japan
Chapter 9 Concluding Remarks
Nobuhiko Kikuchi Research & Development Group, Hitachi Ltd., Kokubunji, Tokyo, Japan
6.2 Application of SDM Technologies in Short-reach Systems
Werner Klaus National Institute of Information and Communications Technology, Koganei, Tokyo, Japan
Chapter 6 Abstract
6.1 Basic technologies for network nodes and network control
Soichi Kobayashi Photonic Science Technology, Inc., Chitose, Hokkaido, Japan
4.5 Preface
4.5.1 1.3-μm Band Bismuth-Doped Amplifier
Masafumi Koga Oita University, Oita, Oita, Japan
5.4.1.3 Optical Diversity Transmission Using MCF

Yasuo Kokubun Institute of Technologists (Formerly with Yokohama National University), Gyoda, Saitama, Japan
2.3.3.2 Ultimate Number of Transmission Channels in Few-Mode Multi-core Fiber with Air-Hole-Assisted Double-Cladding Structure
3.5 Mode multi/demultiplexing technologies for few-mode fibers
Masanori Koshiba Hokkaido University, Sapporo, Hokkaido, Japan
2.3.1 Theory of Crosstalk
Kenji Kurokawa Kitami Institute of Technology, Kitami, Hokkaido, Japan
7.1.4.1 Introduction
7.1.4.3 Fiber Fuse Terminator with HAF
7.1.4.4 Passive Fiber Fuse Terminators and Automatic Power Reduction (APR)
7.1.4.5 Conclusion
Yong Lee Research & Development Group, Hitachi Ltd., Kokubunji, Tokyo, Japan
6.2 Application of SDM Technologies in Short-reach Systems
Hiroji Masuda Shimane University, Matsue, Shimane, Japan
4.3.3.5 Remotely Pumped Multi-core EDFA
4.3.4 State-of-the-Art Multi-core Raman Amplification Technology
4.4.3 State-of-the-Art Few-Mode Raman Amplification Technology
Shoichiro Matsuo Fujikura Ltd., Sakura, Chiba, Japan
2.3.2 Low Crosstalk MCF
2.4.3 Mode coupling
2.5.1 FM-MCFs with Uncoupled Core Design
2.6.2 High-Density MCF Cables
3.2 Fusion splicing technology
4.4.1 Target of Few-Mode EDFA
4.4.2 State-of-the-Art Few-Mode EDFA
Yuichi Matsushima Waseda University, Shinjuku, Tokyo, Japan
Chapter 9 Concluding Remarks
Yutaka Miyamoto NTT Network Innovation Labs., NTT Corporation, Yokosuka, Kanagawa, Japan
1.4 State-of-the-Art Terrestrial Optical Transmission
5.4.1.1 MCF Transmission Technologies for Terrestrial System
5.4.2 Multi-mode Transmission
5.4.3.1 Research on Multi-core Multi-mode Transmission
8.3 i-Action
Tetsuya Miyazaki National Institute of Information and Communications Technology, Koganei, Tokyo, Japan
Chapter 8 Abstract
Takayuki Mizuno NTT Electronics America, San Jose, CA, USA
5.4.2 Multi-mode Transmission
5.4.3 Multi-core Multi-mode Transmission
5.4.3.1 Research on Multi-core Multi-mode Transmission
5.4.3.2 Multi-core Multi-mode Transmission Experiments

Takayoshi Mori NTT Access Network Service Systems Labs, NTT Corporation, Tsukuba, Ibaraki, Japan
2.4.1 Design Trend of FMF
2.4.2 DMD and MIMO
Toshio Morioka Technical University of Denmark, Kongens Lyngby, Denmark
Preface
Chapter 1 Abstract
1.1 Physical Limits and Prospects of Optical Communication Systems
1.2 EXAT Initiative and 3M technologies
Chapter 7 Abstract
Chapter 8 Abstract
8.1 Introduction
Itsuro Morita KDDI Research Inc., Fujimino, Saitama, Japan (presently with Waseda University, Shinjuku, Tokyo, Japan)
1.5 State-of-the-Art Optical Submarine Cable Systems
Chapter 5 Abstract
5.1 Introduction
8.3 i-Action
Kazunori Mukasa Furukawa Electric Co., Ltd., Kameyama, Mie, Japan
7.1.2 Fiber Fuse of Optical Communication Fibers
Ryo Nagase Chiba Institute of Technology, Narashino, Chiba, Japan
Chapter 3 Abstract
3.1 Introduction
3.3 Optical connectors
7.1.2 Fiber Fuse of Optical Communication Fibers
7.1.4.2 Fiber Fuse Terminator with a TEC Fiber
Kazuhide Nakajima NTT Access Network Service Systems Laboratories, NTT Corporation, Tsukuba, Ibaraki, Japan
2.3.4 Reliability of MCF
2.6.1 Prospect for MCF Cables and Their Application Areas
2.7 Future Perspective
8.2 i-FREE
8.4 i-FREE^2
Moriya Nakamura Meiji University, Kawasaki, Kanagawa, Japan
4.2 History of SDM Amplification Technology
Masataka Nakazawa Tohoku University, Sendai, Miyagi, Japan
Preface
2.4.4 Measurement Technology for Mode Coupling
5.3 Multi-level Transmission Technologies
Shu Namiki The National Institute of Advanced Industrial Science and Technology (AIST), Tsukuba, Ibaraki, Japan
4.1 Introduction
Masaharu Ohashi Osaka Metropolitan University, Sakai, Osaka, Japan
Chapter 2 Abstract
2.1 Introduction

2.3.6 Measurement Technology
Yukihiko Okumura R&D Strategy Department, NTT DOCOMO, INC., Yokosuka, Kanagawa, Japan
1.3.2 Optical Network Technologies for Wireless Communication Network
Hirotaka Ono Shonan Institute of Technology, Fujisawa, Kanagawa, Japan
4.3.2 Categorization of Multi-core EDFA
4.4.2 State-of-the-Art Few-Mode EDFA
4.4.3 State-of-the-Art Few-Mode Raman Amplification Technology
4.4.4 Evaluation Methods
4.4.5 Issues and Future Work
Benjamin J. Puttnam National Institute of Information and Communications Technology, Koganei, Tokyo, Japan
5.4.1.1 MCF Transmission Technologies for Terrestrial System
5.5.2 SDM Processing Technologies
6.1 Basic technologies for network nodes and network control
Kunimasa Saitoh Hokkaido University, Sapporo, Hokkaido, Japan
2.3.2 Low Crosstalk MCF
2.5.1 FM-MCFs with Uncoupled Core Design
Jun Sakaguchi National Institute of Information and Communications Technology, Koganei, Tokyo, Japan
5.4.3.2 Multi-core Multi-mode Transmission Experiments
6.1 Basic technologies for network nodes and network control
Taiji Sakamoto NTT Access Network Service Systems Labs, NTT Corporation, Tsukuba, Ibaraki, Japan
2.4.1 Design Trend of FMF
2.4.2 DMD and MIMO
2.5.2 125-μm Cladding 2 LP-mode 6-core Fiber
Akihide Sano Ritsumeikan University, Kusatsu, Shiga, Japan
5.4.1.1 MCF Transmission Technologies for Terrestrial System
Takashi Sasaki Innovation Core SEI, Inc, San Jose, CA, USA
2.2 Recent Progress of Single Core Fiber Characteristics, Loss and A_{eff}
Ryuichi Sugizaki Furukawa Electric Co., Ltd., Ichihara, Chiba, Japan
4.3.3.2 Core-Pumped Multi-core EDFA
4.3.3.3 Multi-element EDFA
4.3.3.4 Cladding-Pumped Multi-core EDFA
4.3.5 Evaluation Methods
4.3.6 Issues and Future Work
Masatoshi Suzuki KDDI Research Inc., Fujimino, Saitama, Japan (presently with Graduate School of Waseda University, Shinjuku, Tokyo, Japan)
Preface
1.5 State-of-the-Art Optical Submarine Cable Systems
Hidehiko Takara National Institute of Technology, Okinawa College, Nago, Okinawa, Japan
1.4 State-of-the-Art Terrestrial Optical Transmission
Chapter 5 Abstract

5.1 Introduction
5.4.1.1 MCF Transmission Technologies for Terrestrial System
5.4.2 Multi-mode Transmission
5.4.3.1 Research on Multi-core Multi-mode Transmission
7.1.2 Fiber Fuse of Optical Communication Fibers
7.2 Safety of Optical Communication Systems (From the Viewpoint of IEC Laser Safety Standardization)
8.3 i-Action

Kenichi Tanaka Research & Development Group, Hitachi Ltd., Kokubunji, Tokyo, Japan
6.2 Application of SDM Technologies in Short-reach Systems

Jun Terada NTT Device Technology Labs., NTT Corporation, Atsugi, Kanagawa, Japan
1.3.2 Optical Network Technologies for Wireless Communication Network

Shin-ichi Todoroki National Institute for Materials Science, Tsukuba, Ibaraki, Japan
7.1.1 Basic Properties

Yukihiro Tsuchida Furukawa Electric Co., Ltd., Ichihara, Chiba, Japan
4.3.3.2 Core-Pumped Multi-core EDFA
4.3.3.3 Multi-element EDFA
4.3.3.4 Cladding-Pumped Multi-core EDFA
4.3.5 Evaluation Methods
4.3.6 Issues and Future Work

Takehiro Tsuritani KDDI Research Inc., Fujimino, Saitama, Japan
5.4.1.2 MCF Transmission Technologies for Submarine Systems
8.4 i-FREE^2

Makoto Yamada Osaka Metropolitan University, Sakai, Osaka, Japan
Chapter 4 Abstract
4.3.1 Target of Multi-core EDFA
4.3.2 Categorization of Multi-core EDFA
4.3.3.1 Bundled Type EDFA
4.5 Preface
4.5.2 Over 1.65-μm Band Amplifier
4.6 Further Progress
7.1.3.3 Detection by Monitoring the Peculiar Luminescence Observed During the Fiber Fuse Propagation
7.1.5 Fiber Fuse-Based Incidence
7.2 Safety of Optical Communication Systems (From the Viewpoint of IEC Laser Safety Standardization)

Kenji Yamamoto Tokushima University, Tokushima, Tokushima, Japan
1.3.1 Ultra-Realistic Communication

Shinji Yamashita The University of Tokyo, Meguro, Tokyo, Japan
5.5.1 State-of-the-Art Optical Signal Processing Technologies

Shuichi Yanagi NTT Advanced Technology Corporation, Musashino, Tokyo, Japan
7.1.4.2 Fiber Fuse Terminator with a TEC Fiber

Contents

Contributors

Yoshiteru Abe NTT Access Network Service Systems Laboratories, NTT Corporation, Tsukuba, Ibaraki, Japan

Kazi S. Abedin LGS Innovation—CACI, Reston, VA, USA

Yoshinari Awaji National Institute of Information and Communications Technology, Koganei, Tokyo, Japan

Kazumasa Enami Tokyo Institute of Technology, Meguro, Tokyo, Japan

Kiyoshi Fukuchi NEC Corporation, Minato, Tokyo, Japan

Kiichi Hamamoto Kyushu University, Kasuga, Fukuoka, Japan

Nobutomo Hanzawa NTT Access Network Service Systems Laboratories, NTT Corporation, Tsukuba, Ibaraki, Japan

Tetsuya Hayashi Sumitomo Electric Industries, Ltd, Yokohama, Kanagawa, Japan

Toshihiko Hirooka Research Institute of Electrical Communication (RIEC), Tohoku University, Sendai, Miyagi, Japan

Koji Igarashi Osaka University, Suita, Osaka, Japan

Katsunori Imamura Furukawa Electric Co., Ltd, Chiyoda, Tokyo, Japan

Takeshi Kamiya The University of Tokyo, Bunkyo, Tokyo, Japan

Nobuhiko Kikuchi Research & Development Group, Hitachi Ltd., Kokubunji, Tokyo, Japan

Werner Klaus National Institute of Information and Communications Technology, Koganei, Tokyo, Japan

Soichi Kobayashi Photonic Science Technology, Inc., Chitose, Hokkaido, Japan

Masafumi Koga Oita University, Oita, Oita, Japan

Yasuo Kokubun Institute of Technologists (Formerly with Yokohama National University), Gyoda, Saitama, Japan

Masanori Koshiba Hokkaido University, Sapporo, Hokkaido, Japan

Kenji Kurokawa Kitami Institute of Technology, Kitami, Hokkaido, Japan

Yong Lee Research & Development Group, Hitachi Ltd., Kokubunji, Tokyo, Japan

Hiroji Masuda Shimane University, Matsue, Shimane, Japan

Shoichiro Matsuo Fujikura Ltd., Sakura, Chiba, Japan

Yuichi Matsushima Waseda University, Shinjuku, Tokyo, Japan

Yutaka Miyamoto NTT Network Innovation Laboratories, NTT Corporation, Yokosuka, Kanagawa, Japan

Tetsuya Miyazaki National Institute of Information and Communications Technology, Koganei, Tokyo, Japan

Takayuki Mizuno NTT Electronics America, San Jose, CA, USA

Takayoshi Mori NTT Access Network Service Systems Labs, NTT Corporation, Tsukuba, Ibaraki, Japan

Toshio Morioka Technical University of Denmark, Kongens Lyngby, Denmark

Itsuro Morita KDDI Research Inc., Fujimino, Saitama, Japan;
Waseda University, Shinjuku, Tokyo, Japan

Kazunori Mukasa Furukawa Electric Co., Ltd., Kameyama, Mie, Japan

Ryo Nagase Chiba Institute of Technology, Narashino, Chiba, Japan

Kazuhide Nakajima NTT Access Network Service Systems Laboratories, NTT Corporation, Tsukuba, Ibaraki, Japan

Moriya Nakamura Meiji University, Kawasaki, Kanagawa, Japan

Masataka Nakazawa Research Institute of Electrical Communication (RIEC), Tohoku University, Sendai, Miyagi, Japan

Shu Namiki The National Institute of Advanced Industrial Science and Technology (AIST), Tsukuba, Ibaraki, Japan

Masaharu Ohashi Osaka Metropolitan University, Sakai, Osaka, Japan

Yukihiko Okumura R&D Strategy Department, NTT DOCOMO, INC., Yokosuka, Kanagawa, Japan

Hirotaka Ono Shonan Institute of Technology, Fujisawa, Kanagawa, Japan

Benjamin J. Puttnam National Institute of Information and Communications Technology, Koganei, Tokyo, Japan

Kunimasa Saitoh Hokkaido University, Sapporo, Hokkaido, Japan

Jun Sakaguchi National Institute of Information and Communications Technology, Koganei, Tokyo, Japan

Taiji Sakamoto NTT Access Network Service Systems Labs, NTT Corporation, Tsukuba, Ibaraki, Japan

Akihide Sano Ritsumeikan University, Kusatsu, Shiga, Japan

Takashi Sasaki Innovation Core SEI, Inc, San Jose, CA, USA

Ryuichi Sugizaki Furukawa Electric Co., Ltd., Ichihara, Chiba, Japan

Masatoshi Suzuki KDDI Research Inc., Fujimino, Saitama, Japan;
Graduate School of Waseda University, Shinjuku, Tokyo, Japan

Hidehiko Takara National Institute of Technology, Okinawa College, Nago, Okinawa, Japan

Kenichi Tanaka Research & Development Group, Hitachi Ltd., Kokubunji, Tokyo, Japan

Jun Terada NTT Device Technology Labs., NTT Corporation, Atsugi, Kanagawa, Japan

Shin-ichi Todoroki National Institute for Materials Science, Tsukuba, Ibaraki, Japan

Yukihiro Tsuchida Furukawa Electric Co., Ltd., Ichihara, Chiba, Japan

Takehiro Tsuritani KDDI Research Inc., Fujimino, Saitama, Japan

Makoto Yamada Osaka Metropolitan University, Sakai, Osaka, Japan

Kenji Yamamoto Tokushima University, Tokushima, Tokushima, Japan

Shinji Yamashita The University of Tokyo, Meguro, Tokyo, Japan

Shuichi Yanagi NTT Advanced Technology Corporation, Musashino, Tokyo, Japan

Abbreviations

3D	Three-dimensional
3M	Multi-core fiber, multi-mode control and multi-level modulation
5GMF	5G Mobile Communications Promotion Forum
5G-PPP	5G Infrastructure Public Private Partnership
ACF	Autocorrelation function
ADC	Analog-to-digital converter
ADSL	Asymmetric digital subscriber line
AEL	Accessible emission limit
AGC	Automatic gain control
ALC	Automatic level control
AoD	Architecture-on-demand
APE	Asymptotic power efficiency
APR	Automatic power reduction
ARF	Anti-resonant fiber
ASE	Amplified spontaneous emission
ATT	Attenuator
AVC	Advanced video coding
AWG	Arrayed waveguide grating/arbitrary waveform generator
BBU	Baseband unit
BCH	Bose–Chaudhuri–Hocquenghem
BDF	Bismuth-doped silica fiber
BER	Bit-error rate
BIF	Bending loss insensitive fiber
BP	Backpropagation
BPSK	Binary phase shift keying
BRF	Band rejection filter
C/U	Control and user
CAPEX	Capital expenditure
CC	Core coding
CCD	Charge-coupled device
CD	Chromatic dispersion/cladding diameter

CDC	Colorless, directionless, contentionless
CDP	Capacity-distance product
CMA	Constant modulus algorithm
CMOS	Complementary metal-oxide semiconductor
CMP	Cyclic mode permutation
CNT	Carbon nanotube
CPC	Core-polarization coding
CPR	Carrier phase recovery
CPRI	Common public radio interface
CPT	Coupled-power theory
C-RAN	Centralized radio access network
CRZ-OOK	Chirped return-to-zero on–off keying
CSF	Cut-off shifted fiber
CT	Cladding thickness
CW	Continuous wave
DA	Double armored
DBA	Dynamic bandwidth allocation
DBM	Double balanced mixer
DBR	Distributed Bragg reflector
DC	Data center/double-clad
DCF	Dispersion-compensating fiber
DDC	D-shaped double-clad
DD-LMS	Decision-directed least-mean square
DFB	Distributed feedback
DGD	Differential group delay
DMA	Differential mode attenuation
DMD	Differential mode delay
DMF	Dispersion-managed fiber
DP	Dual polarization
DPOH	Dual polarization optical hybrid
DPSK	Differential phase-shift-keying
DQPSK	Differential quadrature phase shift keying
DRA	Distributed Raman amplifier
D-RAN	Distributed radio access network
DRoF	Digital radio over fiber
DRS	Dual-ring structure
DSDM	Dense space-division multiplexing
DSF	Dispersion-shifted fiber
DSP	Digital signal processor/processing
EAF	Exponential autocorrelation function
ECTL	External cavity tunable laser
EDF	Erbium-doped fiber
EDFA	Erbium-doped fiber amplifier
EDX	Energy dispersive X-ray spectroscopy
EHF	Extremely high frequency

EPON	Ethernet passive optical network
EVM	Error vector magnitude
EVS	End-view system
EXAT	EXtremely advanced transmission
EYDF	Er/Yb-doped fiber
EYDFA	Er/Yb-doped fiber amplifier
FBA	Fixed bandwidth allocation
FBF	Fiber bundle fan-in/fan-out
FDE	Frequency-domain equalization
FEC	Forward error correction
FFT	Fast Fourier transform/fiber fuse terminator
FI/FO, FIFO	Fan-in/fan-out
FIR	Finite impulse response
FM	Few-mode
FMF	Few-mode fiber
FOM	Figure of merit
FP	Fabry–Perot
FPGA	Field programmable gate array
FRA	Fixed radio access/fiber Raman amplifier
FRPE	Flame retardant polyethylene
FRS-MS	Failure recovery scheme by monitoring signals
FSK	Frequency-shift keying
FSO	Free space optics
FSR	Free spectral range
FTM	Fiber termination module
FTTH	Fiber to the home
FWM	Four-wave mixing
GD	Group delay
GE-PON	Gigabit Ethernet Passive Optical Network
GFF	Gain-flattening filter
GI	Graded-index
GN	Gaussian noise
G-PON	Gigabit Passive Optical Network
GRIN	Gradient Refractive Index
GVD	Group velocity dispersion
HAF	Hole-assisted fiber
HARQ	Hybrid automatic repeat request
HCPS	Hexagonal close-packed structure
HD	Hard-decision
HDC	Hexagonal-shaped double-clad
HDTV	High-definition television
HetNet	Heterogeneous network
HEVC	High-efficiency video coding
HL	Hazard level
HNLF	Highly nonlinear fiber

HR	Highly reflective
ID	Intradyne detection
IEC	International Electrotechnical Commission
IEEE	Institute of Electrical and Electronics Engineers
IEICE	Institute of Electronics, Information and Communication Engineers
IET	Institution of Engineering and Technology
IF	Intermediate frequency
IFFT	Inverse fast Fourier transform
IL	Insertion loss
IM-DD	Intensity modulation and direct detection
IMT-2000	International Mobile Telecommunications-2000
IoT	Internet of Things
IP	Internet Protocol
IPA	Interrelation profile analysis
IQ	In-phase and quadrature
ISI	Inter-symbol interference
IT	Information technology
ITU	International Telecommunication Union
ITU-T	International Telecommunication Union, Telecommunication Standardization Sector
LC	Lucent connector
LCoS	Liquid crystal on silicon
LD	Laser diode
LDPC	Low-density parity check
LG	Laguerre–Gaussian
LIDS	Local injection and detection system
LO	Local oscillator
LP mode	Linearly polarized mode
LSI	Large-scale integration
LTE	Long-term evolution
LW	Lightweight
MBH	Mobile backhaul
MC	Multi-core
MCF	Multi-core fiber
MCS	Multicast switch
MCVD	Modified chemical vapor deposition
MDG	Mode-dependent gain
MDL	Mode-dependent loss
MDM	Mode-division multiplexing
MDPM	Multi-dimensional position modulation
ME	Multi-element
MEF	Multi-element fiber
MEMS	Micro-electro-mechanical system
MFD	Mode field diameter

MFH	Mobile fronthaul
MIC	Ministry of Internal Affairs and Communications of Japan
MIMO	Multiple-input multiple-output
MLSE	Maximum likelihood sequence estimation
MM	Multi-mode
MMF	Multi-mode fiber
MMI	Multi-mode interference
MPE	Maximum permissible exposure of light radiation
MPEG	Moving Picture Experts Group
MPI	Multi-path interference
MPLC	Multi-plane light conversion
MPO	Multi-fiber push-on
MQW	Multiple quantum well
MRC	Maximum-ratio combination
MT connector	Mechanically transferrable connector
MTP™	Multifiber termination push-on
MU connector	Miniature universal connector
MUX/DEMUX	Multiplexer/demultiplexer
MVC	Multiview video coding
MZ	Mach–Zehnder
MZI	Mach–Zehnder interferometer
NA	Numerical aperture
NALM	Nonlinear amplifying loop mirror
NF	Noise figure
NG-PON2	Next-Generation Passive Optical Network 2
NHK	Nippon Hoso Kyokai (Japan Broadcasting Corporation)
NICT	National Institute of Information and Communications Technology
NIMS	National Institute for Materials Science
NLI	Nonlinear interference
NLPS	Nonlinear phase shift
NOHD	Nominal ocular hazard distance
NOLM	Nonlinear optical loop mirror
NOMA	Non-orthogonal multiple access
NPR	Nonlinear polarization rotation
NRZ	Non-return-to-zero
NZDSF	Non-zero dispersion-shifted fiber
OAM	Orbital angular momentum
OAPE	Optical amplifier power efficiency
OBPF	Optical band-pass filter
OBR	Over-cladding bundled rod
OCT	Outer cladding thickness
ODN	Optical distribution network
OEO	Optical-electrical-optical
OFC	Optical frequency shifter

OFDM	Orthogonal frequency-division multiplexing
OFDR	Optical frequency domain reflectometry
OLT	Optical line terminal
ONU	Optical network unit
OOK	On–off keying
OPEX	Operational expenditure
OPLL	Optical phased-locked loop
OPM	Optical performance monitor
ORS	One-ring structure
OSA	Optical spectrum analyzer
OSNR	Optical signal-to-noise ratio
OTDM	Optical time-division multiplexing
OTDR	Optical time-domain reflectometer
OTN	Optical transport network
OXC	Optical cross-connect
PA	Parametric amplification
PAM	Pulse amplitude modulation
PAPR	Peak-to-average power ratio
PAS	Profile alignment system
PBC	Polarization beam combiner
PBGF	Photonic band gap fiber
PBS	Polarization beam splitter
PC	Physical contact
PCC	Power coupling coefficient
PCF	Photonic crystal fiber
PCP	Phase-conjugate light pair
PD	Photodiode
PDF	Probability density functions
PDI	Propagation-direction interleaving
PDL	Polarization-dependent loss
PDM	Polarization-division multiplexing
PFE	Power feeding equipment
PIC	Photonic integrated circuit
PL	Photonic lantern
PLC	Planar lightwave circuit
PLL	Phased-locked loop
PLS	Pump light source
PLZT	Lead lanthanum zirconate titanate
PM	Polarization-maintaining
PME	Polarization multiplexing emulator
PNC	Phase noise cancellation
PON	Passive optical network
POSU	Protection optical switch unit
PPM	Pulse-position modulation
PRBS	Pseudorandom binary sequence

PS	Polarization scrambler/ Polarization-switched
PSCD	Polarization-split segmented coherent detection
PSCF	Pure-silica core fiber
PSD	Power spectrum density
PSK	Phase-shift keying
PT	Pilot-tone
PVC	Polyvinyl chloride
PZT	Lead zirconate titanate
QAM	Quadrature amplitude modulation
QoT	Quality of transport
QPSK	Quadrature phase-shift keying
RAN	Radio access network
RAT	Radio Access Technology
RF	Radio frequency
RIDS	Remote injection and detection system
RIN	Relative intensity noise
RMS	Root mean square
ROADM	Reconfigurable optical add-drop multiplexer
ROPA	Remote optically-pumped amplifier
RP	Remotely pumped
RRH	Remote radio head
RSA	Routing and spectrum assignment
RSE	Relative spatial efficiency
RT	Room temperature
RWA	Routing and wavelength assignment
RX	Receiver
RZ	Return-to-zero
RZ-DQPSK	Return-to-zero differential quadrature phase-shift-keying
S2 imaging	Spatially and spectrally resolved mode imaging
SA	Saturable absorber
SAFARI	Scalable and flexible optical architecture for reconfigurable infrastructure
SBS	Stimulated Brillouin scattering
SC connector	Single fiber coupling connector
SC	Single-core/single-clad/single-channel
SCC	Spatial channel count/spatial core count
SCF	Single-core fiber
SD	Signal detector/soft decision
SDM	Space-division multiplexing/switch database management
SDN	Software-defined networking
SDTV	Standard definition TV
SE	Spectral efficiency/spatial efficiency
SESAM	SEmiconductor saturable absorbing mirror
SFF	Small form factor
SFP	Small form-factor pluggable

SHD	Self-homodyne detection
SHF	Super high frequency
SI	Step-index
SLM	Spatial light modulator
SLS	Square lattice structure
SMF	Single-mode fiber
SMS	Single-multi-single
SNR	Signal-to-noise ratio
SNS	Social networking service
SOA	Semiconductor optical amplifier
SP	Set-partitioning
SPC	Single-parity check
SPIE	Society of Photo-Optical Instrumentation Engineers
SPM	Self-phase modulation
SPPM	Space-polarization-position modulation
SRS	Stimulated Raman scattering
SSB	Single side band
SSC	Spatial super-channel
SSE	Spatial spectral efficiency
SSMF	Standard single-mode fiber
SSS	Spectrum selective switch
SUE	Space utilization efficiency
SW	Switch
TA	Trench-assisted
TDD	Time division duplex
TDE	Time-domain equalization
TDM	Time-division multiplexing
TDMA	Time-division multiple access
TE mode	Transverse electric mode
TEC	Thermally diffused expanded core
TEM	Transverse electromagnetic
TIA	Transimpedance amplifier/Telecommunications Industry Association
TM mode	Transverse magnetic mode
TO	Thermo-optic
TPA	Two photon absorption
TPC	Trans-Pacific Cable
TPC-5CN	Trans-Pacific Cable 5 Cable Network
TWDM-PON	Time- and wavelength- division multiplexing passive optical network
TX	Transmitter
UE	User equipment
UHD	Ultra-high definition/ultra-high density
UHDTV	Ultra-HDTV
USIC	Unreplicated successive interference cancellation

UV	Ultra-violet
VAD	Vapor-phase axial deposition
VCSEL	Vertical-cavity surface-emitting laser
WDM	Wavelength-division multiplexing
WSS	Wavelength selective switch
XFP	10 Gbit/s form-factor pluggable
XG-PON	10 Gigabit-capable passive optical network
XPM	Cross-phase modulation
XT	CrossTalk

Chapter 1
Introduction

Toshio Morioka, Yoshinari Awaji, Kazumasa Enami, Yutaka Miyamoto, Itsuro Morita, Yukihiko Okumura, Masatoshi Suzuki, Hidehiko Takara, Jun Terada, and Kenji Yamamoto

Abstract In this introductory chapter, the physical limits and prospects of the present optical fiber communication systems as well as EXAT initiative and 3M technologies are briefly described. Then, the demands from future applications, namely ultra-realistic communications, and wireless communication networks, such as 5G technologies are described. Lastly, the state-of-the-art terrestrial optical transmission systems and optical submarine cable systems are summarized.

Toshio Morioka is a chapter editor.

T. Morioka (✉)
Technical University of Denmark, Kongens Lyngby, Denmark
e-mail: tomo@fotonik.dtu.dk

Y. Awaji
National Institute of Information and Communications Technology, Koganei, Tokyo, Japan

K. Enami
Tokyo Institute of Technology, Meguro, Tokyo, Japan

Y. Miyamoto
NTT Network Innovation Labs., NTT Corporation, Yokosuka, Kanagawa, Japan

I. Morita · M. Suzuki
KDDI Research Inc., Fujimino, Saitama, Japan

I. Morita
Waseda University, Shinjuku, Tokyo, Japan

M. Suzuki
Graduate School of Waseda University, Shinjuku, Tokyo, Japan

Y. Okumura
R&D Strategy Department, NTT DOCOMO, INC., Yokosuka, Kanagawa, Japan

H. Takara
National Institute of Technology, Okinawa College, Nago, Okinawa, Japan

J. Terada
NTT Device Technology Labs., NTT Corporation, Atsugi, Kanagawa, Japan

K. Yamamoto
Tokushima University, Tokushima, Tokushima, Japan

M. Nakazawa et al. (eds.), *Space-Division Multiplexing in Optical Communication Systems*, Springer Series in Optical Sciences 236,
https://doi.org/10.1007/978-3-030-87619-7_1

1.1 Physical Limits and Prospects of Optical Communication Systems

Figure 1.1 shows the evolution of transmission capacity per fiber of commercial optical fiber communication systems over the last four decades since 80s and future projections toward the next two decades. It is seen that the transmission capacity has been increasing constantly at an annual increase rate of about 1.5 dB per year, which has been enabled by various great inventions and breakthrough technologies. These include ultra-low-loss single-mode optical fibers, longitudinally single-mode laser-diodes, single-mode connection technologies (optical connecters, fusion splicing), Erbium-doped fiber amplifiers (EDFAs), WDM (wavelength division multiplexing) technologies with arrayed waveguide gratings (AWGs) and digital coherent technologies. However, it has been pointed out since the turn of the century that optical communication systems are rapidly approaching their capacity limit [1, 2] of around 100 Tbit/s per fiber [3–14], due to three major physical limiting factors, i.e., optical nonlinear effects in optical fibers, bandwidths of optical amplifiers as well as a destructive phenomenon called fiber fuse [15, 16]. Fiber fuse is described in detail in Chapter 7.

Figure 1.2 depicts how the total transmission capacity per fiber is determined by a product of spectral efficiency (SE), i.e., bit/s/Hz and a signal bandwidth. Firstly, the SE is governed by "Shannon limit" shown as a solid curve in Fig. 1.2a, but is further affected by signal distortions caused by various optical nonlinear effects in optical fibers, leading to a practical curve with a maximum peak called "nonlinear Shannon limit" shown as a dotted curve, an example of which is shown in the same figure. These nonlinear effects include self-phase modulation (SPM), cross-phase modulation (XPM), four-wave mixing (FWM) and stimulated Raman scattering, etc.,

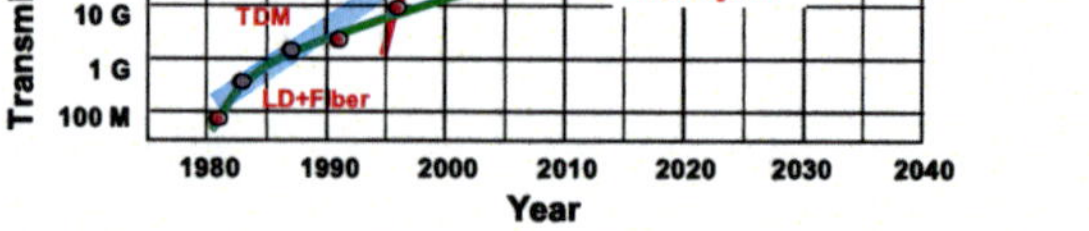

Fig. 1.1 Evolution of transmission capacity per fiber of commercial optical fiber communication systems over the last four decades and the projection for the next two decades. LD: laser diode, EDFA: erbium-doped fiber amplifier, TDM: time division multiplexing, WDM: wavelength division multiplexing, SDM: space-division multiplexing, MDM: mode-division multiplexing, SMF: single-mode fiber (Reprinted with permission from [17]. ©2017 IEICE)

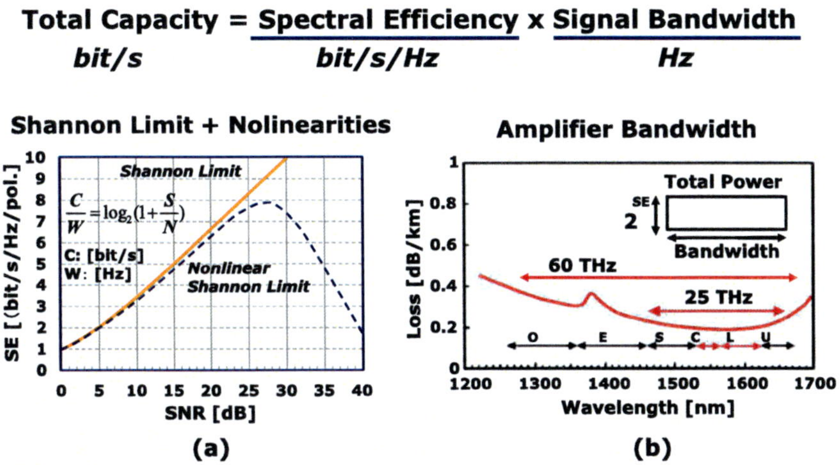

Fig. 1.2 Total capacity and its limiting factors

which are generated by high-power signals in optical fibers. The signal bandwidth given by the amplifier bandwidths is also limited up to around 40 nm per band in the case of rare-earth doped fiber amplifiers and around 100 nm for Raman amplifiers. The total amplifier bandwidth of the low-loss 1.5 μm bands including C-band (1530–1565 nm), L band (1565–1625 nm) and S-band (1460–1530 nm) amounts to 25 THz or combining all the other communication bands, i.e., O-band (1260–1360 nm), E-band (1360–1460 nm), U-band (1625–1675 nm), to around 60 THz. This would result in the maximum potential capacity of around 250 Tbit/s, or 600 Tbit/s, respectively per core if we assume an SE of 5 bit/s/Hz per polarization for all the bands. The ultimate capacity is, however, limited by the maximum allowable optical powers into the fiber, i.e., the fiber fuse propagation threshold optical powers, defined as those at which the fiber fuse stops once it has started. It should also be noted that distributed Raman amplification systems requiring pumping powers of several hundred mW up to W pause a big challenge where their pump powers combined with the signal powers are approaching the fiber fuse propagation threshold power of around 1.4 W for standard single-mode fibers and 1.2 W for dispersion-shifted fibers.

1.2 EXAT Initiative and 3M Technologies

1.2.1 EXAT Initiative

A study group named "EXAT (<u>EX</u>tremely <u>A</u>dvanced <u>T</u>ransmission) Initiative" was organized and initiated in January, 2008 by the National Institute of Information and Communications Technology (NICT), Japan, with 25 members from Japanese industries, academia, and national institutes in order to identify the ultimate limits of the present optical communication systems and to invent new breakthrough technologies which enable to substantially increase the transmission capacity well over Pbit/s per fiber [4, 7, 9, 11, 12, 14, 17]. They focused on identifying ultimate physical limitations of the present optical fiber communication systems, i.e., the maximum optical power that can be transmitted safely in optical fibers, the optical amplification bandwidths, as well as those of optical submarine cables systems, the capacity of which is limited by the electrical power consumed by the optical repeaters. As new R&D directions, the initiative proposed the use of a "space" as the last degree of freedom for multiplexing and the urgent need to develop new optical fibers (multi-core fibers (MCFs) [18–20], few-mode fibers (FMFs)) and both combined [21] as well as new multiplexing schemes, namely space-division multiplexing (SDM) and mode-division multiplexing (MDM), as depicted in Fig. 1.1. It should be pointed out that MCFs and MDM were initially proposed nearly four decades ago as a means to realize high-density optical cables primarily for subscriber lines [22, 23] and a new multiplexing scheme [24] compatible with time division multiplexing (TDM) and WDM, respectively.

After its first investigation period was over in 2008, an international symposium EXAT 2008, the first of its kind in the field of SDM was held in Tokyo in November 2008, where the initiative reported its first study on new optical fibers and related SDM technologies with a clear message that there is an urgent need of developing them in order to overcome the rapidly approaching limit of the optical communication systems. In its second term in 2009, EXAT Initiative discussed specific technological proposals toward the creation of national projects. Technical reports of each term were incorporated into a Japanese book entitled "Innovations in Optical Fiber Communications Technologies" published in 2012. The NICT EXAT then evolved to a new Institute of Electronics, Information and Communication Engineers (IEICE) EXAT study group in 2010, which has continued its vigorous activities, organizing 20 + international workshops, symposia, including EXAT 2013 (Sapporo), EXAT 2015 (Kyoto), EXAT 2017 (Nara), EXAT 2019 (Ise) and EXAT 2021 (virtual).

EXAT Initiative has also led to the creation of a series of pioneering national projects on SDM technologies, such as i-FREE (Innovative Optical Fiber Technologies: 2010–2012), i-ACTION (Innovative Optical Communication Infrastructure: 2011–2015), and i-FREE^2 (Innovative Optical Fiber and Communication Technology for Exa-bit Era with SDM: 2013–2018) which will be described in detail in Chapter 8. Recently, new projects SDM-PN (R&D of Space-Division Multiplexing Photonic Node: 2016–2020), i-FAST (R&D on Innovative Optical Fiber and Communication Technologies Toward Standardization: 2018–2022) and OCEANS (R&D on high-capacity multi-core fiber transmission systems: 2018–2022) were initiated and are ongoing. Furthermore, an EU-Japan coordinated R&D project SAFARI (**S**calable **A**nd **F**lexible optical **A**rchitecture for **R**econfigurable **I**nfrastructure: 2013–2017) was created as the first internationally collaborative project between Japanese EXAT and the European partners, commissioned by the Ministry of Internal Affairs and Communications (MIC) of Japan and European Commission Horizon 2020. These national projects have been leading the SDM technologies in the world, creating many world records such as 1-Pbit/s transmission (2012 [25], 2017 [26]), 1-Ebit/s·km transmission (2013) [27, 28], 2-Pbit/s (2015) [29, 30] and 10-Pbit/s transmission (2017) [31].

1.2.2 3M Technologies

As described in the preceding sections, EXAT Initiative explored new optical fibers making use of "space" dimension, namely MCFs and FMFs or multi-mode fibers (MMFs) as depicted in Fig. 1.3. MCFs have the propagation mode in each core either coupled with those in other cores or un-coupled. FMFs/MMFs, on the other hand, have different modes in a core, which normally couple with each other over some distances and thus multiple-input, multiple-output (MIMO) processing is usually required to separate different modes at the receiver side.

In developing SDM technologies, EXAT Initiative identified three major fundamental research topics, namely "Mutil-core Fiber," "Multi-mode Control" and "Multi-level Modulation," which we call "3M technologies" as depicted in Fig. 1.4 where a factor of 10 in each category should enable a factor of 1000 increase in capacity [4, 11, 17]. In fact, 32-core [26]/37 core [32]/38 core fiber [33] as well as 45-mode [34] transmission has already been demonstrated.

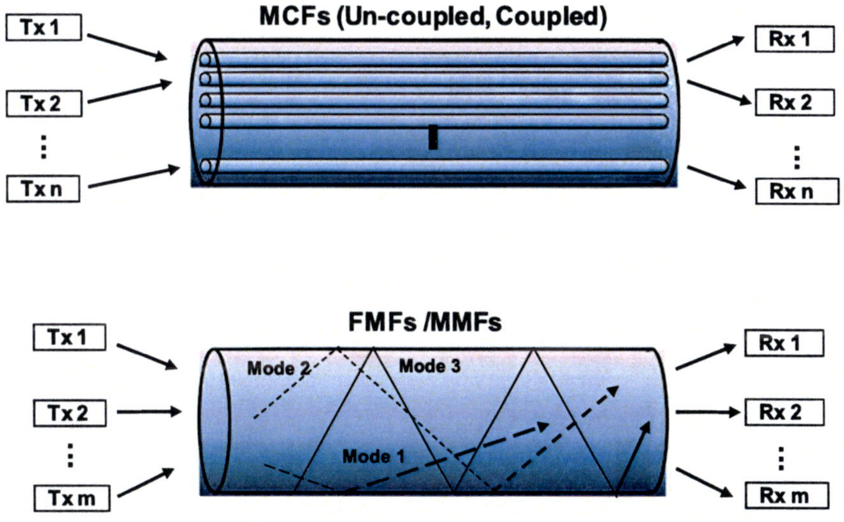

Fig. 1.3 Schematics of multi-core fibers (MCFs) and few-mode fibers (FMFs) or multi-mode fibers (MMFs) (Reprinted with permission from [17]. ©2017 IEICE)

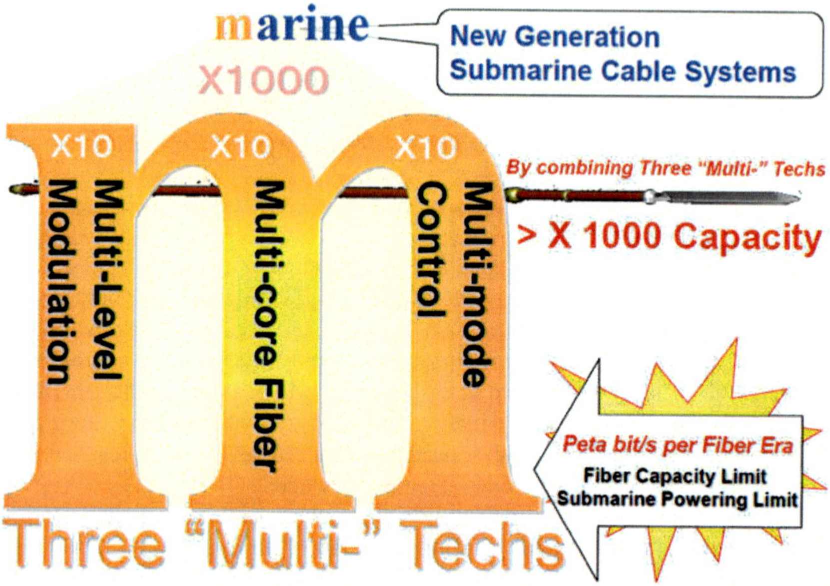

Fig. 1.4 3M technologies (Reprinted with permission from [17]. ©2017 IEICE)

Recently, few-mode, multi-core fibers (FM-MCFs) have been developed to further increase the transmission capacity and used to achieve up to 10-Pbit/s transmission [29, 31, 33, 35]. The major SDM components to be developed are SDM fibers, SDM multiplexers/demultiplexers (SDM-MUXs/DEMUXs), SDM optical amplifiers, optical connectors/splicing as depicted in Fig. 1.5. Important characteristics of the passive components are naturally low insertion loss, low core-/mode-dependent loss, low crosstalk among modes/cores and wide-bandwidths to support as many WDM/SDM channels as possible. SDM optical amplifiers require low core-/mode-/wavelength-dependent gain, wide-bandwidths with high gain and low NFs. All these components will be described in detail in the following chapters.

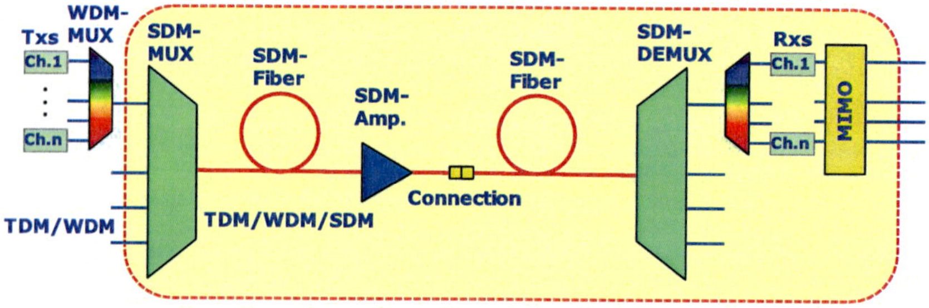

Fig. 1.5 Basic optical components for SDM technologies. Tx: transmitter, Rx: receiver (Reprinted with permission from [17]. ©2017 IEICE)

1.3 Requirements for Future Applications

1.3.1 Ultra-Realistic Communication

To confront various societal issues such as global warming or the aging of the population, a great deal of attention is being paid to ultra-realistic communication, which makes viewers feel as if they were somewhere else without actually leaving their own locations. If meetings could be conducted so that actual face-to-face discussions were held even though the participants were far apart, no energy need be consumed for travel and if there were a system that made it possible for people with physical impairments to take part in social activities, society could make the best use of the knowledge and skills of the elderly.

Although teleconferencing systems already exist, they certainly feel as if the meeting is being conducted through a screen and there is a considerable sense of discomfort compared with a face-to-face meeting. This sense of discomfort would be eliminated if ultra-realistic communication technology was implemented by delivering advanced video conveying the sensations of immersiveness, extremely fine detail, or three-dimensionality; 3D sound imparting even the presence of people nearby or the environmental sounds of a room; or sensations that make it feel as if something in a distant location were close at hand.

This section will introduce ultra-high definition video or 3D video, which are the core technologies of ultra-realistic communication, and will describe the requirements of an optical communication system for transmitting them.

1.3.1.1 Evolution of Video and Their Corresponding Transmission Capacities

The history of the evolution of video, especially television broadcasting, is a history of increasing realism. Figure 1.6 shows the history of this progress from photographs to black and white television, from black and white to color, and from Standard

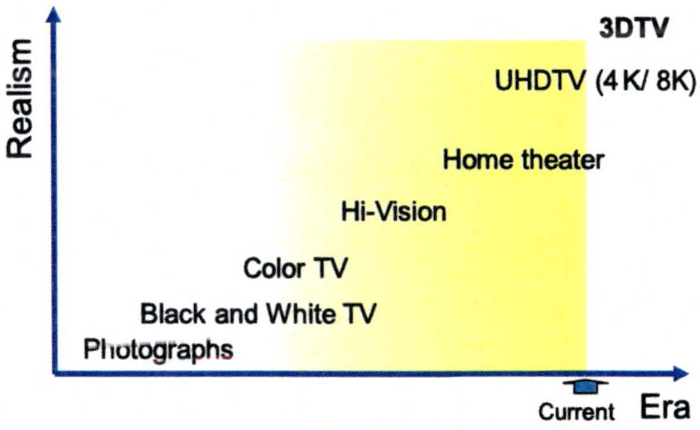

Fig. 1.6 History and future of TV broadcasting advancing toward ultra-realism

Definition TV (SDTV) with 500 or 600 scan lines to High-Definition Television: commonly called Hi-Vision (HDTV) with 1080 scan lines. The marketing of large flat screen TVs for projecting Hi-Vision quality images contributed to this progress. Recently, home theaters with screen sizes of 70 inches or more have become quite popular.

In addition, 4 K Ultra-High Definition (UHD) and 8 K UHD television, which aim to increase high-definition standards even further, were steadily implemented as the latest TV, and research is being conducted on 3D television as the next-generation TV after that.

(1) UHDTV (4 K/8 K)

I have installed a home theater in my home which combines a full HDTV projector and a 90 inch screen so that I can enjoy watching Hi-Vision broadcasts that make me feel as if I were actually present in the scene being broadcast. However, since I am viewing this in the narrow environment of a 6 tatami mat room (approximately 99 square feet), I am unable to maintain the optimum HDTV viewing distance (3 times the screen height) and must watch the picture from too close a distance. Screen display degradation due to the scan line structure or digital compression is conspicuous in full-impact scenes. This increases my desire to be able to view a high-quality picture even when viewing the screen from a close distance.

Ultra-HDTV (UHDTV), which aims to increase high definition standards even further, is an attempt to achieve this goal. Other common names for this technology are 4 K UHDTV and 8 K UHDTV. Specifications for these technologies have already been standardized by the International Telecommunication Union (ITU) [36] and started broadcasting from 2018.

4 K UHDTV is a video system with double the number of scan lines of HDTV at 2160 lines and double the number of horizontal pixels at 3840. Since the number of horizontal pixels is approximately 4000, this has "4 K" in its name. The color reproduction capability is also greater than that of HDTV. Displays and cameras that satisfy 4 K specifications are already on the market. Test terrestrial broadcasts of 4 K UHDTV were conducted in Korea in 2012, and experimental satellite broadcasts were started in Japan in 2014. In June 2014, the FIFA World Cup from Brazil was broadcast in Japan in 4 K video.

Research and development of 8 K UHDTV, which is also known as Super Hi-Vision television, has been conducted with NHK Science and Technology Research Laboratories playing the leading role. Super Hi-Vision television has 4320 vertical scan lines and 7680 horizontal pixels, which are each 4 times those of Hi-Vision television, as shown in Fig. 1.7. The optimum viewing distance for this kind of extremely detailed video can be viewed from a distance much closer to the screen (0.75 times the screen height). The angle of view with which the viewer observes the screen at that time is 100°. The relationship between the visual field and the size of the image projected on the retina indicates that the angle of view of Super Hi-Vision television is almost equivalent to the size of the induced visual field that affects the

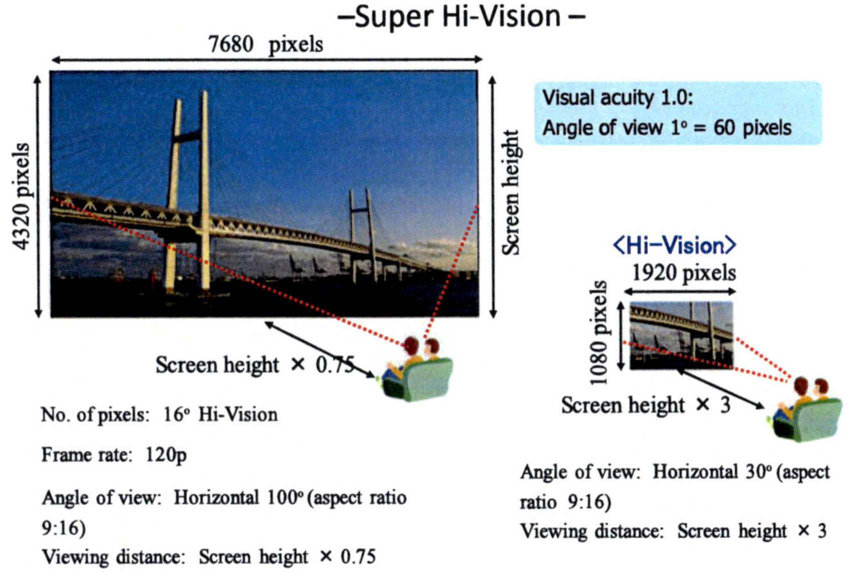

Fig. 1.7 8 K UHDTV

sense of orientation. In other words, a large screen UHD video system can greatly bring about an enhanced sense of reality in which viewers instinctively move their bodies due to motion of the image on the screen. NHK Science and Technology Research Laboratories performed subjective evaluation experiments concerning the visual field angle of view and the sense of reality and obtained results indicating that the sense of reality is saturated for horizontal visual field angles of view ranging from 80 to 100° [37]. In addition, as the screen size increases, motion resolution must also increase. Therefore, 8 K UHDTV also includes standards that increase the number of images per second (frame rate) from the normal 60–120. The color reproduction capability is also greater than that of HDTV as it is for 4 K UHDTV. In addition to this kind of high-definition video, a 22.2 channel multi-channel surround sound system is also provided as part of the Hi-Vision broadcasting service.

While developing prototype equipment such as 8 K cameras, editing systems, transmission devices, broadcasting media (satellite broadcasting, terrestrial broadcasting and broadband networks) and display devices, transmission trials have already been performed. NHK provided public viewings of prototype broadcasts of the 2012 London Olympics at several locations throughout Japan. In addition, experimental broadcasting began in 2016 and actual broadcasting in 2018.

The amount of information (uncompressed signal) carried by Super Hi-Vision television is approximately 144 G bit per second (bps). This number can be calculated from (12 bits per pixel) × (4320 × 7680 pixels) × (120 frames) × (3 primary colors). The amount of information for HDTV, however, is approximately 1.5 Gbps.

Currently, HDTV (Hi-Vision) is being broadcast in Japan by both terrestrial and satellite digital broadcasters, but due to bandwidth constraints, MPEG 2 is being used to compress the signals to approximately 15 Mbps for terrestrial broadcasts and to approximately 20 Mbps for satellite broadcasts (using a compression ratio of approximately 1/100). The compression standards that have been determined for Super Hi-Vision television use High-Efficiency Video Coding (HEVC) [38], which has greater compression efficiency than MPEG 2, so that the drop in image quality is visually permissible even when compressed to approximately 100 Mbps (using a compression ratio of approximately 1/1000).

(2) 3D television (3DTV)

The ultimate video service for ultra-realistic imaging techniques is 3D television, and research and development of 3DTV is being conducted by various organizations including the National Institute of Information and Communications Technologies (NICT). A 3D imaging system can be constructed by understanding the mechanism by which people perceive an external three-dimensional space from two-dimensional images projected on their retinas. Physiological cues that cause us to perceive stereopsis are listed below.

- Binocular parallax: Difference according to the distances to the subject in the images projected on the left and right eyes
- Convergence: The angle determined by both eyes focusing on the object (the eyes will cross when viewing a nearby object)
- Accommodation: Focal adjustment of the eye lens according to the distance to the object
- Motion parallax: Difference in how objects appear due to changes in their relative motion or degree of overlap

In addition to these, there are also graphic clues such as contours and shadows (psychological cues) that enable us to perceive objects three-dimensionally.

The simplest 3D video system for stimulating perception of stereopsis uses the binocular parallax method. In other words, the stereoscopic method presents separate images for the left and right eyes simultaneously on one screen and apportions those images, respectively, to the left and right eyes by using some means such as special glasses. In 2010, Hollywood studios began making many stereoscopic 3D movies which were shown in digital theaters throughout the world. However, although some TV sets that were manufactured for home use also had features that enabled stereoscopic 3D images to be viewed, they did not become very popular because relatively little content was created and because of the following drawbacks of the stereoscopic method. First, special glasses are typically worn to properly view images created by the stereoscopic method, and since the position of the display surface differs from the position where the 3D image is reproduced, the resulting eye adjustments for convergence and accommodation are unbalanced, which often results in eyestrain. In addition, stereopsis can become disrupted when there are more sudden scene changes than can be followed by changes in binocular parallax.

Recently, a great deal of research has been conducted concerning methods of reproducing three-dimensional space in natural light. Some examples include a multiview autostereoscopic display method in which numerous viewpoints are strictly arranged in a fixed alignment [39], a light field method that uses a lens array, etc., to create light in various directions [40] and an electronic holography method that reproduces light with light-wave precision based on the theory of holography [41].

The amount of information for the stereoscopic method is twice the amount of image information for one eye because there are left and right images. For the multiview method, this increases to a multiple corresponding to the number of views. For example, the large screen glasses-free (multiview) 3D display on which we are currently conducting research at NICT uses an ultra-multiview autostereoscopic 3D imaging method in which several hundred Hi-Vision picture-quality projectors are arranged horizontally, and the amount of information in this case is approximately 300–500 Gbps. Figure 1.8 shows the 200-inch, approximately 200-view horizontal parallax, large screen ultra-multiview autostereoscopic display that we developed [42]. The theoretical value for a system with specifications of a 50-inch screen, 30° field of view (range in which stereopsis is possible), Hi-Vision picture quality using the holographic method for generating the most ideal 3D images is 600 Tbps.

The Moving Picture Experts Group (MPEG), which determines international standards for video encoding, discussed encoding for compressing these kinds of multiviewpoint images with massive amounts of data and created Multiview Video Coding (MVC) as the video compression standard for multiview images. This standard encodes data by treating the multiview data as a single base view and some residual non-base views. It uses interview prediction, which references frames contained in

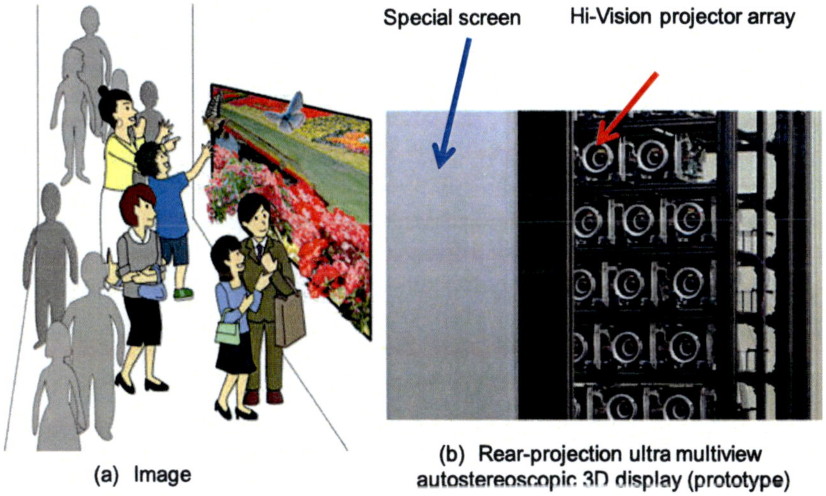

Fig. 1.8 Ultra-multiview autostereoscopic 3D display developed by NICT

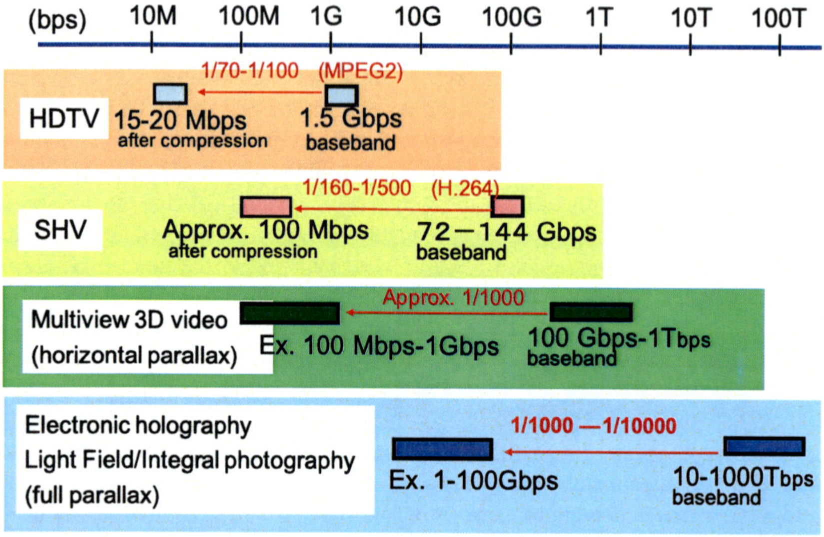

Fig. 1.9 Expected amounts of information before and after compression for various types of video methods

the other view to encode the non-base view. The non-base view can be compressed approximately half as much as the base view.

MPEG was also examining 3D-AVC and 3D-HEVC as separate encoding methods for compressing 3D images. These methods simultaneously encode multiview images and depth data by using interrelations between them. Since 3D image information is based on the projection of an object in 3D space, it makes sense to encode the multiview images and depth data simultaneously. Not only are the compression capabilities of 3D-AVC and 3D-HEVC highly efficient, but the decoder can also generate images at various viewpoints depending on the display system at the decoder side.

Since hologram images or multiview images and depth data are used for 3D video information as described above, it is advantageous that the compression efficiency of these methods theoretically is relatively high for increasing amounts of data. The ultra-realistic multiview 3D imaging methods or holographic methods that NICT will be developing in the future will probably require a transmission capacity of several Gbps to several dozen Gbps per content (Fig. 1.9).

1.3.1.2 Ubiquitous Video

Recently, surveillance cameras have been installed all over for security purposes. Public vehicles such as buses and taxis have also been equipped with cameras. These are connected to closed networks, and each of them is used independently.

On the other hand, live video cam images are publically available from Web pages to inform viewers of traffic conditions, how crowded sightseeing spots are, and how much snow has fallen at ski slopes. Bird's-eye views like those of Google Earth are also available.

Mutually connecting this kind of video, which is flowing throughout the world, to a network and collectively delivering or recording it together with metadata from various types of sensors such as position information, sound, temperature or CO_2 measuring instruments will make it possible for almost any location to be viewed together with data that changes from moment to moment, including even video from the past, at any time from various angles. This is the so-called Internet of Things (IoT) including video information. Of course, policies must be devised for privacy protection, but this certainly can be implemented technically. We have coined the term "ubiquitous video" for this kind of wireless sensor network centered on video. These videos and metadata will be the input information for ultra-realistic video communication.

Implementation of ubiquitous video can give rise to new public services or businesses. For example, this surely can be useful for disaster prevention, security, and elder care, and also can be used to implement monitoring of environment changes (environmental logging), trip navigation, and even life-logging by videoing oneself.

Not only will video be distributed by these kinds of ubiquitous video services, but ordinary Internet users also will continue to distribute video in email attachments and blogs. Although Asymmetric Digital Subscriber Line (ADSL) networks had previously been popular, they were based on the premise that much less information would be uploaded than downloaded. However, networks from now on must guarantee that the amount of uploaded information can also be transmitted.

1.3.1.3 Requirements for Optical Communication Networks

The "killer application" for optical communication networks will surely be video information. To stably distribute the high-capacity video information of the future including UHDTV or 3DTV or to make practical use of video information anywhere throughout the world, the next-generation optical transmission network must satisfy the following conditions.

- Ultra-fast broadband: Able to distribute UHD, 3D video information at rates of several Gbps to several dozen Gbps.
- Low latency: Latency on the order of several dozen milliseconds. Able to provide stress-free interaction, remote control, and negotiation between transmission and reception terminals.
- Variation-free latency: Latency must not differ due to differences in transmission channels or audio and video media.
- Balanced upstream/downstream transmission speeds: Uploading of video information will also increase in the future.

- Smooth access switching: Must be able to execute distributed server transfer and area control without interruption. Must be able to allow smooth channel surfing. Must be able to support changes in the viewing environment (e.g., a viewer may want to continue viewing content that had been viewed through a wired connection from a separate wireless device).
- Plug and play: Video information is expected to be generated by various methods in the future. To support this information with flexibility, any kind of information must be able to be sent and received just by plugging in to a network.
- Robust and highly stable: Must be stable as the core information distribution infrastructure. Must be resilient and durable even during disasters. An unstable network when charges for content are widespread cannot be a core infrastructure.
- Highly reliable and secure: Content must not be easily falsified. Copyrights must be diligently protected. Privacy must be safeguarded.
- Energy conservation: Network and terminals both must operate with low-power requirements.

Conclusions

In this section, we described future video technologies and their transmission capacities as well as requirements for optical communication networks. To increase a sense of realism, information will have to be delivered with as much fidelity as possible. To accomplish this, the some information may have to be delivered directly without compression encoding. Other information such as audio, tactile and vestibular sensation information may also have to be delivered in addition to video information. Delays will not be permitted for interactions with other parties. In this sense, even higher speeds and greater capacities are desirable.

1.3.2 Optical Network Technologies for Wireless Communication Network

Optical networks continue to play an important role in wireless communication systems, especially to link base stations to the core network in the radio access network (RAN).

Two RAN architectures have been developed to accommodate base stations: distributed RAN (D-RAN) and centralized RAN (C-RAN). In the D-RAN architecture, as shown in Fig. 1.10a, base stations are connected to the core network via an IP/Ethernet-based network and connected to other base stations via switches or routers. The link between the base stations and the core network is called the mobile backhaul (MBH). The data rate of MBH from base stations roughly equals that of mobile data traffic. Latency between base stations is categorized as non-ideal backhaul, and latency should be decreased to provide higher system performance. Commercial IP and Ethernet services are currently used as the transmission medium for MBH.

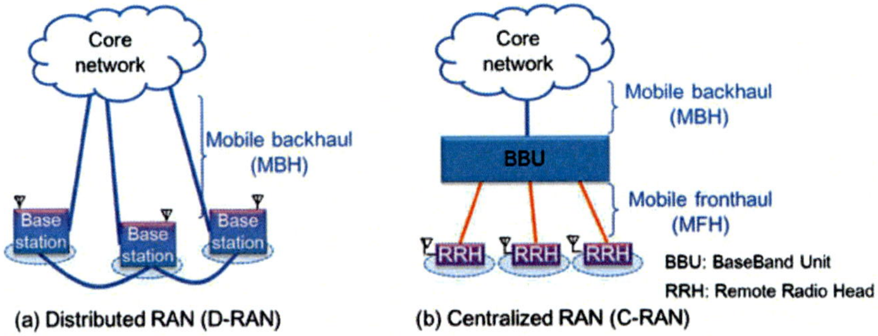

Fig. 1.10 RAN architectures

In the C-RAN architecture, as shown in Fig. 1.10b, the baseband function of base stations are concentrated and remote antennas need offer only transmit and receive functions. The C-RAN configuration realizes advanced cooperation between the cells covered by each base station. In the C-RAN architecture, the link between baseband unit (BBU) and remote radio head (RRH) is called the mobile fronthaul (MFH) and is based on digital radio over fiber (DRoF) for low-cost optics and waveform transparency. Common public radio interface (CPRI) is the de facto standard for MFH transmission [43]. CPRI transmission is based on IQ data transmission and requires high bandwidth (1 Gbit/s for single antenna 20 MHz radio bandwidth transmission). The number of antennas, frequency bands and MIMO layers are increased until the required bandwidth is satisfied. The BBU-RRH response is a severe problem. Since the response time of hybrid automatic repeat request (HARQ) is set to 8 ms, the latency limit for CPRI transmission is about 300 μs in general [44] and the transmission distance is limited to 10–20 km.

Moreover, accurate frequency and time synchronization are required for proper operation of baseband units [45]. The frequency accuracy of 50–200 ppb is required for proper frequency radio transmission. Time accuracy is required for time division duplex (TDD) transmission control, and heterogeneous network (HetNet) operation, etc., and required accuracy is about 1 μs.

1.3.2.1 Prospects of 5G Mobile Communication System

Mobile communication services have demonstrated significant growth over the last few decades. Currently, many operators worldwide have already deployed Long-Term Evolution (LTE) and LTE-Advanced, i.e., fourth generation mobile communications system (4G) to offer faster access with lower latency and higher efficiency than its predecessors, 3G and 3.5G. Given these trends, there is inexorable demand for new mobile communications systems with even further enhance capabilities, namely 5G. Several organizations have established worldwide to study 5G networks, such as the 5G Infrastructure Public Private Partnership (5G-PPP) in Europe [46], IMT-2020 (5G) promotion Group in China [47], 5G Forum in Korea and 5G Mobile Communications Promotion Forum (5GMF) in Japan [48].

The radio access network (RAN) of 5G assumes that small cells will be laid over macro-cells [49]. The small cells offer wide-bandwidth and super high-speed transmission using higher frequency bands from super high frequency (SHF) to extremely high frequency (EHF) in small coverage area, while the macro-cells provide coverage and mobility. 5G RAN will also support extension so that existing small cells as well as various future small cells are accommodated flexibly and appropriately by the networks, and cell selection and radio resource assignment can be performed according to each user's moving situation, service requirements, etc.

The assumed targets of 5G network are as follows:

1. Augmented system capacity is 1000 times higher traffic volume than around 2010 when LTE service started,
2. Supporting the connection of one hundred times more devices than LTE,
3. Higher data rate deployments, e.g.,> 10 Gbit/s,
4. User-plane latency of less than 1 ms over the RAN, (currently 5 ms in LTE).

To meet these 5G targets, we need to address spectrum efficiency, spectrum extension and network densification.

Figure 1.11 shows an assumed 5G RAN [50]. It uses the multi-layered cell architecture with the cooperation of all layer radio access technologies (RATs) to achieve high data throughput and high availability everywhere. As shown in Fig. 1.11, each macro-cell is overlaid by many small cells. The small cells are used for high capacity, while the macro-cell provides coverage and user mobility. Especially in high traffic density areas, small cells should be deployed densely with the cell radius of several

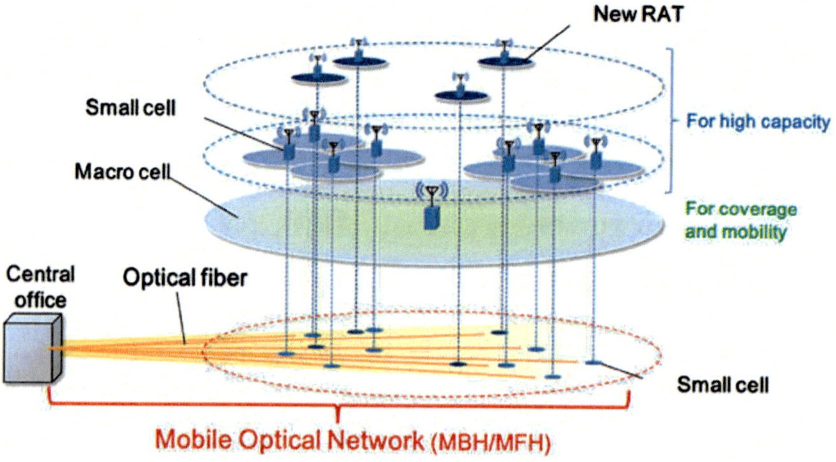

Fig. 1.11 5G RAN

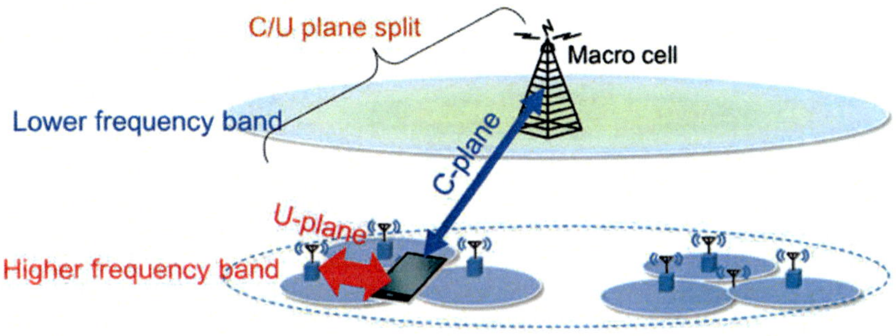

Fig. 1.12 C/U plane split

tens of meters. The RATs of these new small cells are not the same as those of conventional small cells so RAT enhancement is also needed for 5G RAN.

With regard to RAT enhancement, higher order modulation schemes such as 256-QAM and high-frequency utilization schemes such as non-orthogonal multiple access (NOMA) have been studied [51, 52].

To permit the new RATs to offer higher throughput, it is preferable to use of higher frequency bands, such as Super High Frequency (SHF) band, i.e., 3–30 GHz, and Extremely High Frequency (EHF) band, i.e., 30–300 GHz, with large bandwidth (> 100 MHz). Radio waves in the high-frequency range have exceptionally strong line of sight performance so interference between cells is very small. Moreover, higher order MIMO with many antennas can be realized with small antenna size, and this will enhance cell capacity. Therefore, 10-Gbit/s class capacity per cell per RAT can be achieved [53].

For high user mobility, the frequency of handover between macro- and small cells and between small cells will be raised, and control traffic between base station and core network will be increased. Moreover, HetNet operation between macro- and small cells also requires control bandwidth for high mobility. To reduce the control traffic, splitting the control and user planes (C/U plane split), see Fig. 1.12, is a useful technique [50]. The C-plane traffic is transmitted over the macro-cell network, and high connectivity is assured. The high capacity of the U-plane is achieved by small cells, and handover and HetNet operation in a macro-cell area is controlled by the macro-base station. Radio resource management for efficient traffic accommodation is also controlled by the macro-base station. In this case, cooperation between macro- and small cells is very important, and low latency and high-capacity connections between macro- and small base stations are critical.

1.3.2.2 Challenges Imposed by Optical Access Technology for 5G Network

Optical access technology, especially the passive optical network (PON), has been widely deployed in the form of fiber-to-the-home (FTTH) service. In a PON system, one fiber is shared with multiple subscribers and this enables broadband services at low cost. Gigabit class FTTH services are widely offered using the gigabit PON (G-PON) system standardized by ITU-T or the gigabit Ethernet PON (GE-PON) by IEEE. These PON systems are based on the time division multiplexing (TDM) technique. TDM-PON is a cost-effective solution to accommodate small cell clusters; however, there are many technical issues to resolve, such as latency and bandwidth. It is also important for the next optical networks to support the flexible configuration of small cell clusters. Securing the cooperation of mobile and optical networks is one key to resolve these issues, and in the following, some challenges are briefly introduced.

Low latency transmission is very important for the mobile optical network. In downstream TDM-PON, the latency of network equipment is estimated to be several tens of microseconds and 20 km fiber length (100 μs latency) is possible for MFH. In the upstream, due to the dynamic bandwidth allocation (DBA) algorithm, the latency is more than 1 ms, which is too large to support MFH transmission. One approach to reducing DBA latency is using a fixed bandwidth allocation (FBA) scheme. By using FBA, no control signal between OLT and ONU for bandwidth allocation is transmitted, which reduces the latency to several tens of microseconds. However, with FBA operation, the bandwidth of each connection is fixed and efficient bandwidth usage based on statistical multiplexing cannot be achieved.

Recently, a new bandwidth allocation scheme for upstream PON based on mobile bandwidth allocation has been proposed [54]. In the LTE system, the base station allocates uplink bandwidth for each user terminal and also passes scheduling information to UE with 4 ms timing advance. Therefore, as shown in Fig. 1.13, the required upstream bandwidth of the PON can be calculated using scheduling information of the mobile system, and grant data reaches each ONU before uplink signal reception. Thus, the latency of TDM-PON is minimized and total latency, excluding fiber delay, should theoretically be of order of several tens of microseconds. Some measured results of a prototype are shown in Fig. 1.14. Latency under 30 μs is achieved with the proposed scheme. To implement this method, an interface of scheduling information should be placed in the base station.

There are two ways to handle the huge service demands placed on MBH and MFH of 5G RAN, expansion of transmission bandwidth and data bandwidth reduction.

With regard to bandwidth expansion, NG-PON2 can provide 40-Gbit/s class bandwidth, which is enough for current MBH and MFH. However, much higher bandwidth will be needed in future given the vision of 10-Gbit/s class user bandwidth. Digital coherent detection with high level modulation can be used to enhance capacity per wavelength. However, in optical access, cost is a very important factor, and the development of low cost digital coherent systems is critical. WDM-based bandwidth

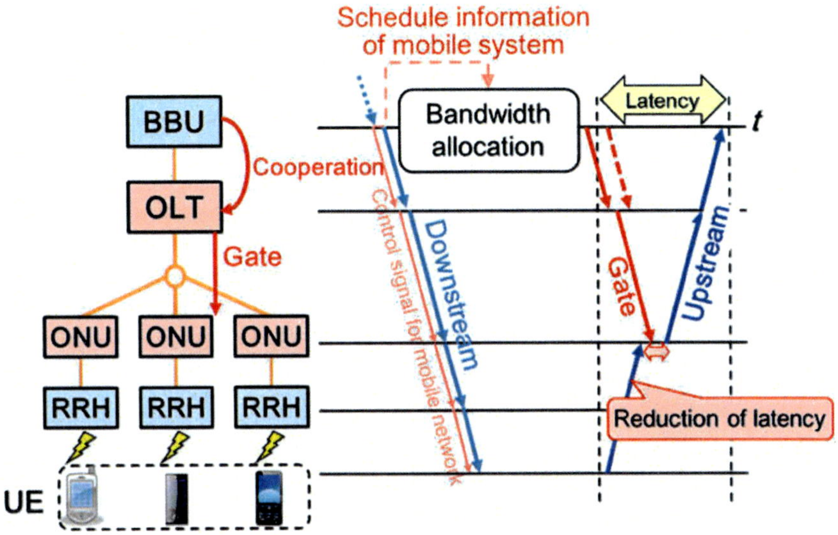

Fig. 1.13 Mobile optical network with TDM-PON

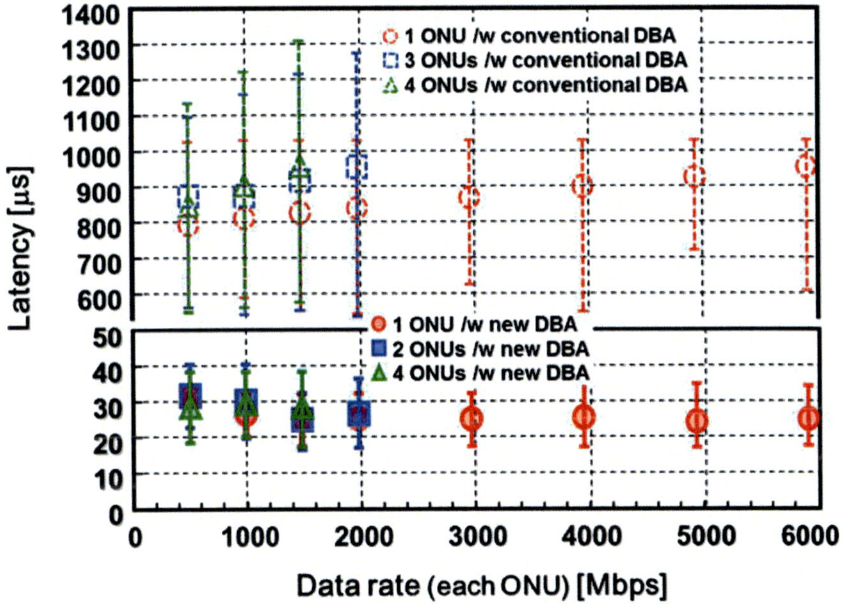

Fig. 1.14 Measured latency with conventional DBA and DBA using schedule information of mobile system (Reprinted with permission from [54]. ©2014 Authors)

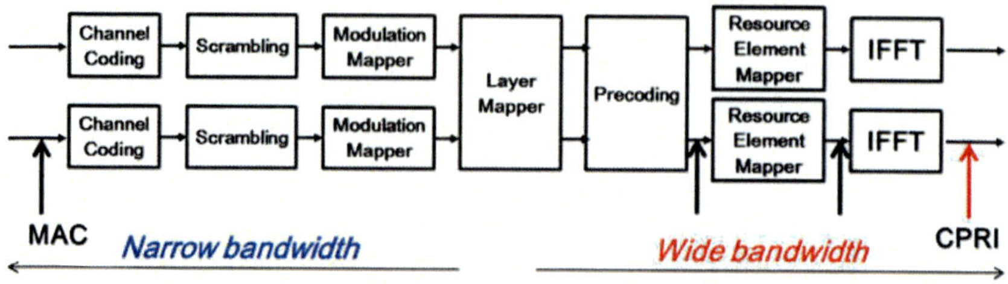

Fig. 1.15 BBU-RRH separation point

expansion is also available, and wavelength allocation for WDM overlay has been discussed in NG-PON2/ITU-T [55, 56].

Bandwidth reduction is very useful for MFH. In the recent CPRI specification [43], a new line coding (64b/66b coding) that offers less redundancy than conventional coding (8b/10b coding) and sampling rate reduced transmission have been proposed. More antennas or bands can be accommodated on the same CPRI link. IQ data compression is also useful. Currently, many compression schemes have been proposed that mainly use sampling rate reduction and quantization bit width reduction; compression ratios of under 1/2 can be realized [57, 58]. However, these methods are lossy compression for low latency and high compression rates, and some distortion in IQ data seems inevitable. In the latest proposals, distortion is set to low enough that signal quality degradation is insignificant (< 1% EVM degradation); however, more accurate transmission of IQ data and high dynamic range is needed for enhanced RAT (256 QAM and NOMA), and the merit of data compression seems to be problematic.

Another approach to bandwidth reduction in MFH is changing the BBU-RRH separation point in the physical layer (Fig. 1.15). Higher sublayer separation reduces the required bandwidth to the mobile data rate [59]. Pre-IFFT data transmission also reduces bandwidth drastically [60, 61]. The MFH data rate of these techniques dynamically changes with the mobile traffic, and in TDM-PON systems, statistical multiplexing gain can be expected. Therefore, a combination of these methods and DBA using scheduling information of the mobile system is very effective for future MFH. These methods need tight integration of the base stations and TDM-PON system, and if the mobile operator is not the FTTH operator, development of an effective control interface between systems is very important.

Conclusions

In the 5G RAN for 2020 and beyond, the role of small cells is very important, and to support the efficient and cost-effective operation of 5G RAN, cooperation of mobile and optical access networks is essential. To accommodate small cell clusters, the TDM-PON-based optical network is useful for its flexibility and cost effectiveness in terms of CAPEX and OPEX, but further enhancement to meet 5G RAN requirements is needed.

1.4 State-of-the-Art Terrestrial Optical Transmission

1.4.1 Expansion of Broadband Services in Japan

There has been a rapid expansion of broadband services in Japan in recent years, spurred by the development of diverse services that use the Internet. According to statistics published by the Ministry of Internal Affairs and Communications, the total number of broadband subscribers in Japan exceeded 40 million and the download traffic of broadband service subscribers reached 19.0 Tbit/s as of May 2020 (Fig. 1.16 and Fig. 1.17) [62]. Fiber to the home (FTTH) has taken over the lead from ADSL in Japan, and Japan now has the highest penetration of optical access services of any

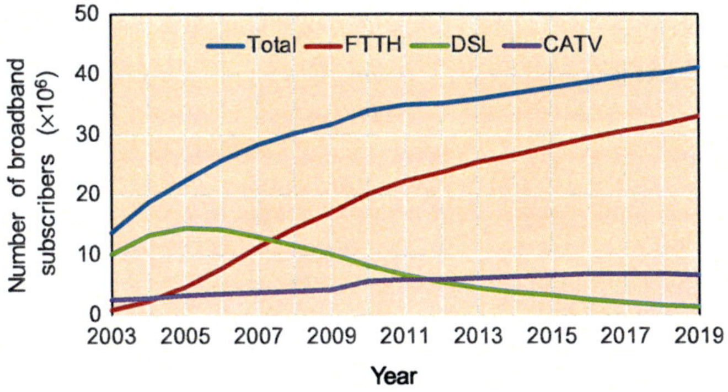

Fig. 1.16 Expansion of the broadband infrastructure

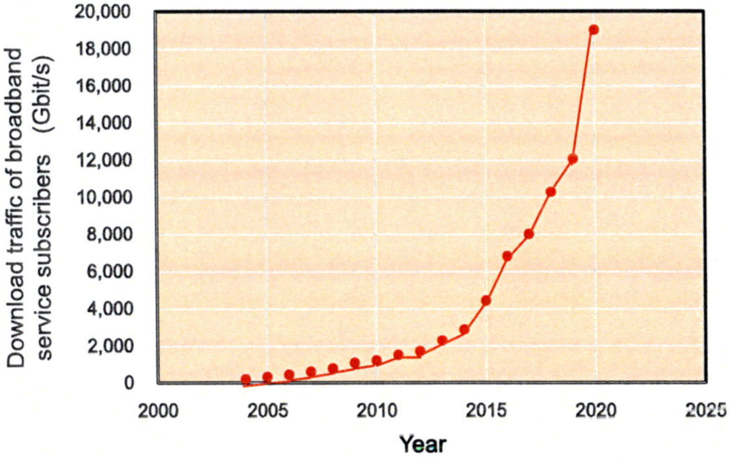

Fig. 1.17 Increase in Internet traffic

country in the world. Furthermore, the data traffic volume on the core network, which carries the high-volume multiplexed access network signals, has been increasing annually by a factor of from 1.3 to 1.4. Research and development on access networks and photonic transport networks that can handle high-volume traffic is indispensable for the growth and further development of the future broadband society. In this section, we introduce the current state of optical access technology and high-capacity optical transmission technology.

1.4.2 Optical Access Technology

The leading optical access technology to support FTTH is Passive Optical Network (PON). It realizes point-to-multipoint communication between Optical Line Terminal (OLT) at the central office and a number of Optical Network Units (ONUs) at the subscribers' side via Optical Distribution Network (ODN). The massive FTTH deployment in recent years has been supported by Gigabit Ethernet PON (GE-PON) which was standardized by IEEE in 2004 as a part of the IEEE Standard 802.3 [63]. GE-PON offers symmetric 1-Gbit/s bandwidth which is shared among up to 32 ONUs through the use of Time-Division Multiple Access (TDMA). The TDMA technique not only allows the shared use of the optical fiber and the OLT optics, but also enables the aggregation of the traffic from the number of ONUs. Thus, it realizes a simple and cost-effective access and aggregation network.

As a successor of GE-PON, IEEE standardized 10G-EPON in 2009 [64]. Its downstream bandwidth is 10 Gbit/s while its upstream bandwidths are 1 Gbit/s and 10 Gbit/s in the asymmetric and the symmetric options, respectively. 10G-EPON OLT can accommodate GE-PON ONUs so that a smooth migration from GE-PON to 10G-EPON is realized.

ITU-T standardized Gigabit-capable PON (G-PON) and 10Gigabit-capable PON (XG-PON) in 2004 and 2010, respectively [65, 66]. For further evolution, ITU-T standardized 40 Gigabit-capable PON in 2015, which has been also known as Next-Generation PON2 (NG-PON2) [67]. NG-PON2 adopts the Time- and Wavelength-Division Multiplexing PON (TWDM-PON) as the primal technology, in which 4–8 wavelength pairs are used to realize the 40-Gbit/s aggregated capacity or over. TDMA is used to share the bandwidth in each upstream wavelength among a number of ONUs. NG-PON2 also adopts Point-to-Point WDM PON (PtP WDM-PON) as the secondary technology. The evolution of PON technologies will continue to support the growth of the broadband access services in the long term.

1.4.3 High-Capacity Optical Transmission Technology

Optical transmission technologies have advanced rapidly over the past thirty years in three main technological innovations (Fig. 1.18): time division multiplexing

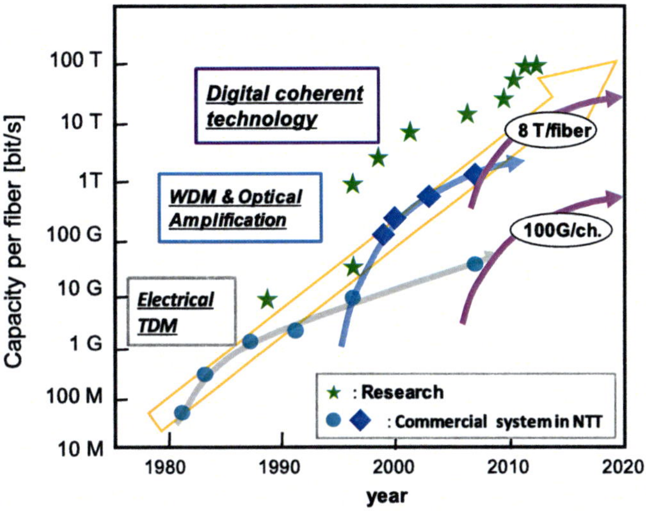

Fig. 1.18 Increase in optical fiber transmission capacity

(TDM) technology based on electrical multiplexing, optical amplification technology/wavelength division multiplexing (WDM) technology, and digital coherent technology. Progress in R&D in the 1980s resulted in a high transmission speed of 10 Gbit/s, and since the 1990s, systems that apply WDM technology to attain terabit-per-second class transmission speeds have been developed [68]. Since 2000, transmission experiments in which capacity up to 100 Tbit/s in the research stage have been reported [68–77].

Ethernet speeds have also been increasing since the 1990s, and 10 GbE has become practical. There has also been development of 100G–class transmission technology, and the appearance of 100G Ethernet as a new client signal has led to continuing discussion of expansion of the OTN. We can expect that Ethernet will continue to drive increases in optical fiber transmission capacity.

The problems associated with such technology include limits on the power input to optical fiber, OSNR, chromatic dispersion and polarization mode dispersion, band restrictions, and nonlinear optical effects. To overcome such problems, research and development on optical amplifier technology, modulation and demodulation technology and ultrafast device technology is important.

The relation between transmission capacity and average optical fiber input power is shown in Fig. 1.19, where the distance per span is 100 km and the signal-to-noise ratio is the same. Increasing the optical power by about 1 W increases the likeliness of damage to the optical fiber or the optical components. One way to avert that problem is to use a Raman amplifier or other distributed amplifier. The optical power of the input signal to the transmission fiber can be reduced by using a hybrid amplifier that includes both an EDFA lumped optical amplifier and a Raman distributed amplifier. The hybrid amplifier prevents damage to the optical fiber and

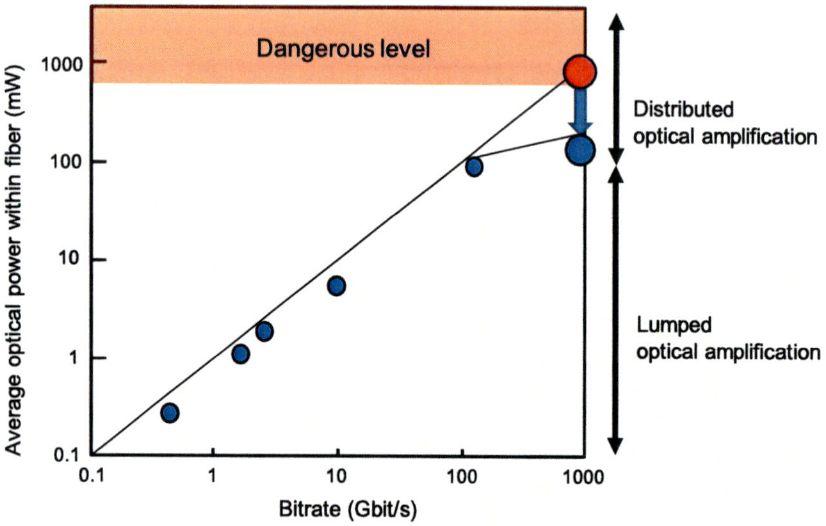

Fig. 1.19 Relationship of transmission capacity and mean optical fiber input power

optical components and reduces the nonlinear optical effect in the transmission path. Studies on increasing bandwidth are also underway, and broadband amplification of 10 THz has been achieved by hybrid amplification using a Raman distributed amplifier and an extended L–band EDFA. That is an increase by a factor of 2.5 compared to the conventional method [75, 77].

Also, incoming signal sensitivity and robustness against wavelength and polarization mode dispersion has increased in recent years, so research and development on phase modulation and demodulation technology and multi-level modulation and demodulation technology has been progressing. RZ–DQPSK modulation, which can increase spectral efficiency by a factor of 2 compared to the conventional OOK scheme, was introduced. Recently, there has been vigorous research on next-generation modulation schemes using polarization-division multiplexing to further improve spectral efficiency.

Digital coherent technology has been attracting attention in recent years, and there are very active works on practical implementation of the research [69–71]. The technology applies ultra-fast digital signal processing achieved through progress in ultra-fast CMOS LSI circuits for highly sensitive coherent detection [70]. Digital coherent technology involves coherent detection using digital signal processing as is done in wireless systems rather than the conventional intensity modulation direct detection, thus enabling coherent detection without physical frequency or phase synchronization of the signal light and the local oscillator light. In addition to greatly increasing receiver sensitivity and spectral efficiency, previously difficult optical fiber dispersion compensation in long-distance transmission can be greatly improved. Digital coherent optical transmission system (Fig. 1.20) is the main method being used in the high-capacity optical transport system for the current core network of

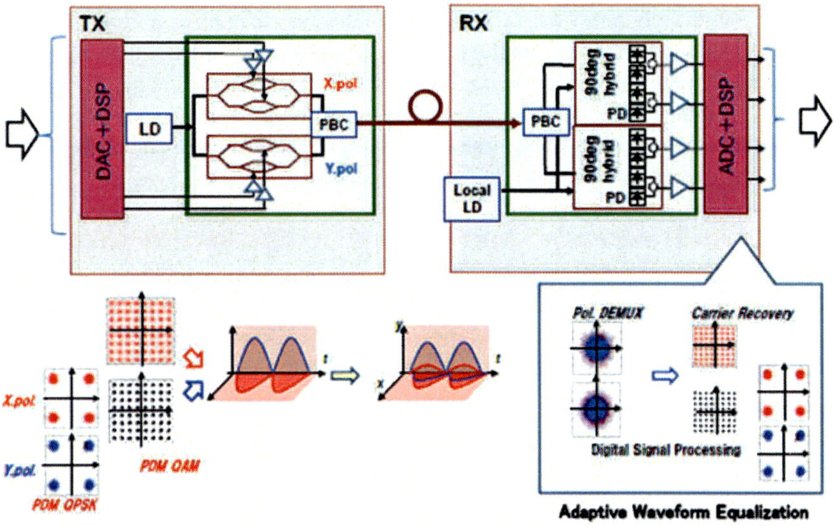

Fig. 1.20 Configuration for digital coherent optical transmission

2020. High-capacity transmission at the 100 Gbit/s per wavelength level has been achieved by using PDM-QPSK to transmit optical signals in optical fibers that have different polarization axes. A practical 10 Tbit/s-class high-capacity optical network using existing optical fiber and 50 GHz channel spacing in the same way as with the conventional wavelength division multiplexing system is possible. With practical digital coherent technology, the main bit rate per wavelength in the core network has shifted from the range between 10 and 40 Gbit/s to 100 Gbit/s. By 2020, continuing reduction of the size and cost of optical transmitter/receiver and miniaturization of digital coherent DSP have lowered power consumption and cost, and 100 Gbit/s per wavelength technology have spread rapidly in the metropolitan network and access network as well.

Conclusion

The total number of broadband subscribers in Japan has increased to over 40 million, and FTTH has recently taken the lead over ADSL. Internet traffic has been increasing by a factor of from 1.3 to 1.4 each year. The evolution of PON technologies has continued to support the growth of the broadband access services. Optical transmission technologies also have advanced rapidly over the past thirty years in three main technological innovations: time division multiplexing (TDM) technology based on electrical multiplexing, optical amplification technology/wavelength division multiplexing (WDM) technology and digital coherent technology. With practical digital coherent technology, the main bit rate per wavelength in the core network have shifted from the range between 10 and 40 Gbit/s to 100 Gbit/s and 400 Gbit/s, and a 19-Tbit/s high-capacity optical network has been achieved. By 2020, 100 Gbit/s per wavelength technology has rapidly come into wide use in metropolitan networks and access networks as well. Implementation of a photonic network that is capable of broadband transport of diverse, huge-volume client signals, including 100 Gigabit Ethernet, is indispensable for the future broadband society. That requires progress in optical access technology and large-capacity optical transmission technology, and future development of these technologies is expected.

1.5 State-of-the-Art Optical Submarine Cable Systems

1.5.1 Main Features of Optical Submarine Cable Systems

Today, optical submarine cable systems connect many countries in the world as shown in Fig. 1.21, and as of 2020, 1.2 million kilometers of optical submarine cable have been installed [78]. The role of satellite systems was taken over by submarine cable systems many years ago, and the majority of international telecommunications is carried on optical submarine cable systems currently. Since the continuously growing of the demand for international telecommunications is foreseen, the role of optical submarine cable systems is becoming larger.

A transmission capacity of optical submarine cable systems has drastically increased since 1980s to satisfy huge demands for international telecommunication services. Table 1.1 and Fig. 1.22 show main parameters and the capacity expansion of major transpacific optical submarine cable systems. In transpacific submarine

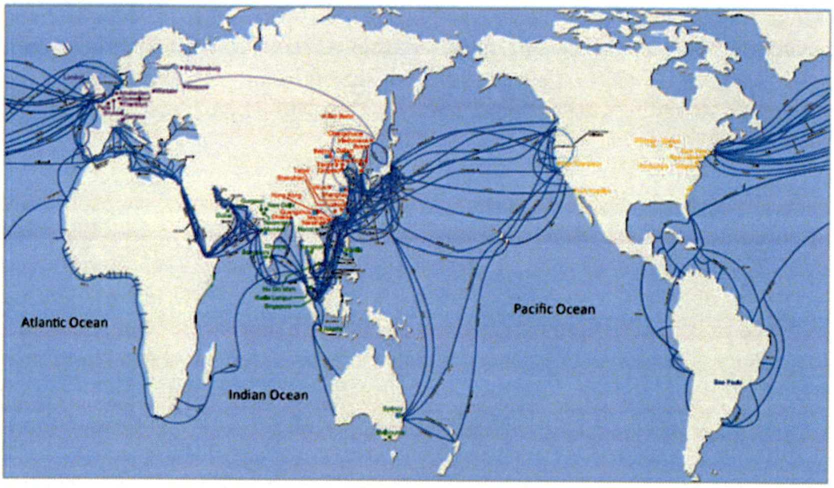

Fig. 1.21 Global network with optical submarine cable systems

Table 1.1 Main parameters of transpacific cable systems

	Ready for service year	Designed fiber capacity (bit/s)	No. of fiber pair	Designed cable capacity (bit/s)	Modulation format	Fiber type	Repeater type
TPC-3	1989	280 M (280 M × 1)	2	560 M	NRZ-OOK	SMF	Digital regenerator
TPC-4	1992	560 M (560 M × 1)	2	1.12G	NRZ-OOK	CSF	Digital regenerator
TPC-5	1996	5G (5G × 1)	2	10G	NRZ-OOK	DSF	Optical amplifier
China-US	1999	20G (2.5G × 8)	4	80G	NRZ-OOK	NZDSF	Optical amplifier
PC1	2000	160G (10G × 16)	4	640G	CRZ-OOK	NZDSF	Optical amplifier
Japan-US	2001	160G (10G × 16)	4	640G	CRZ-OOK	NZDSF	Optical amplifier
VNSL-P (TGN-P)	2002	960G (10G × 96)	8	5.12 T	CRZ-OOK	DMF (+D/−D)	Optical amplifier
Unity	2010	960G (10G × 96)	5	4.8 T	RZ-DPSK	DMF (+D/−D)	Optical amplifier
Faster	2016	10 T (100G × 100)	6	60 T	DP-QPSK	+ D	Optical amplifier

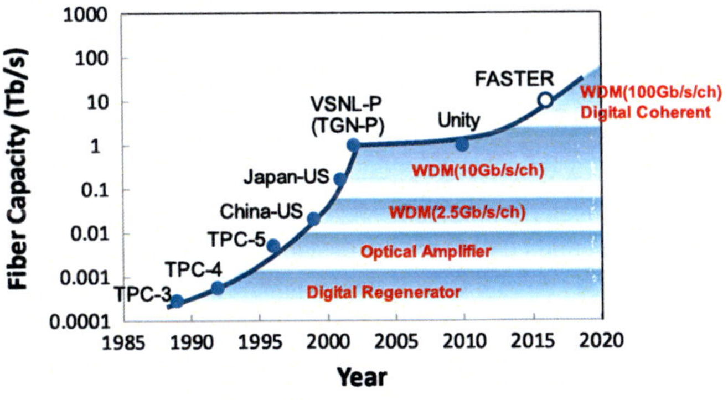

Fig. 1.22 Progress of fiber capacity in transpacific submarine cable systems

cable systems, fiber optic transmission technology was firstly introduced in TPC-3, which was placed into service in 1989. In this system, the digital regenerators were used. In the digital regenerator, the incoming optical signals were converted to electrical signals and then the converted signals were reshaped, retimed and regenerated with high-speed electrical circuits and retransmitted with a local laser. Then, optical amplification technology was introduced in TPC-5CN, which was placed into service in 1996. In China-US cable system, which was placed into service in 1999, wavelength-division-multiplexing (WDM) technology was introduced, and the introduction of WDM accelerated the capacity expansion in optical submarine cable systems. In the earlier generation WDM systems, non-return-to-zero on–off keying (NRZ-OOK) signal with a channel bit rate of 2.5 Gbit/s was used, since it is easy to generate and detect. With introducing chirped return-to-zero on–off keying (CRZ-OOK) signal which is more tolerant against fiber nonlinear effects, the channel bit rate was increased to 10 Gbit/s.

Although the initial designed capacity of transpacific cable shown in Table 1.1 and Fig. 1.22 had not increased drastically from 2003 to 2010, the actual fiber capacity had been upgraded by replacing the transmitter and receiver for higher bit rate by using advanced modulation formats, such as differential phase-shift-keying (DPSK). For example, in 2008, the cable capacity of Japan-US cable was upgraded up to 1.28 Tbit/s, doubling the initial design capacity of 640 Gbit/s.

The submarine cable systems are different from the terrestrial system in many aspects. The main differences are as follows.

- High reliability: system design life over 25 years
- High water pressure resistance: 800 atm
- High tension during laying: 6 t
- Resistance to high voltage: ± 12 kV
- Fewer fiber pairs in cable: less than 10 pairs
- Large accumulation of optical properties: large accumulation of nonlinear effects, chromatic dispersion and noise generated in optical amplifiers in over 6000 km transmission in transoceanic cable systems.

Since the maximum water depth in transpacific cable systems is around 8000 m, submarine cables and repeaters must have high tolerance against water pressure and tension. When a submarine repeater breaks down, it takes a lot of time to repair or exchange it. Thus, each submarine repeater must have very high reliability to minimize the unavailability of the system. In addition, the number of fiber pairs in submarine cable systems is limited because the provided power-by-power feeding equipment (PFE) at landing points and the size of repeater housing have some limitation. Therefore, it is very important to increase the transmission capacity per fiber.

1.5.2 *Main Building Blocks of Optical Submarine Cable Systems*

Figure 1.23 shows main components of optical submarine cable systems. They mainly consist of terminal equipment, PFE, optical fiber cable and optical repeaters.

1.5.2.1 Terminal Equipment and PFE

Terminal equipment is installed at the cable station and sends and receives optical signals and supervises the system. The basic functions of transmitter and receiver for submarine cable systems are identical as those for terrestrial systems. In many submarine cable systems, however, the different modulation formats which are more tolerant to fiber nonlinear effects have been utilized. With the increase of the channel bit rate, the modulation format was changed from intensity modulated signals, NRZ-OOK and CRZ-OOK, to phase-modulated signals, RZ-DPSK and quaternary phase-shift keying (QPSK).

In the latest systems with digital coherent transponders, polarization-division multiplexing (PDM) which can double the spectral efficiency is also introduced.

PFE is installed at the cable station to supply a precisely controlled constant direct current of around 1 A to optical repeaters through the optical cable conductor together with sea ground. PFE generates very high voltage of around 10 kV to drive a few hundreds of repeaters in a transoceanic system.

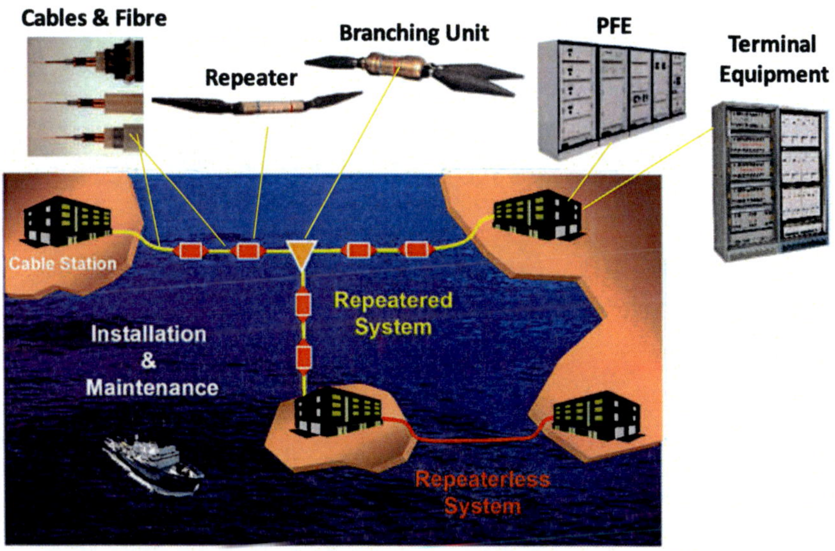

Fig. 1.23 Optical submarine cable system

1.5.2.2 Optical Fiber Cable and Dispersion Management

Optical fiber cables have contained less than ten fiber pairs since their first introduction to submarine cable systems. Recently, the number of fiber pairs is increased to more than 20 in order to increase the cable capacity with the limited supplied power by PFE. The different cable structures are adopted with water depth. The optical fiber cable for deep sea has light structure designed to protect from water pressure. Figure 1.24a shows an example of such cable, which is called lightweight (LW) cable. The optical fiber cable for shallow sea is heavily protected against the cable cut by ships. Figure 1.24b shows the example, which is called double armored (DA) cable. The optical fiber cables mainly consist of an optical fiber unit and power feeding copper conductors which supply electric power to optical repeaters.

In submarine cable systems, new fiber is usually installed when a new system is constructed, which is different from the terrestrial systems where the systems are optimized on the already deployed fiber. Therefore, the optical fiber and dispersion map optimized for each system are utilized in optical submarine cable systems.

For high-speed optical signals, small chromatic dispersion is required to keep a waveform. With the small dispersion, however, signals and noise propagate with a same group velocity, resulting in a long interaction length with a phase-matched condition and a degradation of signal performance even in single-channel transmission through four-wave mixing (FWM) between the signal and noise. For WDM signals, nonlinear effects such as FWM and cross-phase modulation (XPM) between WDM signals become the dominant impairment. To prevent this, a dispersion management was introduced. In the dispersion management, the local dispersion is set to be large to reduce the phase matching while the system average dispersion is close to zero by inserting dispersion compensation fiber. In the first-generation WDM submarine cable systems with a channel bit rate of 2.5 Gbit/s, a fiber with a

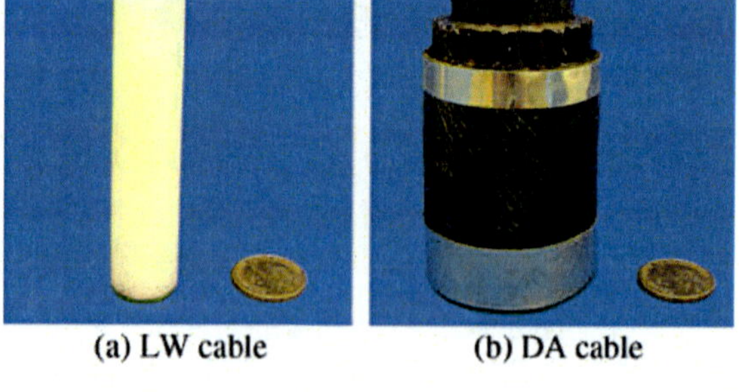

Fig. 1.24 Optical fiber cable for **a** deep sea and **b** shallow sea

large dispersion around 2 ps/nm/km was introduced to reduce inter-channel nonlinear effects. For further reduction of nonlinear effects, new fibers with larger effective area (A_{eff}) around 80 μm^2 or more have been developed for WDM systems with a channel bit rate of 10 Gbit/s. In order to compensate for the accumulated dispersion of the large dispersion fiber, the fiber with opposite-sign chromatic dispersion was inserted periodically as shown in Fig. 1.25a. In this dispersion map, only the second-order dispersion is compensated and the third-order dispersion is not compensated. Therefore, a large amount of dispersion is accumulated at the channels located far away from the system zero dispersion wavelength after long distance transmission. This large accumulated dispersion causes large pulse broadening and overlap with many neighboring pulses. Since the impact of such pulse broadening is more significant in

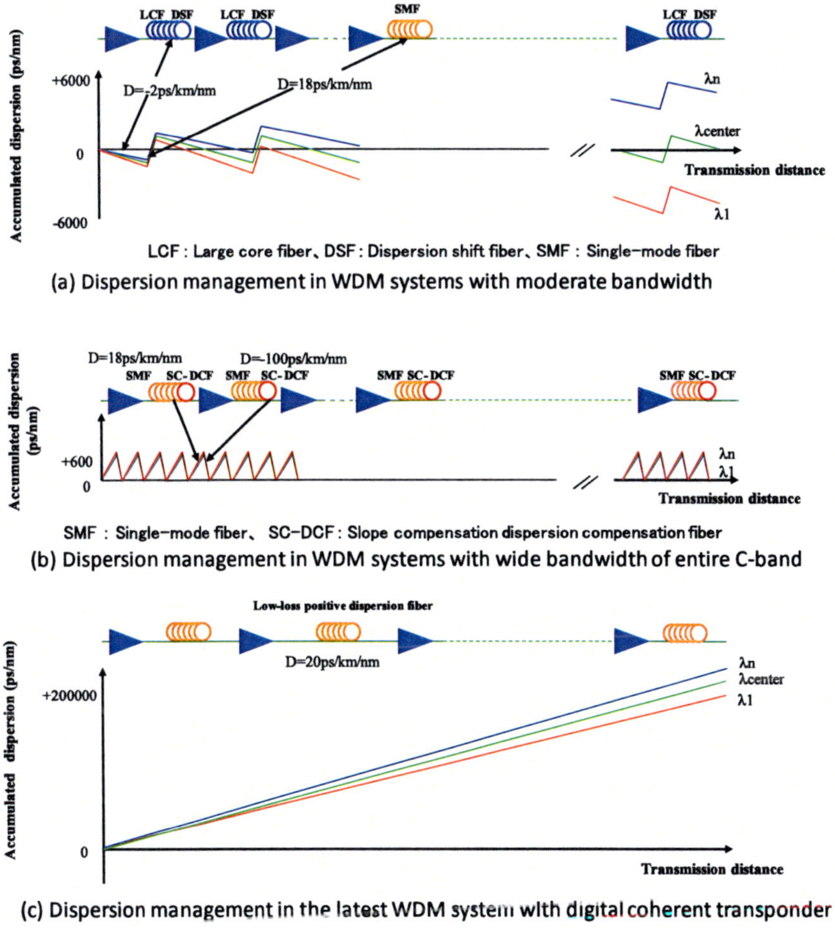

Fig. 1.25 Dispersion management in optical submarine cable systems

WDM systems with higher channel bit rate and wider bandwidth, another dispersion map shown in Fig. 1.25b was introduced in WDM systems with around 1-Tbit/s fiber capacity where the bandwidth was expanded to full C-band around 30 nm. In this dispersion map, each fiber span was constructed with two types of fiber with opposite-sign chromatic dispersion to reduce the span average values of both second- and third-order dispersion. With this configuration, chromatic dispersion characteristic was flattened in around 30 nm bandwidth.

In the latest systems with digital coherent transponders, however, the dispersion management is not utilized. In such systems, the fiber launch power is drastically reduced by using coherent receiver with higher sensitivity and forward error correction (FEC) codes with higher gain. This power reduction together with the use of optical fiber with large dispersion and larger A_{eff} makes the system more linear. Thus, the large amount of accumulated dispersion in the transmission over transoceanic distance can be compensated only at the terminal as shown in Fig. 1.24c and the transmission line is constructed with a single type of fiber, which contributes to reduce the loss of transmission line.

1.5.2.3 Optical Repeater

In the first-generation optical submarine cable systems, digital regenerators, which converted the incoming optical signals to electrical signals, then reshaped, retimed and regenerated the data with high-speed electrical circuits and retransmitted the data with a local laser, was used as repeaters. The regenerator had limitations of its operation speed in electrical circuits, and erbium-doped fiber amplifier (EDFA) was introduced to solve such problems. EDFA has many advantages rather than it can amplify high-speed signals without distortion. One of the advantages of EDFA is its inherent automatic gain control function. Figure 1.26a shows the gain characteristics of EDFA as function of the input power with a constant pump power. When the EDFA is operated in the saturation regime under the nominal operating condition, the EDFA

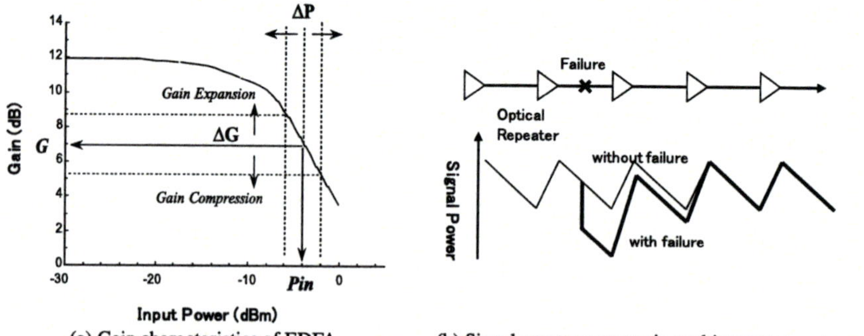

(a) Gain characteristics of EDFA (b) Signal power recovery in multi-repeater systems

Fig. 1.26 Automatic gain control in EDFA repeaters

gain increases by nature and the output power from the EDFA remains constant in the event of the reduction of input power. Figure 1.26b shows the signal power recovery in multi-repeater systems when the loss increases in a span. Owing to the nature of gain-saturated EDFA, the signal power can return to the nominal level after a few optical repeaters with a slight degradation optical signal-to-noise ratio (OSNR). Another advantage of EDFA is that it can amplify multiple signals simultaneously without crosstalk, which enables to introduce WDM in fiber optic communication systems. In submarine cable systems, the repeaters are usually inserted with around 50–100 km spacing, which means more than 100 repeaters are concatenated in transpacific cable systems. Therefore, it is very important to reduce the noise generated in each EDFA as lower as possible. In WDM systems, with wide-bandwidth, the gain equalization is also indispensable to obtain sufficient signal performance with all WDM channels. Usually, the gain equalization is realized with passive gain equalizing filters which are designed to the inverse characteristics of EDFA gain and fiber span loss.

1.5.3 The State-of-the-Art Technologies in Optical Submarine Cable Systems

The transpacific submarine cable system, FASTER, was ready for service in 2016. Table 1.2 summarizes main parameters of FASTER cable system. In FASTER system, the designed system capacity reaches to 60 Tbit/s (10 Tbit/s × 6 fiber pairs) with the state-of-the-art technologies shown in Fig. 1.27.

Table 1.2 FASTER cable system

Partners	KDDI (Japan), China mobile international (China), Chana telecom global (China), Google (USA), Sigtel (Singapore), Global transit, (Malaysia)
System length	about 9000 km
Designed fiber capacity	10 Tbit/s(100 Gbit/s × 100 WDM channels)
No. of fiber pair	6 fiber pairs
Designed cable capacity	60 Tbit/s
Ready for service year	2016

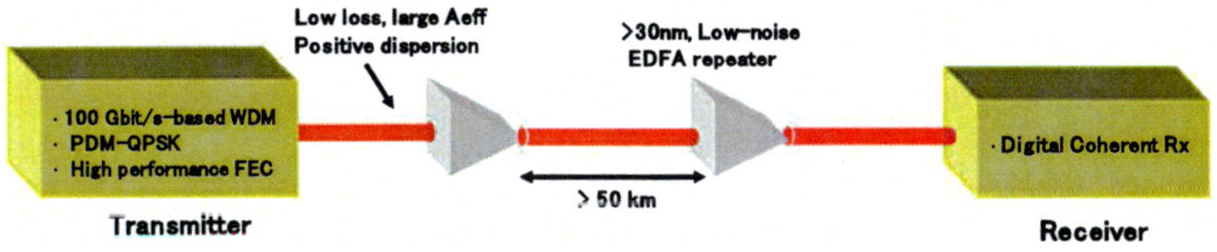

Fig. 1.27 Technologies used in the latest transoceanic submarine cable systems

10-Tbit/s fiber capacity is realized with 100 × 100 Gbit/s WDM transmission, which is enabled with advanced digital signal processing technologies, i.e., digital coherent reception and high-performance FEC. With the latest FEC technology, a net coding gain of higher than 10 dB is achieved with soft-decision FEC. With PDM-QPSK signals, the spectral efficiency reaches to around 2 bit/s/Hz. These technologies used in terminal equipment are similar with those used in the terrestrial systems with channel bit rate of 100 Gbit/s.

The transmission line consists of optical fiber spans whose loss is compensated for by C-band EDFA repeaters with gain bandwidth of larger than 30 nm. The transmission line is constructed with a single type of optical fiber, and the accumulated dispersion is compensated at the receiver in electrical domain. To reduce the nonlinear effects in fiber propagation, the optical fiber with large dispersion around 20 ps/nm/km and large A_{eff} larger than 100 μm^2 is employed. The transmission line with single type of optical fiber has smaller loss, which contribute to obtain a sufficient OSNR after transmission with lower fiber launch power.

The current demand for transpacific traffic can be covered with above-mentioned technologies. In order to expand the fiber capacity furthermore to meet the future explosive traffic demands, the research on disruptive technologies is expected.

References

1. P.P. Mitra, J.B. Stark, Nonlinear limits to the information capacity of optical fibre communications. Nature **411** (6841), 1027–1030 (2001)
2. J.M. Kahn, K.-P. Ho, A bottleneck for optical fibres. Nature **411**(6841), 1007–1009 (2001)
3. E.B. Desurvire, Capacity demand and technology challenges for lightwave systems in the next two decades. J. Lightwave Technol. **24**(12), 4697–4710 (2006)
4. T. Morioka, New generation optical infrastructure technologies: EXAT initiative: towards 2020 and beyond, in *OECC 2009*, FT4 (2009)
5. A.R. Chraplyvy, The coming capacity crunch, in *ECOC 2009* (Plenary Session, 2009)
6. R.-J. Essiambre, G. Kramer, P.J. Winzer, G.J. Foschini, B. Goebel, Capacity limits of optical fiber networks. J. Lightwave Technol. **28**(4), 662–701 (2010)
7. M. Nakazawa et al., Special feature: basic R&D into fiber optic networks in Japan—NICT EXAT study group. New Breeze, the ITU association of Japan, Vol. 22(1), pp 3–15, January; No. 2, pp 1–10 (2010)
8. A.D. Ellis, J. Zhao, D. Cotter, Approaching the non-linear Shannon limit. J. Lightwave Technol. **28**(4), 423–433 (2010)
9. M. Nakazawa, Giant leaps in optical communication technologies towards 2030 and beyond, in *European Conference on Optical Communication (ECOC 2010)* (Plenary Talk, 2010)
10. P.J. Winzer, Energy-efficient optical transport capacity scaling through spatial multiplexing. IEEE Photon. Technol. Lett. **23**(13), 851–853 (2011)
11. M. Nakazawa, Extremely advanced transmission with 3M technologies (multi-level modulation, multi-core and multi-mode), in *OFC/NFOEC 2012*, OTu1D.1 (2012)
12. T. Morioka, Y. Awaji, R. Ryf, P. Winzer, D. Richardson, F. Poletti, Enhancing optical communications with brand new fibers. IEEE Commun. Mag. **50**(2), s31–s42 (2012)
13. D.J. Richardson, J.M. Fini, L.E. Nelson, Space-division multiplexing in optical fibres. Nature Photonics **7**(5), 354–362 (2013)
14. T. Morioka, Recent progress in space-division multiplexed transmission technologies, in *(invited) OFC 2013*, OW4F.2 (2013)

15. IEC Technical Report IEC 61292-4, Optical amplifiers-part 4: maximum permissible optical power for the damage-free and safe use of optical amplifiers, including Raman amplifiers (2004)
16. ITU-T Recommendation G.664, Optical safety procedures and requirements for optical transport systems (2006)
17. T. Morioka, Y. Awaji, Y. Matsushima, T. Kamiya, R&D of 3M technologies towards the realization of exabit/s optical communications. IEICE Trans. Commun. **E100.B**(9), 1707–1715 (2017)
18. M. Koshiba, K. Saitoh, Y. Kokubun, Heterogeneous multi-core fibers: proposal and design principle. IEICE Electron. Express **6**(2), 98–103 (2009)
19. K. Imamura, K. Mukasa, R. Sugizaki, Y. Mimura, T. Yagi, Multi-core holey fibers for ultra large capacity wide-band transmission, in *ECOC 2008*, p. 1.17 (2008)
20. K. Imamura, K. Mukasa, Y. Mimura, T. Yagi, Multi-core holey fibers for the long-distance (> 100 km) ultra large capacity transmission, in *OFC 2009*, OTuC3 (2009)
21. Y. Kokubun, M. Koshiba, Novel multi-core fibers for mode division multiplexing: proposal and design principle. IEICE Electron. Express **6**(8), 522–528 (2009)
22. S. Inao, T. Sato, S. Sentsui, T. Kuroha, Y. Nishimura, Multicore optical fiber, in *OFC 1979*, WB1 (1979)
23. N. Kashima, E. Maekawa, F. Nihei, New type of multicore fiber, in *OFC 1982*, ThAA5 (1982)
24. S. Berdagué, P. Facq, Mode division multiplexing in optical fibers. Appl. Opt. **21**(11), 1950–1955 (1982)
25. H. Takara, A. Sano, T. Kobayashi, H. Kubota, H. Kawakami, A. Matsuura, Y. Miyamoto, Y. Abe, H. Ono, K. Shikama, Y. Goto, K. Tsujikawa, Y. Sasaki, I. Ishida, K. Takenaga, S. Matsuo, K. Saitoh, M. Koshiba, T. Morioka, 1.01-Pb/s (12 SDM/222 WDM/456 Gb/s) crosstalk-managed transmission with 91.4-b/s/Hz aggregate spectral efficiency, in *ECOC 2012*, Th.3.C.1 (2012)
26. T. Kobayashi, M. Nakamura, F. Hamaoka, K. Shibahara, T. Mizuno, A. sano, H. Kawakami, A. Isoda, M. Nagatani, H. Yamazaki, Y. Miyamoto, Y. Amma, Y. Sasaki, K. Takenaga, K. Aikawa, K. Saitoh, Y. Jung, D. J. Richardson, K. Pulverer, M. Bohn, M. Nooruzzaman, T. Morioka, 1-Pb/s (32 SDM/46 WDM/768 Gb/s) C-band dense SDM transmission over 205.6-km of single-mode heterogeneous multi-core fiber using 96-Gbaud PDM-16QAM channels, in *OFC 2017*, Th5B.1 (2017)
27. K. Igarashi, T. Tsuritani, I. Morita, Y. Tsuchida, K. Maeda, M. Tadakuma, T. Saito, K. Watanabe, R. Sugizaki, M. Suzuki, 1.03-Exabit/s-km super-Nyquist-WDM transmission over 7,326-km seven-core fiber, in *ECOC 2013*, Paper PDP3.E.3 (2013)
28. T. Kobayashi, H. Takara, A. Sano, T. Mizuno, H. Kawakami, Y. Miyamoto, K. Hiraga, Y. Abe, H. Ono, M. Wada, Y. Sasaki, I. Ishida, K. Takenaga, S. Matsuo, K. Saitoh, M. Yamada, H. Masuda, T. Morioka, 2 × 344 Tb/s propagation-direction interleaved transmission over 1500-km MCF enhanced by multicarrier full electric-field digital back-propagation, in *European Conference and Exhibition on Optical Communication 2013*, p. PD3.E.4 (2013)
29. D. Soma, K. Igarashi, Y. Wakayama, K. Takeshima, Y. Kawaguchi, N. Yoshikane, T. Tsuritani, I. Morita, M. Suzuki., 2.05 Peta-bit/s super-nyquist-WDM SDM transmission using 9.8-km 6-mode 19-core fiber in full C band, in *ECOC 2015*, PDP.3.2 (2015)
30. B.J. Puttnam, R.S. Luís, W. Klaus, J. Sakaguchi, J.M.D. Mendinueta, Y. Awaji, N. Wada, Y. Tamura, T. Hayashi, M. Hirano, J.R. Marciante, 2.15 Pb/s transmission using a 22 core homogeneous single-mode multi-core fiber and wideband optical comb, in *ECOC 2015*, PDP.3.1 (2015)
31. D. Soma, Y. Wakayama, S. Beppu, S. Sumita, T. Tsuritani, T. Hayashi, T. Nagashima, M. Suzuki, H. Takahashi, K. Igarashi, I. Morita, M. Suzuki, 10.16 Peta-bit/s dense SDM/WDM transmission over low-DMD 6-mode 19-core fibre across C + L band, in *ECOC 2017*, Th.PDP.A.1 (2017)
32. D. Kong, E. Porto da Silva, Y. Sasaki, K. Aikawa, F. Da Ros, M. Galili, T. Morioka, L. Katsuo Oxenløwe, H. Hu, Kramers–Kronig detection with adaptive rates for 909.5 Tbit/s dense SDM and WDM data channels, in *ECOC 2018*, Th3F.5 (2018)

33. G. Rademacher, B.J. Puttnam, R.S. Luís, J. Sakaguchi, W. Klaus, T.A. Eriksson, Y. Awaji, T. Hayashi, T. Nagashima, T. Nakanishi, T. Taru, T. Takahata, T. Kobayashi, H. Furukawa, N. Wada, 10.66 Peta-Bit/s transmission over a 38-core-three-mode fiber, in *Optical Fiber Communication Conference (OFC) 2020*, OSA Technical Digest (Optical Society of America, 2020), paper Th3H.1 (2020)
34. R. Ryf, N.K. Fontaine, S. Wittek, K. Choutagunta, M. Mazur, H. Chen, J.C. Alvarado-Zacarias, R. Amezcua-Correa, M. Capuzzo, R. Kopf, A. Tate, H. Safar, C. Bolle, D.T Neilson, E. Burrows, K. Kim, M. Bigot-Astruc, F. Achten, P. Sillard, A. Amezcua-Correa, J.M Kahn, J. Schröder, J. Carpenter, High-spectral-efficiency mode-multiplexed transmission over graded-index multimode fiber, in *2018 European Conference on Optical Communication (ECOC)*, paper Th3B.1 (2018)
35. T. Mizuno, T. Kobayashi, H. Takara, A. Sano, H. Kawakami, T. Nakagawa, Y. Miyamoto, Y. Abe, T. Goh, M. Oguma, T. Sakamoto, Y. Sasaki, I. Ishida, K. Takenaga, S. Matsuo, K. Saitoh, T. Morioka, 12-core × 3-mode dense space division multiplexed transmission over 40 km employing multi-carrier signals with parallel MIMO equalization, in *OFC 2014*, Th5B.2 (2014)
36. UHDTV Standards: ITU-R Recommendation BT2020-1 (June 2014)
37. K. Masaoka, M. Emoto, M. Sugawara, Y. Nojiri, Contrast effects in evaluating the sense of presence of wide displays. SID **14**(9), 785–791 (2006)
38. HEVC Standards: H.265 (ISO/IEC 23008-2 HEVC) (January 2013)
39. Y. Takaki, Development of super multi-view displays. ITE Trans. MTA **2**(1), 8–14 (2014)
40. J. Arai, F. Okano, M. Kawakita, M. Okui, Y. Haino, M. Yoshimura, M. Furuya, M. Sato, Integral three-dimensional television using a 33-megapixel imaging system. J. Display Technol. **6**(10), 422–430 (2010)
41. T. Senoh, T. Mishina, K. Yamamoto, R. Oi, T. Kurita, Viewing-zone-angle-expanded color electronic holography system using ultra-high-definition liquid crystal displays with undesirable light elimination. J. Display Technol. **7**(7), 382–390 (2011)
42. S. Iwasawa et al., REI: an automultiscopic projection display, in *Three Dimensional Systems and Applications (3DSA)* (2013)
43. Common Public Radio Interface (CPRI); Interface specification, V.6.1 (2014)
44. R. Heron, Heterogeneous access fiber networks enabled by multi-wavelength PONs and virtualization, in *ECOC 2013*, Mo.3.F.1 (2013)
45. SmallCellForumDocument075.02.01, Synchronisation for LTE small cells (2013)
46. https://5g-ppp.eu
47. http://www.imt-2020.cn/en/introduction
48. http://5gmf.jp/en/
49. DOCOMO 5G white paper, https://www.nttdocomo.co.jp/english/binary/pdf/corporate/technology/whitepaper_5g/DOCOMO_5G_White_Paper.pdf
50. Y. Okumura, 5G mobile radio access system using SHF/EHF bands, in *2014 Asia-Pacific Microwave Conference* (2014)
51. 3GPP TR36.872, Small cell enhancements for E-UTRA and E-UTRAN—physical layer aspects (2013)
52. Y. Saito, Y. Kishiyama, A. Benjebbour, T. Nakamura, A. Li, K. Higuchi, Non-orthogonal multiple access (NOMA) for future radio access, in *IEEE VTC-Spring 2013* (2013)
53. S. Suyama, J. Shen, Y. Oda, H. Suzuki, K. Fukawa, 11 GHz band 8 × 16 MIMO-OFDM outdoor transmission experiment for 10 Gbps super high bit rate mobile communications, in *IEEE PIMRC* 2013 (2013)
54. T. Tashiro, S. Kuwano, J. Terada, T. Kawamura, N. Tanaka, S. Shigematsu, N. Yoshimoto, A novel DBA scheme for TDM-PON based mobile fronthaul, in *Optical Fiber Communication Conference, OSA Technical Digest (online)* (Optical Society of America, 2014), paper Tu3F.3
55. http://www.fsan.org/news/
56. ITU-T Recommendation G.989.1, 40-Gigabit-capable passive optical networks (NG-PON2): general requirements (2013)

57. B. Guo, W. Cao, A. Tao, D. Samardzija, LTE/LTE—a signal compression on the CPRI interface. Bell Labs Tech J. **18**(2), 117–133 (2013)
58. ORI Contribution ORI (13) M17009, Compression method for open radio interface (2013)
59. T. Pfeiffer, F. Schaich, Optical architectures for mobileback and fronthauling, in *Workshop of OFC2012* (2012)
60. T. Kubo, T. Asai, Y. Okumura, A study on optical bandwidth reduction for future radio access and optical network, in *IEICE Tech. Report,* CS2013-51 (2013) (In Japanese)
61. K. Miyamoto, S. Kuwano, J. Terada, A. Otaka, Uplink joint reception with LLR forwarding for optical transmission bandwidth reduction in mobile fronthaul, in *Proceedings VTC2015-Spring*, pp. 1–5 (2015)
62. https://www.soumu.go.jp/main_sosiki/joho_tsusin/eng/pressrelease/2020/7/31_3.html
63. IEEE Std 802.3ah-2004, Media access control parameters, physical layers, and management parameters for subscriber access networks (2004)
64. IEEE Std 802.3av-2009, Physical layer specifications and management parameters for 10 Gb/s passive optical networks (2009)
65. ITU-T Recommdandation G.984 series, Gigabit-capable passive optical network (2004)
66. ITU-T Recommdandation G.987 series, 10 Gigabit-capable passive optical network (2010)
67. ITU-T Recommdandation G.989 series, 40 Gigabit-capable passive optical network (2015)
68. K. Hagimoto, NTT Tech. Rev. **3**(6), 20–26 (2005)
69. S. Tsukamoto, D. Ly-Gagnon, K. Katoh, K. Kikuchi, Coherent demodulation of 40-Gbit/s polarization-multiplexed QPSK signals with 16-GHz spacing after 200-km transmission, in *OFC/NFOEC 2005*, paper PDP29 (2005)
70. E. Yamazaki, S. Yamanaka, Y. Kisaka, T. Nakagawa, K. Murata, E. Yoshida, T. Sakano, M. Tomizawa, Y. Miyamoto, S. Matsuoka, J. Matsui, A. Shibayama, J. Abe, Y. Nakamura, H. Noguchi, K. Fukuchi, H. Onaka, K. Fukumitsu, K. Komaki, O. Takeuchi, Y. Sakamoto, H. Nakashima, T. Mizuochi, K. Kubo, Y. Miyata, H. Nishimoto, S. Hirano, K. Onohara, Fast optical channel recovery in field demonstration of 100-Gbit/s Ethernet over OTN using real-time DSP. Opt. Express **19**, 13179–13184 (2011)
71. A. Sano, T. Kobayashi, S. Yamanaka, A. Matsuura, H. Kawakami, Y. Miyamoto, K. Ishihara, H. Masuda, 102.3-Tb/s (224 × 548-Gb/s) C and extended L-band all-Raman transmission over 240 km using PDM-64 QAM single carrier FDM with digital pilot tone, in *OFC/NFOEC 2012*, PDP5C.3 (2012)
72. K. Fukuchi, T. Kasamatsu, M. Morie, R. Ohhira, T. Ito, K. Sekiya, D. Ogasahara, T. Ono, 10.92-Tb/s (273 × 40-Gb/s) triple-band/ultra-dense WDM optical-repeatered transmission experiment, in *OFC 2001*, paper PD24 (2001)
73. Y. Frignac, G. Charlet, W. Idler, R. Dischler, P. Tran, S. Lanne, S. Borne, C. Martinelli, G. Veith, A. Jourdan, J. Hamaide, S. Bigo, Transmission of 256 wavelength-division and polarization- division-multiplexed channels at 42.7 Gb/s (10.2 Tb/s capacity) over 3×100 km of TeraLightTM fiber, in *OFC 2002*, paper FC5 (2002)
74. A.H. Gnauck, G. Charlet, P. Tran, P.J. Winzer, C.R. Doerr, J.C. Centanni, E.C. Burrows, T. Kawanishi, T. Sakamoto, K. Higuma, 25.6-Tb/s C+L-band transmission of polarization-multiplexed RZ-DQPSK signals, in *OFC/NFOEC 2007*, paper PDP19 (2007)
75. H. Masuda, A. Sano, T. Kobayashi, E. Yoshida, Y. Miyamoto, Y. Hibino, K. Hagimoto, T. Yamada, T. Furuta, H. Fukuyama, 20.4-Tb/s (204 × 111 Gb/s) Transmission over 240 km using bandwidth-maximized hybrid Raman/EDFAs, in *OFC/NFOEC 2007*, paper PDP20 (2007)
76. G. Charlet, J. Renaudier, H. Mardoyan, O.B. Pardo, F. Cerou, P. Tran, S. Bigo, 12.8 Tb/s transmission of 160 PDM-QPSK (160 × 2 × 40 Gbit/s) channels with coherent detection over 2550 km, in *ECOC 2007*, paper PD1.6 (2007)
77. H. Masuda, Review of wideband hybrid amplifiers, in *OFC 2000*, paper TuA1 (2000)
78. https://www.submarinecablemap.com/

Chapter 2
Optical Fibers for Space-Division Multiplexing

Masaharu Ohashi, Shoichiro Matsuo, Tetsuya Hayashi, Katsunori Imamura, Yasuo Kokubun, Masanori Koshiba, Takayoshi Mori, Kazuhide Nakajima, Masataka Nakazawa, Kunimasa Saitoh, Taiji Sakamoto, and Takashi Sasaki

Abstract This chapter describes the design, the transmission characteristics, and the measurements technology of multi-core fibers (MCFs), few-mode fibers (FMFs), and few-mode multi-core fibers (FM-MCFs). Moreover, the cabling technology and future perspectives of innovative optical fiber cable technologies are presented.

2.1 Introduction

With the development of various Internet services, the data traffic in the network is rapidly increasing at an annual rate of 20–40% globally. According to the data traffic

---

Masaharu Ohashi and Shoichiro Matsuo are chapter editors.

M. Ohashi
Osaka Metropolitan University, Sakai, Osaka, Japan

S. Matsuo (✉)
Fujikura Ltd, Sakura, Chiba, Japan
e-mail: shoichiro.matsuo@jp.fujikura.com

T. Hayashi
Sumitomo Electric Industries, Ltd, Yokohama, Kanagawa, Japan

K. Imamura
Furukawa Electric Co., Ltd, Chiyoda, Tokyo, Japan

Y. Kokubun
Institute of Technologists (Formerly with Yokohama National University), Gyoda, Saitama, Japan

M. Koshiba · K. Saitoh
Hokkaido University, Sapporo, Hokkaido, Japan

T. Mori · K. Nakajima · T. Sakamoto
NTT Access Network Service Systems Labs, NTT Corporation, Tsukuba, Ibaraki, Japan

M. Nakazawa
Research Institute of Electrical Communication (RIEC), Tohoku University, Sendai, Miyagi, Japan

T. Sasaki
Innovation Core SEI, Inc, San Jose, CA, USA

M. Nakazawa et al. (eds.), *Space-Division Multiplexing in Optical Communication Systems*, Springer Series in Optical Sciences 236,
https://doi.org/10.1007/978-3-030-87619-7_2

increase, the amount of light power launched into the fiber is also rapidly increasing. However, it is recognized that the capacity of the conventional single-mode fiber is approaching its physical limit of about 100 Tbit/s due to the optical nonlinear effects and fiber fuse phenomenon. Therefore, innovative optical fiber cable technologies will be required for exceeding the transmission capacity limit of fibers in the next decades.

To develop a breakthrough technology, which enables this giant leap, an industry–government–academia research group called EXAT (EXtremely Advanced Transmission) was established in Japan in January 2008. We started to investigate the innovative optical communication infrastructure technologies required to address the capacity limitation of the present system. As the core technology of the optical communication infrastructure of the next generation, EXAT initiative proposed that the research and development of 3M technologies as stated in Chap. 1 is clearly necessary. Regarding the 3M technologies, the three major technologies of multi-core fiber (MCF), multi-mode control, and multi-level modulation were clearly identified; in addition, the relevance of 3M technologies to the transmission technologies as well as the processing and node architecture of 3M technologies was discussed. Accordingly, we have intensively investigated innovative optical fiber cable technologies, as one of the main themes since the beginning of EXAT.

In this chapter, our achievements with regard to 3M technologies and the investigation results are described. The chapter begins in Sect. 2.2 with the development goals of optical fibers and describes the relation among effective area A_{eff}, optical loss, and transmission capacity in addition to the recent low-loss optical fiber property. In Sect. 2.3, the crosstalk between cores in the MCF is derived by means of coupled-mode theory and coupled-power theory and the effect of the bending on the crosstalk is described for the various MCFs, such as homogeneous core MCF, heterogeneous core MCF, and ring MCF. A suitable MCF design regarding the effective area A_{eff} and core number is also described from the viewpoint of optical signal-to-noise ratio. Moreover, high-density MCFs, the reliability of MCF, and fabrication technology and measurement technology of MCF are clarified. In Sect. 2.4, the trend of few-mode fiber (FMF) design is described. The design of FMFs, such as low-differential modal delay fibers, coupled MCFs, and photonic band gap fibers, is presented. The design from two-mode fiber to multi-mode fiber is also described, taking into consideration the multiple-input and multiple-output (MIMO) processing. The mode coupling in the FMF and measurement technology for FMFs are also described. In Sect. 2.5, FM-MCFs are described for expanding the capacity, considering the advantages of core and mode numbers, and in Sect. 2.6, cabling technology is described. In particular, the prospects for the application of MCF cables and high-density MCF cables are given. Finally, in Sect. 2.7, future perspectives on innovative optical fiber cable technology are presented. Optical fiber standardization is also discussed.

2.2 Recent Progress of Single-Core Fiber Characteristics, Loss and A_{eff}

The single-core fibers are classified by their structures to standard solid fibers, photonic crystal fibers (PCFs), and photonic bandgap fibers (PBGFs). Since 1970s, as shown in Fig. 2.1, their loss has gradually improved thanks to the development of manufacturing techniques. Obviously, the loss of the optical fiber is the most critical characteristic for supporting the actual long-haul optical transmission system deployment. Also, to suppress the nonlinear effects, expanding effective area (A_{eff}) is of importance.

In this section, the history of loss improvement of several types of fibers is reviewed, and the state-of-the-art low-loss and large Aeff pure-silica-core fibers (PSCFs), and its figure-of-merit are reviewed.

2.2.1 *Progress of Loss Improvement for Various Types of Single-Core Fiber*

Firstly, the all-solid optical fiber was realized by partial doping to the core or cladding and was largely deployed everywhere in terrestrial and submarine communication systems. The state-of-art lowest loss fiber ever reported has the minimum loss of 0.1419 dB/km at 1560 nm [1], enabled by Ge-free silica core with low Rayleigh scattering. The loss improvement and expanding A_{eff} make sense for the system performance, to be described in Sect. 2.2.3, not only for the classical on off keying (OOK) modulation format but also for the phase shift keying (PSK) modulation format by using digital coherent detection.

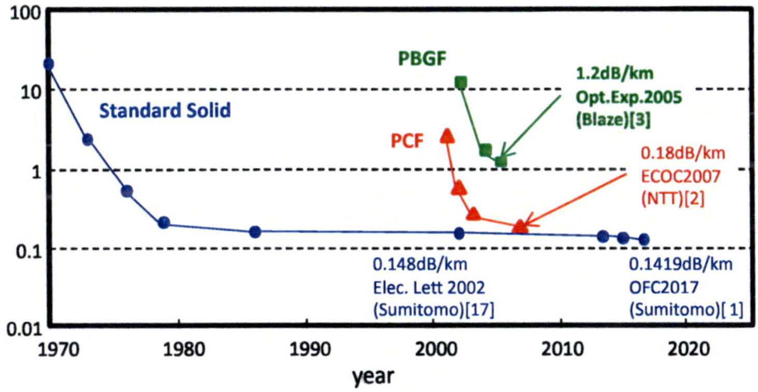

Fig. 2.1 Historical minimum loss improvement for various types of fibers

Next, as for the PCF type, low loss of 0.18 dB/km, which is comparable to standard Ge-doped core single-mode fibers, was demonstrated in 2007 [2], and further loss improvement is a challenge since the core of the PCF is composed of silica glass, the same as that of a standard fiber.

Finally, loss of PBGF was demonstrated to be as low as 1.2 dB/km in 2005 [3], a bit higher than that of previous two types of fibers. It shall be noted that the hollow-core structure does propagate the light in the air core region and is less sensitive to the silica glass absorption, which is always anticipated to work in a low-loss medium. By designing a proper bandgap structure, expanding the manufacturability of hollow-core fiber is also reported [4]. Recently, 10–100× lower power fraction in the glass than that in PBGFs is realized using antiresonant hollow-core fiber (ARF) structure [5]. In addition to the potential for low transmission losses, inherent low latency, low nonlinearity, capability of new wavelength windows would make the hollow-core fibers based on PBGFs with ARF to be important and promising.

2.2.2 *Low-Loss Pure-Silica-Core Fiber (PSCF)*

In this section, low-loss PSCFs to support long-haul, high-capacity systems are reviewed. In order to reduce fiber loss, it is essential to reduce the Rayleigh scattering loss that dominates about 80% of fiber loss at 1550 nm. The Rayleigh scattering results from microscopic non-uniformity of refractive index, due to dopant concentration and glass-density fluctuation. Therefore, the use of pure-silica glass with no dopant as the core material eliminates the dopant concentration fluctuation compared to the Ge-doped standard single-mode fiber (SSMF). In addition, by suppressing the glass-density fluctuation, PSCF with a loss less than 0.149 dB/km at 1550 nm was reported [6], while the loss of SSMF is less than 0.20 dB/km. Its loss spectrum is shown in Fig. 2.2, along with that of a SSMF. The fiber characteristics are summa-

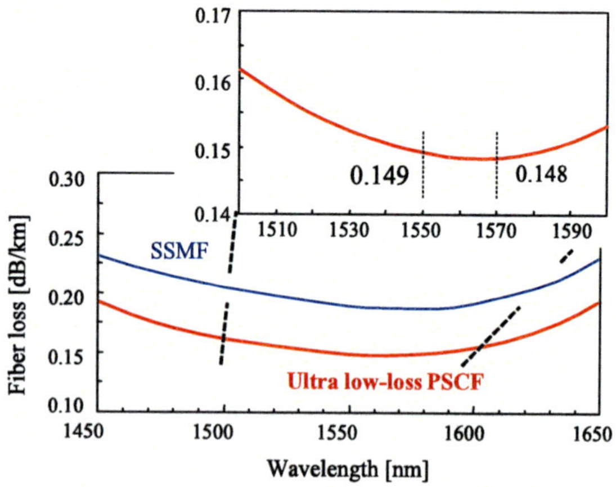

Fig. 2.2 Loss spectra of ultra-low-loss PSCF and SSMF (Reprinted with permission from [7]. © 2015 Sumitomo Electric)

Table 2.1 Characteristics of fabricated PSCF at 1550 nm

	PSCF	SSMF
Fiber loss (dB/km)	0.149	0.190
A_{eff} (μm^2)	135	80
Dispersion (ps/(nm * km))	21.0	16.8

Reprinted with permission from [7]. © 2015 Sumitomo Electric

rized in Table 2.1. To reduce the nonlinearity, larger A_{eff} than SSMF (80 μm^2) is applied.

It is also noted that a high chromatic dispersion is preferable for suppressing nonlinear effects such as cross phase modulation (XPM) and four-wave mixing (FWM) especially for high-capacity digital coherent systems. The chromatic dispersion of the PSCF is as large as 21 ps/(nm*km), which almost reaches the material dispersion of silica of about 22 ps/(nm*km). For large A_{eff} fibers, degradation of the fiber loss due to bending losses is of concern. In Fig. 2.2, a 20-km-long ultra-low-loss PSCF was spooled on a bobbin with a 170-mm-diameter barrel, and there was no obvious degradation due to macro- and micro-bending losses even in a longer wavelength region, where it is sensitive for bending effects. In order to improve the macro-bending performance for a larger A_{eff} fiber, a depressed cladding index profile is applied [8]. As a result, the PSCF shows no significant loss increase by the macro-bending loss over the C + L band wavelength region.

For the ultra-low-loss PSCF, a ring-core refractive index profile having a center core slightly doped with fluorine, surrounded by a pure-silica ring core, is applied as shown in Fig. 2.3. As this profile gives a smaller mode-field diameter while enlarging the A_{eff}, a better dissimilar-splice performance to a SSMF than that of a conventional step-core profile having the same A_{eff} [6] is anticipated. As shown in Fig. 2.4, a lower splicing loss than a step-index core is actually confirmed. For example, with fibers having the same A_{eff} of 130 μm^2, dissimilar-splice loss of the ring-core profile is 0.02 dB/facet-lower than that of a step-core profile.

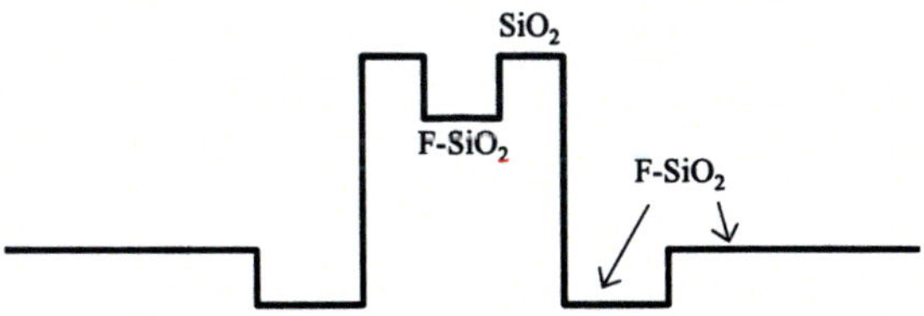

Fig. 2.3 Schematic ring-core refractive index profile (Reprinted with permission from [7]. © 2015 Sumitomo Electric)

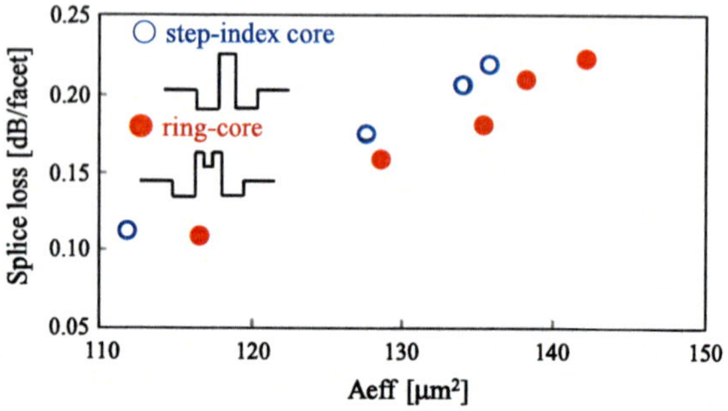

Fig. 2.4 A_{eff} dependence of dissimilar-splice loss to SSMF (Reprinted with permission from [7]. © 2015 Sumitomo Electric)

2.2.3 Optimal Fiber Design Based on Fiber FOM

In this section, an analytical development of fiber figure-of-merit (FOM) is discussed, which can predict the degree of improvement on Q-factor and transmission distance from the fiber characteristics in order to decide an appropriate optical fiber for large capacity and long-haul transmission. Figure 2.5 shows the block diagram of considered point-to-point link. An assumed system is a multi-span digital coherent transmission link composed of a transmission fiber, an erbium-doped fiber amplifier (EDFA), and coupling losses between them [9–11]. Symbols used for this formulation are listed in Table 2.2.

In this link, signal, ASE noise, and nonlinear effects are assumed to have a Gaussian distribution, and they do not interfere with each other, and therefore, the OSNR is expressed as [12]

$$\text{OSNR} = \frac{P_{\text{ch}}}{P_{\text{ASE}} + P_{\text{NLI}}}. \tag{2.1}$$

Using (2.1), optimal launched power (P_{opt}) and maximum Q-factor (Q_{MAX}) can be expressed as [10, 11],

$$P_{\text{opt}}[\text{dBm}] = -\frac{10}{3}\log\left(\gamma^2 L_{\text{eff}}|D|^{-1}\right) + \frac{1}{3}\alpha L + \frac{4}{3}\alpha_{\text{sp}} + C_1, \tag{2.2}$$

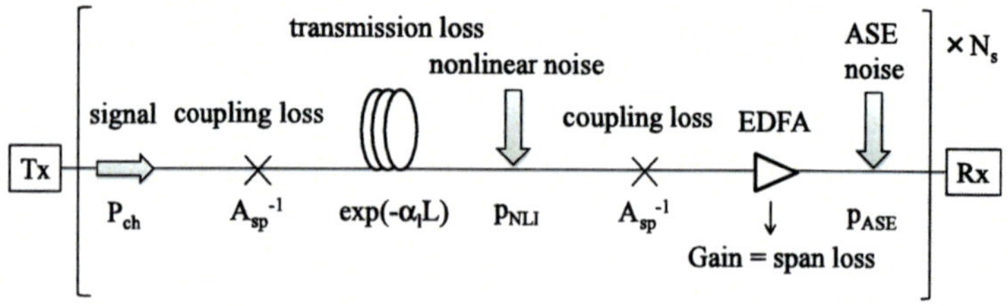

Fig. 2.5 Block diagram of considered link (Reprinted with permission from [7]. © 2015 Sumitomo Electric)

Table 2.2 List of symbols

Symbol	Unit	
α	dB/km	Transmission loss of fiber
α_L	1/km	Transmission loss of fiber $\alpha_L = \alpha/10 \ln 10$
α_{span}	dB	Span loss
L_{eff}	km	Effective length $(=\{1-\exp(-\alpha_L L)\}/\alpha_L)$
n_2	m^2/W	Nonlinear refractive index
γ	1/W/km	Nonlinear coefficient $(=(2\pi/\lambda)\times(n_2/A_{eff}))$
D	ps/(nm km)	Chromatic dispersion
α_{sp}	dB	Coupling loss of fiber to EDFA $\alpha_{sp} = 10\log_{10}Asp$
A_{sp}	–	Coupling loss of fiber to EDFA
L	km	Fiber span length
D_T	km	Total transmission distance
N_s		Number of spans
P_{ch}	W	Launched signal power per channel
P_{ASE}	W	Accumulated ASE noise from EDFAs
P_{NLI}	W	Accumulated nonlinear noise
P_{ASE}	W	ASE noise from an EDFA
P_{NLI}	W	Nonlinear noise per span

Reprinted with permission from [7]. © 2015 Sumitomo Electric

$$Q_{max}[\text{dB}] = -\frac{10}{3}\log\left(\gamma^2 L_{eff}|D|^{-1}\right) - \frac{2}{3}\alpha L - \frac{2}{3}\alpha_{sp} + 10\log(L) - 10\log(D_T) + C_2, \tag{2.3}$$

where C_1 and C_2 are coefficients determined by a transmission system including a back-to-back penalty, noise figure (NF) of EDFA, channel bandwidth, and number of channels. If the FOM is defined as,

$$\text{FOM}[\text{dB}] = -\frac{10}{3}\log\left(\gamma^2 L_{eff}|D|^{-1}\right) - \frac{2}{3}\alpha L + 10\log(L) - \frac{2}{3}\alpha_{sp}. \tag{2.4}$$

P_{opt} and Q_{max} can be expressed as,

$$P_{opt}[\text{dBm}] = \text{FOM} - \alpha_{span} - 10\log(L) + C_1, \tag{2.5}$$

$$Q_{max}[\text{dB}] = \text{FOM} - 10\log(D_T) + C_2. \tag{2.6}$$

Assuming the practical submarine wet-repeaters transmitting 100-channel WDM signal channels, the output power of EDFA is generally limited to +16 dBm to + 18 dBm in total because of a limitation of electric power supply and broad gain bandwidth. As a result, the output power per channel under an actual operation condition will be limited to −2 dBm/ch. Therefore, the actual launched signal power may be less than P_{opt}. At an arbitrary signal power of $P_{ch} = r \cdot P_{opt}$, Q-factor (Q_R) is expressed as [10, 11]

$$\begin{aligned} Q_R &= Q_{max} + 10\log\{3r/(r^3+2)\} \\ &= \mathrm{FOM}_R - 10\log(D_T) + C_2 \end{aligned}, \quad (2.7)$$

where FOM_R is an FOM at an arbitrary P_{ch} and can be written as

$$\mathrm{FOM}_R = \mathrm{FOM} + 10\log\{3r/(r^3+2)\}. \quad (2.8)$$

As can be seen from (2.4) and (2.8), FOM and FOM_R are parameters determined by fiber characteristics, span length, and launched signal power. Furthermore, from (2.6) and (2.7), improvement of Q-factor and/or transmission distance can be predicted easily when using a fiber having different FOM_R in the same system configuration. Alternatively, from (2.4), the higher FOM_R achieves a longer span length at the same Q-factor and transmission distance. For example, a fiber having 1 dB-higher FOM_R will realize a 10 km-longer span length. This can reduce the construction cost of submarine transmission systems.

Figure 2.6 shows iso-FOM_R lines as functions of fiber loss and A_{eff} at L of (a) 80 km and (b) 100 km with solid lines, along with FOM with dashed lines. In this calculation, $D = +21$ ps/(nm km) and $n_2 = 2.2 \times 10^{-20}$ m^2/W were assumed. The coupling loss of a fiber to EDFA was calculated as dissimilar-splice loss from MFD-mismatching between the applied fiber and a SSMF [13, 14]. P_{ch} was set as −2 dBm/ch when P_{opt} was calculated to be more than −2 dBm/ch using (2.2), otherwise, $P_{ch} = P_{opt}$ ($r = 1$). C_1 and C_2 were set as −6.6 dBm/ch and 38.4 dB, respectively, fitted from 100G-QPSK-DWDM transmission experiments in [10, 11, 15]. FOM_R of the ultra-low-loss PSCF shown in the previous Sect. 2.2.2 and reported fibers [13, 16–18] are also plotted in Fig. 2.6. It is clearly found from Fig. 2.6 that the FOM_R improvement mainly depends on the lowering of fiber loss. On the other hand, as for A_{eff}, there is an optimal value in which the FOM_R becomes saturated at around 120–140 μm^2 for L of 80 km and 110–130 μm^2 for L of 100 km.

2.2.4 *Micro-bending Loss Sensitivity*

It is generally known that micro-bending loss sensitivity is degraded for larger A_{eff} fibers. Therefore, reduction of the micro-bending loss is required for installing such a large A_{eff} fiber to submarine cables. Hence, a resin-coating with low Young's

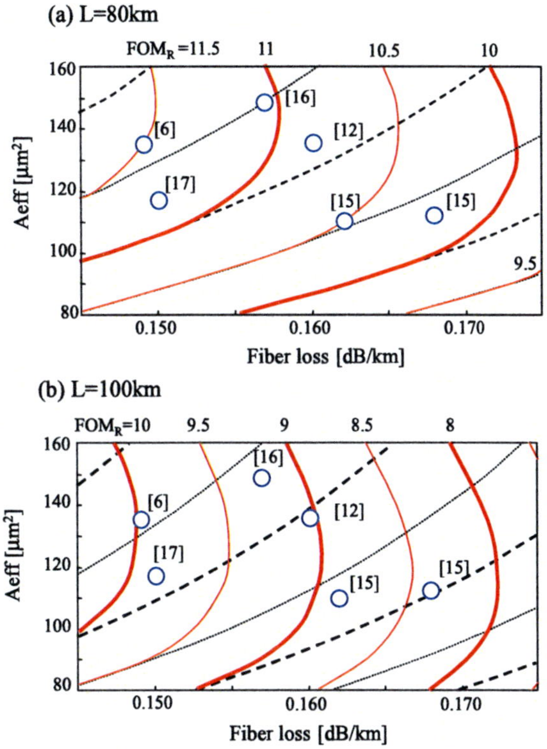

Fig. 2.6 Iso-FOM$_R$ as functions of fiber loss and A_{eff} at span length of **a** 80 km and **b** 100 km (Reprinted with permission from [7]. © 2015 Sumitomo Electric)

modulus in a primary layer was applied on the fabricated PSCFs in order to realize a excellent micro-bending loss performance [19]. Figure 2.7 shows micro-bending losses for fibers having different A_{eff} with conventional and low modulus primary

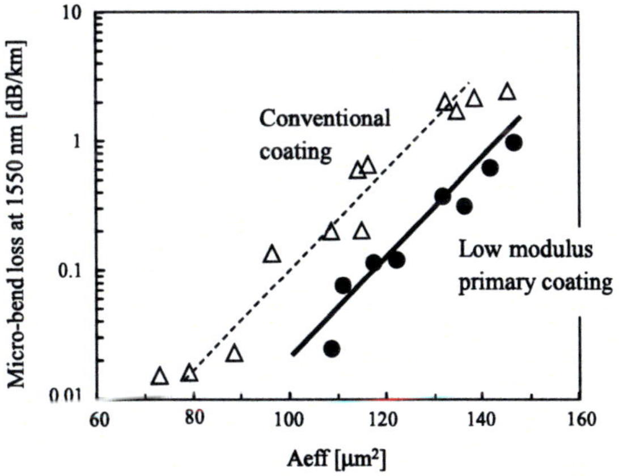

Fig. 2.7 Micro-bending loss for fibers with low Young's modulus primary and conventional coating (Reprinted with permission from [7]. © 2015 Sumitomo Electric)

coatings. The micro-bending loss was characterized by a wire mesh bobbin method at winding tension of 80 gf [20]. As can be seen from Fig. 2.7, micro-bending loss is dramatically reduced by applying the soft primary coating. Micro-bending loss of the large A_{eff} fiber with the soft primary coating is the same level as one of conventional fibers with the conventional coating.

2.2.5 Environmental and Mechanical Performances of Ultra-low-loss PSCF

Environmental and mechanical tests have been conducted according to IEC 60793-2-50 including the damp heat, dry heat, temperature cycling, water immersion, tensile strength, stress corrosion susceptibility, fiber curl, and proof tests. The PSCFs exhibited excellent stabilities for all tests, which show high reliability and durability practicable for a submarine cabling. For example, Fig. 2.8 shows the fiber loss change during damp heat test at a temperature of 85 °C and a relative humidity of 85%, in which measurable degradation was not confirmed.

In addition to the IEC tests, a hydrogen aging test in order to verify a long-term reliability in the submarine environment has been conducted. Figure 2.9 shows a spectrum of loss change after the hydrogen aging test, where the fiber was exposed to hydrogen partial pressure of 1 atm at a room temperature for 4000 h. No Measurable degradation was observed in the wavelength of 1400–1600 nm, and it is found that the PSCFs have the excellent stability against the hydrogen exposure.

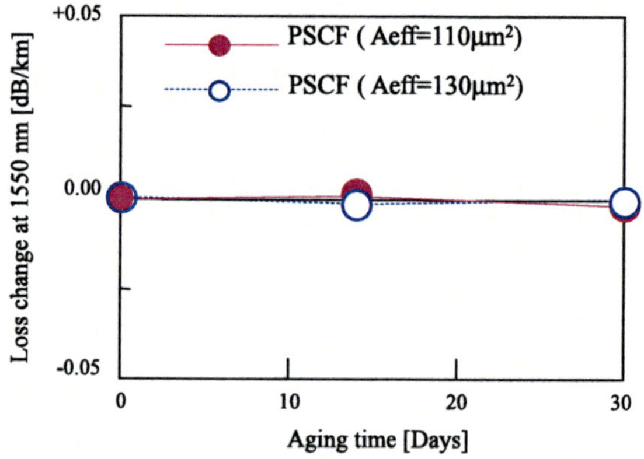

Fig. 2.8 Damp heat test of PSCF at 85 °C and 85% relative humidity (Reprinted with permission from [7]. © 2015 Sumitomo Electric)

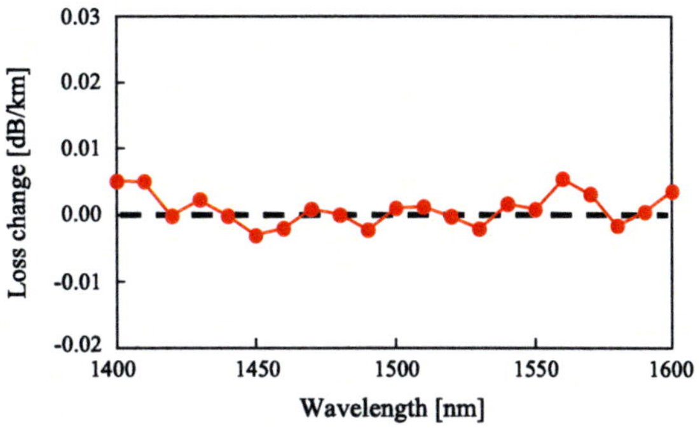

Fig. 2.9 Hydrogen aging test of PSCF (partial pressure = 1 atm at room temperature for 4000 h) (Reprinted with permission from [7]. © 2015 Sumitomo Electric)

2.2.6 *Conclusion*

The state-of-the-art improvement of loss including all-solid single-core fibers, PCFs, and PBGFs has been reviewed, and especially, the steady improvement of all-solid single-core fibers is described. The lowest attenuation loss level of 0.1419 dB/km at 1560 nm has recently been reported. By virtue of ultra-low loss and optimal values of A_{eff}, such PSCFs have the highest fiber FOM among transmission fibers. These fibers will contribute to a dramatic acceleration of capacity growth in submarine systems in the near-term telecommunication systems.

2.3 Multi-core Fiber

2.3.1 *Theory of Crosstalk*

Inter-core crosstalk (XT) is one of the most important properties of uncoupled MCFs, and suppression of the XT has been actively studied by various groups. Since the longitudinal evolutions of the phases of core modes in the MCF are affected by various perturbations, the XT cannot be predicted by the simple coupled-mode equations without considering such perturbations, and a statistical mean of the XT accumulates with obeying a coupled-power equation.

In this section, the power-coupling coefficient between cores is derived from the longitudinally perturbed coupled-mode equation. According to the derivation, the relationship between fiber parameters and the XT is discussed, and XT suppression methods are described.

2.3.1.1 Inter-core Power-Coupling Coefficient of the Multi-core Fibers

By considering longitudinal variations of propagation constants, the coupled-mode equation can be expressed as

$$\frac{\mathrm{d}A_n}{\mathrm{d}z} = -j\kappa_{nm}\exp\left[-j\int_0^z(\beta_m-\beta_n)\mathrm{d}z\right]A_m$$
$$= -j\kappa_{nm}\exp\left[-j(\beta_{\mathrm{c},m}-\beta_{\mathrm{c},n})z - j\int_0^z(\beta_{\mathrm{v},m}-\beta_{\mathrm{v},n})\mathrm{d}z\right]A_m, \quad (2.9)$$

where A_n is the complex amplitude of Core n, κ_{nm} is the mode-coupling coefficient from Core m to Core n, $\beta = 2\pi n_{\mathrm{eff}}/\lambda$ is the propagation constant, n_{eff} is the effective refractive index, and λ is the vacuum wavelength. Subscripts c and v of β_n or β_m in Core n or Core m represent *constant* and *variable* perturbed parts of β_n or β_m, respectively.

Based on (2.9), in case of low crosstalk, the crosstalk in amplitude within the fiber segment $[z_1, z_2]$ can be expressed as

$$\Delta x_{nm} = \frac{\Delta A_n}{A_m} \approx -j\kappa_{nm}\int_{z_1}^{z_2}\exp[-j(\beta_{\mathrm{c},m}-\beta_{\mathrm{c},n})z]f(z)\mathrm{d}z, \quad (2.10)$$

by using

$$f(z) \equiv \exp\left[-j\int_0^z[\beta_{\mathrm{v},m}(z')-\beta_{\mathrm{v},n}(z')]\mathrm{d}z'\right]. \quad (2.11)$$

Accordingly, the average crosstalk increases in power within the segment $[z_1, z_2]$ can be expressed as

$$\Delta X_{nm} = \langle|\Delta x_{nm}|^2\rangle = \langle\Delta x_{nm}\Delta x_{nm}^*\rangle$$
$$\approx \kappa_{nm}^2\int_{z_1}^{z_2}\int_{z_1}^{z_2}\exp[-j(\beta_{\mathrm{c},m}-\beta_{\mathrm{c},n})(z-z')]\langle f(z)f^*(z')\rangle\mathrm{d}z\mathrm{d}z'$$
$$\approx \kappa_{nm}^2\int_{z_1}^{z_2}\int_{z_1-z'}^{z_2-z'}\exp[j\Delta\beta_{\mathrm{c},nm}\zeta]\langle f(z'+\zeta)f^*(z')\rangle\mathrm{d}\zeta\mathrm{d}z'$$

$$\approx \kappa_{nm}^{2}\int_{z_1}^{z_2}\mathrm{d}z'\int_{-\infty}^{\infty}R_{ff}(\zeta)\exp\left(j\Delta\beta_{\mathrm{c},nm}\zeta\right)\mathrm{d}\zeta$$

$$\approx \kappa_{nm}^{2}\Delta z\int_{-\infty}^{\infty}R_{ff}(\zeta)\exp\left(j\Delta\beta_{\mathrm{c},nm}\zeta\right)\mathrm{d}\zeta, \tag{2.12}$$

where the bracket $\langle\cdot\rangle$ represents the ensemble average, R_{ff} is the autocorrelation function (ACF) of $f(z)$, $\Delta z = z_2 - z_1$, $\Delta\beta_{\mathrm{c},nm} = \beta_{\mathrm{c},n} - \beta_{\mathrm{c},m}$ and the correlation length l_{c} of R_{ff} is assumed to be adequately shorter than Δz. $R_{ff}(\zeta)$ can be understood as the correlation between the coupled and non-coupled lights that are propagated for the length of ζ after the coupling. For example, where $\zeta \gg l_{\mathrm{c}}$, the coupled and non-coupled lights become incoherent even if the lights are very coherent. Based on the Wiener–Khinchin theorem, the power spectrum density (PSD) is the Fourier transform of the ACF:

$$S_{ff}^{(\tilde{\nu})}(\tilde{\nu}) = \int_{-\infty}^{\infty}R_{ff}(\zeta)\exp(j2\pi\tilde{\nu}\zeta)\mathrm{d}\zeta, \tag{2.13}$$

where $\tilde{\nu} = n_{\mathrm{eff}}/\lambda = \beta/(2\pi)$ represents the wave number (or spatial frequency) in the medium—whereas the propagation constant β is the angular wave number. Note that $\tilde{\nu}$, n_{eff}, and β have common subscripts, e.g., $\tilde{\nu}_{\mathrm{c}} = n_{\mathrm{eff,c}}/\lambda = \beta/(2\pi)$. To describe the PSDs with respect to $\tilde{\nu}$ and β with common expressions, we would like to define the PSD with respect to β, whose total power is equivalent to (2.13). From the Parseval's theorem, the average power of $f(z)$, or expected value of $|f(z)|^2$, is equivalent to the integral of the PSD over whole, and the following equation holds between $f(z)$ and the PSDs of $f(z)$:

$$\int_{-\infty}^{\infty}S_{ff}^{(\tilde{\nu})}(\tilde{\nu})\mathrm{d}\tilde{\nu} = \int_{-\infty}^{\infty}S_{ff}^{(\tilde{\nu})}(\tilde{\nu})\frac{\mathrm{d}\tilde{\nu}}{\mathrm{d}\beta}\mathrm{d}\beta = \mathrm{E}\left[|f(z)|^2\right] = 1, \tag{2.14}$$

where $\mathrm{E}[\cdot]$ represents the expected value. Therefore, in this section, the PSD $S_{ff}^{(\beta)}(\beta)$ with the scale of the propagation constant β (the angular wave number in the medium) is defined as:

$$S_{ff}^{(\beta)}(\beta) \equiv S_{ff}^{(\tilde{\nu})}(\tilde{\nu})\frac{\mathrm{d}\tilde{\nu}}{\mathrm{d}\beta} = \frac{1}{2\pi}S_{ff}^{(\tilde{\nu})}(\tilde{\nu}) = \frac{1}{2\pi}\int_{-\infty}^{\infty}R_{ff}(\zeta)\exp(j\beta\zeta)\mathrm{d}\zeta. \tag{2.15}$$

From (2.12)–(2.15), the power-coupling coefficient can be expressed as

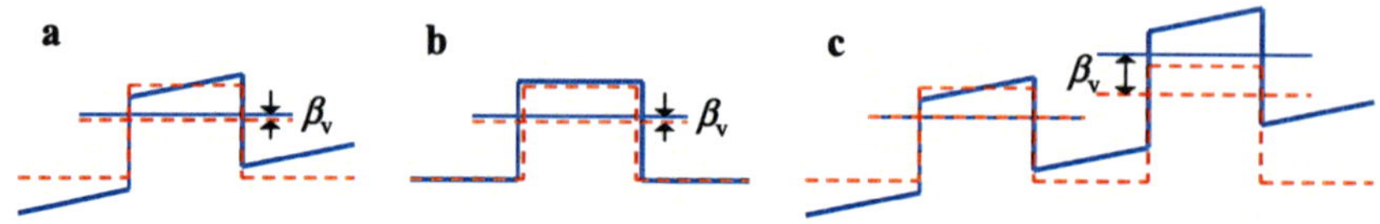

Fig. 2.10 Schematics of perturbations on the propagation constant. **a** A slight change of the propagation constant in a core due to bend, **b** a slight change of the propagation constant in a core due to structure fluctuation, and **c** a bend-induced change of the propagation constant in a core when assuming another core as a reference of the propagation constant

$$h_{nm} = \frac{\Delta X_{nm}}{\Delta z} \approx \kappa_{nm}^2 S_{ff}^{(\tilde{\nu})}\left(\frac{\Delta n_{\mathrm{eff,c},nm}}{\lambda}\right) = \kappa_{nm}^2\left[2\pi S_{ff}^{(\beta)}(\Delta\beta_{\mathrm{c},nm})\right]. \tag{2.16}$$

Figure 2.10 shows the schematics of perturbations on β, or how β_v can vary in the optical fiber. As shown in Fig. 2.10a, b, the bend and the structure fluctuation can induce a slight change in β_v in one core, which can occur either in the single-core fiber or in the MCF. In the single-core fiber, by assuming proper R_{ff} or S_{ff} for the perturbations shown in Fig. 2.10a, b, (2.16) is utilized for analyzing the power coupling between modes in the multi-mode fiber, microbend loss—power coupling from the core modes to the cladding modes, and so on. In the MCF, as shown in Fig. 2.10c, the bend can induce relatively large β_v in a core when assuming another core as a reference of the propagation constant. Fini et al. [21] and Hayashi et al. [22] assumed that β_v in the MCF is induced by the macrobend and twist of the MCF as

$$\beta_{\mathrm{v},n} = \beta_{\mathrm{c},n}\frac{x_n \cos\theta_{\mathrm{f}}(z) - y_n \sin\theta_{\mathrm{f}}(z)}{R_{\mathrm{b}}(z)} = \beta_{\mathrm{c},n}\frac{r_n \cos\theta_n(z)}{R_{\mathrm{b}}(z)}, \tag{2.17}$$

and investigated the crosstalk characteristics of the MCFs. Here, (x_n, y_n) and (r_n, θ_n) are the local Cartesian and polar coordinates of Core n in a fiber cross section, respectively, $\theta_n = 0$ is the radial direction of the macrobend, θ_{f} the angle between the x-axis and the radial direction of the macrobend, and R_{b} the macrobend radius of the MCF—that is, the distance between the center of the macrobend and the origin of the local coordinates.

However, it is not easy to assume a proper β_v, R_{ff}, or S_{ff} that can include the perturbations of both the bend and the structure fluctuation. Therefore, by assuming that R_{ff} includes only the effect of structure fluctuation and does not include that of macrobend and twist, Koshiba et al. investigated the effects of correlation length l_c and of the shape of the ACF R_{ff} on the average crosstalk μ_X [23]. They investigated some types of R_{ff} and found that the exponential ACF (EAF)

$$R_{ff}(\zeta) = \exp\left(-\frac{|\zeta|}{l_{\mathrm{c}}}\right) \tag{2.18}$$

is proper for estimating actual μ_X of the MCFs. The EAF has been introduced to micro-bending loss analysis [24]. Since the PSD of the EAF is the Lorentzian distribution, the power-coupling coefficient was obtained from (2.16) as [23]:

$$\begin{aligned} h_{nm}(z) &= \kappa_{nm}^2 \frac{1}{\pi} \frac{1/(2\pi l_c)}{1/(2\pi l_c)^2 + \left[\Delta n'_{\text{eff,c},nm}(z)/\lambda\right]^2} = \kappa_{nm}^2 2\pi \frac{1}{\pi} \frac{1/l_c}{1/l_c^2 + \left[\Delta\beta'_{c,nm}(z)\right]^2} \\ &= \kappa_{nm}^2 \frac{2l_c}{1 + \left[\Delta\beta'_{c,nm}(z) l_c\right]^2}, \end{aligned} \tag{2.19}$$

where $\Delta\beta'_{c,nm}$ is $\beta'_{c,n} - \beta'_{c,m}$, and β'_c is redefined β_c that includes the effects of macrobend and twist:

$$\beta'_{c,n} = \beta_{c,n}\left(1 + \frac{x_n \cos\theta_f(z) - y_n \sin\theta_f(z)}{R_b(z)}\right) = \beta_{c,n}\left(1 + \frac{r_n \cos\theta_n(z)}{R_b(z)}\right). \tag{2.20}$$

Average crosstalk μ_X estimated using coupled-power equation with the power-coupling coefficient of (2.19) may be valid in cases where changes of R_b and θ_f are gradual enough compared to l_c, since $\Delta\beta'_{c,nm}$—which is variable and includes macrobend and twist—is substituted to $\Delta\beta_{c,nm}$—which is constant—in (2.16).

Practically, the MCF is unintentionally twisted and θ_f may be averaged if the fiber length is adequately long. So, we derive the power-coupling coefficient averaged over θ_f. For simplicity, the center of Core m is taken as the origin of the local coordinate, and accordingly $\Delta\beta'_{c,nm}$ can be written as

$$\Delta\beta'_{c,nm}(R_b, \theta_{nm}) = \Delta\beta_{c,nm} + \Delta\beta_{b,nm}(R_b, \theta_{nm}), \tag{2.21}$$

$$\Delta\beta_{b,nm}(R_b, \theta_{nm}) = \Delta\beta_{b,nm}^{\text{dev}}(R_b) \cos\theta_{nm}, \tag{2.22}$$

$$\Delta\beta_{b,nm}^{\text{dev}}(R_b) = \beta_{c,n} \frac{D_{nm}}{R_b}, \tag{2.23}$$

where θ_{nm} represents the angle between the radial direction of the bend and a line segment from Core m to Core n, $\Delta\beta_{b,nm}$ the difference of β variation between Core m and Core n from the macrobend, and $\Delta\beta_{b,nm}^{\text{dev}}$ the peak deviation of $\Delta\beta_{b,nm}$.

Let $p_\theta(\theta_{nm})$ and $p_{R_b}(R_b)$ be the probability density functions of θ_{nm} and of R_b, respectively, along the MCF; by assuming that $p_\theta(\theta_{nm})$ and $p_{R_b}(R_b)$ are statistically independent, the twist of the MCF is gradual enough, and average crosstalk is adequately low; the average crosstalk $\mu_{X,nm}$ from Core m to Core n can be expressed as

$$\mu_{X,nm}(L) \approx \int_0^L h_{nm}(z)\mathrm{d}z \approx L\left[\frac{1}{L}\int_0^L h_{nm}(z)\mathrm{d}z\right] \approx L\mathrm{E}[h_{nm}]$$

$$\approx L\int_0^\infty p_{R_\text{b}}(R_\text{b})\overline{h}_{nm}(R_\text{b})\text{d}R_\text{b},\tag{2.24}$$

where the average power-coupling coefficient is

$$\overline{h}_{nm}(R_\text{b}) = \int_0^{2\pi} p_{\theta_{nm}}(\theta_{nm})h_{nm}(R_\text{b},\theta_{nm})\text{d}\theta_{nm}.\tag{2.25}$$

By assuming that the twist of the MCF is random enough and the MCF is adequately long, $p_\theta(\theta_{nm})$ can be assumed to be constant (=1/(2π)) over all θ_{nm}; therefore, by substituting $\Delta\beta = \Delta\beta_{\text{b},nm}(R_\text{b},\theta_\text{nm}) = \Delta\beta_{\text{b},nm}^\text{dev}(R_\text{b})\cos\theta_{nm}$ and using (2.16) and $\sin(\arccos x) = \sqrt{1-x^2}$, (2.25) can be rewritten as

$$\begin{aligned}\overline{h}_{nm}(R_\text{b}) &= \int_0^{2\pi}\frac{1}{2\pi}h_{nm}(R_\text{b},\theta_{nm})\text{d}\theta_{nm} = \int_0^{2\pi}\frac{1}{2\pi}\kappa_{nm}^2 2\pi S_{ff}^{(\beta)}\left[\Delta\beta'_{\text{c},nm}(R_\text{b},\theta_{nm})\right]\text{d}\theta_{nm}\\ &= 2\int_0^{\pi}\kappa_{nm}^2 S_{ff}^{(\beta)}\left[\Delta\beta_{\text{c},nm} + \Delta\beta_{\text{b},nm}^\text{dev}(R_\text{b})\cos\theta_{nm}\right]\text{d}\theta_{nm}\\ &= 2\pi\int_{-\Delta\beta_{\text{b},nm}^\text{dev}}^{\Delta\beta_{\text{b},nm}^\text{dev}}\frac{\kappa_{nm}^2}{\pi\sqrt{\left[\Delta\beta_{\text{b},nm}^\text{dev}(R_\text{b})\right]^2-\Delta\beta^2}}S_{ff}^{(\beta)}(\Delta\beta_{\text{c},nm}-\Delta\beta)\text{d}(\Delta\beta),\end{aligned}\tag{2.26}$$

where S_{ff} is the Lorentzian distribution as shown in (2.19). By using the arcsine distribution:

$$p_{\Delta\beta_\text{b}}\left(\Delta\beta_{\text{c},nm}\right) = \begin{cases}\frac{1}{\pi\sqrt{\left(\Delta\beta_{\text{b},nm}^\text{dev}\right)^2-\Delta\beta_{\text{c},nm}^2}}, & \left|\Delta\beta_{\text{c},nm}\right| \le \Delta\beta_{\text{b},nm}^\text{dev},\\ 0, & \text{otherwise},\end{cases}\tag{2.27}$$

which is the probability distribution of $\Delta\beta_\text{b}$, (2.26) can be rewritten as

$$\overline{h}_{nm}(\Delta\beta_{\text{c},nm},R_\text{b}) = \kappa_{nm}^2 2\pi\left(p_{\Delta\beta_\text{b}} * S_{ff}^{(\beta)}\right)_{\Delta\beta}(\Delta\beta_{\text{c},nm}) = \kappa_{nm}^2\left(p_{\Delta\tilde{\nu}_\text{b}} * S_{ff}^{(\tilde{\nu})}\right)_{\Delta\tilde{\nu}}(\Delta\tilde{\nu}_{\text{c},nm}),\tag{2.28}$$

where the expression of $(f * g)_x$ denotes the convolution of f and g with respect to x, and the expression with respect to $\tilde{\nu}$ is also shown for comparison. If we consider the case where PSD S_{ff} in (2.16) includes both the effects of the structure fluctuation and the macrobend, the convolution term in (2.28) may be understood as the PSD S_{ff} in (2.16).

A closed-form expression of the average power-coupling coefficient $\overline{h}$ was analytically derived by Koshiba et al. [25] by assuming constant R_b and twist rate:

$$\overline{h}_{nm} = \kappa_{nm}^2 \sqrt{2} l_c \left[\frac{1}{\sqrt{a\left(b+\sqrt{ac}\right)}} + \frac{1}{\sqrt{c\left(b+\sqrt{ac}\right)}} \right], \tag{2.29}$$

$$a = 1 + \left(\Delta\beta_{c,nm} l_c - \frac{B_{nm} l_c}{R_b} \right)^2 \approx 1 + \left(\Delta\beta_{c,nm} l_c - \Delta\beta_{b,nm}^{dev} l_c \right)^2, \tag{2.30}$$

$$b = 1 + \left(\Delta\beta_{c,nm} l_c \right)^2 - \left(\frac{B_{nm} l_c}{R_b} \right)^2 \approx 1 + \left(\Delta\beta_{c,nm} l_c \right)^2 - \left(\Delta\beta_{b,nm}^{dev} l_c \right)^2, \tag{2.31}$$

$$c = 1 + \left(\Delta\beta_{c,nm} l_c + \frac{B_{nm} l_c}{R_b} \right)^2 \approx 1 + \left(\Delta\beta_{c,nm} l_c + \Delta\beta_{b,nm}^{dev} l_c \right)^2, \tag{2.32}$$

$$B_{nm} = \sqrt{\left(\beta_{c,n} x_n - \beta_{c,m} x_m \right)^2 + \left(\beta_{c,n} y_n - \beta_{c,m} y_m \right)^2}, \tag{2.33}$$

where B_{nm} can be approximated as $\beta_{c,n} D_{nm}$ if $\beta_{c,m}/\beta_{c,n} \approx 1$, D_{nm} is the center-to-center distance between Core m and Core n.

Figure 2.11 shows comparisons between $\overline{h}$ calculated by using (2.28) and $\overline{h}$ calculated by using (2.29)–(2.33). Figure 2.11a, b show the PSDs normalized with respect to the Lorentzian S_{ff} and to the arcsine distribution $p_{\Delta\beta_b}$, respectively. The Lorentzian and arcsine distributions represent the spectra of the perturbations induced by the structure fluctuation and by the macrobend, respectively. Solid lines represent $\overline{h}$

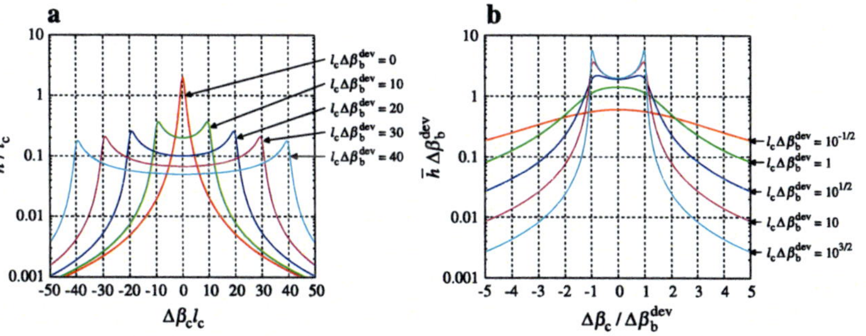

Fig. 2.11 Comparisons between $\overline{h}$ calculated by using (2.28) and $\overline{h}$ calculated by using (2.29)–(2.33). **a** $\overline{h}$ normalized with respect to the Lorentzian, **b** $\overline{h}$ normalized with respect to the arcsine distribution. Solid lines: $\overline{h}$ calculated by using (2.28), dashed lines: $\overline{h}$ calculated by using (2.29)–(2.33). The solid lines and the dashed lines are overlapped (Reprinted with permission from [26]. © 2013 The Optical Society)

calculated by using (2.28) and dashed lines represent $\bar{h}$ calculated by using (2.29)–(2.33); however, the solid lines and the dashed lines are overlapped, and we can only see the solid lines.

In particular cases, we can obtain simpler expressions from (2.28)–(2.33). For example, where $|\Delta\beta_{c,nm}|$ and the bandwidth of $S_{ff}^{(\beta)}$ are adequately smaller than $\Delta\beta_{b,nm}^{dev}$, S_{ff} becomes a narrow delta-function-like distribution and the convolution contains only a gradually varying part of $p_{\Delta\beta_b}(\Delta\beta_{c,nm})$; therefore, (2.28) can be approximated as

$$\bar{h}_{nm}(\Delta\beta_{c,nm}, R_b) \approx \kappa_{nm}^2\left[2\pi p_{\Delta\beta_b}(\Delta\beta_{c,nm})\right] = \kappa_{nm}^2 p_{\Delta\tilde{\nu}_b}(\Delta\tilde{\nu}_{c,nm}) \approx \kappa_{nm}^2 \frac{2}{\sqrt{\left(\frac{\beta_{c,n}D_{nm}}{R_b}\right)^2 - \Delta\beta_{c,nm}^2}} = \kappa_{nm}^2 \frac{\lambda}{\pi\sqrt{\left(\frac{n_{eff,c,n}D_{nm}}{R_b}\right)^2 - \Delta n_{eff,c,nm}^2}}, \quad (2.34)$$

which is also obtained from (2.16) by approximating the PSD $S_{ff}^{(\beta)}$ as the probability distribution of $\Delta\beta_b$—shown in (2.27)—with constant R_b. In case of homogeneous MCFs ($\Delta\beta_{c,nm} = 0$), (2.34) is reduced to

$$\bar{h}_{nm}(R_b) \approx \kappa_{nm}^2 \frac{2R_b}{\beta_{c,n}D_{nm}} = \kappa_{nm}^2 \frac{\lambda R_b}{\pi n_{eff,c,n}D_{nm}}. \quad (2.35)$$

The difference between (2.34) and (2.35) is less than 0.1 dB when $\Delta\beta_c < 0.21\beta_c D/R_b$; therefore, (2.35) may be also used for estimating the crosstalk of a bent heterogeneous MCF with small $\Delta\beta_c$.

The average power-coupling coefficient $\bar{h}$ can predict the XT in actual MCFs.

Koshiba et al. [25] investigated the agreement between measured XT and the calculated XT using (2.29)–(2.33). Figures 2.12 and 2.13 show the relationships between the XT and the bending radius/diameter of MCFs. Figure 2.12 shows the bending-diameter dependence of the XT calculated with (2.21)–(2.33). The correlation length for which we have no reliable information can be predicted by comparing the calculations with the measurement results. The average crosstalk values with the correlation length of 50 mm agree well with measurement results denoted by closed circles.

Hayashi et al. [26] discussed the applicability of (2.29)–(2.33) and (2.28) to microbend-affected MCFs. In the above discussion, S_{ff} only includes structure fluctuation, and $p_{\Delta\beta_b}(\Delta\beta_{c,nm})$ includes macrobend perturbation that gradually varies in longitudinal direction. Based on the assumption that the macrobend perturbation is gradual enough compared to l_c, we may redefine the S_{ff} as the PSD of high-frequency perturbations other than the macrobend perturbation, and thus S_{ff} may include not only the effect of structure fluctuation but also the effect of microbend, in (2.29)–(2.33) and (2.28). In this case, the increase of the microbend can be understood as the decrease of the correlation length l_c of S_{ff}. Figure 2.13 shows the dependences of the

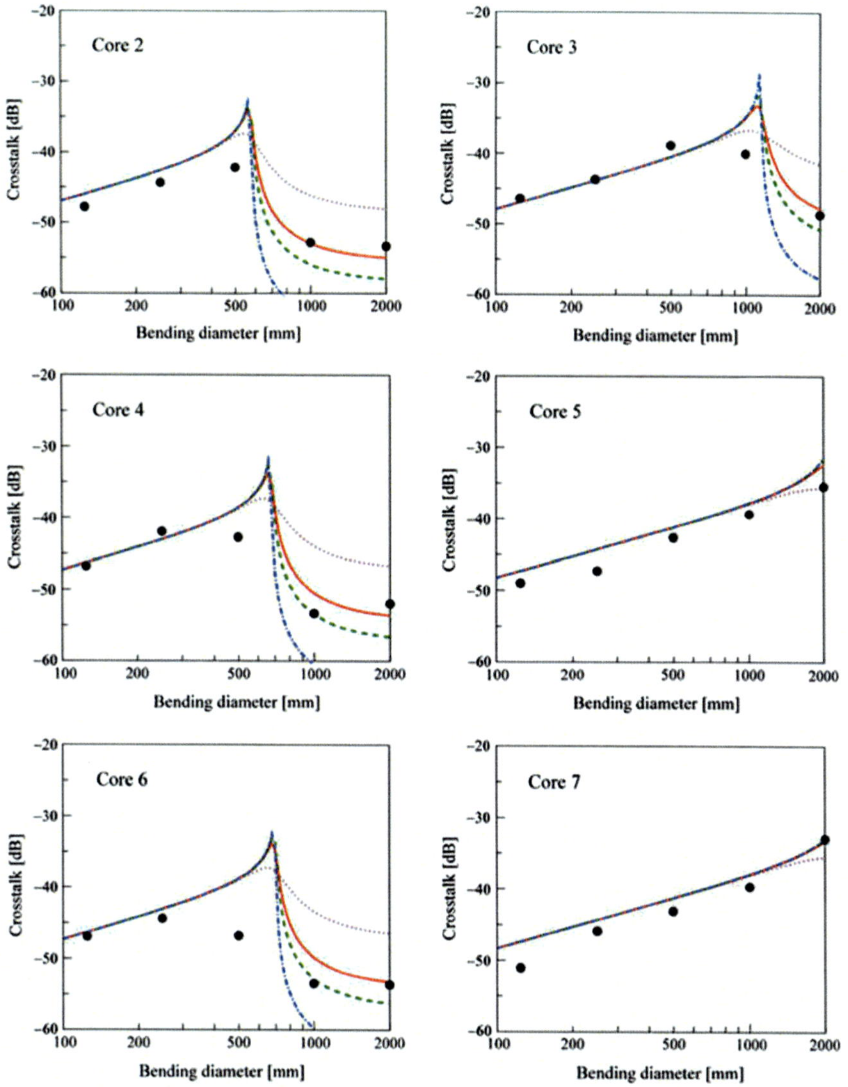

Fig. 2.12 Bending-diameter dependence of crosstalk in a quasi-homogeneous seven-core fiber calculated with (2.29)–(2.33). Dotted line: $d = 10$ mm, solid line: $d = 50$ mm, dashed line: $d = 100$ mm, dashed and dotted line: $d = 500$ mm, and closed circles: measurement data [27] (Reprinted with permission from [25]. © 2012 IEEE)

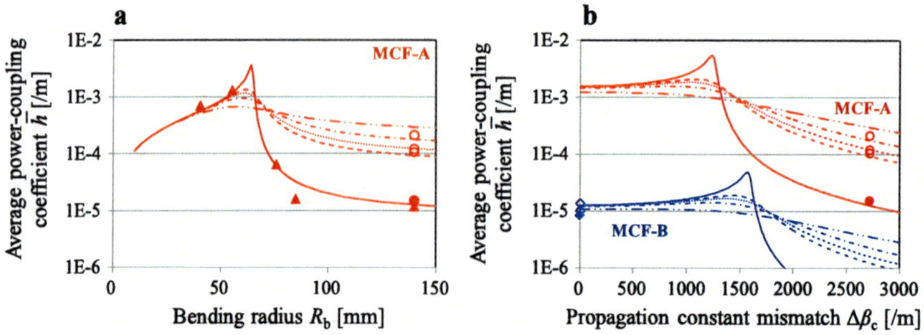

Fig. 2.13 Comparisons of the average power-coupling coefficient $\bar{h}$ obtained from the measurements and from (2.29) to (2.33). **a** The dependences of $\bar{h}$ in a heterogeneous MCF (MCF-A) on the bending radius R_b and on the microbend. **b** The dependence of $\bar{h}$ in MCF-A and a homogeneous MCF (MCF-B) on the propagation-constant mismatch $\Delta\beta_c$ and on the microbend at $R_b = 140$ mm. Closed marks: $\bar{h}$ measured without the microbend, open-marks: $\bar{h}$ measured with the microbend. Solid lines: $\bar{h}$ calculated at $l_c = 3$ cm, dashed lines: $\bar{h}$ at $l_c = 4$ mm, dotted lines: $\bar{h}$ at $l_c = 3$ mm, dashed-dotted lines: $\bar{h}$ at $l_c = 2$ mm, dashed-two dotted lines: $\bar{h}$ at $l_c = 1$ mm (Reprinted with permission from [26]. © 2013 The Optical Society)

average power-coupling coefficient $\bar{h}$ on microbend conditions for a heterogeneous MCF-A [22] and for a homogeneous MCF-B [28]. The microbend was applied by winding the MCFs on a 140-mm-radius bobbin with sandpaper at winding tension T. $\bar{h}$ at $T = 0$ N was measured using a 140-mm-radius bobbin *without* the sandpaper. $\bar{h}$ of the heterogeneous MCF was increased by the microbend, but that of the homogeneous MCF was varied only slightly. These results may be well explained as the shortening of l_c by using (2.29)–(2.33) and (2.28).

2.3.1.2 Crosstalk Suppression Methods Related to Macrobend and Structure Fluctuation

Based on (2.28) and its derivation, it can be understood that the crosstalk is proportional to the power of the mode-coupling coefficient and to the PSD of the perturbations. Of course, the suppression of the mode-coupling coefficient is important and various ways were proposed for confining power into cores such as high-index and small-diameter core structure [29, 30], hole- or trench-assisted core structures [31–35], and photonic crystal structures [36, 37].

The PSD can be intuitively explained as the amount of the phase matching. Accordingly, how to suppress the PSD can be understood as how to suppress the phase matching. In this section, the methods for suppressing the phase matching are described.

The phase-matching suppression methods can be categorized into some types according to how to utilize what kind of the perturbations. Here, three types of suppression methods are explained in the following subsections. A schematic example of $\bar{h}_{nm}(\Delta\beta_{c,nm}, R_b)$ in (2.28) shown in Fig. 2.14 will help with under-

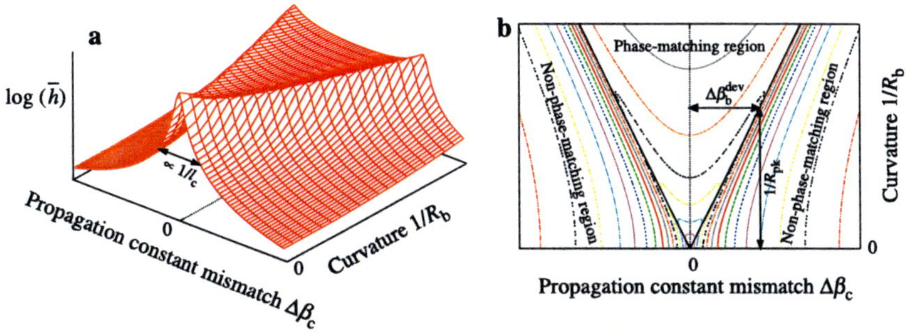

Fig. 2.14 A schematic example of the average power-coupling coefficient $\bar{h}$, as a function of the propagation-constant mismatch $\Delta\beta_c$ and the curvature $1/R_b$, in case that twist of an MCF is gradual and random enough. **a** A three-dimensional plot, **b** a contour map of log($\bar{h}$). Thick solid lines in **b** are the thresholds between the phase-matching region and the non-phase-matching region

standing, along with Fig. 2.11.

Utilization of the Propagation-Constant Mismatch

One is the method utilizing the propagation-constant mismatch $\Delta\beta_c$ to suppress the phase matching [21, 38, 39]. As shown as the non-phase-matching regions in Fig. 2.14, $\Delta\beta_c$ larger than $\Delta\beta_b^{dev}$ can prevent the bend-induced phase-matching between dissimilar cores and can suppress the crosstalk [21, 22]. In other words, for suppressing the crosstalk, the bending radius of the MCF has to be managed to be *adequately* larger than the critical bending radius R_{pk} [22, 23, 25]:

$$R_{pk} = \frac{B_{nm}}{\left|\Delta\beta_{c,nm}\right|} \approx D_{nm}\frac{\beta_{c,n}}{\left|\Delta\beta_{c,nm}\right|} = D_{nm}\frac{n_{eff,c,n}}{\left|\Delta n_{eff,c,nm}\right|}, \tag{2.36}$$

that is, R_b where (2.34) can be infinite, or the maximal R_b where the phase matching due to the macrobend can occur even if there is no structure fluctuation. Some margin from R_{pk} is needed for avoiding the phase matching induced by the spectral broadening of S_{ff}, due to the structural fluctuations. In heterogeneous MCFs, it is preferred if the correlation length l_c of the structural fluctuation can be elongated, because the spectral broadening of S_{ff} can be narrowed and the PSD leakage into the non-phase-matching region can be suppressed, as shown in Figs. 2.11b and 2.14.

If most part of an MCF is deployed in gentle-bend conditions, a slight difference in propagation constants or effective indices may be enough for the phase-matching suppression [21].

Utilization of the Bend-Induced Perturbation

The bend can also be utilized for the phase-matching suppression [32]. As shown in Figs. 2.11a and 2.14, enlargement of the bend-induced perturbation—caused by the increase of the curvature or the decrease of the bending radius—can spread the PSD and suppress the crosstalk even in case of homogeneous MCFs ($\Delta\beta_c = 0$). Identical core structure is rather desirable for suppressing the PSD. The PSD changes gradually with the bend radius, and there is no drastic PSD increase like that around R_{pk} in case of heterogeneous MCFs, since the PSD is suppressed in the phase-matching region. As shown in (2.35), the average crosstalk of a homogeneous MCF is proportional to the average bending radius, where $|\Delta\beta_{c,nm}|$ and the bandwidth of $S_{ff}^{(\beta)}$ are adequately smaller than $\Delta\beta_{b,nm}^{dev}$. Therefore, if the average bending radius of the MCF is managed to be smaller than a certain value, or if the MCF is deployed in bend-challenged conditions, low crosstalk can be achieved with identical cores.

Utilization of the Longitudinal Structural Fluctuation

As shown in Figs. 2.11b and 2.14a, the power spectrum of the perturbations is broadened by the longitudinal structural fluctuations. If the R_{ff} due to the structure fluctuation has a very short correlation length l_c, the power spectrum spreads broadly over the propagation-constant mismatch $\Delta\beta_c$, and thus the PSD may be suppressed even in case of an unbent homogeneous MCF ($\Delta\beta_c = 0$, $1/R_b = 0$). A homogeneous MCF utilizing the longitudinal structural fluctuations was conceptually proposed by Takenaga et al. as "quasi-homogeneous MCF" in [29, 30]. To the author's knowledge, the crosstalk suppression by the structural fluctuation has not been actually observed yet, because the bend-induced perturbations are much larger than the fluctuation-induced perturbations in the measurement conditions. However, the structural fluctuation may work when the MCF is cabled and installed in very-gently bent conditions. For example, $l_c\Delta\beta_{b,nm}^{dev} \cong l_c\beta_c D/R_b$ of 1.3, 1/3, 1.0×10^{-1}, 1.0×10^{-2}, and 1.0×10^{-3} correspond to 1-dB, 5-dB, 10-dB, 20-dB, and 30-dB decreases in $\bar{h}$ from (2.35) at $\Delta\beta_c = 0$, respectively.

2.3.1.3 Statistical Distribution of the Crosstalk

So far, the ensemble average μ_X of the crosstalk X was discussed. In this section, how the crosstalk X is statistically distributed is discussed, especially in the phase-matching case that $R_b \ll R_{pk}$, for simplicity. In the phase-matching case, longitudinal evolution of the crosstalk X_{nm} from Core m to Core n can be approximated as the changes at discrete phase-matching points [32, 40]:

$$A_n(N_{PM}) \approx A_n(N_{PM} - 1) - j\mathrm{K}_{nm}(N_{PM})\exp[-j\varphi_{rnd}(N_{PM})]A_m(N_{PM} - 1)$$

$$\approx A_n(0) - j\sum_{l=1}^{N_{\mathrm{PM}}} \mathrm{K}_{nm}(l)\exp[-j\varphi_{\mathrm{rnd}}(l)]A_m(l-1), \tag{2.37}$$

where $A_n(N_{\mathrm{PM}})$ represents the complex amplitude of Core n after N_{PM}-th phase-matching point, $\phi_{\mathrm{rnd}}(N_{\mathrm{PM}})$ is the phase difference between Cores m and n at N_{PM}-th phase-matching point, and K_{nm} is the coefficient for the discrete changes caused by the coupling from Core m to Core n. ϕ_{rnd} can be regarded as a random sequence, and ϕ_{rnd} significantly varies with slight variations of the bend, twist, and other perturbations. Here, the adequately low crosstalk ($|A_n(N_{\mathrm{PM}})| \ll 1$) is assumed for the approximations that $A_m(0) \approx 1$ and the crosstalk $X_{nm} \approx |A_n(N_{\mathrm{PM}})|^2$. Since $\Re[K_{nm}\exp(j\phi_{\mathrm{rnd}})]$ and $\Im[K_{nm}\exp(j\phi_{\mathrm{rnd}})]$ have the variance $\sigma^2_{2\mathrm{df},nm}$ of $|K_{nm}|^2/2$, probability density functions (PDF) of $\Re A_n(N)$ and $\Im A_n(N)$ converge to Gaussian distributions whose variance $\sigma^2_{4\mathrm{df},nm}$ is $N_{\mathrm{PM}}|K_{nm}|^2/2$ if N_{PM} is adequately large, based on the central limit theorem. When assuming random polarization mode coupling, coupled power can be equally distributed to two polarization modes statistically. Therefore, PDFs of $\Re A_n(N_{\mathrm{PM}})$'s and $\Im A_n(N_{\mathrm{PM}})$'s of two polarization modes converge to the Gaussian distribution with the variance $\sigma^2_{4\mathrm{df},nm}$ of $N_{\mathrm{PM}}|K_{nm}|^2/4$. In this case, the crosstalk X can be represented as a sum of powers of $\Re A_n(N_{\mathrm{PM}})$'s and $\Im A_n(N_{\mathrm{PM}})$'s of two polarization modes. Since the sum χ^2 of powers of four normally distributed random numbers divided by their own variances is chi-squared distributed with 4 degrees of freedom (df):

$$f_{\chi^2,4\mathrm{df}}(x) = \frac{x}{4}\exp\left(-\frac{x}{2}\right). \tag{2.38}$$

$X_{nm}/\sigma^2_{4\mathrm{df},nm}$ can be chi-square distributed with 4 df, and its statistical average $\mu_{X,nm}/\sigma^2_{4\mathrm{df},nm}$ is 4. Therefore, the PDF and the statistical mean $\mu_{X,nm}$ of X_{nm} can be obtained as

$$f_{X,4\mathrm{df}}(X_{nm}) = f_{\chi^2,4\mathrm{df}}\left(\frac{X_{nm}}{\sigma^2_{4\mathrm{df},nm}}\right)\frac{\mathrm{d}}{\mathrm{d}X}\left(\frac{X_{nm}}{\sigma^2_{4\mathrm{df},nm}}\right) = \frac{X_{nm}}{4\sigma^4_{4\mathrm{df},nm}}\exp\left(-\frac{X_{nm}}{2\sigma^2_{4\mathrm{df},nm}}\right), \tag{2.39}$$

$$\mu_{X,nm} = 4\sigma^2_{4\mathrm{df},nm} = N_{\mathrm{PM}}|\mathrm{K}_{nm}|^2. \tag{2.40}$$

This mean XT equals to the XT calculated from (2.35)

Hayashi et al. [32, 41] observed the statistical distribution of the XT X in the actual MCF by utilizing wavelength sweeping for randomizing ϕ_{rnd}. Figure 2.15 shows examples of a spectrum of the XT of the actual MCF and the probability distribution of the XT obtained from the spectrum. The spectrum was measured using a tunable laser whose linewidth is specified at 100 kHz, which was adequately narrow to avoid averaging of the XT over the wavelength. The measured XT distribution is well fitted by the theoretical fitting line of (2.39). So, indeed, the XT in the measured MCF is considered to be normally distributed on the I-Q planes of the two polarizations.

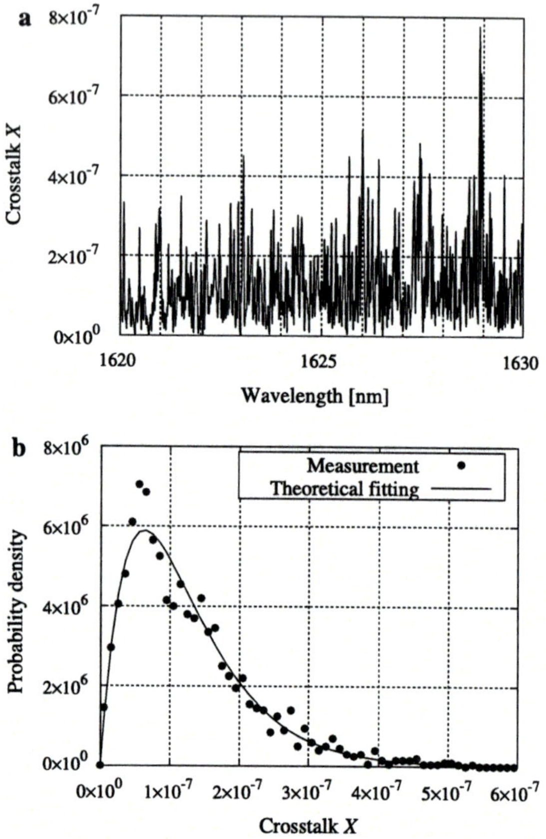

Fig. 2.15 Examples of **a** a crosstalk spectrum of a 17-km-long homogeneous MCF and **b** a probability distribution of the crosstalk obtained from the spectrum (Reprinted with permission from [42]. © 2013 Elsevier)

They also observed that the XT varies with time and its statistical distribution also obeys (2.39) [40]. For organizing the stochasticity of the XT, the variation, averages, and probability distribution of X are illustrated schematically in Fig. 2.16.

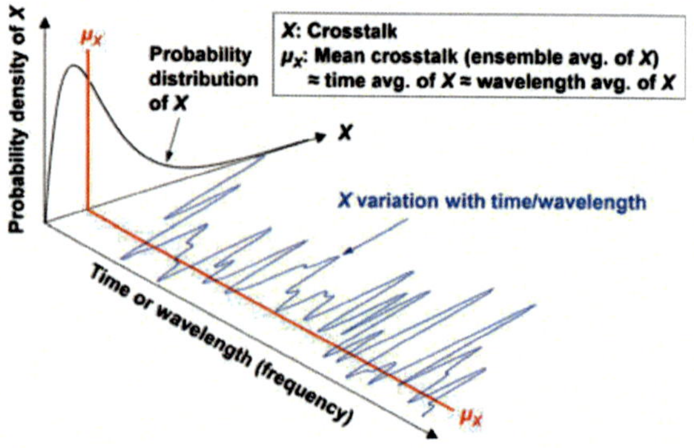

Fig. 2.16 Schematic illustration of the parameters related to the crosstalk (Reprinted with permission (#19SB0033) from [40]. © 2014 IEICE)

Though we discussed the case where the bend perturbation is dominant, the PDF of the XT in the MCF can be considered to converge to normal distribution on the I–Q plane even in cases where the bend perturbation is small if the MCF length is adequately long. This is because the phase correlations between the cores exponentially decay in longitudinal direction with a correlation length of only several centimeters, and the phase differences between the cores are uncorrelated between the adequately separated different longitudinal points.

2.3.2 *Low-Crosstalk MCF*

MCFs for SDM can be fundamentally classified into two different categories as shown in Fig. 2.17. The first type is a weakly coupled MCF, in which each core is used as an independent waveguide with sufficiently low interference (crosstalk) between neighboring cores. In weakly coupled MCFs, the optical crosstalk between adjacent cores is an important problem, since part of the optical power launched into one of the cores is coupled with neighboring cores during the propagation. In order to keep the crosstalk at low levels, the coupling coefficient κ between neighboring cores should be lower than 10^{-2} m^{-1} [43] for long-distance transmission, sacrificing the core density and the typical core pitch is around 40 μm. In this case, multiple-input multiple-output (MIMO) digital signal processing (DSP) is not needed at the receiver side for recovering the signals. The second type is a strongly coupled MCF, in which the crosstalk between cores is intentionally introduced by decreasing the core pitch, resulting in the core density improvement. In theory, the strongly coupled MCF supports several super-modes, and it can be considered to have a form of multi-mode fibers (MMFs). In practice, these super-modes are strongly mixed due to structural parameter fluctuations and/or bending effect, if the effective refractive index difference Δn_{eff} between each mode is relatively small (typically $\Delta n_{\mathrm{eff}} < 10^{-5}$). In MMF

	Weakly-coupled MCF	Strongly-coupled MCF
Examples of schematic cross-sectional view	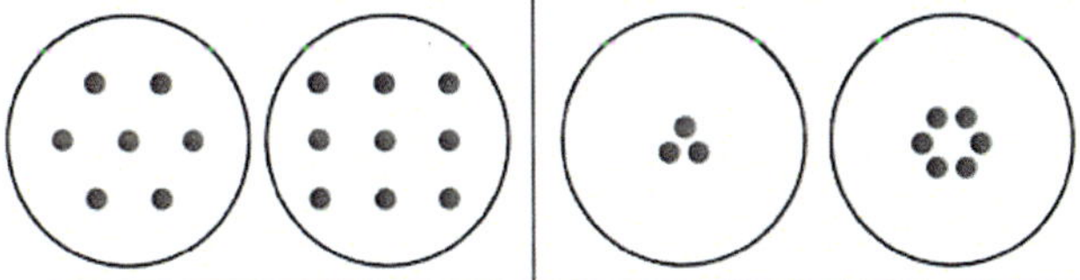	
Coupling coefficient κ [m^{-1}]	κ<0.01	κ>0.1
Core pitch Λ [μm]	Λ>30	Λ<30
MIMO for inter-core crosstalk	No need	Need

Fig. 2.17 Classification of multi-core fibers for SDM (Reprinted with permission from [49]. © 2016 IEEE)

transmission, a group delay spread originating from the modal dispersion is one of the major problems since the magnitude of group delay spread determines the complexity of MIMO receiver (and hence, the system reach) [44]. The strong mode mixing in strongly coupled MCFs is beneficial for reducing the group delay spread, and the group delay spread is proportional to the square root of the transmission distance [45]. Recently, this square root dependence of group delay spread was experimentally reported by using a well-designed coupled three-core fiber up to 4200 km [46] and a coupled six-core fiber up to 1705 km [47]. The coupling coefficient κ between neighboring cores in the strongly coupled MCFs is in the order of 10^{-1} m^{-1} or more, and the typical core pitch is less than 30 μm. In this case, MIMO DSPs with low complexity are needed at the receiver side to recover the signals. It should be noticed that the mode coupling during the propagation between super-modes becomes weak if the coupling between cores is too strong, since the effective refractive index difference Δn_{eff} between each mode becomes large (typically $\Delta n_{eff} > 10^{-4}$). In this case, it can be used as an MMF similar to a single-core MMF with high spatial density [48], although its group delay spread is proportional to the transmission distance and complex MIMO DSP is required as the transmission distance increases. Among these two categories of MCFs, SDM based on weakly coupled MCFs is a simple and promising approach for increasing spatial channel count (SCC), resulting in improvement of transmission capacity per fiber. Various kinds of weakly coupled MCFs have been developed to achieve high-capacity long-distance transmission. Recent single-mode MCF (SM-MCF) transmission experiments have achieved the transmission capacity well beyond the fundamental limit of single-core single-mode fiber [31].

In order to decrease crosstalk in weakly coupled MCFs, the coupling coefficient κ between cores has to be reduced. Trench-assisted MCFs [31, 50, 51] and hole-assisted MCFs [33, 35, 52, 53] have been proposed for reducing the coupling coefficient. Figure 2.18a shows an example of cross-sectional view of a fabricated trench-assisted MCF with 12 cores, where each core has low-index trench layer around the core. In Fig. 2.18b, the schematic diagram of an index profile with trench-assisted structure

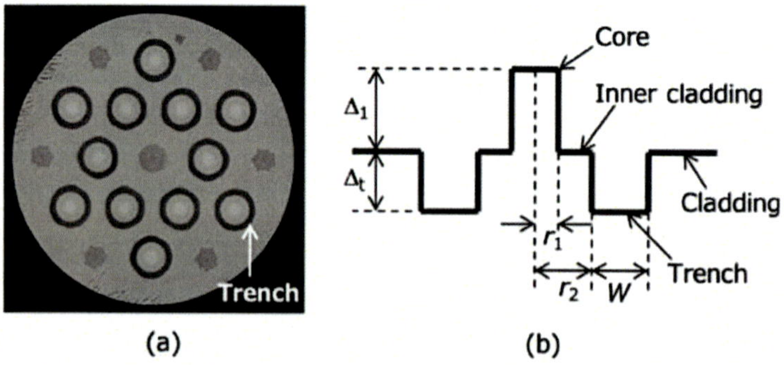

Fig. 2.18 **a** Cross-sectional view of a fabricated trench-assisted MCF and **b** its index profile (Reprinted with permission from [49]. © 2016 IEEE)

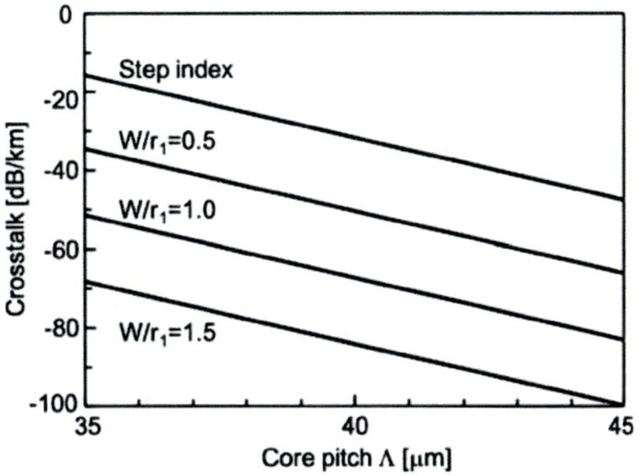

Fig. 2.19 Crosstalk between two cores as a function of the core pitch in a step-index MCF and trench-assisted MCFs with various relative trench widths of W/r_1 at a wavelength of 1550 nm with a bending radius of $R =$ 140 mm, where $r_1 =$ 4.5 μm, $r_2/r_1 = 2.0$, $\Delta_1 =$ 0.35%, and $\Delta_t = -0.70\%$ (Reprinted with permission from [49]. © 2016 IEEE)

is shown. The relative refractive index differences between core and cladding, trench and cladding are Δ_1 and Δ_t, respectively, and r_1, r_2, and W are the core radius, the distance from the core center to the inner edge of trench layer, and the trench width, respectively. Due to the existence of low-index trench layer with the thickness of W, the overlap of the electromagnetic fields between cores can be greatly suppressed, resulting in the suppression of crosstalk compared with that in an MCF without trench. Figure 2.19 shows the crosstalk between two cores as a function of the core pitch in a step-index MCF and trench-assisted MCFs with various relative trench widths of W/r_1 at a wavelength of 1550 nm with a bending radius of $R = 140$ mm, where $r_1 = 4.5$ μm, $r_2/r_1 = 2.0$, $\Delta_1 = 0.35\%$, and $\Delta_t = -0.70\%$, as an example. The crosstalk decreases linearly as increasing the core pitch and the amount of crosstalk reduction in trench-assisted structure relative to step-index structure is core pitch independent [54]. It can be seen that more than 30 dB crosstalk reduction is achievable with a typical trench width of $W/r_1 = 1.0$.

The second approach to suppress the crosstalk is to introduce intrinsic index difference between adjacent cores [38, 55] which is called heterogeneous MCFs. The heterogeneous MCF consists of several kinds of cores whose propagation constants are different from each other, and the bending perturbation plays an important rule for predicting the crosstalk in heterogeneous MCFs [21, 22]. Figure 2.20 shows the averaged power-coupling coefficient h [m^{-1}] as a function of the intrinsic effective index difference Δn_{eff} between cores and the bending radius R at 1550 nm wavelength, where the core pitch is $\Lambda = 40$ μm and the coupling coefficient is assumed to be $\kappa = 0.01$ m^{-1} as an example. It is shown that, when the bending radius is smaller than a critical bending radius R_c, there is a linear relation between a bending radius in a logarithmic scale and a power-coupling coefficient, where $R_c = n_{\text{eff}}\Lambda/\Delta n_{\text{eff}}$ with n_{eff}, Δn_{eff}, and Λ being the effective index of the propagating mode in a core, the intrinsic effective index difference between cores, and the core pitch, respectively. On the other hand, over the bending radius larger than the critical bending radius R_{c}, the power-coupling coefficient decreases rapidly. This is because, if the bending radius is smaller than R_{c}, the bent-induced effective index variation is larger than

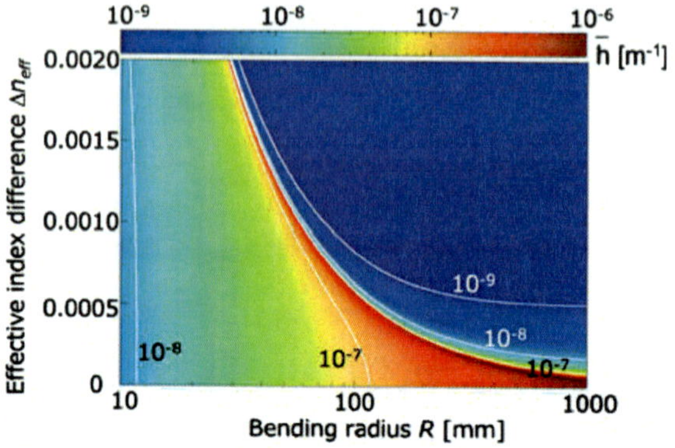

Fig. 2.20 Averaged power-coupling coefficient as a function of the intrinsic effective index difference Δn_{eff} between cores and the bending radius R at 1550 nm wavelength, where the core pitch is $\Lambda = 40\ \mu\text{m}$ and the coupling coefficient is assumed to be $\kappa = 0.01\ \text{m}^{-1}$ (Reprinted with permission from [49]. © 2016 IEEE)

the intrinsic index difference Δn_{eff} between heterogeneous cores, and large crosstalk degradation occurs many times during the propagation at the phase-matching points between cores. On the other hand, in the non-phase-matching region of bending radii $> R_c$, the bent-induced effective index variation is smaller than the intrinsic index difference Δn_{eff}; therefore, the crosstalk is dominated by the statistical properties [23, 26]. In this non-phase-matching region, the heterogeneous MCFs can be used as a bending insensitive fiber in terms of crosstalk. If the effective index difference between cores is sufficiently large, the value of R_c can be pushed toward small range lower than an effective bending radius in MCF cables. Recently, a heterogeneous MCF with 30 cores has been reported [56]. Figure 2.21a, b show a schematic structure and cross-sectional view of the fabricated 30-core fiber with heterogeneous core arrangement, respectively. The fiber structure is based on a hexagonal close-packed structure (HCPS) with 37 cores. Six outermost cores were removed to reduce the cladding diameter. The center core was also removed because the cutoff wavelength lengthening of the center core due to the trench-assisted structure was unavoidable. Four kinds of cores were used to produce a heterogeneous relation for all the adjacent cores under the limitation of the similar effective area (A_{eff}) for all the cores. Cores 1, 2, and 3 employed a trench-assisted index profile shown in Fig. 2.18b to reduce the crosstalk, whereas Core 4 has a step-index profile without trench layer to avoid the cutoff wavelength lengthening of the inner cores surrounded by trench-assisted

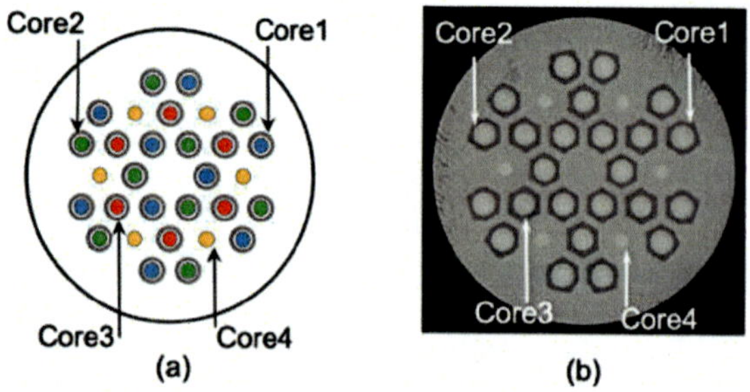

Fig. 2.21 **a** Schematic structure, **b** cross-sectional view, and **c** index profiles for a fabricated 30-core fiber with heterogeneous core arrangement [56] (Reprinted with permission from [49]. © 2016 IEEE)

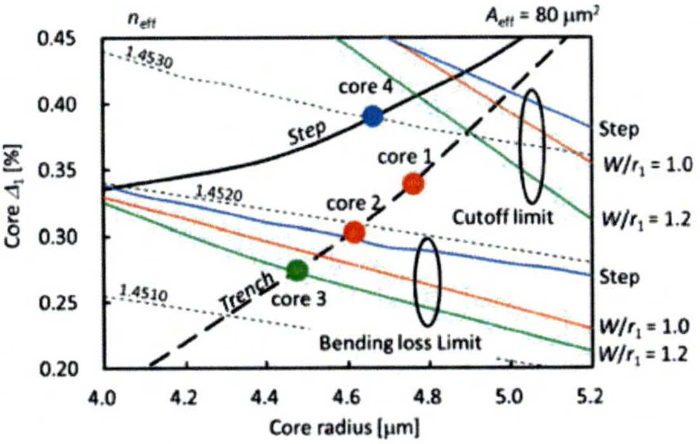

Fig. 2.22 Relationship between core parameters and effective indices in trench-assisted and step-index cores (Reprinted with permission from [49]. © 2016 IEEE)

cores. Their core parameters were determined based on the effective index difference as shown in Fig. 2.22, where the normalized trench position is $r_2/r_1 = 1.7$ and $\Delta_t = -0.7\%$ for trench-assisted cores. All the cores were selected so that the effective index difference between adjacent cores becomes larger than 0.0005 with the similar A_{eff} of 80 μm^2. The fabricated fiber length was 9.6 km, and it had the core pitch of 29.7 μm, the cladding diameter of 228 μm, and the averaged A_{eff} of 77.3 μm^2 at 1550 nm wavelength. In addition, the critical bending radius R_c, where the crosstalk becomes a peak value, was less than 100 mm for all the core combinations. The measured crosstalk was less than −50 dB.

Recently, a 37-core fiber with the HCPS has been reported [57]. The 37-core is composed of three kinds of cores with different propagation constants to suppress crosstalk.

Another option for crosstalk suppression in MCFs is to utilize propagation-direction interleaving (PDI) technique [58, 59], where adjacent cores are assigned to opposite direction. We can reduce the number of adjacent core in which the signal propagates to the same direction to one for all cores by using PDI. Accordingly, effective crosstalk is expected to be reduced. In the unidirectional propagation scheme, the inputted power in excited core is coupled to adjacent cores and the crosstalk is forward propagated crosstalk, XT_f, as shown in Fig. 2.23a. On the other hand, in the bidirectional propagation scheme, the crosstalk from the adjacent cores is generated when the backscattered light is coupled with the opposite direction as shown in Fig. 2.23b, and the dominant factor in back-propagating signals is Rayleigh backscattering [59]. The backward propagated crosstalk, XT_b, is given by

$$\text{XT}_b = \frac{\frac{S\alpha_R}{2\alpha}\overline{h}_{mn}\left[\frac{1-\exp(-2\alpha L)}{\alpha} - 2L\exp(-2\alpha L)\right]}{\exp(-\overline{h}_{mn}L)\cos h(\overline{h}_{mn}L)\exp(-\alpha L)}, \tag{2.41}$$

where S and α_R are the recapture factor of the Rayleigh scattering component into the backward direction and the attenuation coefficient due to Rayleigh scattering, respectively. $\overline{h}_{mn}$ is the averaged power-coupling coefficient. α and L are the fiber

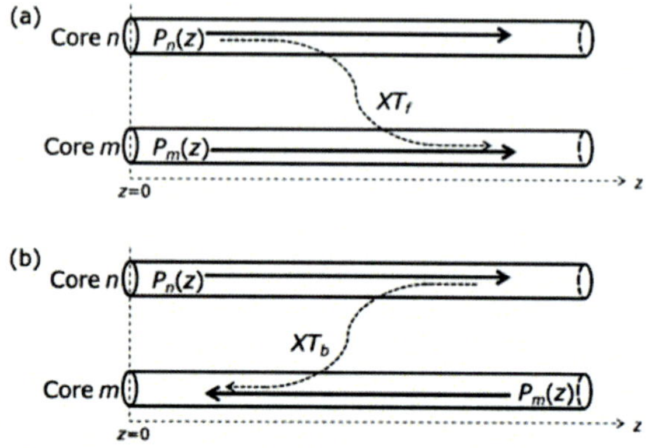

Fig. 2.23 Crosstalk in Core *m* from Core *n* in **a** unidirectional propagation scheme and **b** bidirectional propagation scheme of PDI (Reprinted with permission from [49]. © 2016 IEEE)

attenuation coefficient and fiber length, respectively. Figure 2.24 shows the forward propagated crosstalk, XT_f, and the backward propagated crosstalk, XT_b, as a function of the fiber length in a trench-assisted MCF at a wavelength of 1550 nm with a bending radius of $R = 140$ mm, where $r_1 = 4.5$ µm, $r_2/r_1 = 2.0$, $W/r_1 = 1.0$, $\Delta_1 = 0.35\%$, $\Delta_t = -0.70\%$, and the core pitch is 40 µm, as an example. The Rayleigh backscattering coefficient ($S\alpha_R/2a$) and the fiber attenuation coefficient were assumed to be -32 dB [60] and 0.2 dB/km, respectively. XT_b is smaller than XT_f by more than 18 dB for the fiber length of 100 km or less; however, it can be seen that the crosstalk reduction effect of PDI becomes weaker as the fiber length increases and it is limited by Rayleigh scattering.

By adopting combination of these crosstalk suppression techniques, low-crosstalk and high-core-count MCFs can be designed, such as heterogeneous trench-assisted MCFs [56] and trench-assisted MCFs with PDI [59]. The allowable crosstalk is depending on the transmission distance as well as the modulation format to be used. In [61], the effect of the crosstalk on the optical signal to noise ratio was numerically investigated by assuming the crosstalk as a static coupling, and experimentally by realizing the crosstalk as an effectively static coupling using optical couplers and variable optical attenuators. It has been reported that the crosstalk-induced penalty

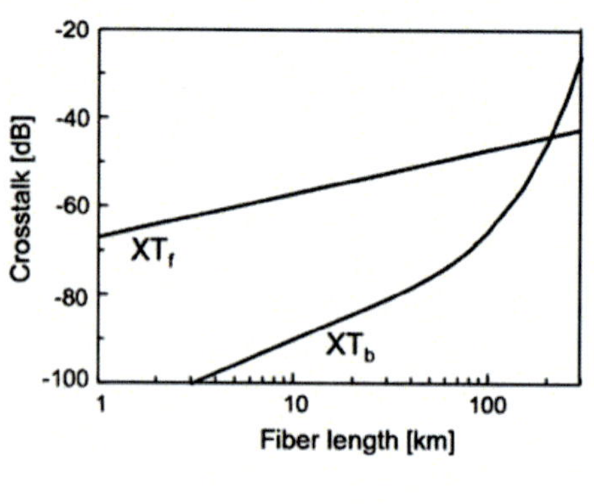

Fig. 2.24 Forward propagated crosstalk and backward propagated crosstalk as a function of the fiber length in a trench-assisted MCF at a wavelength of 1550 nm with a bending radius of $R =$ 140 mm, where $r_1 = 4.5$ µm, $r_2/r_1 = 2.0$, $W/r_1 = 1.0$, $\Delta_1 = 0.35\%$, and $\Delta_t = -0.70\%$ (Reprinted with permission from [49]. © 2016 IEEE)

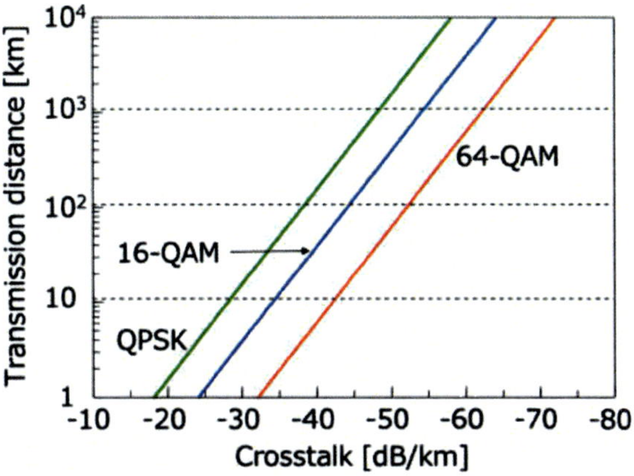

Fig. 2.25 Relation between the achievable transmission distance and the allowable crosstalk level for different modulation format (Reprinted with permission from [49]. © 2016 IEEE)

at the bit-error rate of 10^{-3} is less than 1 dB, when the crosstalk is less than -18 dB for quadrature phase-shift keying (QPSK), -24 dB for 16 quadrature amplitude modulation (QAM), and -32 dB for 64 QAM [61]. Similarly, in [59], the crosstalk penalty in Q-factor for QPSK, 16 QAM, and 32 QAM signals were experimentally measured by using 50-km 12-core fiber. It has been reported that the allowable crosstalk for 0.3-dB Q-penalty was -18 dB for QPSK, -25 dB for 16 QAM, and $-$ 29 dB for 32 QAM. Based on these relations between the allowable crosstalk level and the modulation format, a relation between the achievable transmission distance and the allowable crosstalk level for different modulation format can be plotted as shown in Fig. 2.25. If the modulation format is QPSK, the crosstalk level of -40 dB/km is acceptable for 100 km transmission. On the other hand, if the modulation format is 64 QAM, the crosstalk level of around -55 dB/km is required for 100 km transmission. Therefore, the crosstalk level in MCFs should be carefully designed based on the transmission distance and the modulation format for the transmission experiment.

We should also notice, in actual MCFs, each core generally has several adjacent cores and it is affected by all the nearest cores. If all cores carry equal signal power, the worst-case crosstalk of inner cores is larger than that of outer cores, since the number of nearest neighbor cores is larger for inner cores. For example, in the case of the HCPS, the inner core has six nearest neighbor cores, while the outer cores have three or four nearest neighbor cores. The worst crosstalk, $\mathrm{XT_{worst}}$, in [dB] is given by

$$\mathrm{XT_{worst}} = \mathrm{XT} + 10\log n, \tag{2.42}$$

where XT is the crosstalk between two cores in [dB], and n is the number of nearest neighbor cores. Figure 2.26 shows the crosstalk increment due to nearest neighbor cores (ΔXT) for HCPS, one-ring structure (ORS), and dual-ring structure (DRS), where $\Delta\mathrm{XT}=\mathrm{XT_{worst}} - \mathrm{XT} = 10\log n$. In HCPS, the ΔXT of inner cores is 7.8 dB and that of outer cores is 4.8 dB for three nearest neighbor cores or 6.0 dB for four

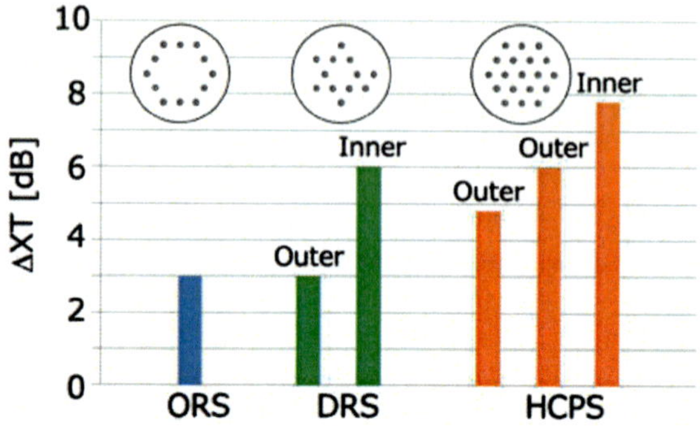

Fig. 2.26 Crosstalk increment due to nearest neighbor cores (ΔXT) for hexagonal close-packed structure (HCPS), one-ring structure (ORS), and dual-ring structure (DRS) (Reprinted with permission from [49]. © 2016 IEEE)

nearest neighbor cores. On the other hand, in ORS, ΔXT is 3 dB for all the cores. The allowable crosstalk level in MCFs should be determined by considering XT_{worst}.

MCFs for SDM applications should have high spatial efficiency compared with a conventional single-core single-mode fiber. The number of cores to be multiplexed in a fiber is related to the core pitch, and the core pitch can be determined by considering the allowable crosstalk based on the transmission distance and the modulation format. The upper limit of the core number is also limited by a cladding diameter (CD), since a small cladding diameter is preferable for maintaining high mechanical reliability for bending. The failure probability of a standard single-mode fiber after 20 years with 125 μm cladding diameter and 30 mm bending diameter is about 10^{-7}, where the proof level is 1% and the number of bending is 100 turns. This failure probability of 10^{-7} would be employed for a threshold also in MCFs, and the failure probability of MCFs should be smaller than 10^{-7} at bending diameter of 60 mm, which is the bending diameter in a storage box. It has been reported that the cladding diameter of MCFs should be smaller than around 250 μm to satisfy the limit of failure probability [62]. In addition, the cladding thickness (CT), which is a distance from a center of outer core to outer cladding edge, should be sufficiently large for reducing micro-/macro-bending losses in outer cores [63]. Therefore, the number of cores as well as the core arrangement has to be determined by taking the allowable crosstalk and the required CT into account under the limitation of the CD.

Table 2.3 summarizes the recently reported weakly coupled SM-MCFs [41, 53, 56, 57, 64–72]. In Table 2.3, in order to compare the core density and crosstalk level of these MCFs, SCC, A_{eff}, CD, CT, and 100-km worst crosstalk are included. In SM-MCFs, SCC corresponds to a core count. Figure 2.27 shows the SCC as functions of the 100-km worst crosstalk and the CD for MCFs listed in Table 2.3. For SM-MCFs, the reported maximum SCC is 37 [56] with the worst crosstalk of −30 dB/100 km in the $C + L$ band. It should be noted that the crosstalk of MCFs increases with increasing the wavelength and the slope of crosstalk over wavelength regimes was measured to be approximately 0.1 dB/nm [73, 74]. Therefore, the worst-case crosstalk at the longer edge of L band is about 10 dB larger than that at the shorter edge of C band.

Table 2.3 Characteristics of recently reported weakly coupled single-mode multi-core fibers

Ref.	SCC	A_{eff} (μm^2) at 1550 nm	CD (μm)	CT (μm)	100-km worst crosstalk (dB) at 1550 nm
[41]	7	80	150	30	−64
[64]	7	~75	186.5	46.4	~−31
[62]	10	116	204.4	43	−23
[65]	7	110	195	48.5	−56
[66]	12	80	225	39	−42
[67]	7	99	200	55	−39.6
[68]	19	72	200	30	−14.3
[53]	6	72.7	125	30.9	−32.5 (at 1625 nm)
[69]	12	105.8	230	35	−45
[70]	19	85	220	34.8	−36.8
[56]	30	77.3	228	33.8	−42
[72]	32	80.3	243	37.2	−34
[57]	37	80.6	248	37	−39.5
[71]	8	~55 (at 1310 nm)	125	22	−42 (at 1310 nm)

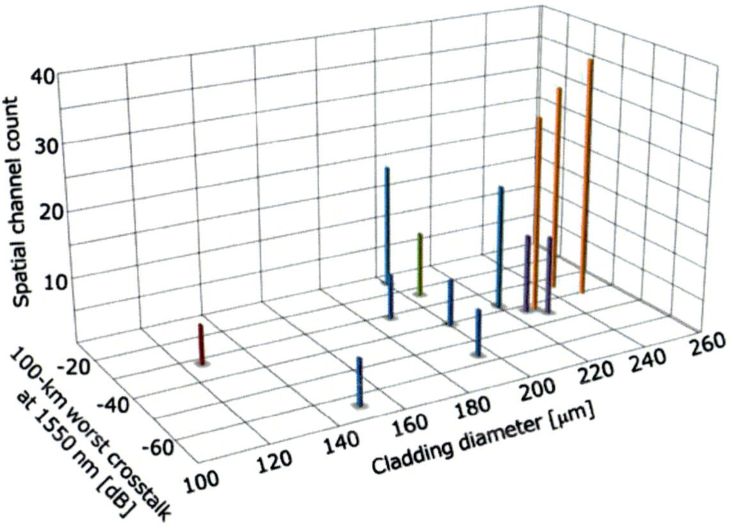

Fig. 2.27 SCC as functions of the 100-km worst crosstalk and the cladding diameter for SM-MCFs listed in Table 2.3 (Reprinted with permission from [49]. © 2016 IEEE)

2.3.3 *High-Density MCF*

To realize high-capacity and long-distance transmission, SNRs in MCF cores are an important factor as with single-core fibers. Of course, the XT suppression is necessary for improving the SNR in the MCF cores. However, there is a trade-off between higher core density, lower XT, larger effective area A_{eff}. In Sect. 2.3.3.1, we discuss the effects of system parameters and fiber parameters including the XT on SNRs in uncoupled MCF systems and investigate how the SNR can be improved in the MCF systems. Then, we discuss how the spatial spectral efficiency (aggregate spectral efficiency per fiber cross-sectional area) can be improved in the uncoupled MCF. In Sect. 2.3.3.2, we review actual MCF designs that can realize ultimately high spatial channel count/density.

2.3.3.1 Preferable Characteristics for Spatial Capacity Enhancement

Spectral Efficiency in MCF Core

In this section, according to [75], the SNR in the MCF core is derived from GN model [12, 76] for the single-mode fiber transmission system by assuming the XT can be regarded as additive white Gaussian noise for simplicity. When assuming an ideal filtering (a noise bandwidth is equivalent to a signal bandwidth) and Nyquist limit of WDM, SNR of dispersion-uncompensated single-core (SC) transmission link can be expressed as [12, 76]:

$$\mathrm{SNR}_{\mathrm{SC}} = \frac{P_{\mathrm{Tx,ch}}}{P_{\mathrm{ASE}}|_{B_{\mathrm{n}}=R_{\mathrm{s}}} + P_{\mathrm{NLI}}|_{B_{\mathrm{n}}=R_{\mathrm{s}}}}, \tag{2.43}$$

$$P_{\mathrm{ASE}} = N_{\mathrm{s}} F\left(e^{\alpha L_{\mathrm{s}}} - 1\right) \mathrm{h}\nu B_{\mathrm{n}}, \tag{2.44}$$

$$P_{\mathrm{NLI}} \approx a_{\mathrm{NLI}} P_{\mathrm{Tx,ch}}^{3}, \tag{2.45}$$

$$a_{\mathrm{NLI}} \approx \left(\frac{2}{3}\right)^{3} N_{\mathrm{s}} \gamma^{2} L_{\mathrm{eff}} \frac{\ln\left(\pi^{2}|\beta_{2}| L_{\mathrm{eff}} B_{\mathrm{WDM}}^{2}\right)}{\pi |\beta_{2}| R_{\mathrm{s}}^{3}} B_{\mathrm{n}}, \tag{2.46}$$

where $P_{\mathrm{Tx,ch}}$, P_{ASE}, P_{NLI} are the average powers of transmitted power per channel, of the ASE noise within the OSNR noise bandwidth B_{n}, and of the nonlinear interference (NLI) within B_{n}, respectively, and where N_{s} is the number of spans, F the EDFA noise figure, α the power loss per unit length, L_{s} the span length, h the Planck's constant, ν the center frequency of WDM comb, $\gamma = 2\pi n_2/(\lambda A_{\mathrm{eff}})$ is the nonlinear coefficient, n_2 the nonlinear refractive index, $L_{\mathrm{eff}} = [1 - \exp(-\alpha L_{\mathrm{s}})]/\alpha$ is the effective length, $\beta_2 = -\lambda^2 D/(2\pi c)$ is the chromatic dispersion in [(unit time)^2/(unit length)], B_{WDM}

$= N_{ch}R_s$ is the WDM bandwidth, N_{ch} the number of WDM channels, and R_s the symbol rate.

Based on the characteristics of XT in MCF shown in Sect. 2.3.1, XT in MCFs may be regarded as a virtual additive white Gaussian noise by assuming adequately broad bandwidth of signal light. Thus, SNR in a certain core in dispersion-uncompensated multi-core (MC) transmission link with uncoupled MCF can be simply expressed as

$$\begin{aligned} \mathrm{SNR}_{\mathrm{MC}} &= \frac{P_{\mathrm{Tx,ch}} - P_{\mathrm{XT}}}{P_{\mathrm{ASE}}|_{B_n=R_s} + P_{\mathrm{NLI}}|_{B_n=R_s} + P_{\mathrm{XT}}} \\ &\approx \frac{P_{\mathrm{Tx,ch}}}{P_{\mathrm{ASE}}|_{B_n=R_s} + P_{\mathrm{NLI}}|_{B_n=R_s} + P_{\mathrm{XT}}}, \end{aligned} \tag{2.47}$$

where P_{XT} is the power of the XT from or to the certain core. In (2.47), $P_{\mathrm{Tx,ch}}$s equivalent between all cores and losses equivalent between all cores are assumed. "$-P_{\mathrm{XT}}$" in the numerator represents the XT from the certain core to other cores, and "$+P_{\mathrm{XT}}$" in the denominator represents the XT from other cores to the certain core. However, "$-P_{\mathrm{XT}}$" is negligible in the case of low XT ($P_{\mathrm{Tx,ch}} \gg P_{\mathrm{XT}}$). Depending on system capability, P_{XT} may have some different definitions. If skew between cores is small and thus pulse broadening due to multipath interference (MPI) can be mitigated by digital signal processing, P_{XT} can simply be the power of light that is input to all cores except the certain core and output from the certain core. If the skew is very large and thus the MPI cannot be mitigated, P_{XT} can be the power of light that is input to all cores, coupled between cores, and output from the certain core. But in either case, if the XT is adequately low (i.e., $\mu_{\mathrm{XT}} <$ ~0.01), P_{XT} can be related to $P_{\mathrm{Tx,ch}}$ by μ_{XT} as

$$P_{\mathrm{XT}} \approx \mu_{\mathrm{XT}} P_{\mathrm{Tx,ch}}. \tag{2.48}$$

In the following discussion, we discuss the case of a core (the worst core) to which total μ_{XT} from other cores is highest. When the XT is adequately low and XT between non-adjacent cores is negligible, the total μ_{XT} ($\mu_{\mathrm{XT,WC}}$) to the worst core can be approximated as

$$\mu_{\mathrm{XT,WC}} \approx \eta_{\mathrm{WC}} N_s L_s, \quad (\mu_{\mathrm{XT,WC}} < \sim 0.01) \tag{2.49}$$

$$\eta_{\mathrm{WC}} \approx N_c \eta, \tag{2.50}$$

where η_{WC} is the aggregate power-coupling coefficient to the worst core, and N_c is the maximal number of the adjacent cores.

(2.43), (2.47)–(2.49) yield

$$\mathrm{SNR}_{\mathrm{MC}}^{-1} \approx \mathrm{SNR}_{\mathrm{SC}}^{-1} + \mu_{\mathrm{XT,WC}}$$

$$\approx \mathrm{SNR}_{\mathrm{SC}}^{-1} + \eta_{\mathrm{WC}} N_s L_s. \quad (2.51)$$

Then SNR penalty due to XT can be expressed as

$$\frac{\mathrm{SNR}_{\mathrm{SC}}}{\mathrm{SNR}_{\mathrm{MC}}} \approx 1 + \mathrm{SNR}_{\mathrm{SC}} \mu_{\mathrm{XT,WC}}. \quad (2.52)$$

Figures of merit of fibers can be evaluated by comparing maximal SNRs achievable in the same system using the different fibers. In both of the SC and MC cases, SNRs are maximized at $P_{\mathrm{Tx,ch}} = [P_{\mathrm{ASE}}/(2a_{\mathrm{NLI}})]^{1/3}$. Therefore, the maximal SNR ($\mathrm{SNR}_{\mathrm{SC,max}}$) achievable in a SC system can be derived as

$$\begin{aligned}
\mathrm{SNR}_{\mathrm{SC,max}} &= \left[3\left(\frac{P_{\mathrm{ASE}}}{2}\right)^{\frac{2}{3}} a_{\mathrm{NLI}}^{\frac{1}{3}} \right]^{-1} \Bigg|_{B_n = R_s} \\
&\approx \left\{ \left[\frac{\left(e^{\alpha L_s} - 1\right)^{\frac{2}{3}} \left(\gamma L_{\mathrm{eff}}\right)^{\frac{2}{3}}}{\left(|\beta_2| L_{\mathrm{eff}}\right)^{\frac{1}{3}}} \right] \left[\ln\left(\pi^2 |\beta_2| L_{\mathrm{eff}} B_{\mathrm{WDM}}^2\right)\right]^{\frac{1}{3}} \right. \\
&\quad \left. \left[\left(\frac{2}{\pi}\right)^{\frac{1}{3}} N_s (Fh\nu)^{\frac{2}{3}} \right] \right\}^{-1} \\
&\approx \left[\frac{\left(e^{\alpha L_s} - 1\right)^{\frac{2}{3}} \left(\gamma L_{\mathrm{eff}}\right)^{\frac{2}{3}}}{\left(|\beta_2| L_{\mathrm{eff}}\right)^{\frac{1}{3}}} \right]^{-1} \frac{C_{\mathrm{system}}}{N_s}, \quad (2.53)
\end{aligned}$$

where C_{system}/N_s includes the second and third bracket terms in the second line. When assuming moderately high $|\beta_2|$ and adequately broad B_{WDM}, C_{system}/N_s can be canceled out by taking a ratio of $\mathrm{SNR}_{\mathrm{SC,max}}$s in the same system between different fibers. Accordingly, a figure-of-merit of a single-core fiber can be expressed only by the bracket term in the last line.

Since the Shannon limit of the SE for a polarization mode can be expressed as

$$\mathrm{SE} = \log_2(1 + \mathrm{SNR}), \quad (2.54)$$

the maximum achievable SE at the optimum input power, hereinafter referred to as SE limit ($\mathrm{SE}_{\mathrm{lim}}$), in the MCF can be calculated from the XT-affected SNR. For example, Fig. 2.28 shows the relationship between the core pitch (Λ), the effective area (A_{eff}), and $\mathrm{SE}_{\mathrm{lim}}$ at 1550 nm for a step-index core surrounded by its six identical cores in hexagonal lattice. The step-index core was assumed to have attenuation of 0.18 dB/km and the cable cutoff wavelength (λ_{cc}) of 1530 nm. The core Δ and radius depend on A_{eff}, since λ_{cc} is fixed. The calculation assumed 6 dB of EDFA noise figure, 10 THz of whole signal bandwidth, 80-km span length, 3200 km of total length, and fiber bending radius R_b of 1 m. A steep change of the $\mathrm{SE}_{\mathrm{lim}}$ in

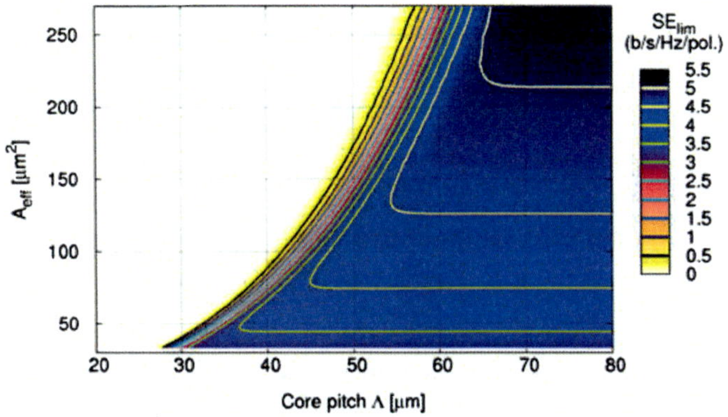

Fig. 2.28 A relationship between Λ, A_{eff}, and XT-affected SE_{lim} at 1550 nm for hexagonally arranged step-index cores whose attenuation is 0.18 dB/km and λ_{cc} is 1530 nm (Reprinted with permission from [77]. © 2013 IEEE)

the horizontal direction is caused by the large dependence of the XT on Λ. We can understand that A_{eff} enlargement improves the SE_{lim} but requires larger Λ for avoiding an XT increase.

Spatial Spectral Efficiency of MCF

Based on the relationship shown in Fig. 2.28, the spatial spectral efficiency (SSE), which was defined as the aggregate SE per cross-sectional area (A_{cs}), can be expressed as:

$$\mathrm{SSE} \equiv \frac{\sum \mathrm{SE}}{A_{cs}}. \tag{2.55}$$

The SSE may be defined for either the bare MCF, the coated MCF, or the MCF cable, by using their aggregate SE and A_{cs}.

Before considering the SSE limit (SSE_{lim})—that is, SSE obtained from SE_{lim}—for the MCF, the potential SSE_{lim} of the MCF core can be defined as the SSE_{lim} where the cores are arranged in the hexagonal lattice on the infinite plane:

$$\mathrm{SSE}_{\mathrm{lim,core}} \equiv \frac{\sum_{\mathrm{core}} \mathrm{SE}_{\mathrm{lim}}}{\left(\frac{\Lambda^2\sqrt{3}}{2}\right)} \tag{2.56}$$

and we can discuss how the SSE of the MCF can be improved.

Figure 2.29 shows the relationship between Λ, A_{eff}, and the ratio of the $SSE_{lim,core}$ of step-index cores, divided by SSE_{lim} ($SSE_{lim,SMF}$) of the SMF whose A_{eff} is 80 μm^2 at 1550 nm, λ_{cc} is 1260 nm, and coating diameter (D_{coat}) is 250 μm. For this graph,

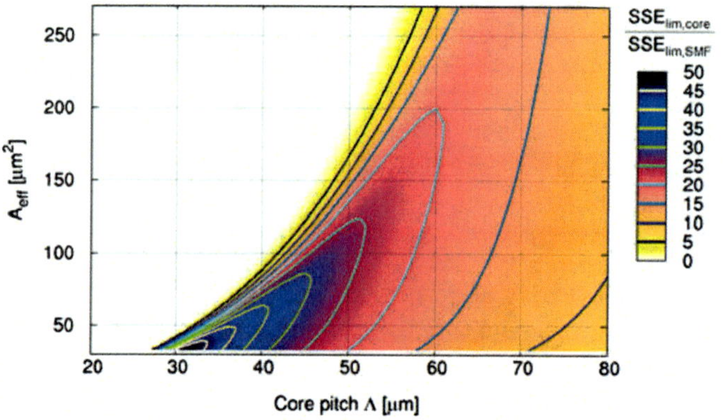

Fig. 2.29 A relationship between Λ, A_{eff}, and a ratio of potential SSE_{lim} ($SSE_{lim,core}$) of the hexagonally arranged step-index cores, divided by SSE_{lim} ($SSE_{lim,SMF}$) of the SSMF whose A_{eff} is 80 μm^2 at 1550 nm and D_{coat} is 250 μm (Reprinted with permission from [77]. © 2013 IEEE)

$SSE_{lim,SMF}$ is also calculated using (2.56) by substituting D_{coat} to Λ. According to the SSE definition in (2.56), Fig. 2.29 can be understood as Fig. 2.28 multiplied by the core density $1/(\Lambda^2\sqrt{3}/2)$. The core density increase with Λ decrease is much larger than SE decrease with Λ decrease. So, the highest $SSE_{lim,core}$ can be observed at A_{eff} less than ~50 μm^2 and Λ around 30 μm in this case and was 40 times larger than $SSE_{lim,SMF}$ despite the simple structure of the cores.

The SSE_{lim} of the MCF is smaller than and different from the potential SSE_{lim} of the MCF core, because the actual MCF cross section is not infinite plane and needs peripheral margin, i.e., outer cladding thickness (OCT) for suppressing coating-leakage loss, and coating layer.

To improve the core density in the MCF cross section, the core layout design is important. Since the hexagonal packing of equal circles is the densest in the plane, one might assume that a proper hexagonal core layout can be the optimal packing; however, there are cases where non-hexagonal layouts can increase the core count more than hexagonal layouts, because the core-arrangeable area is limited.

The core layout problem can be reduced to the problem of packing circles in a circle, once the cladding diameter (D_{clad}), the OCT, and the minimum core distance (d_{min}) are determined—D_{clad}, OCT, and d_{min} are determined by mechanical reliability and ease of handling, by leakage loss to the coating, and by the XT and cutoff wavelength, respectively. Accordingly, we can leverage the achievements in mathematics [78, 79]. Figure 2.30 shows a relationship between the radial position (r_{om}) of the center of the outmost core from the fiber center, normalized by d_{min}, and the number of cores. It can be seen that the densest packing can effectively utilize a slight increase in r_{om} or a slight decrease in d_{min} for increasing the core count, compared to the hexagonal packing.

Based on the results in Figs. 2.28 and 2.30., the SSE_{lim} ($SSE_{lim,MCF}$) of the MCF with the densest core packing was calculated using $A_{cs} = \pi(D_{coat}/2)^2$, and shown

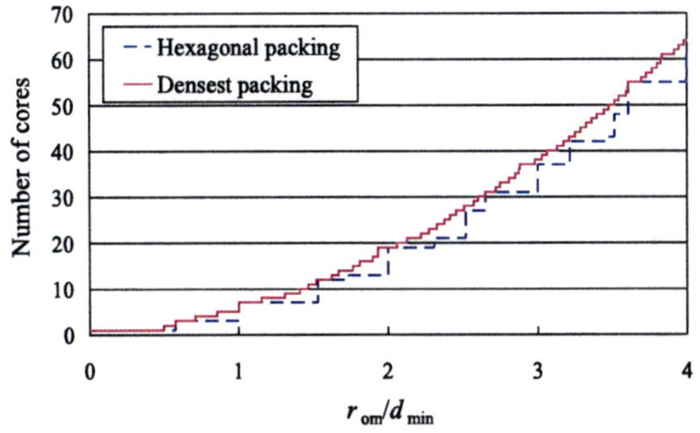

Fig. 2.30 A relationship between the radial position r_{om} of the center of the outmost core from the fiber center, normalized by the minimum core distance d_{min}, and the number of cores. The densest packing was calculated from [5, 6], by reducing the problem to packing equal circles, whose radius is $(d_{min}/2)/(r_{om} + d_{min}/2)$, in a unit circle (Reprinted with permission from [77]. © 2013 IEEE)

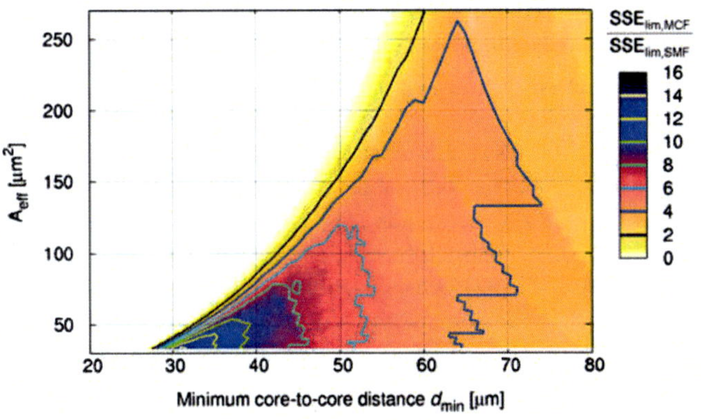

Fig. 2.31 A relationship between the Λ, A_{eff}, and a ratio of the SSE_{lim} ($SSE_{lim,MCF}$) of the MCF divided by that ($SSE_{lim,SMF}$) of the SSMF. The densest core packing was assumed, and D_{clad} and D_{coat} of the MCF were assumed as 225 μm and 350 μm, respectively (Reprinted with permission from [77]. © 2013 IEEE)

in Fig. 2.31. D_{clad} and D_{coat} of the MCF were assumed to be 225 μm and 350 μm, respectively. The OCT of the MCF was determined so that leakage loss from the outmost core to the coating can be no more than 0.001 dB/km at 1625 nm, thus r_{om} becomes larger when A_{eff} becomes smaller—the power confinement becomes stronger. From Fig. 2.31, it can be clearly seen that not large A_{eff} but small A_{eff} with moderately short Λ (moderate XT) can enhance the SSE. Smaller A_{eff}, or strong light confinement is also effective for shortening OCT while avoiding the coating-leakage loss increase.

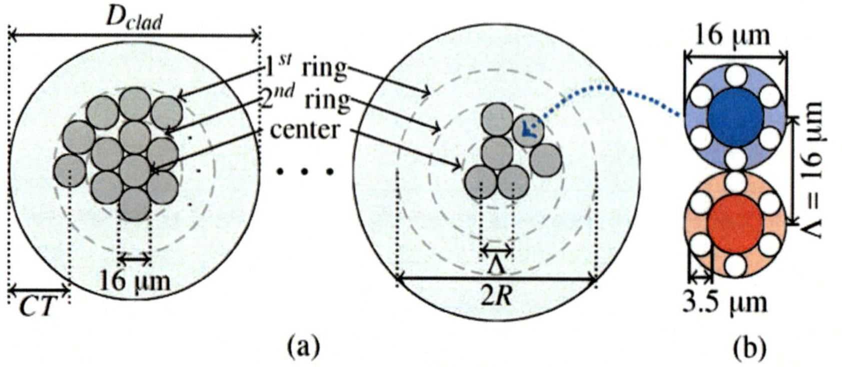

Fig. 2.32 **a** Arrangement of the cores in ring structures. The examples shown here are for the cases with 1 and 3 cores in the center, respectively. **b** Non-identical cores with hole-assisted double-cladding structure [83]. The core diameter is 9 μm, the same as in standard single-mode fibers (Reprinted with permission (#19SB0032) from [82]. © 2013 IEICE)

Chung et al. [80] also reported the investigation on the SSE of the MCF based on numerical simulation of transmission and revealed that smaller A_{eff} is beneficial for improving the SSE.

So far, the homogeneous single-mode MCF was discussed. However, the heterogeneous MCF with small A_{eff} with moderate XT may have possibilities to further enhance the SSE, if the XT increase in the small bending radius can be suppressed by the high index-mismatch structure, e.g., double-cladding structure [81] reviewed in Sect. 2.3.3.2. Ye et al. [82] investigated the total fiber capacity in the cases of MCFs where cores with the hole-assisted double-cladding structure design [83] are arranged on concentric circles in 220-μm-diameter cladding, shown in Fig. 2.32. They calculated the $\mathrm{SE}_{\mathrm{lim}}$ for three cases (A_{eff}: 48 μm^2, 80 μm^2, 120 μm^2) in the same way shown in the item "Spectral Efficiency in MCF Core" of this section and revealed that the smaller A_{eff} reduces $\mathrm{SE}_{\mathrm{lim}}$ of each core but increases core count and the total fiber capacity. Theoretical nonlinear limit of the total fiber capacity can exceed 20 Pbit/s in the case of the A_{eff} of 48 μm^2, as shown in Fig. 2.33.

2.3.3.2 Ultimate Number of Transmission Channels in Few-Mode Multi-core Fiber with Air-Hole-Assisted Double-Cladding Structure

Recent Progress of Air-Hole-Assisted Multi-Core Fiber

Recently, ultra-high bit rate transmission experiments using MCFs have been demonstrated such as 305-Tbit/s transmission using a 19-core fiber [84], 1.01-Pbit/s transmission using a 12-core fiber [85], and 1.05-Pbit/s transmission using a 14-core fiber [86]. However, further increase of the number of cores is difficult due to the

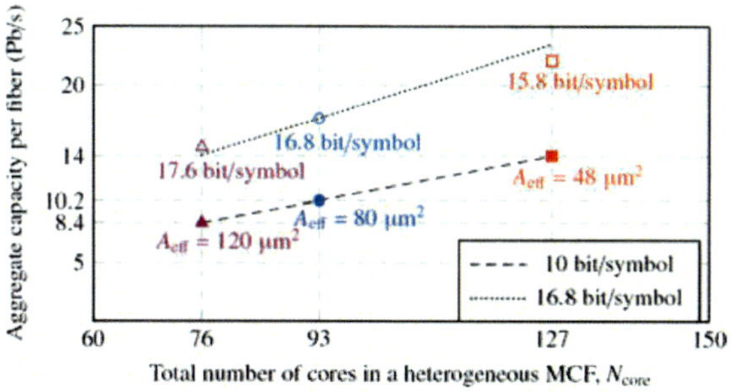

Fig. 2.33 Aggregate capacity per fiber as a function of total number of cores. Solid marks: with PDM-32QAM (10 bit/symbol) for practical application. Hollow marks: theoretical nonlinear maximum spectral (Reprinted with permission (#19SB0032) from [82]. © 2013 IEICE)

thick diameter of core with a trench structure and the combination of homogenous cores. Generally, the minimum core spacing between identical cores to achieve low crosstalk, e.g., −30 dB at 100 km transmission, is much greater than that between non-identical cores, which is the primary factor limiting the core packing density. Thus, the incorporation of many types of heterogeneous cores has been suggested as a method to increase the packing density [38]. However, when employing many types of heterogeneous cores, the difference of V parameter among non-identical cores will be large, resulting in significant differences in the propagation characteristics among the cores. To overcome this problem, a double-cladding structure which creates differences in the equivalent index among several types of cores while maintaining nearly identical propagation characteristics was proposed [81].

Subsequently, a novel heterogeneous MCF with air-hole-assisted double-cladding structure has been proposed, which can drastically suppress the crosstalk between adjacent cores and can achieve extremely high density of cores [83]. Since the adjacent cores almost contact with each other owing to the extremely low crosstalk, further increase of the packing density of cores is almost impossible. Some MCFs with hole-assisted structure have been demonstrated [34, 53]. However, identical cores were used, and so the core density has not reached its limit.

Therefore, to increase the number of transmission channels in the MCF, a few-mode heterogeneous MCF with air-hole-assisted double-cladding structure has been proposed [87]. In this section, firstly the design principle of air-hole-assisted double-cladding structure and the design example of dense heterogeneous MCF are discussed. Next, the optimization of few-mode air-hole-assisted double-cladding structure and the ultimate number of transmission channels in a MCF are shown.

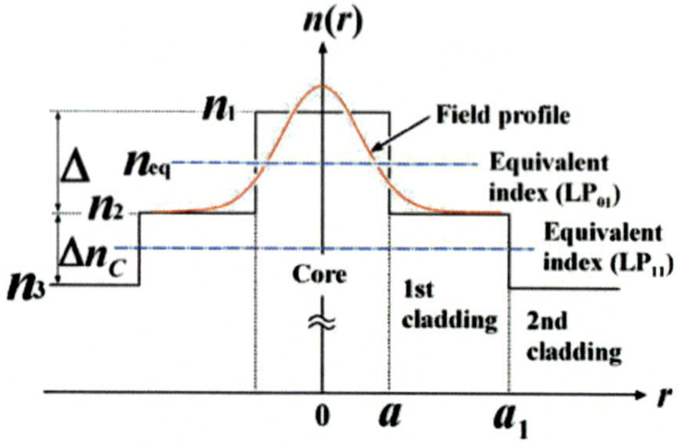

Fig. 2.34 Index and field profile of double-cladding structure (Reprinted with permission from [87]. © 2014 The Optical Society)

Dense Heterogeneous Air-Hole-Assisted Double-Cladding Single-Mode MCF

Let us consider a core surrounded by a double-cladding structure as shown in Fig. 2.34. The field profile of the LP_{01} mode is mostly confined in the core and the first cladding region owing to the difference in the refractive indices between the core and the first cladding, while the equivalent index n_{eq} is regulated so that it is different between adjacent cores by changing the refractive index of the core n_1 and that of the first cladding n_2. If the index difference, $n_1 - n_2$, is kept the same for all cores, propagation characteristics such as the dispersion and the mode-field diameter (MFD) can be kept identical for different cores while the equivalent index n_{eq} can be made different to prevent coupling between adjacent cores by varying the index difference between the first cladding and the second cladding, Δn_c. Using this double-cladding structure with low-index contrast ($\Delta = 0.375\%$), we assumed multiple cores with different pairs of n_1 and n_2 while maintaining the same index difference $n_1 - n_2$. The refractive index of the second cladding, n_3, was assumed to be 1.44402 (pure SiO_2 at $\lambda = 1550$ nm).

To maximize the core packing density, the radius of the first cladding a_1 must be minimized, while keeping the effect of the second cladding on the equivalent index negligibly small. To clarify the radius of the first cladding a_1 required to minimize the effect of the second cladding, the relationship between the equivalent index n_{eq} and the radius of the first cladding a_1 was calculated using the following equation, based on the perturbation theory.

$$n'_{eq} = \sqrt{n_{eq}^2 - I}, \tag{2.57}$$

where n'_{eq} is the perturbed equivalent index due to the existence of the second cladding and n_{eq} is the equivalent index determined by the unperturbed step-index profile consisting of n_1, n_2 and the core radius a. I is the perturbation term given by

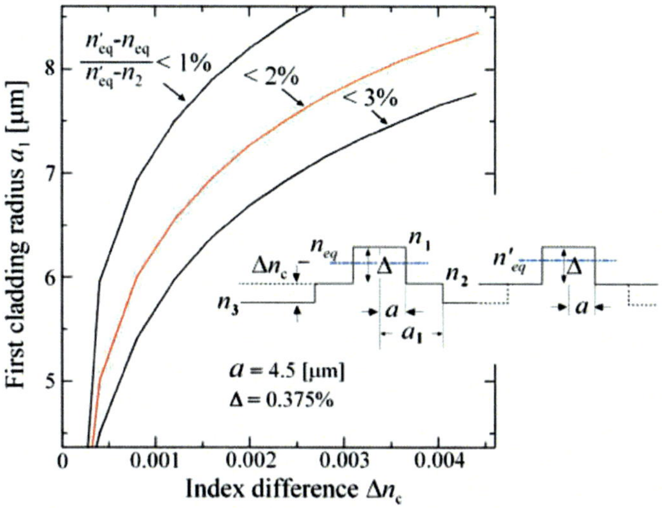

Fig. 2.35 Perturbed equivalent index vs radius of first cladding (Reprinted with permission from [87]. © 2014 The Optical Society)

$$I = \frac{\int_{a_1}^{\infty} \left(n_2^2 - n_3^2\right) \cdot E(r)^2 r \cdot \mathrm{d}r}{\int_0^{\infty} E(r)^2 r \cdot \mathrm{d}r}, \tag{2.58}$$

where $E(r)$ represents the unperturbed field profile for the case that the refractive index of the first cladding is equal to that of the second cladding, $n_2 = n_3$. The calculated result is shown in Fig. 2.35. It can be seen that $a_1 = 8$ μm is sufficient to minimize the effect of the second cladding, because the change in the perturbed equivalent index is only 2% of the equivalent index determined by Δ, even for the case of sufficiently large value of $\Delta n_c = 0.003$ [87].

In the double-cladding structure, not only the LP_{01} mode but also the LP_{11} mode is guided in the first cladding that will result in the crosstalk in the case of single-mode MCF. To suppress the excitation of the LP_{11} mode, air holes are added to the double-cladding structure. The relation between the equivalent index n_{eq} of each mode and the refractive index difference Δn_c between the first cladding and the second cladding was numerically calculated using a mode solver (FemSIM by Rsoft). The calculated results are shown in Fig. 2.36.

It can be seen from this figure that the cutoff Δn_c of LP_{11} mode of the double-cladding structure without air holes is as small as 3×10^{-4}. On the other hand, the cutoff Δn_c of air-hole-assisted double-cladding structure is extended to 3×10^{-3} when the diameter of air hole is 5 μm and the center of the air holes are located on the outer circumference of the first cladding. When the equivalent index is less than 1.44402 (refractive index of SiO_2), it was confirmed that the field profile spread out into the whole area and is no longer confined in the core region. Therefore, this corresponds to the cutoff of LP_{11}. Thus, the index difference Δn_c of non-identical cores

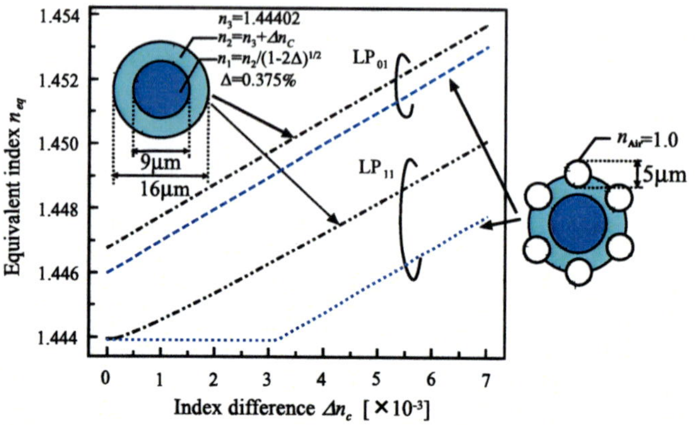

Fig. 2.36 Equivalent index versus index difference Δn_c defined by $(n_2 - n_3)$ (Reprinted with permission from [87]. © 2014 The Optical Society)

can be determined so that the value of Δn_c for different cores is distributed within the range of the single-mode condition. In this study, since the triangle arrangement of non-identical three types of cores was adopted as described in the next section, three different values of Δn_c were determined to be $\Delta n_c = 0.001$, 0.002 and 0.003.

The crosstalk (XT) between adjacent cores was evaluated using the coupled-power theory (CPT) [23]. In this work, the bending radius R was assumed to be sufficiently larger than the critical bending radius, i.e., non-phase-matching region, and the exponential autocorrelation function for the statistical characteristics of the power-coupling coefficients (PCCs) was adopted. Thus, the average PCC [25] between the m-th and n-th core is given by,

$$\bar{h}_{mn}(z) = \frac{2K_{mn}^2 l_c}{1 + (\Delta\beta_{mn}(z) l_c)^2}, \tag{2.59}$$

where K_{mn} is the coupling coefficient calculated by the coupled-mode theory (see (10) in [23]), $\Delta\beta_{mn}(z)$ is the local propagation-constant difference and l_c is the correlation length. The correlation length is defined for the coupling between the fundamental modes of two adjacent cores. These definitions are the same as that in [25]. The correlation length of 0.05 m was hypothetically adopted in this work which was obtained experimentally in [23].

The calculated results of crosstalk are shown in Fig. 2.37. In this calculation, we assumed that Δn_c of one of the two adjacent cores is zero, i.e., $n_2 = n_3$, because the coupling in this case is stronger than the case of $\Delta n_c > 0$. The crosstalk of the double-cladding structure without air holes is about 0 dB in the case of 100 km transmission when the core pitch is 21 μm. When the air holes are located on the outer circumference of the first cladding, the crosstalk for 100 km transmission is as low as −40 dB or less in the range of $0.0005 < \Delta n_c < 0.003$. Here, the core spacing

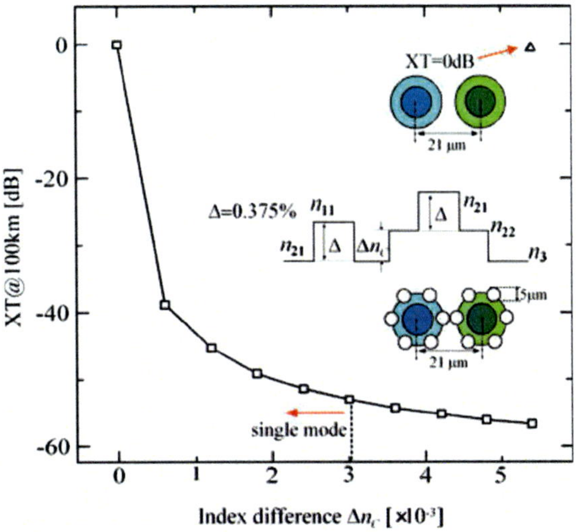

Fig. 2.37 Crosstalk at 100 km transmission versus index difference Δn_c (Reprinted with permission from [87]. © 2014 The Optical Society)

can be greatly reduced to 21 μm, and so the adjacent core regions consisting of the core and the first cladding are in contact with each other. These data show that the crosstalk between adjacent cores can be extremely suppressed by the air holes.

In addition to the crosstalk, another important parameter is the critical bending radius, because the crosstalk increases when the bending radius approaches the critical bending radius. Therefore, the critical bending radius should be as small as possible. The critical bending radius R_c is given by

$$R_c = \frac{\Lambda \cdot n_3}{\delta n_{eq}}, \tag{2.60}$$

where Λ is the adjacent core spacing and δn_{eq} is the difference of the equivalent index between adjacent cores. In this case, since Λ can be greatly reduced to 21 μm and δn_{eq} nearly corresponds to δn_c of 0.001, the critical bending radius can be reduced to 30 mm.

Optimization of Few-Mode Air-Hole-Assisted Double-Cladding Structure

Figure 2.38a shows the schematic view and the definition of structural parameters of air-hole-assisted double-cladding structure. a and a_1 are the core and first cladding radii, respectively, and h is the diameter of the air hole. In this work, a and a_1 were assumed to be 4.5 μm and 8 μm, respectively, the same as the case of single-mode design. The core pitch is Λ, and the gap between circumferences of adjacent first cladding is g. Then, a_{hole}, the distance between the center of the core and the center of the hole, is half of Λ, which means holes are located at the midpoint between adjacent cores. Here, Δ is the refractive index contrast between the core

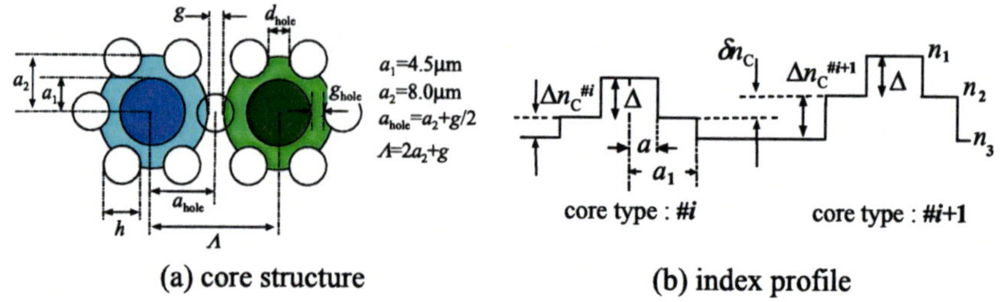

Fig. 2.38 Definition of structural and optical parameters of cores (Reprinted with permission from [87]. © 2014 The Optical Society)

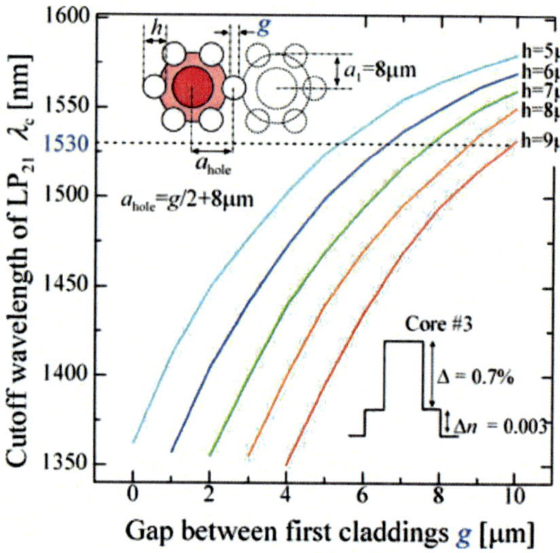

Fig. 2.39 Cutoff wavelength of LP_{21} of core #3 (Reprinted with permission from [87]. © 2014 The Optical Society)

and the first cladding which is kept identical for all cores. To increase the number of modes without deteriorating the core density, Δ was increased. In this design, the index contrast Δ of 0.7% was adopted so that 3 modes, i.e., LP_{01}, $LP_{11}{}^{a}$ and $LP_{11}{}^{b}$ modes can be supported. Then, in this section, since the triangle arrangement of non-identical three types of cores was adopted and numbered as #1, #2, and #3, three different values of Δn_c were determined to be $\Delta n_c = 0.001$ for core #1, 0.002 for core #2, and 0.003 for core #3. Here, δn_c represents the index difference of first cladding between the core #i and the core #$i + 1$ as shown in Fig. 2.38b.

First, the theoretical cutoff wavelength of the LP_{21} for this structure was calculated. Figure 2.39 shows the relationship between g and the theoretical cutoff wavelength of LP_{21} mode against the hole diameter h. Here, the theoretical cutoff wavelength λ_c was calculated for the core #3, because that of core #3 should be longest among the three types of cores due to the large Δn_c. It can be seen that the theoretical cutoff wavelength can be shorter as the overlap area of the air hole to the first cladding

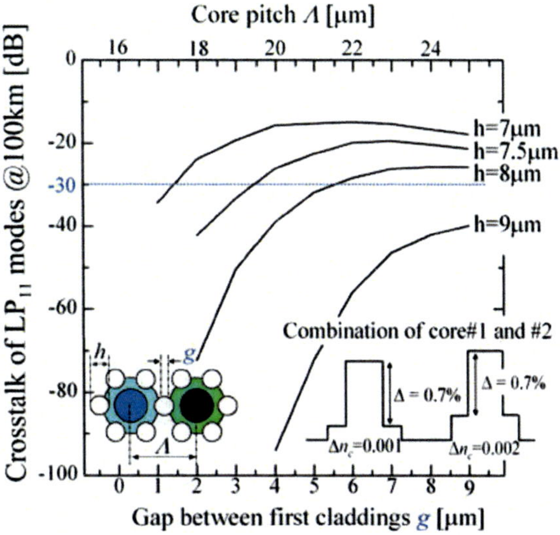

Fig. 2.40 Worst crosstalk of LP_{11} between core #1 and core #2 (Reprinted with permission from [87]. © 2014 The Optical Society)

become larger. The design condition for the cutoff wavelength λ_c was assumed to be 1530 nm in consideration for the use of *C* band and *L* band.

Next, the crosstalk of LP_{11} modes between adjacent cores at the 100 km transmission was calculated using the PCC given by (2.59). Although the correlation length was defined for the calculation of the crosstalk between the fundamental modes in [23], the correlation length l_c was assumed to be 0.05 m [23]. In this work, since the difference between the equivalent index of LP_{11} mode and the refractive index of the first cladding n_2 is almost equal to that in the single-mode case, we assumed this value. Figure 2.40 shows the calculated results. It is seen that the crosstalk is extremely suppressed as the core pitch is reduced, when the hole diameter is constant. Usually, the crosstalk becomes large as the core pitch is reduced. However, in this case, the field is more effectively shielded by the holes as the hole approaches the core with the decrease of core pitch according to the relation $g_{\text{hole}} = (\Lambda - 2a_1 - h)/2$. Therefore, the crosstalk is suppressed with the decrease of the core pitch. In the case of hole diameter of 9 μm, the crosstalk is less than −30 dB for any core pitch as shown in Fig. 2.40.

Lastly, the optimum design condition was obtained using the calculated results of the theoretical cutoff wavelength of LP_{21} and the crosstalk of LP_{11} modes between adjacent cores as shown in Fig. 2.41. Here, g_{hole} indicates the gap between the circumference of the hole and the core, and d_{hole} indicates the gap between the circumferences of the adjacent holes. In the sky-blue region which is surrounded by the lines of $g_{\text{hole}} = 0$ μm and $d_{\text{hole}} = 0$ μm, and the curves of XT = −30 dB and $\lambda_{\text{th}} = 1530$ nm, the air-hole-assisted double-cladding structure can be designed. However, it is not practical that the holes contact each other, i.e., $d_{\text{hole}} = 0$ μm, and the holes and the core contact with each other. Therefore, the deep blue region which is surrounded

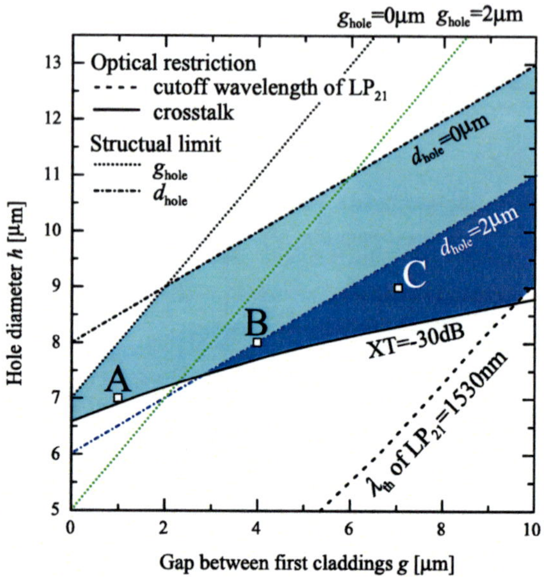

Fig. 2.41 Design restriction on air hole diameter and gap between first cladding (Reprinted with permission from [87]. © 2014 The Optical Society)

by the lines of $g_{\text{hole}} = 2\ \mu\text{m}$ and $d_{\text{hole}} = 2\ \mu\text{m}$ and the curves of XT = −30 dB and $\lambda_{\text{th}} = 1530$ nm represents the more practical design region.

Number of Transmission Channels

Using the optimum design parameters obtained above, ultra-high-density multi-core fiber can be designed. In this study, the triangle arrangement using three types of non-identical cores was adopted since it can realize the highest core density. Once the core pitch, the core arrangement, the number of non-identical core types, and the diameter of core accommodation area are given, the number of cores can be calculated using the lattice algorithm [88]. The algorithm to calculate the maximum core capacity in the triangle lattice arrangement can be briefly summarized as follows.

Let us consider a case #1 that core center is located at the vertex of the triangular lattice as shown in Fig. 2.42. The distance between the center of fiber and the lattice point R_{ik} is given by (2.61). The number of cores existing in the circle with radius R is determined by counting the lattice points satisfying $R_{ik} < R$. This counting procedure is equivalent to the sum of the unit step function $U\ (R - R_{ik})$ with respect to i and k. Thus, the summation of the lattice points in the normalized circle R/D is given by (2.62), where μ and ν are the maximum ranges of summation with respect to i and k and D is the core spacing between identical cores which is equal to $\sqrt{3}\Lambda$.

$$R_{ik} = \frac{1}{2}\Lambda\sqrt{(2i + k)^2 + 3k^2}, \tag{2.61}$$

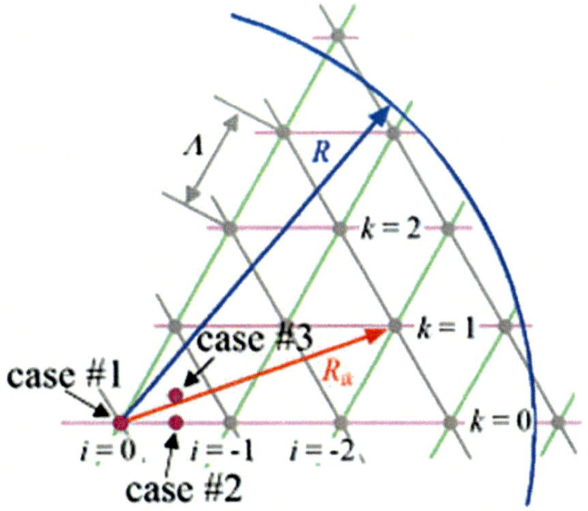

Fig. 2.42 Lattice point in a triangle arrangement and definition of label *i* and *k* (Reprinted with permission from [87]. © 2014 The Optical Society)

$$N_{\#1} = \sum_{i=-\mu}^{\mu} \sum_{k=-\nu}^{\nu} U(R - R_{ik}). \tag{2.62}$$

The maximum core capacity for cases #2 and #3 where the center of fiber is located at the middle point between adjacent lattice points and the center of triangle, respectively, as shown in Fig. 2.42, can be calculated in the same manner. The maximum core capacity for a given *R/D* is determined by comparing these numbers using (2.63).

$$N_{\text{core}} = \max(N_{\#1}, N_{\#2}, N_{\#3}). \tag{2.63}$$

Using the lattice algorithm, the maximum core capacities for triangle lattice were calculated for three cases where the diameter of effective fiber area 2*R* were assumed to be 100, 150, and 200 μm, as shown in Fig. 2.43.

Here, we sampled three designed parameters from Fig. 2.41 and the case of single-mode MCF as described in the second item of this section. The maximum number of cores was evaluated using Fig. 2.43 and is summarized in Table 2.4. Cases *A*, *B*, and *C* correspond to the points *A*, *B*, and *C* in Fig. 2.41, respectively.

In the design of few-mode transmission, three channels, i.e., LP_{01}, $LP_{11}{}^{a}$, and $LP_{11}{}^{b}$ modes, are supported per core. Thus, the number of transmission channels in a fiber will be 3 times the number of the cores. In the case of these three sampled structures, the number of transmission channels ranges from 129 to 387 when the effective fiber area diameter 2*R* is assumed to be 200 mm. Even in the case of 2*R* = 100 μm which is almost equivalent to the conventional single-mode fiber, the number of transmission channels can be increased to 57–93. Figure 2.44 shows the example of

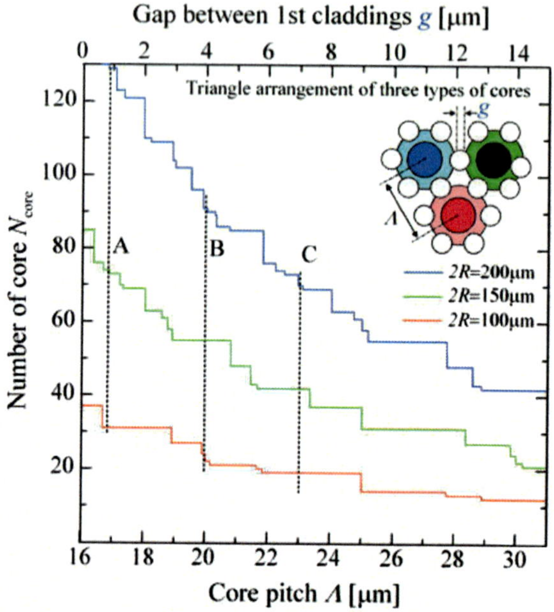

Fig. 2.43 Maximum number of cores versus core pitch against diameter of effective fiber area (Reprinted with permission from [87]. © 2014 The Optical Society)

Table 2.4 Design examples of ultra-large number of channels

Case	Hole dia. h (μm)	Core pitch Λ (μm)	Worst XT (dB)	Number of cores		
				$2R = 100$ μm	$2R = 150$ μm	$2R = 200$ μm
SM	5	21	−37.0	21	48	85
A	7	17	−34.3	31	73	129
B	8	20	−39.0	22	55	91
C	9	23	−46.6	19	42	70

Reprinted with permission from [87]. © 2014 The Optical Society
Note Worst crosstalk (XT) is the evaluated value at 100 km transmission. Cases *A*, *B*, and *C* correspond to the points *A*, *B*, and *C* in Fig. 2.42

case C for $2R = 200$ μm, which corresponds to the number of transmission channels of 210.

In addition, the critical bending radius for the cases *A*, *B*, and *C* were calculated to be 24 mm, 29 mm, and 33 mm, respectively, using (2.60).

As for the fabrication, two fabrication methods, i.e., rod-in-tube method [32] and stack-and-draw method [63], can be used to realize the proposed hole-assisted double-cladding structure. In the case of rod-in-tube method, the air holes are formed by drilling in the preform rod which has double-cladding structure. In the case of stack-and-draw method, the preform rod is prepared by stacking rods which have several types of index and diameter and thin capillary tubes in a silica tube.

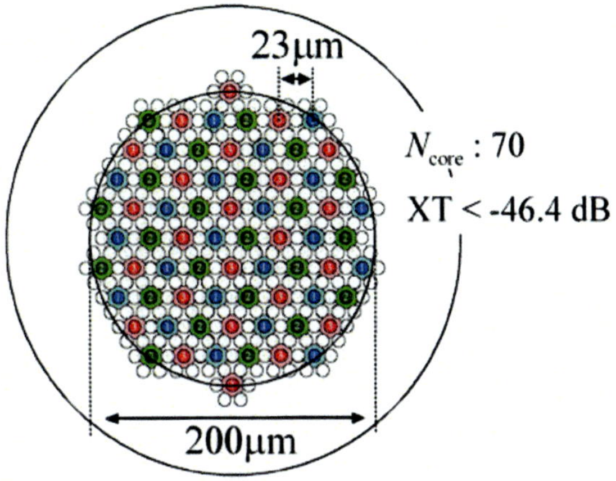

Fig. 2.44 Design example of 70 core few-mode multi-core fiber (Reprinted with permission from [87]. © 2014 The Optical Society)

2.3.4 Reliability of MCF

During the fabrication process of an optical fiber, a tiny crack is remained on the glass surface. These tiny cracks will be growing during its lifetime and finally cause the fiber break. Thus, the mechanical reliability of optical fiber is affected in terms of the strength of glass. The mechanical reliability of an optical fiber can be investigated by using power law theory [89, 90], and the failure probability F can be expressed as (2.64).

$$F = 1 - \exp\left[N_P L\left\{1 - \left(1 + \frac{t_s\sigma_a^n}{t_p\sigma_p^n}\right)^{\frac{m}{n-2}}\right\}\right]. \tag{2.64}$$

N_p, t_p, and σ_p represent the mean breakrate per unit length during prooftesting, effective prooftime, and proofstress, respectively. The reliability of optical fiber can be improved by removing the weak portion of the fabricated fiber prior to the cabling, and it can be achieved utilizing the prooftest with adequate stress level. Generally, one to two percent strain is added during the prooftesting in order to ensure the long-term reliability of optical fiber. m and n in (2.64) represent the Weibull parameter and stress corrosion parameter, respectively. These values can be obtained by examining the failure time of the actual optical fiber statistically as a function of the strain/bending. t_s denotes lifetime. L and σ_a represent the effective fiber length under uniform stress and applied stress after prooftesting, respectively. Here, σ_a can be expressed as a function of bending radius R and fiber diameter D.

Figure 2.45 shows example relationship between failure probability F and bending radius R when the fiber diameter is 125 μm, which corresponds to the standard cladding diameter of a conventional single-mode fiber (SMF). We assumed one percent prooftesting and set the N_p at 0.01 km^{-1}. m and n values were, respectively, set at 2.8 and 18 as example, and we consider 20 years lifetime. Figure 2.45 confirms

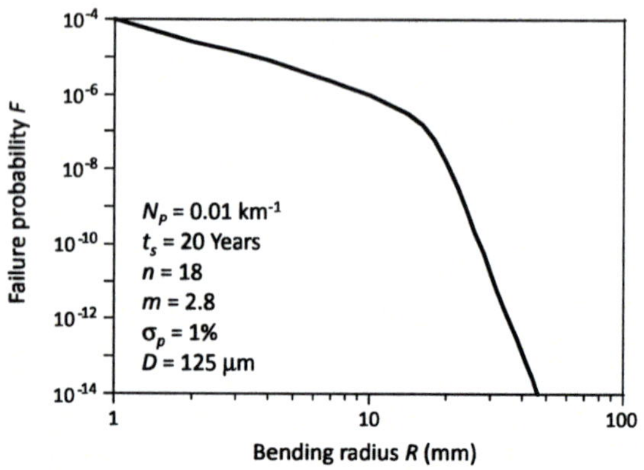

Fig. 2.45 Example relationship between failure probability F and bending radius R when the cladding diameter D is 125 μm

that the failure probability increases as decreasing the bending radius. Here, it would be noted that the change in bending radius dependence found in Fig. 2.45 represents the transition of the failure process. When the bending radius becomes less than 10 mm, the failure characteristics of glass move to its intrinsic regime. Then, a conventional optical fiber with 125 μm cladding diameter breaks instantly when the bending radius is tightened to the high-strength region of the optical fiber, roughly less than 3 mm. Although (2.64) and Fig. 2.45 neglect the high strength region for simplicity, the detail of the intrinsic strength model can be found in [91]. In a conventional SMF, it is recommended to set the allowable bending radius at 30 mm or more. Figure 2.45 then confirms that the failure probability F in the conventional SMF link is expected to be 10^{-11}.

Regarding with a multi-core fiber (MCF), we can also utilize the power law theory to predict the long-term reliability. Generally speaking, the cladding diameter D of MCF tends to be larger than 125 μm as increasing the number of multiple cores. Figure 2.46 shows example relationship between relative failure probability R_F and

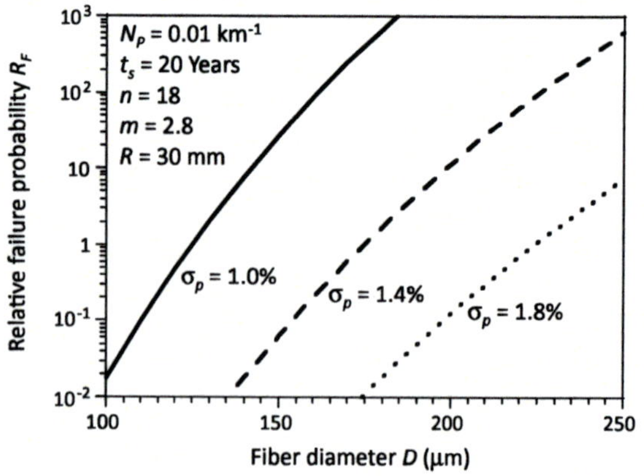

Fig. 2.46 Example relationship between relative failure probability R_F and cladding diameter D as a function of proofstress σ_p

cladding diameter D. In this calculation, we set the bending radius R at 30 mm. The relative failure probability R_F shows the normalized value when we set the failure probability F at $R = 30$ mm and $D = 125$ μm as reference. Solid, dashed, and dotted lines show the results when we set the proofstress σ_p at 1.0%, 1.4%, and 1.8%, respectively. Figure 2.46 confirms that the failure probability is greatly degraded as increasing the cladding diameter D. It can be seen that the relative failure probability R_F becomes more than 10^3 when the cladding diameter D is enlarged from 125 to 180 μm at 1% proofstress. It is also found that we can relax the influence of the enlarged cladding on the failure probability by increasing the proofstress level. Figure 2.46 confirms that the cladding diameter D can be expanded to up to 225 μm while keeping the relative failure probability R_F at one if we can increase the proofstress from 1.0 to 1.8%. However, it would be noted that the higher proofstress generally affects the productivity. Moreover, a 1.8% prooftesting at 200 μm cladding diameter requires a 4.6 times higher strain in its process. These result in the further development/optimization to the existing fabrication process and/or equipment. However, it can be said that we can realize an MCF with feasible reliability by optimizing the cladding diameter and fabrication process simultaneously although further clarification is needed particularly on the relationship with coating conditions.

2.3.5 Fabrication Technology of MCF

MCF is fabricated by drawing MCF preform. To correspond to mass production, the method which is applicable to the mass production is required. There are two types of typical fabrication technology of MCF preform. One of them is rod-in-tube method and the other is stack-and-draw method (Fig. 2.47). In case of the rod-in-tube method, the MCF preform is composed of glass rod with plural holes and core rods, which are inserted into the holes. In case of the stack-and-draw method, the core rods are stacked in the glass tube. To prevent the distortion of the shape of the fiber cladding, it is effective to use the spacer rod to fill the space between core rods and glass tube. The core rods are fabricated by means of Vapor-phase axial deposition (VAD) method, MCVD method, and so on.

There are issues to enlarge the preform size in each case of the rod-in-tube and stack-and-draw method. The main issue of the rod-in-tube method is the length of the preform because there is a limitation of the length of the drill tool. It is difficult to keep the uniformity of the hole position along the preform, and it causes the distortion of the core arrangement. The hole diameter is the other issue of rod-in-tube method. It is difficult to open the many holes with small diameter precisely. Additionally, it should be addressed that the longitudinal uniformity of the glass materials tends to be degraded as the preform becomes long. The degradation of the longitudinal uniformity makes scratches of the glass surface by friction increase so that there is a limitation of enlargement of the MCF preform. By the way, the cost of rod-in-tube method becomes higher as the number of holes (cores) increases because the machine time required for the drilling holes increases. In case of the stack-and-draw method,

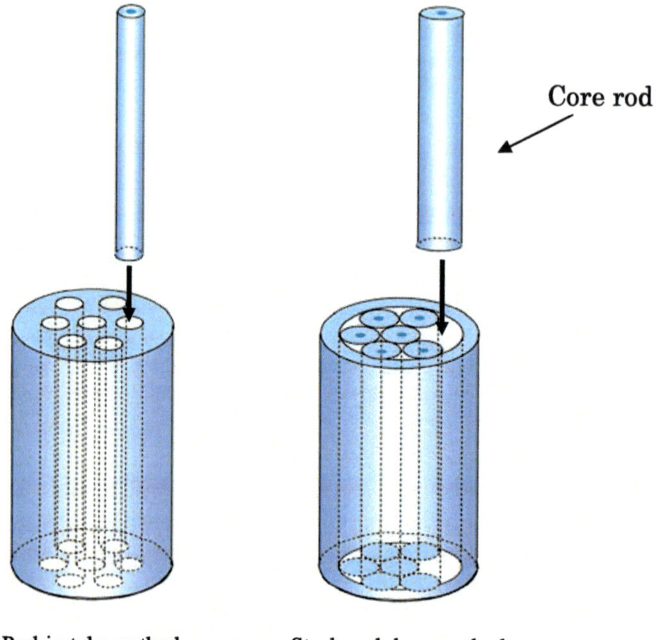

Fig. 2.47 Methods for MCF preform fabrication

the cost also increases as the number of cores increases because the required time for the assembling increases. The optimum method and core numbers depend on the scale of the production.

To overcome the limitation of the enlargement of the MCF preform by means of the rod-in-tube and stack-and-draw methods, the novel method which can surpass these methods should be established. One of the candidates of such method is powder method (Fig. 2.48). By filling the silica powder around the core rods, pressing them, and sintering, the MCF preform could be fabricated. The other candidate is over-cladding method (Fig. 2.49). By bundling plural core rods, depositing silica soot cladding in vapor phase, and sintering, the MCF preform can be made. Each method has possibility of scaling the preform size, so it can be expected that low-cost production could be realized. Each of them has the issues of reducing the attenuation loss and making the precise core arrangement.

Some trials on new fabrication methods have been reported. A seven-core fiber fabricated by slurry casting method with high pure SiO_2 powder showed an attenuation loss of 0.25 dB/km and a crosstalk of −33.4 dB per 100 km at 1550 nm [92]. A four-core fiber fabricated by over-cladding bundled rods (OBR) method showed an attenuation loss of 0.19 dB/km and crosstalk of −63 dB/km at 1550 nm [93]. The OBR was also used to fabricate a 2-LP-mode 4-core fiber with an air-hole structure [94].

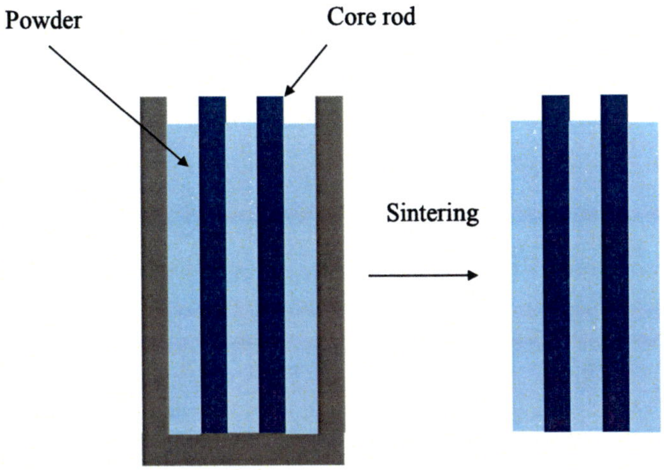

Fig. 2.48 Powder-in-tube method

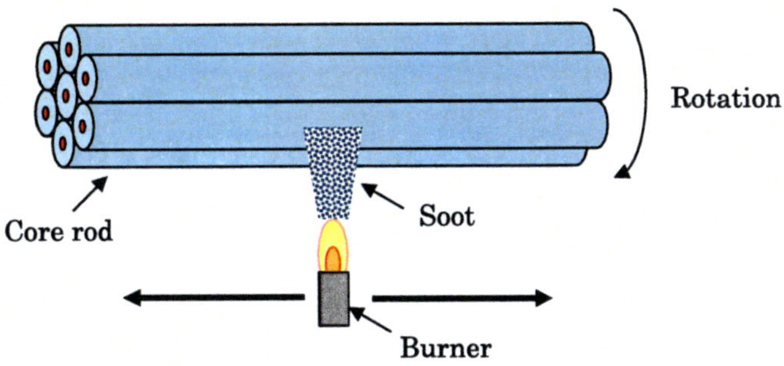

Fig. 2.49 Over-cladding method

In each method, purification is the big issue which must be solved. Because there is assembling process in each method, the purification process to remove the contamination on the surface of core rods is essential. The chemical treatment (e.g., hydrofluoric acid), thermal treatment (oxyhydrogen flame, plasma flame), and chlorine treatment (dehydration) are added to remove the contamination effectively. Additionally, the removing the space around core rods is an issue which should be addressed. Optimization of the condition of collapse process or assembling process with inner blocks (spacer et al.) are necessary (Fig. 2.50). The core-cladding ratio of the core rods should also be investigated. In case of core rod with thin cladding, the attenuation loss tends to be higher due to the impurity around the core rods. So the thicker cladding is suitable from the point of view of attenuation; however, it restricts the design flexibility because the silica bridge tends to be thinner. Additionally, there is the other issue that the amount of glass to be removed tends to increase, and it leads

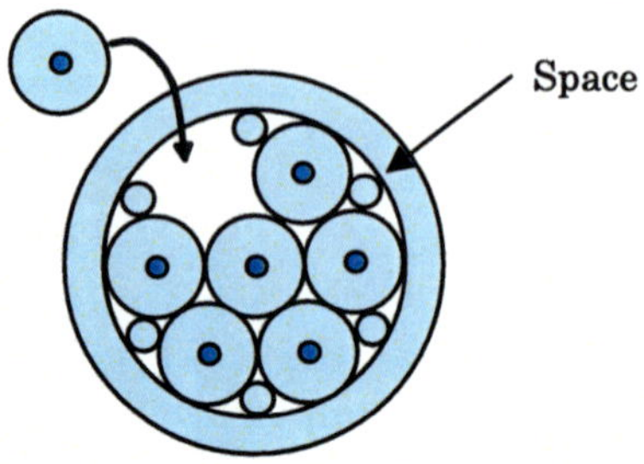

Fig. 2.50 Arrangement of spacer

the cost increase. The optimum core-cladding ratio should be investigated to keep the good productivity.

For the mass production of MCF, gaining the drawing speed is also essential issue. If the drawing speed was increased with keeping the drawing temperature, the drawing tension must be increased. On the other hand, it is necessary to rise the drawing temperature to gain the drawing speed with keeping the drawing tension. Moreover, the increase of the drawing tension causes the increase of the residual stress in the MCF and the increase of the drawing temperature causes the diffusion of the dopants in the cores. The residual stress and the dopant diffusion cause the refractive index distribution change in the fiber cross section, and the distributions of the residual stress or dopant diffusion are possible to have non-uniformity. So, the refractive index distribution change of each core may have different tendency in case of MCF which have plural cores in one cladding (Fig. 2.51).

The cladding diameter of MCF can be larger than the standard 125 μm because the core pitch needs to be sufficiently large to realize desirable crosstalk property. In that case, there is necessary to enlarge the preform size by increasing the preform diameter and/or length to obtain the same fiber length as the fiber with standard cladding diameter (Fig. 2.52). From this point of view, scaling the preform size is

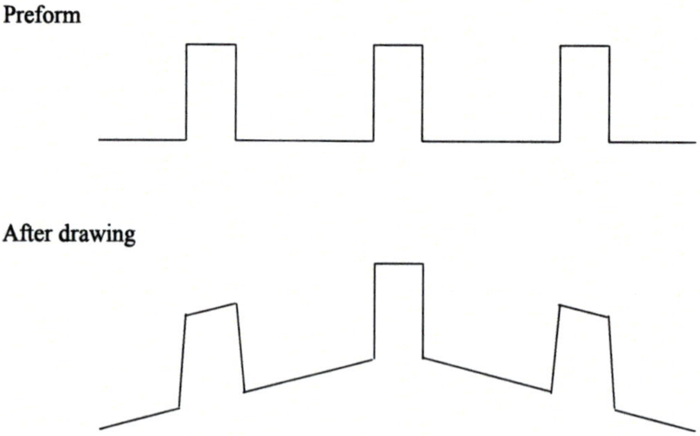

Fig. 2.51 Concept image of the refractive index distribution change in MCF

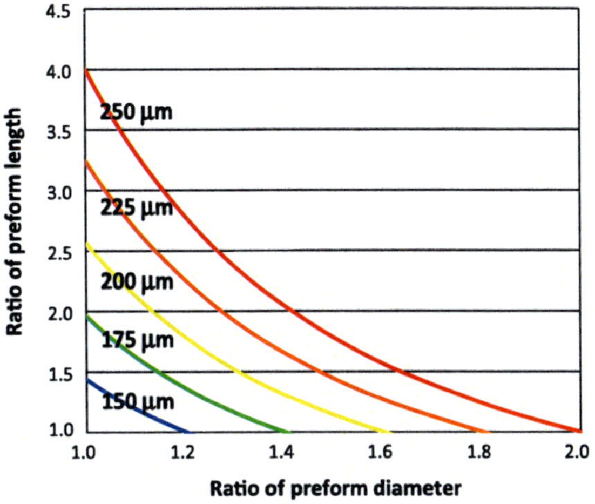

Fig. 2.52 Necessary preform diameter and length depends on cladding diameter

important subject of development of MCF.

2.3.6 Measurement Technology

2.3.6.1 Crosstalk Measurement

The powers of two-core fiber in multi-core fiber (MCF) can be expressed using the following sets of coupled-power equations as

$$\begin{aligned}\frac{\mathrm{d}P_1(z)}{\mathrm{d}z} &= -\alpha_1 P_1(z) + h(z)(P_2(z) - P_1(z))\\ \frac{\mathrm{d}P_2(z)}{\mathrm{d}z} &= -\alpha_2 P_2(z) + h(z)(P_1(z) - P_2(z))\end{aligned}, \tag{2.65}$$

where α_i $(i = 1, 2)$ denotes the attenuation and $h(z)$ is the power-coupling coefficient at position z. For simplicity, it is assumed that $\alpha_1 = \alpha_2 = \alpha$. For the boundary conditions, $P_1(0) = P_0$ and $P_2(0) = 0$, we obtain the following equations:

$$\begin{aligned}P_1(z) &= \frac{P_0}{2}\exp(-\alpha z)\left[1 + \exp\left\{-2\int_0^z h(x)\mathrm{d}x\right\}\right]\\ P_2(z) &= \frac{P_0}{2}\exp(-\alpha z)\left[1 - \exp\left\{-2\int_0^z h(x)\mathrm{d}x\right\}\right]\end{aligned}. \tag{2.66}$$

Therefore, the crosstalk $XT\ (z)$ (in dB) of the two-core fiber at position z can be expressed as

$$XT(z) = 10\log\frac{P_2(z)}{P_1(z)} = 10\log\left[\frac{1-\exp\{-2\int_0^z h(x)\mathrm{d}x\}}{1+\exp\{-2\int_0^z h(x)\mathrm{d}x\}}\right]. \tag{2.67}$$

The crosstalk can be estimated by measuring the output powers of two cores when the input power is launched into the only one core in the MCF. We have two different kinds of crosstalk measurement, direct method and OTDR method.

Direct Method

As for the direct method, the crosstalk can be estimated according to the definition of the crosstalk mentioned above. The direct method for measuring the crosstalk statistically and crosstalk characteristics of the fabricated MCF in [32, 50, 95] has been reported. In particular, the cause of measurement floor (lower limit) of crosstalk measurement system for measuring ultra-low crosstalk has been investigated and eliminated. Reference [41] has proposed a novel measurement method to obtain mean and shape of the statistical crosstalk distribution by utilizing wavelength dependence of the crosstalk.

The measured profiles were observed to have power floors. The floors for standard single-mode fibers (SSMFs) and trench-assisted single-mode fibers (TA-SMFs) were about −60 dB, and less than −90 dB, respectively. The floor for SSMFs is caused by coupling from cladding modes to a core mode during propagation in the receiving SMF, which can be the cause of the crosstalk discrepancy and can be suppressed by a trench layer. TA-SMFs, which have almost the same core profile with cores of the MCF, as a launching fiber and a receiving fiber in measurements have a great effect on the suppression of the floors.

The crosstalk after fiber propagation is a statistic value, because how much the crosstalk increases or decreases at the phase-matching points is varied by variations of the phase offsets between coupled cores at the phase-matching points, and because the phase offset can be easily varied by slight difference of fiber conditions such as fiber bend, twist, and so on. Accordingly, randomization of the phase offsets between the cores along MCFs is needed to obtain statistical distributions of the crosstalk of the MCFs [41].

OTDR Method

(a) ***Principle***

If we detect backscattered power of the adjacent core when the optical pulse is launched into the core, we can estimate the crosstalk by using the OTDR.

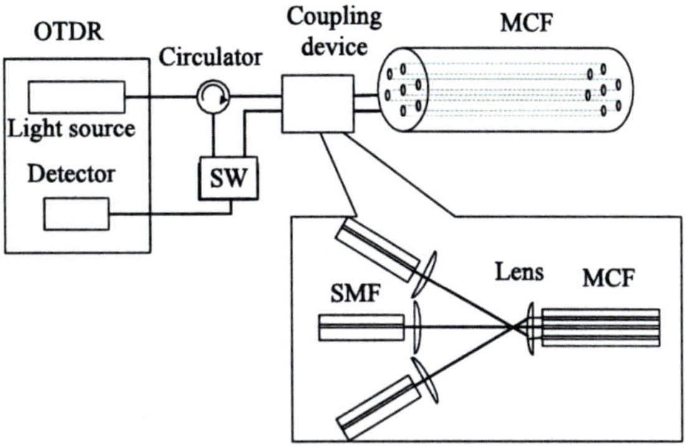

Fig. 2.53 Experimental setup for measuring the crosstalk (Reprinted with permission from [96]. © 2014 The Optical Society)

For simplicity, we consider the backscattered power of each core in the two-core MCF. When the optical pulse is launched into the core 1, the backscattered powers $P_{b1}(z)$ and $P_{b2}(z)$ from a position z in a two-core fiber can be expressed by using (2.66) as

$$P_{b1}(z) = \frac{P_0}{2}\alpha_{s1}B_1\exp(-2\alpha z)\left[1+\exp\left\{-4\int_0^z h(x)\mathrm{d}x\right\}\right], \tag{2.68}$$

$$P_{b2}(z) = \frac{P_0}{2}\alpha_{s2}B_2\exp(-2\alpha z)\left[1-\exp\left\{-4\int_0^z h(x)\mathrm{d}x\right\}\right], \tag{2.69}$$

where α_{si} ($i = 1, 2$) is the local scattering coefficient and B_{i} ($i = 1, 2$) is the backscattered capture fraction.

Here, if we assume that $\alpha_{s1} = \alpha_{s2} = \alpha_s$ and $B_1 = B_2 = B$, the crosstalk of the two-core MCF estimated using OTDR is obtained from (2.68) and (2.69) as

$$XT_{\mathrm{OTDR}}(z) = 10\log\frac{P_{b2}(z)}{P_{b1}(z)} = 10\log\left[\frac{1-\exp\left\{-4\int_0^z h(x)\mathrm{d}x\right\}}{1+\exp\left\{-4\int_0^z h(x)\mathrm{d}x\right\}}\right]$$
$$= 10\log\left[\tan h\left\{2\int_0^z h(x)\mathrm{d}x\right\}\right]. \tag{2.70}$$

As the crosstalk $XT_{\mathrm{OTDR}}(z)$ estimated using OTDR is that for a fiber length of $2z$, the crosstalk at the position z can be evaluated by using (2.70).

Figure 2.53 shows the experimental setup for measuring the crosstalk of a 1-km-long multi-core fiber with seven single-mode cores. Conventional OTDR was used to transmit the pulse and to receive the backscattered power and the powers of the launched core and the adjacent core can be measured by using the optical switch.

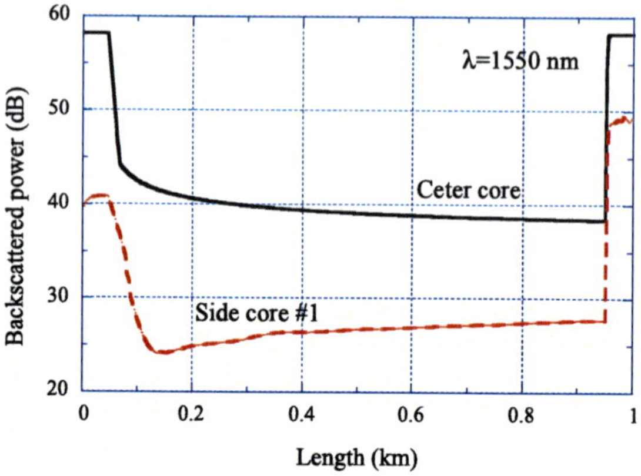

Fig. 2.54 Backscattered powers of the center and side cores at 1.55 μm

The coupling between the conventional single-mode fiber and each core of the multi-core fiber was realized by the space coupling device as shown in Fig. 2.53. The OTDR wavelength was 1550 nm, pulse width was set at 500 ns, and the averaging time was 1 min. The optical pulsed OTDR light was launched into the center core of the multi-core fiber using the space coupling device, and the backscattered powers of the side cores were measured by using the optical switch.

We measured the crosstalk of the multi-core fiber at 1550 nm using the OTDR technique and the conventional direct method. The direct method is a method whereby the crosstalk can be estimated from the ratio of the adjacent core power to the launched core power, which is directly measured with the power meter. A 1-km-long multi-core fiber with seven single-mode cores was used as the sample fiber.

Figure 2.54 shows the backscattered powers of the center and side cores at a wavelength of 1550 nm. The solid line shows the backscattered power of the center core, and the broken line shows that of the side core. It can be seen that the reflections near the input and output portions occurs due to Fresnel reflection, which generated when using the space coupling device. The Fresnel reflection light has been removed from the crosstalk measurement result. It is also seen that the backscattered light of the center core decreases and that of the side core increases due to the power coupling between the cores.

Figure 2.55 shows the length dependence of the crosstalk with this method. The circles show the crosstalk of the three side cores measured by the power meter method. We found that the crosstalk becomes large as the fiber length increases. It is also seen that the crosstalk values at about 1 km measured using OTDR were in good agreement with those obtained by the direct method. Thus, it is clear that the crosstalk distribution can be successfully estimated with this technique.

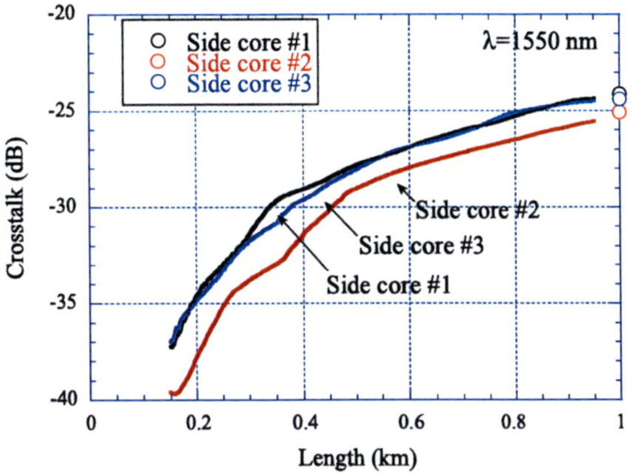

Fig. 2.55 Length dependence of the crosstalk

2.3.6.2 Fiber Parameter Measurement

For single-mode fibers, there are a lot of techniques for measuring the fiber parameters such as the mode-field diameter (MFD), cutoff wavelength, the relative-index difference, and chromatic dispersion. These techniques can be applied to the MCF by using the fan-in/fan-out (FI/FO) devices. Here, the definition and measurement technique for cutoff wavelength of the MCF are introduced. Moreover, a novel technique for measuring the MFD, relative-index difference Δ, and the chromatic dispersion *D* is described by using bidirectional OTDR.

(a) ***Cutoff wavelength measurement***

cutoff wavelength definition of the MCF and its measurement technique has been proposed to reduce the measurement time [97].

In ITU-T Rec. G.650.1 [98], the cutoff wavelength is defined as the wavelength above which the ratio between the total power, including that of the launched higher-order modes, and the fundamental mode power falls to less than 0.1 dB. The cutoff wavelength of single-mode fiber is specified as the fiber parameter that guarantees single-mode operation above a specified wavelength and its specified value is determined in ITU-T fiber series Recommendations. The conventional definition of cutoff wavelength means that we have to measure the cutoff wavelength of each core in a multi-core fiber. However, as MCF has many cores, it takes a lot of time to measure the cutoff wavelength of each core. Therefore, a definition for the MCF cutoff wavelength should be determined taking account of the measurement time and single-mode operation for the MCF.

Reference [97] proposed the following definition of the cutoff wavelength for MCF taking the above items into account.

> The cutoff wavelength of MCF shall be defined as the maximum cutoff wavelength among the cores in the MCF.

Fig. 2.56 Bend reference method for measuring the cutoff wavelength

When we define the cutoff wavelength of MCF as above, we can measure the cutoff wavelength of an MCF by using conventional transmitted power techniques such as the bend reference and multi-mode reference techniques specified in ITU-T Rec. G.650.1. This means that if the fundamental and higher-order modes can be excited into all the cores in an MCF, the maximum cutoff wavelength of an MCF can be measured in principle by using the transmitted power techniques.

The transmitted power technique uses the variation of the power transmitted through a short length of fiber under test with wavelength, under defined conditions, compared with a reference transmitted power. The reference transmitted power can be obtained in two possible ways, (a) by using a test fiber with a loop with a small radius $R = 30$ mm (bend reference method) or (b) by using a short (1–2 m) length of multi-mode fiber (multi-mode reference method).

Figure 2.56 shows the setup for the bend reference method for measuring the cutoff wavelength of conventional single-mode fibers. The bend reference method measures the cutoff wavelength of the LP_{11} mode based on the difference between the mode dependent losses of the LP_{01} and LP_{11} modes. With the bend reference technique, the launch and detection conditions are not changed, but an additional bend is added to the test fiber. The test fiber is bent to a radius of 30 mm or less to suppress the second-order mode at all the scanned wavelengths. The bending radius of 30 mm is specified in ITU-T Rec. G.650.1. The total transmitted power at $z = L$ with and without a small bend near the output portion is measured by sweeping the wavelength while keeping the launch condition at the input fiber unchanged.

The cutoff wavelength is defined as the wavelength above which the ratio between the total power, including that of the launched higher-order modes, and the fundamental mode power falls to less than 0.1 dB [98]. We derive the power ratio r of the total output power at $z = L$ with a bend near the output portion to the total power without the bend based on the model shown in Fig. 2.56. Here, P_{01} (z) and P_{11} (z) represent the average powers of the LP_{01} and LP_{11} modes at the position of z, respectively. α_{01} and α_{11} denote the mode-dependent losses of the LP_{01} and LP_{11} modes, respectively. α_{b0} and α_{b1} are the bending losses with a bending radius R for the LP_{01} and LP_{11} modes, respectively. The power ratio r is a function of fiber length and wavelength and is obtained as [99–101]

$$r(\lambda, z) = \frac{P_{01}(\lambda, z) + P_{11}(\lambda, z)}{P_{01}(\lambda, z)\exp(-\alpha_{b0}L_b) + P_{11}(\lambda, z)\exp(-\alpha_{b1}L_b)}, \quad L_b = 2\pi R, \tag{2.71}$$

where L_b is the length of the bent fiber.

The average powers P_{01} and P_{11} of the LP_{01} and LP_{11} modes at the position of z without a bend can be expressed as

$$P_{01}(\lambda, z) = P_{01}(\lambda, 0)\exp(-\alpha_{01}z), \tag{2.72}$$

$$P_{11}(\lambda, z) = P_{11}(\lambda, 0)\exp(-\alpha_{11}z), \tag{2.73}$$

where $P_{01}(\lambda, 0)$ and $P_{11}(\lambda, 0)$ denote the input powers of the LP_{01} and LP_{11} modes, respectively.

Substituting (2.72) and (2.73) into (2.71), the power ratio r can be obtained as

$$\begin{aligned} r(\lambda, z) \\ &= \frac{P_{01}(\lambda, 0)\exp(-\alpha_{01}z) + P_{11}(\lambda, 0)\exp(-\alpha_{11}z)}{P_{01}(\lambda, 0)\exp(-\alpha_{01}z)\exp(-\alpha_{b0}L_b) + P_{11}(\lambda, 0)\exp(-\alpha_{11}z)\exp(-\alpha_{b1}L_b)} \\ &= \frac{\exp(-\alpha_{01}z) + k\exp(-\alpha_{11}z)}{\exp(-\alpha_{01}z)\exp(-\alpha_{b0}L_b) + k\exp(-\alpha_{11}z)\exp(-\alpha_{b1}L_b)}. \end{aligned} \tag{2.74}$$

Here, k is the ratio P_{11} $(\lambda, 0)/P_{01}$ $(\lambda, 0)$ of the LP_{11} mode power to the LP_{01} mode power at the fiber input end. The cutoff wavelength λ_c is defined as

$$10\log r(\lambda, z) = 0.1(\text{dB}). \tag{2.75}$$

The theoretical analysis of the cutoff wavelength of single-core fiber by the bend reference technique can be applied to MCF mentioned above.

Here, the excitation power ratio γ_i of the i-th core power P_i of MCF to the total input power P_t is expressed as

$$\gamma_i = \frac{P_i}{P_t}, \tag{2.76}$$

$$\sum_i \gamma_i = 1. \tag{2.77}$$

When the fundamental and higher-order modes for all the cores in the MCF are excited, the power ratio r of MCF with and without a bend can be expressed as

$$\begin{aligned} r(\lambda, z) &= \sum_{i=1}^{n} A_i / \sum_{i=1}^{n} B_i \\ A_i &= \gamma_i P_t(\lambda, 0)\left[\exp(-\alpha_{01}^i z) + k\exp(-\alpha_{11}^i z)\right] \\ B_i &= \gamma_i P_t(\lambda, 0)\left[\exp(-\alpha_{01}^i z)\exp(-\alpha_{b0}^i L_b) + k\exp(-\alpha_{11}^i z)\exp(-\alpha_{b1}^i L_b)\right], \end{aligned} \tag{2.78}$$

where A_i and B_i denote the transmitted power of the i-th core with and without a single bend, respectively. n is the number of the cores in the MCF.

We measured the cutoff wavelength of a 2-m-long seven-core MCF using our proposed technique. Figure 2.57 shows the experimental spectral loss of an MCF with seven cores and reveals that our present technique can estimate the cutoff wavelength of MCF according to our proposed definition.

A cutoff wavelength of 1.51 μm was obtained for the MCF from Fig. 2.57. The measurement reproducibility of the cutoff wavelength was about ±3 nm. The cutoff wavelength of each core in the MCF was measured, and the experimental results are shown in Fig. 2.58.

The MCF could operate in a single mode at wavelengths exceeding 1.51 μm. It is confirmed that the maximum cutoff wavelength of the seven-core MCF could successfully estimate in the same way as with two-core MCF.

(a) ***Mode-field diameter, relative-index difference, and chromatic dispersion***

Fig. 2.57 Experimental spectral loss of MCF with seven cores

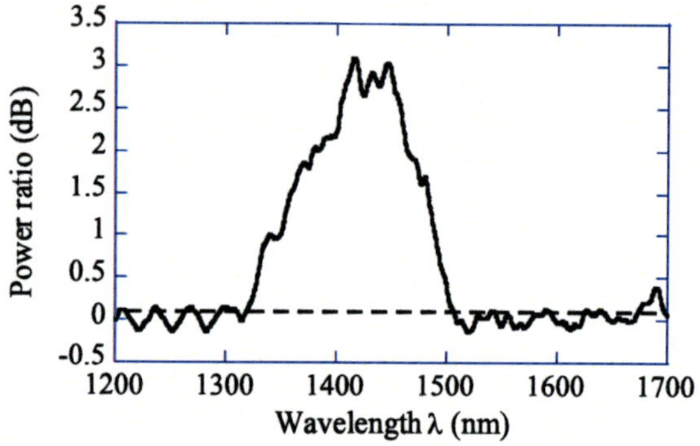

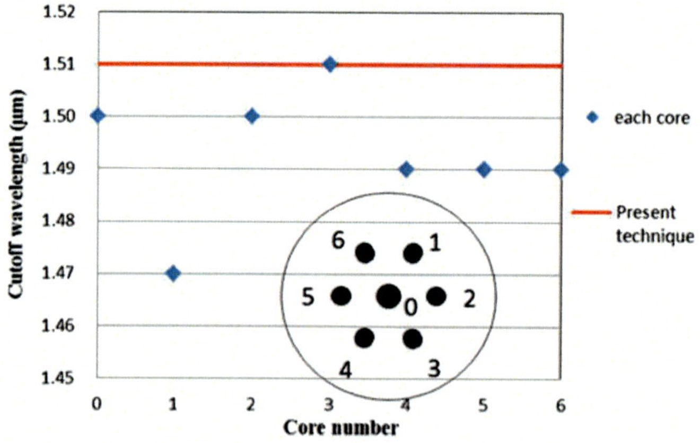

Fig. 2.58 Experimental cutoff wavelengths of individual cores in MCF

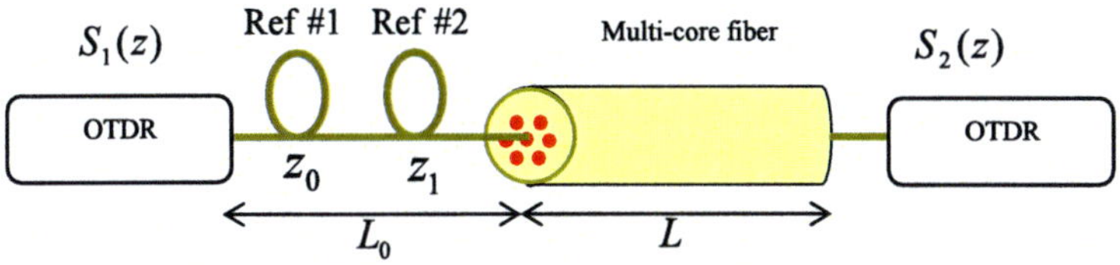

Fig. 2.59 Experimental setup for measuring MFD (Reprinted with permission from [96]. © 2014 The Optical Society)

longitudinal fiber parameters of MCF is shown in Fig. 2.59.

For simplicity, a homogeneous two-core MCF is considered as shown in Fig. 2.59. Consider the longitudinal fiber parameter of the core #1 in MCF. When the input power $P_0(0)$ is launched into the core #1 of MCF, the powers of center and outer cores can be obtained by solving the coupled-power equations with regards to two cores. The powers of the cores #1 and #2 $P_1(z)$ and $P_2(z)$ with respect to fiber length z are given by (2.66).

Two different kinds of reference fiber with a length of L_0 as shown in Fig. 2.59 are employed to determine the absolute value of the longitudinal fiber parameter before an MCF of length L. OTDR signals of the core #1 $S_1(z)$ and $S_2(z)$ (in dB) launched from the opposite ends (subscripts 1 and 2) of a fiber of length $L + L_0$ can be expressed as [96]

$$S_1(z) = \begin{cases} 10\log P_1 + 10\log(\alpha_s(z)) + 10\log(B(z)) - 2\alpha z 10\log e & (0 \le z \le L_0) \\ 10\log P_1 + 10\log(\alpha_s(z)) + 10\log(B(z)) - 2\alpha z 10\log e & \\ +10\log\left[\frac{1+\exp[-4h(z-L_0)]}{2}\right] & (L_0 + L \ge z \ge L_0) \end{cases}, \quad (2.79)$$

$$S_2(z) = \begin{cases} 10\log P_2 + 10\log(\alpha_s(z)) + 10\log(B(z)) - 2\alpha(L + L_0 - z)10\log e & \\ +10\log\left[\frac{1+\exp[-4hL]}{2}\right] & (0 \le z \le L_0) \\ 10\log P_2 + 10\log(\alpha_s(z)) + 10\log(B(z)) - 2\alpha(L + L_0 - z)10\log e & \\ +10\log\left[\frac{1+\exp[-4h(L+L_0-z)]}{2}\right] & (L_0 + L \ge z \ge L_0) \end{cases}, \quad (2.80)$$

where P_1 and P_2 are the input powers, α_s the local scattering coefficient, B the backscattered capture fraction, and α the local attenuation coefficient. h denotes the mode-coupling coefficient between two cores in an MCF.

(2.79) and (2.80) include the optical losses caused by optical power decay, imperfection, and mode coupling. Therefore, the imperfection loss contribution $U(z)$ [102] excluding the mode-coupling loss can be obtained as

$$U(z) = 5\log P_1 P_2 + 10\log(\alpha_s(z)B(z)) - 2\alpha(L + L_0)(10\log e)$$

$$= \begin{cases} \frac{S_1(z)+S_2(z)}{2} - 5\log\left[\frac{1+\exp[-4hL]}{2}\right] & (0 \le z \le L_0) \\ \frac{S_1(z)+S_2(z)}{2} - 5\log\left[\frac{\{1+\exp[-4h(z-L_0)]\}\{1+\exp[-4h(L+L_0-z)]\}}{4}\right] & (L_0 \le z \le L_0 + L) \end{cases} \quad (2.81)$$

The backscatter capture fraction B is expressed using the refractive index of the core n and the MFD $2w(\lambda, z)$ as [102]

$$B(\lambda, z) = \frac{3}{2}\left[\frac{\lambda}{2\pi n w(\lambda, z)}\right]^2. \quad (2.82)$$

The imperfection loss contribution $U_n(z)$ normalized by the value at the first reference point $z = z_0$ can be expressed as

$$U_n(z) = U(z) - U(z_0) = 10\log\left(\frac{\alpha_s(z)n^2(z_0)}{\alpha_s(z_0)n^2(z)}\right) + 20\log\left(\frac{2w(\lambda, z_0)}{2w(\lambda, z)}\right). \quad (2.83)$$

With conventional single-mode fibers, the variation in the local scattering coefficient α_s is negligible compared with that in the MFD. The second reference point $z = z_1$ satisfies (2.83). Therefore, the longitudinal MFD distribution of the core in the MCF can be estimated by using the normalized imperfection loss and MFDs at the two reference points z_0 and z_1 as [103]

$$2w(\lambda, z) = 2w(\lambda, z_0)\left(\frac{2w(\lambda, z_0)}{2w(\lambda, z_1)}\right)^{\frac{U(z)-U(z_0)}{U(z_1)-U(z_0)}}. \quad (2.84)$$

To obtain the pure imperfection loss contribution $U(z)$, the crosstalk between cores have to be estimated. When the OTDR pulse is launched into the core #1 of an MCF with a length L, the backscattered powers $S_3(z)$ (=10log$P_3(z)$) of the core #1 and $S_4(z)$ (=10log$P_4(z)$) of the core #2 in dB from a given position z in an MCF are measured and the crosstalk is measured as shown in Sect. 2.3.6.1. Here, if we assume that $\alpha_{s1} = \alpha_{s2} = \alpha_s$, $\alpha_1 = \alpha_2 = \alpha$, and $B_1 = B_2 = B$, the crosstalk (*XT*) between two cores in an MCF estimated using the OTDR is obtained as the following equation.

$$XT(z) = 10\log\left[\frac{P_4(z)}{P_3(z)}\right] = 10\log\left[\frac{1-\exp(-4hz)}{1+\exp(-4hz)}\right]. \quad (2.85)$$

Note that the crosstalk given by (2.85) estimated using the OTDR is that for a fiber length of $2z$.

The average mode-coupling coefficient $h(z)$ over the fiber length z can be estimated from (2.85). Thus, the imperfection loss contribution $U(z)$ can be estimated by taking the mode-coupling coefficient $h(z)$ into account. The MFD distribution of the core in an MCF can be measured with the present technique.

Once the longitudinal MFD distribution is known, the relative-index difference $\Delta(z)$ (in %) can be estimated as [104]

$$\Delta(z) = \frac{1}{k}\left[(1 + k\Delta(z_0)) \times 10^{\frac{U(\lambda,z) - U(\lambda,z_0) - 20\log\left(\frac{2w(\lambda,z_0)}{2w(\lambda,z)}\right)}{10}} - 1\right], \tag{2.86}$$

where k is a proportional constant of the Rayleigh scattering coefficient to the dopant concentration of GeO_2-doped core fiber [105].

Moreover, we can estimate the longitudinal chromatic dispersion $D(z)$ by using the wavelength dependence of MFD and the relative-index difference Δ. The chromatic dispersion D is given as the sum of the waveguide dispersion D_w and the material dispersion D_m. The material dispersion D_m and waveguide dispersion D_w are expressed as [106]

$$D = D_m + D_w, \tag{2.87}$$

$$D_m = -\frac{\lambda}{c}\frac{d^2n}{d\lambda^2}, \tag{2.88}$$

$$D_w = \frac{\lambda}{2\pi^2 cn}\frac{d}{d\lambda}\left(\frac{\lambda}{w^2}\right), \tag{2.89}$$

where c is the light velocity and n is the refractive index of the core.

The material dispersion D_m can be estimated from the dopant concentration in an optical fiber by using Sellmeier's relation [107]. By contrast, the waveguide dispersion D_w can be estimated by determining the wavelength dependence of the MFD [106].

We approximated the wavelength dependence of the MFD [14] as

$$w(\lambda) = g_0 + g_1\lambda^{1.5}. \tag{2.90}$$

Substituting (2.90) into (2.89), the waveguide dispersion can be expressed as [108]

$$D_w = \frac{\lambda}{2\pi^2 cnw^2}\left(1 - \frac{3g_1\lambda^{1.5}}{w}\right). \tag{2.91}$$

Here, when the MFDs are given at two wavelengths, the coefficient g_1 is determined by solving (2.90) with regard to the two wavelengths.

Figure 2.60 shows the MFD distribution $2w(z)$ of the fiber link at $\lambda = 1.55$ μm. The fiber link consists of two reference fibers and the seven-core MCF. Their fiber parameters are listed in Table 2.5. It is seen that the MFD of MCF decreases a little bit as the fiber length increases. We found that the experimental result for the MFD of the test MCF agreed well with the value measured with the far-field pattern (FFP) technique as shown in Table 2.5. As a result, it was confirmed that the MFD distribution was successfully estimated by the proposed technique.

Figure 2.61 shows the relative-index difference distribution $\Delta(z)$ of the fiber link.

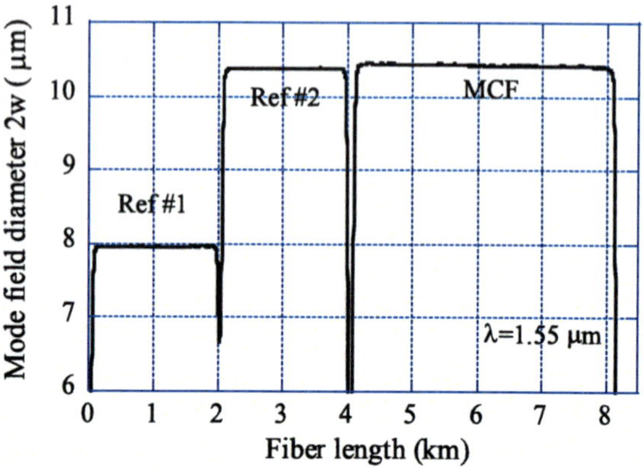

Fig. 2.60 MFD distribution of test fiber link at λ = 1.55 µm (Reprinted with permission from [96]. © 2014 The Optical Society)

Table 2.5 Parameters of test MCF and reference fibers

Parameters	MCF	Ref #1	Ref #2
MFD at 1.55 µm (µm) at 1.31 µm (µm)	10.4 –	7.9 6.4	10.3 9.2
Cutoff wavelength λ_c (µm)	1.21	1.11	1.23
Relative-index difference Δ (%)	–	0.915	0.41
Core pitch Λ (µm)	40.46	–	–
Outer diameter (µm)	179.4	125.1	125.0
Chromatic dispersion at 1.55 µm D (ps/km/nm)	–	0.9	15.5
Crosstalk* at 1.55 µm (dB)	−24.9	–	–

Reprinted with permission from [96]. © 2014 The Optical Society
*Average crosstalk between the center and outer cores

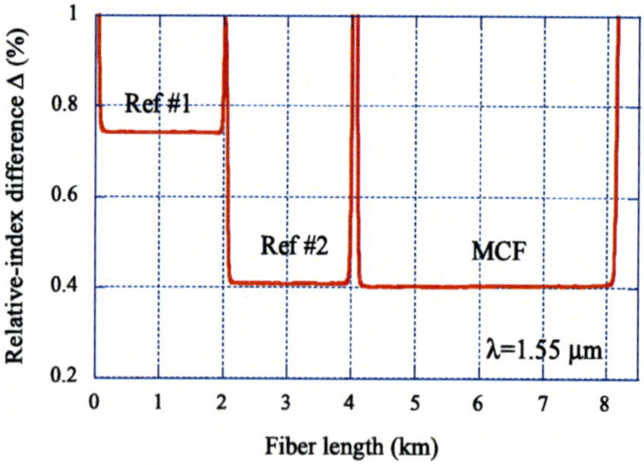

Fig. 2.61 Relative-index difference Δ (%) distribution of fiber link (Reprinted with permission from [96]. © 2014 The Optical Society)

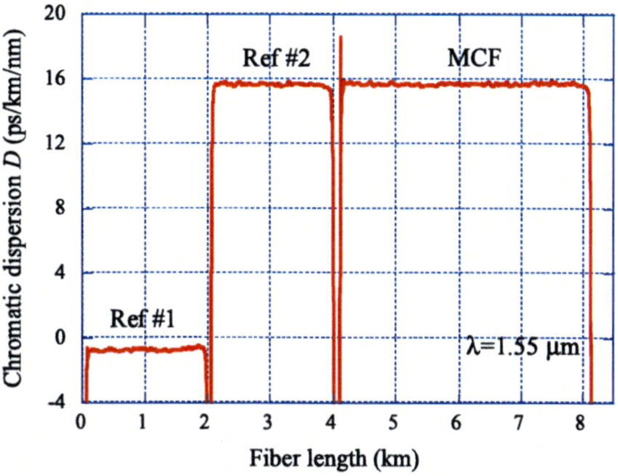

Fig. 2.62 Chromatic dispersion distribution of fiber link at 1.55 μm (Reprinted with permission from [96]. © 2014 The Optical Society)

The relative-index difference Δ was estimated by using (2.86). It is seen that the relative-index difference of the test MCF is almost the same as that of the conventional single-mode fiber, Ref #2. The Δ of Ref #2 was used to estimate the relative-index difference of the MCF. Therefore, the Δ value of Ref #1 is slightly smaller than that shown in Table 2.5. The Δ of MCF becomes large as the fiber length increases. This property of the MCF influences the MFD distribution as shown in Fig. 2.60.

Figure 2.62 shows the chromatic dispersion distribution $D(z)$ of the fiber link at a wavelength of 1.55 μm. The chromatic dispersion D can be estimated by using the estimated relative-index difference Δ and the wavelength dependence of the MFD. We used the experimental MFD results that we obtained at 1.31 and 1.55 μm. Here, a k value of 0.62 was used.

The longitudinal parameters such as MFD $2w$, relative-index difference Δ, and chromatic dispersion D can be successfully estimated with the present OTDR technique. This technique will provide a powerful way of measuring the longitudinal fiber parameter of MCF.

2.4 Few-Mode Fiber

2.4.1 Design Trend of FMF

Few-mode fiber (FMF) is a fiber through which multiple propagation modes can be transmitted. Although in the broad sense it can be categorized as a multi-mode fiber, it is called "few-mode fiber" to distinguish it from conventional graded-index multimode fibers (GI-MMFs). One important difference between FMF and conventional GI-MMF is the number of propagation modes. Conventional GI-MMF typically accommodates several tens of modes. On the other hand, FMF has many fewer modes (typically fewer than ten-LP modes). Another important difference is their application area. GI-MMF has been widely deployed over short-to-medium distances (e.g., in data centers or local area networks). In contrast, FMF has been investigated for use in large capacity networks to overcome the capacity limit of conventional single-mode fiber (SMF).

A new approach for realizing a much higher transmission capacity using MMF has been proposed [109, 110], where electrical signal processing at the receiver is carried out to actively utilize the higher-order modes and thus increase the transmission capacity. This approach mirrors that of the MIMO transmission system in wireless communication. Figure 2.63 shows a conceptual diagram of an optical MIMO transmission system. Multiple signals are multiplexed at the mode multiplexer and transmitted through multiple modes in the FMF. They are then demultiplexed at the mode multiplexer and detected at the coherent receivers. Although each signal can be individually transmitted when there is no modal crosstalk in the transmission line, inter-modal crosstalk can occur in the multi/demultiplexer or FMF, and this causes serious signal degradation. A MIMO equalizer is deployed to compensate for the crosstalk. The modal crosstalk can be correctly compensated for by adaptively adjusting the equalizing coefficient w_{ij} in the MIMO equalizer. However, the computation needed to recover the signals at the receiver becomes more complex as the differential mode delay (DMD) between the propagation modes increases. Thus, it is desirable to reduce the DMD before the signals are received, and low DMD fibers are suitable as transmission lines to reduce the computational complexity of MIMO processing.

Various FMFs have already been proposed as shown in Fig. 2.64. They fall roughly into three categories; weakly coupled FMF without DMD tailoring, low \DMD FMF, and strongly coupled FMF. The common and mandatory requirement for these FMFs is that they can propagate more than n modes including the spatial/polarization mode when the MIMO equalizer has an $n \times n$ configuration. To meet just this requirement, simple step-index (SI) or W shaped-index FMF [111] can be utilized as the transmission line. However, its DMD value is relatively high (e.g., a few ns/km), and thus

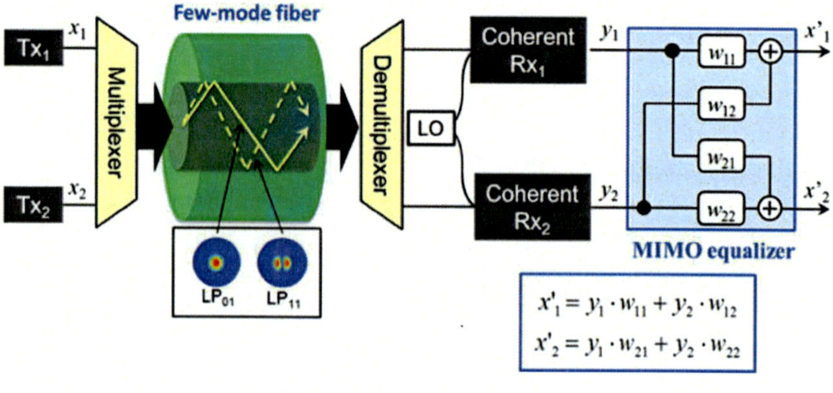

Fig. 2.63 Conceptual diagram of optical MIMO transmission system

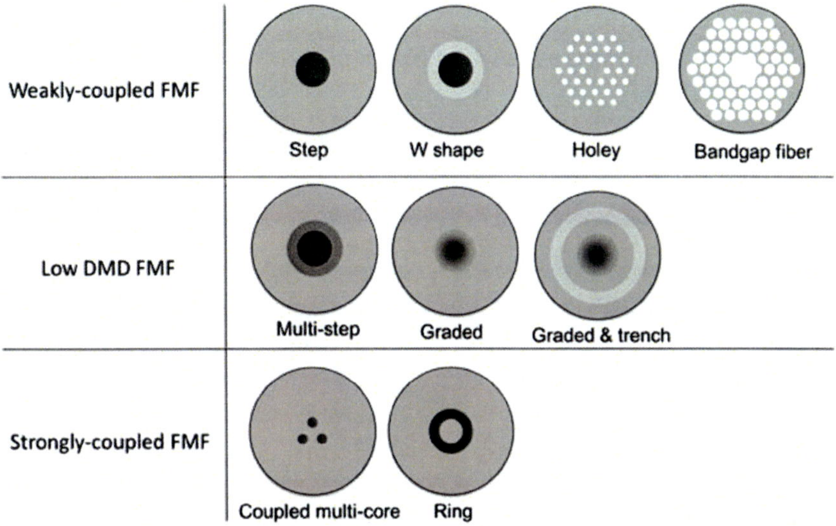

Fig. 2.64 Various types of proposed FMF

the MIMO processing complexity becomes high. Figure 2.65 shows one example of the measured impulse response characteristic of two-mode SI-FMF, where the core radius a is 7.0 μm and the relative-index difference Δ is 0.4% [112]. Two pulses corresponding to the LP_{01} and LP_{11} modes were observed, and a large DMD of about 2 ns/km was obtained. One way to reduce MIMO processing complexity is to reduce the mode crosstalk in the fiber and employ a less complex MIMO equalizer [112]. It is considered that FMF with a large propagation constant difference between modes $\Delta\beta$ is effective in preventing each mode from coupling owing to structural perturbations [113, 114]. Another approach to the MIMO issue is to use true eigenmodes for mode-division multiplexing [115–117].

Fig. 2.65 Measured impulse response of two-mode step-index fiber

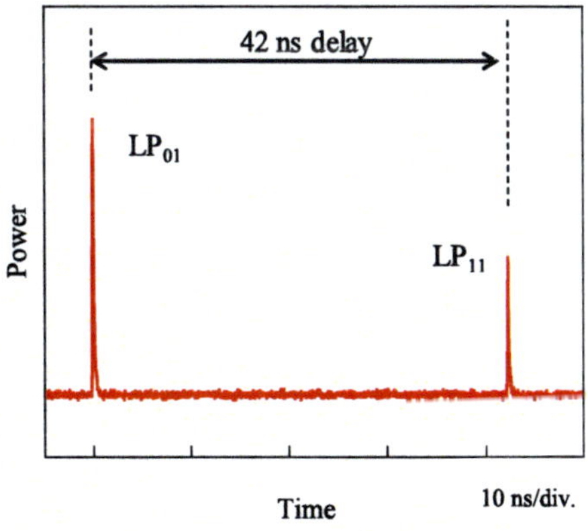

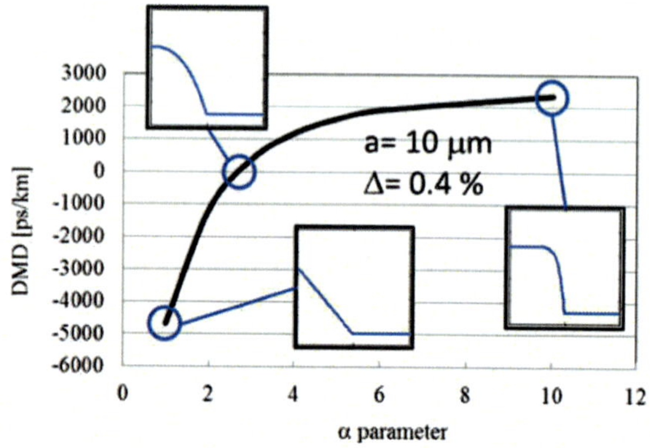

Fig. 2.66 Calculated DMD of 2-mode GI-FMF

Multi-step [118] or graded-index (GI) FMF [119] has been investigated to lower the DMD value. Figure 2.66 shows the calculated DMD value of two-mode GI-FMF where a is 10 μm and Δ is 0.4%. The DMD can be varied from negative to positive by adjusting the parameter α, and a zero DMD value can be obtained by optimizing the core profile. A detailed GI-FMF design for lowering the DMD is discussed in Sect. 2.4.2. Recently, fiber with up to 9-LP modes was proposed [120–122], and a core profile with a graded-index and a trench has been intensively investigated to realize low DMD FMF because of its DMD and mode number controllability.

Although in theory low DMD characteristics can be realized by optimizing the core profile of FMF, the actual DMD of fabricated FMF could be larger than expected owing to manufacturing errors. Figure 2.67 shows a schematic diagram of DMD evolution as a function of transmission distance. As it is very difficult to obtain perfectly "zero" DMD fiber, the total DMD increases as the transmission distance increases. Therefore, DMD-compensated fiber line was then proposed for further reducing the DMD value. It consists of two types of fibers, positive DMD fiber and negative DMD fiber. Here, a positive DMD means that the group velocity of a fundamental mode is higher than that of a higher-order mode. The total DMD can be compensated when the signal passes through positive and negative DMDF in the same manner as chromatic dispersion compensation. Therefore, the total DMD value

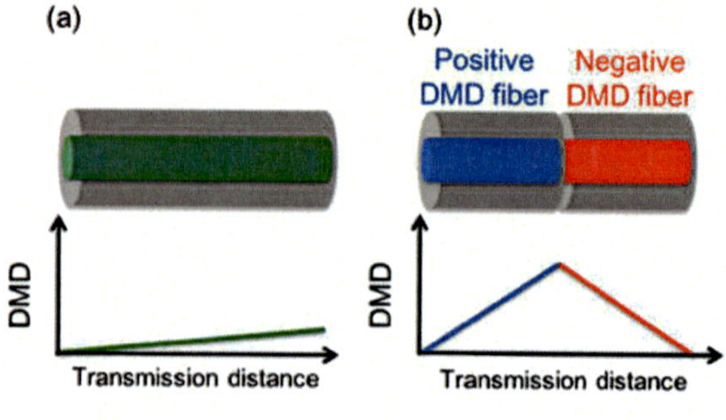

Fig. 2.67 Schematic diagram of DMD-compensated fiber line

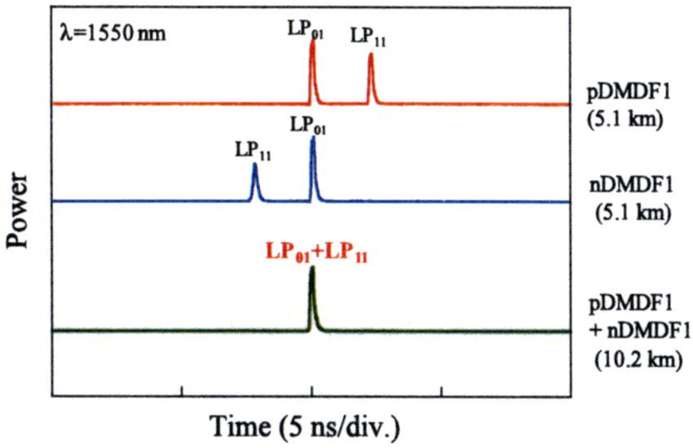

Fig. 2.68 Impulse response of two-mode DMD-compensated fiber line 118] (Reprinted with permission from [118]. © 2014 IEEE)

can be adjusted by changing the length ratio between two FMFs, which enables us to precisely control the total DMD at the end of the transmission line. Figure 2.68 shows the measured impulse response of two-mode DMD-compensated fiber line [118]. Only two pulses can be seen, which correspond to pulses being transmitted in the LP_{01} and LP_{11} modes with each FMF. This figure also shows the measured impulse response of the DMD-compensated fiber line with positive and negative FMF. Although the LP_{11} mode propagation at the end of the fiber line was confirmed as shown in the near field pattern, there was only one pulse, which indicated that the DMD was successfully compensated at the end of the fiber line. It should be noted that the modal crosstalk that was present throughout the transmission line should be suppressed as much as possible because this DMD compensation scheme is based on there being no modal crosstalk.

In contrast to the weakly coupled FMF discussed above, strongly coupled FMFs have recently been investigated. By employing a coupled multi-core [47, 123–127] or ring-core structure [128], the propagation modes couple strongly with each other in the FMF because the $\Delta\beta$ values between modes are small enough to cause mode coupling even when a small perturbation such as micro-bending is applied to the fiber. It has been proposed that distributed mode coupling in the fiber reduces the spread of the impulse response and is effective in reducing the MIMO processing complexity [47], where impulse response growth is in proportion to the square root of a kilometer in the same way as polarization mode dispersion.

Holey FMF such as two-mode photonic crystal fiber (PCF) [129] and photonic bandgap fiber have also been proposed. These fibers have unique characteristics that cannot be obtained with conventional all-solid fibers. Few-mode PCF can realize a larger effective area while maintaining the number of propagation modes over a wide wavelength range. Hollow-core PBGF has been studied with the aim of utilizing the 2 μm band, which is considered a potential transmission window with ultra-low loss. Although transmission loss is still higher than that of solid fiber, high-capacity mode-division multiplexed transmission over 2 μm has been demonstrated [130, 131].

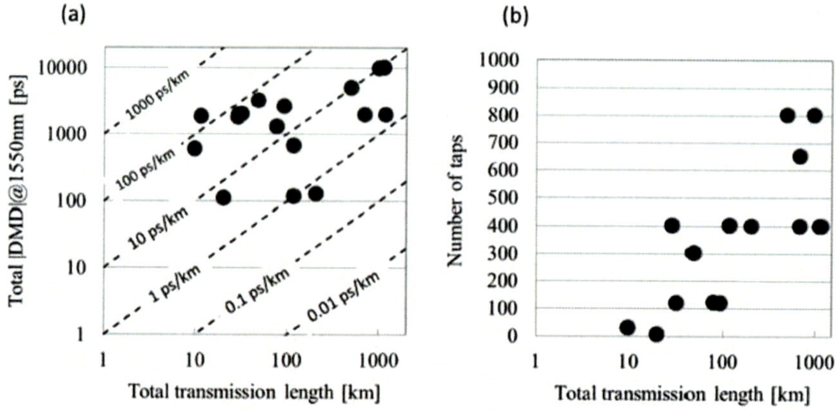

Fig. 2.69 Recently reported transmission experiments with two-mode fiber

Finally, a recently reported mode-division multiplexed transmission experiment using two-mode fiber is summarized in Fig. 2.69. Although a transmission experiment with a very small DMD of less than 1 ps/km was demonstrated, all the experiments were carried out with off-line-based signal processing; thus, a further reduction of MIMO processing complexity is still required if we are to achieve long-haul MIMO transmission with online signal processing.

2.4.2 DMD and MIMO

FMF should be carefully designed taking the following into account;

1. DMD
2. Bending loss of propagation mode
3. Cutoff wavelength for undesirable mode

When assuming two-mode operation for example, the bending losses of the LP_{01} and LP_{11} modes should be enough low to propagate stably through the transmission line, and the undesirable mode (LP_{21} mode) should be cutoff over the operational band while minimizing the DMD value. Of course, a larger effective area is preferable for each mode in terms of reducing the nonlinear effect in the fiber.

This section describes in detail the FMF design where a graded-index with a trench core profile is assumed and shows how to realize few-mode operation while maintaining a low DMD value. Figure 2.70a, b show the relationship between core radius a and normalized propagation constant b at 1550 nm and the relationship between core radius a and DMD at 1550 nm, respectively. The parameter Δ is fixed at 1.0%, and α is fixed at 2.0. As shown in Fig. 2.70a, modes with the same or very similar propagation constants can be grouped together in sets. In general, modes

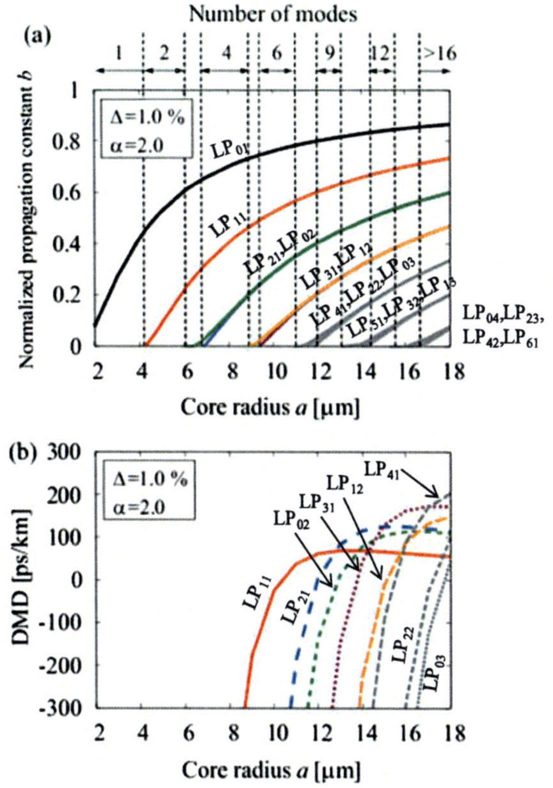

Fig. 2.70 LP-mode characteristics of GI profile, **a** normalized propagation constant *b* versus core radius *a* at 1550 nm, **b** DMD versus core radius *a* at 1550 nm (Reprinted with permission from [133]. © 2014 IEEE)

belong to the same principal mode group *M* if the indices of the $LP_{l,p}$ modes fulfill the condition [132]:

$$M = 2p + l - 1. \tag{2.92}$$

For example, the LP_{21} and LP_{02} modes have an *M* of 3, and the LP_{31} and LP_{12} modes have an *M* of 4. The modes within a certain mode group have very similar propagation constants. The DMD is less than 200 ps/km as a result of tuning the core radius as shown in Fig. 2.70b. On the other hand, Fig. 2.70a, b show that it is not possible to realize *N*-mode operation and a low DMD solely by adjusting the core. Here we define *N*-mode operation as the propagation of *N* specific modes in a fiber for the target wavelength. According to [134], a GI fiber with two-mode operation and a DMD of less than 36 ps/km over the C band can be realized by tuning the structural parameters a, Δ, and α. However, it becomes difficult to realize more than two-mode operation and a low DMD simultaneously with a GI core alone. Therefore, we have to use both a GI core and a trench to realize more than two-mode operation and a low DMD [135, 136].

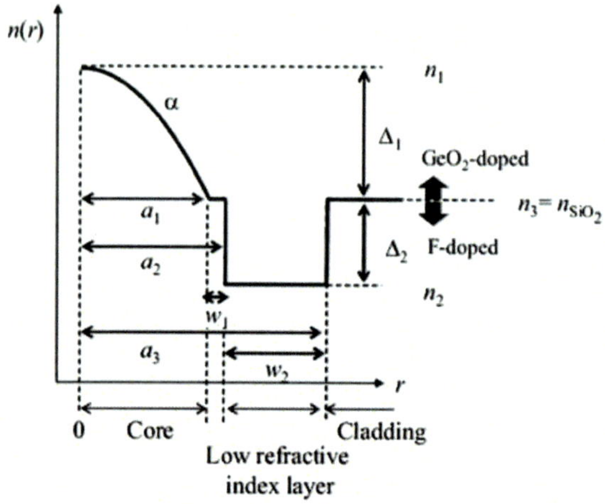

Fig. 2.71 Refractive index profile of GI core with trench fiber (Reprinted with permission from [133]. © 2014 IEEE)

Figure 2.71 shows the refractive index profile of a GI core with a trench fiber. This profile for four-mode operation is investigated in [135, 137].

This profile $n(r)$ is given by

$$n(r) = \begin{cases} n_1\left[1 - 2\Delta_1\left(\frac{r}{a_1}\right)^{\alpha}\right]^{\frac{1}{2}}, & 0 \leq r \leq a_1 \\ n_3, & a_1 \leq r \leq a_2 \\ n_2, & a_2 \leq r \leq a_3 \\ n_3, & r \geq a_3 \end{cases}, \tag{2.93}$$

where Δ_1 is the relative-index difference between the core and the low refractive index layer, which is defined as $(n_1{}^2 - n_3{}^2)/2n_1{}^2$, Δ_2 is the relative-index difference between the cladding and the low refractive index layer, which is defined as $(n_2{}^2 - n_3{}^2)2n_2{}^2$, $w_1 = a_2 - a_1$ is the width between the core and the low refractive index layer, or trench, and $w_2 = a_3 - a_2$ is the width of the low refractive index layer. Until now, the design process has been unclear. In this analysis, we clearly specify the fiber design procedure as regards a GI core with a trench profile after a careful consideration of feasibility.

Figure 2.72 shows the DMD at 1550 nm as function of core radius a_1. The fiber structure parameters were fixed at $w_1 = 1.5$ μm, $w_2 = 11.0$ μm, $\Delta_1 = 0.60\%$, $\Delta_2 = -0.40\%$, and $\alpha = 2.0$, respectively. When 10.4 μm < a_1 < 13.0 μm, we can realize a DMD of less than 200 ps/km and four-mode operation.

As shown in Fig. 2.73, the DMDs of the LP_{21} and LP_{02} modes can be designed so that they are almost the same by adjusting the structural parameters. The Δn_{eff} of the LP_{11} mode and the effective index difference between the LP_{11} mode and the LP_{21} or LP_{02} modes are large. However, the Δn_{eff} values of the LP_{21} and LP_{02} modes are almost the same and mode coupling between these modes is inevitable. Hence, a four-LP-mode fiber with three-mode groups (LP_{01}, LP_{11}, and $LP_{21} + LP_{02}$ mode)

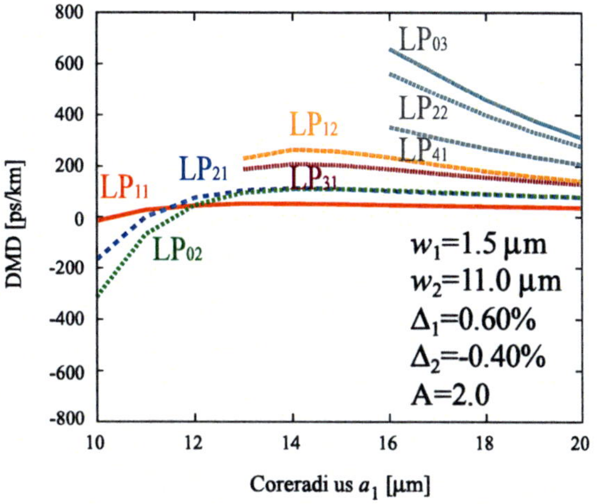

Fig. 2.72 DMD versus core radius a_1 at 1550 nm of proposed GI core with trench fiber (Reprinted with permission from [133]. © 2014 IEEE)

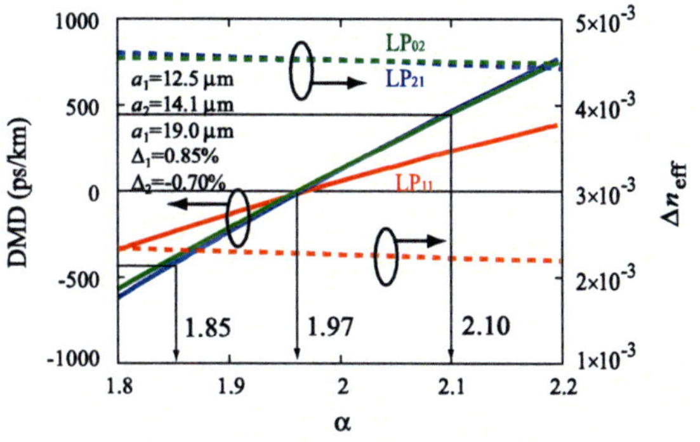

Fig. 2.73 DMD and Δn_{eff} at 1550 nm as a function of α for six-LP-mode fiber (Reprinted with permission from [133]. © 2014 IEEE)

was realized as a DMD compensation transmission line. The DMDs of the LP_{21} and LP_{02} modes can be made the same by adjusting the structural parameters. Moreover, the DMD can be widely tuned from positive to negative by controlling the profile exponent α while maintaining three-mode groups. Thus, even though there is mode coupling between the LP_{21} and LP_{02} modes, a four-mode DMD compensation line can be realized. We can also realize a four-LP-mode DMD compensation line, for example, by combining a positive DMD fiber of $\alpha = 2.10$ and a negative DMD fiber of $\alpha = 1.85$. Although the grouped LP_{21} and LP_{02} modes have very similar DMD and Δn_{eff} values, we can treat them as two independent LP modes in the MIMO system.

Three kinds of four-LP-mode fibers (A, B, and C) and three kinds of six-LP-mode fibers (D, E, and F) were fabricated for use in a $C + L$ band WDM-MDM transmission. The structural parameters were set in consideration of the feasibility of manufacture and the bending loss of the propagating modes. The wavelength dependence of the DMD of the fabricated fibers was obtained by measuring the impulse responses. The solid lines and plots in Fig. 2.74 show the calculated and experimental results, respectively, and these were in good agreement. Fibers B and E had the smallest DMDs of the three types of fiber. Fibers A, D and C, E had positive and negative DMDs, respectively. The maximum DMD of Fiber B was 92 ps/km over the C band and 135 ps/km over the $C + L$ band with four-mode operation and the maximum DMD of Fiber E was 69 ps/km over the C band and 124 ps/km over the $C + L$ band with six-LP-mode operation.

The impulse responses of the fibers were measured to confirm that the DMD compensation performed properly. Figure 2.75a shows the impulse responses of Fibers A and C when we launched pulses with a 100 ps duration and a 100 ns interval at 1550 nm. We can see only three pulses for Fibers A and C, which correspond to pulses being transmitted in the LP_{01}, LP_{11}, and $LP_{21} + LP_{02}$ modes. We also measured the impulse response of the cascaded fiber line with Fibers A and C by using a

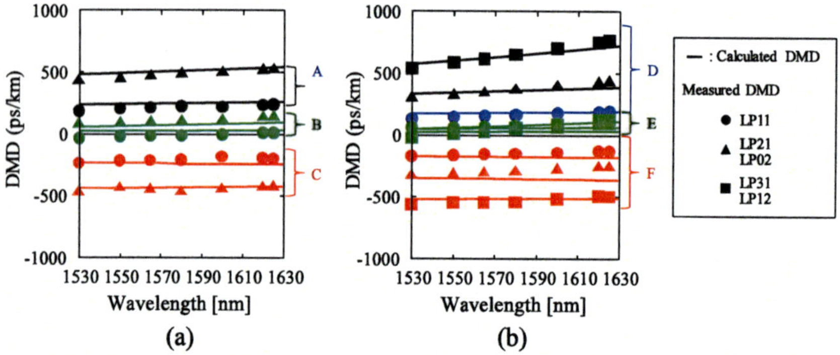

Fig. 2.74 Wavelength dependence of DMD for fabricated fibers (Reprinted with permission from [133]. © 2014 IEEE)

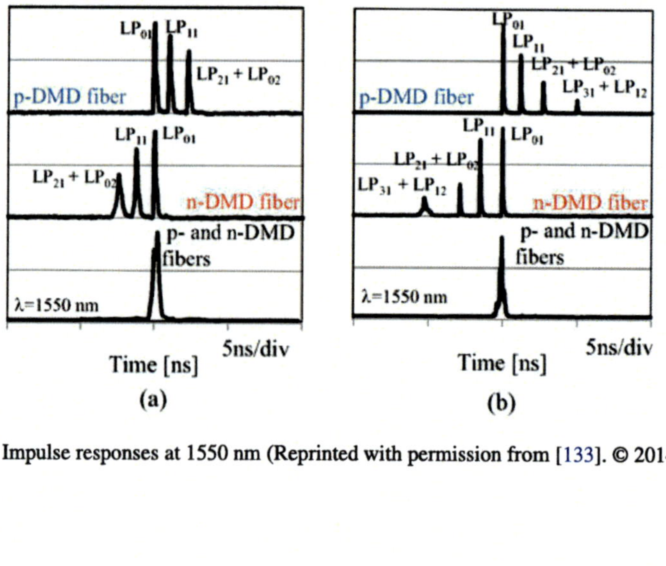

Fig. 2.75 Impulse responses at 1550 nm (Reprinted with permission from [133]. © 2014 IEEE)

fusion splice, and we could observe only one pulse. Figure 2.75b shows the impulse responses of Fibers D and F when 100 ps pulses were launched at 1550 nm. Only four pulses can be seen for Fibers D and F, which correspond to pulses being transmitted in the LP_{01}, LP_{11}, $LP_{21} + LP_{02}$, $LP_{31} + LP_{12}$ modes. We also measured the impulse response of a cascaded fiber line consisting of Fibers D and F by using a fusion splice, and we could observe only one pulse. Therefore, there was almost no modal crosstalk at the fusion-spliced point, and the DMD was successfully compensated for by combining six-LP-mode positive and negative DMD fibers.

Figure 2.76 shows the reported relationship between the DMD and A_{eff} of the LP_{01} mode at 1550 nm in fabricated FMFs with a large A_{eff}. As a reference, the figure also provides the A_{eff} values of recently reported large A_{eff} SMF and conventional SMF. Low DMD fiber led to the suggestion of fibers with a large A_{eff} and a low DMD. It can be seen that two-LP-mode fibers can achieve a larger A_{eff} than the large A_{eff} SMF. On the other hand, four-, six-, and nine-LP-mode fibers tend to have a smaller A_{eff} than two-LP-mode fibers [120, 121, 135, 138]. This is because the higher-order mode is increasingly confined by the high relative-index difference Δ.

Therefore, we have to take account of the nonlinear effect caused by the small A_{eff} when we attempt to realize a large transmission capacity by increasing the number of propagation modes. Figure 2.77 shows that the maximum DMD increases as the

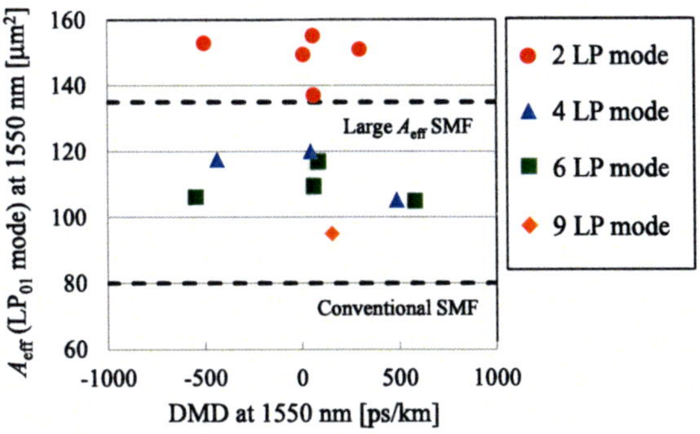

Fig. 2.76 Relationship between DMD and A_{eff} at 1550 nm in reported fabricated FMF with large A_{eff} (Reprinted with permission from [133]. © 2014 IEEE)

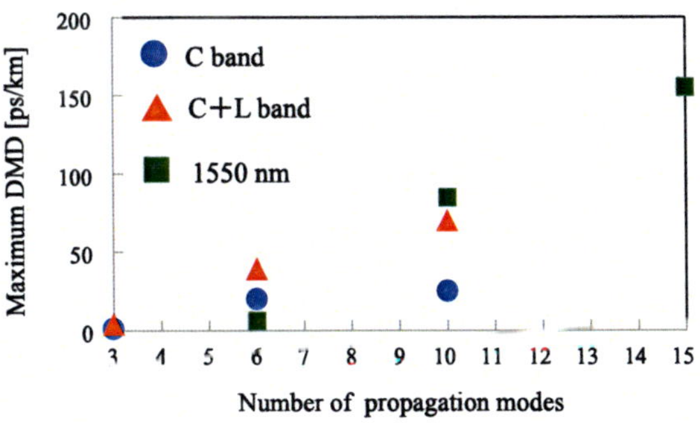

Fig. 2.77 Relationship between number of propagation modes and maximum DMD in reported fabricated FMF

number of propagation modes increases. The concept of DMD compensation by cascading positive and negative fibers requires a higher Δ and stricter fiber fabrication tolerances if we are to realize nine-LP-mode fibers. Furthermore, it becomes necessary to consider the DMD of degenerate modes in the LP modes, and DMD compensation to achieve a lower DMD becomes more difficult.

2.4.3 Mode Coupling

As mentioned in the previous section, a DMD of an FMF is a key factor affecting MIMO complexity. We should also consider the mode-coupling effect to evaluate the MIMO complexity of an MDM system over an FMF. Impulse responses that are monitored at a receiver vary depending not only on the degree of mode coupling but also on coupling location. Mode coupling during signal propagation through a fiber is a well-known phenomenon. In addition, mode coupling caused at splices with misalignment results in characteristic pulses in the impulse response.

In this section, mode coupling in a two-mode fiber, which is the most primitive case of FMFs, is explained. At first, the impulse response with mode coupling over a fiber is theoretically introduced. Impulse responses for different mode-coupling coefficients are presented. Finally, the impact of mode coupling at splice points is discussed.

2.4.3.1 Impulse Response with Mode Coupling Over a Fiber

The impulse response has already been investigated in the early stages of optical transmission development [139, 140]. Here, mode 1 and mode 2 are defined as modes with group velocities of v_1 and v_2, respectively, where $v_1 > v_2$. If an impulse of mode 1 or mode 2 is solely excited in a two-mode fiber, the responses of the modes are described as follows.

$$
\begin{aligned}
P_{11}(z,\tau) = {} & \delta(\tau)\exp\{-(\alpha_1+h)z\} \\
& + \left\{\frac{hn_0}{c\tau_g}\sqrt{\frac{1-\tau}{\tau}}I_1(X')\right\}\exp\{-(\alpha_1+h)z\} \\
& \exp\{(\alpha_1-\alpha_2)\tau z\},
\end{aligned} \tag{2.94}
$$

$$
P_{12}(z,\tau) = \left\{\frac{hn_0}{c\tau_g}I_0\big(X'\big)\right\}\exp\{-(\alpha_1+h)z\}\exp\{(\alpha_1-\alpha_2)\tau z\}, \tag{2.95}
$$

$$
P_{21}(z,\tau) = \left\{h\left(1+\frac{n_0}{c\tau_g}\right)I_0\big(X'\big)\right\}\exp\{-(\alpha_1+h)z\}\exp\{(\alpha_1-\alpha_2)\tau z\}, \tag{2.96}
$$

$$P_{22}(z,\tau) = \delta(1-\tau)\exp\{-(\alpha_2+h)z\}$$
$$+\left\{h\left(1+\frac{n_0}{c\tau_g}\right)\sqrt{\frac{1-\tau}{\tau}}I_1(X')\right\}$$
$$\exp\{-(\alpha_1+h)z\}\exp\{(\alpha_1-\alpha_2)\tau z\}, \tag{2.97}$$

$$\tau \equiv \frac{t - z/v_1}{z(1/v_2 - 1/v_1)}, \tag{2.98}$$

$$X' \equiv 2hz\sqrt{\tau(1-\tau)}, \tag{2.99}$$

where z is fiber length, τ is normalized time; P_{ij} is the power of mode j for mode i excitation ($i, j = 1, 2$) at z and τ; c is light speed in vacuum; n_0 is the refractive index of the core in the fiber, τ_g is DMD, α_i is the transmission loss of mode i; $\delta(x)$ is the delta function; I_0 and I_1 are Bessel functions of the first and second kind, respectively; and h is the mode-coupling coefficient between the two modes.

Figure 2.78 shows an example of simulated impulse response $\alpha_1 = \alpha_2 = 0.2$ dB/km, $\tau_g = 1$ ns/km, $z = 50$ km, and $h = 1.0 \times 10^{-4}$/km. The upper three graphs indicate the result for mode 1 excitation. The peak of P_{11} at $t = 0$ ns corresponds to the power that arrived at the receiver side without mode coupling. The plateau after the peak represents the recoupled power from P_{12}. P_{t-1}, which is the summation of P_{11} and P_{12}, is monitored by a receiver at $z = 50$ km. The lower three

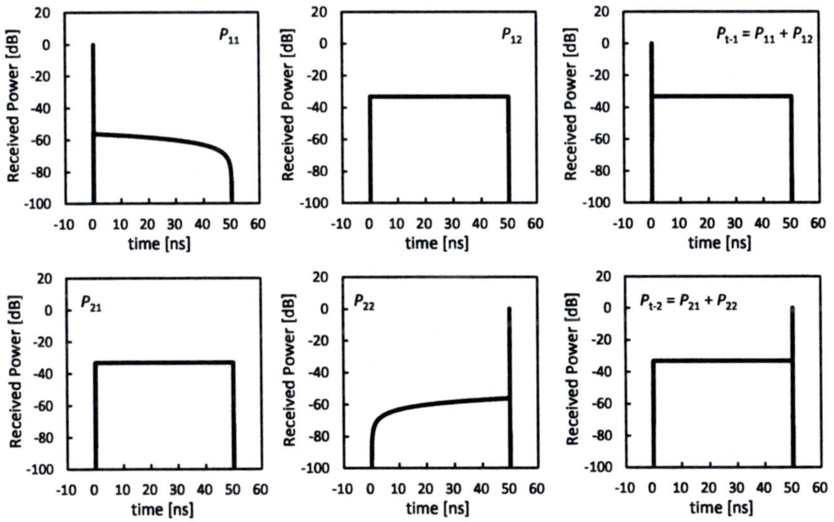

Fig. 2.78 An example of impulse responses of a two-mode fiber: P_{ij} represents the monitored power of mode j for mode i excitation condition. $P_{t\text{-}j}$ represents the monitored power for both mode excitation conditions

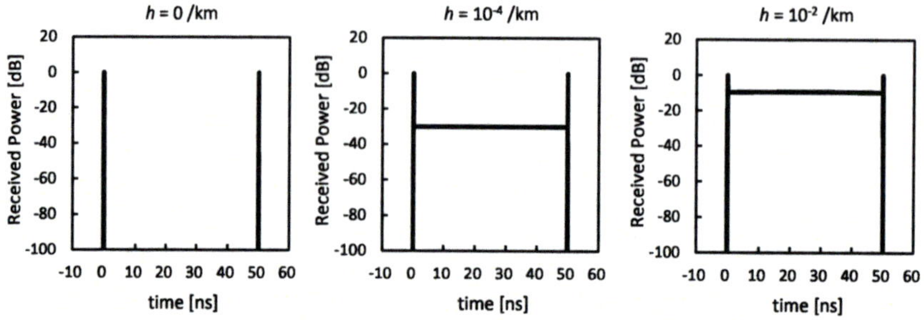

Fig. 2.79 Impulse responses of a two-mode fiber for varying mode-coupling coefficient, *h*

graphs correspond to the impulse response for mode 2 excitation. If both the modes are excited simultaneously, the monitored power will be the summation of $P_{t\text{-}1}$ and P_{t-2}.

Figure 2.79 shows an example of impulse response with both the modes equally excited for different mode-coupling coefficients *h*. The same parameters as for Fig. 2.78 were used. The time difference between two impulses in the figures corresponds to DMD at the length of 50 km. The plateau between the impulses becomes larger with increasing *h*. The level of the plateau has impact on the MIMO complexity. It has been reported that *h* varies with not only Δn_{eff} between the modes but also the fiber tension [114]. The results indicate that we need to take into account disturbances that are imposed on fibers in the field when designing FMFs.

2.4.3.2 Impulse Response with Mode Coupling at Splice Points

If FMFs were perfectly spliced without any misalignment, no mode conversion splice points would be observed. However, some kinds of misalignment are unavoidable for actual splices. The mode conversion at splice points can be observed in an impulse response. Figure 2.80 illustrates the schematics of impulse response with or without a spliced point. Figure 2.80a shows an impulse response without a splice point. Two impulses with a plateau are observed as discussed in the previous section. Figure 2.80b shows an impulse response for the same fibers, now spliced with misalignment. The additional pulses that correspond to the mode coupling at the splice point are observed in the plateau. The position of the additional pulses is estimated with the followed equation [141].

$$t_{01-11} = t_{01} + \Delta t(L - x)/L, \tag{2.100}$$

$$t_{11-01} = t_{01} + \Delta t x/L, \tag{2.101}$$

$$\Delta t = t_{11} - t_{01}, \tag{2.102}$$

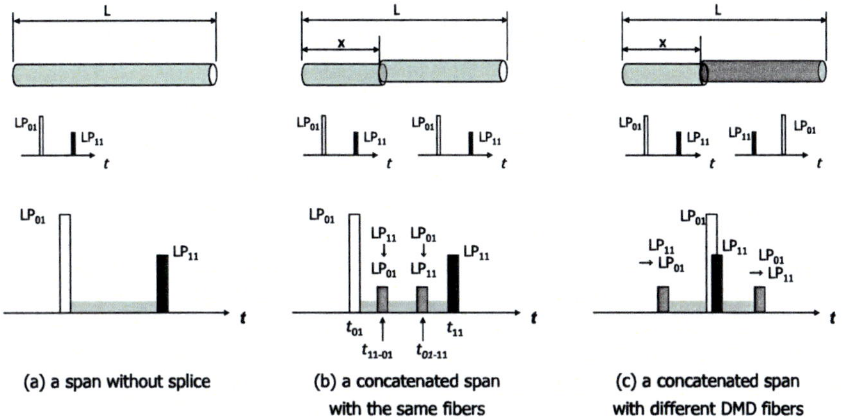

Fig. 2.80 Comparison of impulse responses for different fiber span

where $t_{01\text{-}11}$ and $t_{11\text{-}01}$ are the times of the pulses owing to mode conversion from LP_{01} to $LP_{11}$and from LP_{11} to LP_{01}, respectively; t_{01} and t_{11} is the time of LP_{01} and LP_{11} pulses, respectively; L is the total length of the fibers; and x is the length of the first span of the spliced fibers.

If fibers with different DMD characteristics are spliced, a different impulse response will be observed. Figure 2.80c shows an example of an impulse response over concatenated fibers with different DMD characteristics. LP_{01} mode is faster than LP_{11} mode in fiber 1, and LP_{11} mode is faster than LP_{01} mode in fiber 2. As a result, the time difference between LP_{01} and LP_{11} is compensated. However, additional impulses that originate from mode coupling at the splice points are observed to the either side of the original pulses of the LP_{01} and LP_{11} modes. We should take into account the additional pulse to recover the original signals from the coupled signals using to MIMO technology.

Figure 2.81 shows the results of an experimental demonstration of the effect of a misaligned splice on an impulse response [141]. The impulse responses were monitored on a 25-km two-mode fiber, which is a concatenated span with two fibers of 8.75 and 16.25 km with DMD at 1550 nm of 5.2 ns/km. An impulse which contains both LP_{01} and LP_{11} modes was introduced into the 8.75 km section and was monitored at the end of the 16.25 km section. Figure 2.81a shows an impulse response for a carefully spliced condition; misalignment was reduced as much as possible. Additional pulses that originate from mode coupling at the splice point were not observed between original impulses of LP_{01} and LP_{11}. Figure 2.81b shows an impulse response for a misaligned condition. The offset of 0.7 μm was added from the Fig. 2.81a condition. Two additional pulses were observed between original impulses. The positions of the additional pulses agree with the estimated position from (2.100) to (2.102). It has been reported that large mode coupling larger than −20 dB would be generated for misalignment larger than approximately 0.5 μm [141].

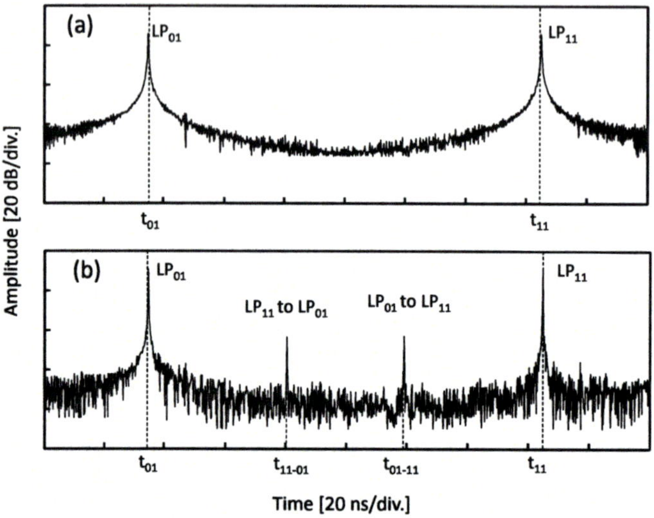

Fig. 2.81 Experimental demonstration of the effect of a misaligned splice on an impulse response (Reprinted with permission from [141]. © 2014 The Optical Society)

Precise splicing may be an issue if mode coupling larger than −20 dB has an impact on MIMO complexity.

2.4.4 Measurement Technology for Mode Coupling

One of the most critical issues in relation to MCF and FMF transmission is the inter-core and inter-modal crosstalk. It has been found that the dominant factor causing the crosstalk is stochastic mode coupling along a fiber resulting from longitudinal perturbations. If we can measure the distribution of the mode-coupling coefficient along the propagation direction, it will give us useful information with which to identify the source of transmission performance degradation in SDM and MDM transmissions.

Recently, a technique for measuring the mode coupling along an MCF and FMF has been proposed that uses synchronous multi-channel optical time domain reflectometry (OTDR) [142, 143]. In this scheme, the OTDR method is extended in such a way that one of the cores/modes is excited with an optical pulse to measure the ratio of the backward Rayleigh scattering between the excited and unexcited cores/modes. This synchronous OTDR scheme is depicted in Fig. 2.82, in which the center core in the MCF is excited and the OTDR signals are received from all the cores. This technique clarifies the non-uniformity of the mode-coupling coefficient along the fiber caused by the fiber's structural irregularity, such as a strong correlation between the

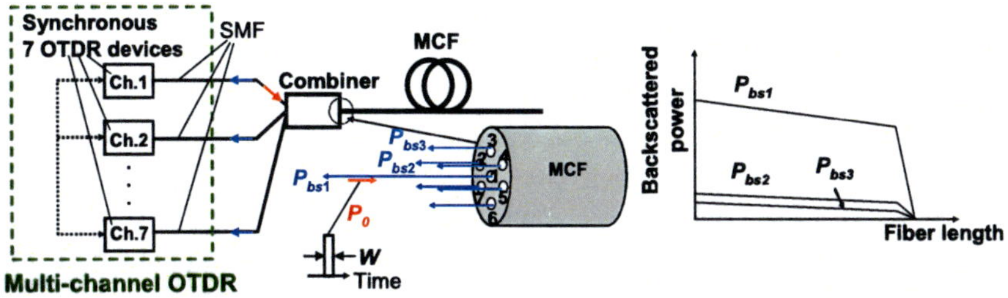

Fig. 2.82 Measurement setup for mode coupling along multi-core fiber (Reprinted with permission from [143]. © 2014 The Optical Society)

change in the mode-coupling ratio and the cladding diameter fluctuation along the fiber.

Figure 2.83 shows the setup for mode-coupling measurement along an FMF using a synchronous multi-channel OTDR [143]. A 100 ns optical pulse yielding a spatial resolution of 10 m was emitted from ch. 1 of the multi-channel OTDR operated at 1550 nm and coupled to the LP_{01} mode of an FMF through the LP_{01} port of a phase-plate-based mode coupler. Here, an optical masking apparatus is installed to remove the Fresnel reflection at the fiber end, which may otherwise distort the

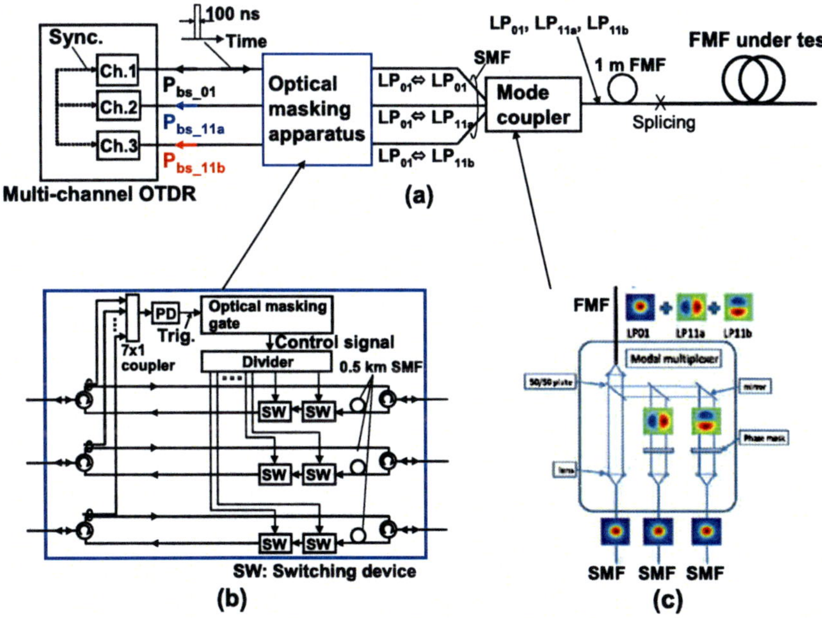

Fig. 2.83 Experimental setup for mode-coupling measurement along an FMF. **a** Complete setup, **b** optical masking apparatus, **c** mode coupler (Reprinted with permission from [143]. © 2014 The Optical Society)

backward Rayleigh scattering waveforms. The backscattered lights from unexcited higher-order modes of the FMF were then detected simultaneously by the three-channel synchronous OTDR. For example, the backscattered power in the LP_{11a} mode, P_{bs_11a}, which resulted from mode coupling with the LP_{01} mode, was detected through the LP_{11a} port of the mode coupler (demultiplexer) by ch. 2 of the multi-channel OTDR. Then the mode-coupling ratio between the LP_{01} and LP_{11} modes along the fiber could be obtained from the power ratio between P_{bs_01} and P_{bs_11}:

$$\eta_{11,01}(L) = \frac{P_{bs\_11}}{P_{bs\_01}} = 2h_{11,01}L + K, \tag{2.103}$$

where $h_{11,01}$ is the mode-coupling coefficient between the LP_{01} and LP_{11} modes, L is the fiber length, and K is a constant determined by the Fourier transformation of the autocorrelation function of the mode-coupling coefficient [144, 145] and the crosstalk caused by the mode coupler. Note that (2.103) is valid when the mode-coupling coefficient h is small.

Figure 2.84 shows measurement results for the mode-coupling distribution along a 5.9 km FMF supporting the LP_{01} and $LP_{11a,b}$ modes with core and cladding diameters of 32.6 and 116.5 μm, respectively. Figure 2.84a shows the backscattered signals from the FMF measured through the LP_{01}, LP_{11a}, and LP_{11b} ports when a 100 ns input optical pulse was coupled into the LP_{01} mode. It can be seen that the backscattered power profile in the LP_{11a} and LP_{11b} modes is not uniform, especially after ~3 km, as shown by the blue and red curves. This deviation is caused by the longitudinal variation in the optical fiber loss difference between the LP_{01} and LP_{11} modes, $\exp[(\alpha_{11} - \alpha_{01})L]$, as plotted in Fig. 2.84b, which was measured with a conventional OTDR for each mode. This indicates that the difference in mode attenuation is not uniform along the FMF, and this results in non-uniformity in the backscattered power profile. This is not the case for mode coupling between single-mode cores in MCF. When evaluating the mode-coupling ratio $\eta_{11,01}$, we take account of the non-uniformity in the attenuation coefficient difference (calibration factor), $\alpha_{11} - \alpha_{01}$, by multiplying the difference in the mode attenuation profile given by Fig. 2.84b, i.e.,

$$\eta_{11,01}(L) = \frac{P_{bs\_11}}{P_{bs\_01}} \exp[(\alpha_{11} - \alpha_{01})L]. \tag{2.104}$$

Mode-coupling ratios calculated from (2.104) are shown in Fig. 2.84c, d, and correspond to the coupling with the LP_{11a} and LP_{11b} modes, respectively. The clearly large offset K is likely to be caused by the crosstalk at the mode coupler, but this does not matter when evaluating the crosstalk along an FMF as the mode-coupling effect is given by the slope of η. From these mode-coupling profiles, the calculated crosstalk values between the two modes over a 5.9 km FMF transmission are −21.1 dB (LP_{01} and LP_{11a}) and −22.4 dB (LP_{01} and LP_{11b}).

When applying this method to the mode coupling between LP_{11a} and LP_{11b} modes, where they are coupled strongly and hL is not smaller than unity, the present

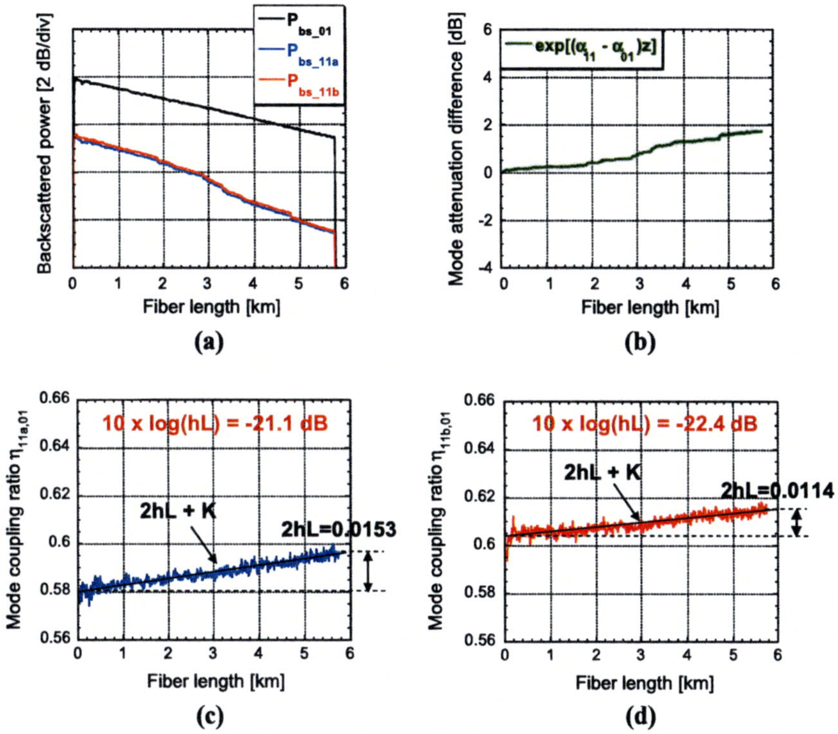

Fig. 2.84 Measurement results for mode coupling along FMF from LP_{01} to $LP_{11a,b}$ mode. **a** Backscattered signals from the LP_{01} and $LP_{11a,b}$ mode channels when a 100 ns optical pulse was coupled into the LP_{01} mode. **b** Difference in attenuation coefficient for LP_{01} and LP_{11a} modes. **c**, **d** Mode-coupling ratio between LP_{01} and LP_{11a} modes and between LP_{01} and LP_{11b} modes, respectively (Reprinted with permission from [143]. © 2014 The Optical Society)

mode-coupling measurement does not provide an accurate result but we can roughly evaluate the way in which the strong mode coupling occurs along the fiber. The results are shown in Fig. 2.85. Here, we show the backscattered waveform that we obtained when we excited the LP_{11a} mode and detected the LP_{11a} and LP_{11b} backward Rayleigh scattering. Approximately 3 dB down LP_{11b} backward Rayleigh scattering was observed as seen in Fig. 2.85(a-1). The details of the OTDR signals are enlarged in Fig. 2.85(a-2). It is interesting to see the waveform changes, in which the power of the backscattered signal changes during propagation. Here we see a completely out-of-phase change with a mode-coupling period of approximately 50 m. This indicates that strong power coupling occurs during propagation. When we take the ratio between them, it becomes almost constant around a mode-coupling ratio of 0.62 as shown in Fig. 2.85(b-1). The initial value of 0.62 comes from the coupling crosstalks at the phase mask mode coupler and the FMF itself. Figure 2.85(b-1) is enlarged in (b-2), in which we observe a similar change to that seen in (a-2). Since P_{bs_11b} was

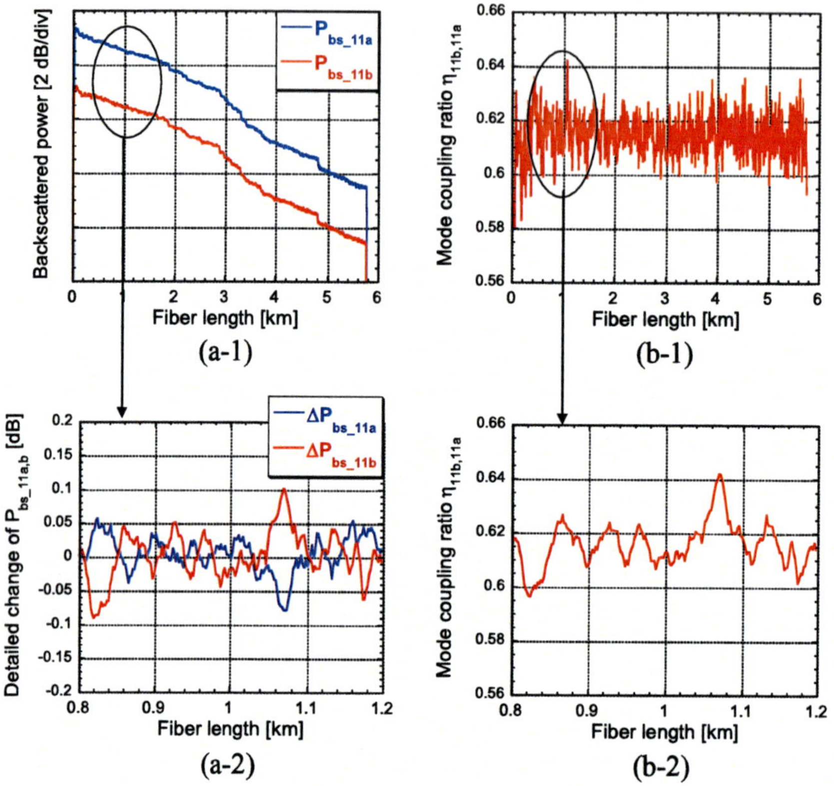

Fig. 2.85 Measurement results for mode coupling along FMF from LP_{11a} to LP_{11b} mode. **a-1, a-2** Backscattered signals. **b-1, b-2** mode-coupling ratios (Reprinted with permission from [143]. © 2014 The Optical Society)

divided with P_{bs_11a}, the ripples at around 50 m have been enhanced. This is due to the strong mode coupling between them.

Spatially and spectrally resolved mode imaging (S^2 imaging) [146] and swept wavelength interferometry [147] have been demonstrated as other methods for characterizing the mode properties of FMF. In general, spectral interference between multiple modes, which can be generated by launching a broadband source into FMF, provides information on the differential group delay through $|DGD| = \lambda^2/(|\Delta\lambda|c)$, where λ is the wavelength, $\Delta\lambda$ is the fringe spacing, and c is the speed of light in a vacuum [148]. In S^2 imaging, the spectral interference pattern is spatially recorded at multiple locations on the end face of a fiber. This enables us to evaluate not only DGD but also the power levels of higher-order modes relative to a fundamental mode. In [146], a tunable laser and a CCD camera were used to resolve the spatial modes, which were obtained by taking Fourier transform of the spectral interference pattern for each location.

The swept wavelength interferometry technique developed in [147], which is an extended form of optical vector network analyzer, can provide information on the phase and intensity response between each N input mode and N output mode ($N \times N$ complex transfer matrix). Specifically, the output of a tunable laser is split into two arms, where one is transmitted through FMF (transmissivity: $H(\omega)$) and the other through a reference fiber ($R(\omega)$), and the phase and intensity of $H(\omega)$ are recovered from their interference pattern. The impulse response $h(t)$ can also be obtained from the inverse Fourier transform of $H(\omega)$. Mode-dependent loss and DGD can be easily evaluated from these data. The polarization-split segmented coherent detection (PSCD) method is a high-speed mode analyses method [149]. The PSCD can obtain a full set of amplitudes, phases, and polarization states of guided modes including degenerate modes in few-mode fibers within the speed limit imposed by the response time of the electronics.

2.5 Few-Mode Multi-core Fiber

As described in the previous sections, MCFs and FMFs have been developed to overcome the capacity limit of a conventional single-mode fiber. However, the possible spatial channel count (SCC) of MCFs and FMFs would be limited by certain factors: crosstalk between cores and cladding diameter for MCFs and DMD and mode coupling for FMFs. A few-mode multi-core fiber (FM-MCF) is the hybrid of the MCF and the FMF: a multi-core fiber whose cores have a few propagating modes. An FM-MCF is expected to overcome the *SCC* limit of the MCF and FMF. In 2012, several organizations proposed FM-MCFs almost simultaneously: four-core [150], seven-core [151], and 19-core [152] FM-MCFs with all-solid structures and a seven-core FM-MCF with a hole-assisted structure [34]. Many kinds of FM-MCFs have been proposed to maximize spatial mode count since then.

In this section, the features of FM-MCFs are presented. At first, the possibility of uncoupled core design is discussed. After overviews of uncoupled FM-MCFs and theoretical discussions, the characteristics of FM-MCFs with spatial channel count of 36 and reduced DMD is presented. Finally, the characteristics of a moderately coupled multi-core fiber are discussed.

2.5.1 FM-MCFs with Uncoupled Core Design

2.5.1.1 Scalability of Spatial Channel Count

Figure 2.86summarizes characteristics of the FM-MCFs presented so far. Certain parameters have been proposed as measure of the performance of FM-MCFs [63]. SCC is redefined for an FM-MCF as follows.

Cross sectional view	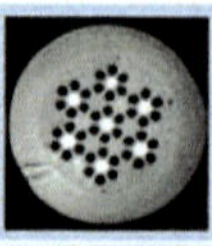	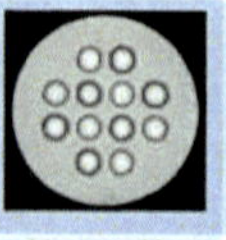	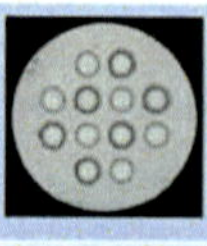		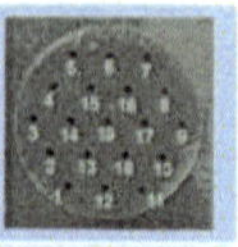
CC	7	12	12	36	19
SC	3 (2LP)	3 (2LP)	3 (2LP)	3 (2LP)	6 (4LP)
SCC	21	36	36	108	114
Cladding Diameter [μm]	192	230	230	306	318
RSE	8.9	10.6	10.6	17.8	17.6
A_{eff} at 1550 nm [μm^2]	113 (LP_{01}) 170*(LP_{11})	97 (LP_{01}) 141*(LP_{11})	110 (LP_{01}) 154*(LP_{11})	76 (LP_{01}) 106*(LP_{11})	
|DMD| [ps/km]	4600 at 1550 nm	< 519 over C+L band	< 96 over C+L band	< 7800 at 1550 nm	~ 1000* at 1550 nm
References	[33]	[156, 165]	[157]	[158]	[159]

* Designed value

Fig. 2.86 Characteristics of FM-MCFs with uncoupled core design. Cross sectional views are provided by authors. Courtesy of Prof. Guifang Li, Dr. Shoichiro Matsuo, Dr. Yoshinari Awaji, and Dr. Itsuro Morita

$$\text{SCC} = \text{CC} \cdot \text{MC}, \tag{2.105}$$

where CC is the core count in a cladding and MC is the mode count in a core. Spatial efficiency (SE) is an SCC density for a cladding.

$$\text{SE} = \frac{\text{SCC}}{(\pi/4)D^2}, \tag{2.106}$$

where D is cladding diameter. Relative spatial efficiency (RSE) is the ratio of SE of an FM-MCF ($SE_{\text{FM-MCF}}$) to that of a conventional single-mode fiber with a single-core and a cladding diameter of 125 μm (SE_{SMF}).

$$\begin{aligned} \text{RSE} &= \frac{\text{SE}_{\text{FM - MCF}}}{\text{SE}_{\text{SMF}}} \\ &= \left(\frac{\text{SCC}}{(\pi/4)D^2}\right) / \left(\frac{1}{(\pi/4)125^2}\right) \\ &= \text{SCC} \cdot (125/D)^2. \end{aligned} \tag{2.107}$$

Figure 2.87 shows a comparison of the SCC of FMFs, SM-MCFs and FM-MCFs as a function of cladding diameter. The contour lines indicate the RSE. The FM-MCFs are categorized into three groups by cladding diameter. The first group contains an FM-MCF with a cladding diameter of 125 μm [153]. The fiber exhibited an SCC of

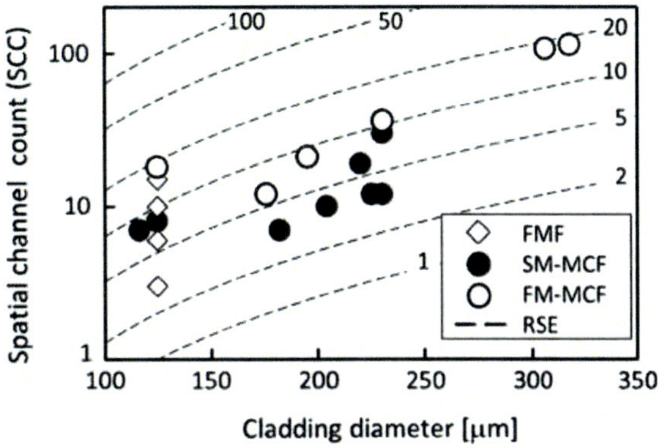

Fig. 2.87 SCC as a function of cladding diameter for FMFs, SM-MCFs and FM-MCFs (Reprinted with permission from [157]. © 2015 IEEE)

18. The RSE of the fiber was 18, which was the maximum RSE. The second group contains FM-MCFs with cladding diameter greater than 300 μm. Two fibers in the second group exhibited an SCC larger than 100 and the highest RSE of approximately 18. There is room for discussion on the practicality of such a large cladding diameter in terms of mechanical reliability and splice loss of the outer cores. The last group comprises FM-MCFs that aim to maximize the SCC and RSE with a cladding diameter smaller than 250 μm, which has been proposed as an allowable cladding diameter for practical condition [62, 154]. A three-mode 12-core fiber with a cladding diameter of 230 μm exhibits an SCC of 36 and RSE of 10.6 [155, 156]. MCFs with SCC greater than or equal to 30 are named MCFs for dense space-division multiplexing (DSDM) [156].

In the case of MCFs, crosstalk between cores is a limiting factor for modulation format, transmission capacity, and transmission distance. Figure 2.88 shows the relationship between SCC and 100-km worst crosstalk of SM-MCFs and FM-MCFs. FM-MCFs with SCC larger than 100 recorded high crosstalk larger than −20 dB. The crosstalk is a limiting factor of the transmission distance over the fibers. By

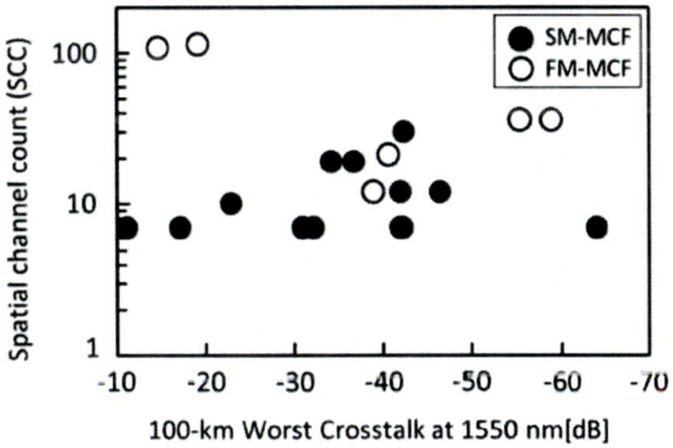

Fig. 2.88 SCC as a function of 100-km worst crosstalk at 1550 nm for SM-MCFs and FM-MCFs (Reprinted with permission from [157]. © 2015 IEEE)

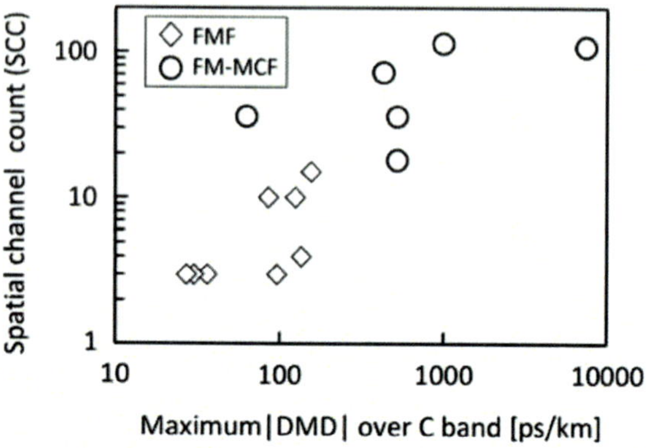

Fig. 2.89 SCC as a function of maximum |DMD| over the C band for FMFs and FM-MCFs (Reprinted with permission from [157]. © 2015 IEEE)

contrast, FM-MCFs with *SCC* of 36 exhibit low crosstalk below −50 dB. High *SCC* combined with low DMD has not been realized by using FM-MCFs.

As described in Sect. 2.4, the suppression of a DMD is indispensable for MDM over FMF. Some FM-MCFs have introduced DMD suppression design to cores. Figure 2.89 shows SCC of FMFs and FM-MCFs as a function of maximum |DMD| over the *C* band. Even for an FMF with a single core, the DMD controllability deteriorates according to the SCC [121]. In the case of an FM-MCF, DMD variation between cores should be controlled in addition to the DMD control in individual cores. The FM-MCF with an SCC of 108 employed a step-index structure, and |DMD| was less than approximately 8000 ps/km at 1550 nm [158]. The FM-MCF with a SCC of 114 utilized cores with a graded-index profile to reduce |DMD|, which is about 1000 ps/km by design [159]. Two types of FM-MCFs with a SCC of 36 have been proposed. Both the fibers successfully reduced the DMD by using a multi-step-index profile or a graded-index profile. The detailed feature and characteristics of the fibers are presented in the following section.

Some FM-MCFs that realized SCC larger than 100 and reduced |DMD| simultaneously [160–162]. A fiber with 12 hexagonally arranged 10-mode cores in a 217-μm cladding realized the maximum SCC of 120 and a relative core multiplicity factor exceeding 100 [161]. A 6-mode 19-core fiber was employed in the transmission experiment that realized transmission capacity over 10 Pbit/s for the first time [162].

2.5.1.2 Design of FM-MCFs

In FM-MCFs, required core pitch becomes large compared with that in SM-MCFs, since the crosstalk between higher-order modes is larger than that between fundamental modes. Figure 2.90 shows the numerically simulated crosstalk between LP_{01} modes ($XT_{01\text{-}01}$), LP_{11} modes ($XT_{11\text{-}11}$), and LP_{02} modes ($XT_{02\text{-}02}$) in a single-mode core, a 2LP-mode core, and a 4LP-mode core, respectively, as a function of core pitch at 1565 nm wavelength, where each core has a trench-assisted index profile shown in Fig. 2.18 with $r_2/r_1 = 2.0$, $W/r_1 = 1.0$, and $\Delta_t = -0.70\%$. All the cores

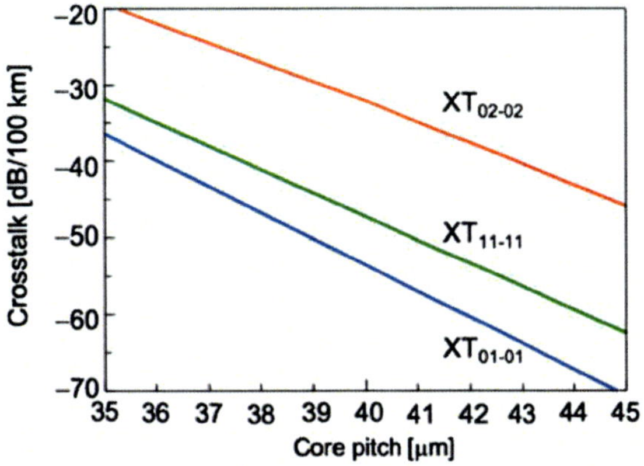

Fig. 2.90 Crosstalk between LP_{01} modes ($XT_{01\text{-}01}$), LP_{11} modes ($XT_{11\text{-}11}$), and LP_{02} modes ($XT_{02\text{-}02}$) in single-mode core, 2-LP-mode core, and 4-LP-mode core, respectively, as a function of core pitch at 1565 nm wavelength (Reprinted with permission from [49]. © 2016 IEEE)

have the same A_{eff} of 80 μm^2 for the fundamental mode, and r_1 and Δ_1 are assumed to be 4.7 μm and 0.38% (Core A), 5.4 μm and 0.56% (Core B), and 5.9 μm and 0.86% (Core C), respectively, for single-mode core, 2LP-mode core, and 4LP-mode core. $XT_{11\text{-}11}$ between the 2LP-mode cores and $XT_{02\text{-}02}$ between the 4LP-mode cores are about 5-dB and 20-dB larger, respectively, compared with $XT_{01\text{-}01}$ between the single-mode cores. Therefore, larger core pitch is required as increasing the number of modes in FM-MCFs for achieving similar crosstalk compared with SM-MCFs. It can be seen from Fig. 2.90 that, if the allowable worst case crosstalk is −30 dB after 100 km propagation, the required core pitches are 35.4 μm, 37.0 μm, and 42.1 μm, for the SM-MCF, the 2LP-mode MCF, and the 4LP-mode MCF with hexagonal close-packed structure (HCPS) core arrangement, whereas they are 34.9, 36.4, and 41.4 μm, for the SM-MCF, the 2LP-mode MCF, and the 4LP-mode MCF with square lattice core arrangement.

In addition, in FM-MCFs, required CT becomes large compared with that in SM-MCFs, since the bending loss of higher-order modes in outer cores is larger than that of fundamental mode. Figure 2.91 shows the calculated bending loss of LP_{01} mode, LP_{11} mode, and LP_{02} mode in the single-mode core (Core A), 2LP-mode core (Core B), and 4LP-mode core (Core C), respectively, as a function of CT at 1565 nm wavelength, where the bending radius is assumed to be $R = 140$ mm. It can be seen from Fig. 2.91 that CT values of 33, 36, and 45 μm or more are needed for the single-mode core, 2LP-mode core, and 4LP-mode core, respectively, to decrease the bending loss below 10^{-3} dB/km at 1565 nm for suppressing the excess loss in outer cores [63].

From these conditions for the core pitch and CT, the scalability of SCC in FM-MCFs with homogeneous core arrangement as a function of CD can be roughly estimated as shown in Fig. 2.92, where the target A_{eff} for the fundamental mode is 80 μm^2 as an example and the allowable worst case crosstalk is assumed to be −30 dB/100 km. For homogeneous SM-MCFs, a realization of DSDM of SCC larger than 30 seems to be difficult with CD < 250 μm. Even in 2LP-mode MCFs, it is a challenging task to achieve SCC over 100 with CD < 300 μm. On the other hand, it

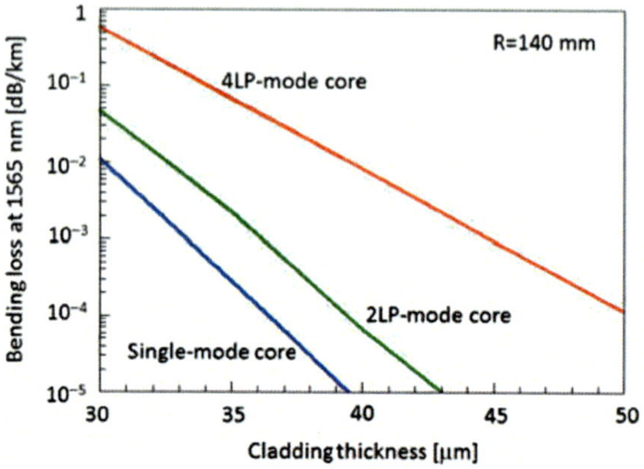

Fig. 2.91 Bending loss of LP_{01} modes, LP_{11} modes, and LP_{02} modes in the single-mode core, 2LP-mode core, and 4LP-mode core, respectively, as a function of cladding thickness at 1565 nm wavelength, where the bending radius is assumed to be $R = 140$ mm

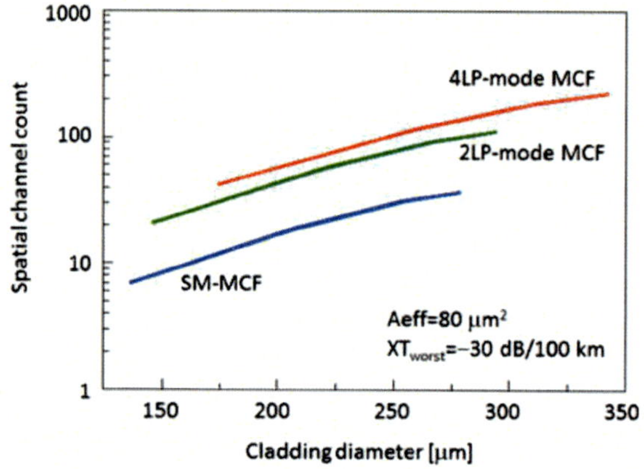

Fig. 2.92 Estimated scalability of spatial channel count in FM-MCFs with homogeneous core arrangement as a function of cladding diameter under the condition of $XT_{worst} = -30$ dB/100 km

is estimated that SCC over 100 would be achievable by using 4LP-mode MCFs with CD of around 250 μm.

The DMD suppression is also a critical issue for designing FM-MCFs. Recently, low crosstalk and low DMD characteristics have been simultaneously achieved by 12-core 2LP-mode MCFs having a multi-step-index profile with trench [156] and a graded-index profile with trench [155], where the heterogeneous MCFs with two different cores and square lattice core arrangement have been used. Figure 2.93a, b show the multi-step-index profile with trench and the graded-index profile with trench, respectively. In the multi-step-index profile, a_1, r_1, r_2, W, Δ_1, Δ_2, and Δ_t stand for inner core radius, outer core radius, the distance between the center of inner core and the inner edge of trench, the thickness of the trench layer, the relative refractive-index difference between inner core and cladding, the relative refractive-index difference between outer core and cladding, and the relative refractive-index difference between trench and cladding, respectively. Similarly, in the graded-index profile, r_1, r_2, W, and Δ_t stand for core radius, the distance between the center of core and the inner edge of trench, the thickness of the trench layer, and the relative refractive-index difference between trench and cladding, respectively. The relative refractive-index difference between core and cladding Δ can be expressed as $\Delta = \Delta_1 [1 - (r/r_1)^\alpha]$, where α represents shape factor of graded-index profile, Δ_1 stands for the peak relative refractive index difference between core and cladding, and r is the radial coordinate. In both cases, it is not simple to determine two different core parameters having sufficiently different effective index difference Δn_{eff} with low DMD and similar A_{eff}. The heterogeneous 2LP-mode core parameters with low DMD and similar A_{eff} are determined as explained below.

The multi-step-index profile is suitable for fabricating large-scale preform compared with the graded-index profile, since the VAD process can be used. DMD in the multi-step-index profile can be mainly controlled by r_2/r_1 and $\Delta_\text{d} = \Delta_1 - \Delta_2$, whereas the suitable value of r_1/a_1 is reported as around 2.0 [118]. Figure 2.94a–c show the DMD at 1550 nm wavelength as function of a_1 and Δ_1 for $r_2/r_1 = 1.5$, $r_2/r_1 = 1.6$, and $r_2/r_1 = 1.7$, respectively, in the multi-step-index core with trench, where we define the target value of A_{eff} of LP_{01} mode to be 110 μm^2. Here, we fixed the value for r_1/a_1 and Δ_t, which are assumed as 2.0 and −0.7%, respectively. As r_2/r_1 shifts, DMD will decrease or increase. So in order to compensate for the decreased or increased DMD, we can alter the Δ_d, and the appropriate values of Δ_d for different r_2/r_1 are (a) $\Delta_d = 0.14\%$, (b) $\Delta_d = 0.13\%$, and (c) $\Delta_d = 0.12\%$, respectively, for achieving low absolute value of DMD around target range of A_{eff}. In Fig. 2.94a–c, the black solid line and black dashed line represent A_{eff} and effective index of LP_{01} mode. The upper and lower white solid lines and white dashed lines represent the cutoff of LP_{21} mode and the limit of LP_{11} mode at W/r_1 of 0.2 and 0.8, respectively. Here, to define the 2LP-mode operation, the bending loss of LP_{21} mode should be > 1 dB/m at a bending radius $R = 140$ mm and we assume the limit value of the bending loss of LP_{11} mode to be 0.5 dB/100 turns at a bending radius $R = 30$ mm. It can be seen that

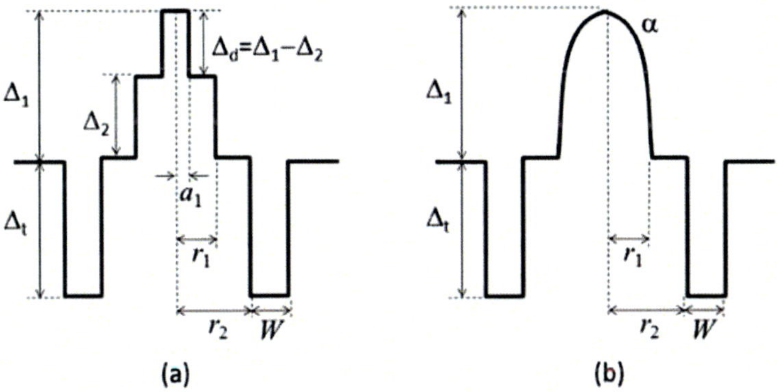

Fig. 2.93 **a** Multi-step-index profile with trench and **b** graded-index profile with trench

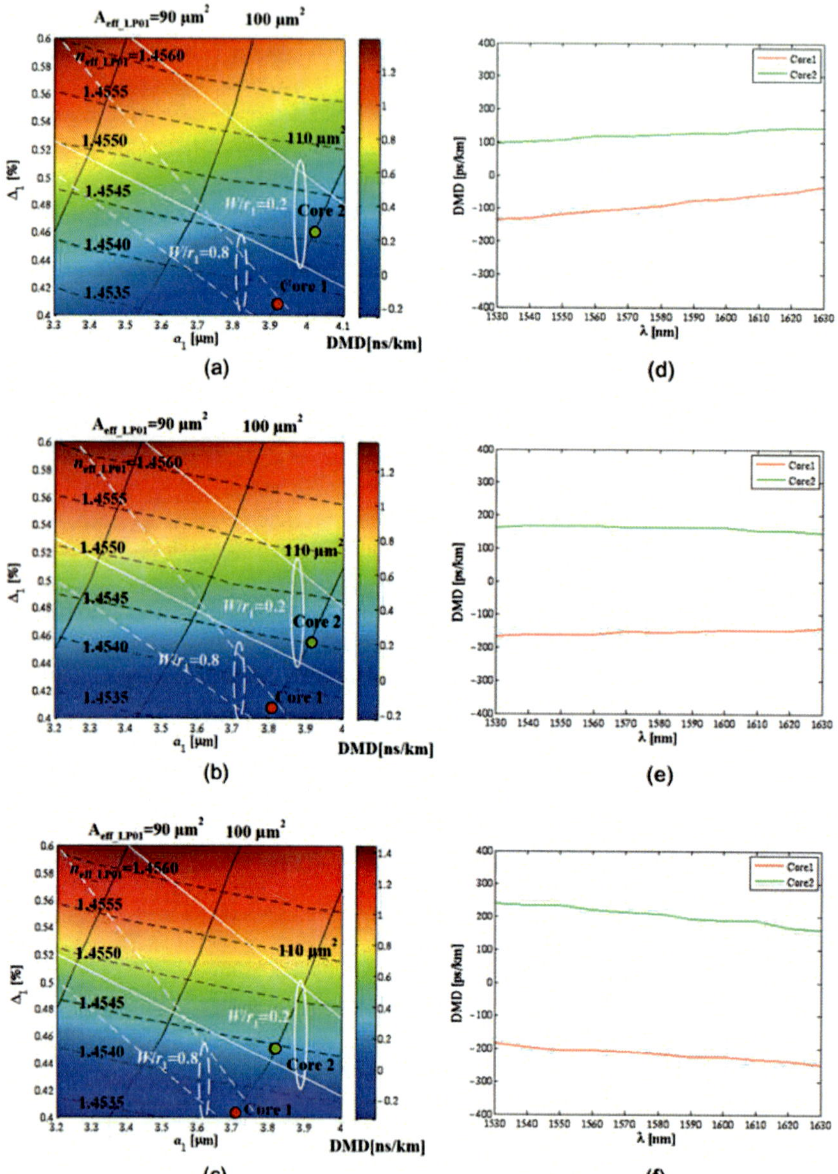

Fig. 2.94 DMD at 1550 nm wavelength as function of a_1 and Δ_1 when $r_1/a_1 = 2.0$ and $\Delta_t = -0.7\%$, where **a** $r_2/r_1 = 1.5$, $\Delta_d = 0.14\%$, **b** $r_2/r_1 = 1.6$, $\Delta_d = 0.13\%$, **c** $r_2/r_1 = 1.7$, $\Delta_d = 0.12\%$. Wavelength dependence of DMD for the selected two cores, where **d** $r_2/r_1 = 1.5$, $\Delta_d = 0.14\%$, **e** $r_2/r_1 = 1.6$, $\Delta_d = 0.13\%$, **f** $r_2/r_1 = 1.7$, $\Delta_d = 0.12\%$ (Reprinted with permission from [163]. © 2014 The Optical Society)

the 2LP-mode operation region will shift as W/r_1 alters. We set W/r_1 to be 0.8 and 0.2 in order to make it probable to choose two sorts of multi-step-index cores with similar A_{eff} of 110 μm^2, low DMD, and relatively large effective index difference Δn_{eff}. To ensure a relative small R_c which is a critical value of bending radius [26], we define the required Δn_{eff} to be about 0.0008. In this case, we can select three pairs of multi-step-index cores with low DMD and DMD slope in the 2LP-mode operation regions at W/r_1 of 0.2 and 0.8, which are shown as the filled circles in green and red in Fig. 2.94. Figure 2.94d–f illustrate the wavelength dependence of DMD for these three pairs of multi-step-index cores for $r_2/r_1 = 1.5$, $r_2/r_1 = 1.6$, and $r_2/r_1 = 1.7$, respectively. It can be observed that as r_2/r_1 increases, the DMD slope is getting smaller and when it equals 1.6, the DMD slope is almost 0 ps/km/nm. Furthermore, when r_2/r_1 increases, the difference between the DMD in both cores becomes larger. From these results, DMD of smaller than several hundreds ps/km can be expected in a 2LP-mode MCF with two different multi-step-index cores [163].

By using a graded-index profile, further decrement of DMD can be achieved. Figure 2.95 illustrates DMD as function of r_2/r_1 and α at 1550 nm wavelength for three different pairs of (r_1, Δ_1) in the graded-index profile core with trench. Three selected pairs of (r_1, Δ_1) are (9.0 μm, 0.4%), (9.5 μm, 0.44%), and (9.75 μm, 0.46%), which do not only achieve A_{eff} of 110 μm^2 but also make sure 2LP-modes are supported in the graded-index cores, where $\Delta_t = -0.7\%$. In Fig. 2.95, the solid, dashed, and dotted lines correspond to the DMD values for (r_1, Δ_1) of (9.0 μm, 0.4%), (9.5 μm, 0.44%), and (9.75 μm, 0.46%), respectively. In addition, the red, green, and blue lines represent DMDs of 100 ps/km, 0 ps/km, and −100 ps/km, respectively. The area filled with pattern in Fig. 2.95 can be regarded as an appropriate design region for r_2/r_1 and α. In this area, we can choose suitable set of r_2/r_1 and α to achieve low DMD and target A_{eff} of 110 μm^2 at the same time. Therefore, we set (r_2/r_1, α) as (1.3,

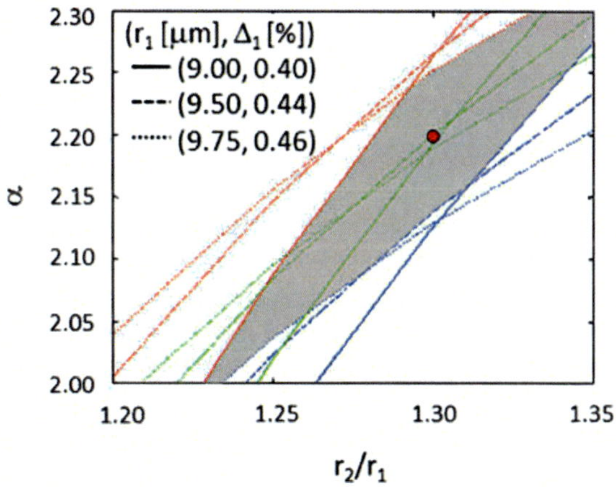

Fig. 2.95 DMD as function of r_2/r_1 and α at 1550 nm wavelength for three different pairs of (r_1, Δ_1) in the graded-index profile core with trench, where the red, green, and blue lines represent DMDs of 100 ps/km, 0 ps/km, and −100 ps/km, respectively

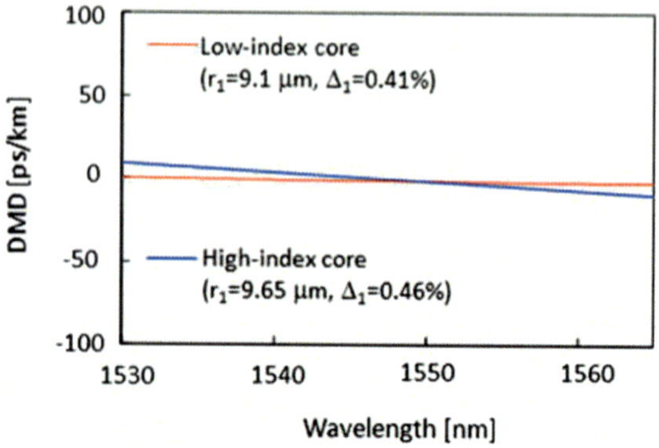

Fig. 2.96 Wavelength dependence of DMD in two different graded-index cores, where $r_2/r_1 = 1.3$, $\alpha = 2.2$, and $\Delta_t = -0.7\%$. The selected pairs of (r_1, Δ_1) are (9.1 μm, 0.41%) and (9.65 μm, 0.46%) for low-index core and high-index core, respectively

2.2) corresponding to the filled circle in red in Fig. 2.95 for relative trench position and graded-index profile. Figure 2.96 shows the wavelength dependence of DMD in two different graded-index cores, where $r_2/r_1 = 1.3$, $\alpha = 2.2$, and $\Delta_t = -0.7\%$. The selected pairs of (r_1, Δ_1) are (9.1 μm, 0.41%) for a low-index core and (9.65 μm, 0.46%) for a high-index core. The final relative trench widths of W/r_1 are determined as 0.7 and 0.3 for the low-index core and the high-index core, respectively, in order to achieve 2LP-mode operation. It can be seen that the absolute DMD values in both cores are smaller than 10 ps/km over C band. From these results, DMD of smaller than several tens ps/km can be expected in a 2LP-mode MCF with two different graded-index cores [164].

2.5.1.3 Characteristics of FM-MCFs with an SCC of 36 and Reduced DMD

As described in Fig. 2.86, two fibers with an SCC of 36 have realized reduced DMD and enlarged SCC under a limited cladding diameter. The fibers have common features: a square lattice structure (SLS) and a heterogeneous core arrangement with the same A_{eff}. A combination of the two approaches is very helpful to maximize the SCC under a limited cladding diameter.

The heterogeneous core arrangement is a well-known method for suppressing inter-core crosstalk [38]. For designing the heterogeneous core arrangement, we need to take care of the homogeneity of A_{eff} among cores with different n_{eff}. The concept of heterogeneous n_{eff} with homogeneous A_{eff} limits the design criteria that satisfy other requirements such as the cutoff wavelength of higher-order modes and bending loss of propagating modes [163, 164]. Figure 2.97 shows the design limitation for two-LP-mode designs with a graded-index profile [164]. For both index profiles, the number of possible core designs is only two. The number of different core design limits the possible core arrangements. Figure 2.98 shows the required number of different cores for various core arrangements [165]. In the case of a hexagonal close-packed structure (Fig. 2.98a) and a dual-ring structure (Fig. 2.98b), three kinds of

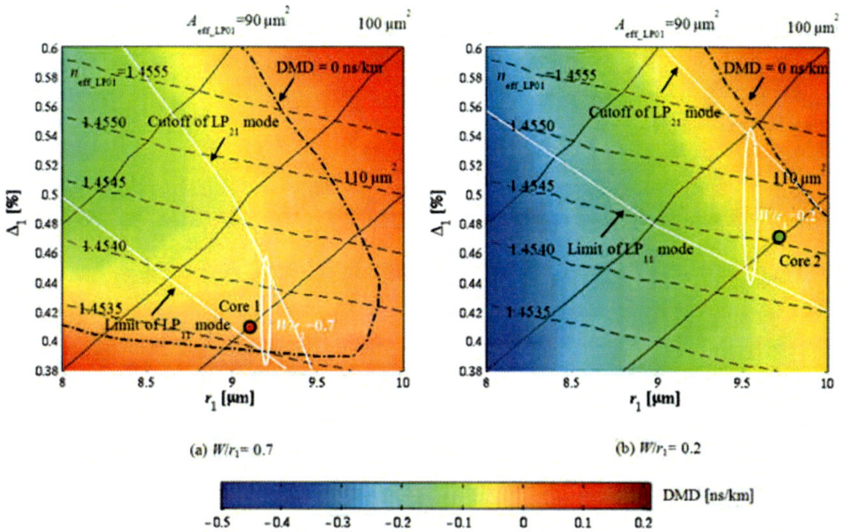

Fig. 2.97 DMD at 1550 nm as function of $r1$ and Δ_1 when $r_2/r_1 = 13$, $\alpha = 2.2$, and $\Delta_1 = -0.7\%$. (Reprinted with permission from [164]. © 2015 The Optical Society)

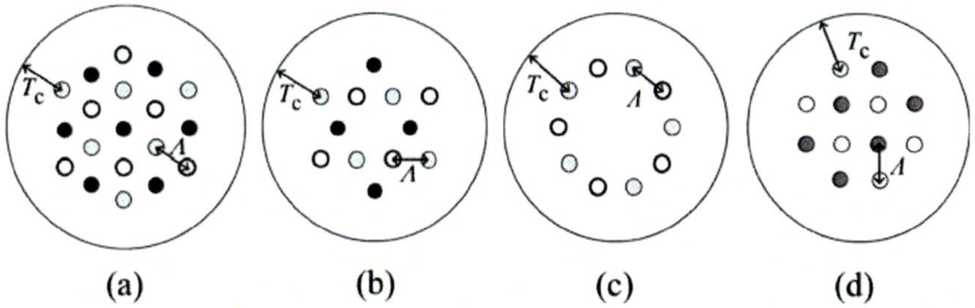

Fig. 2.98 Heterogeneous core arrangement with minimal core types in various core layouts. **a** Hexagonal close-packed structure. **b** Dual-ring structure. **c** One-ring structure. **d** Square lattice structure (Reprinted with permission from [165]. © 2015 IEEE)

cores are required to achieve heterogeneity for all adjacent cores. However, a one-ring structure (Fig. 2.98c) and a square lattice structure (Fig. 2.98d) enable heterogeneous core arrangement with only two kinds of cores. Figure 2.99 shows core count as a function of a cladding diameter for a one-ring structure and a square lattice structure [165]. One measure of the maximum cladding diameter is 250 μm [62]. The square lattice structure is able to maximize the number of cores around the cladding diameter.

Figure 2.100 summarizes DMD characteristics of FM-MCFs with an SCC of 36 [157]. In the case of an FM-MCF with a multi-step-index profile, the maximum |DMD| was 500 ps/km over the C and $C + L$ bands. However, the maximum |DMD| was reduced to 63 and 96 ps/km over the C band and $C + L$ band, respectively,

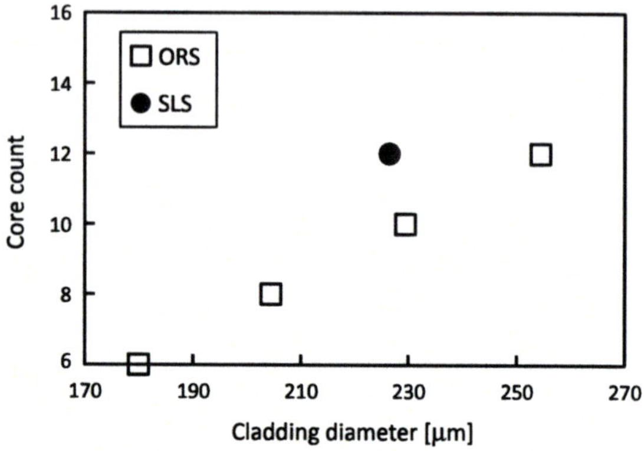

Fig. 2.99 Relationship between minimum cladding diameter and core count in one-ring structure (ORS) and square lattice structure (SLS). Core pitch and outer cladding thickness are assumed to be 40 µm and 50 µm, respectively (Reprinted with permission from [165]. © 2015 IEEE)

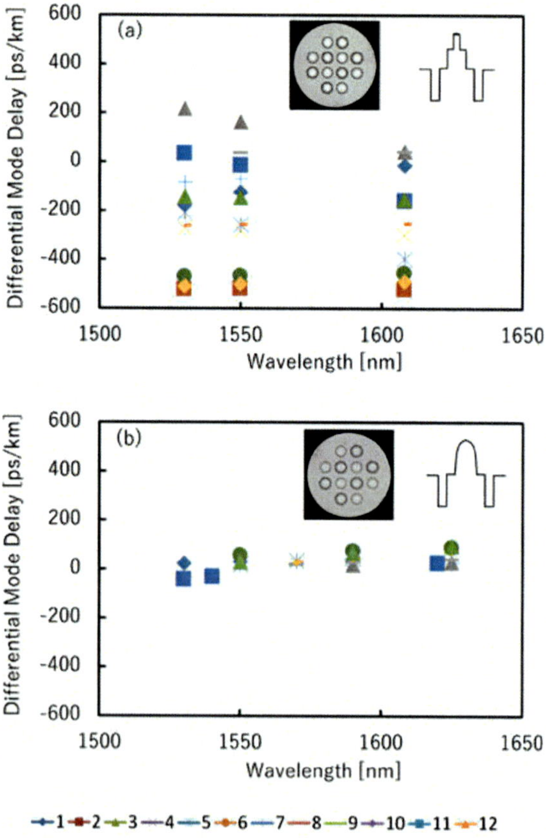

Fig. 2.100 Wavelength dependence of the DMD for two kinds of FM-MCFs with 36 spatial channels (Reprinted with permission from [157]. © 2015 IEEE)

owing to the graded-index profile. As presented in Fig. 2.100 the maximum |DMD| of the FM-MCF with a graded-index profile is close to the maximum |DMD| of single-core FMFs. The low DMD characteristics contributed to 527-km transmission experiments over the FM-MCFs [155].

2.5.2 125-μm Cladding 2 LP-mode 6-core Fiber

This section describes a moderately coupled multi-core fiber (MCF) design [166] for realizing both low inter-core MIMO processing complexity and high space utilization efficiency (SUE). The influence of mode coupling on the inter-core differential mode delay (DMD) in homogeneous core MCF is investigated numerically and experimentally, and moderately coupled MCF design that avoids any increase in the inter-core DMD is revealed, which enables us to compensate for the inter-core crosstalk by MIMO processing with low computational complexity. Finally, the characteristics of fabricated 125 μm-cladding two-LP-mode six-core fiber are described.

The influence of inter-core mode coupling on the DMD characteristic is one important factor for coupled MCF because inter-core crosstalk should be compensated for with the MIMO equalizer deployed at the receiver [47, 167], and the inter-core DMD has a great impact on the MIMO processing complexity. First, homogeneous two-core fiber with a single-mode core is assumed, and the mode-coupling equation is given as

$$\frac{\mathrm{d}}{\mathrm{d}z}\begin{pmatrix} E_1 \\ E_2 \end{pmatrix} = -i\begin{pmatrix} \beta & \kappa \\ \kappa^* & \beta \end{pmatrix}\begin{pmatrix} E_1 \\ E_2 \end{pmatrix}, \tag{2.108}$$

where E_1 or E_2 is the electric field calculated for each core to be isolated, and β or κ is the propagation constant or coupling coefficient. Then the transfer matrix can be diagonalized and we have the transformed equation

$$\frac{\mathrm{d}}{\mathrm{d}z}\begin{pmatrix} E_{\mathrm{even}} \\ E_{\mathrm{odd}} \end{pmatrix} = -i\begin{pmatrix} \beta_{\mathrm{even}} & 0 \\ 0 & \beta_{\mathrm{odd}} \end{pmatrix}\begin{pmatrix} E_{\mathrm{even}} \\ E_{\mathrm{odd}} \end{pmatrix}, \tag{2.109}$$

which means we can treat the propagation modes as super-modes (E_{even} and E_{odd}) without mode coupling. These equations indicate that the propagation constants and the group delays (GD) of the propagation modes are dependent on the κ value.

Figure 2.101 shows the calculated DMD value between super-modes (inter-core DMD) at 1550 nm as a function of coupling coefficient κ when assuming two homogeneous step-index single-mode cores with normalized frequency V values of 1.95 ~ 2.5 separated by a distance D. The electric fields of the even and odd modes are also shown in the inset of this figure. It is found that the inter-core DMD increases as κ increases.

Figure 2.102 also shows the inter-core DMD at 1625 nm as a function of the

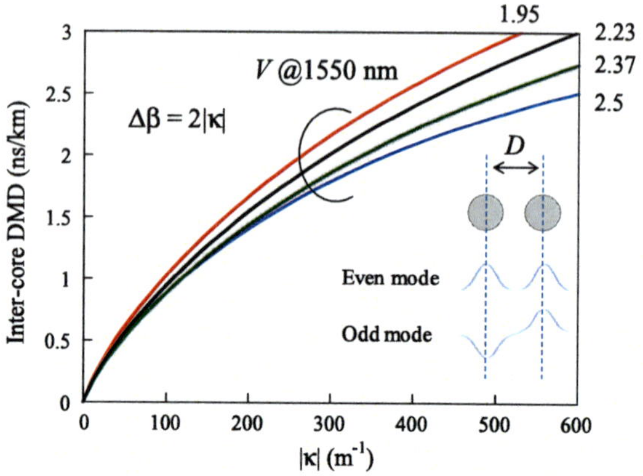

Fig. 2.101 Calculated inter-core DMD variation as a function of coupling coefficient (Reprinted with permission from [166]. © 2015 IEEE)

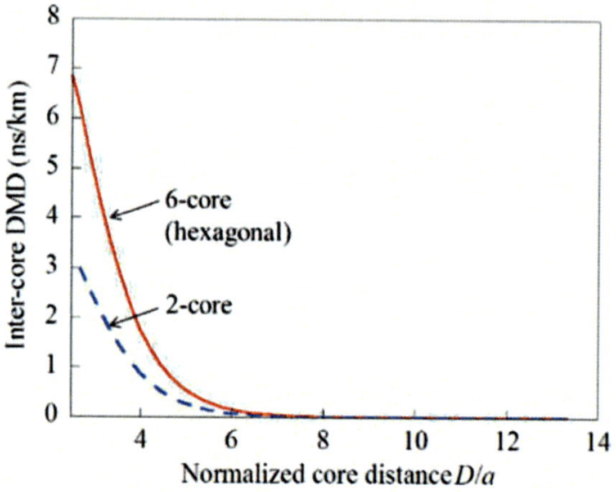

Fig. 2.102 Calculated inter-core DMD variation and crosstalk properties as a function of normalized core distance (Reprinted with permission from [166]. © 2015 IEEE)

core distance normalized by the core radius *a*. For simplicity, a single-mode-based two-core or hexagonally arranged six-core fiber with a step-index profile where *a* is 4.5 μm and the relative-index difference Δ is 0.35% is assumed. The inter-core DMD increases rapidly from a certain core distance in both cases, and the DMD property differs depending on the core number and arrangement. Therefore, there is a lower limit for the core pitch if we are to avoid increasing the inter-core DMD even if we design the coupled MCF.

Based on the previous discussion of the single-mode core based MCF, the inter-core DMD characteristics of two-LP-mode multi-core fiber were investigated next. Figure 2.103 shows the normalized GD values with graded-index two-LP-mode two-core fibers. The full vector finite element method is used for this calculation. Here, *a*, Δ and the alpha parameter of the two cores are 8.0 μm, 0.5% and 3.15, respectively, to obtain a low intra-core DMD (DMD between the LP_{01} and LP_{11} modes). There are six lines in the figure, which correspond to the LP_{01}, LP_{11a}, and LP_{11b} even/odd

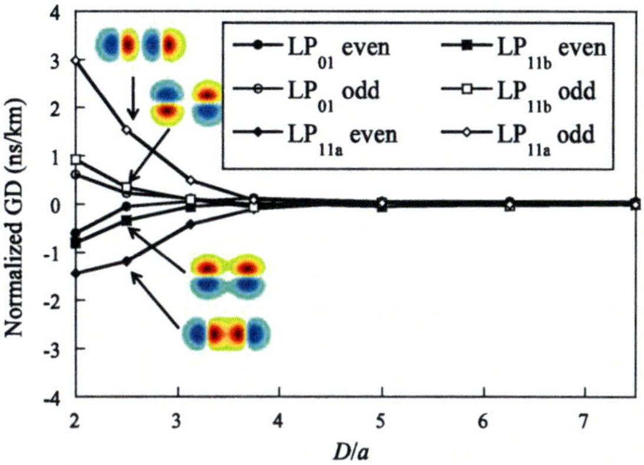

Fig. 2.103 Calculated normalized GD of two-LP-mode two-core graded-index fiber as a function of normalized core distance (Reprinted with permission from [166]. © 2015 IEEE)

modes (the electrical fields of the LP_{11a} and LP_{11b} even/odd modes are shown in the inset of the figure). The inter-core DMD also increases rapidly below a certain core distance, and the DMD between higher-order modes (LP_{11a} even and LP_{11a} odd modes in this case) indicates the maximum DMD value among the six spatial modes. This is because the coupling coefficient between higher-order modes is larger than that between fundamental modes. Therefore, low DMD few-mode MCFs can be designed by appropriately designing the core distance taking account of the coupling level of the higher-order mode.

Finally, the characteristics of fabricated 125 μm-cladding two-LP-mode six-core fiber are described. MCF with 125-μm cladding has the advantages of realizing a high space utilization efficiency while maintaining such properties as good splicing characteristics and mechanical reliability. Each core had a trench-assisted graded-index profile to reduce the DMD value between the LP_{01} and LP_{11} modes within the core, and the core pitch was 30 μm.

Figure 2.104 shows a cross section of the fabricated fiber and near-field patterns of the propagation modes. Clear mode-field patterns of the LP_{01}, LP_{11a}, and LP_{11b} modes are observed without any other higher-order modes.

Figure 2.105 shows the crosstalk characteristics obtained when the LP_{01} or LP_{11} mode was excited on the input side and the transmitted crosstalk power on the output

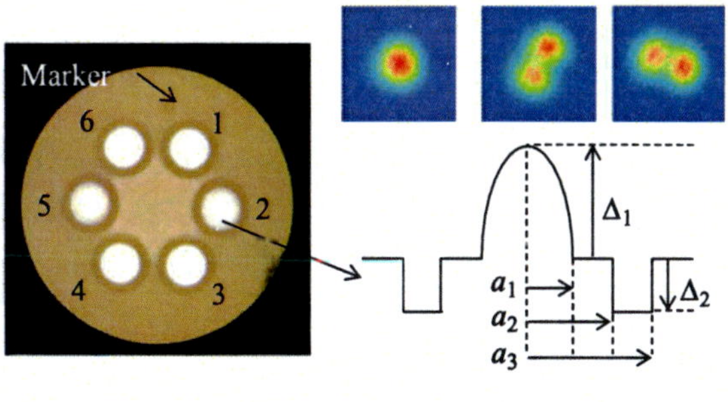

Fig. 2.104 Cross-sectional image of fabricated two-LP-mode six-core fiber and near-field patterns of propagation modes (Reprinted with permission from [166]. © 2015 IEEE)

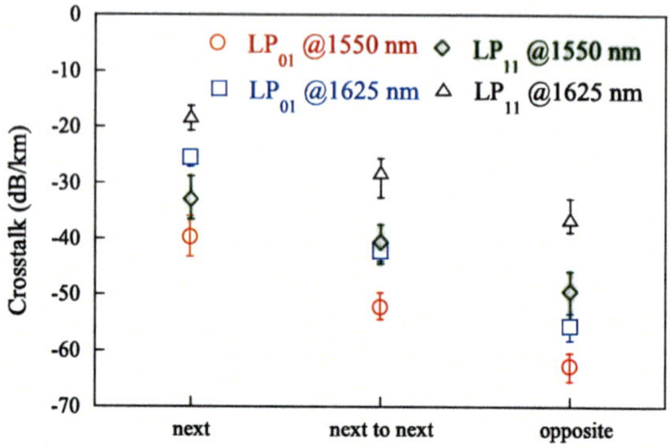

Fig. 2.105 Crosstalk properties of fabricated MCF (Reprinted with permission from [166]. © 2015 IEEE)

side was measured. This fiber realized a moderate coupling level of −20 dB/km, which caused no increase in mode-coupling-induced DMD.

Table 2.6 lists the measured optical properties of the fabricated MCF. A cutoff wavelength of less than 1530 nm for the LP_{21} mode and A_{eff} values of more than 60 and 130 μm^2 for the LP_{01} and LP_{11} modes are obtained. The bending losses of the LP_{11} modes were less than 0.1 dB/turn at a bending radius of 5 mm. There was no noticeable loss increase at 1625 nm, which means that the leakage losses of the LP_{11} modes in all the cores were successfully suppressed. The intra- or inter-core DMD, which means the DMD value between the LP_{01} and LP_{11} modes in the same core or the LP_{01} modes in different cores are also shown. An average intra-core DMD of −0.39 ns/km and an average inter-core DMD of 0.33 ns/km are obtained. These DMD values are assumed to result from core profile manufacturing errors.

The inter-core DMD caused by fabrication error can be mitigated with rotationally spliced fibers. Figure 2.106 shows the measured DMD characteristic based on the GD of the LP_{01} mode in core 1 when a 10 km MCF was split into two 5 km pieces and spliced so as to connect core 1 with core 6. A DMD value of 0.62 ns/km at 1550 nm and a maximum absolute DMD value of less than 0.74 ns/km over the entire *C*-*L* band were realized. The SUE performance of the MCF was finally evaluated. Here one spatial mode is counted as one channel so the LP_{01} and LP_{11} modes were counted as three channels. Fabricated two-LP-mode six-core fiber has a normalized channel multiplicity of 18, which is a considerable improvement compared with that in previously reported non-coupled two-LP-mode MCFs [156, 168].

2.6 Cabling Technology

2.6.1 Prospect for MCF Cables and Their Application Areas

Figure 2.107 schematically shows example structure of existing telecommunication network. Roughly speaking, the current telecommunication network mainly consists

Table 2.6 Optical properties of fabricated MCF

Core	Loss (dB/km)				A_{eff} (μm^2)		Cable cutoff wavelength of LP_{21} mode (nm)	Bending loss of LP_{11} mode @1625 nm (dB/turn)	Intra-core DMD @1550 nm (ns/km)	Inter-core DMD between LP_{01} modes @1550 nm (ns/km)
	LP_{01} mode		LP_{11} mode							
	1550 nm	1625 nm	1550 nm	1625 nm	LP_{01}	LP_{11}				
1	0.47	0.50	0.62	0.65	63	131	1492	0.83	−0.52	0
2	0.74	0.83	0.84	0.92	65	133	1514	0.62	−0.31	0.28
3	0.71	0.77	0.76	0.79	62	130	1520	0.47	−0.17	0.32
4	0.48	0.49	0.55	0.61	62	130	1487	0.56	−0.41	0.41
5	0.51	0.54	0.52	0.58	65	136	1517	0.28	−0.41	0.41
6	0.43	0.47	0.57	0.62	65	136	1522	0.68	−0.48	0.55

*Circulated value

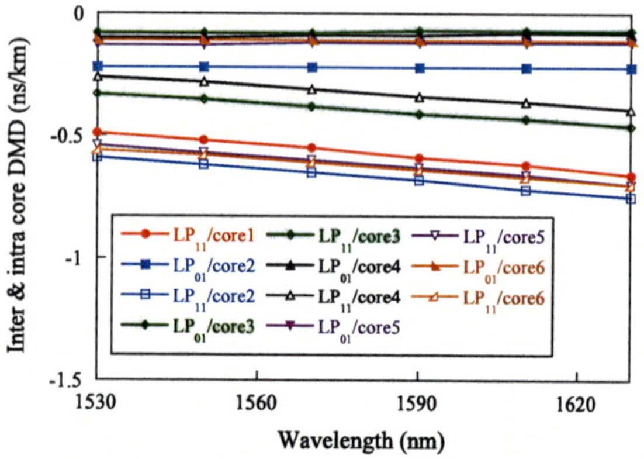

Fig. 2.106 Wavelength properties of intra- and inter-core DMD (Reprinted with permission from [166]. © 2015 IEEE)

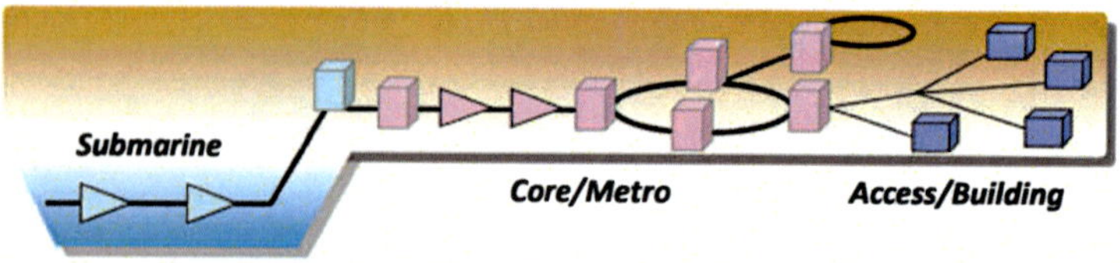

	Submarine	Core/Metro	Access/Building
Example cable structure and fiber number	Fibers ~24	Ribbon Fibers 100 ~ 1000	1 ~ 8
Example link length and number of splicing points	8000 km ~140	500 km ~250	10 km ~5

Fig. 2.107 Schematic image of existing telecommunication network (top), and example cable structure and link condition in each network area (bottom)

of three areas, such as submarine, core/metro and access network. As summarized in the Table in Fig. 2.107, ten to twenty optical fibers are loosely packed in the center region of a submarine cable with a diameter of around 20 mm. Optical submarine cable generally contains appropriate armored structure so that the influence of any external stress on the optical fiber property can be minimized during the lifetime. A maximum link length reached to ten thousand kilometers. However, the number of splicing point is reduced as possible in order to minimize the splicing loss in the link. Concerning the core/metro network, optical fibers are densely packed in the cable. In Fig. 2.107, cross-sectional image of a slot type optical fiber cable is shown as example. In this cable, some slots are allocated radially. Then, we can enhance the optical fiber density in the cable effectively since several ribbon fibers are easily

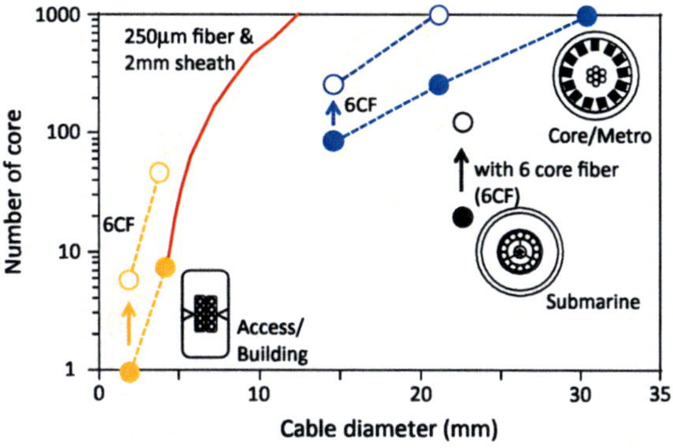

Fig. 2.108 Relationship between number of core and cable diameter. Filled circles show the example of existing optical fiber cable using conventional SMF (single-mode fiber). Open circles show the numerical example when we replace the SMF with 6-core fiber (6CF)

stacked in each slot. The slot type cable can contain 100 to 1000 optical fibers in a few tens millimeter cable diameter. A typical link length of core network becomes several hundred kilometers, and the number of splice point becomes more than 200. This is because a cable piece length is limited to a few kilometers by the allowable install condition in the terrestrial network. In the access network, some cables utilize the slot type structure to achieve the high-density optical fiber distribution. A cable with smaller fiber number, typically one to eight, is also used for optical distribution or indoor wiring. A typical link length is around five kilometers, and number of splice point contained in the cable section becomes less than ten.

If we would like to investigate the applicability of MCF into the existing telecommunication network, everybody can easily imagine its impact on space utilization efficiency. Figure 2.108 shows geometrical relationship between number of core and cable diameter. Filled circles show the example of existing optical fiber cables using conventional SMF. Circles, triangles, and squares represent the optical fiber cables for submarine, core/metro, and access/building network, respectively. The red solid line shows geometrical limit when we assumed that a coated SMF with a 250 μm diameter is packed hexagonally and wrapped with a 2-mm-thick sheath. Open symbols show the numerical example when we assumed to replace the SMF with MCF. Number of cores in an MCF relates to the cladding diameter and allowable core pitch. For simplicity, we set the cladding diameter of MCF at 125 μm which is same with the conventional SMF, although we can increase the number of cores by considering the enlarged cladding diameter. The allowable core pitch is mainly restricted by the crosstalk between neighboring cores, and generally the crosstalk of −30 dB or less is required at target transmission distance to ensure the sufficient transmission performance. If we would like to maintain the mode-field diameter (MFD) characteristic so that comparable to the conventional SMF (about 9 μm at 1310 nm), we can reduce the core pitch to 30–35 μm. Here, we set the target crosstalk at −30 dB after 100 km transmission in 1550 nm wavelength region. These assumptions result in we can allocate 6 SMF cores in one ring using a 125 μmcladding diameter. It would be noted that this prediction is a little bit optimistic since the target distance is not enough particularly for core and submarine network. Moreover, we need additional care to the influence of the reduced cladding thickness since it may degrade the attenuation property of the fiber. Thus, it can be said that Fig. 2.108 provides rough and maximum expectation to the improvement on the core density when we utilized a 6CF instead of the conventional SMF. It can be confirmed from Fig. 2.108 that spatial core density is directly improved in all network areas by simply replacing the conventional SMF with 6CF in the existing optical fiber cable. However, we need further discussion in each network area, if we would like to construct an MCF-based transmission link.

In the optical submarine network, limitation in an optical repeater will be the most important objective. Figure 2.109 shows schematic image of an optical submarine repeater. The optical repeater is composed of cable coupling and housing, and whose typical length is 4600 mm or less. Optical fibers from repeater and submarine cable are spliced in the cable coupling. Typical length and diameter of housing are 4600 and 250 mm as its maximum. Generally, the optical submarine repeater contains amplification, monitoring and power feeding functions in its housing. Therefore, the development of space-division multiplexing (SDM)-based optical amplification technology is mandatory even if we can realize an MCF-based optical submarine cable while keeping the current cable dimension. Moreover, the SDM-based optical amplification function should be realized with lower power consumption because it is difficult to increase the feeding power particularly in an ultra-long-haul optical submarine transmission system using a bidirectional power feeding scheme.

Regarding with a high-density terrestrial optical fiber cable, workability becomes one of the most important factors. As described in before, some terrestrial networks utilize the slot-type optical fiber cable and ribbon fibers (c.f. image and photo on the top of Fig. 2.110) to increase the fiber density effectively. The ribbon fiber can improve not only cable density but also the workability in terms of fiber splicing. Figure 2.110 also shows example working time when we splice 200 fibers [169]. In this figure, purple, orange, yellow, and green blocks represent the preparation, removal of fiber coating, fiber cut, and fusion splice, respectively. Figure 2.110a, c show the results when we splice four and eight mono-coated fibers at once, respectively. In Figure 2.110b, d, respectively we used a four and eight fiber ribbon. It is confirmed that the ribbon fiber is beneficial to improve the splicing workability

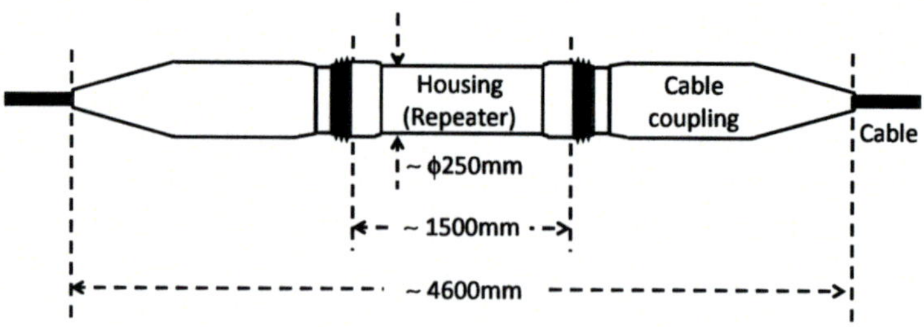

Fig. 2.109 Schematic image of optical submarine repeater

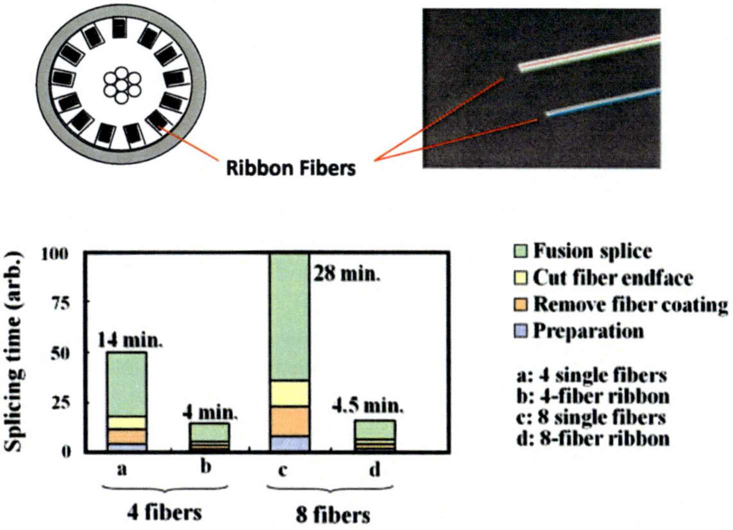

Fig. 2.110 Schematic image of slot-type optical fiber cable used in terrestrial network and photo of ribbon fibers (8 and 4 fiber ribbon). Bottom figure shows example working time when we splice 200 fibers (Reprinted with permission from [169]. © 2008 IEEE)

dramatically. Eight fiber ribbon enables us to reduce splicing time to less than one sixth. As for the MCF, a fusion splicing technique has already been developed for one MCF. Watanabe reported the applicability of side view alignment by employing an MCF with two markers [170]. Yoshida realized a low-loss fusion splice of MCF with thick cladding diameter by employing the end view alignment and swing arc fusion mechanism [171]. However, there is no report on MCF-based fiber ribbon. An MCF with non-circular cladding [172] may relax the difficulty on fiber alignment; however, it is still difficult to fabricate such type of MCF with precise geometrical accuracy along its length. Moreover, it is worth to note that Hogari successfully realized an ultra-high-density optical fiber cable by packing 200 conventional single-core SMF within a 10 mm diameter [169], and whose fiber density much closer to the geometrical limit shown in Fig. 2.108 with red line. In this ultra-high-density optical fiber cable, a rollable fiber ribbon is the most important key technology. In a conventional fiber ribbon, several optical fibers are completely fixed along their length to form a uniform sheet. On the other hand, the neighboring optical fibers are bonded partially along the length in the rollable fiber ribbon, and which enables dense packing similar to that in copper cable while maintaining the applicability to mass fusion splice obtained with conventional fiber ribbon. Therefore, it can be said that it is important to maximize the space utilizing efficiency while maintaining the sufficient workability if we would like to introduce the MCF technology into the terrestrial high-density optical fiber cables.

Concerning the access/building network, the importance of fan-in/fan-out (FI/FO) and connector technologies increases more and more. This is because the feasible access to individual core is mandatory at the end of the network, and pluggable connection is also necessary to realize an effective construction. Intensive research on fan-in/fan-out device and connectors for MCF has already been conducted and Fig. 2.111 shows example photos. Figure 2.111a shows an SC-type connector based pluggable fan-in/fan-out device designed for seven-core MCF, and Fig. 2.111b shows an MU-type connector. These results show that we can realize a pluggable MCF connector by using the existing technology when we assumed the factory-mounted process.

On the other hand, the field assemble splicing technique has also been used in the current access/building network. Figure 2.112 shows photos of existing field assemble splicing technique. Figure 2.112a shows a mechanical splice. The black holder contains V-groove along its length. Two cleaved fibers are butted via refractive index matching oil and hold in the holder. Figure 2.112b, c shows an SC-type-field-mountable connector. A piece of SMF is pre-installed in the ferrule of Fig. 2.112b, c, and it can be connected with conventional SC connector via adaptor. A cleaved

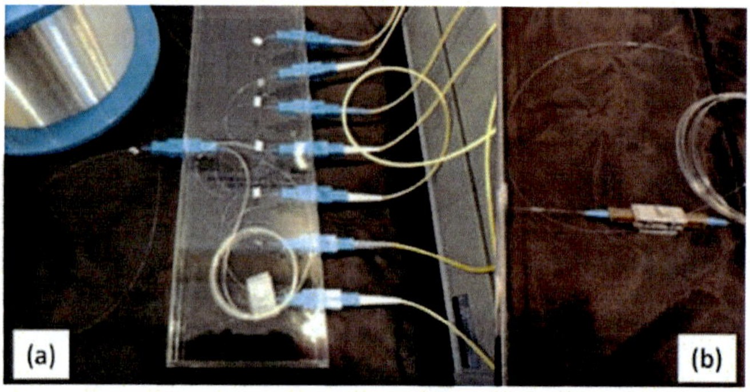

Fig. 2.111 Example of MCF connectors. **a** Photo of pluggable fan-in/fan-out device with SC-type connector. **b** Photo of MU-type MCF connector

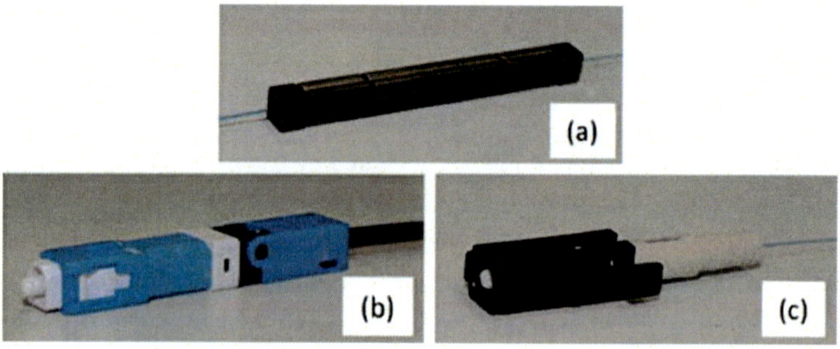

Fig. 2.112 Example of field assemble connectors. **a** Photo of mechanical splice. **b, c** Photo of field-mountable connector

optical fiber is inserted into the adaptor at the field and coupled with the SMF pre-installed ferrule via refractive index matching oil. Figure 2.112b, c respectively correspond to field assemble connector for indoor cable and coated optical fiber. The field assemble splicing techniques provide network operators flexibility to length optimization. However, it is difficult to realize the field-mountable connector for MCF since the MCF intrinsically requires end face alignment in its angular axis.

As the summary of this clause, it can be said that MCF is beneficial enough to improve the spatial density of the existing optical fiber cable directly in all telecom network area. However, it would be noted that an MCF with enlarged cladding diameter requires additional discussion on cabled properties. Moreover, the simultaneous development on various relating technologies, such as SDM in repeater, mass fusion splicing and easy and flexible connector, is mandatory when we will introduce the MCF technology into actual submarine, core/metro, and access/building networks.

2.6.2 High-Density MCF Cables

One purpose of optical fiber cable development has been the realization of high density: a small-diameter cable with sufficient performance. A cable with a reduced diameter is helpful to reduce the construction cost of optical cable network and to realize additional cable construction in the limited space of a cabling duct. Ultra-high-density cables with conventional single-core fibers (SCF) with 125-μm cladding diameter and 250-μm coating diameter (SCF-UHD cable) have been proposed as a solution [173, 174]. SCF-UHD cables have realized a 100-core cable with core density of approximately 2 cores/mm^2, where core density is the ratio of the number of cores and the cross-sectional area of a cable.

The SDM over MCFs or FMFs is a promising technology to overcome the transmission limits of current optical communication systems. Another prospective advantage of MCFs or FMFs is to improve core density and reduce cable diameter. At the end of the 1970s, some trials for high-density cable with MCFs were reported [175]. However, the results could not demonstrate the merit of MCF cables because MCFs used in these trials were not suitable for telecommunication and the cable structures were not optimized for high-density packing.

Figure 2.113 shows an example of core density improvement using an MCF [176]. The triangle symbols are core density of conventional ribbon slot cables using SCFs. The open circle symbols are core density of fabricated SCF-UHD cables [173, 174]. Optimization of cable structure and utilization of a bending-loss-insensitive fiber (BIF) enabled the increase of core density of SCF-UHD cables. The closed circle symbols show an example of core density of MCF-UHD cables. The 12-core fiber shown in the figure is used for the estimation [177, 178]. The same sheath thickness as in [174] was assumed to estimate the cable diameter. The core density of UHD cables can be improved by using MCFs instead of SCFs.

Figure 2.114 shows schematics and measurement results of a fabricated MCF-UHD cable with a length of 2 km [176]. The cable was composed of ten-colored

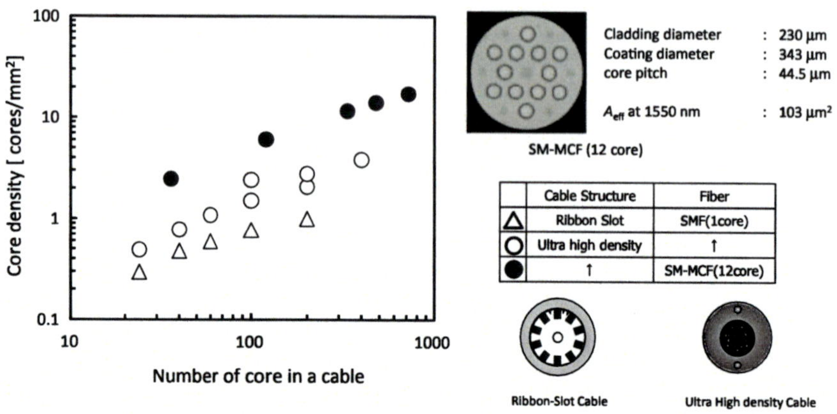

Fig. 2.113 Core density as function of number of cores in a cable (Reprinted with permission from [176]. © 2014 The Optical Society)

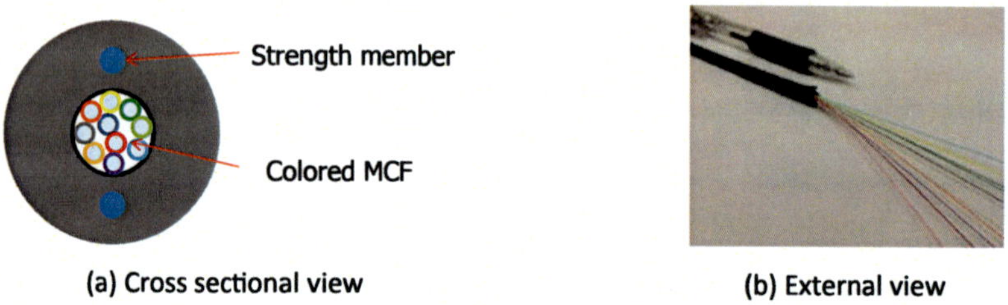

(c) Optical and mechanical test results

Item	Method	Condition	Result
Attenuation	IEC 60793-1-40 Method C	1550 nm	0.20 dB/km
Temperature cycling	IEC 60794-1-2 F1	1550 nm, -30 ~ +70 deg. C, 3 cycles	< 0.15 dB/km
Crush	IEC 60794-1-2 E3	1550 nm, 1960 N/100 m	< 0.05 dB/km
Impact	IEC 60794-1-2 E4	1550 nm, 1 m, 1 kg	< 0.05 dB/km
Repeated bending	IEC 60794-1-2 E6	1550 nm, *R* = 50 mm, 10 cycles	< 0.05 dB/km
Torsion	IEC 60794-1-2 E7	1550 nm, 90 deg./m	< 0.05 dB/km
Bend	IEC 60794-1-2 E11	1550 nm, *R* = 25 mm, 4 turn, 3 cycles	< 0.05 dB/km

Fig. 2.114 Characteristics of a fabricate MCF cable (Reprinted with permission from [176]. © 2014 The Optical Society)

MCFs, strength members, and a polyethylene sheath. The diameter of the cable was designed to be 5.1 mm in consideration of cable mechanical characteristics with a fiber of 230-µm cladding diameter. The averaged attenuation of cores in the cable was 0.20 dB/km, and the attenuation change due to cabling was less than 0.01 dB for all cores. The fabricated cable showed the same optical and mechanical performance with conventional optical cables as shown in the figure. The fabricated MCF-UHD cable achieved a core density of 6 cores/mm^2. The core density of the fabricated MCF-UHD cable is approximately seven times larger than that of a slot cable using

SCFs and approximately three times larger than that of a UHD cable using SCFs with the same number of cores.

Figure 2.115a shows the set up for measurement of crosstalk of the MCF-UHD cable [176]. To improve measurement accuracy, the crosstalk measurement was carried out on a 20-km loop that was made by connecting 2-km MCFs in the cable. Figure 2.115b shows measured crosstalk of the cable as a function of temperature. The cable was wound on a 600-mm-diameter drum. A 3-dB crosstalk change over the range from −50 to 70 °C was observed. Figure 2.115c shows the temperature dependence of the crosstalk of a 3-km freely coiled uncabled MCF. Talcum powder was applied over the MCF to exclude additional stress through the measurement. A 2-dB crosstalk change was observed over the same temperature range. The difference in the temperature dependence originates from macro-bending and micro-bending in

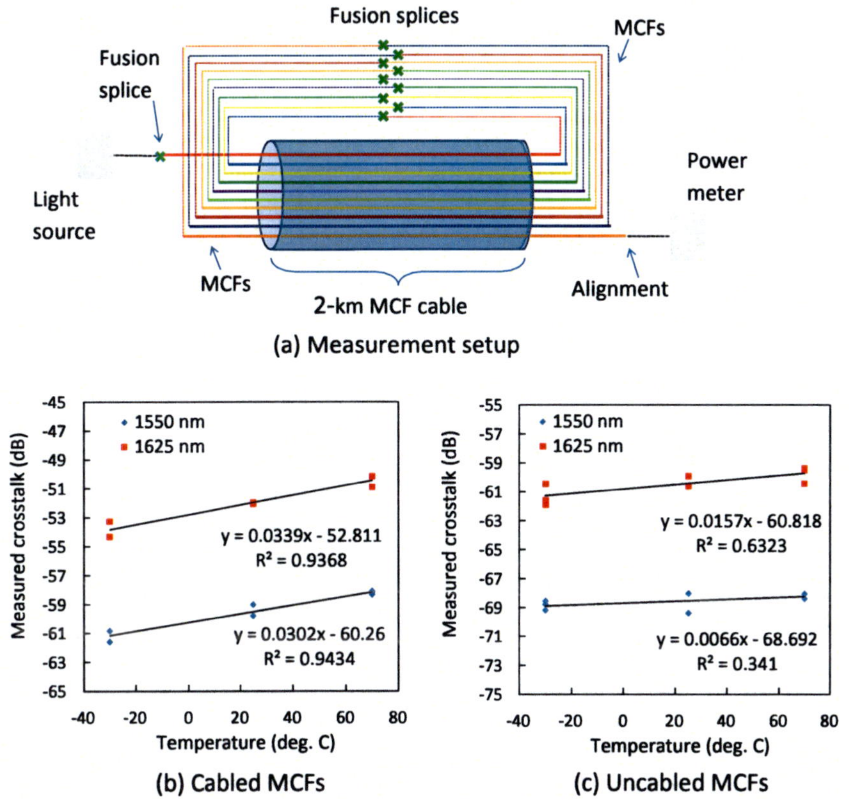

Fig. 2.115 Crosstalk characteristics of a fabricated MCF cable. **a** Measurement setup. **b** Temperature dependence of crosstalk for a 20-km MCF link in a cable. **c** Temperature dependence of crosstalk of a 3-km freely coiled MCF (Reprinted with permission from [176]. © 2014 The Optical Society)

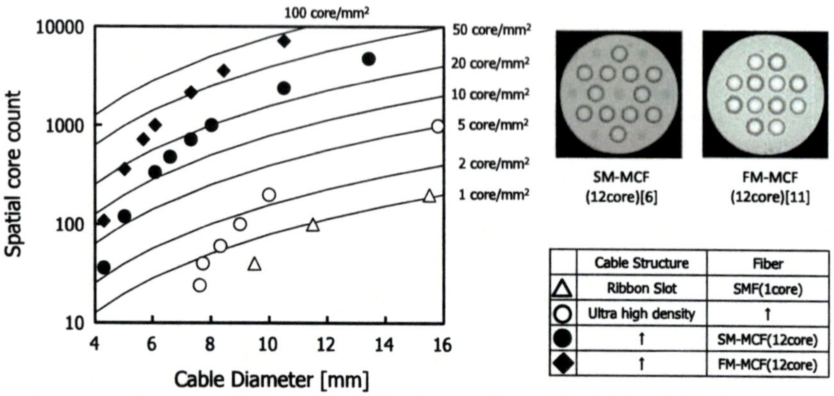

Fig. 2.116 Spatial core count as a function of cable diameter for cables with SCFs and MCFs (Reprinted with permission from [180]. © 2014 The Optical Society)

the cable [26, 179]. The result indicates that we need to design cables in consideration of crosstalk characteristics.

Further improvement of core density will be able to utilize different kinds of MCFs. A core density of 13.6 cores/mm^2 has been reported [71]. The cable utilized 12 MCFs, which involved eight cores that were optimized for 1310-nm transmission, with a cladding diameter of 125 μm. Another approach is a utilizing an FM-MCF. Figure 2.116 shows an estimation of the possibility of a UHD cable with an FM-MCF [176, 180]. Spatial core count (SCC), which is defined as (core count) × (mode count) is used as the vertical axis. An FM-MCF with an SCC of 36 (12 core and 3 spatial modes) was used for the estimation [181]. The cladding diameter and the coating diameter of the FM-MCF were 229 μm and 340 μm, respectively. The diamond symbols indicate the estimated *SCC* as a function of cable diameter for the FM-MCF-UHD cables. A cable with SCC = 1008 can be realized by using only 28 FM-MCFs with SCC = 36. The diameter of an FM-MCF cable with SCC = 1008 was estimated to be approximately 6 mm, which is 60% smaller than that of the SCF-UHD cable with 1000 fibers.

Recently, the characteristics of MCF cables have been reported as investigations toward practical deployment of MCFs [182, 183].

2.7 Future Perspective

2.7.1 Where Will Be the SDM Fibers First Deployed?

In Japan, collaborative research on extremely advanced transmission "EXAT" has been started to realize 1000 times larger transmission capacity by using triple "multi" technologies, multi-core, multi-mode, and multi-level modulation. At the beginning of this clause, we would like to briefly look the impact of SDM optical fiber cable on the expansion of the transmission capacity in terms of the space utilization efficiency (SUE).

Progress on transmission technologies has improved both transmission capacity and cable density. Figure 2.117 shows an example of historical cable density [184]. Here, the dashed line shows the geometrical limit when we assume hexagonally packed 250 μm optical fibers with a 2-mm-thick cable sheath. The black circle corresponds to a 400-pair copper cable whose density was less than 1 mm^{-2}. A slot-type 400 optical fiber cable achieved a cable density six times higher than that of copper cable as shown with an orange circle. The latest 400 optical fiber cable, shown by the light blue plot, realized a cable density of more than 4 mm^{-2} by employing a rollable fiber ribbon [169]. The rollable fiber ribbon enables dense packing similar to that in copper cable while maintaining the operability obtained with conventional fiber ribbon. This is because the neighboring optical fibers in the rollable ribbon are

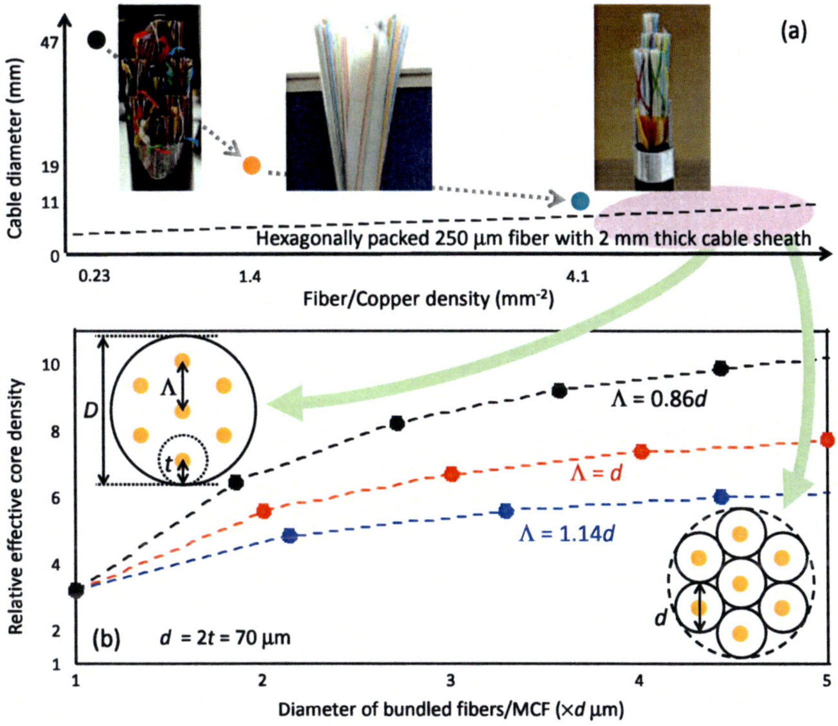

Fig. 2.117 Geometrical relationship of the optical fiber cable. **a** shows the historical relationship between cable diameter and cable density. Inset photos show the cross section of a 400 pair copper, a slot-type 400 optical fiber, and a 400 densely packed optical fiber cables from the left. **b** Shows the relationship between the relative effective core density and the diameter of the bundled single-core fibers or MCF (Reprinted with permission from [184]. © 2017 IEEE)

partially bonded along its length. Figure 2.117a confirms that the latest optical fiber cable has already achieved a cable density close to the geometrical limit.

If we are to overcome the geometrical limit while maintaining a single-core structure, we have to reduce the cladding diameter d to less than 125 μm and bundle the fiber densely. Here, a multi-element fiber [185] can be considered one way of realizing high-density bundled fibers. Alternatively, we can consider an uncoupled-type multi-core fiber (MCF). The d value in a single-core fiber is mainly restricted by the confinement loss characteristic. Roughly speaking, we need a d value of 70–80 μm in order to maintain a typical mode-field diameter (MFD) and feasible attenuation characteristics. Therefore, we can expect to improve the SUE of single-core fiber more than threefold if we can utilize a 70 μm cladding diameter. On the other hand, the structural condition of a hexagonal MCF is determined by the cladding thickness t, core pitch Λ and overall cladding diameter D. The t value is limited by the attenuation property, and it is assumed to be roughly similar to the cladding radius of a single-core fiber, $t = d/2$. Thus, we also need a t value of 35 μm.

Figure 2.117b shows the geometrical relationship between the relative effective core density and the diameter of bundled single-core fibers or MCF. Here, the effective core density is defined as the ratio of the effective area A_{eff} to the unit area. The vertical axis was normalized with the value in conventional single-mode fiber (SMF) with an 80 μm^2 A_{eff} and a 125 μm cladding diameter. We also assumed the d or $2t$ value of a thin cladding fiber or an MCF to be 70 μm for example. Black, red, and blue circles show the relationship when the Λ values were assumed to be 0.86d, d and 1.14d, respectively. We can generally expect a higher SUE by using an MCF, if the Λ value can be reduced below the d value. By contrast, we can improve the SUE by designing a bundled thin cladding fiber when Λ exceeds d. The minimum Λ of a hexagonal MCF would be limited to roughly 35 μm if we were to obtain an MFD similar to that in conventional SMF, although Λ also depends on the required crosstalk level and transmission distance. Thus, it can be said that the improvements in SUE obtained with an MCF and a thin cladding fiber are comparable because the minimum Λ and d values are fairly comparable, although we have to consider the coating thickness in the real application. The maximum D value in an MCF is restricted by mechanical reliability and is estimated to be around 230 μm, which corresponds to 3.3 on the horizontal axis in Fig. 2.117b. Thus, the maximum relative effective core density in bundled fiber/MCF would be improved to around 6.5 for 19 cores. This means that the core multiplexing requires further breakthrough to provide sufficient spatial density for an "EXAT" system.

Convergence of core and mode multiplexing will provide further possibility to realize a SUE of more than 10. The latest research indicates the possibility of an SDM fiber with more than 100 spatial channels [158, 159] by using the multiple few-mode core. However, these fibers employ a cladding diameter of more than 300 μm, and we need additional discussion on the applicability of a larger cladding. These considerations indicate that further research is needed including the transmission technologies to introduce the high-density SDM fiber cable into the telecom backbone network. Another approach is maximizing SUE for a cladding diameter of 125 μm, which is the same diameter with conventional transmission fibers. A transmission experiment over a four-core single-mode fiber with a cladding diameter of 125 μm has been reported [186]. The transmission line consisted of MCFs that were supplied by multiple vendors.

Then, how about the applicability of SDM fiber to short reach application. Worldwide spread of fiber to the home (FTTH) has been increasing the number of optical fiber and connection in a central office rapidly, and it causes conjunction of optical fiber and cable. It is easy to imagine that MCF can reduce the optical fiber conjunction by simply using its spatial multiplexing feature. In these applications, a feasible and reliable connector is mandatory as well as the MCF. There have been several reports on MCF connectors. Nagase realized an MU-type MCF connector by employing Oldham's coupling mechanism to ensure the precise angle alignment and mechanical reliability simultaneously [187]. Watanabe and Shikama individually reported the possibility of multiple MCF connector by using MT-type structure [188, 189]. These results reveal the reality of MCF-based high-density optical connection. When we consider the MCF for short reach in telecommunication network, the compatibility to the existing SMF is one of the most important factors. Recently, Geng reported a dual-core MCF, which has two conventional SMF cores in a 125 μm cladding diameter [190]. Saito also proposed a novel high-speed link which is composed of SMF and MCF [191]. In this study, they used a four-core MCF with a 125 μm cladding diameter as shown in Fig. 2.118a. The inset table reveals that the four cores in MCF maintain the full compliance to the conventional SMF/BIF (bending loss insensitive fiber) specified in ITU-T Recommendations G.652 and G.657 [192, 193].

(a)

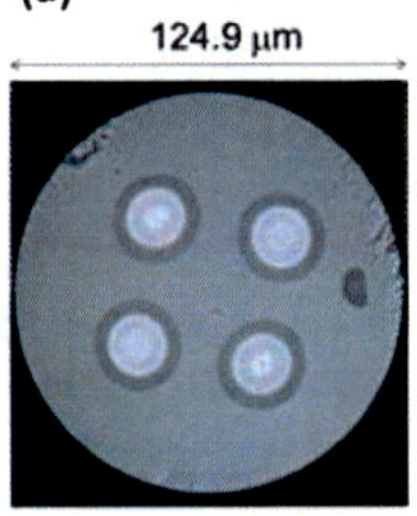

	#1	#2	#3	#4
MFD@1310 nm(μm)	8.60	8.55	8.60	8.40
Cut-off wavelength (nm)	1171	1194	1191	1187
Loss@1310 nm (dB/km) @1550 nm (dB/km)	0.34 0.19	0.33 0.20	0.35 0.21	0.33 0.20
Bending loss* (dB/turn)	0.17	0.19	0.14	0.15
Zero-dispersion wavelength λ_0 (nm)	1317	1318	1318	1319
Zero-dispersion slope S_0 (ps/nm^2·km)	0.091	0.090	0.091	0.091

*) λ = 1625 nm, R = 7.5 mm

(b)

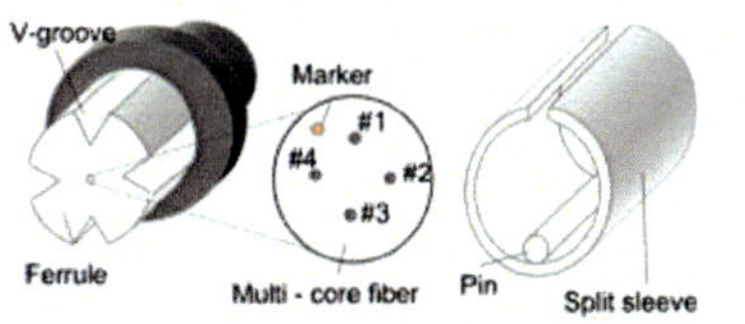

(c)

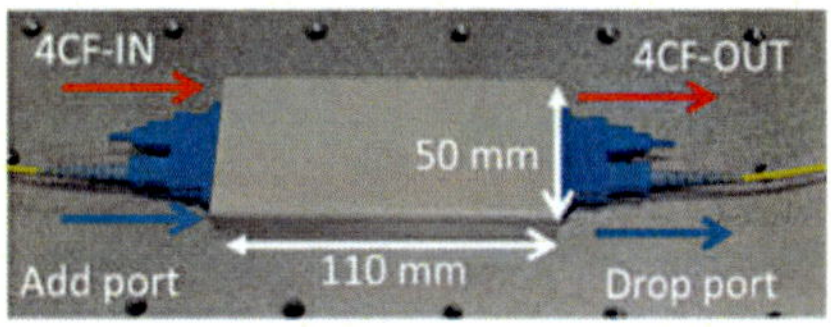

Fig. 2.118 Four-core MCF and optical components used to construct the MCF-based reconfigurable high-speed link [191]. **a** Cross-sectional photo and properties of a four-core MCF with full compliance to conventional SMF. **b** SC-type rotatable connector. **c** Pluggable add/drop module (Reprinted with permission from [184]. © 2017 IEEE)

They also proved the concept of MCF-based reconfigurable link by using a rotatable SC connector and a pluggable add/drop module, respectively shown in Fig. 2.118b, c. These components enable to add/drop signals to/from an arbitrary core of the four-core MCF and enable to provide a connection with the conventional SMF.

Thus, it can be said that the MCF technologies will be a key component for future short reach/high-speed link system including the reduction of the optical fiber and cable conjunction. Moreover, it is worth to note that these MCF technologies will be beneficial not only for telecom network but also for datacom network. This is because the datacom also involves a plenty of optical fiber and cable, and they are increasing day after day. Moreover, the MCF with even core number potentially harmonizes with bidirectional and/or high-speed parallel transmission system. Additional research on transceiver/receiver and optical cross connect for MCF is expected to accelerate the deployment of the SDM fiber.

Optical fiber sensing is another possible application area of SDM fibers. SDM fiber can directly offer additional sensing dimension in terms of spatial channel. One of the most promising application is the shape sensing using the MCF. Moore realized the shape sensing using an MCF with triangular core arrangement as shown in Fig. 2.119a [194]. Luna Innovations Inc. also realized the shape sensing using a four-core MCF shown in Fig. 2.119b [195]. In this MCF, three outer cores consist helical structure along its length, and it enables to sensing the longitudinal curvature and twist by analyzing the longitudinal strain distribution obtained with the OFDR (optical frequency domain reflectometry) technique.

As same as the multi-core, multi-mode can provide additional sensing dimension. SMS sensor, which is simply composed of single-multi-single structure as shown in Fig. 2.120, can be considered as example. At the output end of the sensor, we can observe the optical power fluctuation caused by the modal interference. The modal interference in the SMS sensor is affected by strain, temperature, bending, and so

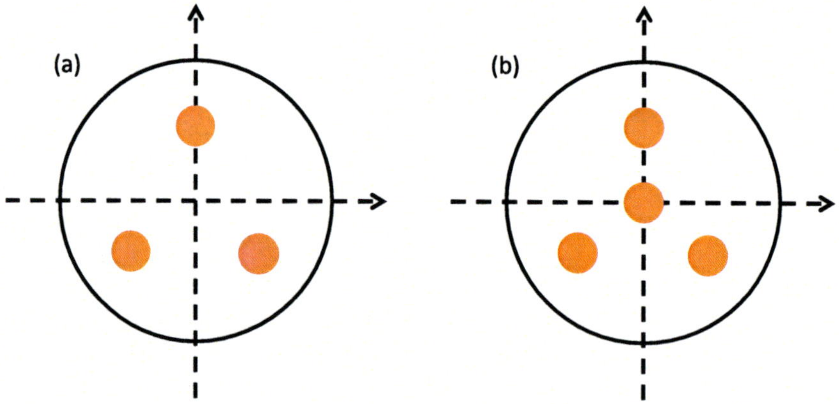

Fig. 2.119 Cross-sectional image of MCF used for shape sensing. **a** MCF with triangular core arrangement [194]. **b** MCF with four-core arrangement, and three outer cores consist helical structure along its length [195]

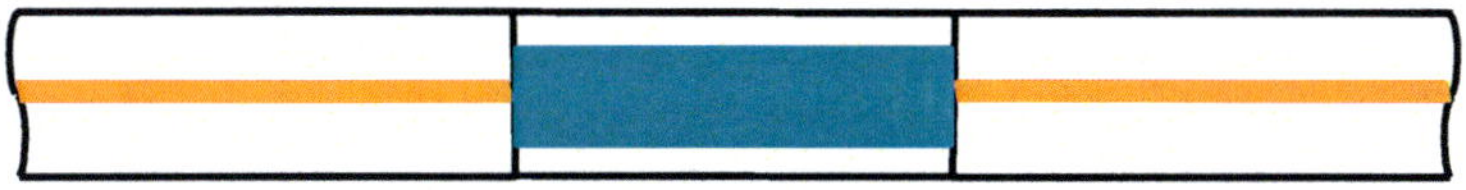

Fig. 2.120 Schematic image of SMS (single-multi-single) sensor

on [196]. Thus, these environmental parameters can be detected by monitoring the variation in the modal interference. The multiple parameter sensing is expected to be achieved if we can control and detect the modal interference obtained with different mode combination. Thus, it can be said that SDM fibers can offer a novel possibility to the optical fiber sensing area.

As summary, open innovative discussion on SDM technology will accelerate the implementation of the SDM fibers since the spatial channel can provide various possibilities not only for telecom application but also for non-telecom application.

2.7.2 Standardization

There are two important international standardization bodies with respect to telecommunication network. One is ITU-T (Telecommunication standardization sector, International Telecommunication Union), and the other is IEC (International Electrotechnical Commission). ITU-T investigates the requirements of telecommunication systems to ensure the inter-connectivity/operability. Historically, international standardization of optical fibers has been developed under responsibility of ITU-T. On the other hand, IEC handles the products specifications, and the international standard for optical cable, components, and subsystems has been mainly specified under IEC responsibility. The cross-reference between ITU-T and IEC has been well organized, and they have conducted harmonized standardization work continuously.

Regarding with the optical fibers, ITU-T has established and maintained seven Recommendations, namely G.65x-series. Table 2.7 summarizes the existing optical fiber standard for telecommunication area, and its relationship between ITU-T and IEC [197]. These optical fibers are categorized into two groups, such as multi- and single-mode fibers. ITU-T Recommendation G.651.1 covers a graded-index multimode fiber (MMF) with a 50 μm diameter core. However, the standardization of MMF is mainly discussed in IEC, since the current telecommunication network does not use MMF except for a specific part of the access network such as multi-dwelling unit and so on. Thus, the Recommendation G.651.1 follows IEC 60793-2-10 and focuses on the category A1a MMF. As for the single-mode fibers (SMFs), ITU-T launched six SMFs, G.652-G.657 fibers, taking the system requirements into account. ITU-T G.652 fiber has been established for a high-speed transmission at 1310 nm wavelength region. However, it is widely used in the current telecommunication network including the 1550 nm wavelength region as a "standard SMF". Following

Table 2.7 International standard of optical fibers, and its relationship between ITU-T (Telecommunication standardization sector, International Telecommunication Union) and IEC (International Electrotechnical Commission)

Fiber type and features	Recommendation/Category			
	ITU-T	IEC		
Multi-mode	50/125 μm graded-index MMF	G.651.1	60793-2-10	A1a
Single-mode	1.3 μm zero-dispersion SMF	G.652	60793-2-50	B1.1 B1.3
	Dispersion shifted SMF	G.653		B2
	Cutoff shift SMF	G.654		B1.2
	Nonzero dispersion SMF	G.655		B4
	Wideband nonzero dispersion SMF	G.656		B5
	Bending loss insensitive SMF	G.657		B6

the ITU-T G.652, ITU-T G.653 and G.654 fibers have been established. ITU-T G.653 fiber is suitable for high-speed and long-haul transmission system since it has zero-dispersion in the 1550 nm wavelength region. ITU-T G.654 fiber is also suitable for the transmission system using 1550 nm wavelength region since it has a longer cutoff wavelength and loss minimized in this transmission window. ITU-T G.655 and G.656 fibers were launched taking account of the progress on the wavelength division multiplexing (WDM) technology. The chromatic dispersions of these fibers were designed and specified so that they have nonzero value in the operation window to reduce the influence of optical nonlinearities. After that, ITU-T established G.657 fiber as a background of worldwide spread of fiber to the x (FTTx) taking account of the importance of the bending loss insensitivity with a viewpoint of optical fiber handling.

Thus, it can be found that ITU-T traditionally established the Recommendations for SMF taking account of their operation window and chromatic dispersion characteristics except for the ITU-T G.657 fiber, which has been developed by focusing on the deployment requirements. Then, how about the standardization of SDM fibers? ITU-T (Question 5/Study Group 15) has already picked up following question as one of considerable work item [198].

What fiber characteristics are needed to support space- and/or mode-division multiplexing applications?

However, there is no active discussion on this topic at present. It can be considered that the further clarification on the requirements of the SDM transmission system is needed as described in the previous clause, if we want to go forward the standardization of SDM transmission technology. The clarification to required core number and allowable cladding diameter is particularly important to discuss the SDM fiber standardization. Moreover, harmonized discussion with SDM components and subsystem is mandatory, even if we consider the short reach application which does not require the use of optical amplification. It can be considered that standardization of both SDM fiber and connector will be key objective in the short reach application. Thus, the harmonized standardization work between ITU-T and IEC is strongly expected to open up a novel telecommunication network using the SDM technology.

References

1. Y. Tamura, H. Sakuma, K. Morita, M. Suzuki, Y. Yamamoto, K. Shimada, Y. Honma, K. Sohma, T. Fujii, T. Hasegawa, Lowest-Ever 0.1419-dB/km Loss Optical Fiber, in *OFC 2017*, Th5D.1 (2017)
2. K. Tajima, Low loss PCF by reduction of hole surface imperfection, in *ECOC2007*, PD2.1 (2007)
3. P.J. Roberts, F. Couny, H. Sabert, B.J. Mangan, D.P. Williams, L. Farr, M.W. Mason, A. Tomlinson, Ultimate low loss of hollow-core photonic crystal fibres. Opt. Express **13**(1), 236–244 (2005). https://doi.org/10.1364/OPEX.13.000236
4. Y. Chen, Z. Liu, S.R. Sandoghchi, G. Jasion, T.D. Bradley, E. Numkam, J.R. Hayes, N.V. Wheeler, D.R. Gray, B.J. Mangan, R. Slavik, F. Poletti, M.N. Petrovich, D.J. Richardson, Demonstration of an 11km Hollow core photonic bandgap fiber for broadband low-latency data transmission, in *OFC2015*, Th5A.1 (2015)
5. T.D. Bradley, J.R. Hayes, Y. Chen, G.T. Jasion, S.R. Sandoghchi, R. Slavik, E.N. Fokoua, S. Bawn, H. Sakr, I.A. Davidson, A. Taranta, J.P. Thomas, M.N. Petrovich, D.J. Richardson, F. Poletti, Record low-loss 1.3dB/km data transmitting antiresonant hollow core fibre, in *ECOC2018*, PDP Th3F.2 (2018)
6. M. Hirano, T. Haruna, Y. Tamura, T. Kawano, S. Ohnuki, Y. Yamamoto, Y. Koyano, T. Sasaki, Record low loss, record high fom optical fiber with manufacturable process, in *OFC/NFOEC 2013*, PDP5A.7 (2013)
7. Y. Kawaguchi, Y. Tamura, T. Haruna, Y. Yamamoto, M. Hirano, Ultra low-loss pure silica core fiber. SEI Techn. Rev. **80**, 51–55 (2015)
8. T. Kato, M. Hirano, M. Onishi, M. Nishimura, Ultra-low nonlinearity low-loss pure silica core fibre for long-haul WDM transmission. Electron. Lett. **35**(19), 1615–1617 (1999). https://doi.org/10.1049/el:19991094
9. V. Carri, A. Carena, G. Bosco, P. Poggiolini, M. Hirano, Y. Yamamoto, F. Forghieri, Fiber figure of merit based on maximum reach. in *OFC/NFOEC2013*, OTh3G.2 (2013)
10. M. Hirano, Y. Yamamoto, V.A.J.M. Sleiffer, T. Sasaki, Analytical OSNR formulation validated with 100G-WDM experiments and optimal subsea fiber proposal, in *OFC/NFOEC2013*, OTu2B.6 (2013)
11. Y. Yamamoto, M. Hirano, V.A.J.M. Sleiffer, T. Sasaki, Analytical OSNR formulation and proposal of optimal fiber for submarine systems. IEICE Techn. Rep. (in Japanese) **113**(156), 23–28 (2013)
12. P. Poggiolini, The GN model of non-linear propagation in uncompensated coherent optical systems. J. Lightwave Technol. **30**(24), 3857–3879 (2012). https://doi.org/10.1109/JLT.2012.2217729
13. M. Hirano, Y. Yamamoto, T. Yoshiaki, T. Haruna, T. Sasaki, Aeff-enlarged pure-silica-core fiber having ring core profile, in *OFC/NFOEC2012*, OTh4I.2 (2012)
14. D. Marcuse, Loss analysis of single-mode fiber splices. Bell Syst. Tech. J. **56**(5), 703–718 (1977). https://doi.org/10.1002/j.1538-7305.1977.tb00534.x
15. V.A.J.M. Sleiffer, D.V. den Borne, M. Kuschnerov, V. Veljanovski, M. Hirano, Y. Yamamoto, T. Sasaki, S. Jansen, H.D. Waardt, A comparison between SSMF and large-Aeff Pure-Silica core fiber for ultra long-haul 100G transmission. Opt. Express **19**(26), B710–B715 (2011). https://doi.org/10.1364/OE.19.00B710

16. S. Ohnuki, K. Kuwahara, K. Morita, Y. Koyano, Further attenuation improvement of a pure silica core fiber with large effective area, in *SubOptic2010*, THU3A03 (2010)
17. S. Bickham, Ultimate limits of effective area and attenuation for high data rate fibers. in *OFC/NFOEC2011*, OWA5 (2011)
18. K. Nagayama, M. Kakui, M. Matsui, T. Saitoh, Y. Chigusa, Ultra-low-loss (0.1484dB/km) pure silica core fiber and extension of transmission distance. Electon. Lett. **38**(20), 1168–1169 (2002). https://doi.org/10.1049/el:20020824
19. Y. Koyano, S. Ohnuki, T. Kawano, M. Hirano, T. Haruna, Y. Yamamoto, Further improvement of linearity-enhanced optical fiber with low attenuation and large effective core area, in *SubOptic2013*, EC11, Paris, France (2013)
20. J.F. Libert, J.L. Lang, J. Chesnoy, The new 160 Gigabit WDM challenge for submarine cable systems, in *IWCS1998*, 375–384 (1998)
21. J.M. Fini, B. Zhu, T.F. Taunay, M.F. Yan, Statistics of crosstalk in bent multicore fibers. Opt. Express **18**(14), 15122–15129 (2010). https://doi.org/10.1364/OE.18.015122
22. T. Hayashi, T. Nagashima, O. Shimakawa, T. Sasaki, E. Sasaoka, Crosstalk variation of multicore fibre due to fibre bend, in *European Conference on Optical Communications (ECOC)*, We.8.F.6. (2010)
23. M. Koshiba, K. Saitoh, K. Takenaga, S. Matsuo, Multi-core fiber design and analysis: coupled-mode theory and coupled-power theory. Opt. Express **19**(26), B102–B111 (2011). https://doi.org/10.1364/OE.19.00B102
24. K. Petermann, Microbending loss in monomode fibers. Electron. Lett. **12**(4), 107–109 (1976). https://doi.org/10.1049/el:19760084
25. M. Koshiba, K. Saitoh, K. Takenaga, S. Matsuo, Analytical expression of average power-coupling coefficients for estimating intercore crosstalk in multicore fibers. IEEE Photonics J. **4**(5), 1987–1995 (2012).https://doi.org/10.1109/JPHOT.2012.2221085
26. T. Hayashi, T. Sasaki, E. Sasaoka, K. Saitoh, M. Koshiba, Physical interpretation of intercore crosstalk in multicore fiber: effects of macrobend, structure fluctuation, and microbend. Opt. Express **21**(5), 5401–5412 (2013). https://doi.org/10.1364/OE.21.005401
27. S. Matsuo, K. Takenaga, Y. Arakawa, Y. Sasaki, S. Tanigawa, K. Saitoh, M. Koshiba, Crosstalk behavior of cores in multi-core fiber under bent condition. IEICE Electron. Express **8**(6), 385–390 (2011). https://doi.org/10.1587/elex.8.385
28. T. Hayashi, T. Sasaki, E. Sasaoka, Multi-core fibers and their crosstalk characteristics, in *IEEE Photonics Society Summer Topical Meeting Series*, Seattle, TuC4.1 (2012)
29. K. Takenaga, S. Tanigawa, N. Guan, S. Matsuo, K. Saitoh, M. Koshiba, Reduction of crosstalk by quasi-homogeneous solid multi-core fiber, in *Optical Fiber Communication Conference (OFC)* 2010, OWK7 (2010)
30. K. Takenaga, Y. Arakawa, S. Tanigawa, N. Guan, S. Matsuo, K. Saitoh, M. Koshiba, An investigation on crosstalk in multi-core fibers by introducing random fluctuation along longitudinal direction. IEICE Trans. Commun. E94.B(2), 409–416 (2011). https://doi.org/10.1587/transcom.E94.B.409
31. K. Takenaga, Y. Arakawa, S. Tanigawa, N. Guan, S. Matsuo, K. Saitoh, M. Koshiba, Reduction of crosstalk by trench-assisted multi-core fiber, in *Optical Fiber Communication Conference* (OFC) 2011, OWJ4 (2011)
32. T. Hayashi, T. Taru, O. Shimakawa, T. Sasaki, E. Sasaoka, Design and fabrication of ultra-low crosstalk and low-loss multi-core fiber. Opt. Express **19**(17), 16576–16592 (2011). https://doi.org/10.1364/OE.19.016576
33. K. Saitoh, T. Matsui, T. Sakamoto, M. Koshiba, S. Tomita, Multi-core hole-assisted fibers for high core density space division multiplexing, in *Optoelectronics and Communications Conference (OECC) 2010*, Sapporo, Japan, 7C2-1 (2010)
34. C. Xia, R. Amezcua-Correa, N. Bai, E. Antonio-Lopez, D. May-Arriojo, A. Schulzgen, M. Richardson, J. Liñares, C. Montero, E. Mateo, X. Zhou, G. Li, Hole-assisted few-mode multi-core fiber for high-density space-division multiplexing. IEEE Photonics Technol. Lett. **24**(21), 1914–1917 (2012). https://doi.org/10.1109/LPT.2012.2218801

35. A. Ziolowicz, M. Szymanski, L. Szostkiewicz, T. Tenderenda, M. Napierala, M. Murawski, Z. Holdynski, L. Ostrowski, P. Mergo, K. Poturaj, M. Makara, M. Slowikowski, K. Pawlik, T. Stanczyk, K. Stepien, K. Wysokinski, M. Broczkowska, T. Nasilowski, Hole-assisted multi-core optical fiber for next generation telecom transmission systems. Appl. Phys. Lett. **105**(8), 081106 (2014). https://doi.org/10.1063/1.4894178
36. K. Imamura, K. Mukasa, R. Sugizaki, Y. Mimura, T. Yagi, Multi-core holey fibers for ultra large capacity wide-band transmission, in *European Conference on Optical Communication (ECOC)* 2008, P.1.17 (2008)
37. D.M. Taylor, C.R. Bennett, T.J. Shepherd, L.F. Michaille, M.D. Nielsen, H.R. Simonsen, Demonstration of multi-core photonic crystal fibre in an optical interconnect. Electron. Lett. **42**(6), 331–332 (2006). https://doi.org/10.1049/el:20064382
38. M. Koshiba, K. Saitoh, Y. Kokubun, Heterogeneous multi-core fibers: proposal and design principle. IEICE Electron. Express 6(2), 98–103 (2009). https://doi.org/10.1587/elex.6.98
39. G. Le Noane, D. Boscher, P. Grosso, J.C. Bizeul, C. Botton, Ultra high density cables using a new concept of bunched multicore monomode fibers: a key for the future FTTH networks, in *International Wire and Cable Symposium (IWCS)* (1994), pp. 203–210
40. T. Hayashi, T. Sasaki, E. Sasaoka, Behavior of inter-core crosstalk as a noise and its effect on Q-factor in multi-core fiber. IEICE Trans. Commun. **E97.B**(5), 936–944 (2014). https://doi.org/10.1587/transcom.E97.B.936
41. T. Hayashi, T. Taru, O. Shimakawa, T. Sasaki, E. Sasaoka, Characterization of crosstalk in ultra-low-crosstalk multi-core fiber. J. Light. Technol. **30**(4), 583–589 (2012). https://doi.org/10.1109/JLT.2011.2177810
42. T. Hayashi, Multi-core optical fibers, in *Optical Fiber Telecommunications*, 6th edn., vol. A, ed. by I. P. Kaminow, T. Li, A.E. Willner (Elsevier, 2013), pp. 321–352
43. K. Saitoh, M. Koshiba, K. Takenaga, S. Matsuo, Crosstalk and core density in uncoupled multi-core fibers. IEEE Photon. Technol. Lett. **24**(21), 1898–1901 (2012). https://doi.org/10.1109/LPT.2012.2217489
44. S.O. Arik, J.M. Kahn, K.-P. Ho, MIMO signal processing for mode-division multiplexing. IEEE Signal Process. Mag. **31**(2), 25–34 (2014). https://doi.org/10.1109/MSP.2013.2290804
45. K.-P. Ho, J.M. Kahn, Statistics of group delays in multimode fiber with strong mode coupling. J. Lightw. Technol. **29**(21), 3119–3128 (2011). https://doi.org/10.1109/JLT.2011.2165316
46. R. Ryf, R.-J. Essiambre, A.H. Gnauck, S. Randel, M.A. Mestre, C. Schmidt, P.J. Winzer, R. Delbue, P. Pupalaikis, A. Sureka, T. Hayashi, T. Taru, T. Sasaki, Space-division multiplexed transmission over 4200-km 3-core microstructured fiber, in *The Optical Fiber Communication Conference*, Los Angeles, CA, USA, PDP5C.2 (2012)
47. R. Ryf, N.K. Fontaine, B. Guan, R.-J. Essiambre, S. Randel, A.H. Gnauck, S. Chandrasekhar, A. Adamiecki, G. Raybon, B. Ercan, R.P. Scott, S.J. Ben Yoo, T. Hayashi, T. Nagashima, T. Sasaki, 1705-km transmission over coupled-core fibre supporting 6 spatial modes, in *The European Conference on Optical Communication*, Cannes, France, PD.3.2 (2014)
48. C. Xia, N. Bai, I. Ozdur, X. Zhou, G. Li, Supermodes for optical transmission. Opt. Express **19**(17), 16653–16664 (2011). https://doi.org/10.1364/OE.19.016653
49. K. Saitoh, S. Matsuo, Multicore fiber technology. IEEE J. Lightw. Technol. **1**, 55–66 (2016). https://doi.org/10.1109/JLT.2015.2466444
50. T. Hayashi, T. Taru, O. Shimakawa, T. Sasaki, E. Sasaoka, Ultra-low-crosstalk multi-core fiber feasible to ultra-long-haul transmission, In *The Optical Fiber Communication Conference*, PDPC2 (2011)
51. K. Imamura, K. Mukasa, R. Sugizaki, Trench assisted multi-core fiber with large Aeff over 100 μm^2 and low attenuation loss, in *The European Conference on Optical Communication*, Geneva, Switzerland, Mo.1.LeCervin.1 (2011)
52. B. Yao, K. Ohsono, N. Shiina, K. Fukuzato, A. Hongo, E. H. Sekiya, K. Saito, Reduction of crosstalk by hole-walled multi-core fibers, in *The Optical Fiber Communication Conference*, Los Angeles, CA, USA, OM2D.5 (2012)
53. T. Sakamoto, K. Saitoh, N. Hanzawa, K. Tsujikawa, L. Ma, M. Koshiba, F. Yamamoto, Crosstalk suppressed hole-assisted 6-core fiber with cladding diameter of 125 μm, in *The European Conference on Optical Communication*, London, U.K., Mo.3.A.3 (2013)

54. F. Ye, J. Tu, K. Saitoh, T. Morioka, A simple analytical expression for crosstalk estimation in homogeneous trench-assisted multi-core fibers. Opt. Express **22**(19), 23007–23018 (2014). https://doi.org/10.1364/OE.22.023007
55. Y. Sasaki, Y. Amma, K. Takenaga, S. Matsuo, K. Saitoh, M. Koshiba, Investigation of crosstalk dependencies on bending radius of heterogeneous multicore fiber, in *The Optical Fiber Communication Conference*, Anaheim, CA, USA, OTh3K.3 (2013)
56. Y. Amma, Y. Sasaki, K. Takenaga, S. Matsuo, J. Tu, K. Saitoh M. Koshiba, T. Morioka, Y. Miyamoto, High-density multicore fiber with heterogeneous core arrangement, in *The Optical Fiber Communication Conference*, Los Angeles, CA, USA, Th4C.4 (2015)
57. Y. Sasaki, K. Takenaga, K. Aikawa, Y. Miyamoto, T. Morioka, Single-mode 37-core fiber with a cladding diameter of 248 μm, in *The Optical Fiber Communication Conference*, Th1H2 (2017)
58. H. Ono, Y. Abe, K. Shikama, T. Takahashi, M. Yamada, K. Takenaga, S. Matsuo, Amplification method for crosstalk reduction in multi-core fibre amplifier. Electron. Lett. **49**(2), 138–140 (2013). https://doi.org/10.1049/el.2012.4307
59. A. Sano, H. Takara, T. Kobayashi, Y. Miyamoto, Crosstalk-managed high capacity long haul multicore fiber transmission with propagation-direction interleaving. IEEE J. Lightw. Technol. **30**(16), 2771–2779 (2014). https://doi.org/10.1109/JLT.2014.2320826
60. M.O. Van Deventer, Polarization properties of Rayleigh backscattering in single-mode fibers. IEEE J. Lightw. Technol. **11**(12), 1895–1899 (1993). https://doi.org/10.1109/50.257947
61. P.J. Winzer, A.H. Gnauck, A. Konczykowska, F. Jorge, J.-Y. Dupuy, Penalties from in-band crosstalk for advanced optical modulation formats, in *The European Conference on Optical Communication*, Geneva, Switzerland, Tu.5.B.7 (2011)
62. S. Matsuo, K. Takenaga, Y. Arakawa, Y. Sasaki, S. Tanigawa, K. Saitoh, M. Koshiba, Large-effective-area ten-core fiber with cladding diameter of about 200 μm. Opt. Lett. **36**(23), 4626–4628 (2011). https://doi.org/10.1364/OL.36.004626
63. K. Takenaga, Y. Arakawa, Y. Sasaki, S. Tanigawa, S. Matsuo, K. Saitoh, M. Koshiba, A large effective area multi-core fiber with an optimized cladding thickness. Opt. Express **19**(26), B542–B550 (2011). https://doi.org/10.1364/OE.19.00B543
64. B. Zhu, T.F. Taunay, M. Fishteyn, X. Liu, S. Chandraskekhar, M.F. Yan, J.M. Fini, E.M. Monberg, F.V. Dimarcello, 112-Tb/s space-division multiplexed DWDM transmission with 14-b/s/Hz aggregate spectral efficiency over a 76.8-km seven-core fiber. Opt. Express **19**(17), 16665–16671 (2011). https://doi.org/10.1364/OE.19.016665
65. H. Takara, H. Ono, Y. Abe, H. Masuda, K. Takenaga, S. Matsuo, H. Kubota, K. Shibahara, T. Kobayashi, Y. Miyamoto, 1000-km 7-core fiber transmission of 10x 96-Gb/s PDM-16QAM using Raman amplification with 6.5W per fiber. Opt. Exp. **20**(9), 10100–10105 (2012). https://doi.org/10.1364/OE.20.010100
66. S. Matsuo, Y. Sasaki, T. Akamatsu, I. Ishida, K. Takenaga, K. Okuyama, K. Saitoh, M. Kosihba, 12-core fiber with one ring structure for extremely large capacity Transmission. Opt. Express **20**(27), 28398–28408 (2012). https://doi.org/10.1364/OE.20.028398
67. H. Takahashi, T. Tsuritani, E.L.T. de Gabory, T. Ito, W.R. Peng, K. Igarashi, K. Takeshima, Y. Kawaguchi, I. Morita, Y. Tsuchida, Y. Mimura, K. Maeda, T. Saito, K. Watanabe, K. Imamura, R. Sugizaki, M. Suzuki, First demonstration of MC-EDFA-repeatered SDM transmission of 40 x 128-Gbit/s PDM-QPSK signals per core over 6,160-km 7-core MCF, in *The European Conference on Optical Communication*, paper Th.3.C.3 (Amsterdam, Netherlands, 2012)
68. J. Sakaguchi, B.J. Puttnam, W. Klaus, Y. Awaji, N. Wada, A. Kanno, T. Kawanishi, K. Imamura, H. Inaba, K. Mukasa, R. Sugizaki, T. Kobayashi, M. Watanabe, 305-Tb/s space division multiplexed transmission using homogeneous 19-core fiber. IEEE J. Lightw. Technol. **31**(4), 554–562 (2013). https://doi.org/10.1109/JLT.2012.2217373
69. A. Sano, H. Takara, T. Kobayashi, H. Kawakami, H. Kisikawa, T. Nakagawa, Y. Miyamoto, Y. Abe, H. Ono, K. Shikama, M. Nagatani, T. Moro, Y. Sasaki, I. Ishida, K. Takenaga, S. Matsuo, K. Saitoh, M. Koshiba, M. Yamada, H. Masuda, T. Morioka, 409-Tb/s + 409-Tb/s crosstalk suppressed bidirectional MCF transmission over 450 km using propagation-direction interleaving. Opt. Express **21**(14), 16777–16783 (2013). https://doi.org/10.1364/OE.21.016777

70. J. Sakaguchi, W. Klaus, B.J. Puttnam, J.M.D. Mendinueta, Y. Awaji, N. Wada, Y. Tsuchida, K. Maeda, M. Tadakuma, K. Imamura, R. Sugizaki, T. Kobayashi, Y. Tottori, M. Watanabe, R.V. Jensen, 19-core MCF transmission system using EDFA with shared core pumping coupled via free-space optics. Opt. Express **22**(1), 90–95 (2013). https://doi.org/10.1364/OE.22.000090
71. T. Hayashi, T. Nakanishi, K. Hirashima, O. Shimakawa, F. Sato, K. Koyama, A. Furuya, Y. Murakami, and T. Sasaki, 125-μm-cladding 8-core multi-core fiber realizing ultra-high-density cable suitable for O-band short-reach optical interconnects, in *The Optical Fiber Communication Conference*, Los Angeles, CA, USA, Th5C.6 (2015)
72. Y. Sasaki, R. Fukumoto, K. Takenaga, K. Aikawa, K. Saitoh, T. Morioka, Y. Miyamoto, Crosstalk-managed heterogeneous single-mode 32-core fibre, in *The European Conference on Optical Communication* , W.2.B.2 (2016)
73. B. Zhu, J.M. Fini, M.F. Yan, X. Liu, S. Chandrasekhar, T.F. Taunay, M. Fishteyn, E.M. Monberg, F.V. Dimarcello, High-capacity space-division-multiplexed DWDM transmissions using multicore fiber. IEEE J. Lightw. Technol. **30**(4), 486–492 (2012). https://doi.org/10.1109/JLT.2011.2173793
74. J. Sakaguchi, Y. Awaji, N. Wada, Fundamental study on new characterization method for crosstalk property on multi-core fibers using long wavelength probe signals, in *The Optical Fiber Communication Conference*, Anaheim, CA, USA, OW1K.1 (2013)
75. T. Hayashi, T. Taru, O. Shimakawa, T. Sasaki, E. Sasaoka, Uncoupled multi-core fiber enhancing signal-to-noise ratio. Opt. Express **20**(26), B94–B103 (2012)
76. A. Carena, V. Curri, G. Bosco, R. Cigliutti, E. Torrengo, P. Poggiolini, A. Nespola, D. Zeolla, F. Forghieri, Novel figure of merit to compare fibers in coherent detection systems with uncompensated links. Opt. Express **20**(1), 339–346 (2012). https://doi.org/10.1364/OE.20.000339
77. T. Hayashi, T. Sasaki, Design strategy of uncoupled multicore fiber enabling high spatial capacity transmission, in *IEEE Photonics Society Summer Topical Meeting Series*, Waikoloa, HI, MC2.4 (2013)
78. R.L. Graham, B.D. Lubachevsky, K.J. Nurmela, P.R. J. Österg\aard, Dense packings of congruent circles in a circle. Discrete Math. **181**(1), 139–154 (1998). https://doi.org/10.1016/S0012-365X(97)00050-2
79. E. Specht, Circles in a circle. www.packomania.com, 21-May-2012. [Online]. Available: http://www.packomania.com/. [Accessed: 20-Apr-2013]
80. J.H. Chang, H.G. Choi, Y.C. Chung, Achievable capacity improvement by using multi-level modulation format in trench-assisted multi-core fiber system. Opt. Express **21**(12), 14262–14271 (2013). https://doi.org/10.1364/OE.21.014262
81. Y. Kokubun, T. Watanabe, Dense heterogeneous uncoupled multi-core fiber using 9 types of cores with double cladding structure, in *Microopics Conference (MOC)*, 17th, Sendai, Japan, K-5 (2011)
82. F. Ye, C. Peucheret, T. Morioka, Capacity of space-division multiplexing with heterogeneous multi-core fibers, in *OptoElectronics and Communication Conference/Photonics in Switching (OECC/PS)*, Kyoto, WR2-3 (2013)
83. T. Watanabe, Y. Kokubun, High density and low cross talk design of heterogeneous multi-core fiber with air hole assisted double cladding, in *OptoElectronics and Communication Conference/Photonics in Switching (OECC/PS)*, Kyoto, MS1-4 (2013)
84. J. Sakaguchi, B. Puttnam, W. Klaus, Y. Awaji, N. Wada, A. Kanno, T. Kawanashi, K. Imamura, H. Inaba, K. Musaka, R. Sugizaki, T. Kobayashi, M. Watanabe, 19-core transmission of 19x100x172-Gb/s SDM-WDM-PDM-QPSK signals at 305Tb/s, in *OFC/NFOEC2012*, Los Angeles, PDP5C.1 (2012)
85. H. Takara, A. Sano, T. Kobayashi, H. Kubota, H. Kawakami, A. Matsuura, Y. Miyamoto, Y. Abe, H. Ono, K. Shikama, Y. Goto, K. Tsujikawa, Y. Sakaki, I. Ishida, K. Takenaga, S. Matsuo, K. Saitoh, M. Koshiba, T. Morioka, 1.01-Pb/s (12SDM/222 WDM/456Gb/s) crosstalk-managed transmission with 91.4-b/s/Hz aggregated spectral efficiency, in *ECOC2012*, Amsterdam, Th.3.C.1 (2012)

86. D. Qian, E. Ip, M.F. Huang, M. Li, A. Dogariu, S. Zhang, Y. Shao, Y.K. Huang, Y. Zhang, X. Cheng, Y. Tian, P. Ji, A. Collier, Y. Geng, J. Linares, C. Montero, V. Moreno, X. Prieto, T. Wang, 1.05Pb/s transmission with 109b/s/Hz spectral efficiency using hybrid single- and few-mode cores, in *FiO2012*, FW6C (2012)
87. T. Watanabe, Y. Kokubun, Ultra-large number of transmission channels in space division multiplexing using few-mode multi-core fiber with optimized air-hole-assisted double-cladding structure. Opt. Express **22**(7), 8309–8319 (2014). https://doi.org/10.1364/OE.22.008309
88. K. Tomozawa, Y. Kokubun, Maximum core capacity of heterogeneous uncoupled multi-core fibers, in *OECC2011*, Sapporo, 7C2-4 (2011)
89. Y. Mitsunaga, Y. Katsuyama, H. Kobayashi, Y. Ishida, Failure prediction for long length optical fiber based on proof testing. J. Appl. Phys. **53**(7), 4847–4853 (1982). https://doi.org/10.1063/1.331316
90. IEC Technical Report IEC 62048, Optical fibres—Reliability—Power law theory (2002)
91. M. Tachikura, Y. Kurosawa, Y. Namekawa, Improved theoretical estimation of mechanical reliability of optical fibers. Proc. SPIE **5623**, 622–629 (2005)
92. J. Yamamoto, T. Yajima, Y. Kinoshita, F. Ishii, M. Yoshida, T. Hirooka, M. Nakazawa, Fabrication of multi core fiber by using slurry casting method, in *The Optical Fiber Communication Conference*, Th1H.5 (2016)
93. S. Nozoe, R. Fukumoto, T. Sakamoto, T. Matsui, Y. Amma, K. Takenaga, K. Tsujikawa, S. Aozasa, K. Aikawa, K. Nakajima, Low crosstalk 125 μm-cladding multi-core fiber with limited air-holes fabricated with over-cladding bundled rods technique, in *The Optical Fiber Communication Conference*, Th1H.6 (2017)
94. S. Nozoe, T. Sakamoto, T. Matsui, Y. Amma, K. Takenaga, Y. Abe, K. Tsujikawa, S. Aozasa, K. Aikawa, K. Nakajima, 125 μm-cladding 2LP-mode and 4-core multi-core fibre with air-hole structure for low crosstalk in C+L Band, in *The European Conference on Optical Communication*, W.1.B.3 (2017)
95. T. Hayashi, T. Taru, O. Shimakawa, T. Sasaki, E. Sasaoka, Lowcrosstalk and low-loss multi-core fiber utilizing fiber bend, in *Optical Fiber Communication Conference*, OWJ3 (2011)
96. M. Ohashi, Y. Miyoshi, H. Kubota, R. Maruyama, N. Kuwaki, Longitudinal fiber parameter measurements of multi-core fiber using OTDR. Opt. Express **22**(24), 30137–30147 (2014). https://doi.org/10.1364/OE.22.030137
97. M. Ohashi, K. Takenaga, S. Matsuo, Y. Miyoshi, Simple technique for measuring cut-off wavelength of multi-core fiber (MCF) and Its definition, in *ACP2012*, AF3A4 (2012)
98. ITU-T Recommendation G. 650.1 (2004)
99. K. Kitayama, M. Ohashi, Y. Ishida, Length dependence of effective cutoff wavelength in single-mode fibers. IEEE J. Lightwave Technol. **LT-2**(5), 629–634 (1983)
100. M. Ohashi, K. Kitayama, T. Kobayashi, Y. Ishida, LP11 mode loss measurements in the two-mode-propagation region of optical fibers. Opt. Lett. **9**(7), 303–305 (1984). https://doi.org/10.1364/OL.9.000303
101. M. Ohashi, N. Shibata, K. Sato, Evaluation of length dependence of cutoff wavelength in a cabled fiber. Opt. Lett. **13**(12), 1123–1125 (1988). https://doi.org/10.1364/OL.13.001123
102. M.S. O'Sullivan, J. Ferner, Interpretation of SM fiber OTDR signatures, in *Proceedings of SPIE'86*, vol. 661, pp. 171–176 (1986)
103. A. Rossaro, M. Schiano, T. Tambosso, D. D'Alessandro, Spatially resolved chromatic dispersion measurement by a bidirectional OTDR technique. IEEE J. Select. Topics Quantum Electron **7**(3), 475–483 (2001). https://doi.org/10.1109/2944.962271
104. M. Ohashi, Novel technique for measuring relative-index difference of fiber links. IEEE Photon. Technol. Lett. **18**(24), 2584–2586 (2006). https://doi.org/10.1109/LPT.2006.887335
105. K. Tsujikawa, M. Ohashi, K. Shiraki, M. Tateda, Effect of thermal treatment on Rayleigh scattering in silica-based glasses. Electron. Lett. **31**, 1940–1941 (1995). https://doi.org/10.1049/el:19951331
106. C. Pask, Physical interpretation of Petermann's strange spot size for single-mode fibres. Electron. Lett. **20**(3), 144–145 (1985). https://doi.org/10.1049/el:19840097

107. N. Shibata, M. Kawachi, T. Edahiro, Optical loss characteristics of high-GeO_2 content silica fibers. IEICE Trans. **E63**(12), 837–841 (1980)
108. K. Nakajima, M. Ohashi, M. Tateda, Chromatic dispersion distribution measurement along a single-mode optical fiber. IEEE J. Lightwave Technol. **15**(7), 1095–1101 (1997). https://doi.org/10.1109/50.596954
109. H. R. Stuart, Dispersive multiplexing in multimode optical fiber. Science **289**(5477), 281–283 (2000). https://doi.org/10.1126/science.289.5477.281
110. A.R. Shah, R.C.J. Hsu, A. Tarighat, A.H. Sayed, B. Jalali, Coherent optical MIMO (COMIMO). IEEE J. Lightwave Technol. **23**(8), 2410–2419 (2005). https://doi.org/10.1109/JLT.2005.850787
111. M. Kasahara, K. Saitoh, T. Sakamoto, N. Hanzawa, T. Matsui, K. Tsujikawa, F. Yamamoto, M. Koshiba, Design of few-mode fibers for mode-division multiplexing transmission. IEEE Photon. J. **5**(6), 7201207–7201207 (2013). https://doi.org/10.1109/JPHOT.2013.2292365
112. T. Sakamoto, T. Mori, T. Yamamoto, N. Hanzawa, S. Tomita, F. Yamamoto, K. Saitoh, M. Koshiba, Mode-division multiplexing transmission system with DMD-independent low complexity MIMO processing. IEEE J. Lightw. Technol. **31**(13), 2192–2199 (2013). https://doi.org/10.1109/JLT.2013.2263495
113. T. Mori, T. Sakamoto, M. Wada, T. Yamamoto, F. Yamamoto, Experimental evaluation of modal crosstalk in two-mode fibre and its impact on optical MIMO transmission, in *ECOC2014*, Th.1.4.4 (2014)
114. R. Maruyama, N. Kuwaki, S. Matsuo, M. Ohashi, Experimental investigation of relation between mode-coupling and fiber characteristics in few-mode fibers, in *OFC2015*, M2C.1 (2015)
115. Y. Kokubun, T. Watanabe, S. Miura, R. Kawata, What is a mode in few mode fibers? Proposal of MIMO-free mode division multiplexing using true eigenmodes. IEICE Electron. Express **13**(18), 20160394 (2016). https://doi.org/10.1587/elex.13.20160394
116. S. Chen, J. Wang, PANDA-type elliptical-core multi-mode fiber with fully lifted eigenmodes for low-crosstalk direct fiber vector eigenmode space-division multiplexing, in *The Optical Fiber Communication Conference*, W4K.4 (2018)
117. S. Chen, J. Wang, Design of PANDA-type elliptical-core multimode fiber supporting 24 fully lifted eigenmodes. Opt. Lett. **43**(15), 3718–3721 (2018). https://doi.org/10.1364/OL.43.003718
118. T. Sakamoto, T. Mori, T. Yamamoto, S. Tomita, Differential mode delay managed transmission line for WDM-MIMO system using multi-step index fiber. IEEE J. Lightw. Technol. **30**(17), 2783–2787 (2012). https://doi.org/10.1109/JLT.2012.2208095
119. L.G-Nielsen, Y. Sun, J.W. Nicholson, D. Jakobsen, R. Lingle, Jr., B. Pálsdóttir, Few mode transmission fiber with low dgd, low mode coupling, and low loss, in *OFC2012*, PDP5A.1 (2012)
120. T. Mori, T. Sakamoto, M. Wada, T. Yamamoto, F. Yamamoto, Six-LP-mode transmission fiber with DMD of less than 70 ps/km over C+L band, in *OFC2014*, paper M3F.3 (2014)
121. P. Sillard, D. Molin, M. BigotAstruc, K. de Jongh, F. Achten, Low-differential-mode-group-delay 9-LP-mode fiber, in *The Optical Fiber Communication Conference*, Los Angeles, CA, USA, paper M2C.2 (2015)
122. F. Feng, G.S.D. Gordon, X.Q. Jin, D.C. O'Brien, F.P. Payne, Y. Jung, Q. Kang, J.K. Sahu, S.U. Alam, D.J. Richardson, T.D. Wilkinson, Experimental characterization of a graded-index ring-core fiber supporting 7 LP mode groups, in *OFC2015*, p. Tu2D.3 (2015)
123. R. Ryf, N.K. Fontaine, M. Montoliu, S. Randel, S.H. Chang, H. Chen, S. Chandrasekhar, A.H. Gnauck, R.-J. Essiambre, P.J. Winzer, T. Taru, T. Hayashi, T. Sasaki, Space-division multiplexed transmission over 3×3 Coupled-Core Multicore Fiber, in *OFC2014*, paperTu2J.4 (2014)
124. T. Hayashi, Y. Tamura, T. Hasegawa, T. Taru, 125-μm-cladding coupled multi-core fiber with ultra-low loss of 0.158 dB/km and record-low spatial mode dispersion of 6.1 ps/km1/2, in *The Optical Fiber Communication Conference*, Th5A.1 (2016)

125. T. Hayashi, Y. Tamura, T. Hasegawa, T. Taru, Record-low spatial mode dispersion and ultra-low loss coupled multi-core fiber for ultra-long-haul transmission. IEEE J. Lightw. Technol. **35**(3,) 450–457 (2016). https://doi.org/10.1109/JLT.2016.2614000
126. T. Sakamoto, S. Aozasa, T. Mori, M. Wada, T. Yamamoto, S. Nozoe, Y. Sagae, K. Tsujikawa, K. Nakajima, Randomly-coupled single-mode 12-core fiber with highest core density, in *The Optical Fiber Communication Conference* Th1H.1 (2017)
127. T. Sakamoto, S. Aozasa, T. Mori, M. Wada, T.Yamamoto, S. Nozoe, Y. Sagae, K. Tsujikawa, K. Nakajima, Twisting-rate-controlled 125 μm cladding randomly coupled single-mode 12-core fiber. IEEE J. Lightw. Technol. **36**(2), 325–330 (2018). https://doi.org/10.1109/JLT.2017.2743205
128. N.K. Fontaine, R. Ryf, M. Hirano, T. Sasaki, Experimental investigation of crosstalk accumulation in a ring-core fiber, in *IEEE summer topical meeting*, TuC4.2 (2013)
129. T. Mori, T. Sakamoto, M. Wada. T. Yamamoto, L. Ma, N. Hanzawa, K. Tsujikawa, S. Tomita, Few-mode photonic crystal fibre for wideband mode division multiplexing transmission, in *ECOC2012*, Tu.1.F.4 (2012)
130. M.N. Petrovich, F. Poletti, J.P. Wooler, A.M. Heidt, N.K. Baddela, Z. Li, D.R. Gray, R. Slavík, F. Parmigiani, N.V. Wheeler, J.R. Hayes, E. Numkam, L. Grűner-Nielsen, B. Pálsdóttir, R. Phelan, B. Kelly, M. Becker, N. MacSuibhne, J. Zhao, F.C. Garcia Gunning, A.D. Ellis, P. Petropoulos, S.U. Alam, D.J. Richardson, First demonstration of 2 mm data transmission in a low-loss hollow core photonic bandgap fiber, in *ECOC2012*, Th.3.A.5 (2012)
131. Y. Jung, V.A.J.M. Sleiffer, N.K. Baddela, M.N. Petrovich, J.R. Hayes, N.V. Wheeler, D.R. Gray, E. Numkam Fokoua, J.P. Wooler, N.H.-L. Wong, F. Parmigiani, S.U. Alam, J. Surof, M. Kuschnerov, V. Veljanovski, H.de Waardt, F. Poletti, D.J. Richardson, First demonstration of a broadband 37-cell hollow core photonic bandgap fiber and its application to high capacity mode division multiplexing, in *OFC2013*, PDP5A.3 (2013)
132. R.E. Freund, C.-A. Bunge, N.N. Ledentsov, D. Molin, and C. Caspar, High-speed transmission in multimode fibers. IEEE J. Lightw. Technol. **28**(4), 569–586 (2010). https://doi.org/10.1109/JLT.2009.2030897
133. T. Mori, T. Sakamoto, M. Wada, T. Yamamoto, F. Yamamoto, Few-mode fibers supporting more than two LP modes for mode-division-multiplexed transmission with MIMO DSP. IEEE J. Lightw. Technol. **32**, 2468–2479 (2014). https://doi.org/10.1109/JLT.2014.2327619
134. K. Sato, R. Maruyama, N. Kuwaki, S. Matsuo, M. Ohashi, Optimized graded index two-mode optical fiber with low DMD, large Aeff and low bending loss. Opt. Express, **21**(14), 16231–16238 (2013). https://doi.org/10.1364/OE.21.016231
135. T. Mori, T. Sakamoto, M. Wada, T. Yamamoto, F. Yamamoto, Low DMD four LP mode transmission fiber for wide-band WDM-MIMO system, in *The Optical Fiber Communications*, Anaheim, CA, 2013, OTh3K.1.R (2013)
136. P. Sillard, D. Molin, A review of few-mode fibers for space-division multiplexed transmissions, in *The 39th European Conference and Exhibition on Optical Communication*, London, U.K., Mo.3.A.1 (2013)
137. F.M. Ferreira, D. Fonseca, H.J.A. da Silva, Design of few-mode fiberswith M-modes and lowdifferential mode delay. IEEE J. Lightw. Technol. **32**(3), 353–360 (2014). https://doi.org/10.1109/JLT.2013.2293066
138. P. Sillard, D. Molin, M. Bigot-Astruc, H. Maerten, D. Van Ras, F. Achten, Low-DMGD 6-LP-mode fiber, in *The Optical Fiber Communication Conference*, San Francisco, CA, p. M3F.2 (2014)
139. D. Marcuse, Pulse propagation in a two-mode waveguide. J. Bell Syst. Tech. **51**, 1785–1791 (1972). https://doi.org/10.1002/j.1538-7305.1972.tb02683.x
140. S. Kawakami, M. Ikeda, Transmission characteristics of a two-mode optical waveguide. IEEE J. Quantum Electron. **QE-14**(8), 608–614 (1978)
141. R. Maruyama, N. Kuwaki, S. Matsuo, K. Sato, M. Ohashi, Experimental evaluation of mode conversion ration at splice point for two-mode fibers and its simulated effect on MIMO transmission, in *OFC 2014*, M3F.6 (2014)

142. M. Nakazawa, M. Yoshida, T. Hirooka, Nondestructive measurement of mode couplings along a multi-core fiber using a synchronous multi-channel OTDR. Opt. Express **20**(11), 12530–12540 (2012). https://doi.org/10.1364/OE.20.012530
143. M. Nakazawa, M. Yoshida, T. Hirooka, Measurement of mode coupling distribution along a few-mode fiber using a synchronous multi-channel OTDR. Opt. Express **22**(25), 31299–31309 (2014). https://doi.org/10.1364/OE.22.031299
144. M. Nakazawa, M. Tokuda, Y. Negishi, Measurement of polarization mode coupling along a polarization-maintaining optical fiber using a backscattering technique. Opt. Lett. **8**(10), 546–548 (1983). https://doi.org/10.1364/OL.8.000546
145. M. Nakazawa, N. Shibata, M. Tokuda, Y. Negishi, Measurements of polarization mode couplings along polarization-maintaining single-mode optical fibers. J. Opt. Soc. Am. A **1**(3), 285–292 (1984). https://doi.org/10.1364/JtheopticalsoA.1.000285
146. K. Jespersen, Z.L.L. Grüner-Nielsen, B. Pálsdóttir, F. Poletti, J.W. Nicholson, Measuring distributed mode scattering in long, few-moded fibers, in *OFC2012*, p. OTh3I.4 (2012)
147. N.K. Fontaine, R. Ryf, M.A. Mestre, B. Guan, X. Palou, S. Randel, Y. Sun, L.G-Nielsen, R.V. Jensen, R. Lingle, Jr., Characterization of space-division multiplexing systems using a swept-wavelength interferometer, in *OFC2013*, OW1K.2 (2013)
148. R. Ryf, S. Randel, A.H. Gnauck, C. Bolle, A. Sierra, S. Mumtaz, M. Esmaeelpour, E.C. Burrows, R.-J Essiambre, P.J. Winzer, D.W. Peckham, A.H. McCurdy, R. Lingle, Mode-division multiplexing over 96 km of few-mode fiber using coherent 6x6 MIMO processing. IEEE J. Lightwave Technol. **30**(24), 521–531 (2012). https://doi.org/10.1109/JLT.2011.2174336
149. R. Kawata, T. Watanabe, Y. Kokubun, Full-set high-speed mode analysis in few-mode fibers by polarization-split segmented coherent detection method: Proposal and simulation of calculation error. IEICE Electron. Express **15**(1), 20171132 (2018). https://doi.org/10.1587/elex.14.20171132
150. Y. Sasaki, K. Takenaga, N. Guan, S. Matsuo, K. Saitoh, M. Koshiba, Large-effective-area uncoupled few-mode multi-core fiber. Opt. Express **20**(26), B77-84 (2012). https://doi.org/10.1364/OE.20.000B77
151. K. Mukasa, K. Imamura, R. Sugizaki, 7-core 2-mode fibers with large Aeff to simultaneously realize 3M, in *The OptoElectronics and Communications Conference*, Busan, Korea, 5C1-1 (2012)
152. K. Mukasa, K Imamura, R. Sugizaki, Multi-core few-mode optical fibers with large Aeff, in *The European Conference on Optical Communication*, P1.08 (Amsterdam, Netherlands, 2012)
153. T. Sakamoto, T. Mori, T. Yamamoto, M. Wada, F. Yamamoto, Moderately coupled 125-μm cladding 2 LP-mode 6-core fiber for realizing low MIMO-DSP and high spatial density, in *The European Conference on Optical Communication*, Cannes, France, Tu4.1.3 (2014)
154. T. Sakamoto, T. Matsui, K. Saitoh, S. Saitoh, K. Takenaga, S. Matsuo, Y. Tobita, N. Hanzawa, K. Nakajima, F. Yamamoto, Few-mode multi-core fibre with highest core multiplicity factor. in *The European Conference on Optical Communication*, Valencia, Spain, We.1.4.3 (2015)
155. K. Shibahara, T. Mizuno, H. Takara, A. Sano, H. Kawakami, D. Lee, Y. Miyamoto, H. Ono, M. Oguma, Y. Abe, T. Kobayashi, T. Matsui, R. Fukumoto, Y. Amma, T. Hosokawa, S. Matsuo, K. Saito, H. Nasu, T. Morioka, Dense SDM (12-core × 3-mode) transmission over 527 km with 33.2-ns mode-dispersion employing low-complexity parallel MIMO frequency-domain equalization, in *The Optical Fiber Communication Conference*, Los Angeles, CA, USA, Th5C.3 (2015)
156. T. Mizuno, T. Kobayashi, H. Takara, A. Sano, H. Kawakami, T. Nakagawa, Y. Miyamoto, Y. Abe, T. Goh, M. Oguma, T. Sakamoto, Y. Sasaki, I. Ishida, K. Takenaga, S. Matsuo, K. Saitoh, and T. Morioka, 12-core × 3-mode dense space division multiplexed transmission over 40 km employing multi-carrier signals with parallel MIMO equalization, in *The Optical Fiber Communication Conference*, San Francisco, CA, USA, Th5B.2 (2014)
157. S. Matsuo, K. Takenaga, K. Saitoh, K. Nakajima, Y. Miyamoto, T. Morioka, High-spatial-multiplicity multi-core fibres for future dense space-division-multiplexing system, in *The European Conference on Optical Communication*, Valencia, Spain., Th.1.2.1 (2015)

158. J. Sakaguchi, W. Klaus, J.-M. D. Mendinueta, B.J. Puttnam1, R.S. Luis, Y. Awaji1, N. Wada, T. Hayashi, T. Nakanishi, T. Watanabe, Y. Kokubun, T. Takahata, T. Kobayashi, Realizing a 36-core, 3-mode fiber with 108 spatial channels, in *The Optical Fiber Communication Conference*, Los Angeles, CA, USA, p. Th5C.2 (2015)
159. K. Igarashi, D. Souma, Y. Wakayama, K. Takeshima, Y. Kawaguchi, T. Tsuritani, I. Morita, M. Suzuki, 114 space-division-multiplexed transmission over 9.8-km weakly-coupled-6-mode uncoupled-19-core fibers, in *The Optical Fiber Communication Conf.*, Los Angeles, CA, USA, Th5C.4 (2015)
160. T. Sakamoto, T. Matsui, K. Saitoh, S. Saitoh, K. Takenaga, T. Mizuno, Y. Abe, K. Shibahara, Y. Tobita, S. Matsuo, K. Aikawa, S. Aozasa, K. Nakajima, Y. Miyamoto, Low-loss and low-DMD few-mode multi-core fiber with highest core multiplicity factor, in *The Optical Fiber Communication Conference*, Th5A.2 (2016)
161. T. Sakamoto, K. Saitoh, S. Saitoh, Y. Abe, K. Takenaga, A. Urushibara, M. Wada, T. Matsui, K. Aikawa, K. Nakajima, 120 Spatial channel few-mode multi-core fibre with relative core multiplicity factor exceeding 100, in *The European Conference on Optical Communication* , We3E.3 (2018)
162. D. Soma, Y. Wakayama, S. Beppu, S. Sumita, T. Tsuritani, T. Hayashi, T. Nagashima, M. Suzuki, M. Yoshida, K. Kasai, M. Nakazawa, H. Takahashi, K. Igarashi, I. Morita, M. Suzuki, 10.16-Peta-B/s dense SDM/WDM transmission over 6-mode 19-core fiber across the C+L Band. J. Lightw. Technol. **36**(6), 1362–1368 (2018). https://doi.org/10.1109/JLT.2018.2799380
163. J. Tu, K. Saitoh, K. Takenaga, S. Matsuo, Heterogeneous trench-assisted few-mode multi-core fiber with low differential mode delay. Opt. Express **22**(4), 4329–4341 (2014). https://doi.org/10.1364/OE.22.004329
164. J. Tu, K. Saitoh, K. Takenaga, S. Matsuo, Heterogeneous trench-assisted few-mode multi-core fiber with graded-index profile and square-lattice layout for low differential mode delay. Opt. Express **23**(14), 17783–17792 (2015). https://doi.org/10.1364/OE.23.017783
165. Y. Sasaki, Y. Amma, K. Takenaga, S. Matsuo, K. Saitoh, M. Koshiba, Few-mode multicore fiber with 36 spatial modes (three modes (LP01 , LP 11a, LP11b) × 12 Cores). IEEE J. Lightw. Technol. **33**(5), 964–970 (2015). https://doi.org/10.1109/JLT.2014.2375876
166. T. Sakamoto, T. Mori, M. Wada, T. Yamamoto, F. Yamamoto, Coupled multicore fiber design with low intercore differential mode delay for high-density space division multiplexing. IEEE J. Lightw. Technol. **33**(6), 1175–1181(2015). https://doi.org/10.1109/JLT.2014.2376526
167. T. Sakamoto, T. Mori, M. Wada, T. Yamamoto, T. Matsui, K. Nakajima, F. Yamamoto, Experimental and numerical evaluation of inter-core differential mode delay characteristic of weakly-coupled multi-core fiber. Opt. Express **22**(26), 31966–31976 (2014). https://doi.org/10.1364/OE.22.031966
168. Y. Sasaki, Y. Amma, K. Takenaga, S. Matsuo, K. Saitoh, M. Koshiba, Trench-assisted low-crosstalk few-mode multicore fiber, in *ECOC2013*, Mo.3.A.5 (2013)
169. K. Hogari, Y. Yamada, K. Toge, Novel optical fiber cables with ultrahigh density. IEEE J. Lightwave Technol. **26**(17), 3104–3109 (2008). https://doi.org/10.1109/JLT.2008.926931
170. K. Watanabe, T. Saito, K. Imamura, Y. Nakayama, and M. Shiino, Study of fusion splice for single-mode multicore fiber. in MOC 2011, Vol. 17, H-8 (2011)
171. K. Yoshida, A. Takahashi, T. Konuma, K. Yoshida, K. Sasaki, Fusion splicer for specialty optical fiber with advanced functions. Fujikura Techn. Rev. **41**, 10–13 (2012)
172. M. Tanaka, M. Hachiwaka, Y. Fujimaki, H. Taniguchi, Butt joint of hexagonal cladding multi-core fiber (in Japanese). *IEICE Technical Report* (in Japanese), OFT2012-68, 77-82 (2013)
173. K. Hogari, Y. Yamada, K. Toge, Design and performance of ultra-high-density optical fiber cable with rollable optical fiber ribbons. OFT **16**(4), 257–263 (2010). https://doi.org/10.1016/j.yofte.2010.05.00
174. M. Yamanaka, K. Osato, K. Tomikawa, D. Takeda, Ultra-high density optical fiber cable with Spider Web Ribbon, in *61st International Wire & Cable Symposium* (2012), pp. 37–41

175. S. Inao, T. Sato, H. Honda, M. Ogai, S. Sentsui, A. Otake, K. Yoshizaki, K. Ishihara, N. Uchida., High density multicore-fiber cable, in *28th International Wire & Cable Symposium* (1979), pp. 370–384
176. I. Ishida, Y. Amma, K. Hirakawa, H. Uemura, Y. Sasaki, K. Takenaga, N. Itou, K. Osato, S. Matsuo, Multicore-fiber cable with core density of 6 cores/mm^2, in *OFC2014*, W4D.3 (2014)
177. T. Kobayashi, H. Takara, A. Sano, T. Mizuno, H. Kawakami, Y. Miyamoto, K. HIraga, Y. Abe,. H. Ono. M.Wada, Y. Sasaki, I. Ishida, K. Takenaga, S. Matsuo, K. Saitoh, Y. Yamada, M. Masuda, T. Morioka, 2×344 Tb/s propagateio-direction interleaved transmission over 1500-km MCF enhanced by multicarrier full electric-fielda digital back-propagation, in *ECOC2013*, PD3.E.4 (2013)
178. K. Takenaga, Multicore Fiber with Dual-Ring Structure. in *OECC/ACOFT 2014*, MO1E-5 (2014)
179. K. Osato, Y. Hashimoto, N. Okada, New design of optical fiber cable for easy mid-span access, in *56th International Wire & Cable Symposium* (2007), pp. 225–229
180. I. Ishida, S. Matsuo, Ultra-high core-density cable with multicore fiber. OECC/ACOF **2015**, 875–877 (2014)
181. Y. Sasaki, Y. Amma, K. Takenaga, S. Matsuo, K. Saitoh, and M. Koshiba, Few-mode multicore fibre with 36 spatial modes (three modes (LP01, LP11a, LP11b) ×12 cores), in *ECOC2014*, Th.1.4.1 (2014)
182. T. Hayashi, T. Nagashima, T. Muramoto, F. Sato, T. Nakanishi, Spatial mode dispersion suppressed randomly-coupled multi-core fiber in straightened loose-tube cable, in *The Optical Fiber Communication Conference*, Th4A.2 (2019)
183. T. Tsuritani, D. Soma, Y. Wakayama, Y. Miyagawa, M. Takahashi, I. Morita, K. Maeda, K. Kawasaki, T. Matsuura, M. Tsukamoto, R. Sugizaki, Field test of installed high-density optical fiber cable with multi-core fibers toward practical deployment, in *The Optical Fiber Communication Conference*, M3J.4 (2019)
184. K. Nakajima, T. Matsui, k. Saito, T. Sakamoto, N. Araki, Multi-core fiber technology: Next generation optical communication strategy. IEEE Commun. Std. Mag. 38–45
185. S. Jain, T.C. May-Smith, V.J.F. Rancano, P. Petropoulos, D.J. Richardson, J.K. Sahu, Multi-element fiber for space-division multiplexed transmission, in *ECOC*, Mo.4.A.2 (2013)
186. T. Matsui, T. Kobayashi, H. Kawahara, E.L.T. de Gabory, T. Nagashima, T. Nakanishi, S. Saitoh, Y. Amma, K. Maeda, S. Arai, R. Nagase, Y. Albe, S. Aozasa, Y. Wakayama, H. Takeshita, T. Tsuritani, H. Ono, T. Sakamoto, I. Morita, Y. Miyamoto, K. Nakajima,118.5 Tbit/s transmission over 316 km-long multi-core fiber with standard cladding diameter, in *OECC/PGC*, PDP2 (2017)
187. R. Nagase, K. Sakaime, K. Watanabe, T. Saito, MU-type multi-core fiber connector. IWCS **61**, 823–827 (2012)
188. K. Watanabe, T. Tsunetoshi, K. Suematsu, R. Nagase, M. Shiino, Development of small MT type 2-multicore fiber connector, in *OFC*2014, W4D.6 (2014)
189. K. Shikama, S. Asakawa, Y. Abe, T. Takahashi, Multiple multicore fiber connector with physical-contact connection. Electron. Lett. **51**(10), 775–777 (2015). https://doi.org/10.1049/el.2015.0624
190. Y. Geng, S. Li, M. J. Li, C.G. Sutton, R.L. MCCollum, R. L. McClure, A.V. Koklyushkin, K.I. Matthews, J.P. Luther, D. L. Butler, High-speed, bi-directional dual-core fiber transmission system for high-density, short-reach optical interconnects. Proc. SPIE **9390** (2015)
191. K. Saito, T. Matsui, K. Nakajima, T. Sakamoto, F. Yamamoto, T. Kurashima, Multi-core fiber based pluggable add/drop link using rotational connector, in *OFC2015*, M2B2 (2015)
192. ITU-T Recommendation G.652, Characteristics of a single-mode optical fibre and cable (2009)
193. ITU-T Recommendation G.657, Characteristics of a bending-loss insensitive single-mode optical fibre and cable for the access network (2012)
194. J.P. Moore, M.D. Rogge, Shape sensing using multi-core fiber optic cable and para-metric curve solutions. Opt. Express **20**(3), 2967–2973 (2012)
195. Fiber optic shape sensing, Luna Innovations Inc., Rev. 3, (2013)

196. Q. Wang, G. Farrell, All-fiber multimode-interference-based refractometer sensor: proposal and design. Opt. Lett. **31**(3), 317–319 (2006). https://doi.org/10.1364/OL.31.000317
197. ITU-T Supplement G.Sup.40, Optical fibre and cable Recommendations and standards guideline. (2010)
198. http://www.itu.int/en/ITU-T/studygroups/2013-2016/15/Pages/q5.aspx.

Chapter 3
Optical Connection Technologies

Ryo Nagase, Kiichi Hamamoto, Yoshiteru Abe, Yasuo Kokubun, and Shoichiro Matsuo

Abstract Optical connection technologies are essential for constructing optical networks. This chapter describes recent progress in splicing technologies, optical connectors, fan-in/fan-out devices and mode multiplexing/demultiplexing technologies for space-division multiplexing optical fibers.

3.1 Introduction

Optical telecommunication networks consist of switching systems, transmission systems and optical fiber cables in a general way. In the central office, various kinds of equipment are installed such as switches, routers, transmission equipment, optical distribution frames, optical line terminals as shown in Fig. 3.1. Huge amount of optical fiber cables with connectors is used for all signal connections between equipment. Long-distance optical fiber cables are installed in service tunnels using splicing technologies and also terminated to the fiber termination modules (FTMs) in central offices. Some long-distance cables are connected to equipment in the central office, while other cables are connected each other at the FTM to transmit signals thorough the central office.

Ryo Nagase and Kiichi Hamamoto are chapter editors.

R. Nagase (✉)
Chiba Institute of Technology, Narashino, Chiba, Japan
e-mail: ryo.nagase@it-chiba.ac.jp

K. Hamamoto
Kyushu University, Kasuga, Fukuoka, Japan

Y. Abe
NTT Access Network Service Systems Laboratories, NTT Corporation, Tsukuba, Ibaraki, Japan

Y. Kokubun
Institute of Technologists (Formerly with Yokohama National University), Gyoda, Saitama, Japan

S. Matsuo
Fujikura Ltd., Sakura, Chiba, Japan

M. Nakazawa et al. (eds.), *Space-Division Multiplexing in Optical Communication Systems*, Springer Series in Optical Sciences 236,
https://doi.org/10.1007/978-3-030-87619-7_3

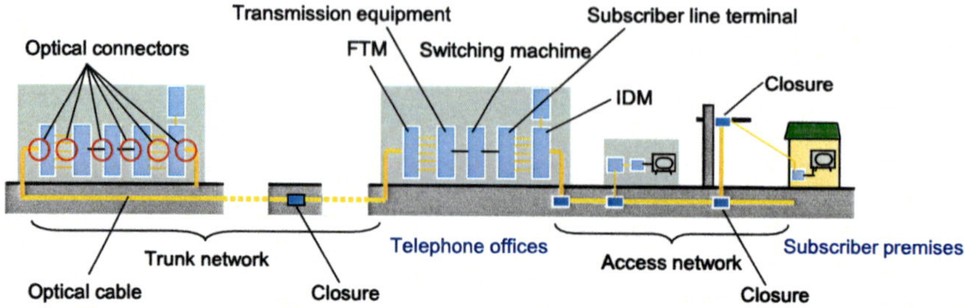

Fig. 3.1 Optical network systems

Optical connection technologies, splicing and connectors are essential for constructing optical networks. In this chapter, optical fiber splicing and connector technologies' fundamentals and applications for multi-core fibers are described.

3.2 Fusion Splicing Technology

Fusion splicing is an indispensable technology to realize permanent splice of optical fibers. Many kinds of fusion splicers for single-core fibers have been commercialized: very compact splicers for field operation, splicers for ribbon fibers and multi-functional splicers for specialty fibers. At present, single-core fibers such as conventional single-mode fibers, polarization-maintaining fibers (PM fibers) and fibers with a large cladding diameter fiber are able to automatically splice with low splice loss.

MCFs have outer cores, which locate away from the center of a cladding. The outer cores cause two issues for splicing MCFs with low attenuation. The first issue is precise alignment of outer cores including makers. The second issue is uniform heating over outer cores. In the followed section, the current solutions for the issues are presented.

3.2.1 Outer Core Alignment

Two technologies are required to realize precise alignment of outer cores. The first is fiber rotation mechanism of alignment stage. Conventional splices for single-core fibers have stages that move *x*-direction and *y*-direction. To align outer cores in a MCF, fiber rotation, that is, θ direction alignment is required. This mechanism has been equipped in fusion splices for PM fibers.

The second is core detection method for multiple cores in a cladding. In addition, core number should be identified through monitoring maker cores. Figure 3.2 summarizes core alignment method presented for single-core fibers [1–3]. The remote injection and detection system (RIDS) is a direct method to realize precise alignment for

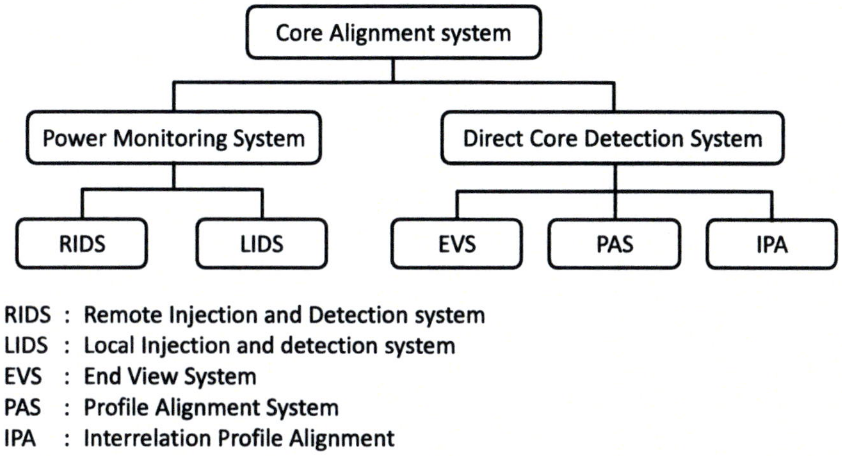

Fig. 3.2 Proposed core alignment system

many kinds of fiber. This method aligns cores to detect optical power at a fiber end which power is injected at another end. A drawback of this method is necessity of operation at both ends of spliced fibers. Accordingly, application of this method is limited to splice of short-length fibers or in laboratories. The local injection and detection system (LIDS) aimed to overcome the drawback of RIDS. The RIDS realized power injection and detection through bending portion of fibers within a fusion splicer. It is very difficult to apply the RIDS to dark-colored fiber and bending loss insensitive fiber. The end-view system (EVS) is developed for aligning PM fibers. Figure 3.3 shows a schematic of end-view function [2]. The light for alignment is introduced through fiber clamp equipment. The image of a fiber end is monitored with a CCD camera through a mirror and lens. The image of both ends of fibers to be spliced can be monitored by changing the direction of a mirror. Core alignment is realized through image processing of the end images. The side-view method is

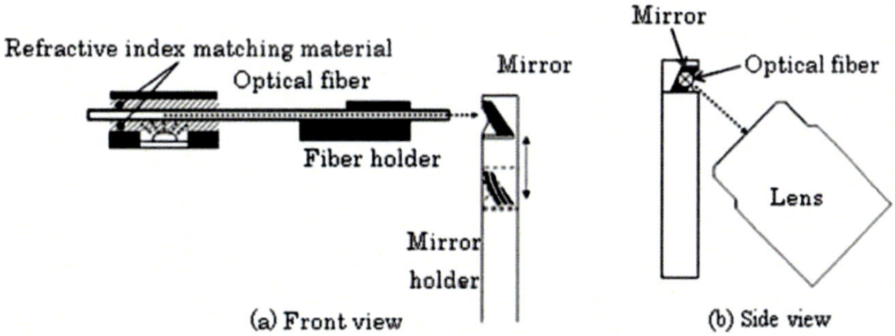

Fig. 3.3 Schematics of end-view system (Reprinted with permission from [2]. © 2012 Fujikura Ltd.)

the most popular alignment method for conventional splicers. The profile alignment system (PAS) identifies a core position by monitoring intensity profile of transmitting light trough an optical fiber. The intensity profile of the PM fibers is more complex than that of the conventional single-core fiber. The PAS enables the core alignment for the most PM fibers through optimizing the position of the focal plane in a cladding. However, a low-contrast PM fiber lacks sufficient refractive index contrast that is to be aligned by the PAS. Interrelation profile analysis (IPA) has been proposed to cope with the splice for the low-contrast PM fiber [3].

At the early stage of the MCF development, the RIDS has been utilized. Recently, automated alignment by the combination of IPA and RIDS has been proposed [4]. The IPA was utilized for rough alignment. Precise core alignment was achieved with the RIDS. Maximum splice loss of side cores was reduced from 1.0 to 0.45 dB by using the precise alignment with the RIDS. Unfortunately, the detection of maker cores has not been realized with the RIDS and IPA.

Automated core alignments including the detection of marker cores have been realized with the EVS [5, 6] of the fusion splicer for specialty fibers [2]. The drawback of the EVS is complex mechanism compared to the side-view system. The development of simplified alignment system is expected for extending the application of MCFs.

3.2.2 *Uniform Heating*

Conventional fusion splicers have two fixed electrodes for discharging. Discharge conditions are optimized through adjusting parameters such as discharge power, discharge time and electrode gap. If optical fibers with a cladding diameter larger 500 μm are spliced with the splices with fixed electrodes, it is difficult to realize optimal heating condition with large discharge path. If the electrode gaps broaden for enlarging discharge path, discharge become unstable and the tip of electrodes are damaged significantly. The swing electrodes system has been proposed to overcome the issue for splicing with a large cladding diameter [2].

The cladding diameter of MCFs is generally about 200 μm. The offset of outer cores tends to be larger than 40 μm from the center of a cladding. Accordingly, wide discharge path is indispensable for uniform heating of all cores. The swing electrode system is effective for splicing MCFs [5, 6]. Figure 3.4 shows a comparison of splice loss with the fixed electrode system and the swing electrode system [5]. Figure 3.4a shows histograms. A seven-core fiber shown as an inset was used for this experiment. Core pitch was 43 μm, and a cladding diameter was 180 μm. A_{eff} was about 110 μm^2 at 1550 nm. Cores were automatically aligned with the EVS. Averaged splice loss is reduced from 0.15 to 0.07 dB by utilizing the swing electrode system. Figure 3.4b shows radar chart of splice losses with both electrode systems. The variations of splice losses that depend on core position were suppressed by using the swing electrode system.

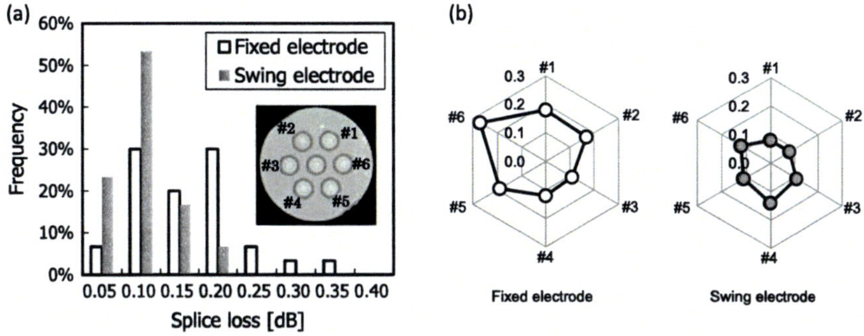

Fig. 3.4 Measured splice loss of a seven-core fiber: **a** histogram of splice loss. **b** splices of outer cores with different electrode system (Reprinted with permission from [5]. © 2013 IEEE)

Splice losses of MCFs strongly depend on a distance from the center of cladding. Figure 3.5 illustrates the effect of the distance on splice losses [6]. If the center of spliced fibers is perfectly aligned, an offset d is determined by a distance from the combination of center (R_{in}, R_{out}) and misalignment of rotational direction (θ_{offset}). If $R_{\text{in}} = 44.5\ \mu\text{m}$, $R_{\text{out}} = 76.9\ \mu\text{m}$ and $w = 5.6\ \mu\text{m}$, the splice loss of outer cores is three times larger than that of inner cores. Figure 3.6 experimentally demonstrated the core position dependence of splice losses [6]. A 12-core fiber with a dual-ring structure was utilized in the experiment. Automated core alignment with the EVS was also used. MFD at 1550 nm was 11.7 μm. Inner cores and outer cores were positioned at R_{in} of 44.5 μm and R_{out} of 76.9 μm, respectively. A cladding diameter was 230 μm. The histogram of splice losses clearly shows the position dependence of splice loss. Average splices loss at 1550 nm was 0.07 dB for inner cores and 0.15 dB for outer cores.

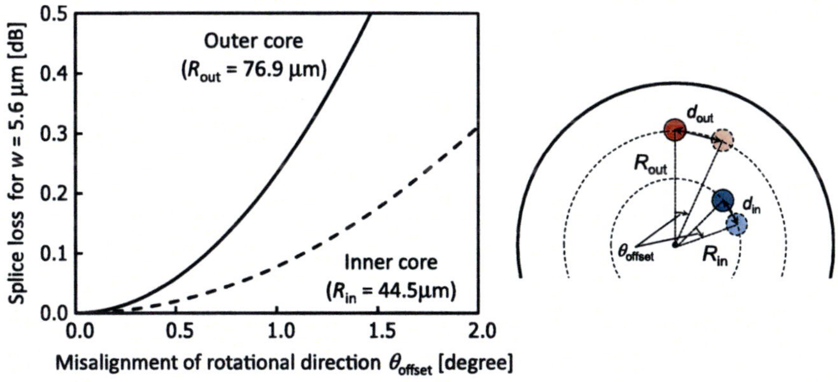

Fig. 3.5 Splice loss as a function of misalignment of rotational direction for different core position (Reprinted with permission from [6]. © 2014 IEEE)

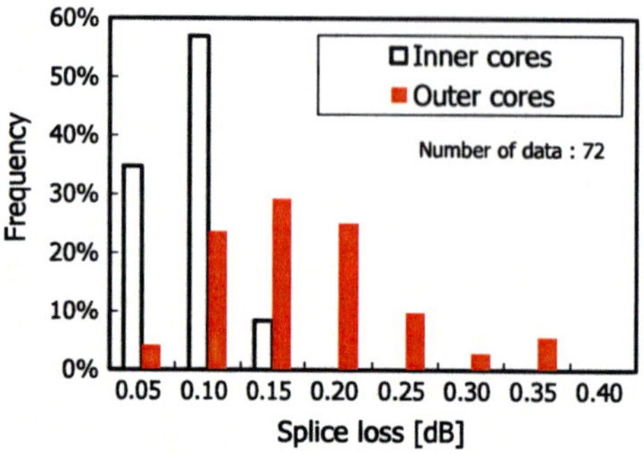

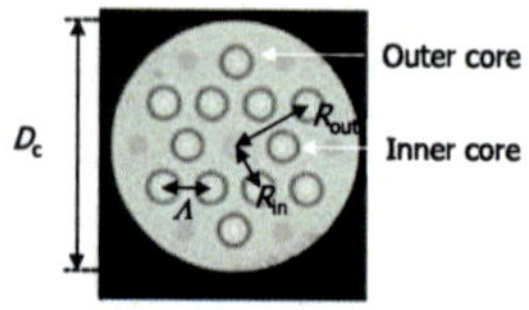

	Splice loss at 1550 nm[dB]		
	Ave.	Max.	Min.
Inner cores	0.07	0.14	0.01
Outer cores	0.15	0.32	0.03

Fig. 3.6 Splice loss of 12-core fiber with a dual-ring structure (Reprinted with permission from [6]. © 2014 IEEE)

Further improvement on alignment accuracy and discharge condition is required for achieving the low splice loss of MCFs. The realization of the low splice loss is projected to be hard if cores have large distance from the center, which is required to realize a high core count fiber with low crosstalk and result in a large cladding diameter. The splicing issue seems to be a limiting factor of the cladding diameter.

3.3 Optical Connectors

Optical connectors are essential for constructing optical networks. They are unique in optical components in that they require resistance from an external force. They should usually allow some component deformation of around tens of microns, and so, accuracy of better than 1 μm is required. Optical connectors used in telecommunication systems have a "floating mechanism" to eliminate the influence of deformation on connection stability. The design difficulty of MCF connector is the precise alignment of each core under a floating mechanism. In this chapter, the design principle and connection characteristics examples of MCF connectors attached to seven-core single-mode fiber are described.

3.3.1 Fundamentals of Optical Connectors

Single-mode optical connectors require an attenuation of less than 0.5 dB and a return loss of 40 dB or more. To realize these requirements with the butt joint mechanism, the fiber core offset should be less than 1 μm. There are four conditions for achieving this requirement.

(a) Precise fiber fixing to the center of the ferrule
(b) Precise ferrule alignment with good repeatability

(c) Suppression of Fresnel reflection at the connection point
(d) Suppression of the influence of housing deformation caused by external force.

Low-cost zirconia ferrule manufacturing technology has been established for single-mode simplex connectors. This is a way of achieving condition (a). Zirconia ferrule can be used for the MCF connector if there is precise alignment of the rotation angle around the ferrule axis.

A precise ferrule alignment technique with a split sleeve was also established for condition (b). We can use the same technique for the MCF connector.

For condition (c), physical contact (PC) connection technology has been established to achieve a return loss of 50 dB or more. We can use PC technology for the MCF connector; however, the specifications of the ferrule end-face geometry must be reconsidered because some cores are not located in the center of the fiber.

Optical connectors are usually operated by hand, so we sometimes have to consider an external force of tens of Newton interacting with the fiber cables. If such a force interacts with a small-sized optical connector, the plug housing will be deformed by more than 10 μm, which is far greater than the alignment tolerance. To solve this problem (condition (d)), a ferrule "floating mechanism" has been widely used.

3.3.2 Butt Joint-type MCF Connector

Butt joint-type single-mode optical connectors are widely used, e.g., SC (IEC 61754-4), LC (IEC 61754-20) and MU (IEC 61754-6) connectors. It is recommended that these existing mechanical interfaces can be used to realize a new MCF connector because the first priority with optical connectors is intermateability. Considerable investment and a long period of development are needed to develop a new mechanical interface that achieves intermateability between the devices supplied by different vendors.

The MU-type MCF connector was proposed based on the following design principles [7].

Small size

The new MCF connector should have the same or greater packaging density than conventional optical connectors. On the other hand, there is some possibility of using an MCF connector for high-speed optical transceivers. In this regard, small form factor (SFF) connectors are recommended.

Symmetrical structure

The plug housing will be deformed not only via interaction with an external force but also via interaction with the internal force that is produced by the spring used for PC connection. To realize stable connection for all the cores in an MCF, uniform pressure is recommended for every outer core.

There are two types of SFF connectors, one type has only one cantilever for a latch (e.g., the LC connector), and the other type has two cantilevers that are arranged symmetrically (e.g., the MU connector). To realize stable MCF connection, the latter type is recommended because it will deform symmetrically when mated.

Floating mechanism

An MU connector has a 0.1 mm gap in each direction between the ferrule flange and plug housing. This gap allows a ferrule rotation of ± 10°, which does not satisfy the allowable tolerance of ± 0.5° to achieve attenuation of less than 0.5 dB. To realize an angle tolerance of ± 0.5°, Oldham's coupling mechanism was used. Figure 3.7a shows the design principle of the MU-type MCF connector, which incorporates

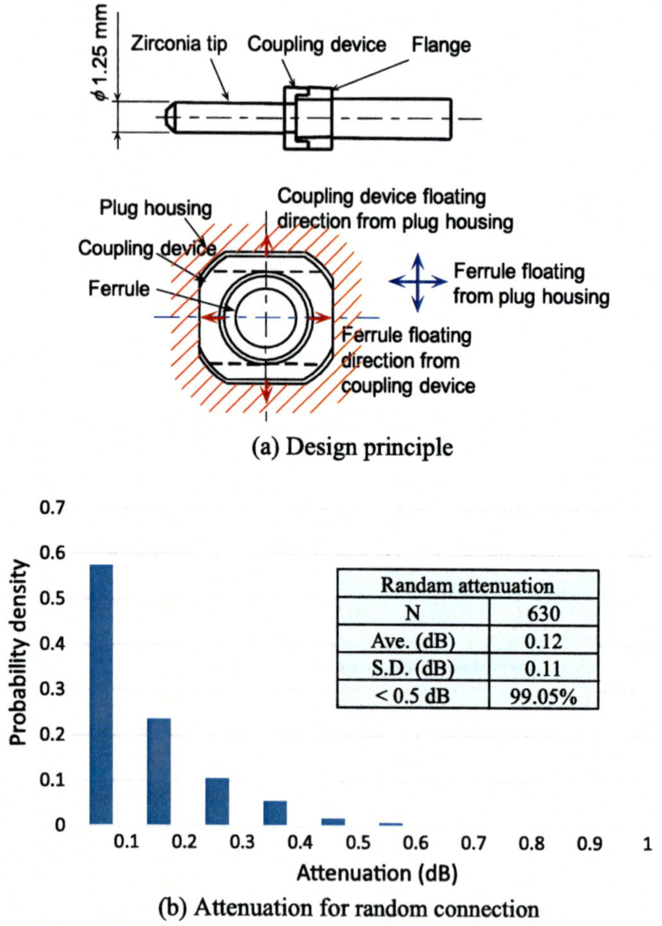

Fig. 3.7 MU-type multi-core fiber connector, **a** design principle and **b** attenuation

Oldham's coupling mechanism [7]. Figure 3.7b shows the attenuation distribution of randomly connected MU-type seven-core MCF connectors [8].

Another butt joint-type MCF connector, SC-type MCF connector with pressurization springs, was proposed. It uses pressurization springs which hold the ferrule and provide both ferrule floating and precise angular alignment [9, 10].

3.3.3 *Lens Coupling Type MCF Connector*

The other technique for achieving the abovementioned condition (d) is the lens coupling technique. Expanded beams allow larger misalignment via interaction with an external force at a connection point than butt joint-type connectors. If each core has a pair located in a symmetrical position, e.g., typical seven-core MCF, it can be coupled using only one lens pair [11].

3.3.4 *Few-Mode Fibers Connection*

Mode-division multiplexing transmission experiments with FMF have advanced rapidly. When using FMF, we need to consider several additional fiber parameters for inter-mode effects, e.g., differential mode attenuation (DMA), differential group delay (DGD), linear cross-mode coupling and inter-mode nonlinearity [12]. Any core offset at the connection point of the FMF would influence the DMA or linear cross-mode coupling. Greater accuracy is needed to achieve an FMF connector than a single-mode fiber connector. Finding a way to design an FMF connector with low DMA and multiple connection points in a practical network is the subject for a future study.

3.4 Fan-In/Fan-Out Device for Multi-core Fiber

When constructing a space-division multiplexing transmission system with multi-core fiber (MCF), it is essential to realize fan-in/fan-out (FI/FO) devices that optically couple each core of the MCF with individual single-core fibers (SCFs). The core arrangement of the FI/FO device must correspond to the core arrangement of the MCF. However, it is impossible to connect the MCF with SCFs simply by bundling conventional SCFs with a cladding diameter of 125 μm because the core pitch of the MCF is tens of microns. Several types of FI/FO devices for MCF transmission have been devised and used in high-capacity MCF transmission experiments. Figure 3.8 shows the basic configuration of typical FI/FO devices including a bundled fiber type, a fused fiber type, a free-space coupling type and a three-dimensional waveguide type. This section describes the features of these FI/FO devices as shown in Fig. 3.8.

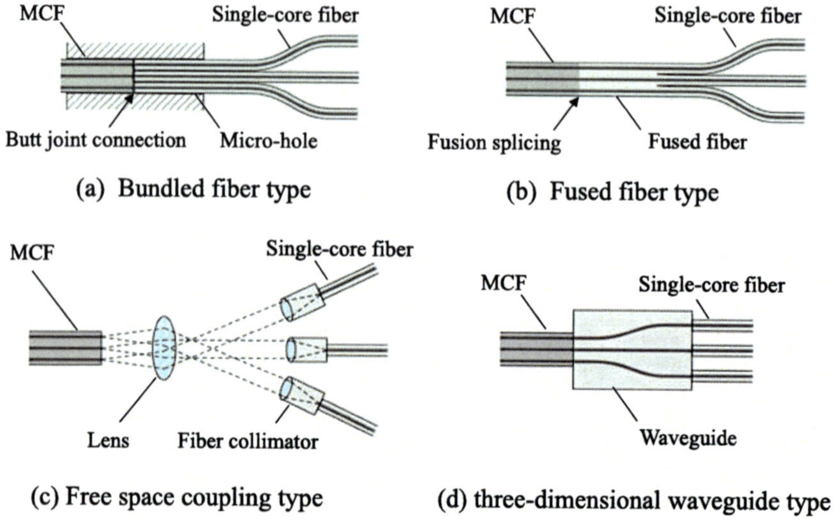

Fig. 3.8 Configuration of fan-in/fan-out device

3.4.1 Bundled Fiber-Type Fan-In/Fan-Out Device

The bundled fiber-type FI/FO device realizes a closely packed core arrangement of an MCF by utilizing individual SCFs. Figure 3.9 shows the basic configuration of the bundled fiber-type FI/FO device, which employs two ferrules and a split sleeve. The MCF is inserted into the micro-hole of one of the ferrules. The SCFs required for MCF core arrangement are inserted into the micro-hole of the other ferrule. The inserted SCFs are then fixed in place with adhesive. The portion of the SCFs inserted into the ferrule has the same cladding diameter as the core pitch of the MCF. The

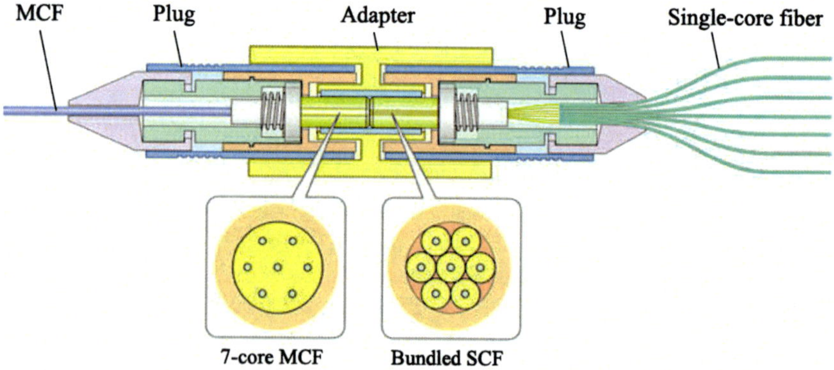

Fig. 3.9 Configuration of bundled fiber-type fan-in/fan-out device

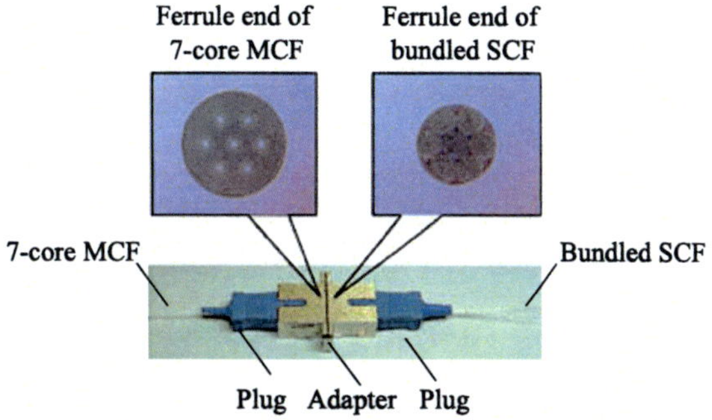

Fig. 3.10 Bundled fiber-type fan-in/fan-out device with SC connector housings

core arrangement of the bundled fibers at the ferrule end is made to correspond to the hexagonal closely packed core arrangement of the MCF by inserting individual SCFs into the circular hole, as shown in Fig. 3.10. Bundled fiber-type FI/FO devices using a ferrule with a hexagonal hole [13] or a square hole [14] have also been reported with the aim of realizing an accurate closely packed fiber arrangement for MCF with more than 12 cores. The bundled fiber-type FI/FO device enables us to realize a pluggable connection by accommodating the bundled fibers in connector housings. To realize a pluggable connection for MCF, accurate rotational angle alignment is required between the coupled cores. There have been several reports of pluggable MCF connection including an SC connector type with a rotational angle alignment structure [15] and an MU connector type with Oldham's coupling mechanism [7]. Figure 3.10 shows an SC connector-type FI/FO device for single-mode seven-core MCF [16]. The connector enables us to adjust the rotational angle of the MCF or SCFs by rotating the ferrule. In a long-haul MCF transmission employing an amplification system with a high-power input, it is desirable to achieve physical contact (PC) connection between the bundled SCFs and MCF. The PC conditions between the SCFs and MCF are calculated by using a finite element simulation [17]. FI/FO devices with a PC connection between the SCFs and MCF have been reported for seven-core [16, 17], 12-core [13, 14] and 19-core [18] MCFs.

3.4.2 *Fused Fiber-Type Fan-In/Fan-Out Device*

The fused fiber-type FI/FO device realizes a closely packed core arrangement of MCF by fusing the SCFs through elongation. Figure 3.11 shows a schematic view of process used to fabricate the fused fiber-type FI/FO device. SCFs are inserted into a glass capillary with a hexagonal closely packed hole arrangement. The inserted SCFs and the capillary are fused by using an elongation process. The core pitch value of the FI/FO device is controlled by the elongation ratio. The elongated fibers are cut at a point that has the same core pitch as the connected MCF. The fused fiber-type FI/FO device is connected to the MCF by a fusion splice technique. It is important that the

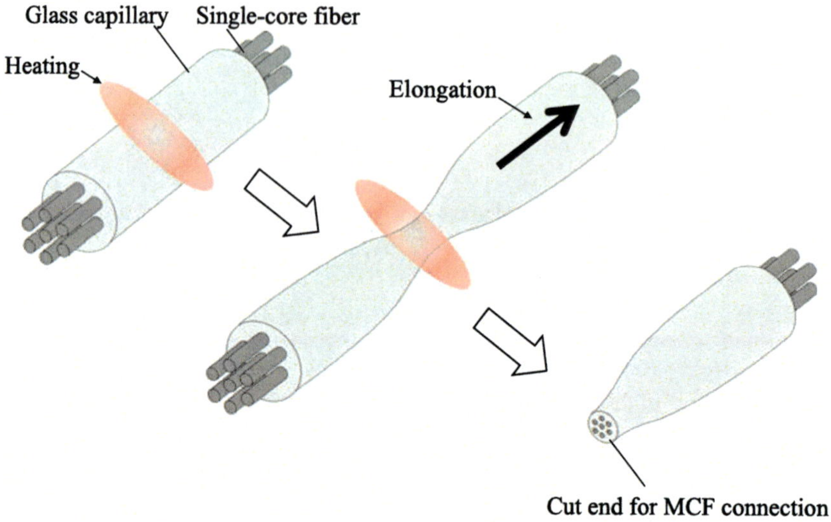

Fig. 3.11 Fabrication process of fused fiber-type fan-in/fan-out device

mode field diameter (MFD) of the FI/FO device corresponds with that of the MCF if we are to obtain a low-loss connection. The MFD at the elongated end of the FI/FO device is determined by the elongation ratio, the relative index difference and the core radius of the SCFs. Therefore, the core profile of the SCFs before the elongation process must be appropriately designed to achieve a low-loss fused fiber-type FI/FO device. Fused fiber-type FI/FO devices have been reported for seven-core [19, 20] and 12-core [21] MCFs.

3.4.3 Free-Space Coupling-Type Fan-In/Fan-Out Device

The free-space coupling-type FI/FO device optically couples each core of the MCF with individual SCFs by utilizing a lens. The basic configuration of the free-space coupling-type FI/FO device is shown in Fig. 3.8c. The beams emerging from the MCF facet propagate in slightly different directions and are more spatially separated by passing through the lens. The separated beams are coupled to the individual SCFs with a fiber collimator. This configuration with only a single lens requires sufficient length to enable the beams in the FI/FO device to be separated. Therefore, in an early report of MCF transmission experiments using this type FI/FO device, the device size for a 19-core MCF was as large as 300 mm in diameter and 430 mm high [22]. In attempts to reduce the size of the FI/FO device using a lens, configurations have been reported with a GRIN lens and prisms [11] and a GRIN lens and micro-lens array [23]. By designing the lens system, a free-space coupling-type FI/FO device based

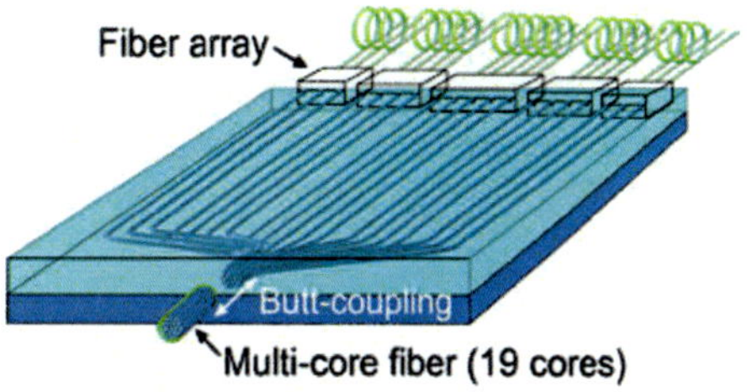

Fig. 3.12 Fan-in/fan-out device (Reprinted with permission from [26]. © 2015 IET)

on lens optics has a potential that corresponding to that of various MCFs, including those with a non-hexagonal core arrangement.

3.4.4 Three-Dimensional Waveguide-Type Fan-In/Fan-Out Device

A three-dimensional waveguide-type FI/FO device provides a compact connection from an MCF to an SCF array. Three-dimensional waveguide-type FI/FO devices fabricated by using an ultra-short pulse laser for four-core MCF [24] or a laminated polymer waveguide for seven-core [25] and 19-core [26] MCFs have been reported. Figure 3.12 shows the laminated polymer waveguide for 19-core MCF. The device was fabricated by using polymer waveguide process that enables multi-layer stacking on a wafer. The waveguide and fibers are connected with adhesive by using an active alignment technique. This type of FI/FO device has the attractive potential for integration with functional waveguide devices such as a wavelength filter for realizing a future space-division multiplexing transmission system.

3.5 Mode Multiplexing/Demultiplexing Technologies for Few-Mode Fibers

3.5.1 Overview

Photonic devices for communication systems have been developed based on transverse single-mode condition so far. Therefore, "transverse single mode" has been a fundamental waveguide-design rule for the photonic devices including laser diodes, modulators, switches and others. Most of the commercially implemented photonic devices are single-mode ones. On the other hand, multi-mode-related devices have been researched for mode-division multiplexing recently [27–48]. For this reason, it is required to realize mode converters, mode couplers, mode multiplexing/demultiplexing devices and the other multi-mode-related devices.

Table 3.1 Features of mode MUX/DEMUX devices

	Need of beam splitter[a]	Wavelength dependence	Applicability to degenerate modes	Integration
Phase plate	Yes	Weak	Yes	Not suited
Spot-based	Yes	None	Yes	Not suited
Photonic lantern	Partially yes	Weak	Yes, but additional device is needed	Yes (waveguide type)
Si photonics	No but large loss	To some extent	Yes	Suited
Fiber directional coupler	None	Weak	Need of 90° rotation	Not suited
Waveguide directional coupler	None	Weak	Need of additional device	Suited
Mode evolutional type	None	Weak	Yes	Suited
MMI type	None	Weak	Yes	Suited

[a]One mode to one port correspondence is not guaranteed by the beam splitter itself and suffers branching loss

In the following sections, we review the recent progress of mode converters, combiner/dividers, mode multiplexing/demultiplexing devices and the other related devices. The features of several types of mode MUX/DEMUX devices are summarized in Table 3.1 (up to the point of 2014).

3.5.2 *Mode Converter, Mode Coupler and the Other Multi-mode-Related Devices Based on Optics*

To realize mode-division multiplexing using few-mode fibers, the mode multiplexer/demultiplexer (MUX/DEMUX) which can selectively excite (multiplex) and separate (demultiplex) spatial modes is inevitable. Thus far, various types of MUX/DEMUXs have been proposed, and these can be classified into two types, one is based on the difference of propagation constants of guided modes, and the other is based on the difference of electromagnetic field profile in the cross-sectional area of few-mode fibers. In general, it is difficult for the former type to demultiplex the degenerate modes, and the latter one is not easily realized by the planar optical waveguide structure. The silicon waveguide mode MUX/DEMUX device can multiplex and demultiplex degenerate guided modes using a grating coupler.

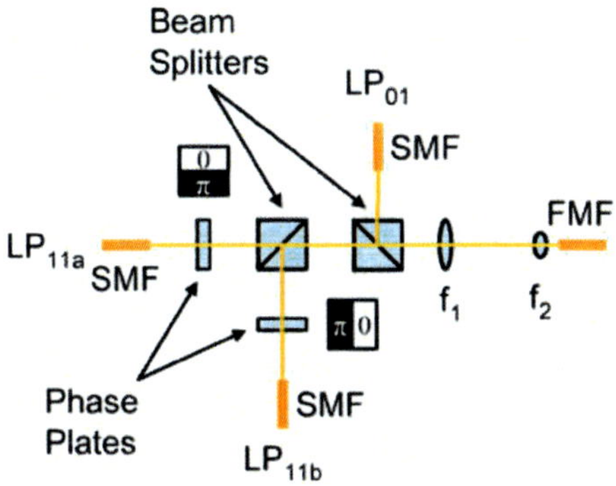

Fig. 3.13 Phase plate-type mode MUX/DEMUX device (Reprinted with permission from [27]. © 2011 The Optical Society)

3.5.2.1 Phase Plate Type

This type of mode MUX/DEMUX is based on the combination of beam splitter, which has no dependence on the modal field profile and polarization, and the phase plate which has the phase pattern depending on the modal filed profile, as shown in Fig. 3.13 [27]. Therefore, this type of mode MUX/DEMUX device suffers the splitting loss in principle, and the insertion loss increases with the increase of the number of modes. However, the degenerate modes can be separated. The scalability of the number of modes is limited by the number of beam splitters. However, the use of programmable spatial light modulator (SLM) can expand the flexibility of modal electromagnetic field pattern [28].

3.5.2.2 Spot-Based Type

The electromagnetic field pattern of higher-order modes can be decomposed into several light spots with opposite phase between neighboring spots. Therefore, the selective mode excitation can be realized by combining and arranging several light spots with the control of phase so that the phase is opposite between neighboring spots. Based on this principle, some spot-based mode multiplexers have been proposed and demonstrated [30, 31]. To combine the spot-like light beams, a set of micro-mirror with different directional angle [30] or a kind of corner mirror [31] is needed together with the lens system to accept all the beam spots as shown in Fig. 3.14. This principle can be utilized to realize the mode multiplexer, but it is difficult to be utilized for the mode demultiplexer using the combination of mirror and lens system, because a precise phase control between branched arms is needed. In addition, although this kind of mode multiplexer can excite selectively a definite order of mode, it requires the beam splitter if different modes are excited simultaneously (multiplexing function).

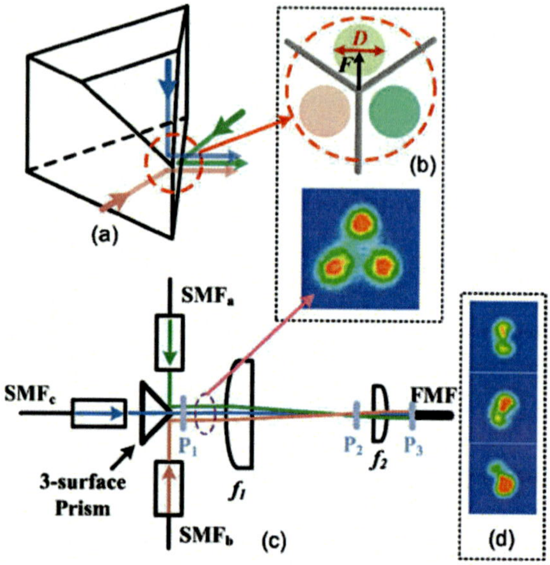

Fig. 3.14 Spot-based-type mode MUX/DEMUX device (Reprinted with permission from [31]. © 2013 IEEE)

3.5.2.3 Photonic Lantern

The photonic lantern was introduced in the field of astrophotonics to connect a multi-mode fiber to single-mode fibers [36], because the low-loss connection can be realized when the number of guided modes in a multi-mode fiber is equal to the number of single-mode cores. In the mode-division multiplexing transmission, the photonic lantern was first utilized to form a mode multiplexer, using a bundle of identical single-mode fibers connected to a multi-mode fiber with tapering structure [37]. When a bundle of identical single-mode fibers is used, the label of guided mode excited in a multi-mode fiber corresponds to a geometrical position of single-mode cores which are launched [37]. Therefore, this type of photonic lantern with identical single-mode cores is a selective mode exciter and is difficult to be used as the mode demultiplexer. If the core size or the V parameter of single-mode cores is dissimilar, however, the correspondence between the mode order and the single-mode core is determined based on the magnitude relation of propagation constants of single-mode cores. Therefore, the mode selectivity can be obtained, and this type of photonic lantern can be utilized as a mode demultiplexer as shown in Fig. 3.15 [38, 39]. The photonic lantern is formed using the tapered bundle of single-mode fibers as well as the laser inscribed waveguide [40].

3.5.2.4 Fiber Coupler Type

Fused fiber coupler can be used as a mode MUX/DEMUX device when dissimilar cores are coupled to each other and the propagation constant of target guided mode

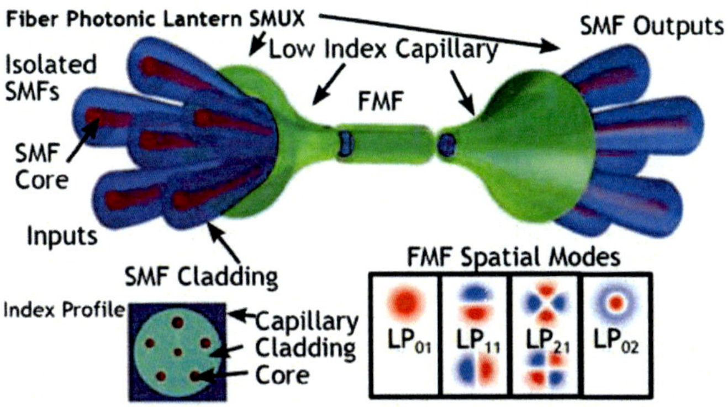

Fig. 3.15 Photonic lantern-type mode MUX/DEMUX device (Reprinted with permission from [38]. © 2013 IET)

in a multi-mode core (bus line) is equal to that of the single-mode fiber which serves as the add or drop line core [43, 44]. Since the principle is based on the synchronous coupling between identical cores with the same propagation constant, the length of coupler region should be properly designed, and the crosstalk has a weak wavelength dependence. In addition, since the dispersion characteristics of fundamental mode in the single-mode (add/drop line) core and higher-order mode of multi-mode core (bus line) are generally non-identical, the wavelength dependence is larger than that in the coupling between identical cores. If a simple fused fiber coupler is used, two couplers should be serially connected with 90° rotation to separate degenerate LP_{11} modes, i.e., LP_{11}^{even} and LP_{11}^{odd} modes [44]. If a specially designed three dissimilar core fiber is used, three mode (LP_{01}, LP_{11}^{even} and LP_{11}^{odd}) multiplexer/demultiplexer can be realized [45] as shown in Fig. 3.16.

3.5.2.5 Silicon Waveguide Type

Since silicon waveguide can achieve very compact size for wiring and bending, it is suitable for a complex photonic circuit like mode MUX/DEMUX device involving many interconnects corresponding to individual mode orders and polarizations. Utilizing this feature, some mode multiplexing devices have been realized using silicon waveguide [46–48]. Other feature of silicon waveguide is the integration of quasi-planar-type grating coupler from the few-mode core to silicon thin film waveguide, which is connected directly to silicon single-mode waveguide by tapering structure. Since the grating coupler has polarization and wavelength dependences, cross-type grating is used for the separation of polarization [46, 47] as shown in Fig. 3.17. To eliminate the polarization dependence, an InP 45° micro-mirror was developed to convert the optical axis from the waveguide surface to the vertical direction, which is the direction of few-mode core. In this device, micro-mirrors were arranged to form the mode pattern, like the spot-based mode MUX/DEMUX [48].

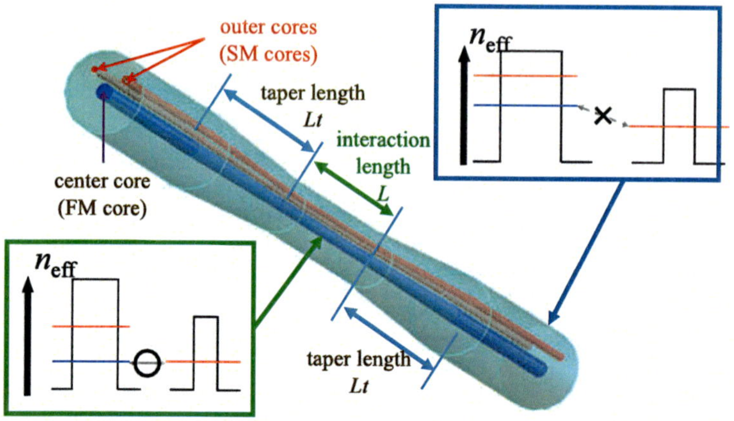

Fig. 3.16 Fiber directional coupler-type mode MUX/DEMUX device (Reprinted with permission from [45]. © 2014 IEEE)

Fig. 3.17 Silicon waveguide-type mode MUX/DEMUX device (Reprinted with permission from [46]. © 2012 IEEE)

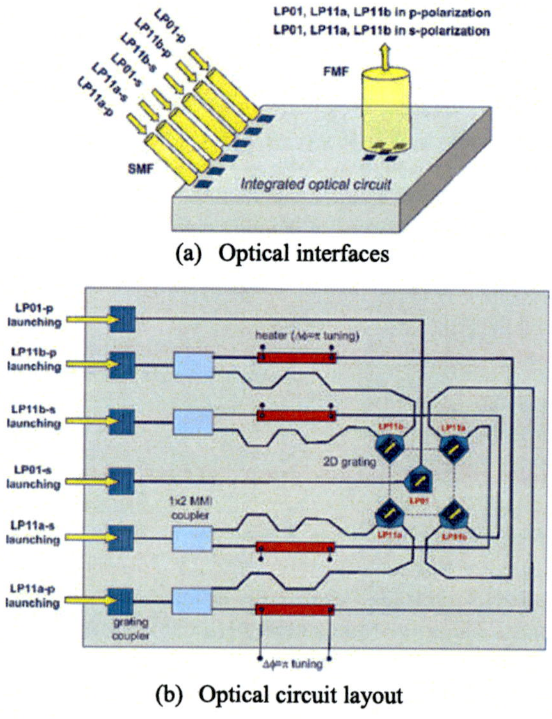

3.5.3 *Mode Converter, Mode Coupler and the Other Multi-mode-Related Devices Based on Waveguide Technology*

In this section, we review several types of photonic-integrated waveguide devices. In the following sections, the proposed and demonstrated device principles are reviewed; however, the technology mentioned here is still not matured due to insufficient research term; therefore, further improvement and development are expected in the future.

3.5.3.1 Mode Coupling Device

When two adjacent waveguides, having identical propagation constant, are placed each other in parallel, light propagating in one waveguide couples to the other. Mode coupling device utilizes this scheme. Generally speaking, modal propagation constant depends on waveguide width, in addition to the mode order itself. By choosing proper waveguide width and mode order, mode coupling condition is secured among two width-different waveguides. The device is designed based on this principle: The modal propagation constant is designed to have exact the same value by choosing waveguide width for each propagation mode. When the both propagation constants match each other while the width itself is different, mode couples as shown in Fig. 3.18 [49].

In the case of waveguide device, however, one difficulty exits in scaling up toward vertical direction, while this is needed for LP modes. To realize LP_{21} mode especially, there are two schemes proposed so far; One is to "cut and paste" (see Fig. 3.19) [50], and the other is to utilize "mode rotation" (see Fig. 3.20) [51, 52].

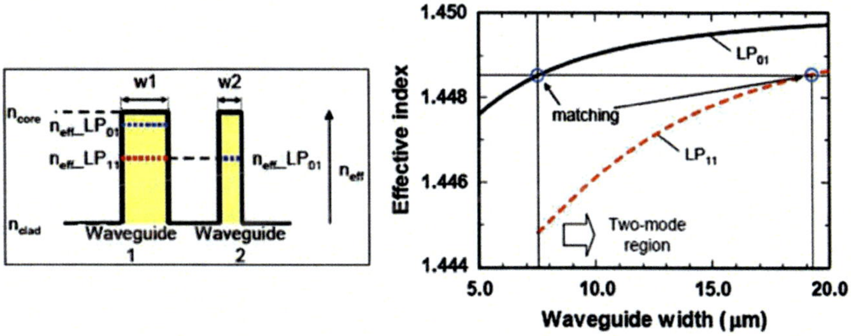

Fig. 3.18 Principle of mode coupling (Reprinted with permission from [49]. © 2012 Authors)

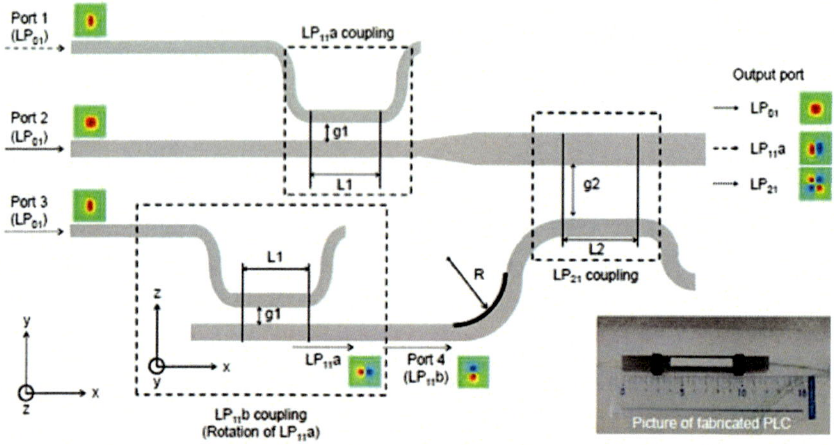

Fig. 3.19 Mode coupling device (Reprinted with permission from [50]. © 2013 IET)

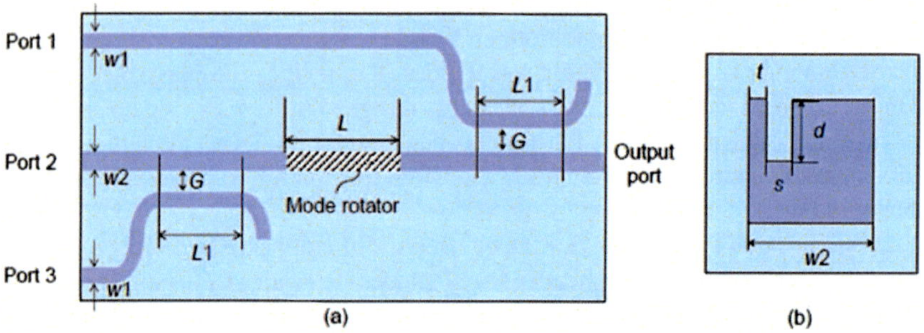

Fig. 3.20 Mode coupling device (Reprinted with permission from [51]. © 2014 The Optical Society)

In the "cut and paste" configuration[1], once LP_{11} mode device is fabricated and detached from the wafer. The device is rotated toward vertical direction by 90°, and it is physically attached to the other LP_{11} mode integrated device. LP_{21} mode converter is realized in such a way. This configuration proves the waveguide device's possibility for LP_{21} mode; on the other hand, it was not possible to fabricate without physical detachment and attachment process. Later, "mode ration" scheme has been proposed that is free from this detachment and attachment process. Narrow slit formed on the top surface of the waveguide core rotates the optical field in 90°. The integrated device

[1] LP mode is one of the two-dimensional space-mode-sets, and the mode-evolution extends in two dimensional direction (horizontal, and vertical). Meanwhile, planar lightwave circuit (PLC) is hard to extend the space dimension toward vertical direction as it is a plane circuit normally. To overcome this problem, one way of realizing LP-mode conversion is to utilize the "cut and paste" configuration. This is, namely, a part of PLC is once cut, rotated 90 deg, and attached to the other PLC. In such a way, higher-evolution in LP mode is realized.

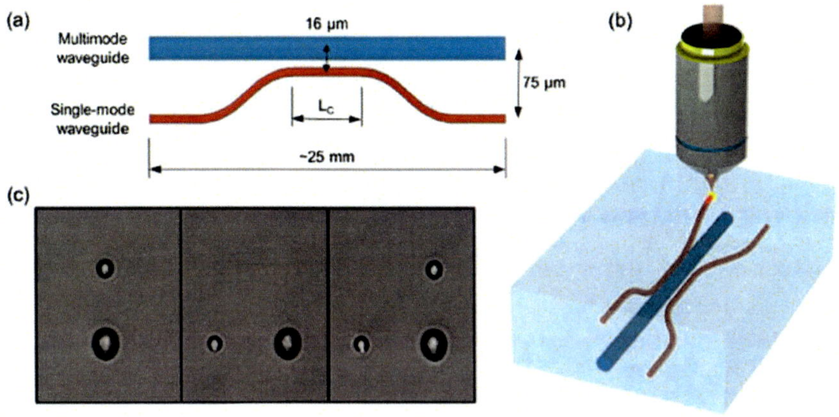

Fig. 3.21 Mode coupling device (Reprinted with permission from [53]. © 2015 Authors)

with this slit waveguide realizes LP_{21}. In addition to these activities, mode coupling device based on 3D waveguide was also demonstrated recently (see Fig. 3.21) [53].

3.5.3.2 Adiabatic Mode Evolution Device

Adiabatic mode evolution is another principle to enable mode multiplexing/demultiplexing [54, 55]. Note that, the working principle is different from directional coupler. When two non-identical cores are separated with somewhat large spacing, even and odd modes are created, and the electric field of even mode has its main peak in the wider (larger V value) core and has very tiny sub-peak in the narrower (smaller V value) core. On the other hand, the electric field of odd mode has its main peak in the narrower (smaller V value) core and has very tiny sub-peak in the wider (larger V value) core. When the two waveguides are tapered in the opposite way, i.e., the wider waveguide at the input end is tapered down toward the output end and the narrower waveguide at the input end is tapered up toward the output end, the magnitude relation of propagation constants is inverted. Then, the main peak of even mode in the wider waveguide at the input end shifts to the other waveguide along with the propagation and is extracted from the other waveguide at the output end. As the transition is adiabatic, it has a benefit that the mode evolution does not have significant wavelength dependency. This device principle was demonstrated for few-mode fiber (see Fig. 3.22) [54]. As is shown in this figure, the highest-order mode is dropped toward drop line. Note that, dropped signal is exchanged into single mode. Later, the device for strongly coupled multi-core fiber was also demonstrated by the same group (see Fig. 3.23) [55].

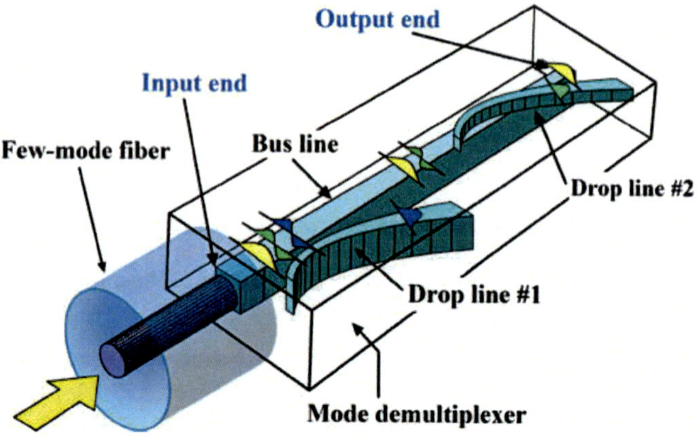

Fig. 3.22 Adiabatic mode evolution for few-mode fiber (Reprinted with permission from [54]. © 2015 IEEE)

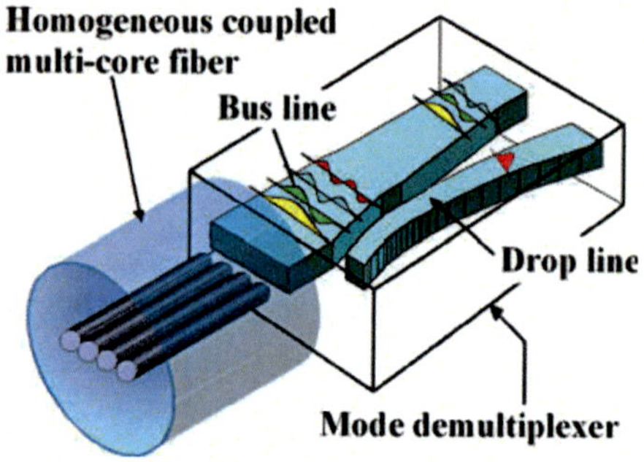

Fig. 3.23 Adiabatic mode evolution for super-mode using strongly coupled multi-core fiber (Reprinted with permission No. 19RA045 from [55]. © 2016 IEICE)

3.5.3.3 OAM Mode Device

Orbital angular momentum (OAM) mode converter has been proposed and demonstrated so far. As the OAM mode is a ring-shaped intensity profile with cylindrical phase distribution of 2π multiple integer per one round, the phase distribution profiles are linear if the optical field is deployed in horizontal direction. Utilizing this principle, the proposed device consists of slab waveguide with waveguide array like arrayed waveguide grating (AWG) (see Fig. 3.24) [56, 57]. In addition, LP mode expression by using linear combination of OAM modes was also proposed and demonstrated [56]. In this case, possible LP mode is limited due to the different mode configuration between LP and OAM modes. This OAM device offers a merit that it also realizes mode add and drop functions simultaneously.

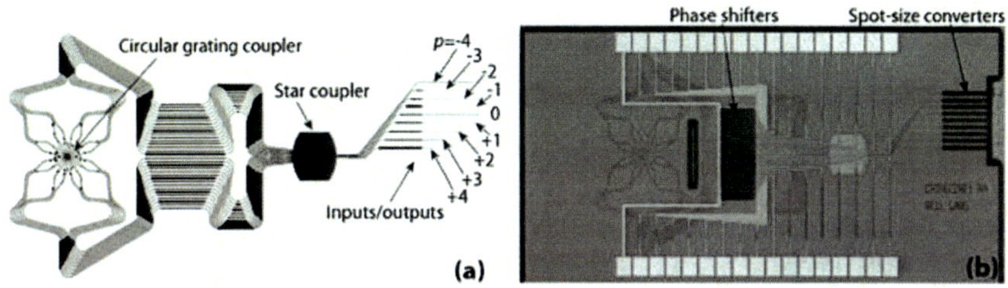

Fig. 3.24 OAM mode device (Reprinted with permission from [56]. © 2011 Authors)

3.5.3.4 MMI Mode Converter

Multi-mode interference (MMI) coupler has been proposed and demonstrated to realize higher-order mode output. In a MMI waveguide, multi-mode interference occurs that leads to LP_{11} mode (see Fig. 3.25) [58].

The device required two MMIs (see Fig. 3.25c). The first MMI section works as 1×2 power splitter. And, the second MMI is designed as 50% mode converter (see Fig. 3.25a). In this MMI, the injected single mode is converted into first-order mode

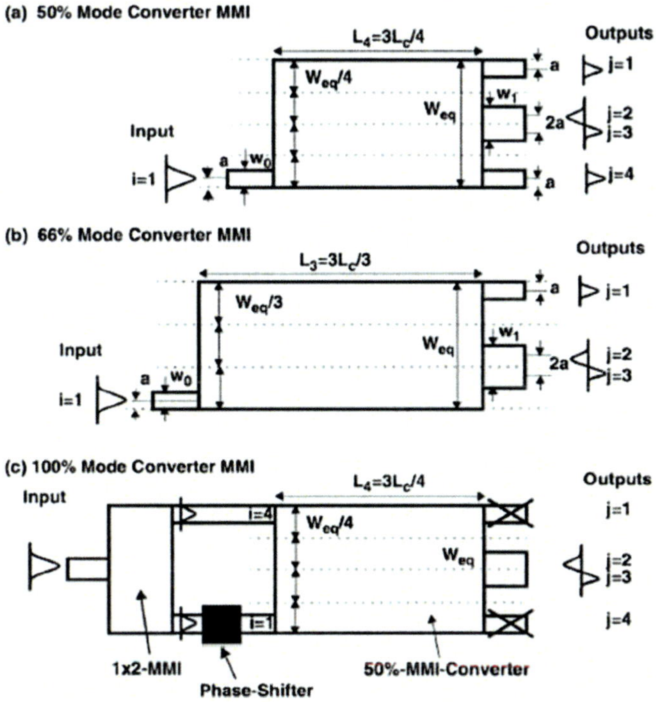

Fig. 3.25 MMI mode converter (Reprinted with permission from [58]. © 1998 IEEE)

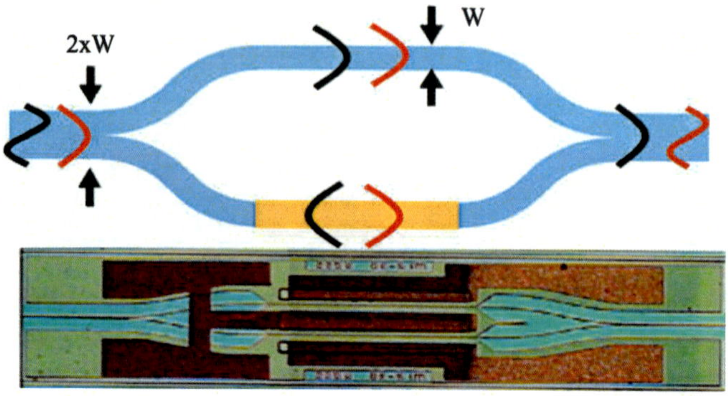

Fig. 3.26 Optical mode switch (Reprinted with permission from [63]. © 2015 Authors)

toward center waveguide, while 25% single mode is remaining at two-side waveguides. The access waveguide is connected into lower port in Fig. 3.25a; however, the upper port connection is indeed possible. To extract the two 25% single mode to achieve 100% mode conversion toward first-order mode, phase shifter is integrated (see Fig. 3.25c). As one of the access waveguide is shifted with π, the resulted output at the 25% single-mode output ports goes zero, while the output power goes 100% into the center waveguide. Note that, MMI itself is wavelength dependent in principle; however, the design scheme to suppress the wavelength dependency has been proposed [59].

3.5.3.5 Optical Mode Switch

Optical mode switch is a device that realizes, not only mode conversion, mode exchange among different propagation mode order [60–63]. This device has a function of mode conversion and mode exchange. The device for two modes (see Fig. 3.26) seems similar with Mach–Zehnder interferometer; however, there is a difference in the width of access waveguide (both input and output). The injected light is injected into input waveguide, which allows first-order mode, and then divided into two single-mode waveguides. When one of the arms works as a phase shifter, the resulted modes after the second *Y*-junction switch each other (cross-state), while they keep same mode at the output in case phase is not changed. Scalable design has been also proposed so far. As it directly switches optical space mode, it is expected to be used in space mode reconfigurable optical add-drop multiplexer (ROADM). Note that, this device switches modes in lateral direction (TE_{XX} and TM_{XX} modes); therefore, it does not switch LP mode itself at the present stage.

References

1. K. Inada, O. Wtanabe, H. Taya, Splicing of fibers by fusion method. IEEE J. Sel. Area Commun. **SAC-4**(5), 706–712 (1986)
2. K. Yoshida, A. Takahashi, T. Konuma, K. Yoshida, K. Sasaki, Fusion splicer for specialty optical fiber with advanced functions. Fujikura Tech. Rev. (41), 10–13 (2012)
3. W. Zheng, D. Duke, T. Kubo, B. Malinsky, Interrelation profile analysis method for alignment for polarization-maintaining fiber, in *OFC/NFOEC 2010*, JThA61 (2010)
4. W. Zheng, Automated alignment and splicing for multicore fibers, in *OFC/NFOEC 2013*, OM3I.4 (2013)
5. Y. Amma, Y. Arakawa, A. Takahashi, K. Takenaga, S. Matsuo, Low-loss fusion splice technique for multicore fiber with a large cladding diameter, in *IEEE Summer Topicals 2013*, MC2.2 (2013), pp. 74–75
6. Y. Amma, A. Takahashi, K. Takenaga, S. Matsuo, Accuracy of core alignment with end-view function for multicore fiber, in *IEEE Summer Topicals 2014* (2014), pp. 170–171
7. R. Nagase, K. Sakaime, K. Watanabe, T. Saito, MU-type multicore fiber connector. IEICE Trans. Electron. **E96-C**(9), 1173–1177 (2013)
8. R. Nagase, K. Sakaime, K. Watanabe, T. Saito, Characteristics of MU-type multicore fiber connector, in *IEICE General Conference 2014* (Mar 2014) (in Japanese)
9. T. Kobayashi, H. Endo, C. Ndiaye, Optical connector for multi-core fibers with axial rotation prevention mechanism, in *Proceedings of the 2014 IEICE General Conference*, paper B-10-1 (Mar 2014) (in Japanese)
10. E. Nomoto, K. Hiruma, T. Sugawara, K. Tanaka, Y. Lee, SC-type multi-core optical-fiber connectors using a pressurization spring. Opt. Rev. **22**(5), 679–685 (2015)
11. Y. Tottori, T. Kobayashi, M. Watanabe, Low loss optical connection module for seven-core multicore fiber and seven single-mode fibers. IEEE Photon. Technol. Lett. **24**(21), 1926–1928 (2012)
12. Y. Sun, R. Lingle Jr., A. McCurdy, D. Peckham, R. Jensen, L. Gruner-Nielsen, Few-mode fibers for mode-division multiplexing, in *Proceedings of the IEEE Summer Topicals 2013*, MC3.1 (2013)
13. Y. Abe, K. Shikama, S. Yanagi, T. Takahashi, Low-loss physical-contact-type fan-out device for 12-core multicore fiber, in *39th European Conference and Exhibition on Optical Communication (ECOC 2013)*, P.1.7 (2013)
14. T. Mizuno, T. Kobayashi, H. Takara, A.Sano, H. Kawakami, T. Nakagawa, Y. Miyamoto, Y. Abe, T. Goh, M. Oguma, T. Sakamoto, Y. Sasaki, I. Ishida, K. Takenaga, S. Matsuo, K. Saitoh, T. Morioka, 12-core $\times$ 3-mode dense space division multiplexed transmission over 40 km employing multi-carrier signals with parallel MIMO equalization, in *OFC 2014*, Th5B.2 (2014)
15. K. Shikama, Y. Abe, S. Asakawa, S. Yanagi, T. Takahashi, Multicore fiber connector with physical-contact connection. IEICE Trans. Electron. **E99-C**(2), 242–249 (2016)
16. H. Takara, T. Mizuno, H. Kawakami, Y. Miyamoto, H. Masuda, K. Kitamura, H. Ono, S. Asakawa, Y. Amma, K. Hirakawa, S. Matsuo, K. Tsujikawa, M. Yamada, 120.7-Tb/s MCF-ROPA unrepeatered transmission of PDM-32QAM channels over 204 km. IEEE J. Lightwave Technol. **33**(7), 1473–1478 (2015)
17. O. Shimakawa, M. Shiozaki, T. Sano, A. Inoue, Pluggable fan-out realizing physical-contact and low coupling loss for multi-core fiber, in *Optical Fiber Communication Conference and Exposition and the National Fiber Optic Engineers Conference (OFC/NFOEC) 2013*, OM3I.2 (2013)
18. K. Shikama, Y. Abe, H. Ono, A. Aratake, Low-loss fiber-bundle-type fan-in/fan-out device for 6-mode 19-core fiber, in *Optical Fiber Communications Conference and Exhibition (OFC) 2017*, M3G.3 (2017)
19. B. Zhu, T.F. Taunay, M. Fishteyn, X. Liu, S. Chamdrasekhar, M.F. Yan, J.M. Fini, E.M. Monberg, F.V. Dimarcello, 112-Tbps Space-division multiplexed DWDM transmission with 14-b/s/Hz aggregate spectral efficiency over a 76.8-km seven-core fiber. Opt. Express **19**, 166665–16671 (2011)

20. H. Uemura, K. Takenaga, T. Ori, S. Matsuo, K. Saitoh, M. Koshiba, Fused taper type fan-in/fan-out device for multicore EDF, in *2013 18th OptoElectronics and Communications Conference held jointly with 2013 International Conference on Photonics in Switching (OECC/PS)*, TuS1-4 (2013)
21. H. Uemura, K. Omichi, K. Takenaga, S. Matsuo, K. Saitoh, M. Koshiba, Fused taper type fan-in/fan-out device for 12 core multi-core fiber, in *2014 OptoElectronics and Communication Conference and Australian Conference on Optical Fiber Technology* (2014), pp. 49–50
22. W. Klaus, J. Sakaguchi, B.J. Puttnam, Y. Awaji, N. Wada, T. Kobayashi, M. Watanabe, Free-space coupling optics for multicore fibers. IEEE Photon. Technol. Lett. **24**(11), 1902–1905 (2012)
23. O. Shimakawa, H. Arao, M. Harumoto, T. Sano, A. Inoue, Compact multi-core fiber fan-out with GRIN-lens and micro-lens array, in *Optical Fiber Communications Conference and Exhibition (OFC) 2014*, M3K.1 (2014)
24. R.R. Thomson, H.T. Bookey, N.D. Psaila, A. Fender, S. Campbell, W.N. MacPherson, J.S. Barton, D.T. Reid, A.K. Kar, Ultrafast-laser inscription of a three dimensional fan-out device for multicore fiber coupling applications. Opt. Express **15**(18), 11691–11697 (2007)
25. T. Watanabe, Y. Kokubun, Laminated polymer waveguide fan-out device for uncoupled multi-core fiber. Opt. Express **20**(24), 26317–26325 (2012)
26. T. Watanabe, Y. Kokubun, Stacked polymer waveguide type fan-in/fan-out device for dense multi-core fiber. IET Optoelectron. 1–5 (2015). https://doi.org/10.1049/iet-opt.2014.0137
27. S. Randel, R. Ryf, A. Sierra, P.J. Winzer, A.H. Gnauck, C.A. Bolle, R.-J. Essiambre, D.W. Peckham, A. McCurdy, R. Lingle Jr., 6 × 56-Gb/s mode-division multiplexed transmission over 33-km few-mode fiber enabled by 6 × 6 MIMO equalization. Opt. Express **19**(17), 16697–16707 (2011)
28. H. Chen, T. Koonen, Single multi-mode mask for multi-channel mode division demultiplexing, in *OFC/NFOEC2013*, Anaheim, OTh1B.4 (2013)
29. K. Igarashi, D. Souma, K. Takeshima, T. Tsuritani, Selective mode multiplexer based on phase plates and Mach-Zehnder interferometer with image inversion function. Opt. Express **23**(1), 183–194 (2015)
30. R. Ryf, M.A. Mestre, A.H. Gnauck, S. Randel, C. Schmidt, R.-J. Essiambre, P.J. Winzer, R. Delbue, P. Pupalaikis, A. Sureka, Y. Sun, X. Jiang, D.W. Peckham, A. McCurdy, R. Lingle, Jr., Low-loss mode coupler for mode-multiplexed transmission in few-mode fiber, in *OFC/NFOEC 2012*, Los Angeles, PDP5B (2012)
31. H. Chen, V. Sleiffer, F. Huijskens, R. van Uden, C. Okonkwo, P. Leoni, M. Kuschnerov, L. Grüner-Nielsen, Y. Sun, H. de Waardt, T. Koonen, Employing prism-based three-spot mode couplers for high capacity MDM/WDM transmission. Photon. Technol. Lett. **25**(24), 2474–2477 (2013)
32. G. Labroille, B. Denolle, P. Jian, P. Genevaux, N. Treps, J.-F. Morizur, Efficient and mode selective spatial mode multiplexer based on multi-plane light conversion. Opt. Express **22**(12), 15599–15607 (2014)
33. G. Labroille, N. Barre, P. Jian, J-F. Morizur, Multi-wavelength multiplexer with independent mode control based on multi-plane light conversion, in *ECOC 2016*, Dusseldorf, W4P1SC4.41 (Proceedings) (2016), pp. 833–835
34. N.K. Fontaine, R. Ryf, H. Chen, S. Wittek, J. Li, J.C. Alvarado, J.E. Antonio Lopez, M. Cappuzzo, R. Kopf, Al Tate, H. Safar, C. Bolle, D.T. Neilson, E. Burrows, K.W. Kim, P. Sillard, F. Achten, M. Bigot, A. Amezcua-Correa, R. Amezcua Correa, J. Du, Z. He, J. Carpenter, Packaged 45-mode multiplexers for a 50-μm graded index fiber, in *ECOC 2018*, Rome, Mo4E.1 (2018)
35. N.K. Fontaine, H. Chen, R. Ryf, D. Neilson, J.C. Alvarado, J. van Weerdenburg, R. Amezcua-Correa, C. Okonkwo, J. Carpenter, Programmable vector mode multiplexer, in *ECC 2017*, Gothenburg, Th.PDP.C.6 (2017)
36. S.G. Leon-Saval, T.A. Birks, J. Bland-Hawthorn, M. Englund, Multimode fiber devices with single-mode performance. Opt. Lett. **30**(19), 2545–2547 (2005)

37. N.K. Fontaine, R. Ryf, J. Bland-Hawthorn, S.G. Leon-Saval, Geometric requirements for photonic lanterns in space division multiplexing. Opt. Express **20**(24), 27123–27132 (2012)
38. N.K. Fontaine, S.G. Leon-Saval, R. Ryf, J.R.S. Gil, B. Ercan, J. Bland-Hawthorn, Mode-selective dissimilar fiber photonic-lantern spatial multiplexers for few-mode fiber, in *ECOC 2013*, London, PD1.C.3 (2013)
39. S.G. Leon-Saval, N.K. Fontaine, J.R. Salazar-Gil, B. Ercan, R. Ryf, J. Bland-Hawthorn, Mode-selective photonic lanterns for space-division multiplexing. Opt. Express **22**(1), 1036–1044 (2014)
40. N.K. Fontaine, R. Ryf, Characterization of mode-dependent loss of laser inscribed photonic lanterns for space division multiplexing systems, in *OECC/PS 2013*, Kyoto, MR202 (2013)
41. Z. Sanjabi Eznaveh, J.E. Antonio-Lopez, J.C. Alvarado Zacarias, A. Schulzgen, C.M. Okonkwo, R. Amezcua, All-fiber few-mode multicore photonic lantern mode multiplexer. Opt. Express **25**(1), 16701–16707 (2017)
42. S. Gross, M. Ams, S.G. Leon-Saval, M.J. Withford, Ultrafast laser inscribed mode-group-selective 6-mode photonic lanterns for mode-division multiplexing, in *CLEO-Europe 2017* (2017). https://doi.org/10.1109/CLEO-E-EQEC.2017.8087236
43. F. Saitoh, K. Saitoh, M. Koshiba, A design method of a fiber-based mode multi/demultiplexer for mode-division multiplexing. Opt. Express **18**(5), 4709–4716 (2010)
44. A. Li, X. Chen, A. Al Amin, W. Shieh, Fused fiber mode couplers for few-mode transmission. Photon. Technol. Lett. **24**(21), 1953–1956 (2012)
45. K. Takenaga, H. Uemura, Y. Sasaki, S. Nishimoto, T. Uematsu, K. Omichi, R. Goto, S. Matsuo, K. Saitoh, Multicore fibre-based mode multiplexer/demultiplexer for three-mode operation of LP_{01}, $LP_{11}{}^{a}$, and $LP_{11}{}^{b}$, in *ECOC 2014*, Cannes, Tu.4.1.4 (2014)
46. A.M.J. Koonen, H. Chen, H.P.A. van den Boom, O. Raz, Silicon photonic integrated mode multiplexer and demultiplexer. Photon. Technol. Lett. **24**(21), 1961–1964 (2012)
47. Y. Ding, H. Ou, J. Xu, C. Peucheret, Silicon photonic integrated circuit mode multiplexer. Photon. Technol. Lett. **25**(7), 648–651 (2013)
48. H. Chen, R. van Uden, C. Okonkwo, T. Koonen, Compact spatial multiplexers for mode division multiplexing. Opt. Express **22**(26), 31582–31594 (2014)
49. N. Hanzawa, K. Saitoh, T. Sakamoto, K. Tsujikawa, T. Uematsu, M. Koshiba, F. Yamamoto, Asymmetric parallel waveguide with mode conversion, in *Tech. Dig. OFC*, OTu1I.4 (2012)
50. N. Hanzawa, K. Saitoh, T. Sakamoto, K. Tsujikawa, T. Uematsu, M.Koshiba, F. Yamamoto, Three-mode PLC-type multi/demultiplexer for mode-division multiplexing transmission, in *Tech. Dig. ECOC*, Tu.1.B.3 (2013)
51. N. Hanzawa, K. Saitoh, T. Sakamoto, T. Matsui, K. Tsujikawa, M. Koshiba, F. Yamamoto, Mode multi/demultiplexing with parallel waveguide for mode division multiplexed transmission. Opt. Express **22**(24), 29321–29330 (2014)
52. Y. Yamashita, Y. Ishizaka, N. Hanzawa, T. Fujisawa, T. Sakamoto, T. Matsui, K. Tsujikawa, F. Yamamoto, K. Nakajima, K. Saitoh, Excitation of LP21b and LP02 modes with PLC-based tapered waveguide for mode-division multiplexing, in *OECC/PS-2016*
53. S. Gross, N. Riesen, J.D. Love, M.J. Withford, C-band mode-selective couplers fabricated by the femtosecond laser direct-write technique, in *Tech. Dig. OFC*, W3B.2 (2015)
54. T. Watanabe, Y. Kokubun, Demonstration of mode-evolutional multiplexer for few-mode fibers using stacked polymer waveguide. IEEE Photon. J. **7**(6), 7103311 (2015)
55. T. Watanabe, K. Kojima, Y. Kokubun, Serial branching mode multi/demultiplexer for homogeneous multi-core fibers. IEICE Electron. Express **13**(1), 20150961 (2016)
56. C.R. Doerr, N.K. Fontaine, M. Hirano, T. Sasaki, L.L. Buhl, P.J. Winzer, Silicon photonic integrated circuit for coupling to a ring-core multimode fiber for space-division multiplexing, in *Tech. Dig. ECOC*, Th.13.A.3 (2011)
57. T. Su, R.P. Scott, S.S. Djordjevic, N.K. Fontaine, D.J. Geisler, X. Cai, S.J.B. Yoo, Demonstration of free space coherent optical communication using integrated silicon photonic orbital angular momentum devices. Opt. Express **20**(9), 9396–9402 (2012)
58. J. Leuthold, J. Eckner, E. Gamper, P.A. Besse, H. Melchior, Multimode interference couplers for the conversion and combining of zero and first order modes. J. Lightwave Technol. **16**(7), 1228–1239 (1998)

59. Y. Chaen, K. Tanabe, H. Jiang, K. Hamamoto, Low wavelength dependency design for MMI (multi-mode interference) mode converter. IEICE Electron. Express (ELEX) **12**(20), 20150727 (2015)
60. K. Hamamoto, Optical mode switch for future optical router, in *Tech. Dig. ECIO-MOC 2014*, Th 4b I1 (2014)
61. R. Takakura, M. Jizodo, A. Fujino, T. Tanaka, K. Hamamoto, Proposal of optical mode switch. Jpn. J. Appl. Phys. **53**, 08MB10 (2014)
62. R. Imansyah, T. Tanaka, L. Himbele, H. Jiang, K. Hamamoto, Mode crosstalk evaluation on optical mode switch, in *Proceedings of the IEEE IBP Conference*, IBP2015-P3.2, vol. 114, no. 453 (2015), pp. 133–136
63. R. Imansyah, L. Himbele, H. Jiang, K. Hamamoto, First demonstration of electrically controlled mode switching, in *Tech. Dig. 20th Microoptics Conference*, D3 (2015)

Chapter 4
Optical Amplification Technologies

Makoto Yamada, Soichi Kobayashi, Hiroji Masuda, Shoichiro Matsuo, Moriya Nakamura, Shu Namiki, Hirotaka Ono, Ryuichi Sugizaki, and Yukihiro Tsuchida

Abstract This chapter provides a comprehensive review of the optical amplification technologies for space division multiplexing (SDM) transmission that has been studied extensively. It consists of the motivation, history, targets, categories, evaluation methods, and ongoing issues of SDM amplifiers. The amplification technologies discussed in this chapter are mostly erbium-doped fiber amplifiers (EDFAs) and Raman-based SDM amplifiers. It also covers amplification techniques based on bismuth and rare earths that are expected to expand the transmission window from the conventional WDM transmission bands.

Makoto Yamada is a chapter editor.

M. Yamada (✉)
Osaka Metropolitan University, Sakai, Osaka, Japan
e-mail: yamada.makoto@omu.ac.jp

S. Kobayashi
Photonic Science Technology, Inc., Chitose, Hokkaido, Japan

H. Masuda
Shimane University, Matsue, Shimane, Japan

S. Matsuo
Fujikura Ltd., Sakura, Chiba, Japan

M. Nakamura
Meiji University, Kawasaki, Kanagawa, Japan

S. Namiki
The National Institute of Advanced Industrial Science and Technology (AIST), Tsukuba, Ibaraki, Japan

H. Ono
Shonan Institute of Technology, Fujisawa, Kanagawa, Japan

R. Sugizaki · Y. Tsuchida
Furukawa Electric Co., Ltd., Ichihara, Chiba, Japan

M. Nakazawa et al. (eds.), *Space-Division Multiplexing in Optical Communication Systems*, Springer Series in Optical Sciences 236,
https://doi.org/10.1007/978-3-030-87619-7_4

4.1 Introduction

The ever-evolving history of optical fiber communications has been partly but most importantly constructed by the advent and progress of optical amplification technologies. The optical amplification has enabled global-scale communication networks now indispensable for the Internet and smartphone services. In particular, erbium-doped fiber amplifiers (EDFAs) were discovered and invented to find their destined application in the wavelength division multiplexing (WDM) transmission. From the historical point of view, in the dawn of optical fiber communications till early 90s, optical amplifiers were expected for other purposes. One was to extend the unrepeated reach of transmission because their gain was significantly larger than that of the regeneration by an optical-to-electrical (O/E) converter followed by an electrical-to-optical (E/O) converter [1]. Another expected application of EDFAs was the video distribution system in which EDFAs compensated the power loss due to the splitters for broadcasting video signals [2].

When the Internet began to be disseminated in the mid-90s, the world information traffic started increasing relentlessly. Then, owing to the wide operating wavelength range of EDFAs, the WDM transmission quickly met the urgent need to increase the communication bandwidth, coping with the "explosion" of the Internet. The rapid progress of EDFA technology had also nourished the pump laser technology to enable fiber Raman amplifiers, a powerful assist of EDFAs for wider bandwidth and lower noise [3]. Thus, WDM technologies along with EDFAs and Raman amplifiers have since been deployed all over the world, leading to the broadband-connected world as of today.

Figure 4.1 illustrates how the optical fiber communication has evolved through the introduction of the optical amplifier. It is intuitive that quite a few transmitters (Tx) and receivers (Rx) could be omitted by introducing the WDM transmission assisted with optical amplifiers. One of the virtues of the WDM transmission was the ease of upgrading capacity of a transmission line without increasing the amount of facility in the repeater locations, which was decisively cost-beneficial.

Now, we already discussed in the previous chapters that the capacity limit of the WDM transmission is being reached. The space division multiplexing (SDM) will be the last degree of freedom that is exploitable for the purpose of further capacity increase within the transmission band limited by the amplification band. It is indeed an inevitable consequence that because all the fundamental attributes of the light wave, i.e., time, frequency, state of polarization, and quadrature components, have been utilized to achieve the maximum capacity within a single-mode fiber set by the non-linear Shannon limit [4]; further, multiple capacity can only be obtained through the spatial parallelization of transmission lines.

However, any of the upgraded systems must be more efficient than the just multiple installations of the conventional technologies. In this sense, even though the optical fibers became multi-core or multi-mode to carry a multiple of transmissions, the system would not be efficient or attractive enough until the optical amplifiers would be able to efficiently amplify the SDM signals. This is exactly analogous to the

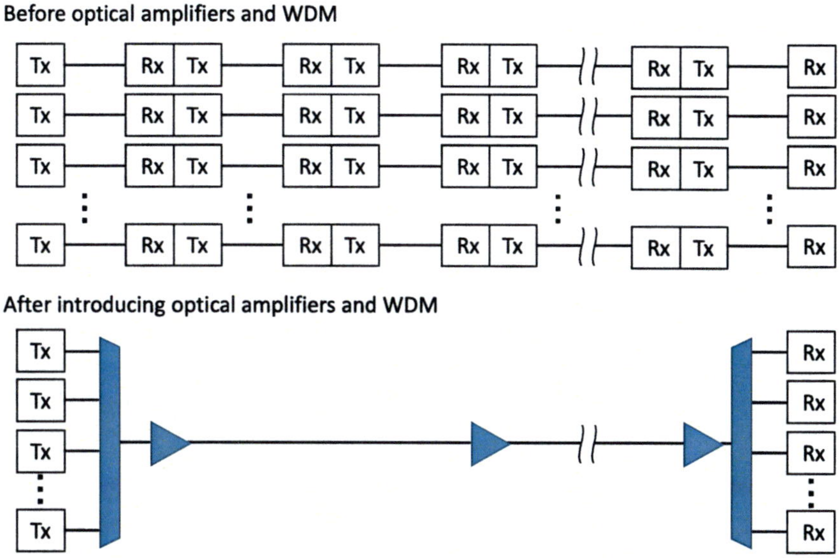

Fig. 4.1 Evolution of optical fiber communications by optical amplification

success of the WDM transmission; for example, the SDM signals need to be simultaneously amplified by a single optical amplification apparatus in order to make the system efficient and cost-beneficial. Figure 4.2 illustrates this analogy from WDM to SDM transmission. The history of WDM readily necessitates a certain form of

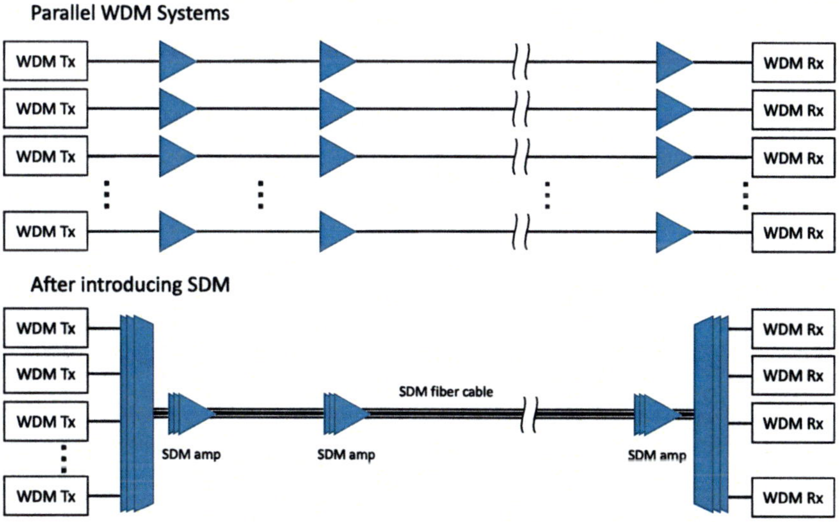

Fig. 4.2 Evolution of optical fiber communications from WDM to SDM

optical amplification tailored to SDM transmission. It should be pointed out that even though the transmission fibers were of neither multi-core nor multi-mode but just bundled as independent fibers in a cable, integration of optical amplifiers into one subsystem to efficiently deal with such multiple transmission lines would still make sense if overall size, cost, and energy consumption became remarkably smaller. Thus, the challenges are how to realize optical amplifiers for SDM, how to satisfy the performance specifications required by the system, and ultimately how to make those optical amplifiers more efficient and cost-beneficial than a multiple of conventional EDFAs.

This chapter is subjected to explore the feasibility and R&D status of the optical amplification technologies for SDM. As optical amplification technologies, only doped fiber amplifiers and fiber Raman amplifiers are taken up, and other technologies such as semiconductor optical amplifiers (SOAs) and fiber optical parametric amplifiers (FOPAs) are not covered in this book. Although it does not mean to deny the future potential of SOAs and FOPAs applicable to SDM, doped fiber amplifiers and fiber Raman amplifiers have shown much more advancement for SDM applications up to date. This may be partly because EDFAs and Raman amplifiers have outperformed the other optical amplification technologies for high-capacity systems and been widely used in various commercial systems.

Apart from SDM optical amplification technologies, this chapter will also cover efforts on amplification techniques that will expand transmission bands, pursuing further scaling of the transmission capacity exploiting the wider low-loss window of the optical fiber. This direction will set a good benchmark before SDM is committed, while it is actually complementary with SDM in the long run.

The structure of this chapter is as follows. First, the historical background for the SDM amplification technology is reviewed in Sect. 4.2. Then, as major promising approaches for SDM, two different kinds of SDM amplification technologies, i.e., multi-core fiber amplification and few-mode fiber amplification technologies, will be referred to in Sects. 4.3 and 4.4, respectively. In Sect. 4.3, 4.3.1 clarifies the target of multi-core EDFA, then Sect. 4.3.2 categorizes their schemes in terms of amplifying media and pumping methods. Section 4.3.3 develops the detailed technical explanations and status updates on each of the multi-core EDFA technologies categorized in Sect. 4.3.2: bundled type EDFA (Sect. 4.3.3.1); core-pumped multi-core EDFA (4.3.3.2); multi-element EDFA (Sect. 4.3.3.3); cladding-pumped multi-core EDFA (Sect. 4.3.3.4); remotely pumped multi-core EDFA (Sect. 4.3.3.5). Section 4.3.4 refers to the state-of-the-art multi-core Raman amplification technology. Section 4.3.5 describes the evaluation methods inherent in the multi-core EDFAs. Future issues on multi-core EDFAs are discussed in Sect. 4.3.6.

Section 4.4 is on few-mode fiber amplification technologies, where Sect. 4.4.1 specifies the target of few-mode EDFAs, followed by the review of state-of-the-art few-mode EDFAs in Sect. 4.4.2. The few-mode fiber Raman amplification technology is discussed in Sect. 4.4.3. Section 4.4.4 then reviews the evaluation methods for few-mode EDFAs. Future issues on few-mode EDFAs are discussed in Sect. 4.4.5. Section 4.5 explores the possibility of amplification outside the conventional WDM transmission bands: 1.3-μm band bismuth-doped fiber amplifier (Sect. 4.5.1) and over 1.65-μm band rare earth-doped fiber amplifiers (Sect. 4.5.2). Section 4.6 concludes this chapter by making remarks for further progress expected for future.

4.2 History of SDM Amplification Technology

SDM amplification technologies, including multi-core (MC) and multi-mode (MM) amplification, have been investigated in various fields of fiber-optic technologies. Some of them are not used for increasing the capacity of optical transmission systems. Here, we review the investigated MC and MM amplification technologies and their background in the primary stage of them and examine how to expand the application of SDM amplification technology to long-haul optical transmission.

MC amplification was needed in the field of optical interconnects to amplify two-dimensional (2-D) optical signals. The optical interconnects have to realize parallel data links, for example, 32-bit or 64-bit parallel data of connected digital signal processors without parallel-to-serial data conversion. 2-D parallel techniques can increase the number of transmission channels dramatically in comparison with conventional one-dimensional parallel interconnects using fiber ribbons. Additionally, massively parallel processing systems with 2-D optoelectronic devices require 2-D parallel optical interconnects [5, 6]. 2-D parallel optical interconnects are also needed in optical computing systems in which 2-D optical information, including image signals, has to be processed [7]. In these systems, however, branching loss is one of the most serious problems because the 2-D optical signals have to be branched repeatedly to enable data links between many devices. The 2-D parallel optical data has to be amplified after transmission and branching functions. Figure 4.3 schematically shows an example of a 2-D parallel optical interconnect system with 2-D optical amplifier. 2-D optical amplifiers are needed to compensate for the transmission loss and branching loss. The 2-D optical signals can be 32- or 64-bit binary parallel signals, as well as analog image signals. In any case, the 2-D parallel optical signals must be amplified while retaining the 2-D geometrical configuration of the signals.

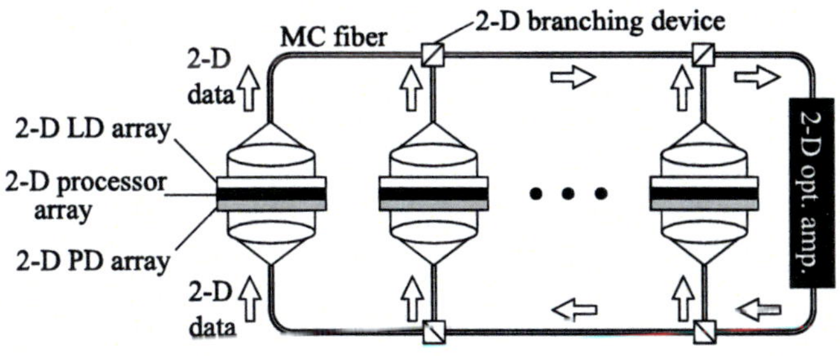

Fig. 4.3 MC-fiber-based 2-D parallel optical interconnect system with 2-D optical amplifier (Reprinted with permission from [11]. © 2001 IEEE)

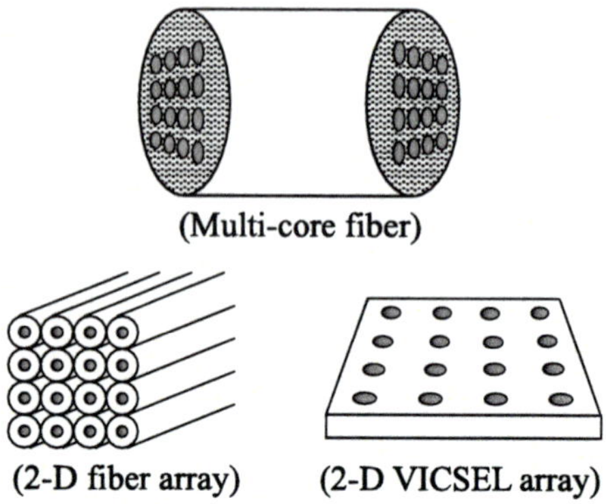

Fig. 4.4 2-D parallel optical amplifiers

Figure 4.4 shows some candidates for 2-D parallel optical amplifiers. A semiconductor optical amplifier (SOA) with a vertical-cavity surface-emitting laser (VCSEL) structure is one possible candidate [8, 9]. A VCSEL-based SOA itself has advantages, including minimal insertion loss and polarization insensitivity. Its high suitability for 2-D integration is also attractive from the viewpoint of 2-D optical systems. However, SOAs have some drawbacks, including a low-output saturation power and low gain compared with fiber-based optical amplifiers. Fiber-based 2-D parallel optical amplifiers can be realized simply by bundling normal EDFAs, forming packages of 2-D fiber arrays [10]. In this case, however, the cost of the amplifier increases in proportion to the number of bundled fibers. In particular, it is difficult to realize a compact assembly of the overall amplifier system, including pumping components. The other candidate is an MC-fiber-based amplifier. An MC fiber has multiple cores in a common cladding. When it is used as an endoscope, it is also called an image fiber. 2-D parallel optical amplifier can be realized by doping the cores with Er. This MC Er-doped fiber (MC-EDF) has all the cores in a common cladding. Therefore, a compact and cost-effective assembly can be realized by sharing the pumping components among the multiple cores.

To the best of the author's knowledge, the first experimental demonstration of MC amplifiers dates back to 2001 [11]. Figure 4.5 shows specifications and cross-sectional views of the first fabricated MC amplifier. It was designed to amplify parallel 1.55-μm optical signals with Er-doped cores. The number of cores was as large as 3000 to enable the amplification of infrared images, as well as binary optical signals. The signal gain characteristics are shown in Fig. 4.6. The characteristics were measured by changing the power of the signal power, P_s, and pump power. The length of the MC-EDF was 3.0 m. The measured gain versus pump power characteristic agreed well with the theoretical saturation curve. However, the gain saturated at less than 1.2 dB because the pumping efficiency was low due to the multi-mode cores. In case where MC-EDF is used only for long-haul data transmission, without considering infrared image amplification, the number of cores can be decreased, and the pumping efficiency can be improved by optimizing the fiber parameters.

Parameter	Specification
Number of cores	3,000
Fiber diameter	630 μm
Cladding diameter	590 μm
Core diameter	4.5 μm
NA	0.25

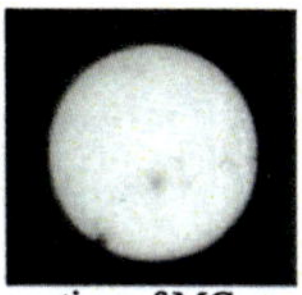
Cross section of MC amplifier

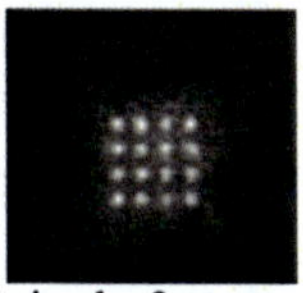
Output signals after propagation through the MC amplifier

Fig. 4.5 Experimentally fabricated MC amplifier (Reprinted with permission from [11]. © 2001 IEEE)

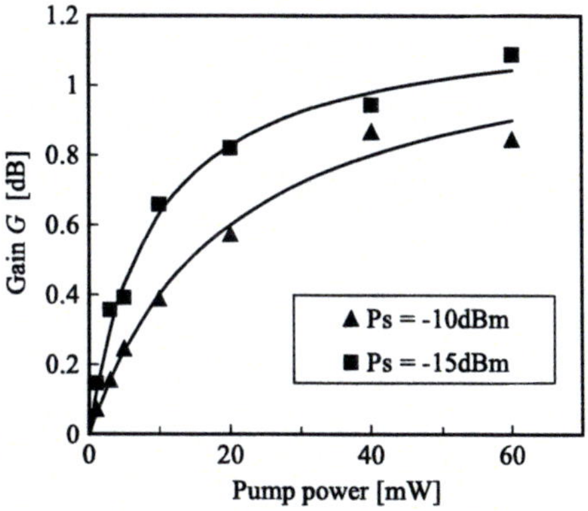

Fig. 4.6 Signal gain characteristics as a function of pump power (Reprinted with permission from [11]. © 2001 IEEE)

To the best of the author's knowledge, the first experimental demonstration of Er-doped MM fiber amplification was reported by Nykolak et al. [12]. In this report, however, the targeted application was not the amplification of mode-division multiplexed signals. The MM amplifier was studied for its application to low-cost and/or robust MM systems, including local area networks (LAN's), avionic distribution systems, and fiber sensors systems. The theoretical number of propagation modes was 22. Measurement was performed using a mode-scrambler, which was used to excite as many propagating signal and pump modes as possible. The measured gain was 16 dB at a signal wavelength of 1543 nm, with a pump power of approximately 100 mW at a wavelength of 980 nm. EDFAs with MM cores are also commonly used in high-power applications to decrease the spatial power density in the cores. A high-power super fluorescent source and high-power amplification are examples of the applications [13–15]. In these applications, however, only the fundamental mode in the MM cores is used.

4.3 Multi-core Fiber Amplification Technology

4.3.1 Target of Multi-core EDFA

Multi-core fiber (MCF) transmission has attracted attention because it has the potential to overcome the capacity crunch and to achieve an ultra-high exabit/s class capacity. A long-haul transmission system utilizing an MCF requires an EDFA to maintain a high optical signal power level, in the same way as a single-core fiber transmission system.

The EDFA of the MCF transmission system may be constructed by simply arranging a large number of conventional EDFAs (arrayed EDFA), but multi-core EDFAs (MC-EDFAs) using new amplification mediums are been developed. The aim is to realize superior performance through the integration of both the optical components and the cores of erbium-doped fiber (EDF) as shown in Fig. 4.7, without any degradation in such amplification properties as the crosstalk (XT) between optical signals, which propagates in each core, and the amplification efficiency. The XT characteristic is a particularly important issue because several multi-core EDFA cores must be incorporated with high density. Furthermore, it is also important in terms of achieving the same amplification characteristics for each core.

With a multi-core EDFA, the aims are to realize a smaller size, lower cost, and lower power consumption than an arrayed EDFA [16–19].

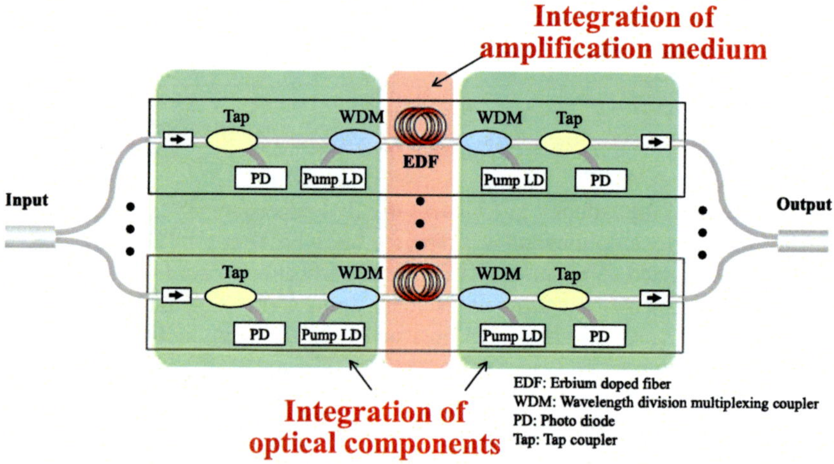

Fig. 4.7 Concept of multi-core EDFA

4.3.2 Categorization of Multi-core EDFA

Table 4.1 categorizes the amplification medium and pump method of an MC-EDFA [17–19]. There are three types of amplification medium, bundled EDF [20–22], multi-core EDF [23–33], and multi-element EDF [34–36].

The advantages of bundled EDF are that a mature EDF fabrication technique can be used, and the length of each EDF can be adjusted to achieve the same gain. Moreover, a combination of different EDFs, such as single-mode and multi-mode EDF, C-band and L-band EDF can be easily achieved; although, we have to develop an additional fabrication process, namely a way of bundling the fiber and reducing the cladding diameter to achieve downsizing with a low XT. With multi-core EDF, a multi-core fiber fabrication technique would be applicable, and cost reduction and downsizing made possible by manufacturing several cores in one fiber fabrication operation. However, we have to achieve a uniform amplification characteristic for each core and reduce the cladding diameter while maintaining a low XT for practical use.

Multi-element EDF can be prepared by improving the fiber drawing process of the mature EDF fabrication technique. The issue is the increase in the number of multi-elements. On the other hand, there are two pumping methods, namely discrete or core pumping [23–27] and cladding pumping [23–36]. Discrete pumping can be applied to the conventional EDF pumping method and its optical components and

Table 4.1 Categorization of amplification medium and pump method for MC-EDFA

Amp. medium	Bundled EDF	Multi-core EDF	Multi-core EDF: Double-cladding structure	Multi-element EDF
	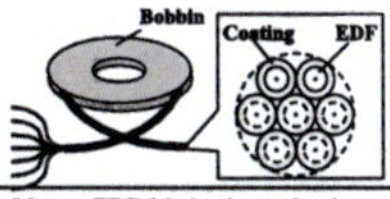	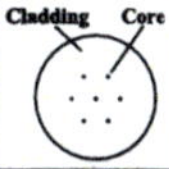	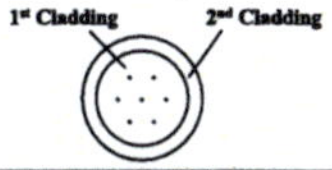	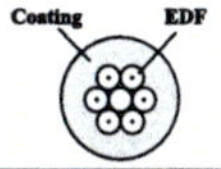
Features	·Mature EDF fabrication techunique can be used. ·Each EDF length can be adjusted to achieve same gain. (Combination of different EDFs).	·Multicore fiber fabrication technique would be applicable. ·Cost reduction and downsizing possible by manufacturing several cores in one fiber fabrication operation.		·Multi-element EDF can be prepared by improving fiber drawing process of mature EDF fabrication technique.
Issues	·Development of bundling technique .	·Development of practical multicore EDF fabrication technique.		·Increase in number of elements.
		Amplification characteristic of each core must be same.		
	·Reduce cladding diameter while maintaining low XT.			
Pump method	Discrete pumping (core pumping)		Cladding pumping	
	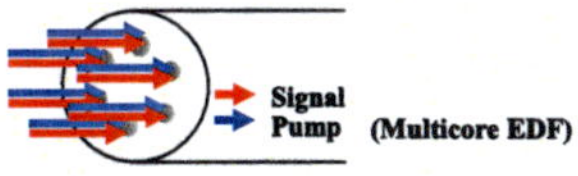		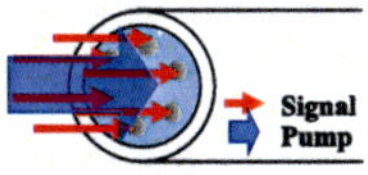	
Features	·Pump method and optical components used for conventional EDF can be adapted. ·Highly efficient pumping and high-speed control.		·Lower power consumption and downsizing possible by decreasing number of pump LDs.	
Issues	·Downsizing, price reduction, low power consumption by integrating optical components.		·Improved pumping efficiency. ·New pump combiner for cladding pump. ·Technique for adjusting technique for pump power of several cores ·High-speed control.	
References	20 ~ 22	23 ~ 27	28 ~ 33	34 ~ 36

Reprinted with permission from [18]. © 2014 SPIE

provides highly efficient pumping and high-speed control. The problems to be solved involve integrating the optical components to reduce the total amplifier size, cost, and power consumption. Cladding pumping, which is employed for multi-core EDF with a double-cladding structure, has the potential to realize both a lower power consumption and downsizing by reducing the number of pump LDs and improving pumping efficiency. We must also develop new optical components for the cladding pumping and a technique for adjusting the gain of several cores to achieve a pump power with high-speed control. Recently, a hybrid pumping technique, which employs both core and cladding pumping, has been proposed for use with multi-core EDF with a double-cladding structure. It has been confirmed that the gain of several cores can be adjusted individually, and the potential for high-speed control and low power consumption has been revealed [37].

Another way to categorize the MC-EDFA is in terms of lumped and distributed amplification. Remotely pumped amplification, which is a distributed amplification technique, is an important approach for realizing long-haul transmission systems [27, 38], in the same way as the multi-core Raman amplification technique [39].

4.3.3 State-of-the-Art Multi-core EDFA Development Technology

In this section, recent results on bundled type EDFA, core-pumped multi-core EDFAs, multi-element EDFAs, cladding-pumped multi-core EDFAs and remotely pumped multi-core EDFA are described in detail.

4.3.3.1 Bundled Type EDFA

The bundled type EDFA is constructed with bundled EDF, which consists of a bundle of identical single-core EDFs. Bundled EDF has the following advantages [20, 21].

1. Mature fiber fabrication technology: Other than the need for an additional fabrication process, namely bundling, bundled EDF can be prepared with a high yield ratio.
2. Amplifier design flexibility: Various configurations are available by combining EDFs with different lengths and/or dopant concentrations. Then, the amplification of different wavelength bands such as the C- and L-bands can be achieved. The combination of single- and multi-mode EDFs in a bundle is also possible.
3. Lower crosstalk (XT) characteristics: This is an important issue regarding the MC-EDFA, and there is the possibility of reducing the XT by changing the environmental material surrounding the EDFs.
4. Possibility of compact size: As mentioned above, the MC-EDFA amplification medium requires low XT and a certain distance between the Er-doped cores of the multi-core EDF, which is another candidate amplification medium for

the MC-EDFA, and this leads to a cladding diameter increase. To allow us to construct a practical compact EDFA unit, we store the fiber in the unit by coiling it on a bobbin with a certain bending diameter that achieves a low failure probability. We may be able to achieve compact storage with a bundled EDF consisting of reduced cladding EDFs by coiling them together on one bobbin.

5. Efficient EDF control technology can be used: A practical MC-EDFA unit for WDM transmission systems requires complicated EDF controls such as automatic level control (ALC) or automatic gain control (AGC), and gain tilt correction. High-speed control may also be required. An EDFA unit constructed with bundled EDF can realize the benefit of current control technologies, which have been developed for a unit with single-core EDF.
6. Elimination of fan-out devices: Each EDF can be directly connected to a signal and pump multiplexer.

The disadvantage of using bundled EDF is the additional fabrication process required for bundling, as mentioned above. Of course, multi-core EDF has the advantages of being an amplification medium that can be prepared by drawing one fiber and of having the potential for cladding pumping with one pump LD, and thus achieving a cost-effective MC-EDFA.

Figure 4.8 shows the configuration of a bundled EDFA [20, 21]. Seven bundled reduced cladding EDFs with a cladding diameter of 60 μm were used to construct the bundled and compactly coiled EDF. The EDF was 7.3 m long. We used a silica-based planar lightwave circuit (PLC) type WDM coupler array and a V groove connection between the bundled EDF and the array to realize a compact MC-EDFA. Figure 4.9 shows the amplification characteristics. A practical average gain of about 23 dB and

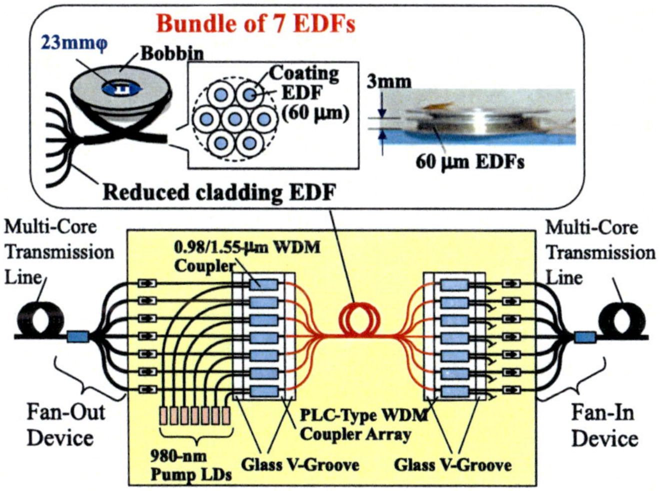

Fig. 4.8 Configuration of bundled EDFA using reduced cladding bundled EDF (Reprinted with permission from [21]. © 2012 IEEE)

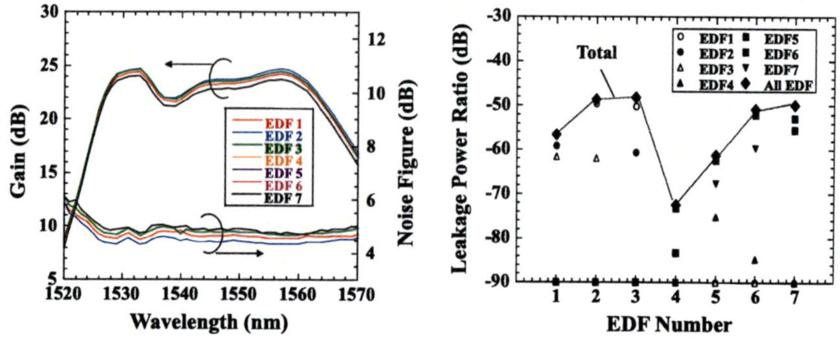

(a) Signal gain and noise figure characteristics (b) XT characteristics

Fig. 4.9 Amplification characteristics of bundled EDFA using reduced cladding bundled type EDFA (Reprinted with permission from [21]. © 2012 IEEE)

a noise figure of less than 5.1 dB were achieved as shown in Fig. 4.9a [21], and it is confirmed that the same amplification characteristic can be realized for every core by employing bundled EDF. The differences between the average gains and noise figures are due to the different splicing losses of the EDF and PLC coupler array. The total XT was less than − 48.2 dB as shown in Fig. 4.9b [21], and we found that the XT characteristics may degrade depending on the way the EDF is wound on the bobbin.

Recently, a technique where the EDF configuration in a bundled EDF is rearranged along the fiber, as shown in Fig. 4.10, has been proposed to avoid an important XT issue with bundled and multi-core EDF, namely that the total/XT values of the cores are not uniform and a specific core has the highest total XT value because of the fixed configuration of the cores along the fiber [22].

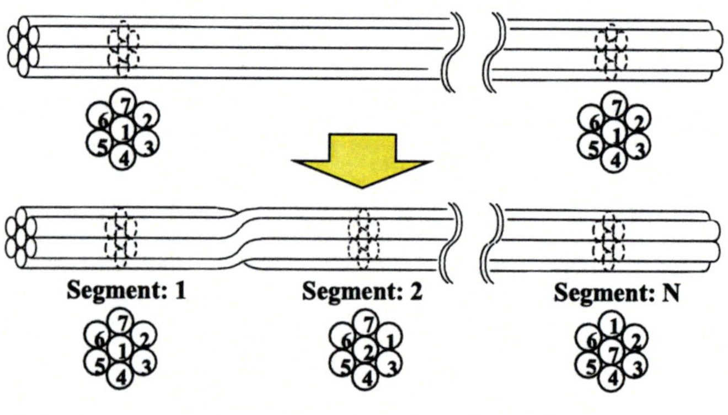

Fig. 4.10 XT improvement technique of bundled EDF

4.3.3.2 Core-Pumped Multi-core EDFA

Amplification scheme of core-pumped MC-EDFA is almost same as standard single core EDFA. The most important difference between them is using MC-EDF for amplification medium for MC-EDFA. Also, input/output devices for MC-EDF are needed only for MC-EDFA. In this section, components including MC-EDF and characteristics of core-pumped MC-EDFA are introduced.

MC-EDF is a MCF which has erbium-doped core. In general, a core optimized for standard EDF is used for the MC-EDF. The key parameters for the design of MC-EDF are core pitch and cladding diameter. Reported MC-EDF with shortest core pitch and smallest cladding diameter are 30 μm and 100 μm, respectively [24]. The merit for this fiber is three cores are included in the smaller cladding area than standard EDF. Crosstalk is one of the most important issues for MC-EDFA. For MC-EDF, crosstalk coefficient between each core is designed to have larger value because length of MC-EDF used in MC-EDFA is very short. For example, adequate length of three-core EDF described in [1] is 12 m. In the transmission experiment, any degradation by crosstalk is not detected. Crosstalk of MC-EDFA with three-core MC-EDF is certified less than − 33 dB at 1550 nm.

Fan-in/out devices are also important for core-pumped MC-EDFA. So far, spatial coupling types and fiber device type fan-in/out devices are mainly reported. One of the requirements for fan-in/out for MC-EDFA is to have low insertion loss because insertion loss of the components determines noise figure value. For high-power amplifiers, reflection from splicing points should be avoided because it causes increase of noise in the amplifier. To satisfy these demands, fiber bundle type fan-in/fan-out device (FBF) is the most promising candidate due to its low loss and low reflection characteristics. Figure 4.11 shows FBF for MC-EDF [40]. FBF is fabricated with thin cladding fibers. Cladding diameter of thin cladding fiber is selected to be the same value of core pitch of MCF. Difference between requirements to thin cladding fibers for FBF for MC-EDF and standard MCF is size of MFD. Because butt joint technique is used for connection between FBF and MC-EDF, it is better to set MFD for both fibers are similar value to reduce splice loss between MC-EDF and FBF. MFD of EDF is set to smaller than the ones of transmission fibers in order to get higher power conversion efficiency. That means MFD of thin cladding fibers for FBF for MC-EDF

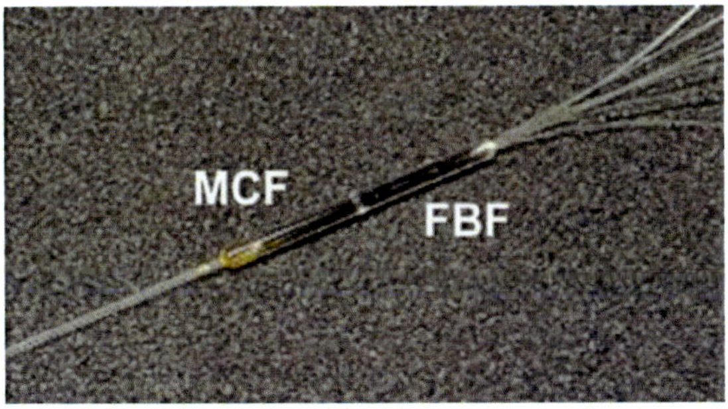

Fig. 4.11 Fiber bundle-type fan-out (FBF)

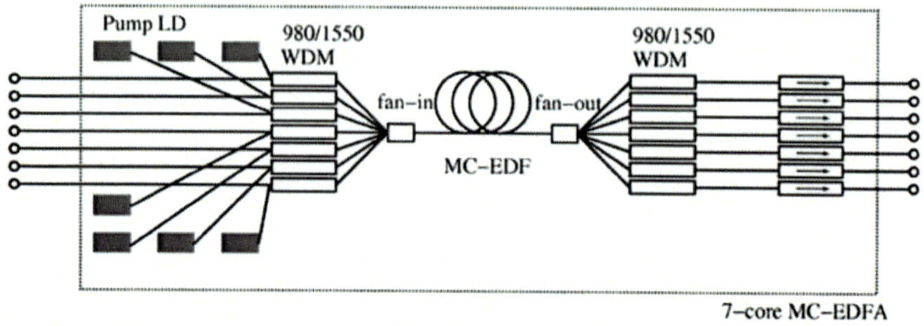

Fig. 4.12 Configuration of core-pumped MC-EDFA (Reprinted with permission from [42]. © 2012 Authors)

should have smaller value. On the other hand, another end of fan-out fiber must be connected with standard fiber. At this splice point, MFD conversion technique may be adopted. By improved FBF reducing both insertion loss and reflection, NF of MC-EDF is regarded as almost same as the one of regular EDFA.

Figure 4.12 shows configuration example of core-pumped MC-EDFA with forward pumping. In this setup, seven-core EDF with cladding diameter of 180 μm, core-to-core pitch of 45 μm, mode field diameter of 7.3 μm at 1580 nm, core absorption coefficient of 3.4 dB/m at 1550 nm is used [41]. Length of MC-EDF is set to 8 m to obtain sufficient gain over entire C-band region. Each signal and pumping power is combined with WDM coupler. Combined light is inputted into each core of MC-EDF with FBF.

Figure 4.13 represents gain and noise figure spectrum of seven-core EDFA where input signal power is set to 0 dBm. From this figure, flat gain and low NF between 1530 and 1560 nm wavelength range for all seven cores are shown. Crosstalk properties between two different cores are shown in Fig. 4.14. Crosstalk component of less than − 50 dB is achieved by using well-developed FBF component.

Based on above consideration of seven-core MC-EDFA with core-pumped scheme, core-pumped MC-EDFAs have the same amplification characteristics as

Fig. 4.13 Gain and noise figure spectrum with core-pumped MC-EDFA (Reprinted with permission from [41]. © 2013 Authors)

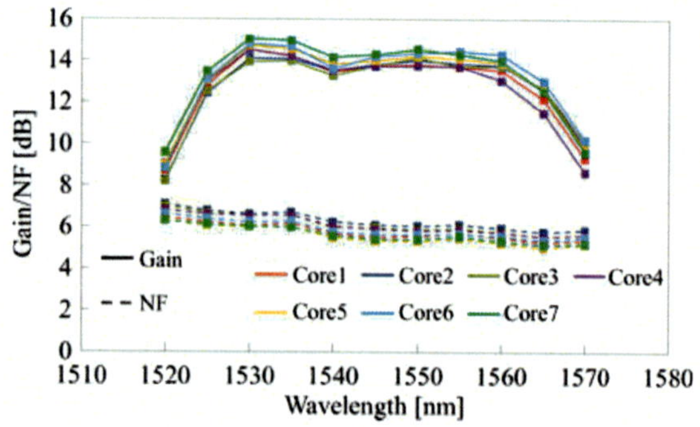

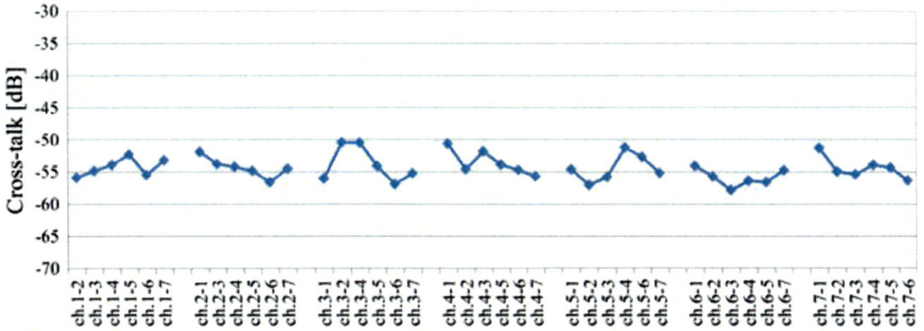

Fig. 4.14 Crosstalk properties between two different cores (Reprinted with permission from [41]. © 2013 Authors)

standard single core EDFAs and have achieved low crosstalk properties with well-developed SDM Mux/DeMux components.

4.3.3.3 Multi-element EDFA

Multi-element fiber (MEF) has multiple fibers in the single coating layer. Each element is thin cladding fiber which is made of glass. Multi-element EDFA (ME-EDFA) is the amplifier whose gain medium is MEF. The shape of MEF is almost same as standard fibers. Since both of them have multiple cores in the fiber, MEF can be used in the same application area for multi-core fibers (MCFs). Configuration of MEF and MCF is shown in Fig. 4.15. Difference between each ones is structure of cladding. MCF has only one cladding in the fiber and multiple cores are located in the common cladding. On the other hand, each element in MEF can work as individual fiber which has core and individual cladding.

First experiment utilizing Multi-element EDF(ME-EDF) which has the same structure as MEF was reported in 2013 by University of Southampton [43]. Jain

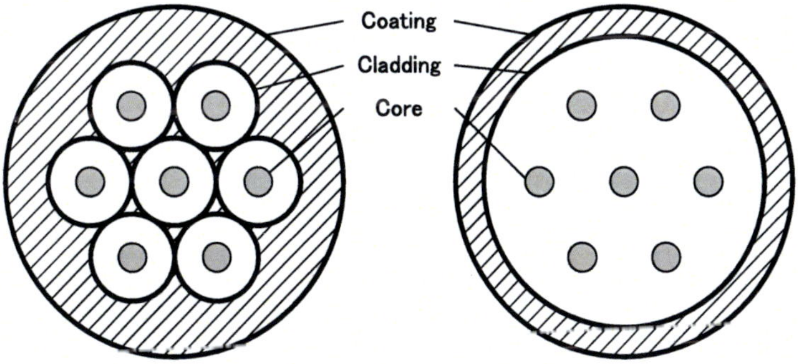

Fig. 4.15 Comparison of MEF and MCF

et al. reported 3-elements and 7-elements ME-EDF in this manuscript. ME-EDFs which have three cores [24] and seven cores [42] had already been reported. Differences between reported ME-EDF and MC-EDF is their fiber outer diameter and core pitches. Reported three-core multi-core fiber has cladding outer diameter of 100 μm, core-pitch of 30 μm, and coating outer diameter of 240 μm. Seven-core multi-core fiber has cladding outer diameter of 180 μm, core-pitch of 45 μm, and coating outer diameter of 220 μm. Furthermore, 12-core Er/Yb-doped fiber (EYDF) [31], whose cladding outer diameter is 216 μm, core-pitch is 37.2 μm, and coating outer diameter is 356 μm, is also reported. Comparing MC-EDF, coating outer diameters of each ME-EDF are relatively larger because outer diameter of each element in reported ME-EDF is 125 μm. 5-element fiber [35, 44] which is reported as Multi-element Er/Yb-doped fiber (ME-EYDF) has single coreless element located center of it and four Erbium-Ytterbium-doped elements are located around the coreless fiber. Figure 4.16 shows structure of 5-element fiber reported in [44] and [35]. Though outer diameter of each element was set to 80 μm, outer coating diameter of this fiber is still larger than seven-core MC-EDF. Parameters for reported ME-EDF, ME-EYDF and MC-EDF are summarized in Table 4.2.

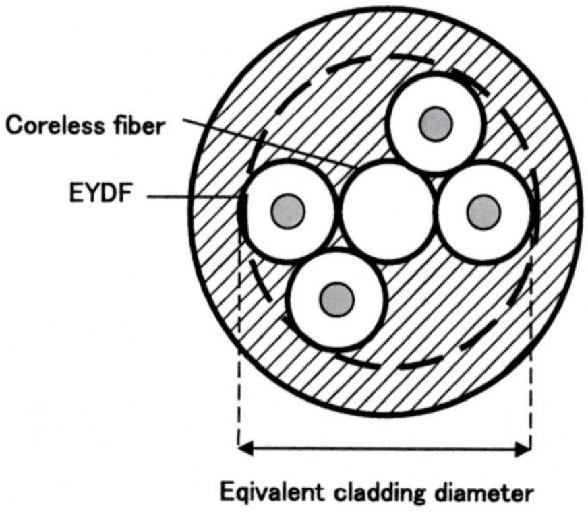

Fig. 4.16 Structure of 5-element EYDF

Table 4.2 Parameters for reported ME-EDF, ME-EYDF, and MC-EDF

Items	Units	Multi-element			Multi-core		
Rare earth	–	Er	Er	Er/Yb	Er	Er	Er/Yb
Core#	–	3	7	5	3	7	12
Core# w/rare earth	–	3	7	4	3	7	12
O.D. of element/core pitch for MCF	μm	125	125	80	30	45	37.2
O.D. of coating	μm	310	460	305	240	220	356

One of the important characteristics for MCF/MEF is crosstalk between each core. Mechanism of crosstalk generation is mode coupling between cores. Mode coupling is occurred through common cladding. In general, crosstalk is mostly determined by distance between each core and fiber length. Length of MC-EDF in the MC-EDFA is relatively shorter than MCF for transmission line, crosstalk of MC-EDF can be set to be larger value comparing other MCFs. Distance between cores for MCF is core pitch. Minimum core pitch of reported MC-EDF is 30 μm due to keeping adequate crosstalk value.

Distance between cores for MEF is determined by diameter of each element. Crosstalk is not a problem for MEF because MEF does not have common cladding. Mode coupling between cores may occur through contact portion of each element but contribution for crosstalk is negligible. That means distance between cores for MEF is not determined by crosstalk. Conversely each element should work as independent fiber. Minimum outer diameter of each element is determined by leaky mode loss of individual fiber elements. Reported minimum outer diameter of ME-EDF is 80 μm. Equivalent cladding radius can be defined to distance from center of cladding to outer glass edge. Equivalent cladding diameter of reported 5-element ME-EYDF is 240 μm. This value is quite larger than reported cladding diameter of MC-EDF. This is the reason why outer coating diameter of ME-EDF is larger than the one of MC-EDF.

Two types of configurations for Multi-element Erbium-doped fiber amplifier (ME-EDFA) has reported. One is core-pumped ME-EDFA. The other is cladding-pumped one. For core-pumped one, configuration is almost same as MC-EDFA. Components for ME-EDFA are almost same as standard single-core EDFA except for fan-in/fan-out devices located at the both ends of ME-EDF. For cladding-pumped one, special pumping devices, which are different from the ones for MC-EDFA, are needed for inputting pump power.

In [44], ME-EYDF for cladding pumping is reported. Pumping light is inputted through coreless fiber located in the center of ME-EYDF. Pumping power is guided by polymer cladding whose refractive index is lower than silica glass and coupled to other Er/Yb-doped elements through contact portion of each element. This is mechanism for exciting four Er/Yb-doped elements by single pumping light inputted through coreless element.

In [35], characteristics of ME-EDFA using 5-element EYDF are reported. Gain of over 20 dB and noise figure (NF) of less than 11 dB were achieved, respectively, for wavelength range between 1530 and 1560 nm where input signal power was set to – 23 dBm. Pumping power launched into coreless element was set to 6.4 W by single multi-mode LD at 960 nm. Since minimum NF is 4.7 dB in longer wavelength region, NF of this setup may be reduced in the future by improvement of fan-in/fan-out devices. Though this ME-EYDF has coreless element in the center, it is possible to set the doped element for increasing total amplified power. But it will be difficult to manage variation of output power level because pumping power intensity in the center core would be quite higher than others. Also, making signal/pump combiner for this setup would be difficult. That means reported 5-element EYDF is one of the best solutions.

The merit of ME-EDF against MC-EDF is in its manufacturing method described in [43]. Comparing the process of standard fibers, attaching assembled preforms to handle is the only process. Assembled preform is drawn as single fiber using furnace and coated inline. By this method, multiple elements are included in a single coated fiber. On the other hand, fabrication method of MC-EDF is complicated and difficult for cost reduction. That means ME-EDF has a potential for low-cost SDM amplifiers.

Studies for ME-EDF have just started. Much potential for possibilities would be existed. Application area for ME-EDF would be the same as MC-EDF. Both MC-EDF and ME-EDF will be used for each adequate application areas.

4.3.3.4 Cladding-Pumped Multi-core EDFA

Cladding pumping technique is used for high output power EDFA which is mainly used for CATV systems. The difference between regular core-pumped EDFA and cladding-pumped one is propagation area of pumping light in EDF. EDF for cladding-pumped one has double-cladding structure, which is consist of core, inner cladding, and outer cladding. Signal light is inputted to core of double-cladding EDF, which is designed to guide signal light. Pumping light is coupled to inner cladding of cladding-pumped EDF. Since inner cladding area is much larger than core area, pumping light to be inputted into double-cladding EDF can be set to be much higher power than regular scheme. This is the reason why higher power can be obtained by cladding-pumped EDFA. In addition, high-power multi-mode LD (MM-LD) has been improving for getting higher output power. Both double-cladding EDF and high-power MM-LD make output power of cladding-pumped EDFA higher.

Cladding pumping scheme is applicable for double-cladding MC-EDF. Only one MM-LD is needed for exciting all cores embedded in MC-EDF. The structure of MC-EDF is almost same as the one for core-pumped one. Design of core and its allocation of MC-EDF for cladding pumping can be adopt as same as core-pumped one. The only difference is its cladding structure. In order to guide pumping light inputted into inner cladding, low index polymer is used as coating layer. The roles of low index polymer layer are both guiding pumping light in inner cladding layer which is made of silica glass and protection for glass layer.

One of the key technology for cladding-pumped MC-EDFA is pump light combiner which introduces pump light into MC-EDF with signal light from the MCF. So far, two types of pump light combiner for MC-EDFA have been proposed. One is based on a space-coupling technique. The other is based on a fiber coupling technique with fiber bundle or fiber taper.

Adoption of free-space optics for signal/pump combiner for MC-EDFA is simple way to couple among any kinds MCF and MC-EDF. Mimura et al. firstly reported MC-EDFA with free-space pump light coupling in 2012 [29]. In this report, two cores in MCF are excited by single MM-LD. Since common objective lens was used to couple both signal lights and pumping light, pump power efficiency was remained low value. To solve this problem, Takasaka et al. proposed a new multi-mode pump light coupler that does not include a shared objective lens to avoid chromatic aberration

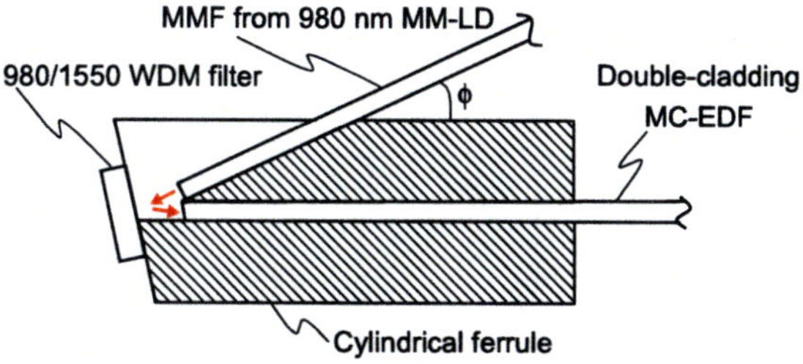

Fig. 4.17 Longitudinal profile of pump coupler based on space-coupling technique

[30]. A longitudinal section profile of the multi-mode pump coupler is shown in Fig. 4.17. The coupler consists of a seven-core double-cladding MC-EDF as a gain medium, a multi-mode fiber for 980 nm multi-mode pump light, a 980/1550 WDM filter that reflects 980 nm light and transmits 1550 nm light, and a cylindrical ferrule to fix all components. The multi-mode fiber has core and cladding diameter of 105 μm and 125 μm, respectively, and numerical aperture of 0.22. Multi-mode pump light emitted from the multi-mode fiber is reflected by the WDM filter and is input to the inner cladding of the double-cladding MC-EDF. To make the coupling loss of 980 nm pump light as low as possible, the optical path from the end of the multi-mode fiber to the end of the double-cladding MC-EDF is reduced less than hundreds micro meters. Incident angle ϕ shown in Fig. 4.17 is optimized to decrease the coupling loss. The insertion loss of the multi-mode pump coupler is less than 3.4 dB. This value consists of the loss of the multi-mode pump coupler and the absorption of the MC-EDF. In addition, distribution of pump power in the inner cladding at the MC-EDF output end was almost flat though that at the multi-mode fiber end was Gaussian shape. That means gain variation caused by imbalanced pump power distribution does not occur by this scheme.

Coupling loss of pumping power should be reduced to achieve higher pump power efficiency and to avoid the damage of combiner by heat from uncoupled power. To reduce the coupling loss of signal light, fiber type combiner is advantageous. Abedin et al. reported cladding-pumped MC-EDFA with fiber bundle pump combiner in 2012 [28]. Fiber bundle pump combiner has close-packed hexagonal lattice with 7 thin cladding fibers, where the centrally located optical fiber is for launching pump light into inner cladding of MC-EDF. Using this technology, six cores can be excited in MC-EDF. For exciting seven cores in MC-EDF, Abedin et al. reported side-coupling method using tapered multi-mode fiber [44]. The coupling efficiency of the combiner for pump light was reported as 67%. Fiber type pump/signal combiner is promising to improve pump power efficiency and to solve heat problem.

MC-EDFA utilized double-cladding MC-EDF and pump/signal combiner have been reported. For C-band amplification, utilizing space-coupling combiner was reported by Takasaka et al. [30]. Figure 4.18 shows the setup to measure amplification characteristics. Length of the MC-EDF is set to be 10 m to obtain the wide-band gain in C-band. The multi-mode fiber of the multi-mode pump coupler is connected with a 980 nm multi-mode LD. At input side, optical signal output from the tunable laser source (TLS) is connected to one of 7 SMFs and is collimated by following lens

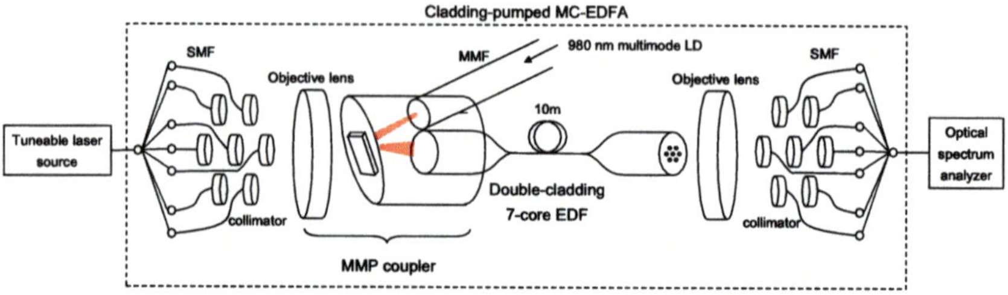

Fig. 4.18 Experimental setup of cladding-pumped MC-EDFA with space-coupling technique (Reprinted with permission from [30]. © 2013 IET)

Fig. 4.19 Gain and noise figure spectrum with cladding-pumped MC-EDFA (Reprinted with permission from [30]. © 2013 IET)

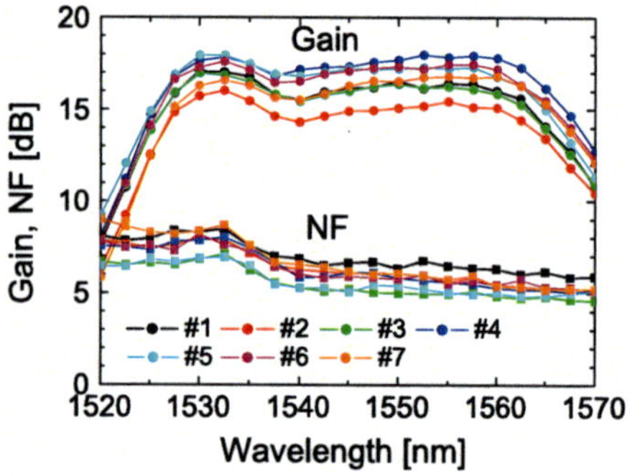

and is focused on the end of the MC-EDF by a shared objective lens. At output side, optical signal output from the MC-EDF is coupled with a SMF by a shared objective lens and one of seven collimators. Amplified signal is measured by an optical spectrum analyzer. Figure 4.19 shows gain and noise figure spectrum for all 7 cores, where pump power and signal power was held at 39.45 dBm (8.8 W) and − 18 dBm, respectively. Gain shape correspond to that for conventional EDFA. In C-band, all cores have gain > 14 dB and NF < 9 dB. At pump power of 39.45 dBm (8.8 W), the maximum output power was 8.7 dBm. The average slope of the saturated power to the pump power is 2.2 dB/dB. The saturation power is too small to amplify WDM signals. Though higher NF and lower output power comparing conventional EDFA is the problem, this can be solved by improved pump/signal combiners.

Seven-core EDFA with higher output power is reported by Maeda et al. using fiber type pump/signal combiner [45]. To obtain higher output power, higher pumping power inputted into MC-EDF and longer MC-EDF are adopted for setup of MC-EDFA. Insertion losses of the pump combiner for pump light and for signal light are 0.17 dB and less than 0.5 dB, respectively. Since insertion loss of pump combiner for pumping power is reduced, higher pumping power can be inputted into MC-EDF without damage of pump combiner. Since higher pumping power is inputted in this setup, reflected light should be cared at the splicing points. At the input end of pump/signal combiner, fiber bundle type fan-in/out (FBF) devices as SDM

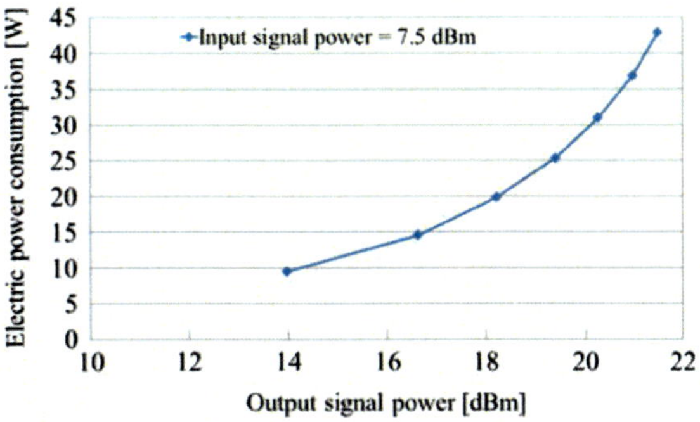

Fig. 4.20 Electric power consumption in MM-LD (Reprinted with permission from [45]. © 2015 IEEE)

Mux/DeMux are used. By applying angled polishing technique to both end faces, reflection less than − 70 dB at splice points for all cores is achieved. Since the length of MC-EDF is set to 107 m to obtain higher output power, amplification region is shifted to L-band. In this report, maximum output power from 20.6 to 21.9 dBm is obtained for 7 cores in L-band. Gain about 11 dB is obtained from 1578.4 to 1608.1 nm in L-band for all the seven cores when total inputted 51ch WDM signal power of 7.5 dBm NF spectrum for each core is flat with deviation less than 1 dB. XTs are measured by launching two signals of which wavelength are 15913 and 1594 nm simultaneously with power of 0 dBm into two different cores, each other. The MM-LD is set to output power with 40.0 dBm during the measurement of the XT. The XTs between all cores is less than − 40 dB in this report. Since all core is excited by only one MM-LD, cladding-pumped MC-EDFA is expected to realize EDFA with low power consumption. Figure 4.20 shows the result when the input signal power is 7.5 dBm. From this result, it is confirmed that the electric power consumption about 33 W when the output signal power is 20 dBm.

Recently, 32-core MC-EYDFA was also reported with a minimum gain of > 17 dB and an average noise figure (NF) of 6.5 dB over all cores in the wavelength range 1534–1561 nm for − 4 dBm input signal power [46].

4.3.3.5 Remotely Pumped Multi-core EDFA

Remotely pumped EDFAs (RP-EDFAs) or remote optically pumped amplifiers have been studied since the 1990s in order to enhance the transmission distances of unrepeatered ultra-long-haul undersea transmission systems using single-core fibers [47, 48]. Transmission distances of up to about 500 km have been reported to date [49]. The RP-EDFA system has an advantage in that there is no need for electrical power supply cable equipment in the transmission line, which optically amplified repeatered transmission systems require. In a conventional RP-EDFA system, the transmission distance is designed to be larger than that of a distributed Raman amplifier (DRA) system; although, the former has a more complicated structure because of the additional RP-EDFA module. Pump powers of up to several watts were used for both

RP-EDFA and DRA systems for undersea system applications, where fusion splicing was used instead of optical connector splicing, if needed, in order to prevent damage to the optical connector due to the high pump power. Furthermore, several studies using RP-EDFAs on terrestrial systems applications were reported [48, 50].

Several studies of RP-EDFA systems using multi-core fibers (MCFs; RP-MC-EDFA systems) have been reported since 2011 [27, 38, 51, 52]. It was theoretically and experimentally shown that the RP-MC-EDFA system required a significantly lower pump power to achieve a predetermined improvement in the optical signal-to-noise ratio (ΔOSNR) than the multi-core DRA (MC-DRA) system [27, 51, 52]. ΔOSNR is defined as the difference between the OSNR in the RP-MC-EDFA system ($OSNR_{RP}$) and that in the conventional optically amplified transmission system using EDFAs ($OSNR_0$): $\Delta OSNR = OSNR_{RP} - OSNR_0$. The predetermined improvements in the ΔOSNR were set to more than ~ 5 dB. The RP-MC-EDFA system reported to date (called the pump power saving type of RP-MC-EDFA) used a significantly smaller pump power per core than the single-core fiber transmission system using the conventional RP-EDFA mentioned above (called the high pump power type of RP-EDFA).

The basic configuration of the RP-MC-EDFA system is schematically shown in Fig. 4.21. The system has two spans of MCFs (MCF-1 and -2), the RP-MC-EDFA module (RP-EDFA), fan-in and fan-out devices (FI and FO), a pump light source (PLS), an optical coupler for the signal and pump lights, an inline EDFA, an optical transmitter (Tx) section, and an optical receiver (Rx) section. Photographs of the cross-sections of the MCF and the multi-core EDF (MC-EDF) used in the experiment in [27] are also shown in Fig. 4.21 as an example of the MCF and MC-EDF used in the RP-MC-EDFA system. MCF-1 and -2 were fusion spliced to the MC-EDF. The RP-EDFA consists of the MC-EDF and two fusion splicing points.

The optical amplification characteristics of the RP-MC-EDFA system operating in the C-band were analyzed by a numerical simulation [51]. Simultaneous equations

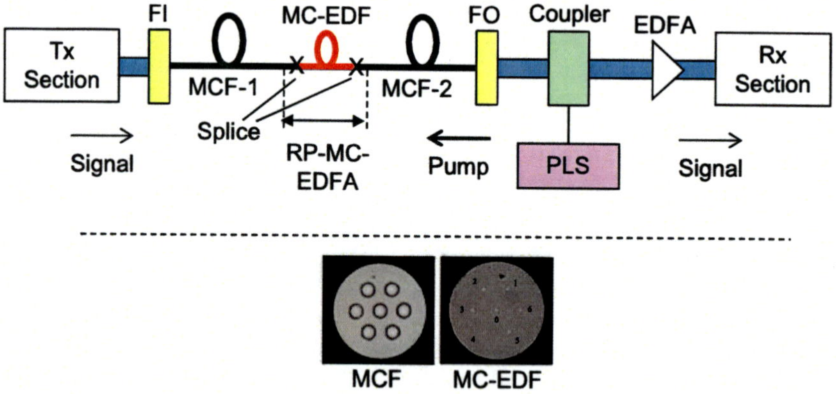

Fig. 4.21 The basic configuration of the RP-MC-EDFA system (Reprinted with permission from [27]. © 2013 IEEE)

for the signal, pump, and amplified spontaneous scattering light were numerically solved. In the simulation, the mode field diameters of the MCFs (MCF-1 and -2) and the EDF were set to 9 and 4.5 μm, respectively. The total length of the MCFs was set to 160 km. The wavelength-dependent coefficients of the loss and the Raman gain of the MCFs were assumed to be the same as those of the G.652 fiber. The number of wavelength-division multiplexed (WDM) channels and the signal power launched into MCF-1 were set to 30 and 0 dBm/channel, respectively. The optimal pump wavelength for the RP-MC-EDFA system ranged from approximately 1460 to 1480 nm depending on the pump power. The OSNR improvement ΔOSNR became maximal at a Raman gain of ~ 15 dB. However, the Raman gain for system operation was set to 10 dB, taking into account the impairment due to the multipath interference effect. ΔOSNR as a function of the pump power for the RP-MC-EDFA and MC-DRA systems was calculated, and the RP-MC-EDFA system yielded much higher pumping efficiency than the MC-DRA system. Moreover, the level diagrams for the RP-MC-EDFA and MC-DRA systems were derived and compared.

An experimental study of the optical amplification characteristics of the RP-MC-EDFA system was reported in [52]. Single-core transmission fibers and a piece of single-core EDF were used instead of MCFs and an MC-EDF for simplicity in the experiment. The OSNR improvements as a function of the pump power were measured for the RP-MC-EDFA and MC-DRA systems. It was confirmed that the RP-MC-EDFA system yielded a pumping efficiency ~ 5 times larger than that of the MC-DRA system.

In another experimental study of the optical amplification and transmission characteristics of the RP-MC-EDFA system [27], two seven-core fibers (MCF-1 and -2) and a piece of seven-core EDF (MC-MCF) were used. The lengths of MCF-1 and -2 were 75 and 46 km, respectively, so the total length was 121 km. The core distance and mode field diameter at 1550 nm of the MCFs (the MC-EDF) were 49.7 μm (49.5 μm) and 11.8 μm (6.0 μm), respectively. The average, minimum, and maximum splice losses between the MC-EDF and the MCF were 0.42, 0.32, and 0.57 dB/point at a wavelength of 1310 nm, respectively. The pump light was launched into MCF-2 via a 1.46/1.55-μm coupler array integrated on a silica-based planar lightwave circuit (PLC) and the FO. The pump wavelength (λ_p) was 1452 nm for the RP-EDFA system, and the λ_p values for the DRA system tested for comparison were 1424 and 1452 nm. Four WDM signals were used, and their frequencies were 193.0–193.3 THz on the ITU-T 100-GHz grid. The OSNRs and bit error rates (BERs) were measured at a signal frequency of 193.3 THz (wavelength of 1551.32 nm). The WDM signals were modulated at 10.7 Gb/s by an $LiNbO_3$ external modulator in a non-return-to-zero modulation format with a pseudorandom binary sequence having a length of a $2^{31}-1$ in the transmitter.

The measured OSNRs for the MC-DRA system as a function of the pump power (P_p) are shown in Fig. 4.22a. P_p was measured at the input port of the PLC coupler. The center core is denoted as Core-0, and the outer cores are denoted as Core-1 to -6. The OSNR without pumping ranged from 20.3 to 22.6 dB, whereas that at the maximum pump power of 1660 mW ranged from 27.5 to 30.8 dB. On the other hand, Fig. 4.22b shows the measured OSNRs for the RP-MC-EDFA system. The

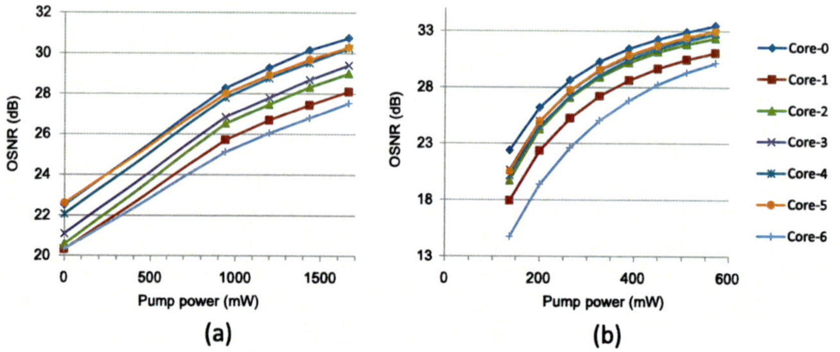

Fig. 4.22 Pump power dependence of OSNR (Reprinted with permission from [27]. © 2013 IEEE)

OSNR at the maximum pump power of 570 mW ranged from 30.1 to 33.5 dB. The OSNR differences among the cores are attributed mainly to differences in the coupling losses between the PLC coupler input point and the MCF-2 input points. The pumping efficiencies of the DRA and RP-EDFA systems are compared in Fig. 4.23. The measured OSNR improvement (ΔOSNR) of Core-0 as a function of P_p are shown. The RP-MC-EDFA gain was ~ 10 dB at $P_p = 480$ mW. The pump powers required to achieve ΔOSNR values of 6, 7, and 8 dB for the MC-DRA and RP-MC-EDFA systems (P_{pDRA} and P_{pEDFA}) were $P_{pDRA} = 265$, 305, and 360 mW and $P_{pEDFA} = 1000$, 1250, and 1550 mW, respectively. Therefore, the pump power ratios of P_{pDRA} to P_{pEDFA} (R) were 3.8, 4.1, and 4.3 at ΔOSNR = 6, 7, and 8 dB, respectively. The ratio R was ~ 4 on average for the three values of ΔOSNR. The OSNR penalty characteristics of the RP-MC-EDFA system were also investigated in the transmission experiment. Seven cores were simultaneously pumped at a P_p value of ~ 330 mW. There was no OSNR penalty in the measured BER plots.

The feasibility of the RP-MC-EDFA system was confirmed in an unrepeatered transmission experiment [38]. A forward distributed Raman amplification scheme

Fig. 4.23 Pumping efficiency characteristics (Reprinted with permission from [27]. © 2013 IEEE)

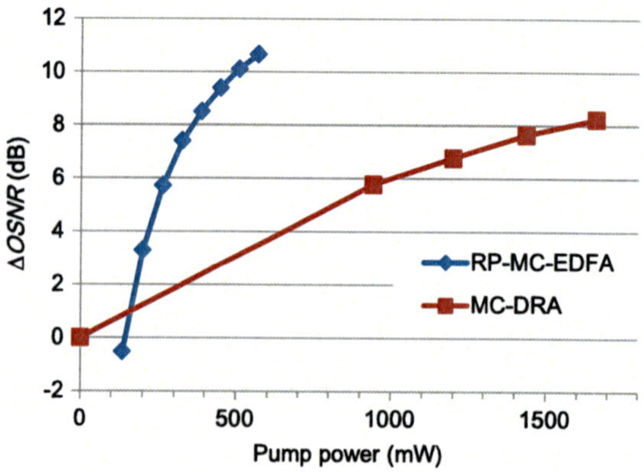

was also employed in addition to the RP-MC-EDFA scheme in the experiment. Space division multiplexed signal channels with an aggregate capacity of 120.7 Tb/s were successfully transmitted over 204 km of seven-core fiber. The capacity was achieved by using 180 WDM channels of polarization-division multiplexed (PDM) 32-QAM signals (95.8 Gb/s) per core: [7 cores] × [180 WDM channels] × [PDM 32-QAM signals of 95.8 Gb/s] = 120.7 Tb/s. A record capacity per fiber and the highest spectral efficiency (53.6 b/s/Hz) for unrepeatered transmission were achieved. The transmission line consisted of two sections of seven-core fibers (MCF-1 and -2) with an effective area of about 110 μm^2. MCF-1 had a total length of 144 km and an average loss of 0.207 dB/km. MCF-2 had a length of 60 km and an average loss of 0.195 dB/km. Low splice loss values ranging from 0.01 to 0.28 dB were realized. The losses of the FI and FO, including the fusion splice losses, were 0.4–0.6 dB and 0.3–0.9 dB, respectively. The total losses including those of MCF-1 and MCF-2 ranged from 42.9 to 43.6 dB, and loss variations between cores were suppressed to less than 0.7 dB. An MC-EDF with high gain efficiency was placed between MCF-1 and MCF-2 [53]. The pump light source for forward DRA had wavelengths of 1435 and 1460 nm, at which the average pump powers were 529 and 514 mW, respectively. The pump light source for backward pumping had wavelengths of 1450 and 1480 nm, at which the average pump powers were 288 and 292 mW, respectively. The total forward/backward pump powers were 7.3 and 4.1 W, respectively. The Q-factor penalty due to intercore crosstalk was negligible (within 0.1 dB). As described above, long-haul and high-capacity unrepeatered transmission was successfully demonstrated using state-of-the-art optical components in the transmission line of the RP-MC-EDFA system, the optical components of which were MCFs, an MC-EDF, an FI, and an FO.

4.3.4 State-of-the-Art Multi-core Raman Amplification Technology

Distributed Raman amplifiers (DRAs) in single-mode fiber transmission systems have been studied since the late 1980s in order to achieve higher optical signal-to-noise ratios (OSNRs) than schemes using only discrete or lumped amplifiers [54, 55]. Each DRA, which uses distributed Raman amplification in a transmission fiber more than several tens of kilometers long, consists of a transmission fiber, a Raman pump light source, and a WDM optical coupler for the signal and pump light. In most cases, the DRA has been used together with a discrete amplifier, that is, an EDFA or a discrete Raman amplifier. The combination of amplifiers was called a hybrid amplifier, for example, a Raman/EDFA hybrid amplifier and so on. All-Raman amplifiers, which use no discrete amplifier, have also been reported. Transmission systems using hybrid amplifiers reportedly achieved OSNR improvements of more than about 5 dB compared with transmission systems using EDFAs [56–59]. Hybrid

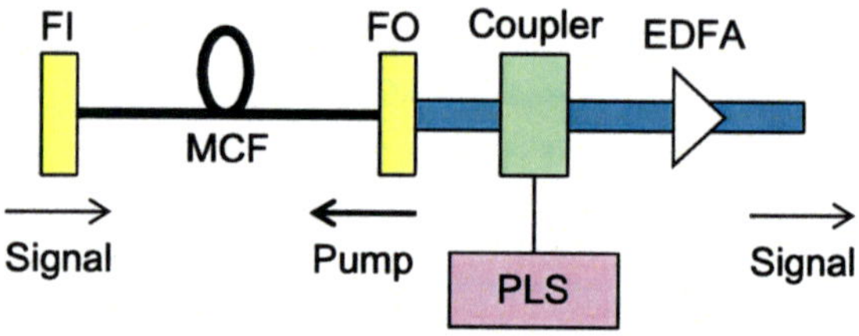

Fig. 4.24 Basic configuration of optical amplification system using MC-DRA

amplifiers have been used in some ultra-long-haul transmission experiments in the laboratory and also in some practical systems [58–61].

An experimental study of distributed Raman amplification in a multi-core fiber transmission system was reported in 2012 [39]. The multi-core DRA (MC-DRA) in the experiment consisted of a seven-core transmission fiber 75 km in length (MCF), a pump light source (PLS), a signal/pump WDM coupler, a fan-in device (FI), and a fan-out device (FO). Note that the FI and FO are not used in the single-core fiber transmission system. The basic configuration of a single span of the optically amplified transmission system using the MC-DRA is schematically shown in Fig. 4.24. A multi-core EDFA is placed after the MC-DRA, as in the case of a hybrid amplifier in a single-core transmission system.

The optical amplification and transmission characteristics of the MC-DRA were clarified, and the feasibility of the transmission system using MC-DRAs was confirmed experimentally [39]. The MCF used in the experiment had a trench-assisted structure. The core pitch and the effective area of the cores were 49 μm and 110 μm^2, respectively. The attenuation coefficient and dispersion for the seven cores at a wavelength of 1550 nm were 0.190–0.199 dB/km and 20.5–20.8 ps/nm/km, respectively. The 75-km MCF was fusion spliced with the FI and FO. The losses of the FI and FO, including the losses at the fusion splice, were 0.5–1.9 dB and 0.7–1.6 dB, respectively. The crosstalk between the cores of the 75-km MCF was − 65 dB at 1550 nm on average. The total losses between the FI input point and the FO output point, including the loss of the MCF, ranged from 16.0 to 18.1 dB. The measured crosstalk of the FI from the center core to the outer cores was − 61 dB. The pump light source consisted of two sets of laser diode modules, which had pump wavelengths of 1424 and 1452 nm and used a polarization multiplexing scheme, and couplers based on a fiber or planar lightwave circuit. The Raman gain at a pump power launched into the FO of 1.1 W ranged from 9.2 to 12.9 dB for the seven cores. The effective noise figure of the Raman/EDFA hybrid amplifier used in the experiment ranged from − 1 to 1 dB at a Raman gain of 12 dB. The dependence of the effective noise figure on the Raman gain was similar to that for the Raman/EDFA hybrid amplifier in the single-core fiber transmission system, as expected.

A transmission experiment using a recirculating loop was reported in the study. Ten signal lights with optical frequencies of 193.00 to 193.45 THz on the ITU-T grid were modulated at a channel bit rate of 96 Gb/s with a PDM 16-QAM format. Error-free operation with Q factors larger than a forward-error correction limit of 8.3 dB was confirmed at a transmission distance of 1050 km. The aggregate pump power for the seven cores was 6.5 W. The optical components used in the experiment were not damaged. As described above, the MC-DRA used in the MCF transmission system had low noise and high OSNR characteristics that were essentially the same as those of the single-core fiber transmission system. The greater pump power for the MC-DRA compared to the single-core DRA was due mainly to the insertion losses of the FI and FO. The cross-talk between the cores was negligible because a state-of-the-art MCF, FI, and FO were used.

Moreover, a study of the crosstalk characteristics of an MC-DRA system was reported in [62]. The crosstalk between the cores was shown to depend on the Raman gains if the Raman gains of the cores were not uniform. A hybrid transmission system was proposed in which the Raman gains in the cores are intentionally designed to have different values. The difference in the crosstalk increases with the difference in the Raman gains.

4.3.5 Evaluation Methods

Because multi-core EDFA which has several cores embedded in a common cladding suffer from crosstalk component penetrated from neighboring cores, evaluation method is required to measure crosstalk component precisely in multi-core EDFA with an excited state. In the case of non-doped multi-core transmission fiber, it is relatively simple to measure a crosstalk component because input signal is directly detected from neighboring cores. However, in the case of rare-earth-doped multi-core fiber, it is difficult to apply direct detection method of non-doped multi-core transmission fiber. This means that not only main signal light but also crosstalk signal light draw influence from absorption effect. On the other hand, when we consider cladding-pumped multi-core EDFA, all rare-earth-doped cores are excited simultaneously and amplified spontaneous emission (ASE) light is generated in all cores. Therefore, it is difficult to distinguish crosstalk component from ASE light. In this section, evaluation methods of crosstalk component between cores in multi-core EDF with an excited state are introduced.

One approach is that we can directly measure crosstalk component with an operating wavelength at which absorption coefficient of rare-earth dopant is low. In [30], Takasaka et al. measure crosstalk component of multi-core EDF with operating wavelength of 1640 nm. In this experiment, multi-core EDF has absorption coefficient of 0.01 dB/m at 1630 nm even though this multi-core EDF has absorption coefficient of 5 dB/m at 1530 nm. Therefore, the crosstalk components are directly measured with the same evaluation method of multi-core fiber. Figure 4.25 shows the experimental setup for measurement of crosstalk component with direct detection. Here, $P_{1,\mathrm{in}}$, $P_{1,\mathrm{out}}$, and $P_{2,\mathrm{out}}$ represent the optical input power to core 1, optical output power from core 1, and optical output power from core 2, respectively. This evaluation method is relatively simple; however, we estimate a crosstalk value of amplified wavelength band from the result of non-amplified wavelength band of 1640 nm.

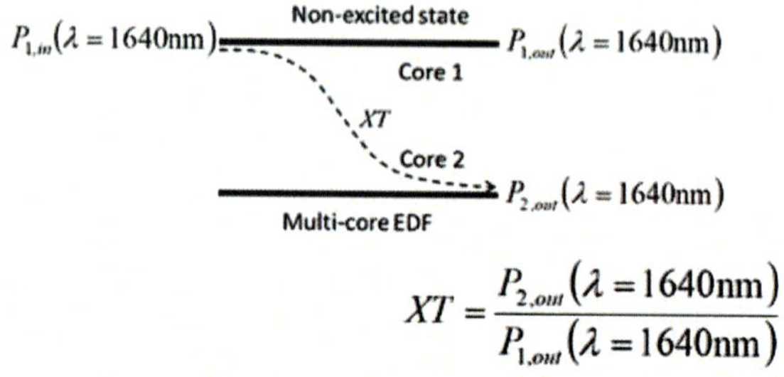

Fig. 4.25 Directly detected crosstalk measurement at an operating wavelength with low absorption

More realistic method of crosstalk evaluation is that we can measure crosstalk component with an excited state. Not only core that main signal light propagates in, but also core that crosstalk component penetrates to, are under excited state simultaneously in the case of cladding-pumped scheme. Therefore, there is a need to input signal light into both cores in order to measure the crosstalk component clearly without any ASE floor level. However, if we launch signal light at the same wavelength into both cores, it is impossible to distinguish amplified main signal light from crosstalk component in optical spectrum result. To solve above issue, it is reported from [24] that two signals of which the wavelength is different between cores that main signal propagates in and that crosstalk component penetrates to. Figure 4.26 shows the experimental setup. Here, it is assumed that gain property and coupling coefficient are almost the same tendency between two wavelength of λ_1 and λ_2. By employing this evaluation method, we can measure crosstalk component directly at amplified wavelength band in optical domain.

Ono et al. report evaluation method of crosstalk component in multi-core EDF with an excited state in modulation frequency domain as shown in Fig. 4.27 [63].

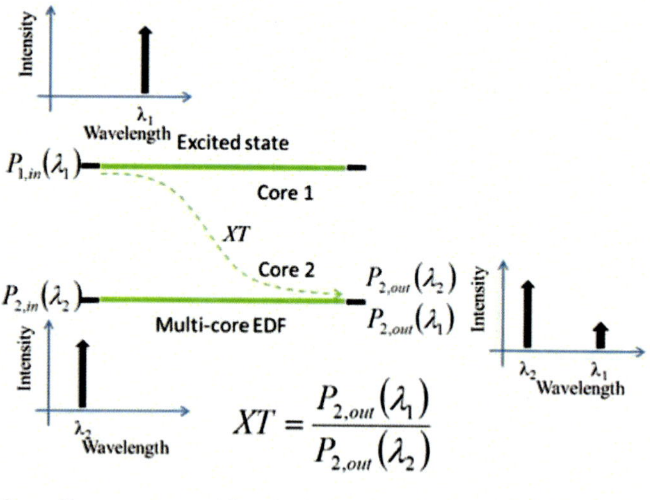

Fig. 4.26 Crosstalk measurement with excited state by using two different wavelength

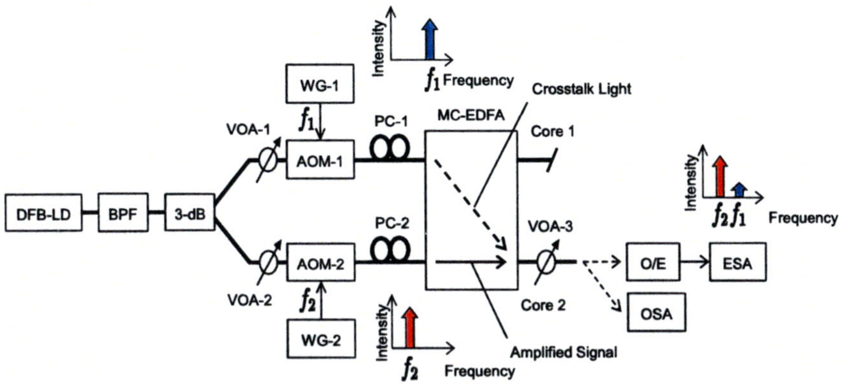

Fig. 4.27 Inter-core crosstalk measurement in multi-core fiber amplifier by using multiple intensity tones (Reprinted with permission from [63]. © 2014 IET)

After dividing signal light by 3-dB coupler, two individual modulators generate intensity-modulated signal light with different modulation frequency. In this case, signal light of intensity-modulated frequency f_1 is input into core 1 and signal light of intensity-modulated frequency f_2 is input into core 2. Then, we measure output signal through electrical spectrum analyzer after O–E conversion. In this work, it is confirmed that crosstalk component is independent on the intensity-modulated frequency and input signal power.

4.3.6 *Issues and Future Work*

SDM amplification technology is reviewed in this section. In this stage, requirements for SDM amplifiers are existed for both amplifiers for SDM systems and amplifiers not for SDM systems.

Firstly, requirements for SDM systems are described. For this application, harmonization for MCF for transmission line is important. Future development for MCF would be improvement for high capacity and applying to longer transmission distance. From this point of view, one solution of MCF for future transmission line is to have lower nonlinearity and lower crosstalk. Similar requirement would be applied for SDM amplifier.

One approach to solve this requirement would be applying distributed amplifiers and/or remote pumped EDFA. Common technologies for them are delivering the high power pumping in the core of the MCF for the transmission line. MCF and peripheral devices should be designed to endure input power of several watts. Especially, fan-in devices for high input power would be higher priority for developing these systems.

The other approach is optimization of lumped MC-EDFA. Main issues for this approach are improvement of gain fiber and components. The optimal design for gain fibers are strongly related with excitation method for MC-EDFA. For core pumping, bundled EDF is the simplest way to make SDM amplifier. It needs no additional development of peripheral components. To emphasize the merits applying bundled EDF, development of small packaged one would be the way to optimization. For cladding-pumped EDFA the most important thing would be developing double-cladding fibers with high cladding absorption for high output power amplifiers.

Though merits of using common cladding for multiple cores are limited for core-pumped amplifiers, it is a big advantage for cladding-pumped amplifiers. In order to simplify the system configuration, ideal approach for amplifiers would be using the same MCF as transmission line as a wiring fiber in amplifiers. Also, fibers and components used in the amplifiers would be applied to have multi-core approach. Components adequate for multi-core approach would be signal/pump combiner, isolator, gain-flattening filter (GFF), and so on. Each component works by different principles requirements for them should be set to reasonable value. Insertion loss of these components affects the NF of amplifiers directly. That means development of low-loss optical multi-core components would be the one of the keys for SDM amplifier.

Technologies of SDM amplifiers can be utilized for outside of SDM systems are also important to enlarge the market of SDM amplifiers. Common requirements for all amplifiers are reducing package size and power consumption. For reducing package size, reducing the number of components by applying multi-core components are effective. This is the same approach to SDM amplifiers. Reducing the number of pumping LD is effective for both reduced package size and low power consumption. Techniques used for pumping of gain fibers, shared pumped EDFA and batch amplification by cladding pumping have already reported.

The merit of low power consumption is reported in [64] using batch amplification by applying single multi-mode LD for excitation for all cores. Using this scheme, long-haul transmission experiment and possibilities to adapt the MC-EDFA to the transoceanic system is reported.

Shared pumped MC-EDFA as shown in Fig. 4.28 was reported in [65]. 19-core EDF is used as gain fiber and ten LD is used. By adopting this technique, advantage of reducing space is also achieved. In this report, all multi-core approach is also tried. All cores of MCF and MC-EDF are set to same position. Also, multi-core isolator is used. This report would be the approach to the ideal SDM amplifier. The next try for SDM amplifier may be reducing the number of LD.

The last issue for SDM amplifiers is technique for gain control. Generally, gain of EDFA is determined by product of concentration of rare earth in the gain fiber and fiber length. To obtain the target gain profile, fiber is set to adequate length. However, length of each core cannot be changed in one MC-EDF. That means gain variation between cores in MC-EDF may be occurred. Furthermore, power distribution of pumping light in the inner cladding may affect the gain variation for cladding-pumped EDFA. For applying SDM amplifier to application area of outside SDM, dynamic gain control may be required. If amplifiers with low power consumption are achieved by batch amplification technique, total power consumption including gain control should be considered. For future development of SDM amplifier, novel gain control

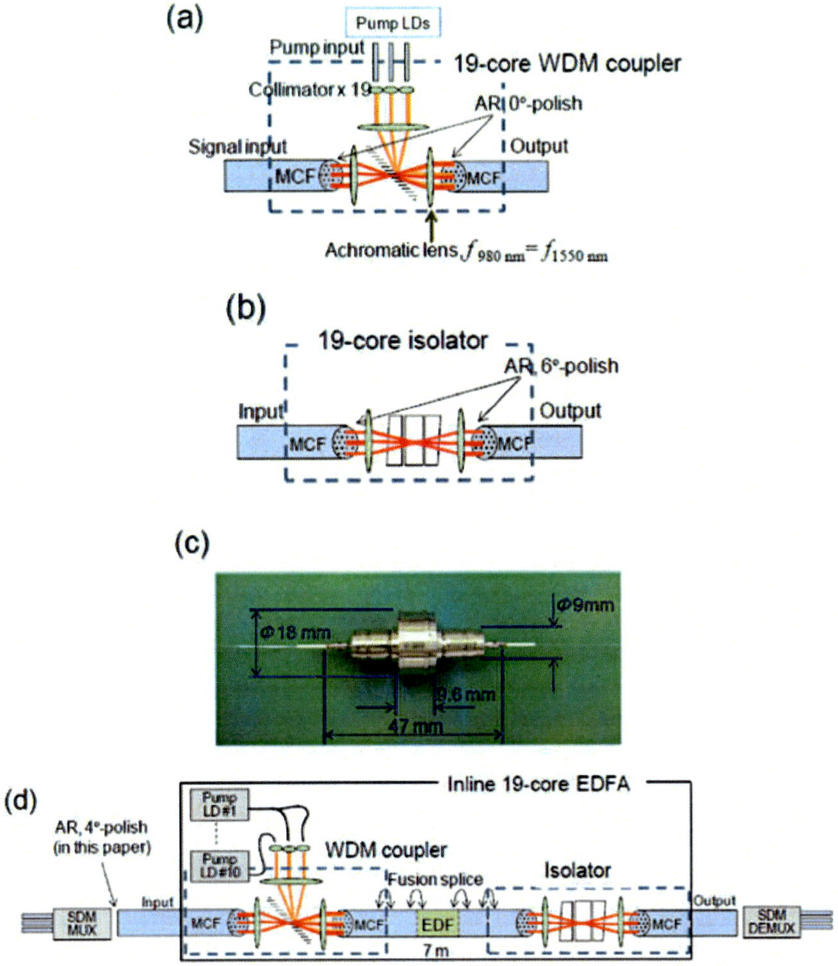

Fig. 4.28 SDM devices with shared optics. **a** WDM coupler configuration, **b** isolator configuration, **c** outlook of the isolator, **d** inline EDFA and characterization setup (Reprinted with permission from [65]. © 2014 The Optical Society)

technique would be required. In addition, recently to solve this problem, a hybrid pumping technology that combines core pumping and cladding pumping has been proposed, and an amplifier technology aiming at both low power consumption and gain control is being studied [66].

In the investigation of the future issues for SDM amplifiers, it is considered that the most attractive merits for SDM amplifiers are possibility to reduce the power consumption. Other merits of SDM amplifier are considered as comparable merits for standard amplifiers but reducing power consumption is the merit to defeat the standard amplifiers. To emphasize this merit, further improvement for gain fibers and peripheral components should be achieved.

For gain fibers, a new approach setting cut-off wavelength between pump wavelength and signal wavelength is proposed [67–69]. The pump power conversion efficiency is improved by expanded core area. For components, the pump power recycling technique utilizing residual pump power after amplification is proposed [70, 71]. Power consumption of MC-EDFA will be reduced by these techniques.

4.4 Few-Mode Fiber Amplification Technology

4.4.1 Target of Few-Mode EDFA

The most mode-division multiplexing (MDM) system over FMFs has utilized MIMO (Multi-Input Multi-output) techniques to restore deteriorated signals due to mode crosstalk to the original signals. It has been pointed out that the performance of MIMO system is affected by the existence of mode-dependent loss (MDL) or mode-dependent gain (MDG) [72]. Accordingly, optical amplifiers for MDM system is required to control MDG in addition to satisfy the basic requirement for amplifiers such as gain and noise figure.

Figure 4.29 shows an example of gain profile of few-mode EDFA [73]. A two-mode EDF with a conventional step-like core structure was used in the EDFA. The EDF was pumped with the light of the LP_{01} mode. A 10-dB gain difference was observed between the LP_{01} mode and the LP_{11} mode. Figure 4.30a illustrates the reason of the huge MDG. For conventional EDFs have the same Er-doping profile with an optical index profile. However, the power distribution of the two modes is quite different. If the EDF is pumped with the LP_{01}-mode light, an activated Er-ion profile will be close to the pumping light. The gain of an EDF depends on the overlap integral of a signal power profile and an activated Er-ion profile. Accordingly, the huge MDG is observed in few-mode EDFAs with conventional EDFs.

Many approaches have been proposed to reduce the MDG of a few-mode EDFA. One approach is to use a higher-order mode for pumping. Actively amplified region in a core vary with the pumping power distribution. By using the LP_{21} mode for

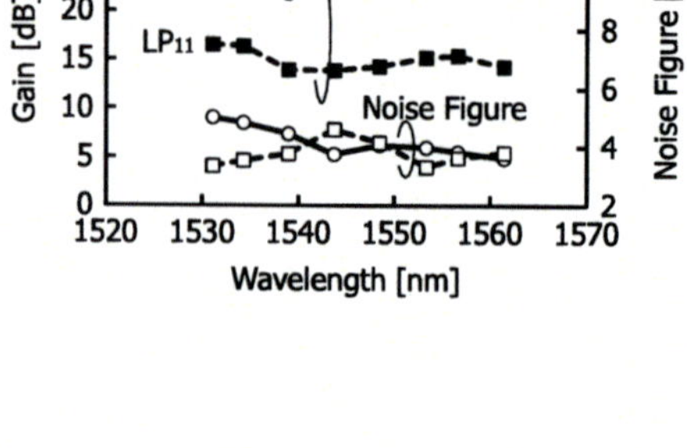

Fig. 4.29 An example of gain and NF of two-mode EDFA with LP_{01} pump: EDFA is composed of an EDF with a step-index profile(Reprinted with permission from [73]. © 2014 IET)

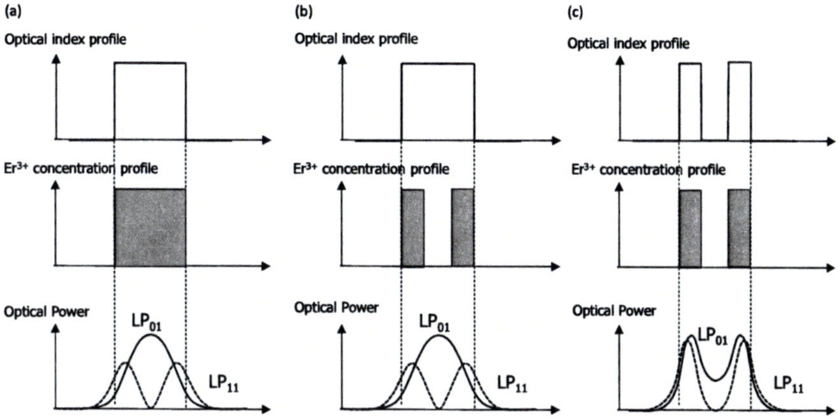

Fig. 4.30 Schematics of optical index profile, Er concentration and optical power distribution for different types of EDF

pumping, the gain of the LP_{11} mode was allowed to be larger than that of the LP_{01} mode [74]. In [3], a free-space optics with phase plates was used to generate the LP_{21} pumping light from the LP_{01} source. Another way to generate the LP_{21} pumping light is to use a PLC-based coupler and a mechanical long-period fiber grating [75].

Other approaches to reduce the MDG are change fiber structure. One way is to adjust Er-ion profile to the power distribution of higher-order modes as shown in Fig. 4.30b. In this way, the optical index profiles intended to keep a step-like profile. Another way to reduce the MDG is to utilize a ring-like profile for both optical index profile and Er-ion profile as shown in Fig. 4.30c. By using a ring-like optical index profile, the power distribution of the LP_{01} mode changes from Gaussian to ring shape, which is similar to the LP_{11} mode. As a result, the gain of the modes is equalized.

DMD-reduced EDFAs have been realized by the combination of the approaches presented above. The detailed setup and characteristics of the EDFAs are presented in the following section.

4.4.2 *State-of-the-Art Few-Mode EDFA*

As described in the previous section, reducing the MDG is a specific issue in a few-mode EDFA, and there are mainly three approaches for achieving it.

The first approach is adaptive modal gain equalization or reconfigurable modal gain. The concept of adaptive modal gain equalization is shown in Fig. 4.31, which illustrates two different methods for achieving it. One is to control the modal content of the pump. Several pump modes with different powers are launched into a multimode EDF (MM-EDF) with multiple modal signals, and the MDG is minimized by adjusting the power of the pump modes. With this approach, several theoretical

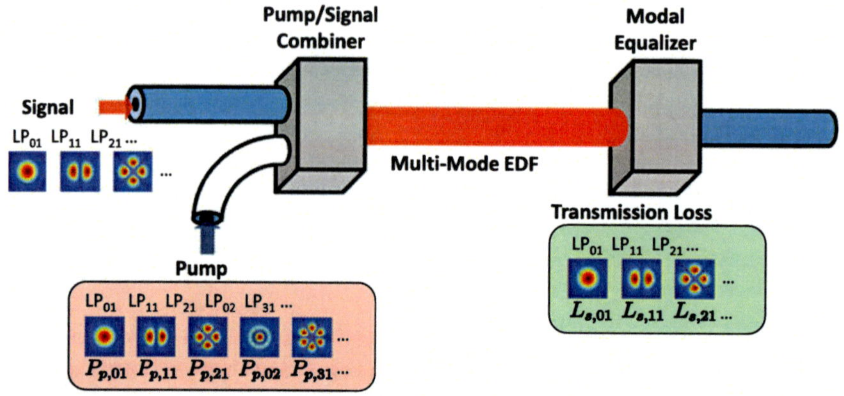

Fig. 4.31 Concept of adaptive modal gain equalization

investigations have been done. Using mode converters has been proposed to transform the spatial mode of the pump source into the N spatial modes with variable attenuators, which enables N-degree control over the mode content of the pump and thereby minimizes the MDG [74]. The gains of LP_{01} and LP_{11} signal modes can be tuned over a wide dynamic range by adjusting the relative amount of LP_{01} and LP_{21}. A relative gain between the two spatially degenerate LP_{11} signal modes (LP_{11}o and LP_{11}e) can also be adjusted by adding a small amount of the azimuth angle of LP_{11} of the pump mode. Using a spatial light modulator (SLM) to minimize the MDG has also been proposed to adjust the power of the pump modes [76]. It is insisted that this method is more power-efficient than controlling the modal content of the pump [74] because it only requires one pump laser per pump polarization, independent of the number of signal modes. The other approach for adaptive modal gain equalization is to adjust the powers of the different amplified modal signal. With this approach, an SLM is placed immediately after the MM-EDF to equalize the MDG in the signal.

Using a higher-order mode as a pump to reduce the MDG has also been experimentally investigated. A demonstration has been reported for 2-LP mode amplification which intended to use a step-index profile fiber with a uniformly doped core [77]. The MDGs of 2.5 and 1 dB for backward LP_{11} and forward LP_{21} mode pumping were obtained, respectively. In another experiment [78], the EDFA was able to control LP_{01} and degenerate LP_{11} modal gain by using a long-period fiber grating. The MDG can be controlled in the − 7 to 5 dB range by varying the ratio of the LP_{01} to LP_{11} pump modes.

The second approach for reducing the MDG is to adjust the doping profile of Er ions to the power distribution of higher-order modes by using, for instance, ring-doping of Er ions as shown in Fig. 4.30b. Two kinds of fiber structures have been investigated for this approach: a conventional solid glass fiber and a microstructured fiber. With the former, a theoretical study has been reported for a 4-LP mode amplifier [79]. This study proposed a fiber design incorporating two separate doped sections and showed that the design provides less than 1-dB MDG in a 4-LP mode EDFA for a relatively simplified pump arrangement using only two pump modes of LP_{01} and LP_{41}. Conventional solid glass fibers have also been experimentally investigated for adjusting the Er-ion profile. A 2-LP mode amplifier in which the Er ions were substantially confined within a ring in the fiber core has been demonstrated [80]. For an optimum offset condition of pump launching, it was observed that the differential modal gain becomes zero (virtually no mode-dependent gain) at a moderate pump power. Another 2-LP mode amplifier employed a ring-doped EDF, and it enabled modal gain to be equalized to 1 dB at a gain of 12 dB per mode [81]. For a microstructured fiber, a multi-mode fiber amplifier that employed an Er-doped fiber with a microstructure has been investigated both theoretically and experimentally [82]. According to this work, MCVD-based fibers with an Er-ion dopant profile in which the MDG is as small as 3.1 dB could be theoretically hard to be fabricated. This is because of Er diffusion that leads to a deterioration of gain equalization over modes and spectra. Therefore, an alternative, based on micro-structuring the 19-cell core of an air/silica PCF, has been proposed. Such a fiber design has been investigated and differential gain over spectra and modes down to 3.2 dB for 21-dB average gain has been predicted.

Another investigation using both a higher-order mode as a pump and adjusted the Er-ion profile to the power distribution of high-order modes [83]. A 4-LP mode amplifier was studied by using a mode-selective bi-directional pumping scheme in con-junction with an EDF with a specially tailored Er-doping profile. The signal gain of over 20 dB was obtained for all guided modes with an MDG of less than 2 dB.

The third approach for a small MDG is to utilize a ring-like profile for both the optical index and Er-ion doping area as shown in Fig. 4.32, which is called a ring-core EDF. The intensities of the fundamental and higher-order modes are similar in a ring-core EDF. Therefore, the difference between two overlap integrals, namely, that for the excited Er-ion area and the intensity profile of the fundamental mode signal and that for the excited erbium ion area and the intensity profile of higher-order signals, is reduced. A small MDG is then achieved by just employing a ring-core EDF for a few-mode EDFA (FM-EDFA). A theoretical study has shown that nearly identical gains for 6-LP modes could be obtained by using a ring-core EDF, and the performance is robust to macro-bending down to 5-cm radius when cladding-pumped or core-pumped by higher-order pump modes [84]. The ring-core EDF has

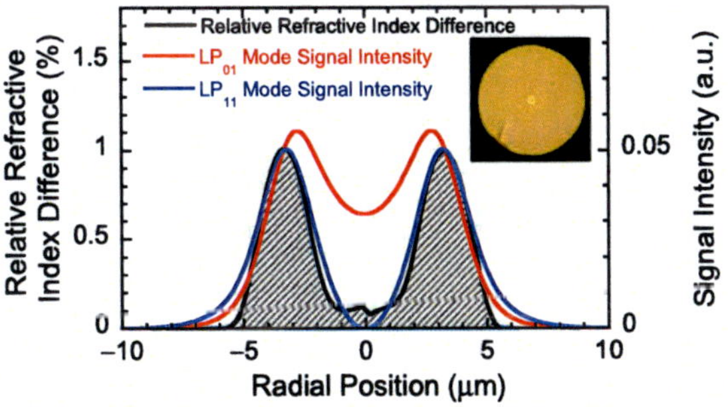

Fig. 4.32 Refractive index profile and intensity profiles of LP_{01} and LP_{11} mode signals of ring-core EDF (Reprinted with permission from [85]. © 2015 The Optical Society)

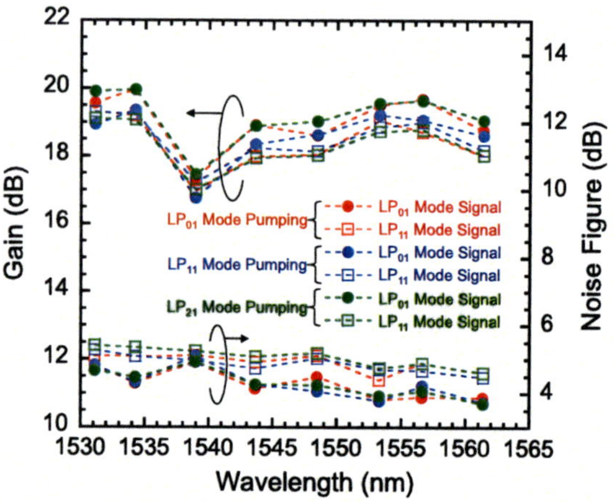

Fig. 4.33 Gain and noise figure of FM-EDFA employing ring-core EDF (Reprinted with permission from [85]. © 2015 The Optical Society)

been proven to reduce the MDG experimentally for a 2-LP mode EDFA [85]. The 2-LP mode EDFA with a ring-core EDF exhibited smaller MDGs of less than 1.0 dB between LP_{01} and LP_{11} mode signals for the LP_{01}, LP_{11}, and LP_{21} mode-pumping as shown in Fig. 4.33. A ring-core EDF was also applied to an L-band, 2-LP mode EDFA and the MDG reduction was demonstrated [86].

4.4.3 State-of-the-Art Few-Mode Raman Amplification Technology

A distributed Raman amplification in an FMF requires a small MDG as well as a relatively flat gain over the signal band in the same way as an FM-EDFA.

2-LP mode distributed Raman amplification has been demonstrated in an FMF transmission experiment [87]. In this experiment, the depolarized LP_{11} mode pump light was launched, and no LP_{01} mode pump light was input into the FMF. This pump condition produces nearly the same gain for both the LP_{01} and LP_{11} modes. Further theoretical studies have been done to minimize the gain difference between 12 polarization- and spatial-mode [88]. This study is based on real waveguide modes of an FMF, not on LP modes. It showed that the MDG could be minimized by optimizing the modal pump power distribution. Another theoretical study introduces a systematic optimization approach to achieve equalized gain for different modes with a method that can achieve simultaneous optimization of the MDG and wavelength-dependent gain [89]. The study revealed an optimization algorithm to minimize MGD with the small wavelength-dependent gain, and the algorithm is used to optimize a 4-LP mode distributed Raman amplification in an FMF with fifty signal wavelengths and four pump wavelengths.

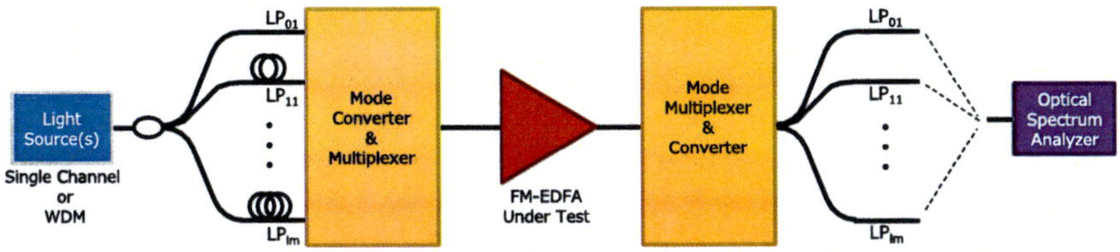

Fig. 4.34 Schematic of a setup for measuring the gain and noise figure of a multi-mode EDFA

4.4.4 Evaluation Methods

In the gain and noise figure (NF) measurement of an FM-EDFA, the input and output signals are the mixture of different mode signals. Figure 4.34 shows a simplified schematic of a setup for measuring the gain and the NF of an FM-EDFA. Different mode signals usually share the same laser sources. The laser lights used as input signals are split to the same number of the modes, and then the signals of different modes are multiplexed with a mode-multiplexer, sometimes with different delays to avoid undesired power fluctuation caused by a mode-coupling. When polarization-multiplexed signals are used as input signals, modulated signals can be used. The mode-multiplexed signal is fed into an FM-EDFA under test. The output signals from the FM-EDFA need to be mode-demultiplexed with a demultiplexer. The mode-multiplexed output signals of the FM-EDFA are fed into a mode demultiplexer, and measured with an optical spectrum analyzer. Here, it should be noted that the two (nominally degenerate) orthogonal modes of the higher-order mode (e.g., LP_{11} mode) experience continuous mode-mixing along the EDF and the spatial orientation of the lobes can also be altered by external perturbations of the fiber. Therefore, the gain of the higher-order mode should be measured in two directions, perpendicular to each other, and the summation or average of the measured power of the two orthogonal modes is then used to estimate the modal gain [81, 85].

4.4.5 Issues and Future Work

This section describes several issues to overcome in order to put FM-EDFAs into practical use.

Several FM-EDFAs have been demonstrated for MDM transmission experiments. Some of them exhibit good gain and noise characteristics with a small MDG. However, better MDG performance may be needed in order to install an FM-EDFA into a practical MDM system. From this viewpoint, it is important to reduce the MDG further over all the supported signal modes in designing a few-mode EDF (FM-EDF). Gain flatness is also required for the entire signal band for all the signal modes, and the required pump power should be as small as possible. Therefore, parameters of an FM-EDF need to be optimized. An optimization algorithm is useful for finding the best FM-EDF parameters, and some methods have already been reported. One employs a genetic algorithm [90]. With this algorithm, an Er-ion-doping area of a cladding-pumping EDF is optimized and the DMGs for 6-LP modes are minimized. Another method utilizes a gradient descent optimization algorithm [91]. This algorithm is applied to find the best Er-ion doping profile, which is achievable with the 19 elements of the core of a microstructure fiber for 6-LP mode amplification with a small DMG. Further studies will reveal the best way to optimize FM-EDF parameters.

In the aspect of constructing an FM-EDFA, small optical components, such as an isolator, WDM coupler, GFF, must be realized. A fiber handling technique may also be required in order to house few-mode fiber pigtails in a small module package.

In future MDM networks, input signal power will change according to channels added or dropped. Therefore, controlling modal gain is a prerequisite for adopting FM-EDFA in such networks. An investigation of inter-modal cross gain modulation and the gain transient [80] indicated that all spatial modes experience roughly similar responses under a range of different add/drop conditions. However, some evidence of mode-dependent sensitivity was observed and the mode-dependent gain dynamics was enhanced further when operating at higher levels of amplifier saturation.

It is also important to figure out how various effects that have been studied only theoretically affect the performance of an actual FM-EDFA. Those effects include rotation dependence, vector mode effects, and a mode beating. Experimental studies of those effects as well as further theoretical studies are expected to clarify whether they cause serious problems or not.

4.5 Amplification Techniques for Expanding Transmission Bands

Expanding the transmission band is an important approach for increasing the capacity of optical communication systems with high performance, along with a multi-core and/or multi-mode transmission technique. Optical fiber amplifiers based on rare-earth-doped fiber play an important role in transmission systems because they compensate for the insertion loss of the optical fiber used for transmission lines and for the loss of the optical components used for optical nodes. Research on expanding the transmission band began immediately after the development of the C-band EDFA, and O-band Pr-doped fiber amplifiers [120–122], S-band or U-band Tm-doped fiber amplifiers [109, 123–125], L-band EDFAs [126], C and L-band EDFAs [127–129], and S-band EDFAs [108, 130, 131] have been developed.

Recently, two types of optical fiber amplifiers, namely the bismuth doped fiber amplifier, which operates in the 1.3-μm band, and the Tm-doped fiber amplifier, which operates in the 1.7 μm to over 2 μm wavelength region, have been actively studied.

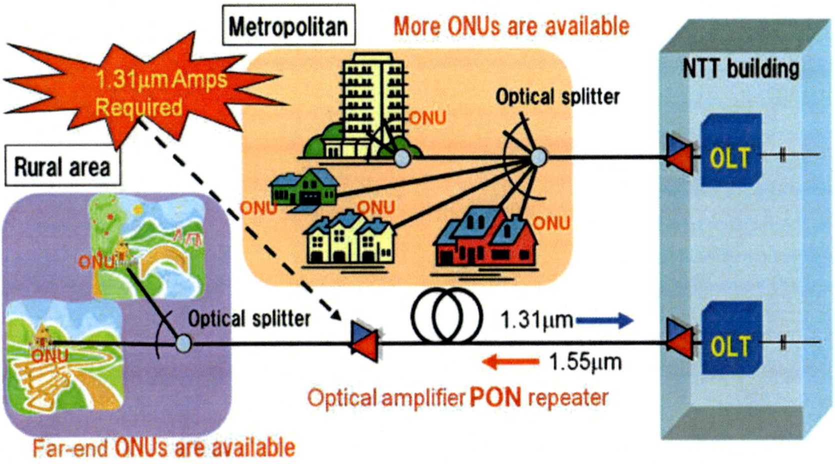

Fig. 4.35 1.3-μm amplifiers in subscriber networks. *ONU* Optical network unit, *OLT* optical line terminal and *PON* passive optical network

The 1.3-μm band bismuth-doped fiber amplifier has the potential for application to subscriber networks as well as for increasing the capacity of long haul transmission networks. As shown in Fig. 4.35, 1.55-μm band optical signals are used in communication networks connecting telephone companies to their subscribers, but a 1.3-μm band optical signal is used for communication between subscribers and the telephone companies. With conventional subscriber lines, the average distance between a subscriber and a telephone company has typically been less than 10 km. However, the recent spread of optical communication networks into rural areas has required optical fibers of 20 km or longer. Furthermore, the number of subscriber networks is expected to increase because of the demand for the high-speed delivery of video in urban areas. 1.3-μm optical amplifiers are attractive for use in future upstream long-distance access lines in telecommunication networks [92].

On the other hand, optical fiber amplifiers that can operate over the 1.65-μm band, which employ hollow core photonic bandgap fiber have been studied for application to optical fiber communication in the mid-infrared wavelength band up to around 2 μm [93, 94]. Future optical communications beyond 1.6 μm will employ photonic bandgap fiber to overcome the capacity limit of conventional systems [95].

The following sections describe recent results for 1.3-μm band bismuth-doped amplifiers and an over 1.65-μm band amplifier.

4.5.1 1.3-μm Band Bismuth-Doped Amplifier

Bismuth-doped silica glass was proposed as an amplification and a laser oscillation at 1.3-μm by Fujimoto and Nakatsuka [96]. Bismuth-doped fiber lasers and amplifiers

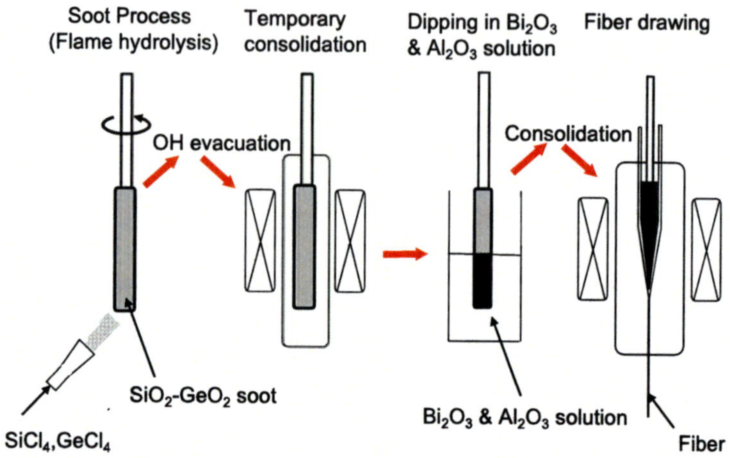

Fig. 4.36 BDF fabrication by VAD method

fabricated by the modified chemical deposition (MCVD) were discussed by the Dianov group [97–99]. Clad pumping is a very effective method of increasing the gain of optical fiber amplifiers. In this section, a new flat-gain amplification method at 1.3 μm based on bismuth-doped silica fiber (BDF) made by the vertical axial vapor-phase deposition (VAD) method is presented. The gain characteristics of VAD-fabricated single- and double-clad bismuth-doped silica fibers with D-shaped and hexagonal first claddings are discussed.

Figure 4.36 shows BDF fabrication using the VAD method. First, Ge-doped silica soot was fabricated by flame hydrolysis. Second, a Bi and Al solvent penetrated into the soot after temporary consolidation by an OH evacuation process. Third, a Bi-doped preform was fabricated by a final consolidation process. The measured preform diameter and length were 14.6 mm and 127 mm, respectively. Finally, the preform was drawn at 1950 °C to form a BDF with a core diameter of several micrometers.

The BDF preform consisted of Si, Ge, Al, and Bi oxides. Bismuth and aluminum ions were solution-doped into a porous germanosilicate layer. The core color was pink because of the Bi doping. The preform was drawn into a BDF with a core diameter of either 4 or 10 μm and a cladding diameter of 125 μm. The concentration of Bi in the core glass was measured to be 0.07–1 mol% by energy dispersive x-ray spectroscopy (EDX). The relative refractive index difference between the core and the cladding of the preform was measured to be 0.5–0.95% using an optical fiber preform analyzer.

The fluorescence spectra of BDF pumped by an 808 nm laser diode (LD) with a pumping input power between 4.83 and 85.4 mW is shown in Fig. 4.37. The spectra showed flat gain between 1200 and 1350 nm. In Fig. 4.37, the dips at 1250 and 1390 nm can be explained as OH ion absorption loss. These dips can be mitigated by annealing at high temperature to evacuate OH ions from the porous germanosilicate glass. The OH ion concentration in the annealed sample, however, was estimated to

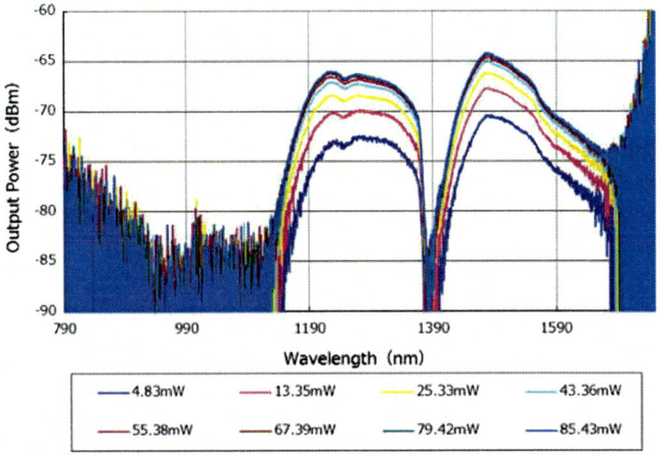

Fig. 4.37 Fluorescence spectra with 4.83–85.43 mW pumping power

be above 40 ppm by comparison with the OH absorption spectra of previous optical fibers fabricated by VAD. Figure 4.38 shows fluorescence spectra measured for pump wavelengths of 808, 980, and 1064 nm. The spectra generated by pumping with the 808 nm LD was very broad and flat from 1150 to 1380 nm.

Figure 4.39 shows the output signal gain versus fiber length in the single-clad bismuth-doped silica fiber (SC-BDF) with a 4 μm core diameter after annealing to evacuate OH ions, measured using the cutback method [100]. The maximum small signal gain was 8 dB for a 3-m-long fiber pumped at 11 mW and 808 nm with a – 68 dBm output signal power at 1310 nm on the display of the optical spectrum analyzer (OSA). The gain slope in Fig. 4.35 is 1.83 dB/m. The propagation loss of the SC-BDF measured by launching the 1310 nm light was − 2 dB/m. In this

Fig. 4.38 Fluorescence spectra of BDF pumped at 808, 980, and 1064 nm

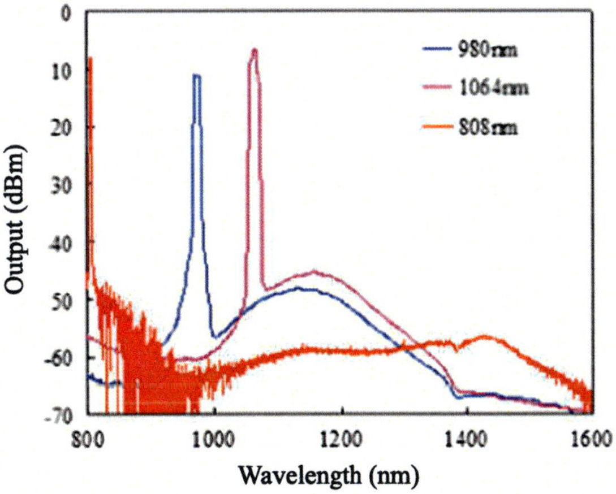

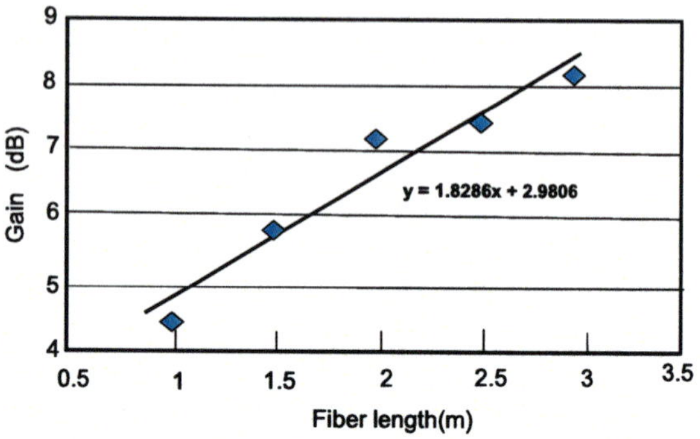

Fig. 4.39 1310 nm signal gain of SC-BDF pumped with 808 nm LD by cutback method (Reprinted with permission from [100]. © 2015 IEICE)

work, the propagation loss in SC-BDF was mainly due to OH ions, but the bismuth itself also absorbed some of the emitted light in mutual band interactions [101]. The amplified gain and the propagation loss of the SC-BDF with 1 mol% Bi concentration and a 4 μm core diameter were 1.41 dB/m and − 3.7 dB/m, respectively. An SC-BDF sample was fabricated with 0.5 mol% Bi. A comparison of SC-BDFs with 0.5 and 1 mol% Bi indicated that increased Bi doping resulted in a larger propagation loss. In order to obtain a high gain in the SC-BDF, the most important parameter is the confinement factor of the amplifier. However, the bismuth oxide was located at the periphery of the core, as was the aluminum oxide. Each oxide concentration distribution was measured by EDX. When comparing the gain values for SC-BDF core diameters of 4 and 10 μm, the 4 μm core gain was larger because of the higher confinement factor related to the overlap integral of the bismuth oxide concentration distribution and the Gaussian laser power distribution.

In the SC-BDF gain saturation was observed along the fiber by fluorescence measurements with core pumping by the fusion splice technique. To eliminate this effect, a saturation double-clad Bi-doped silica fiber (DC-BDF) structure was designed. Bi-doped silica fibers with D-shaped and hexagonal first claddings were fabricated with a polymer second clad layer that has a lower refractive index than silica. The refractive index of the plastic second clad was measured as 1.407.

Figure 4.40a shows the cross-sectional view of a double-clad Bi-doped silica fiber with a D-shaped double-cladding (DDC-BDF) [102]. The core diameter of the DDC-BDF was measured to be 8.4 μm. The longer and the shorter diameters of the D-shaped first-clad were 140 μm and 125 μm, respectively. The relative index difference was Δ~0.5%, measured between the core and the first clad. Fiber pigtails from the LDs, multi-mode fibers (MMF), and the circular DC-BDF were fusion spliced together, except for the asymmetric first cladding in the DDC-BDF. The gain characteristics were measured from the fluorescence spectra of BDFs pumped by LDs using the cutback method. The fluorescence power as a function of fiber length at 1310 nm was measured as the DDC-BDF was cutback in 1 m increments, starting from 4 m. A gain of 0.891 dB/m was obtained from the slope of the gain measurement results.

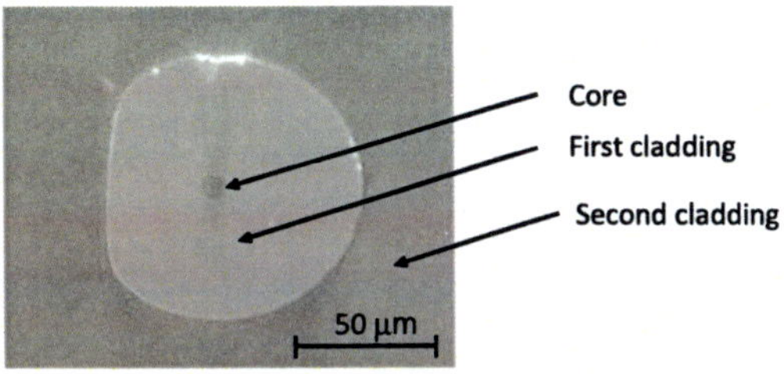

(a) D-shaped core double-clad fiber

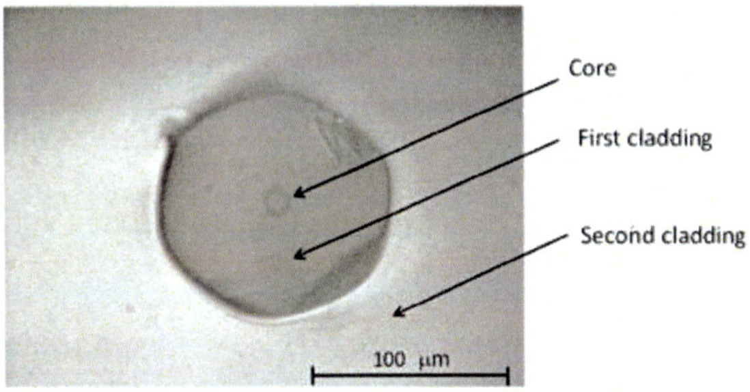

(b) Hexagonal-core double-clad fiber

Fig. 4.40 Cross sections of double-clad Bi-doped fibers (Reprinted with permission from [102]. © 2015 SPIE)

Figure 4.40b shows a cross-sectional view of the hexagonal-shaped double-clad BDF (HDC-BDF). The core diameter of the HDC-BDF was measured to be 7 μm, and the distance between opposite vertices in the hexagon was 100 μm. Side pumping was performed with 808 nm light using a 70° polished multi-mode fiber with a matching gel with the same refractive index as silica. A 1310 nm CW laser light signal was introduced into the HDC-BDF through a fusion splice with the single-mode fiber pigtail. The 1310 nm signal output was – 67 dBm, and the 808 nm LD pumping output power was – 25 dBm on the measured by the optical spectral analyzer after 4 m HDC-BDF. The measured amplified gain was 2.4 dB/m for 4 m of HDC-BDF. The amplified gain was 0.5 dB/m when the 1310 nm output was – 38 dBm. The propagation loss was measured to be – 1.9 dB/m using the cutback method at 1310 nm.

4.5.2 Over 1.65-μm Band Amplifier

A Tm-doped optical fiber amplifier was developed to make it possible to realize a long-haul 2-μm transmission system that employs hollow core photonic bandgap fiber [95]. However, the amplification band of the Tm-doped optical fiber amplifier was limited to 1910–2000 nm. To reduce the transmission potential of the photonic bandgap fiber and increase the transmission capacity, it is necessary to expand the amplification band around 2 μm.

Ho-doped fiber amplifiers are candidates for achieving an amplification band in a wavelength region beyond 2 μm [103–106]. An amplification band in the 2050–2130 nm wavelength region with a gain of over 28 dB has been realized for optical fiber communications [105]. On the other hand, there are two approaches for achieving a shorter amplification band. One is to insert amplified spontaneous emission (ASE) rejection filters along the Tm-doped optical fiber [107], which is the same method used to realize an S-band EDFA [108, 131], and another is doping the cladding of Tm-doped optical fiber with Tb [109, 110]. Although the gain spectrum of Tm-doped fiber has been expanded by increasing the pump power launched into it, the technique only increases the intensity of the tail of the spectrum, thus maintaining the gain spectrum peak wavelength of ~ 2 μm, and it means that the gain of a short wavelength region does not increase due to the gain saturation effect caused by the increasing ASE. The ASE rejection filter and the absorption of Tb can effectively suppress any increase in ASE around 2 μm.

As an example, Fig. 4.41a, b show the configuration of a 1.7-μm band Tm-Tb-doped optical fiber amplifier and the measurement setup, and the amplification characteristics of the fiber amplifiers, respectively [110]. The amplifier employed a forward pump configuration with a 1.2/1.7-μm band WDM coupler and two 1.21-μm pump LDs polarization-multiplexed with a polarization beam combiner (PBC) module. The 14 m long Tm-Tb-doped fluoride fiber was used as an active fiber, and the Tm concentration in the core, the Tb concentration in the cladding, the relative refractive index difference and the cut-off wavelength of the Tm-Tb-doped fluoride fiber were 2000 ppm, 1000 ppm, 2.5% and 1.0 μm, respectively. A 31 nm amplification band (1678–1709 nm) was achieved with low noise characteristics.

These optical amplifiers will also allow us to construct future seamless broadband optical communication systems with wavelengths exceeding 2.1 μm.

4.6 Further Progress

Various multi-core and few-mode EDFAs are currently being developed and studied to improve their characteristics. The most basic amplification media and optical components for these EDFAs have already been successfully developed. Figures 4.42 and 4.43 show the progress made on SDM fiber transmission experiments with multi-core and/or few-mode amplification technologies [32, 38, 111–117]. The transmission distance was increased by employing multi-core and/or few-mode EDFAs, and it is confirmed that the technologies are important as regards devices for constructing a practical long-haul SDM fiber transmission system.

Further development is required for each technology as listed below.

1. Improvement in the characteristics of optical components for constructing practical multi-core and few-mode EDFAs
2. Development of control technology including high-speed response characteristics

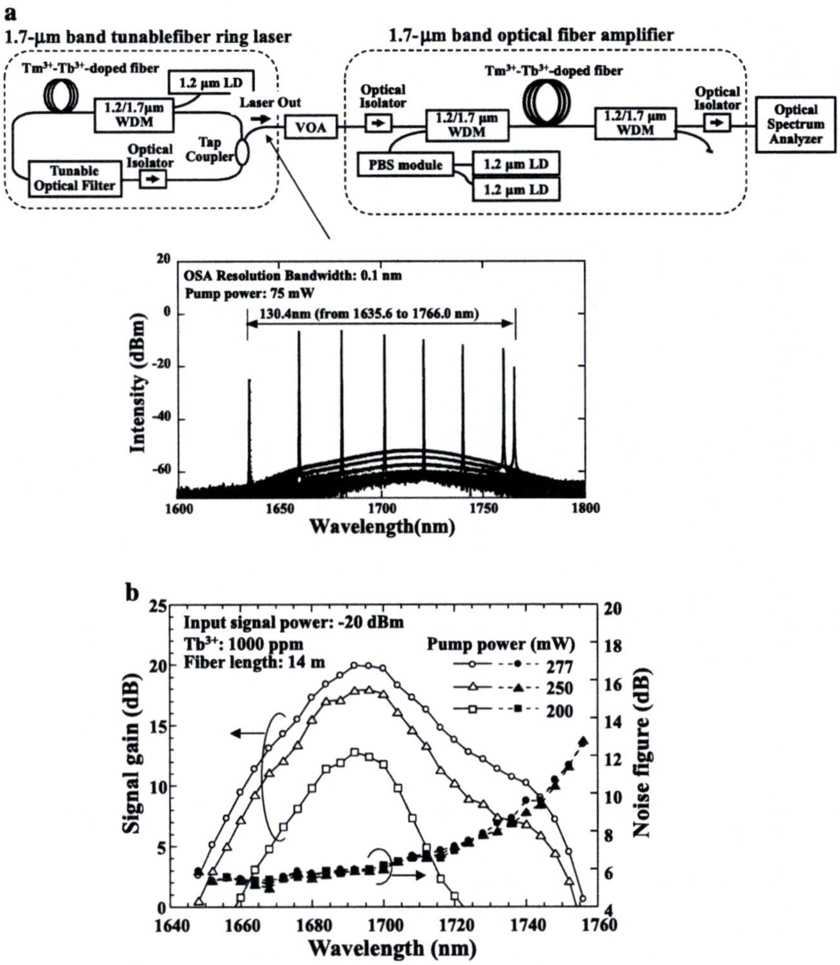

Fig. 4.41 **a** Amplification characteristics of Tm-Tb-doped fiber amplifier. Reprinted with permission from [110]. © 2014 Authors. **b** Amplification characteristics of Tm-Tb-doped fiber amplifier (Reprinted with permission from [110]. © 2014 Authors)

3. Integration of optical components to reduce total amplifier size and cost
4. Effective technique for reducing power consumption.

It is important that we improve the insertion loss and crosstalk characteristics of optical components. A reduction of insertion loss is especially necessary in order to improve the noise and output characteristics, and we must also realize a low power consumption. The realization of an EDFA that has an AGC or ALC function is required for actual SDM fiber transmission systems. It should be noted that, as a gain control technology for multi-core EDFA, a hybrid pumping technology

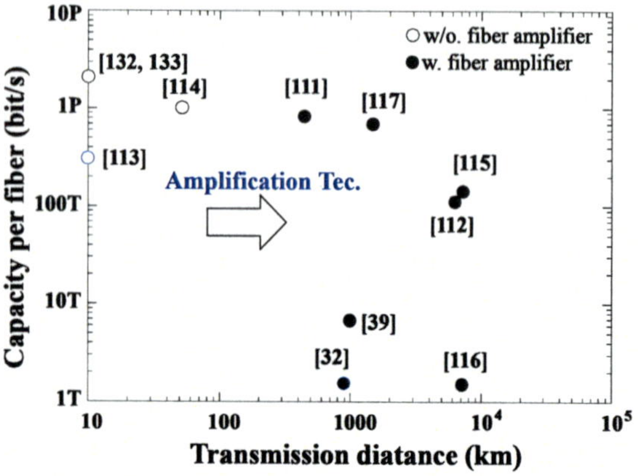

Fig. 4.42 SDM WDM transmission distance versus spatial multiplicity

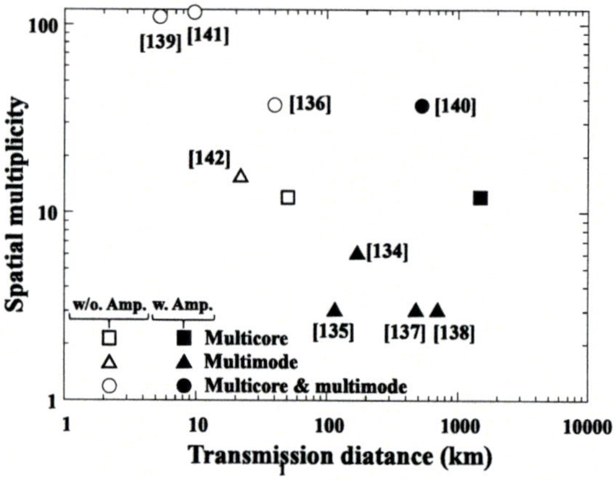

Fig. 4.43 SDM WDM transmission distance versus spatial multiplicity

that combines core pumping and cladding pumping has been proposed [66], and the feasibility of AGC and ALC is increasing. In general, a two-stage configuration of multi-core and few-mode EDFA will be employed as well as the conventional single-core and single-mode EDFA. Since we are at a stage where only the basic techniques for realizing individual optical components for multi-core and few-mode EDFA have been realized, we can expect the development of integrated optical components such as a pump/signal combiner with an optical isolator, a tap coupler with monitor photo-diodes, a pump source where a number of pump lasers are packed into one, and a gain equalization filter in which level equalization is possible.

The most difficult issue is the realization of low power consumption. In multi-core amplification, optical amplifier power efficiency (OAPE) has begun to be used for evaluating the power consumption. Figure 4.44 shows the number of cores versus OAPE for some of the previously reported cladding-pumped multi-core EDFAs [118]. The cladding-pumped multi-core EDFA with a single laser diode pumping

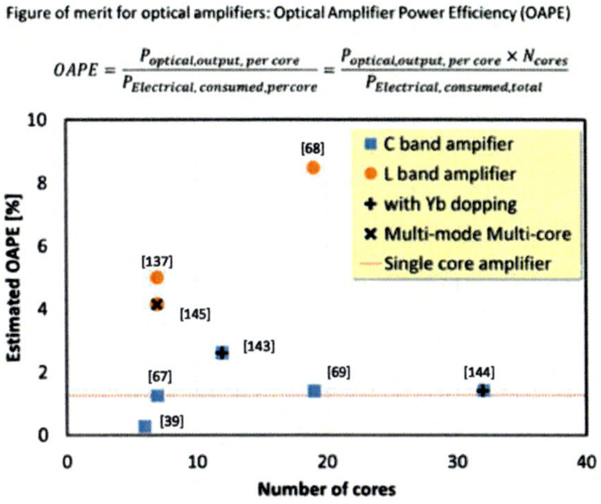

Fig. 4.44 Number of cores versus OAPE for some of the previously reported cladding-pumped multi-core EDFAs (Reprinted with permission from [118]. © 2020 EXAT)

multiple cores within the same cladding is attractive in the L-band operation. It has been found that the C-band operation is more limited than the L-band operation due to the lower maximum output power. In addition, although there was a report that a few-mode EDFA is superior to a multi-core EDFA as regards power consumption, low power consumption is only achieved for small signal amplification, which is not performed with practical amplifiers. We would like to expect future progress in reducing the power consumption of the few-mode amplification.

Furthermore, to achieve a transmission capacity exceeding exabit/s, a combination of optical multi-core and multi- mode fiber transmission is indispensable. Therefore, the development of multi-core and few-mode EDFA, which integrates multi-core and few-mode amplification technology, has also started [119].

Finally, Fig. 4.45 shows the scenarios for SDM amplifiers shown in "EXAT Roadmap ver.2" [118], which describes the roadmap of SDM transmission technology compiled by "Technical Committee on Extremely Advanced Optical Transmission Technologies (EXAT)" group in Japan. It is expected to be applied to phase 1 short distance transmission, Phase 2 long-haul transmission, Phase 3 terrestrial and flexible network, and the progress of power consumption is also shown. We would like to expect further research and development toward these applications.

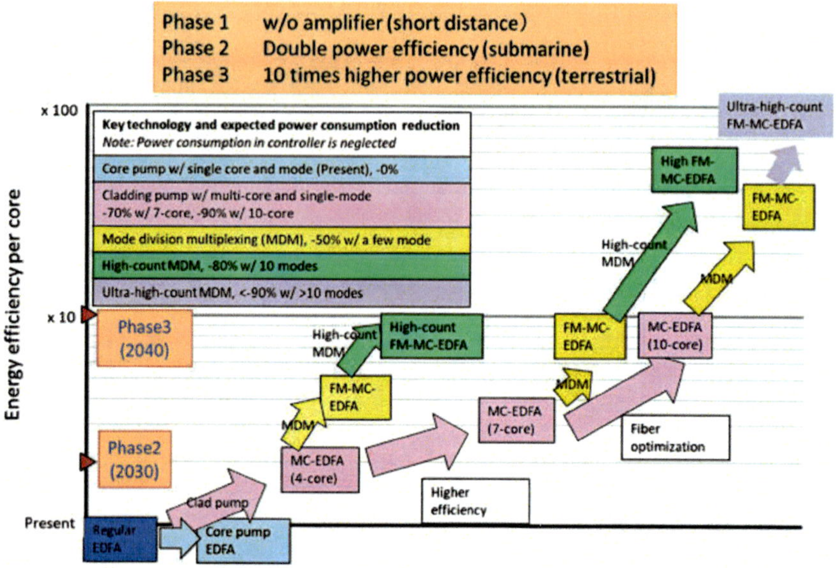

Fig. 4.45 Scenarios for SDM amplifiers (Reprinted with permission from [118]. © 2020 EXAT)

References

1. K. Hagimoto, S. Nishi, K. Nakagawa, An optical bit-rate flexible transmission system with 5 Tb/s km capacity employing multiple in-line erbium-doped fiber amplifiers. IEEE J. Lightwave Technol. **8**
2. E. Yoneda, K. Suto, K. Kikushima, H. Yoshinaga, All-fiber video distribution (AFVD) systems using SCM and EDFA techniques. IEEE J. Lightwave Technol. **11**(1), 128–137 (1993)
3. A. Dutta, N. Dutta, M. Fujiwara, *WDM Technologies: Passive Optical Components*, Chapters 9–10 (Academic, San Diego, CA, 2003)
4. R.J. Essiambre, G. Kramer, P.J. Winzer, G.J. Foschini, B. Goebel, Capacity limits of optical fiber networks. IEEE J. Lightwave Technol. **28**(4), 662–701 (2010)
5. K. Kitayama, M. Nakamura, Y. Igasaki, K. Kaneda, Image fiber-optic two-dimensional parallel links based upon optical space-CDMA. IEEE J. Lightwave Technol. **15**(2), 202–202 (1997)
6. D.T. Neilson, E. Schenfeld, Plastic modules for free-space optical interconnects. Appl. Opt. **37**(14), 2944–2952 (1998)
7. M. Ishikawa, Optoelectronic parallel computing system with reconfigurable optical interconnection, in *SPIE*, vol. CR62 (1996), pp. 156–175
8. C. Tombling, T. Saitoh, T. Mukai, Performance predictions for vertical-cavity semiconductor laser amplifiers. IEEE J. Quantum Electron. **30**(11), 2491–2499 (1994)
9. R. Lewen, K. Streubel, A. Karlsson, S. Rapp, Experimental demonstration of a multifunctional long-wavelength vertical-cavity laser amplifier-detector. IEEE Photon. Technol. Lett. **10**(8), 1067–1069 (1998)
10. K. Koyabu, F. Ohira, T. Yamamoto, Fabrication of two-dimensional fiber arrays using microferrules. IEEE Trans. Comp. Packag. Technol. **21**(1), 11–19 (1998)
11. M. Nakamura, K. Kitayama, N. Shamoto, K. Kaneda, Two-dimensional Erbium-doped image fiber amplifier (EDIFA). IEEE J. Select. Topics Quantum Electron. **7**(3), 434–438 (2001)

12. G. Nykolak, S.A. Kramer, J.R. Simpson, D.J. DiGiovanni, C.R. Giles, H.M. Presby, An erbium-doped multimode optical fiber amplifier. IEEE Photon. Technol. Lett. **3**(12), 1079–1081 (1991)
13. O.G. Okhotnikov, J.M. Sousa, High power superfluorescent source with stable single-transverse-mode output using a multimode Er-doped fibre. Electron. Lett. **33**(20), 1727–1728 (1997)
14. O.G. Okhotnikov, J.M. Sousa, High power single-transverse-mode amplifier based on a multi-mode Er-doped fibre, in *24th European Conference on Optical Communication (ECOC1998)* (1998), pp. 237–238
15. C.D. Stancey, R.M. Jenkins, Demonstration of fundamental mode propagation in highly multimode fibre for high power EDFAs, in *Conference of Lasers and Electro-Optics Europe (CLEO2005)*, Munich, Germany (2005), p. 558
16. M. Yamada, Amplification technologies for multi-core fibers, in Korea-Japan Workshop on Beyond 100G, ThC8 (2011), pp. 45–46
17. M. Yamada, S. Matsuo, H. Masuda, S. Kobayashi, K. Tsujikawa, H. Ono, Recent progress on multicore amplification technologies, in *International Symposium on Extremely Advanced Transmission Technology (EXAT)'13*, Invite Talk in Session III: Multicore Fiber and Devices (2013), pp. 55–59
18. M. Yamada, H. Ono, S. Matsuo, Multicore EDFA for long distance transmission, in Photonics West 2014: Next-Generation Optical Communication: Components, Sub-Systems and Systems III, 9009-21 (2014)
19. H. Ono, K. Takenaga, K. Ichii, M. Yamada, Amplification technology for multi-core fiber transmission, in SUM 2014 IEEE Photonics Society Summer Topical Meeting on Space Division Multiplexing for Optical Systems and Networks, ME2.1 (2014)
20. K. Tsujikawa, L. Ma, K. Ichii, S. Matsuo, M. Yamada, N. Hanzawa, H. Ono, Optical fiber amplifier employing a bundle of reduced cladding erbium-doped fibers for multi-core fiber transmission, in SUM 2012 IEEE Photonics Society Summer Topical Meeting on SDM for Optical Systems and Networks, WC3 (2012)
21. M. Yamada, K.T. Sujikawa, L. Ma, K. Ichii, S. Matsuo, N. Hanzawa, H. Ono, Optical fiber amplifier employing a bundle of reduced cladding erbium-doped fibers. IEEE Photon. Technol. Lett. **24**(21), 1910–1913 (2012)
22. K. Yamaoka, M. Yamada, K. Tsujikawa, K. Ichii, S. Matsuo, H. Ono, Crosstalk reduction in multi-core fiber amplifier with bundled EDF, in *International Symposium on Extremely Advanced Transmission Technology (EXAT)'13*, P07 (2013)
23. K.S. Abedin, T.F. Taunay, M. Fishteyn, M.F. Yan, B. Zhu, J.M. Fini, E.M. Monberg, F.V. Dimarcello, P.W. Wisk, Amplification and noise properties of an erbium-doped multicore fiber amplifier. Opt. Express **19**(17), 16715–16721 (2011)
24. Y. Tsuchida, K. Maeda, Y. Mimura, H. Matsuura, R. Miyabe, K. Aiso, R. Sugizaki, Amplification characteristics of a multi-core erbium-doped fiber amplifier, in *Optical Fiber Communication Conference and Exposition (OFC 2012)*, OM3C.3 (2012)
25. H. Ono, Y. Abe, K. Shikama, T. Takahashi, M. Yamada, K. Takenaga, S. Matsuo, Amplification method for crosstalk reduction in a multi-core fibre amplifier. Electron. Lett. **49**(2), 138–140 (2013)
26. H. Ono, H. Takara, M. Yamada, K. Ichii, K. Takenaga, S. Mastuo, Multi-core EDFA for 12-core fiber transmission, in *International Symposium on Extremely Advanced Transmission Technology (EXAT)'13*, P02 (2013)
27. H. Masuda, H. Ono, H. Takara, Y. Miyamoto, K. Ichii, K. Takenaga, S. Matsuo, K. Kitamura, Y. Abe, M. Yamada, Remotely pumped multicore erbium-doped fiber amplifier system with high pumping efficiency, in SUM 2013 IEEE Photonics Society Summer Topical Meeting on SDM for Optical Communication, WC3.3 (2013)
28. K.S. Abedin, T.F. Taunay, M. Fishteyn, D.J. DiGiovanni, V.R. Supradeepa, J.M. Fini, M.F. Yan, B. Zhu, E.M. Monberg, F.V. Dimarcello, Cladding-pumped erbium-doped multicore fiber amplifier. Opt. Express **20**(18), 20191–20200 (2012)

29. Y. Mimura, Y. Tsuchida, K. Maeda, R. Miyabe, K. Aiso, H. Matsuura, R. Sugizaki, Batch multicore amplification with cladding-pumped multicore EDF, in *38th European Conference on Optical Communication (ECOC2012)*, Tu.4.F.1 (2012)
30. S. Takasaka, H. Matsuura, W. Kumagai, M. Tadakuma, Y. Mimura, Y. Tsuchida, K. Maeda, R. Miyabe, K. Aiso, K. Doi, R. Sugizaki, Cladding-pumped seven-core EDFA using a multimode pump light coupler, in *39th European Conference on Optical Communication (ECOC2013)*, We.4.A.5 (2013)
31. H. Ono, K. Takenaga, K. Ichii, S. Matsuo, T. Takahashi, H. Masuda, M. Yamada, 12-core double-clad Er/Yb-doped fiber amplifier employing pump/signal combiner modules with multi-core fiber pigtails, in *39th European Conference on Optical Communication (ECOC2013)*, We.4.A.4 (2013)
32. J. Sakaguchi, W. Klaus, B. J. Puttnam, J.-M.D. Mendinueta, Y. Awaji, N. Wada, Y. Tsuchida, K. Maeda, M. Tadakuma, K. Imamura, R. Sugizaki, T. Kobayashi, Y. Tottori, M. Watanabe, R.V. Jensen, 19-core MCF transmission system using EDFA with shared core pumping coupled in free-space optics, in *39th European Conference on Optical Communication (ECOC2013)*, Th.1.C.6 (2013)
33. K. Ichii, K. Takenaga, S. Matsuo, H. Ono, M. Yamada, Double-clad multi-core Er/Yb co-doped fiber for cladding pumped amplifier, in *International Symposium on Extremely Advanced Transmission Technology (EXAT)'13*, P08 (2013)
34. V.J.F. Rancaño, S. Jain, T.C. May-Smith, J.K. Sahu, P. Petropoulos, D.J. Richardson, First demonstration of an amplified transmission line based on multi-element fibre technology, in *39th European Conference on Optical Communication (ECOC2013)*, PD1.C.2 (2013)
35. S. Jain, V.J.F. Rancaño, T.C. May-Smith, P. Petropoulos, J.K. Sahu, D.J. Richardson, Multi-element fiber technology for space-division multiplexing applications. Opt. Express **22**(4), 3787–3796 (2014)
36. J.K. Sahu, S. Jain, V.J.F. Rancaño, T.C. May-Smith, A. Webb, P. Petropoulos, D.J. Richardson, Multi-element fiber for space-division multiplexing, in SPIE Photonics West 2014: Next-Generation Optical Communication: Components, Sub-Systems and Systems III, 9009-11 (2014)
37. H. Ono, M. Yamada, T. Hosokawa, K. Ichii, H. Masuda, S. Kobayashi, Study on gain control of MC-EDFA by using hybrid pumping of cladding- and core-direct-pumping schemes. IEICE Technical Report, OCS2014-116 (2015) (in Japanese)
38. H. Takara, T. Mizuno, H. Kawakami, Y. Miyamoto, H. Masuda, K. Kitamura, H. Ono, S. Asakawa, Y. Amma, K. Hirakawa, S. Matsuo, K. Tsujikawa, M. Yamada, 120.7-Tb/s (7 SDM/180 WDM/95.8 Gb/s) MCF-ROPA unrepeatered transmission of PDM-32QAM channels over 204 km. IEEE J. Lightwave Technol. **33**(7), 1473–1478 (2015)
39. H. Takara, H. Ono, Y. Abe, H. Masuda, K. Takenaga, S. Matsuo, H. Kubota, K. Shibahara, T. Kobayashi, Y. Miaymoto, 1000-km 7-core fiber transmission of 10 × 96-Gb/s PDM-16QAM using Raman amplification with 6.5 W per fiber. Opt. Express **20**(9), 10100–10105 (2012)
40. K. Watanabe, T. Saito, Y. Tsuchida, K. Maeda, M. Shiino, Fiber bundle type fan-out for multicore Er doped fiber amplifier, in *18th OptoElectronics and Communications Conference (OECC2013)*, TuS1–5 (2013)
41. Y. Tsuchida, K. Maeda, K. Watanabe, T. Ito, K. Fukuchi, M. Yoshida, Y. Mimura, R. Sugizaki, M. Nakazawa, Multicore EDFA for DWDM transmission in full C-band, in *Optical Fiber Communication Conference and Exposition (OFC 2013)*, JW2A.16 (2013)
42. Y. Tshuchida, K. Maeda, K. Watanbe, T. Saito, S. Matsumoto, K. Aiso, Y. Mimura, R. Sugizaki, Simultaneous 7-core pumped amplification in multicore EDF through fiber based fan-in/out, in *38th European Conference on Optical Communication (ECOC2012)*, Tu.4F.2 (2012)
43. S. Jain, T.C. May-Smith, A.S. Webb, M. Belal, D.J. Richardson, J.K. Sahu, D.N. Payne, Erbium-doped multi-element fiber amplifiers for space-division multiplexing operations. Opt. Lett. **38**(4), 582–584 (2013)
44. K.S. Abedin, J.M. Fini, T.F. Thierry, B. Zhu, M.F. Yan, L. Bansal, F.V. Dimarcello, E.M. Monberg, D.J. DiGiovanni, Seven-core erbium-doped double-clad fiber amplifier pumped simultaneously by side-coupled multimode fiber. Opt. Lett. **39**(4), 993–996 (2014)

45. K. Maeda, Y. Tsuchida, S. Takasaka, T. Saito K. Watanabe, T. Sasa, R. Sugizaki, K. Takeshima, T. Tsuritani, Cladding pumped multicore EDFA with output power over 20 dBm using a fiber based pump combiner, in *20th OptoElectronics and Communications Conference (OECC2015)*, PWe.30 (2015)
46. S. Jain, C. Castro, Y. Jung, J. Hayes, R. Sandoghchi, T. Mizuno, Y. Sasaki, Y. Amma, Y. Miyamoto, M. Bohn, K. Pulverer, Md. Nooruzzaman, T. Morioka, S. Alam, D.J. Richardson, 32-core erbium/ytterbium-doped multicore fiber amplifier for next generation space-division multiplexed transmission system. Opt. Express **25**, 32887–32896 (2017)
47. K. Aida, S. Nishi, Y. Sato, K. Hagimoto, Design and performance of a long-span IM/DD optical transmission system using remotely pumped optical amplifiers. IEE Proc. J. **137**(4), 225–229 (1990)
48. H. Kidorf, M. Nissov, D. Foursa, Ultra-long haul submarine and terrestrial applications, in *Raman Amplifiers for Telecommunications*, ed. by M.N. Islam, 2, Chapter 17 (Springer, Belin, 2004)
49. D. Chang, P. Patki, S. Burtsev, W. Pelouch, 8 × 120 Gb/s transmission over 80.8 dB/480.4 km unrepeatered span, in *Optical Fiber Communication Conference and Exposition (OFC2013)*, JTh2A.42 (2013)
50. H. Masuda, H. Kawakami, S. Kuwahara, A. Hirano, K. Sato, Y. Miyamoto, 1.28-Tbit/s (32 × 43 Gbit/s) field trial over 528-km (6 × 88 km) DSF using L-band remotely-pumped EDF/distributed Raman hybrid inline-amplifiers. Electron. Lett. **39**(23), 1668–1670 (2003)
51. H. Masuda, H. Nagaoka, K. Tayama, Optical SNR characteristics of a space-division-multiplexed optically-amplified transmission system using remotely pumped multi-core EDFAs, in *16th OptoElectronics and Communications Conference (OECC2011)*, 6C2–4 (2011)
52. Y. Yamauchi, H. Masuda, T. Nobukawa, H. Nagaoka, A highly-efficient remotely-pumped multi-core EDFA transmission system with a novel hybrid wavelength-/space-division multiplexing scheme, in *17th OptoElectronics and Communications Conference (OECC2012)*, 5C1–4 (2012)
53. K. Takenaga, K. Ichii, S. Matsuo, Y. Abe, H. Ono, M. Yamada, H. Masuda, Multicore EDF optimized for remotely pumped amplification system over multicore fiber, in *17th OptoElectronics and Communications Conference (OECC2013)*, TuS1–2 (2013)
54. M.N. Islam, Overview of Raman amplification in telecommunications, in *Raman Amplifiers for Telecommunications*, ed. by M.N. Islam, 1, Chapter 1 (Springer, Berlin, 2004)
55. K. Rottwitt, Distributed Raman amplifiers, in *Raman Amplification in Fiber Optical Communication Systems*, ed. by C. Headley, G.P. Agrawal, Chapter 3 (Elsevier Academic Press, Cambridge, 2005)
56. H. Masuda, M. Tomizawa, Y. Miyamoto, K. Hagimoto, High-performance distributed Raman amplification systems with limited pump power. IEICE Trans. Commun. **E89-B**(3), 715–723 (2006)
57. H. Masuda, Advanced transmission systems using distributed Raman amplification technologies, in *SPIE OpticsEast*, vol. 6012, 601204 (2005), pp. 1–13
58. H. Masuda, M. Tomizawa, Y. Miyamoto, K. Hagimoto, Impacts of distributed Raman amplification transmission technologies on terrestrial large-capacity WDM systems. IEICE **J89-B**(3), 307–315 (2006)
59. H. Masuda, M. Tomizawa, Y. Miyamoto, High-performance distributed Raman amplification systems: practical aspects and field trial results, in *Optical Fiber Communication Conference and Exposition (OFC2005)*, OThF5 (2005)
60. H. Masuda, E. Yamazaki, A. Sano, T. Yoshimatsu, T. Kobayashi, E. Yoshida, Y. Miyamoto, S. Matsuoka, Y. Takatori, M. Mizoguchi, K. Okada, K. Hagimoto, T. Yamada, S. Kamei, 13.5-Tb/s (135 × 111-Gb/s/ch) no-guard-interval coherent OFDM transmission over 6,248 km using SNR maximized second-order DRA in the extended L-band, in *Optical Fiber Communication Conference and Exposition (OFC2009)*, PDPB5 (2009)
61. G. Charlet, M. Salsi, P. Tran, M. Bertolini, H. Mardoyan, J. Renaudier, O. Bertran-Pardo, S. Bigo, 72 × 100Gb/s transmission over transoceanic distance, using large effective area fiber,

hybrid Raman-erbium amplification and coherent detection, in *Optical Fiber Communication Conference and Exposition (OFC2009)*, PDPB6 (2009)
62. K. Kitamura, K. Tayama, H. Masuda, Cross-talk characteristics of a hybrid multi-core fiber transmission system using distributed Raman amplification, in *17th OptoElectronics and Communications Conference (OECC2013)*, TuS1-3 (2013)
63. H. Ono, T. Mizuno, H. Takara, K. Ichii, K. Takenaga, S. Matsuo, M. Yamada, Inter-core crosstalk measurement in multicore fibre amplifier using multiple intensity tones. Electron. Lett. **50**(14), 1009–1010 (2014)
64. Y. Tsuchida, K. Maeda, K. Watanabe, T. Saito, S. Takasaka, M. Tadakuma, R. Sugizaki K. Doi, H. Matsuura, K. Takeshima, T. Tsuritani, I. Morita, H. Ogoshi, Cladding-pumped L-band multicore EDFA with reduced power consumption, in SUM 2014 IEEE Photonics Society Summer Topical Meeting on Space Division Multiplexing for Optical Systems and Networks, ME2.2 (2014)
65. J. Sakaguchi, W. Klaus, B.J. Puttnam, J.M.D. Mendinueta, Y. Awaji, N. Wada, Y. Tsuchida, K. Maeda, M. Tadakuma, K. Imamura, R. Sugizaki, T. Kobayashi, Y. Tottori, M. Watanabe, R.V. Jensen, 19-core MCF transmission system using EDFA with shared core pumping coupled via free-space optics. Opt. Express **22**(1), 90–95 (2014)
66. H. Ono, Y. Miyamoto, M. Yamada, Gain control in multi-core erbium-doped fiber amplifier with cladding and core hybrid pumping. IEEE J. Lightwave Technol. **37**(13), 3365–3372 (2019)
67. Y. Tsuchida, K. Maeda, K. Watanabe, K. Takeshima, T. Sasa, T. Saito, S. Takasaka, Y. Kawaguchi, T. Tsuritani, R. Sugizaki, Cladding pumped seven-core EDFA using an absorption-enhanced erbium doped fibre, in *Proceedings of European Conference on Optical Communication 2016*, Düsseldorf, Germany, M.2.A.2 (2016)
68. S. Takasaka, K. Maeda, K. Kawasaki, K. Yoshioka, H. Oshio, R. Sugizaki, Y. Kawaguchi, H. Takahashi, T. Tsuritani, M. Shiino, Increase of cladding pump power efficiency by a 19-core erbium doped fibre amplifier, in *Proceedings of European Conference on Optical Communication 2017*, Gothenburg, Sweden, September Th.2.D (2017)
69. S. Takasaka, K. Maeda, K. Kawasaki et al., EDF Length dependence of amplification characteristics of cladding pumped 19-core EDFA, in *Proceedings of Optical Fiber Communications Conference and Exposition 2018*, San Diego, USA, Th1K.2 (2018)
70. S. Takasaka, K. Maeda, K. Kawasaki et al., Cladding pump recycling in 7-core EDFA, in *Proceedings of European Conference on Optical Communication 2018*, Rome, Italy, We1E.5 (2018)
71. H. Takeshita, K. Matsumoto, S. Yanagimachi et al., Improvement of the pump recycling ratio of turbo cladding pumped MC-EDFA with paired spatial pump combiner and splitter, in *Proceedings of Optical Fiber Communications Conference and Exposition 2019*, San Diego, USA, Th1B.2 (2019)
72. P.J. Winzer, G.J. Foschini, MIMO capacities and outage probabilities in spatially multiplexed optical transport systems. Opt. Express **19**(17), 16680–16696 (2011)
73. H. Ono, T. Hosokawa, K. Ichii, S. Matsuo, M. Yamada, Improvement of differential modal gain in few-mode fibre amplifier by employing ring-core erbium-doped fibre. Electron. Lett. **51**(2), 172–173 (2014)
74. N. Bai, E. Ip, T. Wang, G. Li, Multimode fiber amplifier with tunable modal gain using a reconfigurable multimode pump. Opt. Express **19**(17), 16601–16611 (2011)
75. M. Wada, T. Sakamoto, T. Mori, T. Yamamoto, N. Hanzawa, F. Yamamoto, Modal gain controllable 2-LP-mode fiber amplifier using PLC type coupler and long-period grating. IEEE J. Lightwave Technol. **32**(24), 4092–4098 (2014)
76. R.N. Mahalati, D. Askarov, J.M. Kahn, Adaptive modal gain equalization techniques in multimode erbium-doped fiber amplifiers. IEEE J. Lightwave Technol. **32**(11), 2133–2143 (2014)
77. N. Bai, E. Ip, Y.-K. Huang, E. Mateo, F. Yaman, M.-J. Li, S. Bickham, S. Ten, J. Liñares, C. Montero, V. Moreno, X. Prieto, V. Tse, K. M. Chung, A. Pak T. Lau, H.-Y. Tam, C. Lu, Y. Luo, G. Peng, G. Li, T. Wang, Mode-division multiplexed transmission with inline few-mode fiber amplifier. Opt. Express **20**(3), 2668–2680 (2012)

78. M. Wada, T. Sakamoto, T. Mori, N. Hanzawa, T. Yamamoto, F. Yamamoto, Modal gain controllable all-fiber type multimode fiber amplifier, in *18th OptoElectronics and Communications Conference (OECC2013)*, TuS4-5 (2013)
79. Q. Kang, E. Lim, Y. Jung, F. Poletti, S.-U. Alam, D.J. Richardson, Design of four-mode erbium doped fiber amplifier with low differential modal gain for modal division multiplexed transmissions, in *Optical Fiber Communication Conference and Exposition (OFC2013)*, OTu3G.3 (2013)
80. Y. Jung, Q. Kang, V.A.J.M. Sleiffer, B. Inan, M. Kuschnerov, V. Veljanovski, B. Corbett, R. Winfield, Z. Li, P.S. Teh, A. Dhar, J. Sahu, F. Poletti, S.U. Alam, D.J. Richardson, Three mode Er ring-doped fiber amplifier for mode-division multiplexed transmission. Opt. Express **21**(8), 10383–10392 (2013)
81. E. Ip, M.-J. Li, K. Bennett, A. Korolev, K. Koreshkov, W. Wood, C. Montero, J. Liñares, Experimental characterization of a ring-profile few-mode erbium-doped fiber amplifier enabling gain equalization, in *Optical Fiber Communication Conference and Exposition (OFC2013)*, JTh2A.18 (2013)
82. G.L. Cocq, Y. Quiquempois, A.L. Rouge, G. Bouwmans, H.E. Hamzaoui, K. Delplace, M. Bouazaoui, L. Bigot, Few mode Er-doped fiber with micro-structured core for mode division multiplexing in the C-band. Opt. Express **21**(25), 31647–31659 (2013)
83. Y. Jung, Q. Kang, J.K. Sahu, B. Corbett, J. O'Callagham, F. Poletti, S.U. Alam, D.J. Richardson, Reconfigurable modal gain control of a few-mode EDFA supporting six spatial modes. IEEE Photon. Technol. Lett. **26**(11), 1100–1103 (2014)
84. Q. Kang, E. Lim, Y. Jun, X. Jin, F.P. Payne, S.U. Alam, D.J. Richardson, Gain equalization of a six-mode-group ring core multimode EDFA, in *40th European Conference on Optical Communication (ECOC2014)*, P.1.14 (2014)
85. H. Ono, T. Hosokawa, K. Ichii, S. Matsuo, H. Nasu, M. Yamada, 2-LP mode few-mode fiber amplifier employing ring-core erbium-doped fiber. Opt. Express **23**(21), 27405–27418 (2015)
86. M. Wada, T. Sakamoto, S. Aozasa, T. Mori, T. Yamamoto, N. Hanzawa, F. Yamamoto, L-band 2-LP mode EDFA with low modal dependent gain, in *Optical Fiber Communication Conference and Exposition (OFC2015)*, Tu3C.3 (2015)
87. R. Ryf, A. Sierra, R. Essiambre, S. Randel, A. Gnauck, C. A. Bolle, M. Esmaeelpour, P. J. Winzer, R. Delbue, P. Pupalaikis, A. Sureka, D. Peckham, A. McCurdy, R. Lingle, Mode-equalized distributed Raman amplification in 137-km few-mode fiber, in *37th European Conference on Optical Communications (ECOC2011)*, Th.13.K.5 (2011)
88. R. Ryf, R. Essiambre, J. Hoyningen-Huene, P. Winzer, Analysis of mode-dependent gain in Raman amplified few-mode fiber, in *Optical Fiber Communication Conference and Exposition (OFC2012)*, OW1D.2 (2012)
89. Z. Zhou, An analytical approach for gain optimization in multimode fiber Raman amplifiers. Opt. Express **22**(18), 21393–21402 (2014)
90. Q. Kang, E.-L. Lim, F. Poletti, Y. Jung, C. Baskiotis, S.U. Alam, D.J. Richardson, Minimizing differential modal gain in cladding-pumped EDFAs supporting four and six mode groups. Opt. Express **22**(18), 21499–21507 (2014)
91. G.L. Cocq, Y. Quiquempois, L. Bigot, Optimization algorithm applied to the design of few-mode erbium doped fiber amplifier for modal and spectral gain equalization. IEEE J. Lightwave Technol. **33**(1), 100–108 (2015)
92. K.I. Suzuki, Y. Fukada, N. Yoshimoto, K. Kumozaki, M. Tsubokawa, Automatic level controlled burst-mode optical fiber amplifier for 10 Gbit/s PON application, in *Optical Fiber Communication Conference and Exposition (OFC2010)*, OTuH1 (2010)
93. A.D. Ellis, Current capacity limits and activities within the EU project MODE-GAP to overcome them, in SUM 2012 IEEE Photonics Society Summer Topical Meeting on SDM for Optical Systems and Networks, MC1.1 (2012), pp. 169–170
94. D.J. Richardson, Fiber amplifiers for SDM systems, in *Optical Fiber Communication Conference and Exposition (OFC2013)*, OTu3G.1 (2013)
95. Z. Li, A.M. Heidt, J.M.O. Daniel, Y. Jung, S.-U. Alam, D.J. Richardson, Thulium-doped fiber amplifier for optical communications at 2 μm. Opt. Express **21**(8), 9289–9297 (2013)

96. Y. Fujimoto, M. Nakatsuka, Infrared luminescence from bismuth-doped silica glass, Japan. J. Appl. Phys. **40**(3B), Part 2, L279–L281 (2001)
97. E.M. Dianov, V.V. Dvoryn, V.M. Mashinsky, A.A. Umnikov, M.V. Yashkov, A.N. Guryanov, CW bismuth fiber laser. Quantum Electron. **35**(12), 1082–1084 (2005)
98. V.V. Dvoyrin, V.M. Mashinsky, L.I. Bulatov, I.A. Bufetov, A.V. Shubin, M.A. Melkumov, E.F. Kustov, E.M. Dianov, Bismuth-doped-glass optical fibers—a new active medium for lasers and amplifiers. Opt. Lett. **31**(20), 2966–2968 (2006)
99. V.V. Dvoyrin, E.M. Dianov, Efficient bismuth-doped fiber laser. IEEE J. Quantum Electron. **44**(9), 834–840 (2008)
100. M. Takahashi, T. Fujii, Y. Saito, Y. Fujii, S. Kobayashi, Optical amplification at 1.3 μm with Bi doped fiber fabricated by VAD method, in *18th OptoElectronics and Communications Conference (OECC2013)*, TuPS-9 (2013)
101. E.M. Dianov, Bismuth-doped optical fibers: a new active medium for NIR lasers and amplifiers, in *Proceedings of SPIE*, vol. 8601 (2013), pp. 86010H-1-10
102. M. Takahashi, T. Fujii, Y. Saito, Y. Fujii, S. Kobayashi, 1.3 μm optical amplification with double-clad Bi doped silica fiber, in *SPIE Photonics West 2015* (2015), pp. 9344–9388
103. A.S. Kurkov, V.A. Kamynin, E.M. Sholokhov, A.V. Marakulin, Mid-IR supercontinuum generation in Ho-doped fiber amplifier. Laser Phys. Lett. **8**(10), 754–757 (2011)
104. Y. Tang, X. Li, Z. Yan, X. Yu, Y. Zhang, Q.J. Wang, 50-W 2-μm nanosecond all-fiber-based thulium-doped fiber amplifier. IEEE J. Select. Topics Quantum Electron. **20**(5), 310077 (2014)
105. N. Simakov, Z. Li, S.U. Alam, P.C. Shardlow, J.M.O. Daniel, D. Jain, J.K. Sahu, A. Hemming, W.A. Clarkson, D.J. Richardson, Holmium-doped fiber amplifier for optical communications at 2.05–2.13 μm, in *Optical Fiber Communication Conference and Exposition (OFC2015)*, Tu2C.6 (2015)
106. V.A. Kamynin, S.O. Antipov, A.V. Baranikov, A.S. Kurkovadb, Holmium-doped fibre amplifier operating at 2.1 μm. Quantum Electron. **44**(2), 161–162 (2014)
107. Z. Li1, Y. Jung, J.M.O. Daniel, N. Simakov, P.C. Shardlow, A.M. Heidt, W.A. Clarkson, S.U. Alam, D.J. Richardson, Extreme short wavelength operation (1.65–1.7 μm) of silica based thulium doped fiber amplifier, in *Optical Fiber Communication Conference and Exposition (OFC2015)*, Tu2C (2015)
108. H. Ono, M. Yamada, M. Shimizu, An S-band erbium-doped silica fiber amplifier with a flattened-gain of over 21 dB. Electron. Lett. **38**(19), 1084–1086 (2002)
109. T. Sakamoto, M. Shimizu, M. Yamada, T. Kanamori, Y. Ohishi, Y. Terunuma, S. Sudo, 35-dB gain Tm-doped ZBLYAN fiber amplifier operating at 1.65 μm. IEEE Photon. Technol. Lett. **8**(3), 349–351 (1996)
110. M. Yamada, H. Ono, K. Ohta, S. Aozasa, T. Tanaka, K. Senda, Y. Maeda, O. Koyama, J. Ono, 1.7 μm band optical fiber amplifier, in *Optical Fiber Communication Conference and Exposition (OFC2014)*, Tu2D.3 (2014)
111. A. Sano, H. Takara, T. Kobayashi, H. Kawakami, H. Kishikawa, T. Nakagawa, Y. Miyamoto, Y. Abe, H. Ono, K. Shikama, M. Nagatani, T. Mori, Y. Sasaki, I. Ishida, K. Takenaga, S. Matsuo, K. Saitoh, M. Koshiba, M. Yamada, H. Masuda, T. Morioka, 409-Tb/s + 409-Tb/s crosstalk suppressed bidirectional MCF transmission over 450 km using propagation-direction interleaving. Opt. Express **21**(14), 16777–16783 (2013)
112. T. Kobayashi et al., 2 × 344 Tb/s propagation-direction interleaved transmission over 1500-km MCF enhanced by multicarrier full electric-field digital back-propagation, in *39th European Conference on Optical Communication (ECOC2013)*, PD3.E.4 (2013)
113. J. Sakaguchi, B.J. Puttnam, W. Klaus, Y. Awaji, N. Wada, A. Kanno, T. Kawanishi, K. Imamura, H. Inaba, K. Mukasa, R. Sugizaki, T. Kobayashi, M. Watanabe, 19-core fiber transmission of 19 × 100 × 172-Gb/s SDM-WDM-PDM-QPSK signals at 305 Tb/s, in *Optical Fiber Communication Conference and Exposition (OFC2012)*, PDP5C (2012)
114. H. Takara, A. Sano, T. Kobayashi, H. Kubota, H. Kawakami, A. Matsuura, Y. Miyamoto, Y. Abe, H.Ono, K. Shikama, Y. Goto, K. Tsujikawa, Y. Sasaki, I. Ishida, K. Takenaga, S. Matsuo, K. Saitoh, M. Koshiba, T. Morioka, 1.01-Pb/s (12 SDM/222 WDM/456 Gb/s) crosstalk-managed transmission with 91.4-b/s/Hz aggregate spectral efficiency, in *38th European Conference on Optical Communication (ECOC2012)*, Th.3.C.1 (2012)

115. K. Igarashi, K. Takeshima, T. Tsuritani, H. Takahashi, S. Sumita, I. Morita, Y. Tsuchida, M. Tadakuma, K. Maeda, T. Saito, K. Watanabe, K. Imamura, R. Sugizaki, M. Suzuki, 110.9-Tbit/s SDM transmission over 6,370 km using a full C-band seven-core EDFA. Opt. Express **21**(15), 18053–18060 (2013)
116. K. Igarashi, T. Tsuritani, I. Morita, Y. Tsuchida, K. Maeda, M. Tadakuma, T. Saito, K. Watanabe, K. Imamura, R. Sugizaki, M. Suzuki, 1.03-exabit/s km super-Nyquist-WDM transmission over 7,326-km seven-core fiber, in *39th European Conference on Optical Communication (ECOC2013)*, PD3.E.3 (2013)
117. T. Kobayashi, H. Takara, A. Sano, T. Mizuno, H. Kawakami, Y. Miyamoto, K. Hiraga, Y. Abe, H. Ono, M. Wada, Y. Sasaki, I. Ishida, K. Takenaga, S. Matsuo, K. Saitoh, M. Yamada, H. Masuda, T. Morioka, 2 × 344 Tb/s propagation-direction interleaved transmission over 1500-km MCF enhanced by multicarrier full electric-field digital back-propagation, in *39th European Conference on Optical Communication (ECOC2013)*, PD3.E.4 (2013)
118. Technical Committee on Extremely Advanced Optical Transmission Technologies (EXAT), EXAT Roadmap ver.2. https://www.ieice.org/~exat/Roadmap/EXAT_Roadmap_ver2.pdf (2020)
119. Y. Amma, T. Hosokawa, H. Ono, K. Ichii, K. Takenaga, S. Matsuo, M. Yamada, Ring-core multicore few-mode erbium-doped fiber amplifier. IEEE Photon. Technol. Lett. **29**(24), 2163–2166 (2017)
120. Y. Ohishi, T. Kanamori, T. Kitagawa, S. Takahashi, E. Snitzer, G.H. Sigel, Jr., Pr^{3+}-doped fluoride fiber amplifier operating at 1.31 μm. Opt. Lett. **16**
121. M. Yamada, M. Shimizu, T. Kanamori, Y. Ohishi, Y. Terunuma, K. Oikawa, H. Yoshinaga, K. Kikushima, Y. Miyamoto, S. Sudo, Low-noise and high-power Pr-doped fluoride fiber amplifier. IEEE Photon. Technol. Lett. **7**(8), 869–871 (1995)
122. Y. Nishida, M. Yamada, J. Temmyo, T. Kanamori, Y. Ohishi, Plug-in type 1.3-μm fiber amplifier module for rack-mounted shelves. IEEE Photon. Technol. Lett. **9**(8), 1096–1098 (1997)
123. T. Sakamoto, M. Shimizu, T. Kanamori, Y. Terunuma, Y. Ohishi, M. Yamada, S. Sudo, 1.4 μm-band gain characteristics of a Tm-Ho-doped ZBLYAN fiber amplifier pumped in the 0.8 μm band. IEEE Photon. Technol. Lett. **7**(9), 983–985 (1995)
124. T. komukai, Y. Yamamoto, T. Sugawa, Y. Miyajima, Upconversion pumped thulium-doped fluoride fiber amplifier and laser operating at 1.47 μm. IEEE J. Quantum Electron. **31**(11), 1880–1889 (1995)
125. S. Aozasa, H. Masuda, M. Shimizu, M. Yamada, Highly efficient S-band thulium-doped fiber amplifier employing high-thulium-concentration doping technique. IEEE J. Lightwave Technol. **25**(8), 2108–2114 (2007)
126. H. Ono, M. Yamada, Y. Ohishi, Gain-flattened Er^{3+}-doped fiber amplifier for WDM signal in the 1.57–1.60-μm wavelength region. IEEE Photon. Technol. Lett. **9**(5), 596–598 (1997)
127. M. Yamada, H. Ono, T. Kanamori, S. Sudo, Y. Ohishi, Broadband and gain-flattened amplifier composed of a 1.55 μm-band and a 1.58 μm-band Er^{3+}-doped fiber amplifier in a parallel configuration. Electron. Lett. **33**(8), 710–711 (1997)
128. A. Mori, K. Kobayashi, M. Yamada, T. Kanamori, K. Oikawa, Y. Nishida, Y. Ohishi, Low noise broadband tellurite-based Er^{3+}-doped fiber amplifiers. Electron. Lett. **34**(9), 887–888 (1998)
129. M. Yamada, A. Mori, K. Kobayashi, H. Ono, T. Kanamori, K. Oikawa, Y. Nishida, Y. Ohishi, Gain-flattened tellurite-based EDFA with a flat amplification bandwidth of 76 nm. IEEE Photon. Technol. Lett. **10**(9), 1244–1246 (1998)
130. E. Ishikawa, M. Nishihara, Y. Sato, C. Ohshima, Y. Sugaya, J. Kumasaka, Novel 1500 nm-band EDFA with discrete Raman amplifier, in *27th European Conference on Optical Communication (ECOC2001)*, vol. 6 (2001), pp. 48–49
131. H. Ono, M. Yamada, M. Shimizu, S-band erbium-doped fiber amplifiers with a multi-stage configuration-design, characterization and gain tilt compensation. IEEE J. Lightwave Technol. **21**(10), 2240–2246 (2003)

132. B.J. Puttnam, R.S. Luís, W. Klaus, J. Sakaguchi, J.M.D. Mendinueta, Y. Awaji, N. Wada, Y. Tamura, T. Hayashi, M. Hirano, J. Marciante, 2.15 Pb/s Transmission using a 22 core homogeneous single-mode multi-core fiber and wideband optical comb, in *41th European Conference on Optical Communication (ECOC2015)*, PDP3.1 (2015)
133. D. Soma, K. Igarashi, Y. Wakayama, K. Takeshima, Y. Kawaguchi, N. Yoshikane, T. Tsuritani, I. Morita, M. Suzuki, 2.05 Peta-bit/s super-Nyquist-WDM SDM transmission using 9.8-km 6-mode 19-core fiber in full C band, in *41th European Conference on Optical Communication (ECOC2015)*, PDP3.2 (2015)
134. R. Ryf, S. Randel, N.K. Fontaine, M. Montoliu, E. Burrows, S. Corteselli, S. Chandrasekhar, A.H. Gnauck, C. Xie, R.J. Essiambre, P.J. Winzer, R. Delbue, P. Pupalaikis, A. Sureka, Y. Sun, L. Gruner-Nielsen, R.V. Jensen, R. Lingle, 32-bit/s/Hz spectral efficiency WDM transmission over 177-km few-mode fiber, in *Optical Fiber Communication Conference and Exposition (OFC2013)*, PDP5A.1 (2013)
135. V.A.J.M. Sleiffer, Y. Jung, V. Veljanovski, R.G.H. van Uden, M. Kuschnerov, H. Chen, B. Inan, L. Grüner Nielsen, Y. Sun, D.J. Richardson, S.U. Alam, F. Poletti, J.K. Sahu, A. Dhar, A.M.J. Koonen, B. Corbett, R. Winfield, A.D. Ellis, H. de Waardt, 73.7 Tb/s (96 × 3 × 256-Gb/s) mode-division-multiplexed DP-16QAM transmission with inline MM-EDFA, in *38th European Conference on Optical Communication (ECOC2012)*, Th.3.C.4 (2012)
136. T. Mizuno, T. Kobayashi, H. Takara, A. Sano, H. Kawakami, T. Nakagawa, Y. Miyamoto, Y. Abe, T. Goh, M. Oguma, T. Sakamoto, Y. Sasaki, I. Ishida, K. Takenaga, S. Matsuo, K. Saitoh, T. Morioka, 12-core × 3-mode dense space division multiplexed transmission over 40 km employing multi-carrier signals with parallel MIMO equalization, in *Optical Fiber Communication Conference and Exposition (OFC2014)*, Th5B.2 (2014)
137. R. Ryf, N.K. Fontaine, M. Montoliu, S. Randel, B. Ercan, H. Chen, S. Chandrasekhar, A. Gnauck, S.G. Leon-Saval, J. Bland-Hawthorn, J.R.S. Gil, Y. Sun, R. Lingle, Photonic-lantern-based mode multiplexers for few-mode-fiber transmission, in *Optical Fiber Communication Conference and Exposition (OFC2014)*, W4J.2 (2014)
138. E. Ip, M. Li, Y. Huang, A. Tanaka, E. Mateo, W. Wood, J. Hu, Y. Yano, K. Koreshkov, 146λ × 6 × 19-Gbaud wavelength- and mode-division multiplexed transmission over 10 × 50-km spans of few-mode fiber with a gain-equalized few-mode EDFA, in *Optical Fiber Communication Conference and Exposition (OFC2013)*, PDP5A.2 (2013)
139. J. Sakaguchi, W. Klaus, J.M.D. Mendinueta, B.J. Puttnam, R.S. Luis, Y. Awaji, N. Wada, T. Hayashi, T. Nakanishi, T. Watanabe, Y. Kokubun, T. Takahata, T. Kobayashi, Realizing a 36-core, 3-mode fiber with 108 spatial channels, in *Optical Fiber Communication Conference and Exposition (OFC2015)*, Th5C.2 (2015)
140. K. Shibahara, T. Mizuno, H. Takara, A. Sano, H. Kawakami, D. Lee, Y. Miyamoto, H. Ono, M. Oguma, Y. Abe, T. Kobayashi, T. Matsui, R. Fukumoto, Y. Amma, T. Hosokawa, S. Matsuo, K. Saitoh, H. Nasu, T. Morioka, Dense SDM (12-core × 3-mode) transmission over 527 km with 33.2-ns mode-dispersion employing low-complexity parallel MIMO frequency-domain equalization, in *Optical Fiber Communication Conference and Exposition (OFC2015)*, Th5C.3 (2015)
141. K. Igarashi, D. Souma, Y. Wakayama, K. Takeshima, Y. Kawaguchi, T. Tsuritani, I. Morita, M. Suzuki, 114 Space-division-multiplexed transmission over 9.8-km weakly-coupled-6-mode uncoupled-19-core fibers, in *Optical Fiber Communication Conference and Exposition (OFC2015)*, Th5C.4 (2015)
142. N.K. Fontaine, R. Ryf, H. Chen, A.V. Benitez, B. Guan, R. Scott, B. Ercan, S.J.B. Yoo, L.E. Grüner-Nielsen, Y. Sun, R. Lingle, E. Antonio-Lopez, R. Amezcua-Correa, 30 × 30 MIMO transmission over 15 spatial modes, in *Optical Fiber Communication Conference and Exposition (OFC2015)*, Th5C-1 (2015)
143. H. Ono, K. Takenaga, K. Ichii, S. Matsuo, T. Takahashi, H. Masuda, M. Yamada, 12-core double-clad Er/Yb-doped fiber amplifier employing free-space coupling pump/signal combiner module, in *39th European Conference on Optical Communication (ECOC2013)*, We.4.A.4 (2013)

144. S. Jain, T. Mizuno, Y. Jung, A. Isoda, K. Shibahara, J. Hayes, Y. Sasaki, K. Takenaga, Y. Miyamoto, S. Alam, D.J. Richardson, Improved cladding-pumped 32-core multicore fiber amplifier, in *Proceedings of European Conference on Optical Communication 2017*, Gothenburg, Sweden, September, Th.2.D (2017)
145. M. Wada, S. Jain, K.I. Shibahara, Y. Jung, I.A. Davidson, P. Barua, J.R. Hayes, T. Sakamoto, T. Mizuno, D.J. Richardson, Y. Sasaki, K. Saitoh, K. Nakajima, High density few-mode multicore fibre amplifier for energy efficient SDM transmission, in *Proceedings of European Conference on Optical Communication 2019*, Dublin, Ireland, September, W.2.C.1 (2019)

Chapter 5
Optical Transmission Technologies

Itsuro Morita, Toshihiko Hirooka, Hidehiko Takara, Yoshinari Awaji, Kiyoshi Fukuchi, Koji Igarashi, Masafumi Koga, Yutaka Miyamoto, Takayuki Mizuno, Masataka Nakazawa, Benjamin J. Puttnam, Jun Sakaguchi, Akihide Sano, Takehiro Tsuritani, and Shinji Yamashita

Abstract For a drastic increase of the optical fiber capacity, extremely advanced optical transmission technologies are indispensable. This chapter reviews such technologies including advanced modulation formants for higher spectral efficiency, transmission technologies using new types of fiber for space division multiplexing,

---

Itsuro Morita, Toshihiko Hirooka, and Hidehiko Takara are chapter editors.

I. Morita (✉) · T. Tsuritani
KDDI Research Inc., Fujimino, Saitama, Japan
e-mail: it-morita@waseda.jp

I. Morita
Waseda University, Shinjuku, Tokyo, Japan

T. Hirooka · M. Nakazawa
Research Institute of Electrical Communication (RIEC), Tohoku University, Sendai, Miyagi, Japan

H. Takara
National Institute of Technology, Okinawa College, Nago, Okinawa, Japan

Y. Awaji · B. J. Puttnam · J. Sakaguchi
National Institute of Information and Communications Technology, Koganei, Tokyo, Japan

K. Fukuchi
NEC Corporation, Minato, Tokyo, Japan

K. Igarashi
Osaka University, Suita, Osaka, Japan

M. Koga
Oita University, Oita, Oita, Japan

Y. Miyamoto
NTT Network Innovation Laboratories, NTT Corporation, Yokosuka, Kanagawa, Japan

T. Mizuno
NTT Electronics America, San Jose, CA, USA

A. Sano
Ritsumeikan University, Kusatsu, Shiga, Japan

S. Yamashita
The University of Tokyo, Meguro, Tokyo, Japan

M. Nakazawa et al. (eds.), *Space-Division Multiplexing in Optical Communication Systems*, Springer Series in Optical Sciences 236,
https://doi.org/10.1007/978-3-030-87619-7_5

and signal processing technologies providing a variety of functions in high-speed optical transmission systems.

5.1 Introduction

To support the continuously growing network traffic, the capacity carried with an optical fiber has increased by introducing various kinds of novel technologies, including new types of optical fiber, advanced modulation formats, multiplexing technologies in time, wavelength, and polarization domain together with coherent detection. It is foreseen, however, the fundamental capacity limit of conventional fiber will come soon. For further increase of the capacity beyond such limit, we need some breakthrough optical transmission technologies. The EXAT research group in Japan established in 2008 claimed the 3M technologies, namely multi-core fiber, mode control in optical fiber, and multi-level modulation, as key technologies to overcome such limit. The research on the 3M technologies has rapidly progressed, and many record-breaking experiments including ultrahigh-capacity transmission beyond petabit/s and long-haul transmission with exabit/s·km (Ebit/s·km) capacity distance product were demonstrated. This chapter explains these transmission technologies together with signal processing technologies which enhance the transmission performance and functionalities in optical network.

In this chapter, after the overview of transmission technologies required for a drastic increase of the optical fiber capacity in Sect. 5.2, multi-level transmission technologies to achieving an ultrahigh spectral efficiency toward the Shannon limit are described in Sect. 5.3. Section 5.4 explains space division multiplexed (SDM) transmission technologies using three types of fibers, i.e., multi-core fiber, multi-mode fiber, and multi-core multi-mode fiber. The recent high-capacity transmission experiments using these fibers are introduced in this section. In Sect. 5.5, signal processing technologies to provide a variety of functions in optical networks are explained. SDM processing technologies and orbital angular momentum are also described in Sect. 5.5.

5.2 Overview of Transmission Technologies

The main object of this section is to introduce overview of technology trends and requirements for future exabit network which is required in the following sections. For this purpose, high-capacity optical link technologies are summarized.

The high-capacity link technology is needed to enable transmission of high-capacity information between two places. It is one of the important technologies for realization of future exabit photonic network.

A progress of communication network has been driven by a long-distance transmission of high-speed optical signal employing a low-loss optical fiber transmission line. The loss of the optical fiber transmission line became far below 0.2 dB per km at the 1.5-μm wavelength window where the silica optical fiber exhibits lowest loss.

This excellent low-loss characteristic is enabled by a high-tech production technology of silica-based optical fiber.

In a viewpoint of transmission capacity, very high carrier frequency of 200 THz is an advantage of the light wave having 1.5-μm wavelength. By utilizing it, it is possible to realize high-capacity systems with very wideband transmission. Figure 5.1 shows a trend of capacity increase in transmission system research and development employing the high-capacity link technologies.

In an early stage of the optical network progress up to around 1995, a key enabling technology was a time division multiplexing (TDM). The speed of an aggregated signal by the TDM increased rapidly year after year. It started from several-hundred Mbit/s in the 1980s, then it increased to 2.5 Gbit/s, and finally it reached to 10 Gbit/s in the early 1990s. In high-speed systems using the TDM technology, signal to be transmitted is generated from multiple low-speed signal by electrical signal first. And then, the aggregated signal was used as a driver input of high-speed optical modulator. In this stage, most system used an intensity modulation and direct detection (IM-DD) scheme. A highly efficient transmission system is realized with the IM-DD technology, in which highly integrated electronic circuit technology is utilized to achieve small size and low system cost. However, as the speed of the aggregated signal increased year by year, it is becoming difficult to modulate the light wave with the high-speed signal and to transmit it over long distance. To overcome this problem, a lot of technologies were developed such as high-speed external modulation, high-speed transceiver devices, and dispersion compensation devices. The advance of these technologies made possible the optical network being practical.

After the first stage up to around 1990 employing high-speed TDM technologies, there was a rapid capacity increase by wavelength division multiplexing (WDM) technology. In the optical transmission systems using the WDM technology, multiple optical signals having different carrier frequency (wavelength) transmit multiple signals independently. The multiple optical carriers are modulated by a different

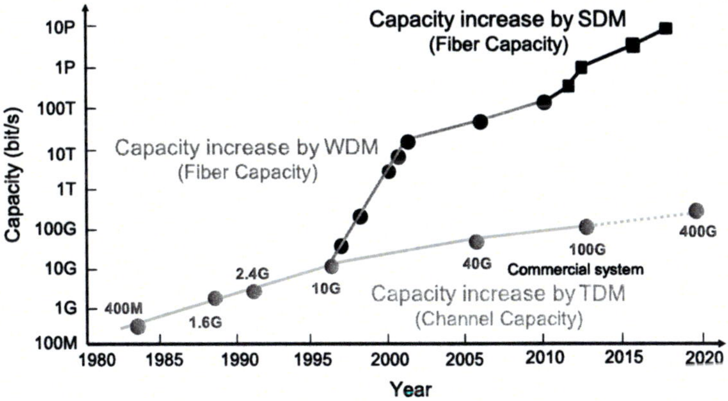

Fig. 5.1 Trend of network capacity increase by optical link technology evolution

information, then they are multiplexed, and finally the carriers are transmitted over optical fiber transmission line. The primal factor why WDM technology progress took over the TDM technology was the speed limitation of the TDM aggregated signal. A long-distance transmission is found to be difficult when the signal speed exceeds 10 Gbit/s because the waveform distortion during the transmission became so severe at this high speed.

Another factor is an appearance of a practical optical amplifier which is an erbium-doped fiber amplifier (EDFA). The EDFA exhibits very wide amplification bandwidth. It exhibited several-tenth nanometer wavelength range. Moreover, the amplification wavelength region of the EDFA coincided with the band at which the silica-based optical fiber exhibited the lowest transmission loss. With these EDFAs, simultaneous amplification of multiple carriers having different wavelength each other became possible so that a cost of the signal repeater equipment was greatly reduced.

In the 1990s, there were many experimental demonstrations exploring limitation of highest capacity in optical fiber transmission. As for a research and development level, three 1-Tbit/s transmission experiments were reported in 1996 [1–3]. In the late 1990s, because a speed of TDM signal successfully increased from 10 to 40 Gbit/s, further capacity expansion was seen. There were two 10-Tbit/s transmission experiments in 2001 [4, 5]. In the late 2000s, use of multi-level modulation format has begun which leaded to "over 10-Tbit/s" capacity era. There was an experimental demonstration of 69.1 Tbit/s capacity in 2011 [6] and 101-Tbit/s experiments in 2012 [7].

The TDM technology and the WDM technology are two important technologies in realizing high-capacity optical link. The total capacity of the system employing these technologies is given by a product of two parameters which are "frequency spectral efficiency" and "wavelength bandwidth."

Frequency spectral efficiency represents the information capacity to be transmitted per unit frequency bandwidth. The value of the spectral efficiency is given by dividing signal bitrate by channel spacing. In order to increase the spectral efficiency value, there are two options; one is to increase the bitrate by more multiplexing by TDM, and the other is to decrease the channel spacing without compromising crosstalk. In both cases, information density gets increased in the frequency spectrum which was called as "ultra-dense multiplexing" in other words.

Wavelength bandwidth is the same as the transmission bandwidth of the optical fiber network. It is limited by the amplification band of the optical amplifiers and low-loss band of the transmission line. In order to increase the capacity, bandwidth extension technology is required that enables to extend the wavelength bandwidth.

Utilizing other dimensions for multiplexing is an alternative way for effectively increasing capacity. One promising dimension, which is other than time and frequency, is a space dimension. Utilizing the space dimension, which is called as "space division multiplexing (SDM) technology," means to use spatially separated multiple transmission channels. Systems using the SDM technology include a strand of multiple fiber in a transmission cable or multiple core inside one optical fiber.

It is also useful to use other dimensions such as polarizations in an optical fiber or modes in a multi-mode optical fiber. Code division multiplexing is one of possible options, too. Among them, polarization-division multiplexing (PDM) is already used in a practical network system.

To achieve "capacity increase," the three parameters contribute which are frequency spectrum efficiency for higher density information multiplexing, bandwidth for providing wider information packing area, and space division or other multiplexing to increase dimensions. These parameters contribute the higher capacity increase; however, there are several different system design directions depending on highlighted parameter. We will discuss this relation using Fig. 5.2.

In order to investigate it, we use a three-dimensional expression with a rectangular solid as total capacity, as shown in Fig. 5.2. In the figure, three rectangular solids are drawn to demonstrate three examples having same capacity but different designs. The first example of high-capacity system, which is tagged by "ultra-dense multiplexed transmission," employs very high spectral efficiency. This configuration is effective when a system has a limitation of wavelength bandwidth or difficulties to use other dimension such as space or polarization.

The second design, which is tagged as "ultra-broadband transmission" in the figure, provides the higher capacity by mostly extending the wavelength bandwidth. In this case, it is possible to configure a system using simpler modulation format signals but less spectral efficiency. This system avoids using advanced modulation format signals which have usually complicated configurations.

These two design directions are so effective that many research and development activities have already proved the efficiency of combined use of the TDM technology and the WDM technology for increase in both high spectral efficiency and very wide bandwidth. Thus, it is not easy to improve the efficiency and the bandwidth

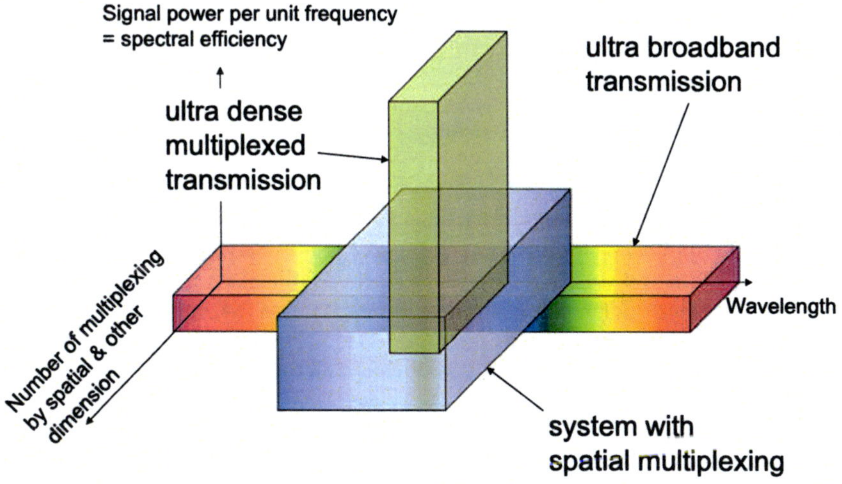

Fig. 5.2 Relationship between total capacity and three key parameters

further. In addition to that, use of signals with higher spectral efficiency or wider bandwidth will eventually lead to a shortening of transmission distance. To avoid this distance reduction, applying other additional technologies is needed. The third system configuration in Fig. 5.2, which has a notation of "system with spatial multiplexing," is another and practical option considering this distance problem in capacity increase. The spatial multiplexing configuration provides good balance in increase of three parameter values. In this option, dimensions of space or polarization are utilized for drastic capacity increase without losing other important parameter value in a system such as transmission distance or repeater spacing.

Based on the above discussion, there are three key parameters to determine transmission system capacity which are spectral efficiency, wavelength bandwidth, and number of multiplications using polarization and/or spatial dimensions.

As for the wavelength bandwidth, use of very wide bandwidth wider than 30 nm is realized in commercial system using either C band or L band. It is possible to make the bandwidth double using two bands simultaneously.

As for other multiplexing such as using space division multiplexing or mode division multiplexing, the number of multiplexing is still small or even one in current practical systems. But recent technology advancement enabled the polarization-division multiplexing as practical in commercial systems. It is mostly because the appearance of commercial-level coherent system with digital signal processing technology is established.

In this chapter, optical link technologies are summarized in the viewpoint of progress how the high-capacity optical transmission system technologies have evolved to improve their performance. The discussion is done in the aspects of highly dense multiplexing, wide bandwidth transmission, and utilization of space division multiplexing and other new dimension in multiplexing.

5.3 Multi-level Transmission Technologies

Achieving an ultrahigh spectral efficiency toward the Shannon limit is one of the targets of our 3M technologies since it will allow us to increase the total WDM capacity within a finite optical amplification bandwidth. Of the various available formats, M-ary quadrature amplitude modulation (QAM) is capable of approaching the Shannon limit most closely by increasing the multiplicity M. A 2^N QAM signal represents N bits, and so, it has N times the spectral efficiency of OOK.

Since the first demonstration of coherent optical QAM [8], the multiplicity level has been increased from 64 [9] to 128 [10], 256 [11, 12], 512 [13, 14], 1024 [15, 16], 2048 [17], and 4096 [18]. Recent QAM transmission experiments with a high spectral efficiency are summarized in Fig. 5.3. Higher-order multi-level modulation also enables us to realize a high-speed system with low-speed/narrow bandwidth devices and therefore helps to enhance tolerance to chromatic dispersion (CD) and polarization mode dispersion (PMD) as well as to reduce power consumption. Orthogonal frequency-division multiplexing (OFDM) is another approach that can realize highly

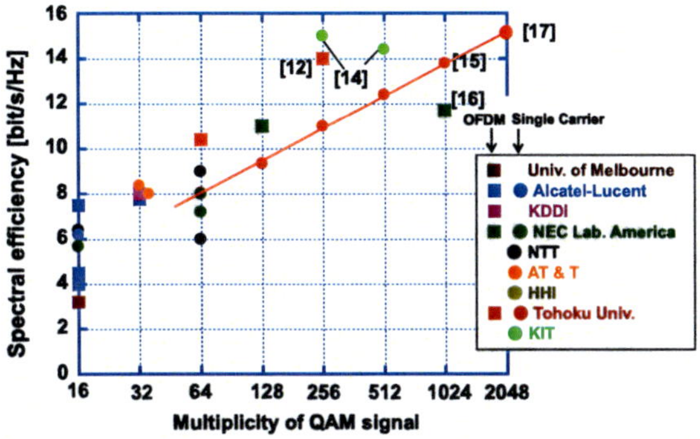

Fig. 5.3 Recent QAM transmission experiments with a high spectral efficiency

spectrally efficient transmission. In OFDM transmission, the multi-carrier transmission of low-speed orthogonal subcarriers enables us to improve both spectral efficiency and dispersion tolerance by adopting high multi-level subcarrier modulation formats and employing coherent detection.

By taking advantage of the high spectral efficiency, QAM has been applied to WDM transmission with a record-breaking capacity exceeding 100 Tbit/s per single core [7, 19]. QAM has also been adopted for transport systems operating beyond 100 Gbit/s such as Pol-Mux 16 QAM transmission at 448 Gbit/s/ch [20] and 856 Gbit/s/ch [21] at symbol rates of 56 and 107 Gsymbol/s, respectively. Recently, Pol-Mux 120 ~ 864 Gbit/s/ch transmissions with 64 QAM at 10 ~ 72 Gsymbol/s have also been reported [22–26].

This section describes fundamental technologies and recent progress on coherent QAM transmission, with a special focus on attempts to realize much higher-order QAM of, for example, 256 ~ 4096 levels.

5.3.1 Spectral Efficiency of QAM Signal and Shannon Limit

QAM is a modulation format that combines two carriers whose amplitudes are modulated independently with the same optical frequency and whose phases are 90° apart. These carriers are called in-phase carriers (I) and quadrature phase carriers (Q). QAM can assign 2^N states by using I and Q, and this is called 2^N QAM. Figure 5.4 shows constellation maps for 16 (2^4) QAM and conventional OOK. As shown in Fig. 5.4, a 2^N QAM signal processes N bits in a single channel, and so, it can realize N times the spectral efficiency of OOK.

The BER of M-ary QAM signals, P_b, can be analytically calculated, and the result for $M = 2^N$ is given by [27]

$$P_b = \frac{\sqrt{M}-1}{\sqrt{M}\log_2\sqrt{M}}\mathrm{erfc}\sqrt{\left[\frac{3\log_2 M}{2(M-1)}\right]E_b/N_0} \tag{5.2.1}$$

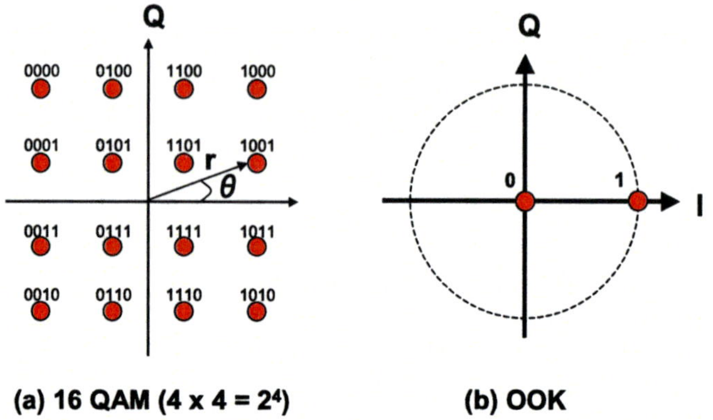

Fig. 5.4 Constellation maps for 16 QAM (**a**) and OOK (**b**)

Here, E_b and N_0 are the energy and noise power per bit; thus, E_b/N_0 corresponds to the signal-to-noise ratio (SNR) per bit. The specific forms of (5.2.1) for $M = 16, 64, 256$, and 1024 are as follows:

$$P_b = \frac{3}{8}\mathrm{erfc}\sqrt{\frac{2}{5}E_b/N_0} \quad \text{(16QAM)} \tag{5.2.2}$$

$$P_b = \frac{7}{24}\mathrm{erfc}\sqrt{\frac{1}{7}E_b/N_0} \quad \text{(64QAM)} \tag{5.2.3}$$

$$P_b = \frac{15}{64}\mathrm{erfc}\sqrt{\frac{4}{85}E_b/N_0} \quad \text{(256QAM)} \tag{5.2.4}$$

$$P_b = \frac{31}{160}\mathrm{erfc}\sqrt{\frac{5}{341}E_b/N_0} \quad \text{(1024QAM)} \tag{5.2.5}$$

Equations (5.2.2) ~ (5.2.5) are plotted in Fig. 5.5 as a function of E_b/N_0.

The spectral efficiency of an M-ary QAM signal is shown in Fig. 5.6a as a function of E_b/N_0. Here, the ultimate spectral efficiency is given by the Shannon limit:

$$\frac{C}{W} = \log_2\left(1 + \frac{E_b}{N_0}\frac{C}{W}\right), \tag{5.2.6}$$

which is well known as the Shannon–Hartley theorem [28]. This figure indicates that, as the multiplicity M increases, the spectral efficiency of M-QAM moves closer to the Shannon limit than other advanced modulation formats such as M-PSK or M-FSK.

However, the increase in M requires a larger E_b/N_0 value under the same BER as shown in Fig. 5.5. To realize a better BER performance with a lower E_b/N_0, the

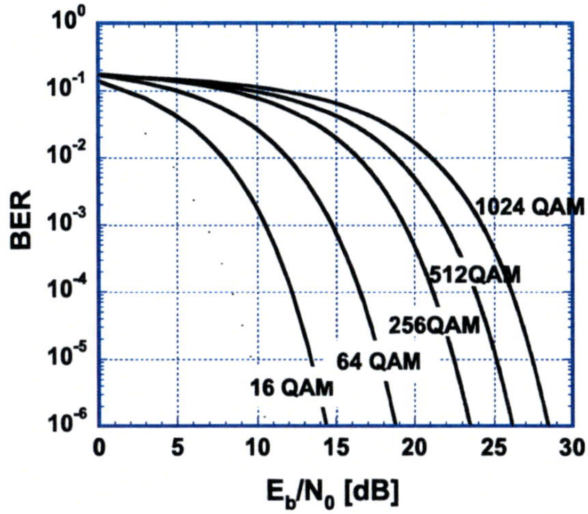

Fig. 5.5 BER of 16 ~ 1024 QAM as a function of E_b/N_0

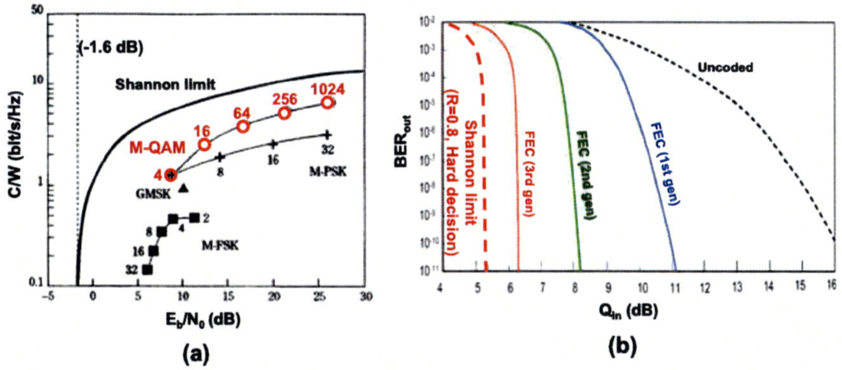

Fig. 5.6 **a** Spectral efficiency of M-ary QAM signal and the Shannon limit. E_b/N_0 at BER = 10^{-4} is shown assuming synchronous detection. **b** Relationship between BER after FEC and Q-value without FEC

forward error correction (FEC) technique has been developed. Figure 5.6b shows the BER after applying FEC versus the input Q-value without FEC, Q_{in} [29]. Q_{in} is the SNR given by

$$Q_{in} = \frac{I_1 - I_0}{\sigma_1 + \sigma_0} \tag{5.2.7}$$

where I_1 and I_0 are the mean values, and σ_1 and σ_0 are the standard deviations of the bits corresponding to 1 and 0, respectively. Here, for a certain code rate R, the

Shannon limit gives the lowest Q_{in} value needed to achieve error-free operation by employing FEC. When the code rate is R, the maximum BER_{in} that can be made error free by using FEC is given by

$$R = 1 + BER_{in} \log_2 BER_{in} + (1 - BER_{in}) \log_2 (1 - BER_{in}) \quad (5.2.8)$$

We can then obtain the lowest Q_{in} value from

$$BER_{in} = \frac{1}{2} \mathrm{erfc}\left(\frac{Q_{in}}{\sqrt{2}}\right) \quad (5.2.9)$$

Equations (5.2.8) and (5.2.9) are known as Shannon's second theorem or the noisy channel coding theorem [28]. This provides the ultimate limit for the minimum Q-value needed to achieve an infinitely low BER. Recently, third-generation FEC, namely the turbo block code with a soft decision, has been developed, which enables us to realize BER performance very close to the Shannon limit. This indicates the distinct possibility of realizing ultrahigh spectral efficiency by combining QAM and FEC.

5.3.2 Fundamental Configuration and Key Components of QAM Coherent Optical Transmission

5.3.2.1 Coherent Light Source

As a coherent light source, a narrow linewidth CW fiber laser or laser diode such as an external cavity laser is employed to meet the narrow linewidth requirement for high-order QAM. In addition, a stable optical frequency in the 1.5 μm region is also very important for achieving higher QAM multiplicity. C_2H_2 and HCN molecules have been utilized as frequency standards to stabilize the frequencies of semiconductor and fiber lasers at 1.55 μm. Figure 5.7 shows the configuration of a $^{13}C_2H_2$ frequency-stabilized fiber ring laser [30]. The frequency deviation of a fiber laser from the $^{13}C_2H_2$ absorption line is detected with a phase sensitive detection circuit and fed back to a PZT to control the laser frequency. The linewidth is only 4 kHz, and the frequency stability is 2.5×10^{-11} for an integration time, τ, of 1 s and 6.3×10^{-12} for a τ of 100 s, which indicates excellent short- and long-term stabilities.

Recently, various types of narrow linewidth semiconductor lasers have been developed. Several examples are shown in Fig. 5.8. The one shown in Fig. 5.8a involves the adoption of waveguide ring resonators [31]. This makes it possible to realize a linewidth as narrow as 15 kHz as well as wide wavelength tunability. Figure 5.8b shows a commercially available external cavity DBR laser with a linewidth of 4 kHz [32]. Figure 5.8c shows a DFB-LD with an external fiber ring cavity [33]. Here, both

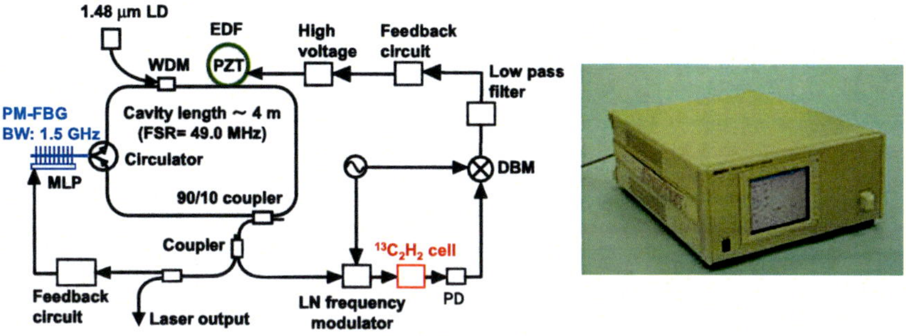

Fig. 5.7 Configuration and overview of a $^{13}C_2H_2$ frequency-stabilized fiber laser

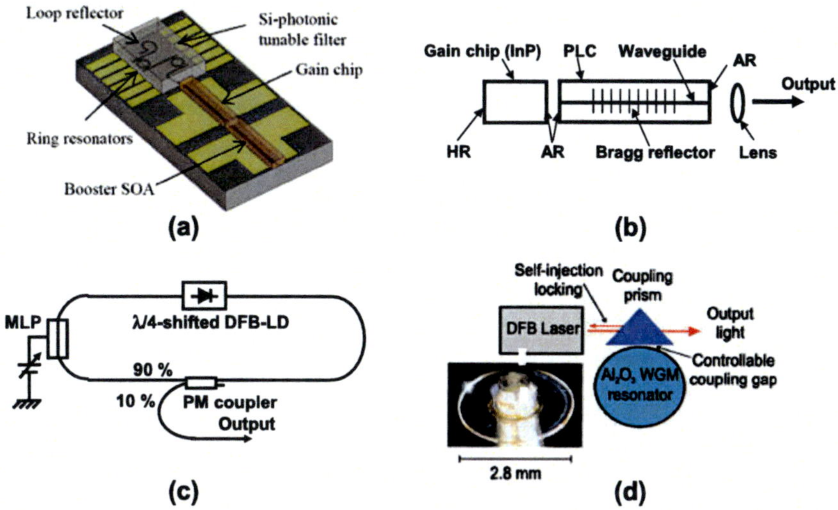

Fig. 5.8 Narrow linewidth semiconductor lasers (Reprinted with permission from [31]. © 2014 IEEE)

edges of the DFB-LD are AR-coated and a fiber ring cavity is attached to form a long external cavity. The DFB-LD features both the narrow linewidth associated with a fiber ring cavity and a low RIN value. Figure 5.8d shows a DFB-LD that employs the resonant optical feedback from a high-Q crystalline whispering-gallery-mode resonator with a linewidth of <1 kHz [34]. In general, however, the LD has a broader linewidth in the tail than fiber lasers.

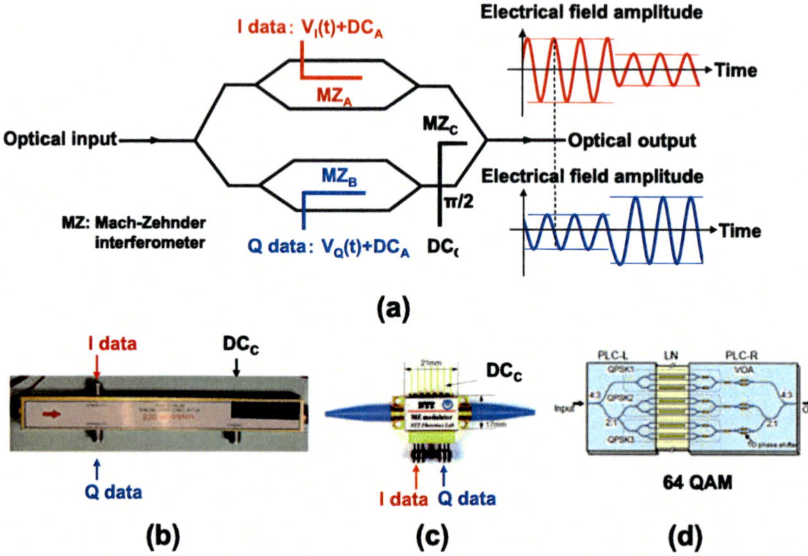

Fig. 5.9 Configuration of IQ modulator (**a**) and overview of LN (**b**), InP (**c**), and PLC-LN (**d**) modulators

5.3.2.2 Optical IQ Modulator

The optical QAM signal can be easily generated with an IQ modulator, consisting of two nested MZ modulators and a 90° phase shifter, driven by QAM baseband signals. The configuration of an optical IQ modulator is shown in Fig. 5.9a [35]. It is composed of three Mach–Zehnder interferometers (MZIs), where sub-MZIs, MZ_A and MZ_B, are installed in each arm of the main MZI (MZ_C). I and Q optical data are generated individually with MZ_A and MZ_B, respectively, using the I and Q components of the QAM baseband signal. They are combined with MZ_C with a DC bias DC_C so that a 90° phase shift is introduced between the two signals.

An optical IQ modulator has been realized not only with an LN modulator as shown in Fig. 5.9b but also with a monolithic InP modulator as shown in Fig. 5.9c [36], which is beneficial for compact integration. A more complicated IQ modulator can be fabricated using PLC-LN hybrid integration as shown in Fig. 5.9d [37]. In this case, three IQ modulators are integrated, and by employing QPSK modulation at each modulator with a power ratio of 4:2:1, a 64 QAM signal can be generated from binary data.

5.3.2.3 Coherent Optical Receiver and Optical PLL

Precise optical carrier phase recovery is a fundamental functionality in coherent optical transmission. Recent advances in digital signal processing (DSP) technologies have enabled intradyne detection with a digital coherent receiver, in which the phase fluctuation between a signal and a local oscillator (LO) is eliminated with a DSP [38]. This technique does not require an optical phased-locked loop (OPLL) for phase synchronization and therefore enables us to realize coherent transmission using free-running lasers as a light source and an LO. Digital coherent receivers relax the requirements as regards the linewidth and frequency stability of a coherent light source, and the IF fluctuation can be reduced in the DSP without employing an OPLL. However, the IF stability thus obtained may not be sufficient for higher QAM multiplicity.

On the other hand, to realize an ultrahigh multiplicity, heterodyne or homodyne detection with an analog optical PLL circuit is more advantageous, since it can automatically produce a stable IF frequency. In heterodyne or homodyne detection, the optical frequency difference between a transmitter and an LO must be kept constant to obtain a stable IF signal. Here, the use of a high-speed OPLL is a key technique for automatic frequency control. The linewidth of the IF signal is evaluated as

$$\sigma_{\phi}^{2} = \frac{\delta f_{\mathrm{T}} + \delta f_{\mathrm{L}}}{2 f_{\mathrm{c}}} \tag{5.2.10}$$

where δf_{T} and δf_{L} are the linewidths of the transmitter and LO, and f_{c} is the bandwidth of the feedback circuit [39]. This indicates that the reduction of the phase noise (linewidth) of the two lasers and the large bandwidth of the feedback circuit are very important factors as regards realizing a precise OPLL. Of the many available lasers, the fiber laser is the most suitable for an OPLL because of its low phase noise (narrow linewidth).

Figure 5.10a shows the OPLL configuration [40]. Here, a free-running fiber laser is used as an LO whose configuration is almost the same as that of the transmitter except that an LN modulator is adopted in the laser cavity for the high-speed tracking of the IF signal. The LO linewidth is also approximately 4 kHz. The signal from the transmitter is heterodyne-detected with the LO signal. The phase of the beat signal (IF signal: $f_{\mathrm{IF}} = |f_{\mathrm{S}} - f_{\mathrm{L}}|$) is compared with the phase of the reference signal from the synthesizer (f_{syn}) by the DBM, and the difference between them is fed back to the LO through the feedback circuits. The phase noise of the OPLL is mainly dominated by the loop bandwidth. With this configuration, we could obtain an FM bandwidth for the LO of up to 1 GHz. The OPLL circuit contains two feedback circuits with different loop filter bandwidths. One is a broadband filter (~1 MHz) for fast frequency tuning with the LN modulator, and the other is a narrowband filter (~10 kHz) for slow frequency tuning with the PZT.

Figure 5.10b, c shows the IF spectrum and the single sideband (SSB) phase noise spectrum. The linewidth of the spectrum was less than 10 Hz, which was below the measurement resolution. The phase noise variance (RMS) of the IF signal estimated by integrating the SSB noise spectrum was as low as 0.3°, which indicates that stable OPLL operation is successfully achieved under a low phase noise condition. This low

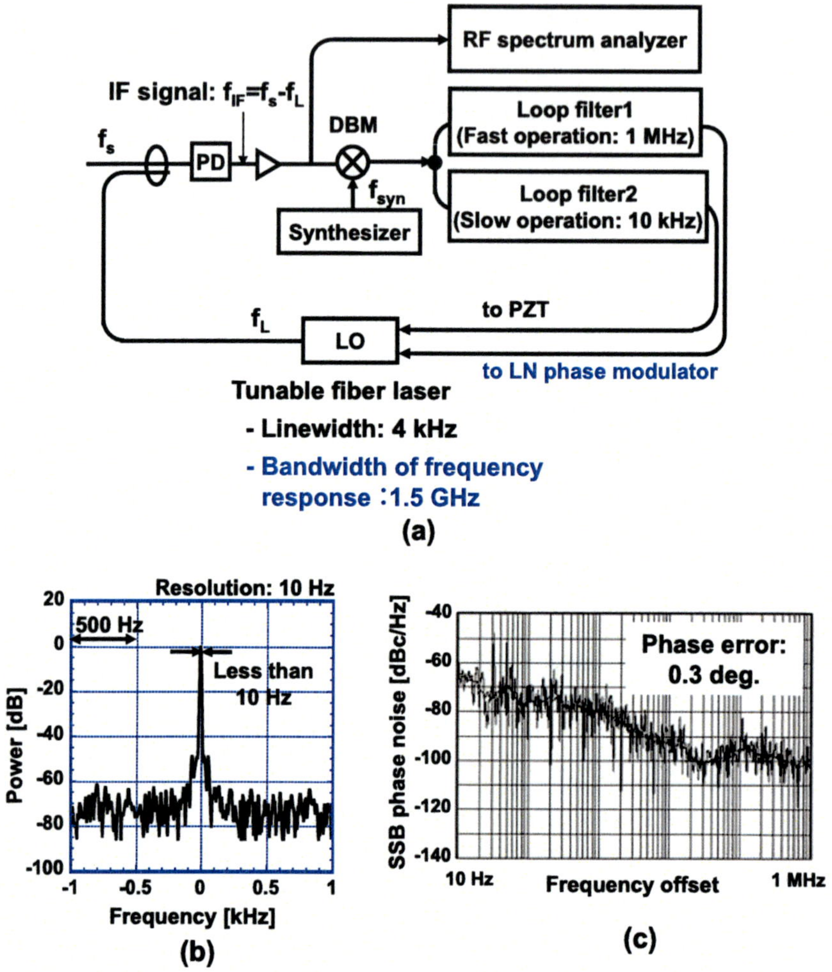

Fig. 5.10 Configuration of OPLL for coherent transmission (**a**), IF signal spectrum (**b**), and SSB phase noise spectrum (**c**)

phase noise of the IF signal despite the relatively large PLL bandwidth is attributed to the narrow linewidth of the fiber laser.

The tolerance of the phase noise for 64, 256 and 1024 QAM signals can be estimated from the constellation maps. As shown in Fig. 5.11, the angle between the two closest symbols is $2\delta\phi = 4.7$, 2.0, and 0.95° for 64, 256, and 1024 QAM, respectively, which correspond to the tolerable phase noise. Therefore, an RMS phase noise of 0.3° is sufficiently small for demodulating even a 1024 QAM signal.

The QAM signal and LO are coupled into a coherent receiver for demodulation. Figure 5.12 shows the configuration of a coherent receiver used for homodyne detec-

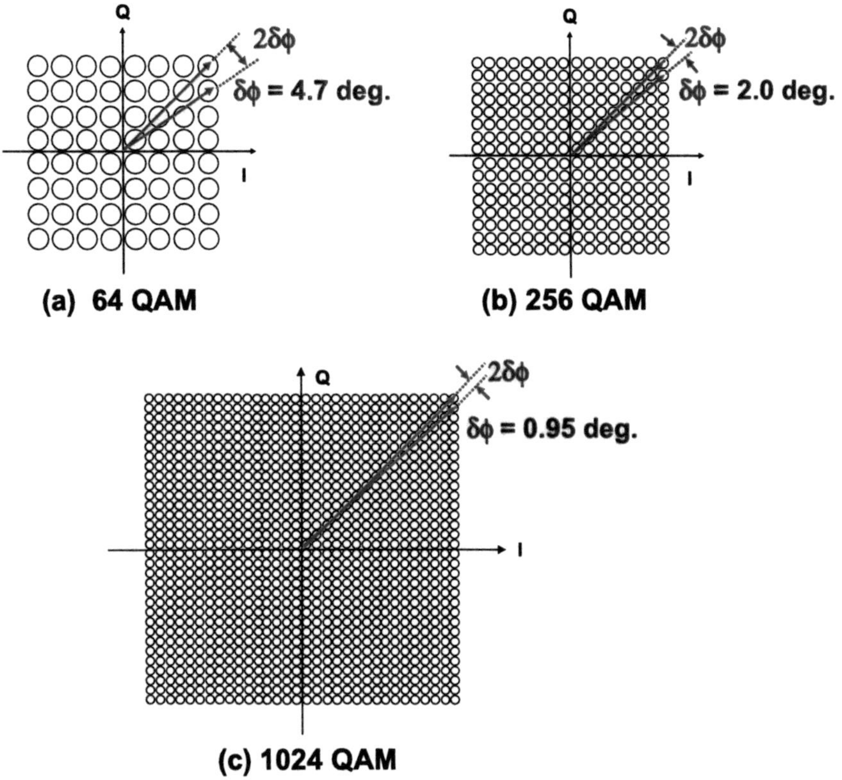

Fig. 5.11 Constellation maps of 64, 256, and 1024 QAM signals and a comparison of the tolerable phase noise

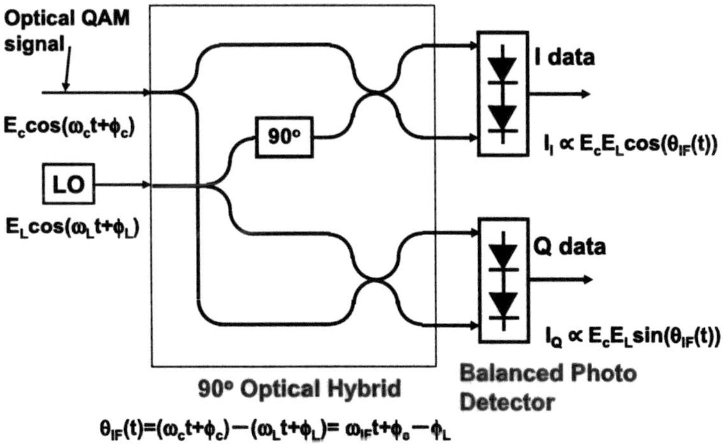

Fig. 5.12 Schematic diagram of coherent receiver

tion. The LO output is divided into two signals with a 90° phase shift and coupled with the QAM signal in a 90° optical hybrid circuit so that the cosine (I) and sine (Q) components are obtained after balanced detection. A compact 90° optical hybrid circuit and balanced photodiodes are commercially available, and the integration of these components is under development [41].

5.3.2.4 Digital Demodulator and Equalizer

The IF signal is finally A/D-converted and demodulated in a digital signal processor (DSP). Figure 5.13 shows a schematic diagram of a digital demodulator. Here, a polarization diversity configuration with a dual-polarization optical hybrid circuit is adopted to receive a polarization-multiplexed signal. The QAM data are first digitized with four A/D converters that are operated synchronously and then fed into the DSP where I and Q data are demodulated. In the DSP, adaptive equalization is employed to compensate for waveform distortions caused by chromatic dispersion and fiber nonlinearity during transmission, or hardware imperfections in individual components such as a non-ideal frequency response or skew in optical or electrical devices. Polarization demultiplexing is also carried out in the equalizer. Finally, the demodulated data are converted into binary data in the decoder.

Recent advances in DSP technologies have enabled not only the coherent detection of high-speed multi-level optical signals with high precision, but also compensation for linear and nonlinear transmission impairments in the electrical domain. A number of adaptive equalization techniques have been demonstrated to compensate for linear impairments, which are categorized mainly into time- and frequency-domain approaches. A finite impulse response (FIR) filter is a well-known time-domain equalizer [42]. In FIR filters, the frequency resolution is determined mainly by the finite number of filter taps, which cannot be increased greatly because this would lead to complex calculations. Insufficient resolution in the equalizers becomes disadvantageous for higher-order QAM, as these signals typically contain non-negligible frequency components in the lower frequency regime, such as below 10 MHz. To improve the frequency resolution without increasing the calculation complexity, frequency-domain equalization (FDE) schemes have been developed [43, 44] and have recently received attention as regards digital coherent transmission

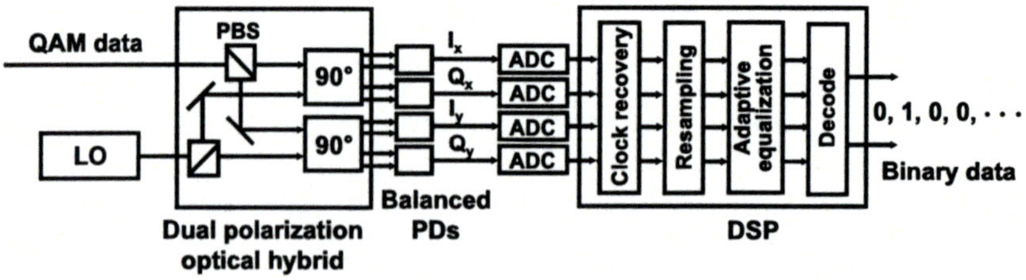

Fig. 5.13 Schematic diagram of digital demodulator

[45]. FDE features an equalization capability with less computational complexity than FIR filters by virtue of the fast Fourier transformation (FFT) operation.

In addition to the linear impairments, fiber nonlinearity also becomes a major limiting factor especially for higher-order QAM transmission because of the power-dependent phase rotation caused by self-phase modulation (SPM) and cross-phase modulation (XPM) [46–48]. As the symbol rate increases, the interplay between optical nonlinearities and dispersion becomes non-negligible, and hence, the individual average compensation of dispersion and SPM is insufficient to eliminate distortions. To compensate for linear and nonlinear impairments simultaneously, the back-propagation (BP) method [49, 50] has recently been proposed, which computes the inverse fiber propagation using the nonlinear Schrödinger equation by using a split-step Fourier method. In a polarization-multiplexed transmission, we employ a split-step Fourier analysis of the Manakov equation, which describes pulse propagation in the presence of dispersion, SPM, and XPM between the two orthogonal polarizations under a randomly varying birefringence [51],

$$\begin{cases} -i\frac{\partial A_x}{\partial z} = -\frac{\beta_2}{2}\frac{\partial^2 A_x}{\partial t^2} + \frac{8}{9}\gamma|A_x|^2 A_x + \frac{8}{9}\gamma\left|A_y\right|^2 A_x + i\frac{\alpha}{2}A_x \\ -i\frac{\partial A_y}{\partial z} = -\frac{\beta_2}{2}\frac{\partial^2 A_y}{\partial t^2} + \frac{8}{9}\gamma\left|A_y\right|^2 A_y + \frac{8}{9}\gamma|A_x|^2 A_y + i\frac{\alpha}{2}A_y \end{cases} \tag{5.2.11}$$

where A_x and A_y represent the amplitude of the x and y polarization components, and α, β_2, and γ are the loss, dispersion, and nonlinear coefficients, respectively. In BP, we solve (5.2.11) with the reversed sign of α, β_2, and γ. BP has the potential to provide significantly improved performance as a result of enhanced OSNR by allowing a higher transmission power.

5.3.3 *Higher-Order QAM Transmission Experiments*

In this section, we present recent demonstrations of (a) 4096 QAM single-carrier transmission, which is currently the highest QAM multiplicity and (b) 256 QAM-OFDM transmission with a spectral efficiency as high as 14 bit/s/Hz. A novel coherent optical Nyquist pulse transmission for realizing highly spectrally efficient, ultrahigh-speed transmission beyond the limit of DSP is also described in (c), where the result of a 7.7-Tbit/s, 64 QAM transmission is presented in detail.

5.3.3.1 4096 QAM Single-Carrier Coherent Transmission

As a first example of much higher-order QAM transmission, here we describe a recent demonstration of 4096 QAM single-carrier coherent optical transmission, in which a 72-Gbit/s polarization-multiplexed signal (3 Gsymbol/s) was transmitted over 160 km within an optical bandwidth of 3.8 GHz [18].

Figure 5.14 shows the experimental setup. At the transmitter, coherent CW light emitted from an C_2H_2 frequency-stabilized fiber laser was modulated at an IQ modulator. It is driven with a 3 Gsymbol/s, 4096 QAM baseband signal, which was generated by an arbitrary waveform generator (AWG) operating at 9 Gsample/s. We employed a raised-cosine Nyquist filter at the AWG using a software program to reduce the bandwidth of the QAM signal [52]. It is well known in the microwave communication field that a Nyquist filter is very useful for reducing the bandwidth of a data signal without introducing intersymbol interference [53]. Figure 5.15 shows

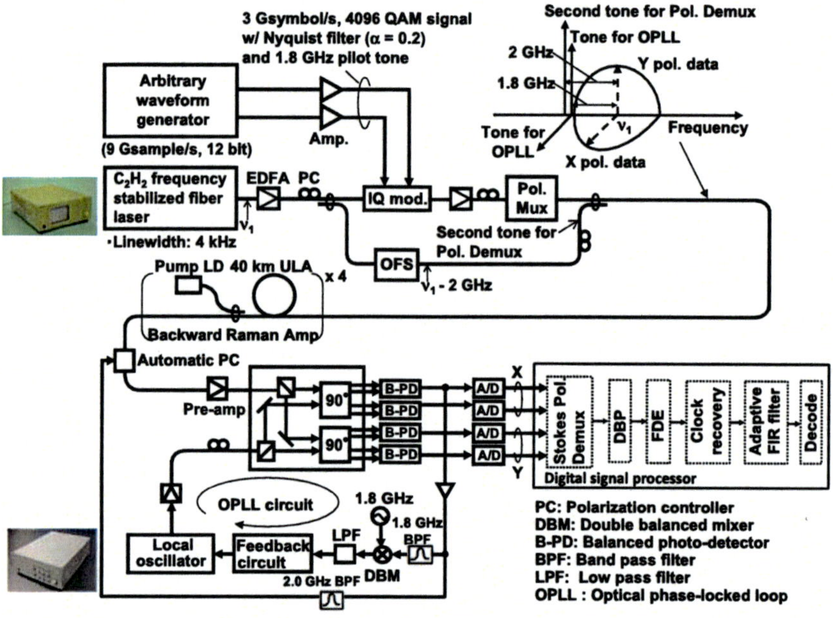

Fig. 5.14 Experimental setup for 4096 QAM (72-Gbit/s) coherent transmission over 160 km (Reprinted with permission from [18]. © 2018 Authors)

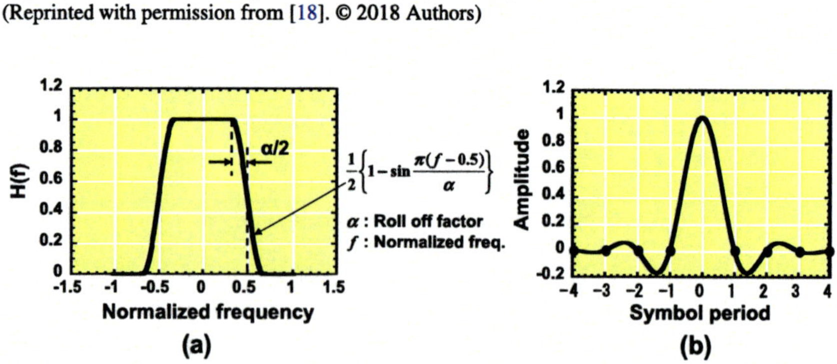

Fig. 5.15 Transfer function (**a**) and impulse response (**b**) of a raised-cosine Nyquist filter

the transfer function and impulse response of the raised-cosine Nyquist filter. The transfer function is given by

$$H(f) = \begin{cases} \frac{1}{2}\left\{1 - \sin\frac{\pi(f-0.5)}{\alpha}\right\}, & 0.5 - \frac{\alpha}{2} \le |f| < 0.5 + \frac{\alpha}{2} \\ 1, & |f| < 0.5 - \frac{\alpha}{2} \\ 0, & |f| \ge 0.5 + \frac{\alpha}{2} \end{cases} \tag{5.2.12}$$

where α is called a roll-off factor. As shown in Fig. 5.15b, the impulse response becomes zero at the location of adjacent symbols. This indicates that the bandwidth can be reduced with the Nyquist filter while avoiding the intersymbol interference. Here, we employed a root raised-cosine Nyquist filter with a roll-off factor $\alpha =$ 0.2, so that the bandwidth of the QAM signal was reduced to 3.8 GHz. In addition, a pre-equalization process based on FDE was adopted to provide high-resolution compensation for the distortions caused by individual components such as the AWG and the IQ modulator. In parallel to these processes, a part of the laser output was divided in front of the IQ modulator, and its frequency was down-shifted by 2 GHz against the carrier frequency. This signal was combined with the QAM data and used as a pilot tone in the receiver for the OPLL as described in Sect. 5.3.2.3.

The optical QAM signal was then orthogonally polarization-multiplexed and launched into a 160-km fiber link. At the receiver, the QAM signal was homodyne-detected at a 90° optical hybrid. As an LO, we used a frequency-tunable fiber laser whose phase was locked to the pilot tone transmitted with the data signal via the OPLL, which enabled low phase noise coherent detection. After detection with balanced photodiodes, the QAM data were A/D-converted and processed with a DSP in an offline condition. In the DSP, a digital back-propagation method using the Manakov equation was adopted to compensate for fiber nonlinearities and dispersion simultaneously. Finally, the compensated QAM signal was demodulated into binary data, and the bit error rate was evaluated.

The experimental results are shown in Fig. 5.16. In this experiment, 72-Gbit/s data were transmitted within an optical bandwidth of only 3.8 GHz. This indicates a

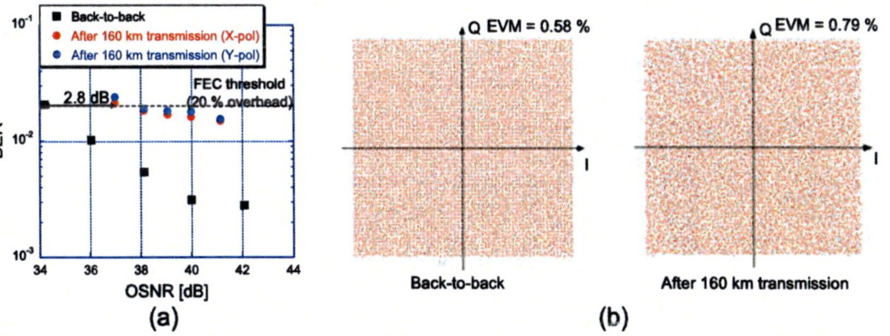

Fig. 5.16 Experimental results for 72-Gbit/s 4096 QAM transmission. **a** BER characteristics. **b** Constellations before and after transmission (Reprinted with permission from [18]. © 2018 Authors)

net spectral efficiency as high as 15.8 bit/s/Hz in a multi-channel transmission, even when taking the 20% FEC overhead into account.

5.3.3.2 256 QAM-OFDM Coherent Transmission

Another attractive candidate for transmission with a high spectral efficiency is orthogonal frequency-division multiplexing (OFDM), where the multi-carrier transmission of low-speed orthogonal subcarriers enables us to improve the spectral efficiency [54]. A typical example of an OFDM signal spectrum and waveform is shown in Fig. 5.17. Although the spectrum of each subcarrier overlaps, the oscillating tail of each subcarrier spectrum intersects at the zero level. The data can be easily demodulated in spite of the spectral overlap by extracting a subcarrier using the orthogonality.

Subcarrier multiplexing with a spectral overlap allows us to improve the spectral efficiency and thus the dispersion tolerance. By virtue of these features, a 1-Tbit/s per channel OFDM transmission has been demonstrated [55]. With this scheme, 1.2-Tbit/s transmission over 7200 km [56] and a back-to-back demonstration at 26 Tbit/s [57] have been reported. However, it should be noted that OFDM is more sensitive to nonlinear effects than single-carrier transmission owing to the large peak-to-average power ratio (PAPR). This occurs as a result of phase coherence among subcarrier signals.

Figure 5.18 shows a block diagram of baseband OFDM signal generation and demodulation. At the transmitter, as shown in Fig. 5.18a, the binary data are first encoded with a multi-level format such as QPSK or QAM. A training symbol used for the amplitude and phase equalization of subcarriers is then added, and the data are converted into a parallel sequence and divided into N subcarriers using inverse fast Fourier transformation (IFFT). After adding a guard interval and employing D/A conversion, a baseband OFDM signal is obtained. At the receiver, as shown in Fig. 5.18b, the IF signal is synchronously detected, and after removing the guard interval and extracting the training symbol, the OFDM signal is demodulated via the fast Fourier transformation (FFT) of the subcarriers. The demodulated signal is finally converted to binary data.

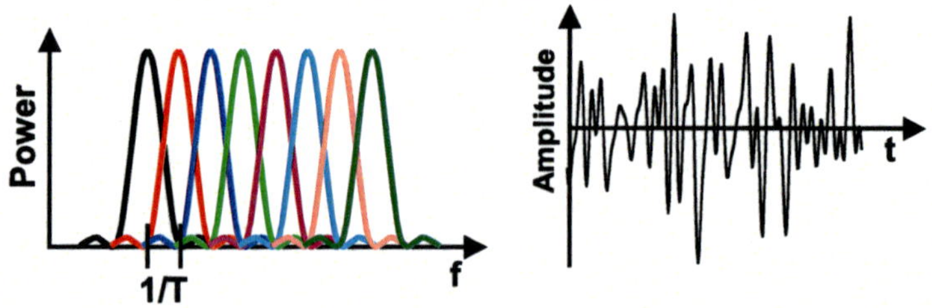

Fig. 5.17 Typical example of an OFDM signal spectrum and waveform

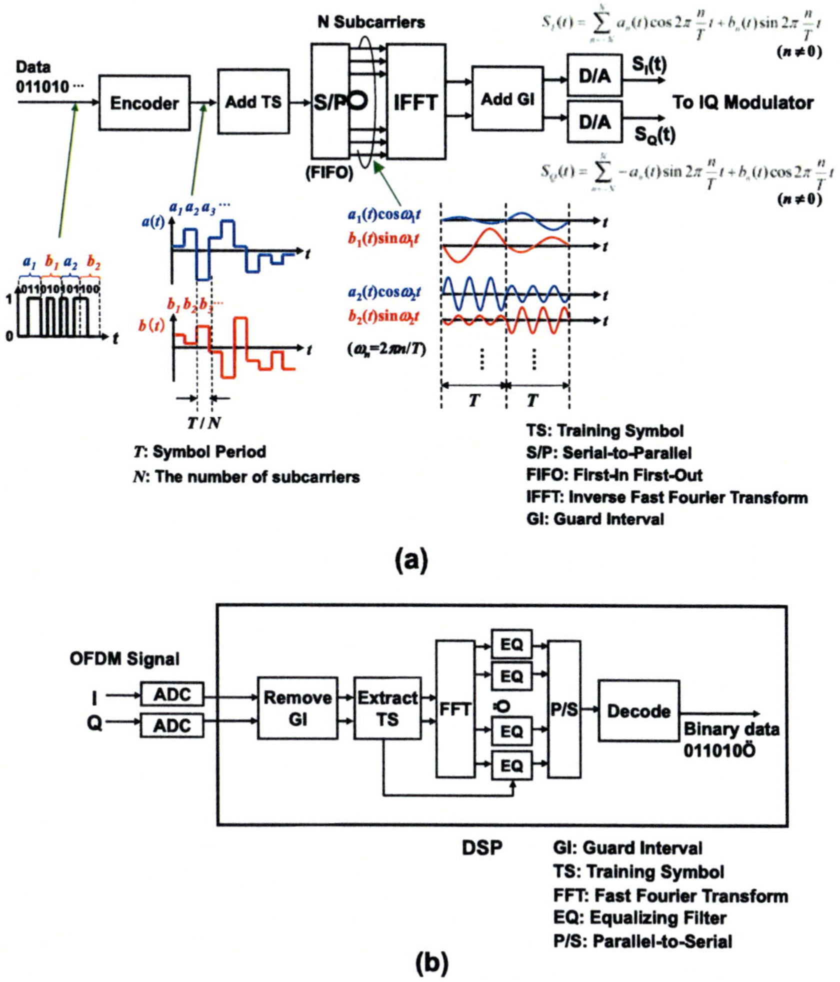

Fig. 5.18 Block diagrams of baseband OFDM signal generation (**a**) and demodulation (**b**)

To achieve higher spectral efficiency in OFDM, high-level subcarrier modulation formats and coherent detection play a very important role. As is shown in Fig. 5.3, a spectral efficiency of 8.0 bit/s/Hz has been realized with 32 QAM subcarrier modulation, where a 400-Gbit/s polarization-division multiplexing (PDM)-OFDM signal was transmitted with a channel spacing of 50 GHz [58]. By increasing the QAM multiplicity level further, a 101.7-Tbit/s WDM capacity has been demonstrated using a Pol-Mux 128 QAM-OFDM with a spectral efficiency of 11 bit/s/Hz [7]. More recently, 1024 QAM subcarrier modulation at 1 Msymbol/s has also been reported [16]. As an alternative format to QAM, 256-level iterative polar modulation has been applied to subcarrier modulation and was found to reduce the required OSNR by 1.2 dB compared with rectangular 256 QAM [59].

Here, we describe a 1-Tbit/s 256 QAM-OFDM transmission that we achieved by adopting high-resolution FDE and digital back-propagation (DBP) [12]. The experimental setup is shown in Fig. 5.19a. Two optical sidebands were first generated from an C_2H_2 frequency-stabilized CW fiber laser using a Mach–Zehnder modulator (MZM1), which were converted into 20 optical carriers (3.3-GHz spacing) using an optical frequency shifter (OFS) and Mach–Zehnder modulators (MZM2). They were modulated using an IQ modulator with a 3.2-Gsymbol/s 256 QAM-OFDM baseband signal generated with an AWG using FFT and high-resolution FDE. After polarization multiplexing, a 1.02-Tbit/s 256 QAM-OFDM signal (2 pol. $\times$ 20 ch $\times$ 3.2 Gsymbol/s) was obtained, whose spectrum is shown in Fig. 5.19b. After 560-km transmission, the OFDM signal was homodyne-detected with a dual-polarization 90° optical hybrid using an LO signal from a frequency-tunable tracking fiber laser, whose phase was locked to the pilot signal via OPLL. After detection with four balanced PDs, the data were A/D-converted and post-processed with a DSP including DBP. Finally, each OFDM band was extracted electronically by a digital filter. The OFDM signal was demodulated with an FFT and converted to a binary data sequence. In this transmission, 1-Tbit/s OFDM signal was transmitted with twenty 3.3-GHz-spaced carriers, and the spectral efficiency reached 14.3 bit/s/Hz after taking a 7% FEC overhead into account. The BER below the FEC limit (2×10^{-3}) was achieved after a 560-km transmission.

5.3.3.3 Ultrahigh-Speed and Highly Spectrally Efficient Coherent Nyquist Pulse Transmission

Along with the aim of higher multiplicity, it is very important to explore new ways of increasing the symbol rate, which is currently limited by the speed and bandwidth of analog-to-digital (A/D) and digital-to-analog (D/A) converters. To overcome these limitations, coherent optical time division multiplexing (OTDM) has been demonstrated that employs multi-level QAM for ultrashort optical pulses [60–62]. However, typical pulse waveforms such as those with Gaussian or sech profiles generally occupy a large bandwidth in the frequency domain and thus may not be appropriate in terms of SE. We recently proposed a new type of optical pulse, which we call an "optical Nyquist pulse" [63], whose shape is given by the sinc-function-like impulse response of the Nyquist filter shown in Fig. 5.15b. In contrast to ordinary RZ pulses, the tail of a Nyquist pulse does not undergo exponential decay but approaches zero slowly accompanied by a periodic oscillation. Here, in spite of a strong overlap, no ISI occurs due to the zero crossing property of the Nyquist pulse at every symbol interval. The fundamental configuration of the ultrahigh-speed Nyquist TDM transmission is shown in Fig. 5.20. The overlapped pulse interleaving without ISI results in a significant bandwidth reduction in OTDM transmission, which is ultimately as low as the TDM symbol rate. This feature is justified by the orthogonality relationship in the time interleaving of Nyquist pulses [63]. Because of these outstanding

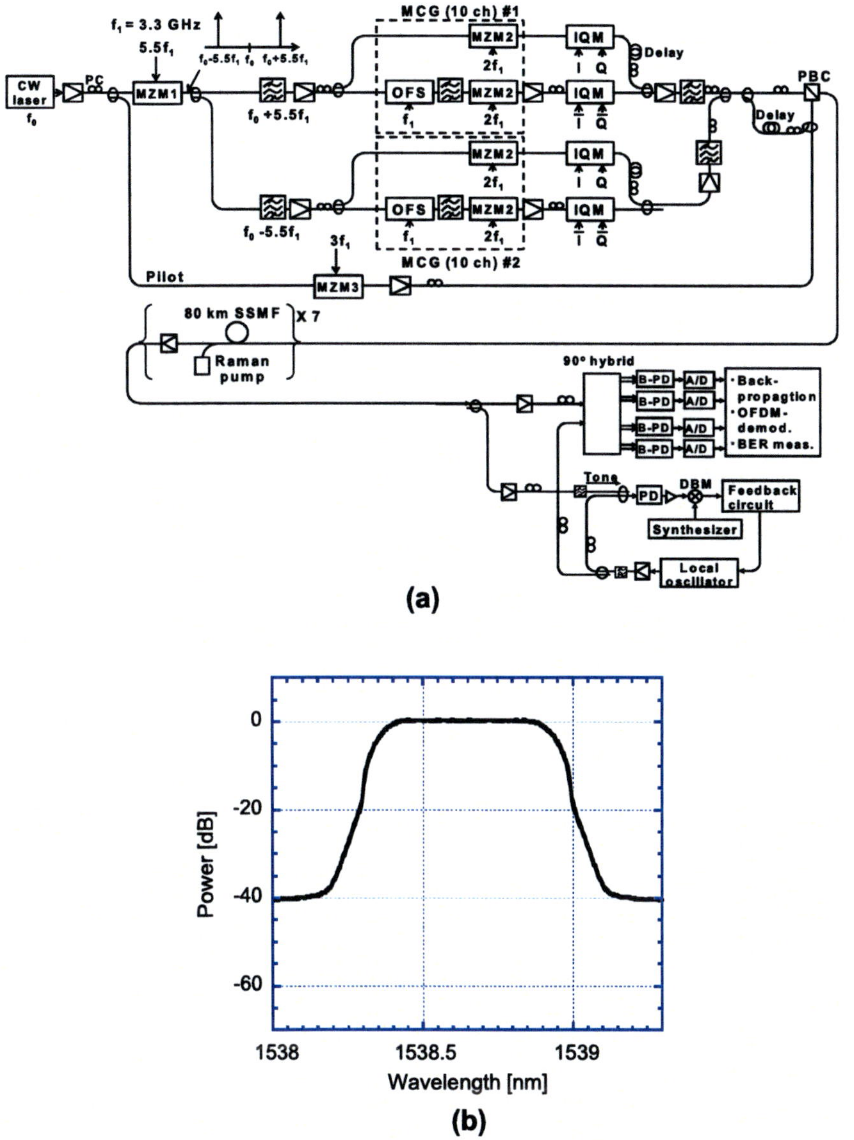

Fig. 5.19 Experimental setup for 1-Tbit/s 256 QAM-OFDM transmission (**a**) and the optical spectrum of the OFDM signal (**b**)

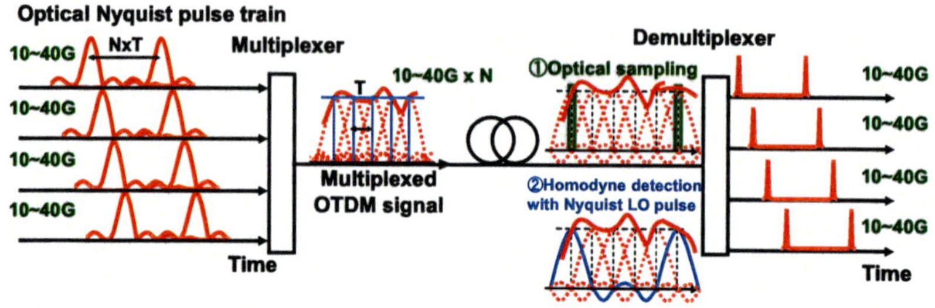

Fig. 5.20 Basic configuration for Nyquist OTDM transmission

properties, optical Nyquist pulses have attracted a lot of attention as ideal pulses for high-speed and highly spectrally efficient OTDM transmission [64–66].

Here, we describe the generation of an optical Nyquist pulse and its high-speed orthogonal time division multiplexing and demultiplexing, and present an ultrafast, spectrally efficient (7.7-Tbit/s/ch, PDM-64 QAM, 9.7 bit/s/Hz) transmission that we demonstrated with coherent Nyquist OTDM [67]. Figure 5.21a shows the experimental setup. On the transmitter side, a 10-GHz coherent Nyquist pulse train with $\alpha = 0$ was generated from a mode-locked fiber laser and a programmable optical filter. The pulse waveform and optical spectrum of the generated Nyquist pulses are shown in Fig. 5.21b, c, respectively. The Nyquist pulse was then data modulated with a 10-Gbaud, 64 QAM signal using an IQ modulator, where we pre-compensated for the nonlinear phase rotation induced by SPM during transmission. The Nyquist pulse was then time domain multiplexed to 640 Gbaud using a delay line bit interleaver, and after polarization multiplexing and combining the pilot tone, the 7.7-Tbit/s PDM-64 QAM signal was launched into a 150-km transmission link. The optical bandwidth including the pilot tone was as narrow as 660 GHz. After transmission, the Nyquist OTDM signal was polarization demultiplexed and synchronously homodyne-detected with a Nyquist LO pulse, in which the demultiplexed 10 Gbaud I and Q data can be obtained after balanced detection. This scheme utilizes the time-domain orthogonality of time-interleaved optical Nyquist pulses $\phi_n(t) = \text{sinc}(t - nT)$ [68]. The data g_n at $t = nT$ can be obtained from the TDM Nyquist data train $u(t) = \sum g_n \phi_n(t)$ as the product of $u(t)$ and $\phi_n(t)$. This indicates that the photo-mixing of $u(t)$ with a Nyquist LO pulse $\phi_n(t)$ on a PD can easily extract a tributary from an overlapping data sequence with a high SNR, and the leakage from other channels can be completely suppressed. This is analogous to OFDM, in which $\phi_n(t)$ is regarded as $\cos(2\pi nt/T)$, where each subcarrier is multiplexed in the frequency domain with a mutual spectral overlap as we described in 5.3.3.2. After A/D conversion with an 80-Gsample/s (66 GHz) digital oscilloscope, the 10G-baud data were demodulated offline in a digital signal processor (DSP) with a 10-GHz demodulation bandwidth.

The BER characteristics are shown in Fig. 5.21d. A BER below the forward error correction (FEC) threshold 2×10^{-2} was achieved after 150 km with a back-to-back power penalty of 3.3 dB. The present result shows the potential for realizing a net spectral efficiency as high as 9.7 bit/s/Hz even when taking account of the 20%

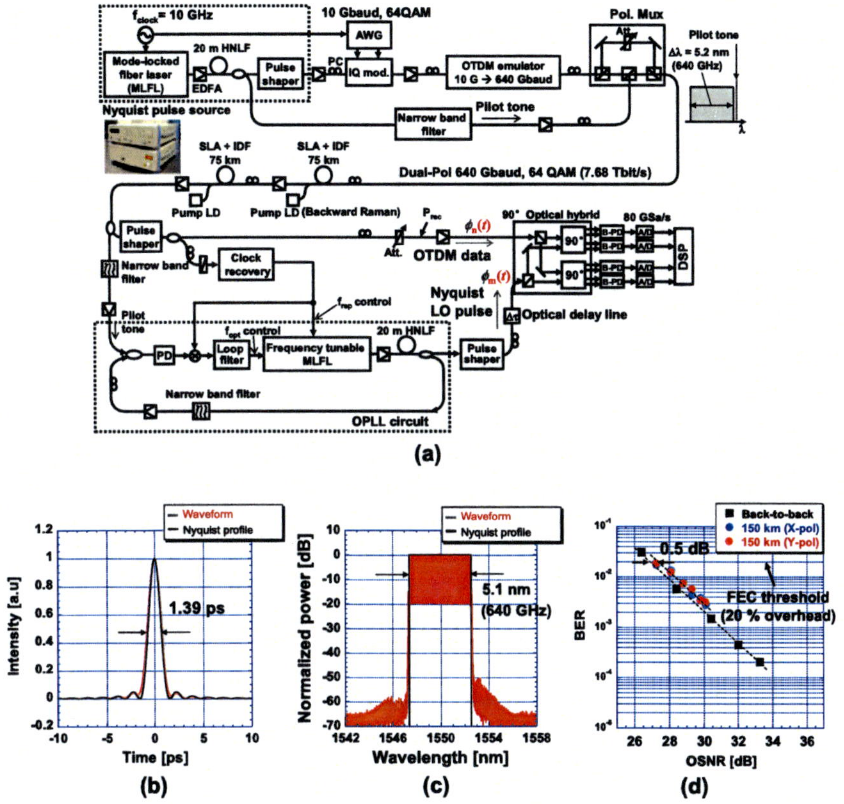

Fig. 5.21 **a** Experimental setup for 7.7-Tbit/s, 64 QAM coherent orthogonal Nyquist TDM transmission over 150 km. **b**, **c** Waveform and spectrum of a generated optical Nyquist sinc pulse. **d** BER performance

FEC overhead. This is to the best of our knowledge the highest spectral efficiency yet realized in a >1-Tbit/s/ch transmission. Recently, a single-channel 10-Tbit/s DQPSK Nyquist pulse transmission has also been achieved at 2.56 Tsymbol/s [69].

5.4 Space Division Multiplexed Transmission Technologies

5.4.1 Multi-core Transmission

5.4.1.1 MCF Transmission Technologies for Terrestrial Systems

In terrestrial optical transport networks, total capacity per fiber has been continuously increased with the growth rate of about 1.4 times per year by employing various innovative technologies. Current WDM transmission systems based on single-core single-mode fibers (SMFs) can support about 10-Tbit/s total capacity by employing 100-Gbit/s polarization-division multiplexed (PDM) QPSK modulation and coherent detection with the aid of digital signal processing [70]. Further improvement of total capacity and spectral efficiency (SE) are investigated by using higher-order multi-level modulation and wideband optical amplification technologies. So far, 102.3-Tbit/s transmission over 240 km was realized by employing PDM-64QAM format and C and extended L band ultra-wideband optical amplification [19]. Total capacity in SMFs, however, is approaching fundamental limit imposed by fiber fuse [71] and nonlinear signal distortion in the transmission fibers [48]. Space division multiplexing using multi-core fibers (MCFs) is very promising to overcome this limitation [72], and rapid growth in total capacity has been realized by using this technique.

Figure 5.22 plots the total capacity and attainable distance of recent single-mode MCF transmission experiments. Large capacity transmission experiments exceeding 100 Tbit/s were first demonstrated by using hexagonally arranged 7-core MCFs: 109-Tbit/s transmission over 17 km [73] and 112-Tbit/s transmission over 77 km [74] were achieved by employing PDM-QPSK modulation. Transmission experiments over higher number of cores have been investigated, and 305-Tbit/s transmission was achieved by using a 10-km 19-core fiber [75]. A total capacity exceeding 1 Pbit/s was first achieved by employing 52-km 12-core fiber with one-ring structure (ORS)

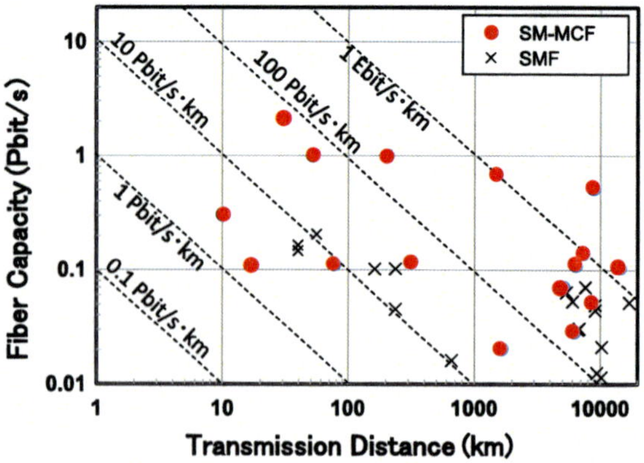

Fig. 5.22 Capacity and transmission distance of recent transmission experiments using single-mode (SM) multi-core fiber (MCF) and conventional single-core single-mode fiber (SMF)

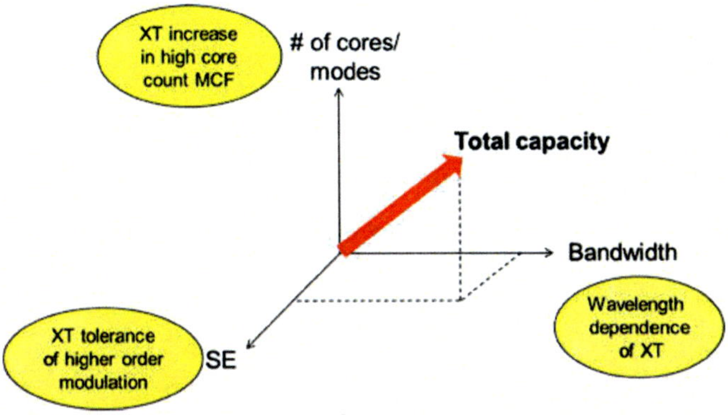

Fig. 5.23 Relation between inter-core crosstalk and the number of cores, SE, and signal bandwidth

and PDM-32QAM signals [76]. Recently, 2.15-Pbit/s transmission over 31-km 22-core fiber was demonstrated [77]. In terms of capacity distance product, transmission experiments exceeding 1 Ebit/s·km were realized [78, 79].

In this section, we describe MCF transmission technologies targeting on terrestrial transport networks. First, we discuss crosstalk management that is indispensable to realize ultrahigh-capacity MCF transmission and show how it is applied in 1-Pbit/s transmission experiments. Next, we discuss crosstalk suppression technique employing bidirectional transmission in MCF: Propagation direction interleaving is very promising in long-distance transmission of higher-order multi-level signals over higher core-count MCFs.

Crosstalk Management

Total capacity of MCF transmission systems is given by the product of the number of cores, SE per core, and the total signal bandwidth, as shown in Fig. 5.23. Inter-core crosstalk is closely related to these factors: Higher-order multi-level modulation is indispensable in order to increase the SE per core. Due to the reduced distance in signal constellation diagrams, however, higher-order modulation format is vulnerable to crosstalk [80]. Figure 5.24 shows an example of the measured penalties in Q-factors as a function of crosstalk in PDM-4QAM, 16QAM, and 32QAM signals after 50-km MCF transmission. The allowable crosstalk for 0.3-dB penalty in Q-factor was −18 dB, but increased to −25 dB for PDM-16QAM and −29 dB for PDM-32-QAM. On the other hand, hexagonal close-packed structure (HCPS) MCF reduces the inter-core gap and thus results in crosstalk increase. Therefore, the number of cores and SE per core has a trade-off relation. Moreover, crosstalk has wavelength dependence and thus is related to the available signal bandwidth [81].

Basically, two schemes have been proposed to cope with the crosstalk problem in MCF transmission: One is the use of multiple-input and multiple-output (MIMO)

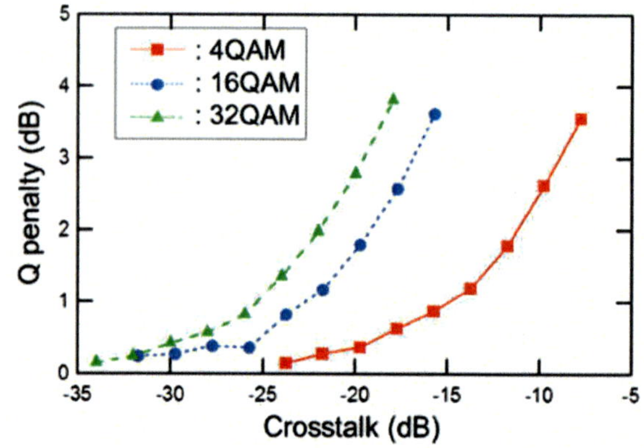

Fig. 5.24 *Q*-penalty as a function of crosstalk in 50-km MCF transmission

digital signal processing (DSP) at the receiver to undo crosstalk generated during the propagation through MCF. This approach utilizes strongly coupled MCF and allows mode coupling between cores. Based on this concept, 4200-km transmission over 3-core strongly coupled MCF was demonstrated [82]. The light propagation in the coupled-core fiber is described in super-modes, and thus, the coupled-core fiber transmission is very similar to multi-mode transmission described in the following section. The other approach, which this section focuses on, is to use weakly coupled MCF and crosstalk management; MIMO processing between different cores is not used, and the modulation format is optimized to maximize the aggregate SE given by the product of the core counts and the SE per core. In this case, the same DSP configuration can be applied as that for conventional SMF. The total capacity and the attainable distance are limited by the crosstalk between adjacent cores.

Total capacity exceeding 1 Pbit/s was achieved by employing weakly coupled MCF and crosstalk management. Figure 5.25 shows the 1.01-Pbit/s transmission of PDM-32 QAM signals over 52 km of 12-core fibers [76]. In this experiment, 12-core fiber with one-ring structure was used. In this core arrangement, the number of adjacent cores was only two. Since the coupling from the adjacent core is the most dominant components in total crosstalk, the total crosstalk of one-ring structure was effectively suppressed below −32 dB within the entire C and extended L bands compared with hexagonal core arrangements. This crosstalk level was sufficiently low for 32-QAM transmission, and the SE per core was 7.6 bit/s/Hz. This resulted in the aggregate SE of 91.4 bit/s/Hz and the total capacity of 1.01 Tbit/s within 11.1-THz total signal bandwidth.

Multi-Pbit/s Transmission in Homogeneous 22-Core MCF

Transmission capacity exceeding 2 Pbit/s was demonstrated in a single-mode, 22-core, homogeneous MCF, by exploiting an extended wavelength range, enabled by a

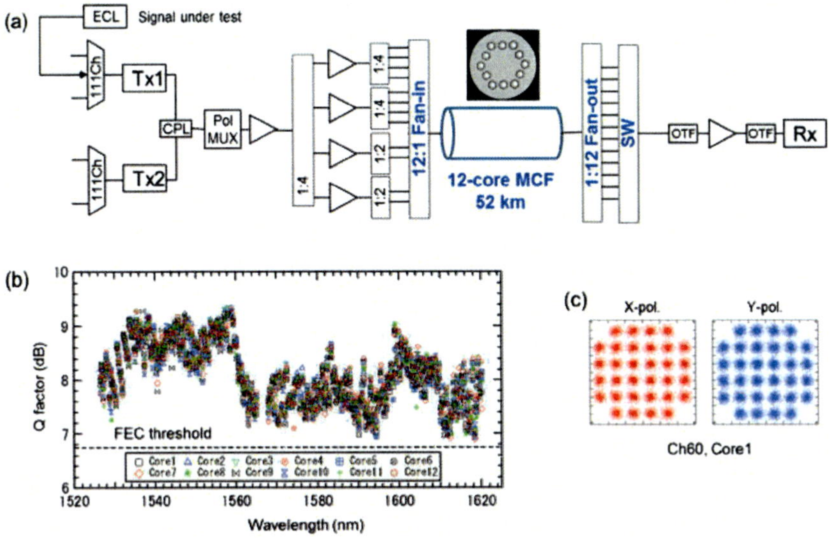

Fig. 5.25 1-Pbit/s transmission over 52-km one-ring structure 12-core fiber. **a** Experimental setup, **b** measured *Q*-factors, **c** example of received constellation

wideband optical comb source in combination with high spectral efficiency modulation of 22-core spatial super-channels (SSCs) [77]. The homogeneous 22-core MCF was based on a new 3-layer design with a two-pitch layout and total cladding diameter of 260 μm, as shown in the inset of Fig. 5.26. The 31-km span was spliced from 5 separately drawn sub-spans, giving rise to total link crosstalk of −37.5 dB at the comb seed wavelength of 1559 nm. The custom-designed frequency comb source consisted of a narrow linewidth (5 kHz) seed laser modulated with a low noise 25-GHz oscillator with the resulting 25 GHz-spaced comb spectrally broadened in a dispersion-engineered fiber mixer. 399 comb lines acted as 22 sub-channels of an SSC, each carrying 64 QAM at 24.5 GBaud (>10 THz bandwidth) to transmit a total data rate of 2.15 Pbit/s confirmed by BER measurements summarized in Fig. 5.26. The low dynamic skew [83], homogeneous fiber, measured to vary over a few picoseconds in 24 h in laboratory conditions, is crucial for SSC transmission, allowing sharing of both transmitter hardware and digital signal processing resources [84, 85]. Furthermore, at the cost of some coding overhead, the wavelength dependence evident in the lower half of Fig. 5.26 may be equalized by optical coding across each SSC with short optical codes of 10s–100s bits [86]. With the lowest BER channels remaining uncoded, different coded modulation schemes may be applied across each SSC according to the average sub-channel BER with the strongest codes being used to reduce the SSCs with highest average BER at the input to the serially coded electronic forward error correction using 1000s bits to minimize overall throughput [87].

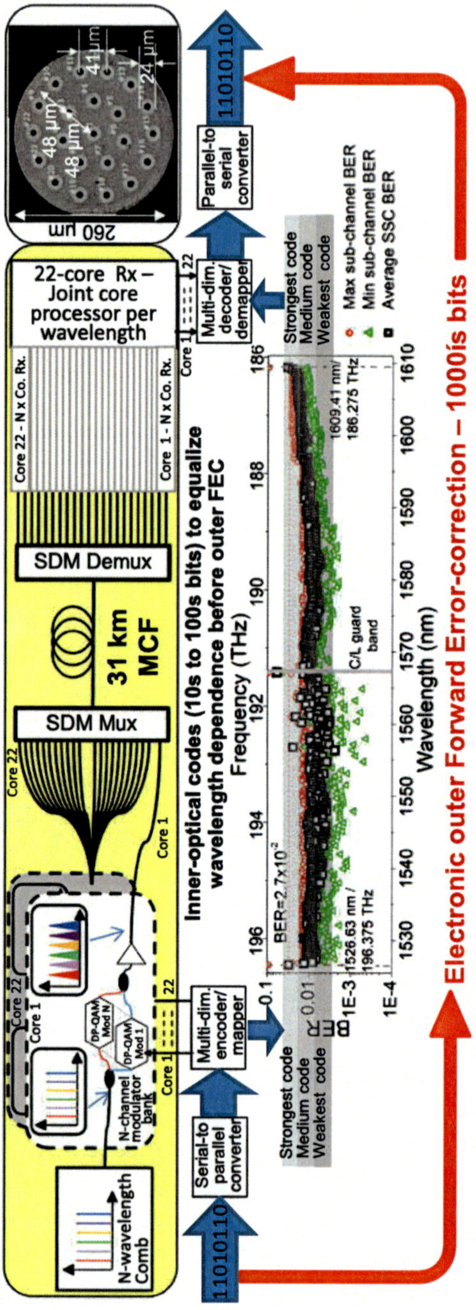

Fig. 5.26 Schematic of >2 Pb/s transmission system based on a 22-core fiber with inner spatial super-channel coding and outer forward error correction (FEC)

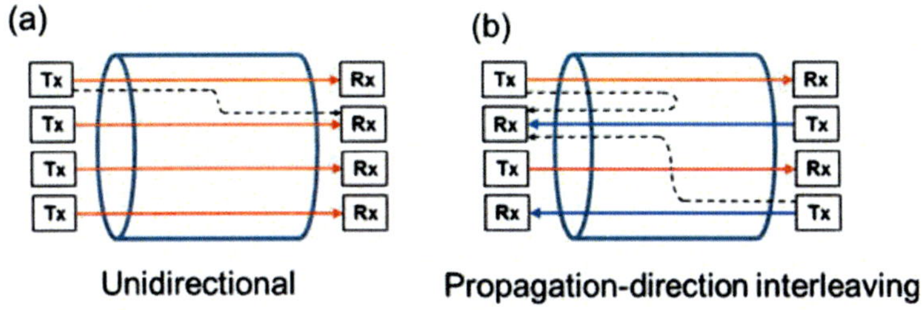

Fig. 5.27 Core assignment of unidirectional (**a**) and bidirectional (**b**) MCF transmission

Large Capacity Long-Haul MCF Transmission Using Propagation Direction Interleaving

Crosstalk management is indispensable to maximize the total capacity of MCF transmission systems. Crosstalk reduction is essential in order to extend the attainable distance of MCF transmission systems. In addition to the core arrangement consideration, bidirectional transmission is also very attractive in reducing the inter-core crosstalk. In bidirectional transmission over MCF, the propagation directions of the adjacent cores are assigned with opposite direction as shown in Fig. 5.27. This scheme is called as propagation direction interleaving (PDI). In conventional unidirectional core assignment, all signals propagate through MCF with the same direction (Fig. 5.27a). In this case, coupling between the co-propagating signals occurs in a distributed manner and causes crosstalk. In PDI, on the other hand, the propagation direction of the adjacent cores is altered (Fig. 5.27b). In this case, crosstalk from the adjacent cores arises when the back-scattered light is coupled with the cores with opposite direction. Therefore, if the splices and connectors yield only slight back-reflection, the dominant factor in back-propagating signals is Rayleigh backscattering, which can be sufficiently low, and so, crosstalk suppression is expected [88].

Figure 5.28a shows an example of the core assignment using PDI in the case of 12-core fiber with dual-ring structure [89]. The number of the closest cores with the same propagation direction is reduced from 4 to 1 by using PDI. Figure 5.28b shows the measured total crosstalk as a function of wavelength when all cores are excited. We can confirm that 3–5 dB crosstalk suppression is obtained by employing PDI.

Next, we describe long-haul transmission of higher-order QAM signals over 12-core DRS fibers with PDI. Figure 5.29 shows the Q-factor as a function of transmission distance in 20-Gbaud PDM-32QAM signals. In the unidirectional case in which measured core and the surrounding 4 cores were excited with the same direction, Q-factor rapidly degraded as the transmission distance increased due to the XT from the neighboring cores, and a 0.6-dB Q-penalty was observed with 500-km transmission. However, by using PDI, this penalty was successfully suppressed to 0.1 dB since only 1 core was excited with the same direction and the other 3 cores were in opposite direction. Large capacity long-haul transmission of PDM-16QAM signals employing PDI was also demonstrated. In this experiment, 748-channel 11.5-Gbaud

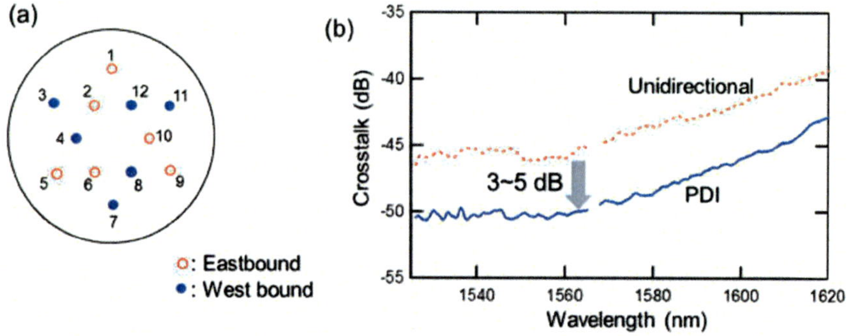

Fig. 5.28 Core assignment (**a**) and measured crosstalk (**b**) of 50-km 12-core fiber with dual-ring structure

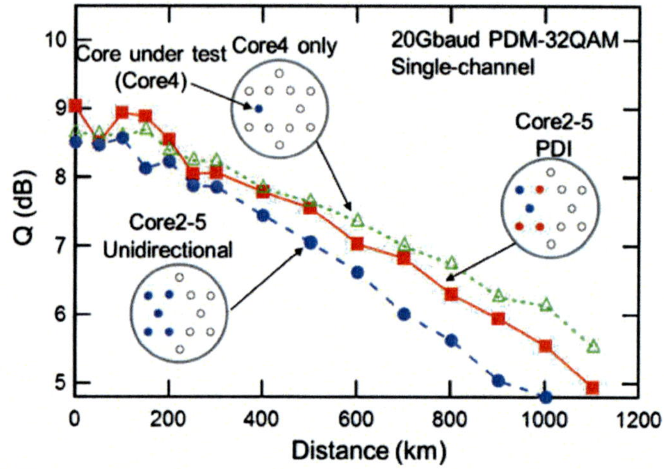

Fig. 5.29 Q-factors as a function of distance in 20-Gbaud PDM-32QAM transmission

PDM-16QAM signals with 12.5-GHz spacing were generated in C and extended L band. The transmission line consisted of a recirculating loop of a 50-km 12-core DRS fiber with PDI. Figure 5.30 shows the measured optical spectra, Q-factors, and an example of constellation diagrams after 1500-km transmission. The total capacity was 2×344 Tbit/s, and thus, the capacity distance product was 1.032 Ebit/s·km, the highest record in MCF transmission.

In this section, MCF transmission technologies targeting on terrestrial transport networks are described. In weakly coupled MCF transmission, total capacity and attainable distance is mainly limited by crosstalk between adjacent cores, and thus, crosstalk management, which optimizes the core multiplicity and the order of multi-level modulation, is indispensable in order to maximize the capacity and transmission distance. Ultrahigh-capacity exceeding 1 Pbit/s was achieved by using

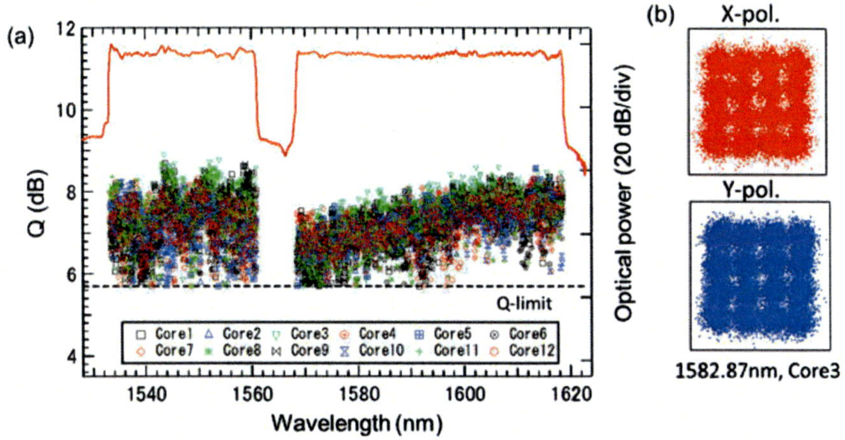

Fig. 5.30 2 × 344-Tbit/s transmission experiment over 1500-km 12-core MCF with dual-ring structure. **a** Measured *Q*-factors. **b** Example of received constellation

crosstalk management. Bidirectional MCF transmission using propagation direction interleaving is promising in order to extend the attainable distance of crosstalk-managed MCF transmission. Capacity distance product exceeding 1 Ebit/s·km was achieved by using this scheme.

5.4.1.2 MCF Transmission Technologies for Submarine Systems

In order to satisfy both requirements of higher capacity and longer distance for submarine systems, highly spectral-efficient modulation format with lower required optical signal-to-noise ratio (OSNR) as well as multi-core fibers (MCFs) with low core-to-core crosstalk is required. In addition, for increasing the number of fiber pairs by the use of MCFs for the submarine systems, reducing power consumption and size of optical repeaters is also required due to the limits of capacity of power feeding equipment. In this section, we introduce two ultra-long-haul transmission experiments using seven-core fibers with ultra-low core-to-core crosstalk. Super-Nyquist-WDM technique based on quadrature phase-shift keying (QPSK) was employed in the experiment with the highest capacity and distance product of 1 Ebit/s·km over 7300 km. Cladding pumped multi-core erbium-doped fiber amplifier (MC-EDFA) was investigated in long-haul MCF transmission experiment at the first time.

Exabit/s·km Ultra-Long-Haul MCF Transmission Experiment

One straightforward approach for improving spectral efficiency is based on an ultra-dense WDM technique packing optical signals as closely as possible in the frequency

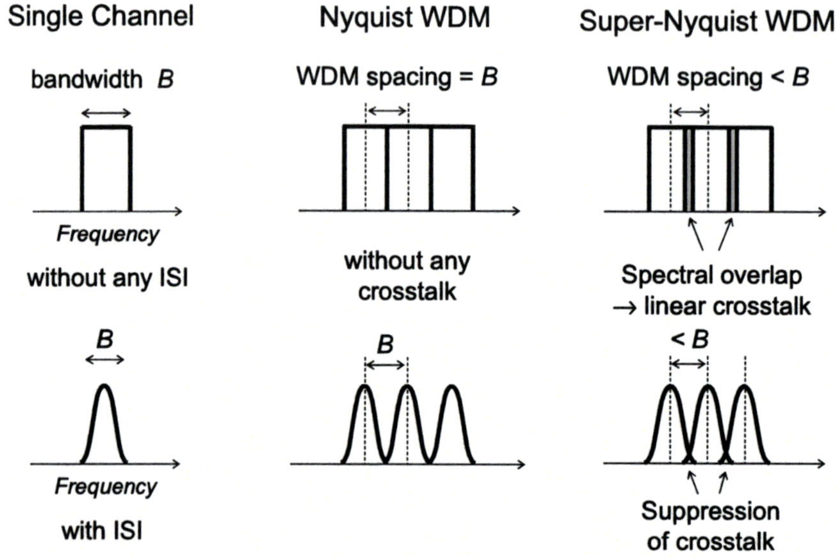

Fig. 5.31 Relationship between the WDM spacing and baud rate of the signal. **a** Nyquist pulse shaping with rectangular-shaped spectrum, **b** pulse shaping over Nyquist criteria. B: baud rate of signal, Δf: WDM spacing (Reprinted with permission from [79]. © 2015 IEEE)

domain. For ultra-dense WDM systems, spectral narrowing is required so that the crosstalk between WDM channels is sufficiently suppressed. In contrast, excessive spectral narrowing, which corresponds to excess limitation of the signal bandwidth, gives rise to the signal degradation due to the intersymbol interference (ISI).

Nyquist pulse shaping is a well-known technique for suppression of ISI [90]. In particular, rectangular-shaped Nyquist pulse shaping with bandwidth of the signal baud rate B is preferable for ultra-dense WDM systems with frequency spacing Δf equal to B because the spectral overlap between WDM channels could be suppressed. In this case as shown in Fig. 5.31a, not only ISI penalty but also WDM crosstalk can be entirely suppressed. This is called Nyquist-WDM technique [91]. For beyond Nyquist-WDM with $\Delta f < B$, which is called super-Nyquist-WDM or faster-than-Nyquist-WDM [92, 93], the spectral overlap is not negligible even if rectangular-shaped Nyquist pulse shaping is used.

For the super-Nyquist-WDM condition with $\Delta f < B$, we consider pulse shaping for tight spectral filtering over the Nyquist criteria, as shown in Fig. 5.31b. In this case, the spectral overlap can be suppressed, compared with the Nyquist pulse shaping. On the other hand, such tight limitation of signal bandwidth gives rise to a remarkable ISI penalty. It is notable that there is a trade-off relationship between ISI and WDM crosstalk in the super-Nyquist-WDM systems. Although the WDM crosstalk suppression is possible by using decoded symbols of adjacent WDM channels in principle, the multi-channel receiver is required. For practical application, a technique

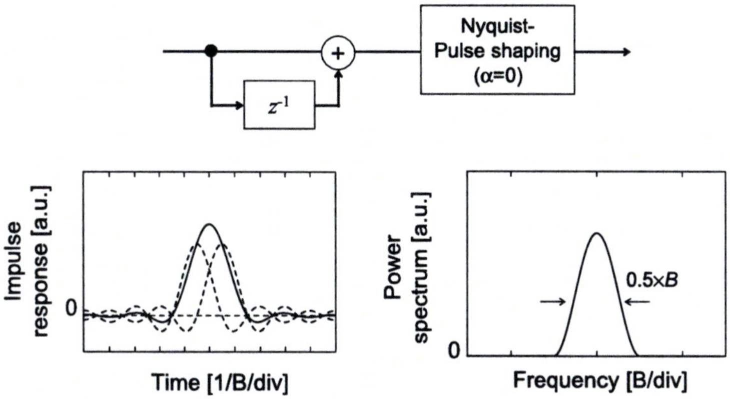

Fig. 5.32 **a** Model of duobinary-pulse shaping. **b** Impulse responses and power spectra of duobinary-pulse shaping (Reprinted with permission from [79]. © 2015 IEEE)

for suppressing the ISI penalty is essential to optimize the trade-off relationship in the super-Nyquist-WDM condition.

In the presence of ISI due to the pulse shaping over Nyquist criteria, the waveform of the received signal would be remarkably distorted. Even in this case, the transmitted symbol with the highest probability can be expected using the sequence received before [94]. This signal processing is called maximum likelihood sequence estimation (MLSE). In the presence of ISI spanning $L + 1$ symbols, the transmitted symbol at time k can be estimated for the received sequence from $k - L$ to k. If the symbols are M-ary, there are $M{\cdot}L$ cases as the possible sequence, which are called states. The trellis diagram describes the possible paths interconnecting between states, and the Viterbi algorithm may be used to determine the most probable path based on the metrics calculated by the decision error of the received symbol [90].

Note that the calculation complexity of MLSE is determined by the span L of ISI due to the pulse shaping, since the number of states of the trellis is $M{\cdot}L$. The duobinary-pulse shaping is most preferable for MLSE, since the symbol span of the ISI is limited to be two [90]. Figure 5.32a shows the model of the duobinary-pulse shaping. It consists of the finite impulse response (FIR) filter with two taps followed by rectangular-shaped Nyquist pulse shaping with the bandwidth of the baud rate. In this case, the state number of trellis is limited to be just M. Figure 5.32b shows the impulse response and power spectrum of duobinary-pulse shaping. The impulse response is composed of two delayed Nyquist pulse shaping indicated by dashed lines. Note that the bandwidth (full width at half maximum) of the power spectrum is half the signal baud rate. Such tight spectrum has advantage over suppression of the WDM crosstalk in the super-Nyquist-WDM condition. A technique based on MLSE with duobinary-pulse shaping was proposed in the 1970s [94], and an application to the super-Nyquist-WDM transmission experiments has recently been reported [92, 93].

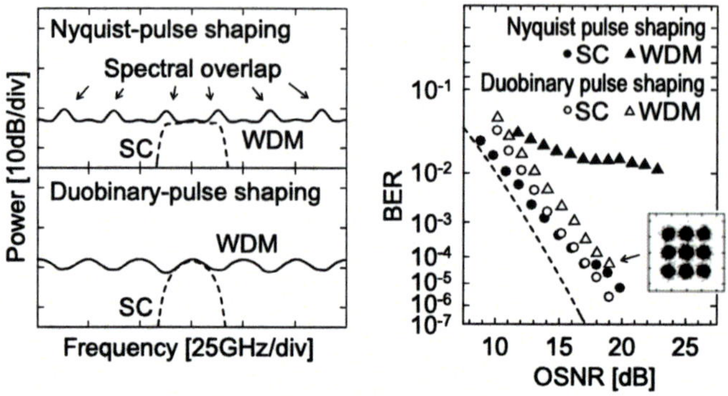

Fig. 5.33 **a** Measured optical spectra of Nyquist-pulse-shaped and duobinary-pulse-shaped PDM-QPSK signals. Dashed lines: single channel (SC) cases, solid lines: WDM cases. **b** Measured bit error rates of 30-Gbaud duobinary-pulse-shaped and Nyquist-pulse-shaped PDM-QPSK signals in the single channel (SC) and WDM cases. Dot: single-channel Nyquist-pulse-shaped QPSK signals, closed triangle: super-Nyquist-WDM Nyquist-pulse-shaped QPSK signals, open circle: single-channel duobinary-pulse-shaped QPSK signals, open triangle: super-Nyquist-WDM duobinary-pulse-shaped QPSK signals (Reprinted with permission from [79]. © 2015 IEEE)

We measured the back-to-back bit error rate (BER) performance of super-Nyquist-WDM PDM-QPSK signals with duobinary-pulse shaping. Here, we multiplexed 201-channel 30-Gbaud duobinary-pulse-shaped PDM-QPSK signals with WDM spacing of 25 GHz. Measured optical spectra with Nyquist pulse shaping and duobinary-pulse shaping are shown in Fig. 5.33a. For the Nyquist pulse shaping, we found the spectral peaks due to the spectral overlap between WDM channels.

On the other hand, such spectral overlap components were sufficiently suppressed by using the duobinary-pulse shaping. The measured BERs of the single channel (SC) and 25-GHz-spaced WDM cases of the duobinary-pulse-shaped PDM-QPSK signals are plotted by dots and closed triangles shown in Fig. 5.33b, respectively. For comparison, the results of those cases of Nyquist pulse shaping are indicated by open circles and open triangles, respectively. Dashed line is the theoretical curve of the SC PDM-QPSK signals. In the 25-GHz-spaced WDM case of Nyquist pulse shaping (closed triangles), we found a remarkable penalty due to WDM crosstalk from adjacent WDM channels. Using duobinary-pulse shaping (open triangles), the penalty was suppressed to below 2 dB compared with the SC PDM-QPSK signals. These results suggest that duobinary-pulse shaping with MLSE is effective for suppression of the ISI penalty in the super-Nyquist-WDM condition.

Here, we introduce a demonstration of 1-Ebit/s·km super-Nyquist-WDM transmission experiments using seven-core fibers with seven-core EDFA. The details have been reported in [79]. Figure 5.34 shows the experimental setup. The transmitter was based on three-rail configuration. For two rails, eight even and eight odd channels were independently modulated by IQ modulators (IQMs) driven by 30-Gbaud duobinary-pulse-shaped binary signals. For the third rail, 201 WDM channels with

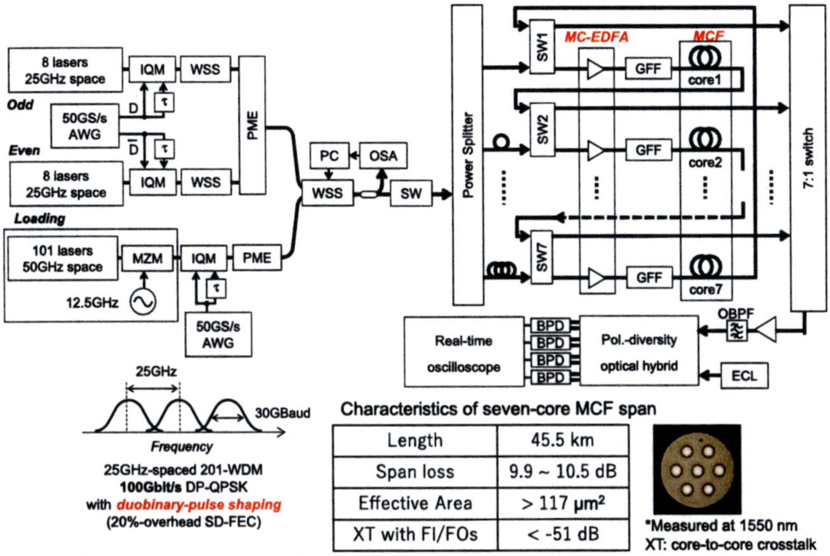

Fig. 5.34 Experimental setup for MCF transmission. IQM: IQ modulator, WSS: wavelength-selective switch, PME: polarization-multiplexed emulator, PC: personal computer, OSA: optical spectrum analyzer, SW, optical switch, GFF: gain-flattening filter, OBPF: optical band-pass filter, ECL: external cavity laser, BPD: balanced photodetector (Reprinted with permission from [79]. © 2015 IEEE)

a 25-GHz spacing ranging from 191.2625 to 196.2875 THz were generated in order to maintain not only an OSNR but also nonlinear effects.

After the modulated WDM signals from three rails were fed into the polarization multiplexing emulators (PMEs), they were combined by a wavelength-selective switch (WSS). Consequently, we obtained 201-channel 25-GHz-spaced super-Nyquist-WDM 30-Gbaud duobinary-pulse-shaped PDM-QPSK signals, resulting in a spectral efficiency of 4 bit/s/Hz assuming the use of a LDPC-based SD-FEC with 20% overhead [95].

For seven-core fiber transmission, the 201-channel WDM signals were launched into a specially configured sevenfold recirculating loop, which was composed of a 45.5-km seven-core fiber span with low core-to-core crosstalk as shown in Fig. 5.34, external gain-flattening filters (GFFs), and optical switches (SWs). The seven loops were synchronously operated as reported in [96]. The seven recirculating loops shared a common load switch that launched identical copies of the WDM signals into each of the loops. After amplification by the seven-core EDFA with external GFFs, the signals were launched into the cores of the 45.5-km seven-core fiber. The output signals from one core were sent to the recirculating loop input of the next core, in a cyclic fashion. The core-to-core configuration can average out variations in span loss, dispersion, and other component imperfections [96]. Two sets of WSS with single-core EDFA were inserted at input of the fourth and seventh cores for gain

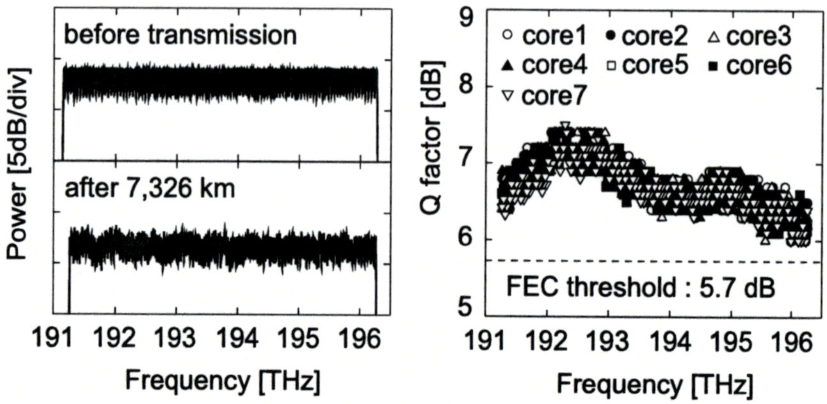

Fig. 5.35 **a** Optical spectra before and after 7326-km transmission. **b** Measured Q-factors of all 201 WDM channels of seven cores of MCF (Reprinted with permission from [79]. © 2015 IEEE)

equalization, and the gain flatness was automatically managed by the WSS at input of the seventh core. A polarization scrambler was inserted at the input of the fifth core.

The transmitted signals were received by a digital coherent receiver after pre-amplification and channel selection with an optical band-pass filter (OBPF) with a bandwidth of 1 nm. Electrical signals from the receiver were stored in sets of 1 M samples by using a four-channel digital oscilloscope operating at 50 Gsample/s. The stored data were processed offline by DSP as follows: The received signals were re-sampled to 2 sample/symbol. After Nyquist pulse shaping and dispersion compensation in the frequency domain, polarization demultiplexing and signal equalization were performed by half-symbol-spaced FIR filters with 80 taps, which were adapted by the decision-directed least mean square (DD-LMS) algorithm [97]. After MLSE, the symbols were decoded.

The optical spectra before and after 7326-km transmission are shown in Fig. 5.35a. After transmission, the power differences of all channels were maintained to be less than ± 2 dB thanks to precise gain equalization. The measured Q-factors after 7326-km transmission are shown in Fig. 5.35b. The Q-factors of all 201 channels for seven cores of MCF were confirmed to be over 6.0 dB, which exceeds the limit of the 20%-overhead LDPC-based SD-FEC with the maximum number of iterations of 12, namely 5.7 dB [95].

Long-Haul MCF Transmission Experiment Using Cladding-Pumped MC-EDFAs

SDM provides a major advantage that an integration of not only transmission fibers but also fiber amplifiers can reduce the number of optical components in an entire optical transmission system [98, 99]. In particular, a multi-core erbium-doped fiber

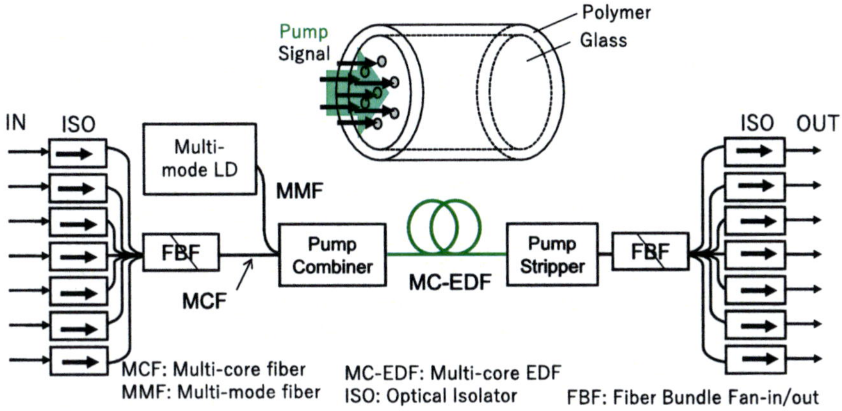

Fig. 5.36 Configuration of cladding-pumped seven-core EDFA

Multi-core EDF (MC-EDF)	
inner cladding diameter	180 μm
core-to-core pitch	45 μm
numerical aperture (NA)	0.36
mode field diameter (MFD)	7.3 μm
core absorption coefficient	7 dB/m
Length of MC-EDF	107 m

(a)

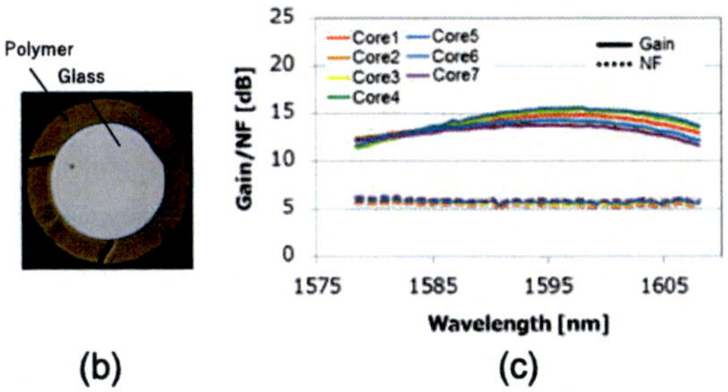

Fig. 5.37 **a** Characteristics of seven-core EDF, **b** cross section of seven-core EDF, **c** gain and NF versus wavelength

amplifier (MC-EDFA) using cladding pumping scheme composed of a single erbium-doped fiber (EDF) with multiple cores and a multi-mode laser diode [100, 101] is an attractive technique to realize all multi-core fiber (MCF) transmission spans without fan-in/fan-out devices. In this paragraph, we introduce a demonstration of long-haul transmission over 4200-km MCF using cladding-pumped MC-EDFAs [102].

Figure 5.36 shows a schematic diagram of cladding-pumped seven-core EDFA. The cladding-pumped MC-EDFA mainly consists of a multi-core EDF with double cladding structure, multi-mode LD as a pump laser, a pump combiner in which signals and pump light are combined and a pump stripper used to eliminate the pump light. Fiber bundle fan-in/fan-out devices and isolators were attached to both edges of seven-core EDF in this demonstration [102]. Since the cladding pumping scheme is possible to simultaneously pump all Er-doped multi-cores by using a single high-power multi-mode LD [103–105] without the dependence on the number of cores, it would hopefully reduce the number of optical components and power consumption in the amplifier. Figure 5.37a, b shows characteristics and a cross section of seven-core

EDF used in the demonstration, respectively. As shown in Fig. 5.37b, seven cores are arranged in a hexagonal array with a core-to-core pitch of 45 μm. We employed a longer length (107 m) of seven-core EDF to achieve required gain and output power in L band amplification. This is because overlap integral of the cladding pump field and the cores in the seven-core EDF is small. Figure 5.37c shows gain profiles and noise figures (NFs) of the cladding-pumped seven-core EDFA. 51 WDM channels with 70-GHz spacing from 1578.333 to 1608.264 nm were used to measure them. Total signal input power and the pump power were set to +7.5 and +42.3 dBm into the amplifier, respectively. The gain over 11 dB was obtained over a wide bandwidth of 30 nm, and the gain difference between cores was about 2 dB. The NF was below 6 dB in all cores, and the NF spectrum for each core is flat with a deviation of less than 1 dB.

We demonstrate a long-haul MCF transmission using cladding-pumped multi-core EDFAs. The experimental configuration is almost same as shown in Fig. 5.34 [106]. The transmitter was based on three-rail configuration. For two rails, five even and five odd channels were independently modulated for quadrature phase-shift keying (QPSK) modulation by IQ modulators (IQMs) driven by 32-Gbaud electrical signals with rectangular-shaped spectra, which were generated by an arbitrary waveform generator (AWG) with a sampling rate of 50 Gsample/s. For the third rail, 76 WDM channels with a 35-GHz spacing ranging from 1578.333 to 1600.451 nm were generated. After the modulated WDM signals from three rails were fed into the polarization multiplexing emulators (PMEs), they were combined by a coupler. In the experiment, we disabled 10 consecutive channels on the third rail and suppressed amplified spontaneous emission (ASE) noise on 10 disabled channels by a tunable optical band rejection filter (BRF), and then, the 10 channels from the two rails were tuned to the corresponding wavelength for the measurement. Consequently, we obtained 76-channel 35-GHz-spaced 32-Gbaud Nyquist-shaped dual-polarization QPSK (PDM-QPSK) signals.

For MCF transmission, the 76-channel WDM signals were launched into a specially configured sevenfold recirculating loop consisting of the cladding-pumped seven-core EDFA, 40-km-length seven-core MCF including external gain-flattening filters (GFFs) and optical switches (SWs). In the cladding-pumped seven-core EDFA, the signal power launched into each core was set to −2 dBm/ch by the multi-mode LD with a pump power of 41.2 dBm. After amplification, the gain differences of WDM signals were equalized within ±0.9 dB through GFFs. The 40-km seven-core fiber has an insertion loss of about 9 dB and the crosstalk of less than −55 dB including the fan-in/fan-out devices. The transmission loss of the fiber at a wavelength of 1600 nm was 0.2 dB/km [107]. The output signals from one core were sent to the recirculating loop input of the next core, in a cyclic fashion. A single-core EDFA and wavelength-selective switch (WSS) were inserted at the output of the seventh core for gain equalization, and the gain flatness was automatically managed by the WSS with an optical spectrum analyzer (OSA). A low-speed polarization scrambler (PS) was inserted at the output of the seventh core.

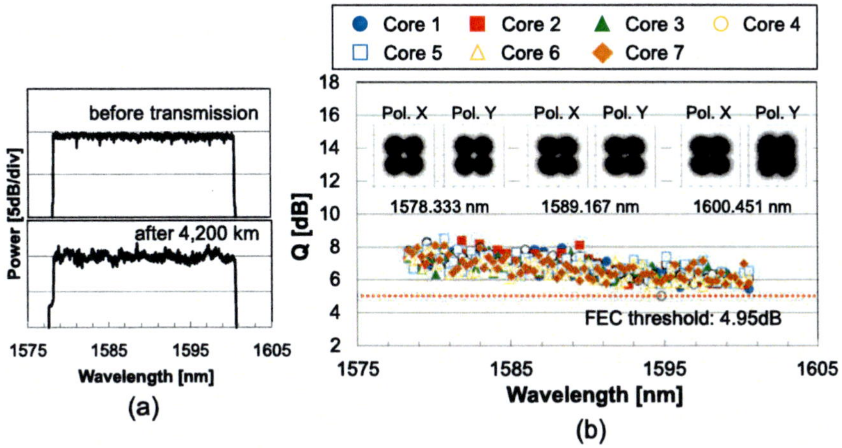

Fig. 5.38 **a** Optical spectrum before and after transmission. **b** Measured Q-factors for all the channels (Reprinted with permission from [102]. © 2015 IEEE)

The transmitted signals were received by a digital coherent receiver after preamplification and channel selection with an optical band-pass filter (OBPF). Electrical signals from the receiver were stored in sets of 1 M samples by using a 4-channel digital oscilloscope with a sampling rate of 50 Gsample/s. The received signals were demodulated by offline digital signal processing as follows: (1) resampling at two samples/symbol, (2) Nyquist filtering and chromatic dispersion compensation in the frequency domain, (3) polarization demultiplexing and signal equalization by half-symbol-spaced adaptive butterfly finite impulse response (FIR) filters with constant modulus algorithm (CMA), (4) decoding and bit error ratio (BER) calculation.

Figure 5.38a, b summarizes measured optical spectra before/after transmission and Q-factors for all 76 WDM channels for seven cores after 4200-km transmission via 105 seven-core EDFAs. The Q-factor is calculated from the measured BER. As shown in Fig. 5.38b, the Q-factors of all 76 channels for seven cores of MCF were confirmed to be over 5.0 dB, which exceeds the limit of assumed forward error correction (FEC) with 25.5% overhead (spatially coupled-type irregular low-density parity check code-based soft decision FEC (LDPC-based SD-FEC) with BCH code) [106, 107]. As a result, we operated the CP-7C-EDFA in L band so as to obtain the higher power and appropriate gain for long-haul transmissions and confirmed that the 76-channel 35-GHz-spaced Nyquist-shaped dual-polarization quadrature phase-shift keying (PDM-QPSK) signals were successfully transmitted over 4200 km using seven-core fibers with cladding-pumped seven-core EDFAs. Note that the transmission distance was limited to 4200 km due to relatively large inter-core crosstalk of the pump combiner in cladding-pumped seven-core EDFAs [102].

To reduce the inter-core crosstalk in the cladding-pumped EDFA, EDF with higher core absorption coefficient which halved the EDF length and a pump combiner with the optimized side pumping scheme were introduced, which enabled 5040-km transmission of 50 WDM × 256 Gbit/s in 7-core MCF [108].

5.4.1.3 Optical Diversity Transmission Using MCF

Internet traffic has continually grown over the past few decades, but its growth rate has been rising rapidly in the last few years (2013–2015), which has been spurred by the explosive spread of smartphones and/or mobile devices. The resulting loads are compelling network operators to optimize their networks with regard to cost-effectiveness, which means minimizing capital and operational expenditures. The innovations expected should enhance the capacity and distance of fiber transmission, and many research engineers have tackled the key issues. However, single-core, single-mode fiber transmission faces the optical power density limit imposed by fiber fuse [109] and fiber nonlinearity, and the limit might set the ceiling on transmission capacity and bitrate-distance product at around 100 Tbit/s [7, 19] and 400 Pbit/s·km [110], respectively.

Space division multiplexing (SDM) in multi-core fiber (MCF) is one solution that can dramatically increase the transmission capacity per fiber [111]. It can enhance the space density of transmission capacity, and several reports have already exceeded 1 Pbit/s [76, 77, 112]. However, the simple SDM technique cannot overcome the issues of the power limit per core and fiber nonlinearity. The fiber nonlinearity limits, as is well known, the Q-value for one channel in single-mode fiber transmission, and overcoming fiber nonlinearity would allow us to raise the fiber input optical power per channel, resulting in higher Q-values. Unfortunately, the total fiber input power is limited by the optical power density limit. This limited Q-value restricts the transmission distance and/or the M-ary level for QAM.

The most intensively discussed nonlinearity mitigation technique is digital backward propagation with nonlinear perturbation pre-distortion [50, 113, 114]. However, it incurs excessive digital processing cost. The backward propagation approach can be realized by utilizing a phase-conjugate light. Several reports have described cascade transmission with mid-span phase reversal [115–118]. Optical diversity transmission with mutually phase-conjugate light pair (PCP) and maximum ratio combination (MRC) is another simple approach that can counter fiber nonlinearity [119, 120]. The MRC of PCP cancels the nonlinear phase shift (NLPS) induced by cross-phase modulation (XPM). This scheme also enables the optical power density limit to be, in effect, doubled. One challenge for MCF transmission is to discover other possibilities [119–121].

This section briefly provides the MRC condition for optical diversity transmission across multi-core fiber and theoretically introduces the transfer function approximation of NLPS cancelation. After that, their impact is demonstrated in a 2-core diversity experiment on 5 channel WDM transmission using 25-Gbit/s QPSK PCP over 3 spans of 60-km MCF. Superior NLPS cancelation is achieved with CD span-by-span compensation rather than its lumped equivalent at the receiver. The optical diversity transmission with PCP and MRC offers Q-value enhancement of better than

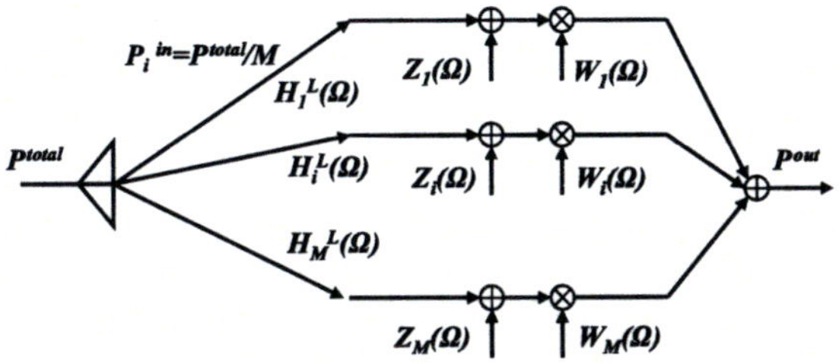

Fig. 5.39 Optical diversity transmission and maximum ratio combining model; P_i^{in}: fiber input power, H_i^{L}: channel #i linear transfer function, Z_i: Fourier transform of white Gaussian noise, W_i: weighting function (Reprinted with permission from [120]. © 2016 The Optical Society)

4 dB, which yields superior bitrate-distance product and optical power density limit performance, compared to twice the single-core transmission in MCF.

Signal-to-Noise Ratio in Optical Diversity Transmission

MRC based on diversity transmission was originally designed for microwave transmission and radar systems [122], but basically it can be applied to the linear transmission region. We have expanded this theory to cover optical fiber transmission that exhibits nonlinear phenomena.

A conceptual diagram of the proposal is depicted in Fig. 5.39. The optical total power, denoted as P^{total}, is split M ways. Each component, $P_i^{\text{in}} = P^{\text{total}}/M$, is modulated by some signal code and launched into a different core of the same multi-core fiber; the number of cores is M_{core} ($M_{\text{core}} \geq M$). We can write the optical signal electric field at core #i input as:

$$e_i(0,t) = \sqrt{2P_i^{\text{in}}}A_i'(0,t)\cos(\omega_0 t) = (1/2)A_i(0,t)\exp(j\omega_0 t) + \text{c.c.} \tag{5.3.1}$$

where $A_i(0, t)$ represents the complex signal amplitude at transmission distance $z = 0$, ω_0 does its carrier angular frequency, and c.c. stands for complex conjugate. The Fourier transform of the complex amplitude $A_i(0, t)$ is $A_i(\Omega)$ with the relationship of

$$A_i(0,t) = \frac{1}{2\pi}\int A_i(\Omega)e^{j\Omega t}d\Omega \tag{5.3.2}$$

where Ω is optical carrier modulation angular frequency. The time average of square of the electric field amplitude, $\overline{|A_i(0,t)|^2}$, is set at core #i with input optical average power P^{in} as $\overline{|A_i(0,t)|^2} = P_i^{\text{in}}$. This split optical signal is transmitted through the core. The electrical field after propagating a distance of z, $e_i(z,t)$, can be obtained from its attenuation and propagation phase delay factor $\exp\left[-(\alpha/2)z + j\beta(\omega_0 + \Omega)z\right]$

$$e_i(z,t) = e^{j\omega_0 t}\frac{1}{4\pi}\int_{-\infty}^{\infty} A_i(\Omega)e^{-\{(\alpha_i/2)+j\beta_i(\omega_0+\Omega)\}z}e^{j\Omega t}d\Omega + \text{c.c.} \tag{5.3.3}$$

where α is the attenuation constant and $\beta_i(\omega_0+\Omega)$ the signal carrier phase parameter at around carrier frequency ω_0. The phase, $\beta_i(\omega_0+\Omega)$, linearly shifts in the fiber linear transmission region, and we can expand $\beta_i(\omega_0+\Omega)$ b near the carrier frequency into a Taylor series

$$\beta_i(\omega = \omega_0 + \Omega) = \beta_i(\omega_0) + \left.\frac{\partial\beta_i}{\partial\omega}\right|_{\omega_0}\Omega + \frac{1}{2}\left.\frac{\partial^2\beta_i}{\partial\omega^2}\right|_{\omega_0}\Omega^2 + \cdots \tag{5.3.4}$$

and obtain

$$\begin{aligned} e_i(z,t) &= (1/2)\exp\left[j(\omega_0 t - \beta_{0i}z)\right] \\ &\times\left[\frac{1}{2\pi}\int_{-\infty}^{\infty} A_i(\Omega)\exp\left\{-(\alpha_i/2)+j\left(\Omega t - \frac{\Omega z}{v_{gi}} - \frac{1}{2}\frac{d}{d\omega}\left(\frac{1}{v_{gi}}\right)\Omega^2 z\right)\right\}d\Omega\right] \\ &+ \text{c.c.} \end{aligned} \tag{5.3.5}$$

as the approximation up to the second-order terms, where $\beta_{0,i} = \beta_i(\omega_0)$, $d\beta_i/d\omega|_{\omega_0} = 1/v_{gi}$ (group velocity). Here, we denote the linear transfer function of core #i ($i = 1, 2, \ldots, M$) branch transmission channel as

$$H_i^L(\Omega) = \exp\left\{-j\Omega\left(1/v_{gi} + b_i\Omega\right)z\right\} \tag{5.3.6}$$

where b_i is the group velocity dispersion (GVD) defined as $b = \left(d^2\beta/d\omega^2\right)/2$. The output electrical field of core #i is given by

$$\begin{aligned} e_i(z,t) &= \exp\left[-(\alpha_i/2) + j(\omega_0 t - \beta_{0i}z)\right] \\ &\frac{1}{4\pi}\int_{-\infty}^{\infty} A_i(\Omega)H_i^L(\Omega)\exp\{j(\Omega t)\}d\Omega + \text{c.c.} \end{aligned} \tag{5.3.7}$$

The M divided signals are combined after weighting by $w_i(t)$. A white Gaussian noise z_i(t) such as amplified spontaneous emission noise is added to the optical signal in each channel. Here, we define Fourier spectrum $Z_i(\Omega)$ and $W_i(\Omega)$ to be the Fourier transform of $z_i(t)$ and $w_i(t)$, respectively.

After combining the M split optical signals with weighting, we obtain

$$A^{\text{out}}(\Omega) = \sum_{i=1}^{M}\left\{W_i(\Omega)\left(H_i^L(\Omega)A_i(\Omega) + Z_i(\Omega)\right)\right\}$$

$$
\begin{aligned}
&= \sum_{i=1}^{M} \left\{ W_i(\Omega) H_i^{\mathrm{L}}(\Omega) A_i(\Omega) \right\} + \sum_{i=1}^{M} \{ W_i(\Omega) Z_i(\Omega) \} \\
&= \left(\mathbf{W}^{\mathrm{T}} \mathbf{H} \right) \mathbf{A} + \mathbf{W}^{\mathrm{T}} \mathbf{Z}
\end{aligned} \tag{5.3.8}
$$

where $\mathbf{W}^{\mathrm{T}} = [W_1(\Omega), W_2(\Omega), \ldots, W_M(\Omega)]$, $\mathbf{H}^{\mathrm{T}} = \left[H_1^{\mathrm{L}}(\Omega), H_2^{\mathrm{L}}(\Omega), \ldots, H_M^{\mathrm{L}}(\Omega)\right]$, and $\mathbf{A}^{\mathrm{T}} = [A_1(\Omega), A_2(\Omega), \ldots, A_M(\Omega)]$. The output signal power, except noise, becomes

$$
\begin{aligned}
P^{\mathrm{out}} &= E\left[\left(\mathbf{W}^{\mathrm{T}} \mathbf{H} \mathbf{A} \right) \left(\mathbf{W}^{\mathrm{T}} \mathbf{H} \mathbf{A} \right)^* \right] = \left(\mathbf{W}^{\mathrm{T}} \mathbf{H} \right) E\left[\mathbf{A} \mathbf{A}^* \right] \left(\mathbf{W}^{\mathrm{T}} \mathbf{H} \right)^* \\
&= \left(\mathbf{W}^{\mathrm{T}} \mathbf{H} \right) \left(\mathbf{W}^{\mathrm{T}} \mathbf{H} \right)^* E\left[\mathbf{A} \mathbf{A}^* \right] = \left(\mathbf{W}^{\mathrm{T}} \mathbf{H} \right) \left(\mathbf{W}^{\mathrm{T}} \mathbf{H} \right)^* P_{\mathrm{in}}
\end{aligned} \tag{5.3.9}
$$

where $E[]$ denotes ensemble average. The total noise power Z^{total} is written as follows, assuming linear transmission region as $z_k(t)$ terms are mutually independent:

$$
\begin{aligned}
Z^{\mathrm{total}} &= E\left[\left(\sum_{i=1}^{M} W_i Z_i \right) \left(\sum_{i=1}^{M} W_i Z_i \right)^* \right] = \left(\mathbf{W}^{\mathrm{T}} \mathbf{Z} \right) \left(\mathbf{W}^{\mathrm{T}} \mathbf{Z} \right)^* \\
&= \mathbf{W}^* \mathbf{W} N_i = \sum_{i=1}^{M} |W_i|^2 N_i
\end{aligned} \tag{5.3.10}
$$

where N_i indicates instantaneous noise power added in core #*i*. Taking the ratio of (5.3.9) to (5.3.10) yields SNR Γ as:

$$
\Gamma = \frac{P^{\mathrm{out}}}{Z^{\mathrm{total}}} = \frac{\left(\mathbf{W}^{\mathrm{T}} \mathbf{H} \right) \left(\mathbf{W}^{\mathrm{T}} \mathbf{H} \right)^* P_{\mathrm{in}}}{\mathbf{W}^* \mathbf{W} N_i} = \frac{\left| \sum_i^M W_i H_i^{\mathrm{L}} \right|^2}{\sum_i |W_i|^2 N_i} P_{\mathrm{in}} \tag{5.3.11}
$$

By applying the Schwarz inequality for complex-valued numbers, this SNR is maximized to

$$
\Gamma_{\max} = P_{\mathrm{in}} \sum_{i}^{M} \left| H_i^L \right|^2 / N_i \tag{5.3.12}
$$

if

$$
W_i = a H_i^{\mathrm{L}^*}(\Omega) / N_i \tag{5.3.13}
$$

is satisfied, where a is some arbitrary complex constant. This condition means that the optimum weight for each core branch is inversely proportional to the branch noise power and proportional to the complex conjugate of the fiber transfer function. In other words, to achieve optimum SNR, the channel with worst noise figure should be most lightly weighted. Considering the phase term, the GVD, $\exp(-jb_i \Omega^2 z)$, should be compensation and the propagation phase delay $\exp(-j\Omega z/v_{gi})$ should also be adjusted in some manner.

Nonlinear Phase-Shift Cancelation by WDM Phase-Conjugate Light Pair

We rewrite (5.3.1) for the kth channel (ch.) optical electrical field in WDM transmission through core #i by adding subscript k as:

$$e_{i,k}(0,t) = (1/2)A_{i,k}(0,t)\exp(j\omega_k t) + \text{c.c.} \tag{5.3.14}$$

where ω_k denotes the kth ch. optical carrier angular frequency among WDM signals. In WDM transmission, XPM is the major factor distorting the waveform via nonlinear phase shift. Based on (5.3.3), the electrical field that explicitly exhibits the perturbation of nonlinear phase shift can be written as

$$e_i(z,t) = \frac{1}{2}e^{j(\omega_k t - \beta_{0i}z)} \cdot \left\{ \frac{1}{2\pi} \int_{-\infty}^{\infty} A(\Omega)e^{-(\alpha_i/2)z} e^{j\{\Omega t - (\Omega(1/v_{gi,k} + b_{i,k}\Omega) + \beta_{i,k}^{\text{NL}}(\Omega))z\}} d\Omega \right\} + \text{c.c} \tag{5.3.15}$$

and the nonlinear phase parameter, $\beta_{i,k}^{\text{NL}}(\Omega)$, is defined as:

$$\beta_{i,k}^{\text{NL}}(\Omega) \equiv 2\gamma_i \sum_{j=1, j\neq k}^{J} \left|A_{i,j}(\Omega)\right|^2 \tag{5.3.16}$$

where γ_i is the nonlinearity coefficient of core #i. Here, we introduce the transfer function approximation for nonlinear phase shift term due to XPM as

$$H_{i,k}^{\text{NL}}(\Omega) = \exp\left\{-j\beta_{i,k}^{\text{NL}}(\Omega)z\right\} \tag{5.3.17}$$

Since $\beta_{i,k}^{\text{NL}}(\Omega)z$ varies with time because of its optical power dependency, it represents the time-averaged nonlinear phase shift. Then, the kth ch. fiber transfer function for core #i becomes $H_{i,k}(\Omega)$ as the product of the linear transfer function $H_{i,k}^{\text{L}}(\Omega)$ and the nonlinear transfer function $H_{i,k}^{\text{NL}}(\Omega)$ as follows:

$$H_{i,k}(\Omega) = H_{i,k}^{\text{L}}(\Omega)H_{i,k}^{\text{NL}}(\Omega) \tag{5.3.18}$$

Then, $H_i^{\text{L}*}(\Omega)$ in (5.3.13) is replaced with

$$H_{i,k}^{*}(\Omega) = \exp\left\{j\Omega\left(1/v_{gi,k} + b_{i,k}\Omega\right)z\right\}\exp\left\{j\beta_{i,k}^{\text{NL}}(\Omega)z\right\} \tag{5.3.19}$$

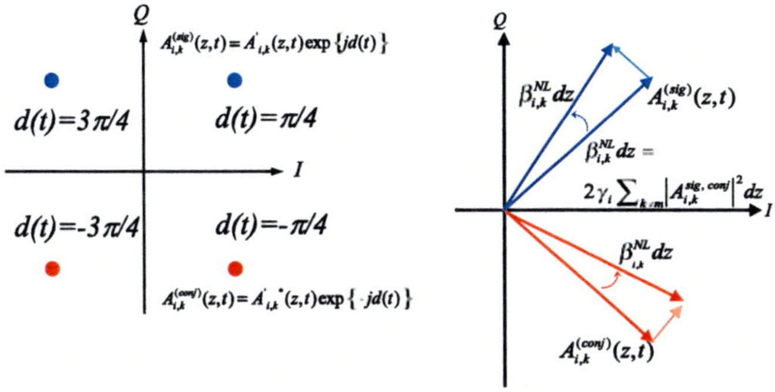

Fig. 5.40 Vector relationship for phase-conjugate signal pair in **a** and for nonlinear phase-shifted phase-conjugate signal pair in **b.** $A_k^{(\mathrm{sig})}(z, t)$ and $A_k^{(\mathrm{conj})}(z, t)$ symbol points for $d = \pi/4$ and $3\pi/4$ are plotted for QPSK modulation (Reprinted with permission from [120]. © 2016 The Optical Society)

Generally speaking, the compensation of $\beta_{i,k}^{\mathrm{NL}}(\Omega)z$ is not so easy, and there are several approaches to mitigate nonlinear signal interaction. The most intensively discussed mitigation technique up to now is digital backward propagation with nonlinear perturbation pre-distortion. However, it incurs excessive digital processing overheads. The other way is the backward propagation by mid-span phase reversal which uses a phase-conjugate light. The other approach, described below, is optical diversity transmission with maximum ratio combining of the WDM signal and its conjugate lights pair.

When we transmit quadrature phase-shift keying (QPSK) and/or M_{L}-ary quadrature amplitude modulation (QAM) signals, the phase modulation component can, in complex amplitude, be expressed by, after explicitly denoting data phase $d(t)$,

$$A_{i,k}^{(\mathrm{sig})}(z, t) \equiv A_{i,k}(z, t) = A_{i,k}^{'}(z, t) \exp\{\mathrm{j}d(t)\} \tag{5.3.20}$$

where $d(t)$ takes the value of $m(\pi/4)$ where $m = 1, 3, 5$, or 7. The corresponding phase-conjugate signal is expressed as

$$A_{i,k}^{(\mathrm{conj})}(z, t) \equiv A_{i,k}^{*}(z, t) = A_{i,k}^{'*}(z, t) \exp\{-\mathrm{j}d(t)\} \tag{5.3.21}$$

Figure 5.40a shows a better understanding of the QPSK signal and its phase-conjugate equivalent vector relationship. At the transmitter, the two red symbols correspond to the phase-conjugate of the original symbols, shown in blue. The conjugate pair has reverse sign along the I-axis.

When propagating MCF, field evolution is governed by the nonlinear Schrödinger equation. Each PCP experiences NLPS in the same direction by $2\gamma_i \sum_{k\neq j} \left|A_{i,k}^{\mathrm{sig,conj}}\right|^2 \mathrm{d}z$ due to the Kerr effect, while the signal envelope broadens due to GVD, as shown in Fig. 5.40b. Note that the GVD signs are reversed relative to

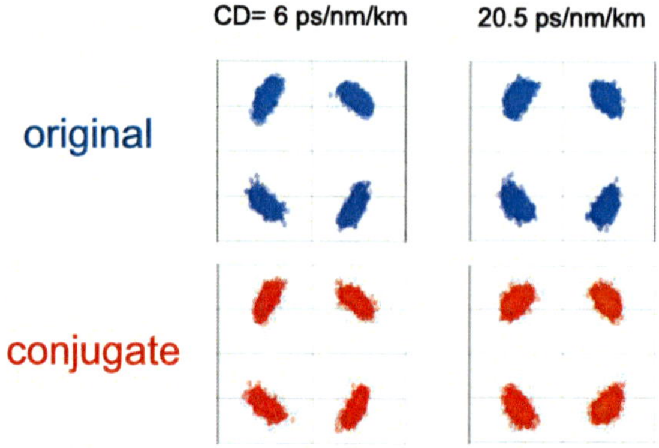

Fig. 5.41 Simulated constellation maps for phase-conjugate signal pair. Those for chromatic dispersion (CD) values of 6 and 20.5 ps/nm/km are simulated (Reprinted with permission from [120]. © 2016 The Optical Society)

each other because of spectral inversion. Since GVD causes power exchange between the XPM and AM components, phase noise is converted into amplitude noise, the quantity of which increases with CD. Figure 5.41 shows the simulated constellation maps for signal and conjugate lights under the condition of 5-ch WDM transmission over 16 spans, each of 60 km, with CD values of 6 and 20.5 ps/nm/km. They indicate that the symbol distribution spreads in the direction of the phase rotation, while the amplitude noise suggests a chromatic dispersion of 20.5 ps/nm/km, because we can see that the radial thickness of the symbol increases when CD is 20.5.

The PCP is combined after weighting. The combined Fourier component is given by

$$A_{i,k}^{\mathrm{out}}(\Omega) = W_{i,k}(\Omega) H_{i,k}^{\mathrm{L}}(\Omega) \left\{ \tilde{A}_{i,k}(\Omega) H_{i,k}^{\mathrm{NL}}(\Omega) + \left(\tilde{A}_{i,k}^{*}(\Omega) H_{i,k}^{\mathrm{NL}}(\Omega) \right)^{*} \right\} \quad (5.3.22)$$

where the Fourier transform of $A_{i,k}^{(\mathrm{sig,\ conj})}(z,t)$ is $A_{i,k}^{(\mathrm{sig})}(\Omega) \equiv \tilde{A}_{i,k}(\Omega)$ and $A_{i,k}^{(\mathrm{conj})}(\Omega) \equiv \tilde{A}_{i,k}^{*}(\Omega)$ with the relationships of:

$$A_{i,k}^{(\mathrm{sig})}(z,t) = \frac{1}{2\pi} \int \tilde{A}_{i,k}(\Omega) \mathrm{e}^{\mathrm{j}\Omega t} \mathrm{d}\Omega \quad (5.3.23)$$

and

$$A_{i,k}^{(\mathrm{conj})}(z,t) = \frac{1}{2\pi} \int \tilde{A}_{i,k}^{*}(\Omega) \mathrm{e}^{\mathrm{j}(-\Omega t)} \mathrm{d}\Omega \quad (5.3.24)$$

When we focus on the nonlinear term in the braces of (5.3.22), substituting (5.3.17) into it yields

$$\tilde{A}_{i,k}(\Omega) H_{i,k}^{\mathrm{NL}}(\Omega) + \left(\tilde{A}_{i,k}^{*}(\Omega) H_{i,k}^{\mathrm{NL}}(\Omega) \right)^{*} = 2\tilde{A}_{i,k}(\Omega) \cos\left\{ \beta_{i,k}^{\mathrm{NL}}(\Omega) z \right\} \quad (5.3.25)$$

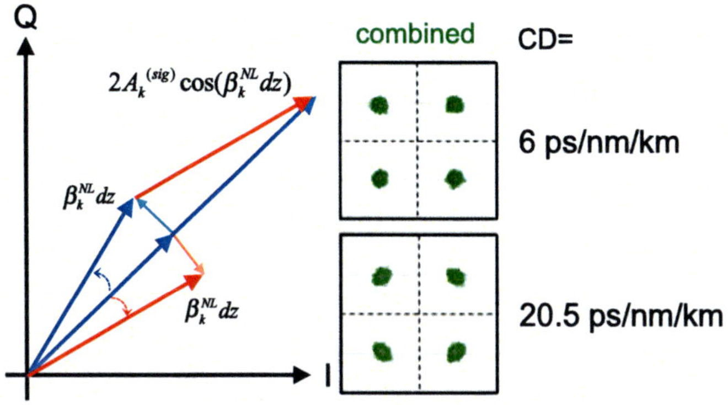

Fig. 5.42 Vector relationship for nonlinear phase shift cancelation and simulated combined constellation maps (Reprinted with permission from [120]. © 2016 The Optical Society)

We can automatically obtain the cosine component of double the signal amplitude by the simple PCP combining process. Figure 5.42 shows the vector relationship and simulated combined constellation maps to help understanding of the cancelation implied in (5.3.25). The constellation maps are inset for CD values of 6 and 20.5 ps/nm/km. We can see the symbol distributions narrow, compared to the original and phase-conjugate symbols depicted in Fig. 5.42. The smaller CD symbol distributions are narrower than those for larger distributions. This cancelation is imperfect; however, decreasing GVD is more effective if the interference noise induced by four-wave mixing (FWM) can be ignored.

With this approach, the degree of approximation rises if nonlinear length *dz* shortens. As a consequence, span-by-span GVD compensation (S-by-S) would be expected to offer greater cancelation effect than lumped compensation (Lump) by a DSP at the receiver.

WDM Signal Phase-Conjugate Pair Diversity Transmission Experiment

The cancelation of nonlinear phase shift was experimentally confirmed, and the *Q*-value enhancement achieved by MRC was evaluated in the 2-core diversity transmission of mutually phase-conjugated lights through an MCF. The experimental setup is shown in Fig. 5.43. Five optical carriers, with frequency spacing of 50 GHz anchored at 193.25 THz, emitted from 5 tunable laser sources, were modulated by 25-Gbit/s QPSK code with a $2^{15}-1$ pseudo-random binary sequence (PRBS) created by an arbitrary waveform generator (AWGen). The tunable laser spectral line widths were ~60 kHz. They were wavelength-multiplexed, split into two, and input into two IQ modulators. The I- and Q-arms of one were driven by positive logic, while Q-arm of the other was driven by negative logic in order to generate the PCP. The modulated signal traversed a 20-km SMF to randomize the bit sequences among the 5 channels. At the output facet, the period difference of 1.5 symbols was set between

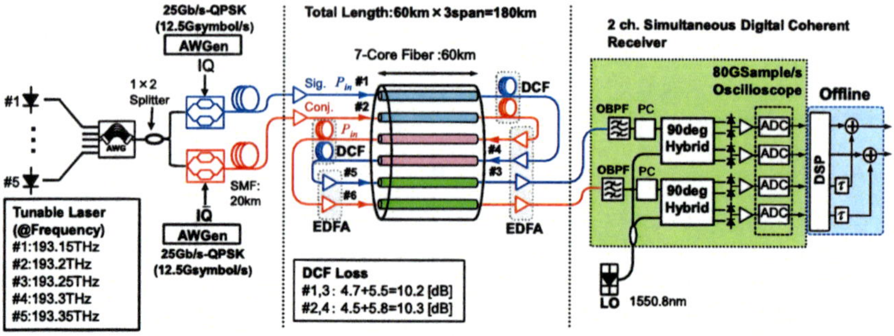

Fig. 5.43 Experimental setup for WDM phase-conjugate pair diversity transmission. DCF was removed in case of lumped compensation. EDFA, erbium-doped fiber amplifier; AWGen, arbitrary waveform generator; OBPF, optical band-pass filter; LO, local oscillator; PC, polarization controller; ADC, analog-to-digital converter

adjacent channels. The original and phase-conjugated signals were launched into #1 and #2 cores of a 60-km 7-core fiber after being amplified by two erbium-doped fiber amplifiers (EDFAs). The launched power to one core is P_{in}. The signal outputs from cores #1 and #2 were optically amplified by two EDFAs and launched into #3 and #4; this was repeated for #5 and #6 to create a 3-span transmission system. In S-by-S, dispersion-compensation fibers (DCFs) were inserted in each span. Their insertion losses are shown in the delineated box in Fig. 5.43. Two of them showed a loss of about 15 dB, so Q-value was degraded. Polarization controllers adjusted the polarization state of the original and phase-conjugated signals before the receiver. The 7-core fiber used had a CD of 20.5 ps/nm/km and an effective core area of 109.5 μm^2.

Demodulation was post-processed offline. The PCP at the frequency of 193.25 THz was coherently detected and evaluated. They were compensated (weighted) for CD and optical frequency offset. The Q-values were estimated from symbol distributions on the constellation maps.

The Q-value versus fiber launched power P^{total} is shown in Fig. 5.44 for QPSK coded signals when compensated by S-by-S and Lump. P^{total} corresponds to $P^{\text{in}} = P^{\text{total}}/(M = 1)$ for single-core transmission and $2 \times P^{\text{in}}$ for 2-core diversity transmission, following the definition in Fig. 5.39. Blue indicates the Q-value for single-core transmission and green that for the MRC of PCP signals. Solid circles plot the results for Lump and hollow circles for S-by-S. The blue circles peaked at 23.7 and 22 dB at $P^{\text{total}} = -2$ dBm for Lump and S-by-S, respectively, yielding a 1.7-dB difference since DCF loss was large in this experiment. The combined Q-values (green) peaked at the same value, 26 dB, at 1 and 4 dBm for Lump and S-by-S, resulting in Lump and S-by-S enhancements of 2.3 and 4 dB, respectively. This confirms that S-by-S is superior to Lump at NLPS cancelation as is expected. The 4 dB enhancement means that the transmission distance can be more than doubled, so this approach raises the distance-capacity product.

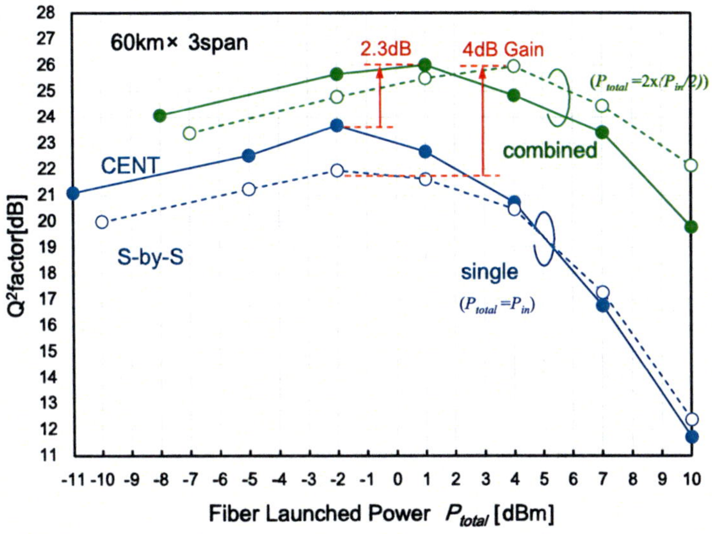

Fig. 5.44 Q-value versus fiber launched total power P^{total}. $P^{\text{total}} = P^{\text{in}}$ for single-core transmission and $P^{\text{total}} = 2 \times P^{\text{in}}$ for diversity

Figure 5.45 shows the measured constellation maps. The data after MRC also confirm the tighter distribution than single transmission map. The tighter distribution verifies the cancelation of nonlinear phase shift both S-by-S and Lump. Lump compensation yields a larger radial distribution that corresponds to the amplitude noise component. The MRC process cannot cancel this component.

The XPM cancelation for smaller GVD was also confirmed by simulations. Figure 5.46 shows the Q-value (blue and green) and its enhancement (black) versus P^{total} for CD = 6 and 20.5 ps/nm/km. The Q-value for CD = 6 ps/nm/km in single-core transmission peaked at 18.4 dB at −5 dBm, while that for CD = 20.5 was 18 dB at −3 dBm. The MRC Q-value for CD = 6 (solid green) peaked at 24 dB at $P^{\text{total}} = 0$ dBm, 5.6 dB enhancement from the Q-value of 18.4 dB for single. The enhancement strengthens with nonlinearity and reaches 14 dB when CD = 6 ps/nm/km. We can see that smaller CD is more effective than larger CD.

The optical diversity transmission with PCP and MRC has demonstrated Q-value enhancement of better than 4 dB, which yielded superior bitrate-distance product, compared to twice the single-core transmission in MCF. And this scheme can improve optical power density limit performance.

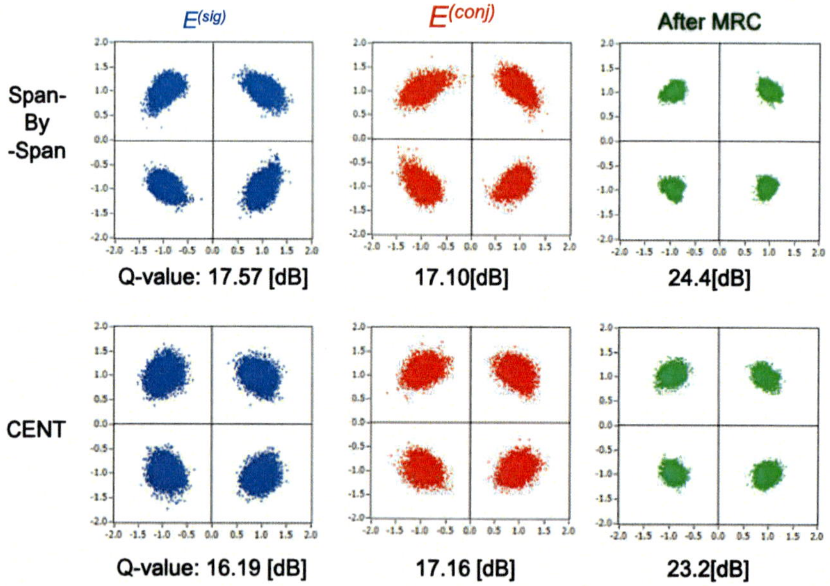

Fig. 5.45 Measured constellation maps for original and its phase-conjugate signal pair. Fiber input power P_{total} was set at 7 dBm (Reprinted with permission from [120]. © 2016 The Optical Society)

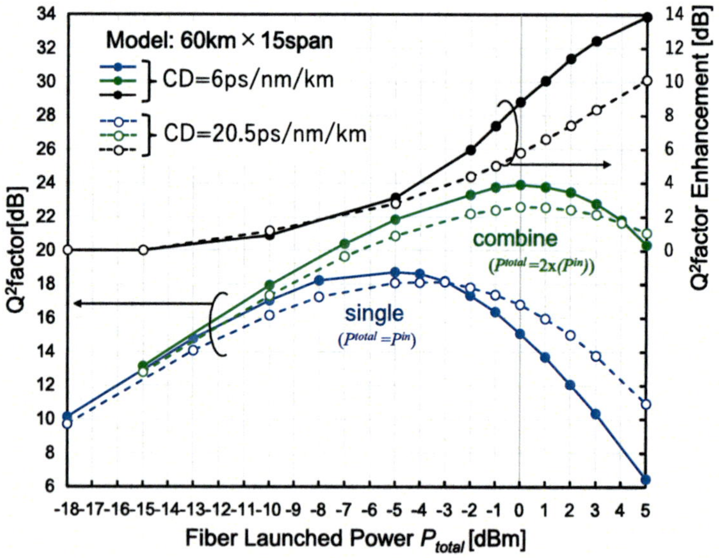

Fig. 5.46 Measured constellation maps for original and its phase-conjugate signal pair. Fiber input power P_{total} was set at 7 dBm (Reprinted with permission from [120]. © 2016 The Optical Society)

5.4.2 *Multi-mode Transmission*

5.4.2.1 Background

Multi-mode transmission based on mode division multiplexing (MDM) had been studied and used in short-distance optical fiber communications. It is realized by multiplexing signals in multiple fiber-guided modes and using a few-mode fiber (FMF) or a multi-mode fiber (MMF) as the transmission media. Major research on space division multiplexing (SDM) using either multi-core or multi-mode in transmission for long-distance core networks started in 2009 [111]. In early multi-mode transmission studies, the reach was short, and transmission capacity and spectral efficiency were low. Owing to the advancement of digital coherent technology and SDM technologies from various fields including signal processing, optical fibers, and optical devices, it became possible to improve transmission performance in recent years.

In the following sections, latest technological advances are described in multi-mode transmission over FMF or MMF.

5.4.2.2 Multi-mode Transmission Schemes

MDM in early studies used an MMF as the transmission media. MDM in most recent studies uses a few-mode fiber (FMF), which is an MMF that limits the number of propagation modes to few lower-order modes. The first three lower-order modes are commonly used in multi-mode transmission studies [123–125], namely LP_{01}, LP_{11a}, and LP_{11b} modes. FMF transmission experiments using six modes, namely LP_{01}, LP_{11a}, LP_{11b}, LP_{21a}, LP_{21b}, and LP_{02} modes, were reported to increase spatial multiplicity [126, 127]. In transmission over a FMF, all propagation modes are fully used, and it is easier to control the modes than an MMF and achieve good transmission performance.

5.4.2.3 Multi-mode Fiber Transmission Technologies

In order to improve performance in multi-mode transmission, advances are required in fundamental technologies including mode multiplexer/demultiplexers (MUX/DEMUXs), SDM fibers, integrated devices, and digital signal processing (DSP) technologies.

Up until now, various types of mode MUX/DEMUX have been reported. They are based on the following four types [128]: mode MUX/DEMUX based on (1) LP mode converters and combiners, (2) index matching by asymmetric mode couplers, (3) simultaneous mode conversion by a phase plate or gratings, and (4) a spot-based mode coupler with arrayed cores. The important characteristics for these devices are low loss, easy handling, and good stability. FMF transmission experiments mostly used mode MUX/DEMUX based on type (1) with phase plates and mode combiners [123, 124] since it is easy to compose the device simply using free-space optics. However, the insertion loss is large among the four types of mode MUX/DEMUX. Moreover, it increases with the number of mode-multiplexed since theoretical loss arises each time it combines a mode. Silica planar light wave circuit (PLC)-type mode MUX/DEMUX based on type (2) with low insertion loss and high mode conversion efficiency was developed and employed in multi-mode transmission experiments [128]. Mode MUX/DEMUX based on type (3) with appropriate design and fabrication can also exhibit low insertion loss and high mode conversion efficiency. Photonic lantern mode MUX/DEMUX based on type (4) also offers low insertion loss and is used in experiments [126, 127]. This type of mode MUX/DEMUX is designed to generate super-modes and require signal processing to uncouple the modes at the receiver.

Modal properties that affect transmission performance include modal crosstalk, differential mode delay (DMD), and mode-dependent loss (MDL). Signals may not be separated and demultiplexed when the modal crosstalk is high. By using mode MUX/DEMUXs with good mode conversion efficiency and non-coupled few-mode optical fiber and components, the modes can be demultiplexed and recovered after transmission. In practice however, mode coupling arises in transmission lines, caused by optical axis misalignments at connection points and in optical components, and in a transmission fiber that increases with distance.

One promising solution for recovering data even with the presence of modal crosstalk is to use multiple-input and multiple-output (MIMO) signal processing technique. It is a commercially used technique used to separate mixed signals at the receiver and is originally from the field of wireless communications. Reducing signal processing complexity is essential to implement such MIMO signal processing function into DSPs used in current optical communications. The computational resources needed for MIMO are given as $2M \times 2\,M \times T$, where M is the number of multiplexed modes, 2 corresponds to polarization-division multiplexing (PDM), and T is the number of taps. The complexity increases with the square of the number of modes. To increase M to achieve higher capacity, suppressing T is essential. In MIMO time-domain equalization (TDE), T increases proportionally with the maximum DMD, as shown in Fig. 5.47. The number of taps used in recent MDM is typically around

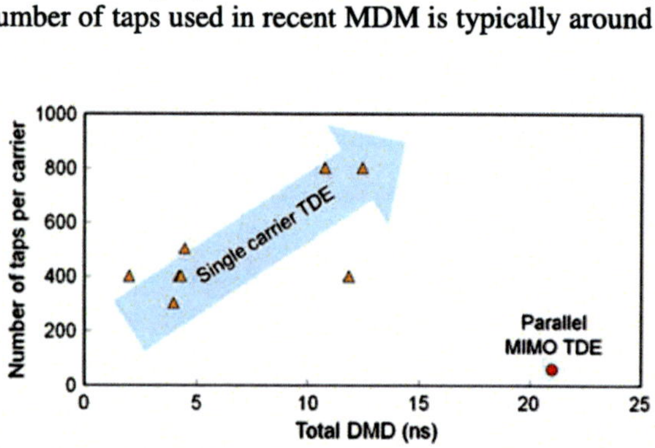

Fig. 5.47 Number of TDE taps required per carrier versus total DMD

50–100 taps/ns. Since the DMD of a step-index (SI)-type FMF is typically 3 ns/km, T will reach 100 in less than a kilometer. Efforts have been made to reduce this signal processing complexity. One can reduce DMD by optimizing the refractive index design. An FMF with multi-step-index refractive index profile can reduce the DMD to around 0.5 ns/km. Further, an FMF with graded-index (GI) profile offers the lowest DMD with <25 ps/km in the C band [129]. Another common approach is to combine spools of FMFs with opposite DMD signs [130] to reduce the maximum absolute DMD within the span.

A different approach to reduce signal process complexity is a novel DSP technique based on parallel MIMO processing with low-baud rate multi-carrier signals [131, 132]. Since each subcarrier is handled as an independent signal, MIMO can be performed for each subcarrier with small number of taps to reduce calculation size and ease convergence. A parallel MIMO TDE was proposed and demonstrated. With 20 subcarriers, the T required for 21-ns maximum DMD was only 61 taps per subcarrier [131]. Further, frequency-domain equalization (FDE) is known to offer lower complexity than TDE since the computational resources needed increase moderately with DMD. A parallel MIMO FDE with 10 subcarriers was used to process signals having 33-ns DMD with 128 taps per subcarrier [132].

Another important modal impairment is the differences of loss and gain in the transmission line depending on mode, that is, MDL and differential modal gain (DMG), respectively. Since DSP cannot fully compensate for MDL, it is important to suppress the absolute value of the MDL in the transmission line. The suggested specification of MDL in long-haul transmission is ±2 dB if we assume a penalty of 0.5 dB [133]. A low MDL of < ±0.002 dB/km is possible in the latest transmission fiber [134], and a multi-mode EDFA with DMG of <1.8 dB is reported [135]. While reducing MDL in transmission line is critical, the use of MDL equalization is effective in extending transmission distance [132, 133].

Transmission based on weak and strong mode coupling regimes has different characteristics. In the weak coupling scheme, modes are weakly coupled during transmission. Using mode MUX/DEMUXs with high conversion efficiency and minimizing mode coupling in the transmission line, MIMO can be simplified or even unnecessary in short reach transmission. In the strong coupling scheme [136], increase of modal impairments such as DMD and MDL is proportional to the square root of distance and is smaller compared to the case with weak coupling. Modes can be strongly coupled by mode scrambler, coupling in a fiber, tight bent in a fiber, and optical axis misalignment. Signal processing is essential to uncouple the signals at the receiver.

5.4.2.4 Multi-mode Transmission

Figure 5.48 shows a typical experimental setup for multi-mode transmission. The setup consists of transmitters, mode MUX, FMF, mode DEMUX, receivers, and MIMO DSP. The number of modes used in recent FMF transmission experiments is mostly three [123–125] or six [126, 127], and MDM using 10 [137] and 15 [138]

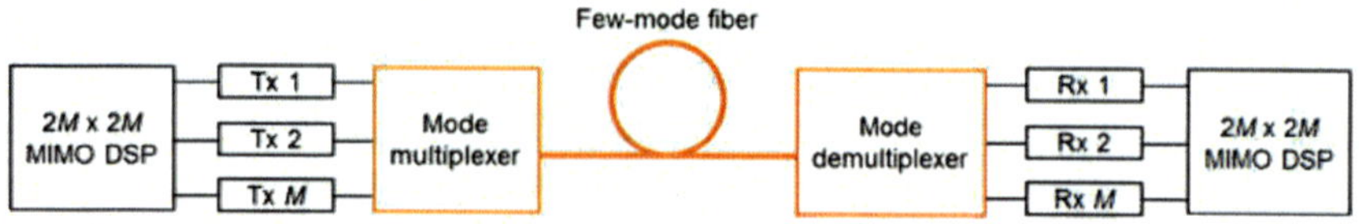

Fig. 5.48 Typical configuration of multi-mode transmission experiment

modes has been reported. Various multi-mode technologies and technique to mitigate modal impairments described above are employed in these experiments.

Transmission performances of some of the recent MDM transmission experiments are summarized in Table 5.1. Several transmission experiments using a multi-core few-mode fiber (MC-FMF) [112, 131, 132] are also shown for reference.

Figure 5.49 shows the spectral efficiency η_{spectral} versus distance of recent WDM-MDM transmission experiments from Table 5.1. The longest transmission distance is 1000 km with 16 wavelength channels and PDM-QPSK modulation using a 3-mode low-DMD GI-type FMF and a low DMG few-mode EDFA [124], and 1050 km with 60 wavelength channels and PDM-QPSK modulation using a 3-mode DMD compensated FMF and Raman amplification [125]. Using higher-order modulation format is essential to increase η_{spectral}. The maximum transmission distance over a FMF for PDM-16QAM in 2014 was 177 km with a η_{spectral} of 5.33 bit/s/Hz [126]. An experimental demonstration showed that a transmission distance using PDM-16QAM can increase to 316.2 km [133] by low MDL transmission line with MDL compensation. The maximum capacity in FMF transmission was 57.6 Tbit/s with a η_{spectral} of 4 bit/s/Hz using three LP modes over 117 km with PDM-16QAM [123]. It was raised to 115.2 Tbit/s with a η_{spectral} of 2.9 bit/s/Hz using ten LP modes over 87 km with PDM-QPSK [137]. Studies on MDM transmission are being conducted for further increasing transmission capacity and distance over a FMF.

In the latest studies, a novel cyclic mode permutation (CMP) technique and advanced DSP technique were proposed with the aim of extending the multi-mode transmission reach [139–141]. A 3-mode ultra-long-haul transmission over a transoceanic distance of 6300 km was demonstrated by employing the CMP scheme in combination with unreplicated successive interference cancelation (USIC) technique [139]. Further works on 3-mode long-distance transmission systems are reported including a 3-mode long-haul transmission of PDM-32QAM signals over 1000 km using the CMP and hybrid multi-core EDFA/Raman amplification schemes [140], and a 3-mode high-capacity long-haul transmission of 40.2 Tbit/s over 3060 km using the full C band [141]. The CMP scheme induced quasi-strongly coupling between multiple modes and allowed long-distance transmission even with DMD unmanaged FMF transmission lines [139–141].

Table 5.1 Performances of recent multi-mode transmission experiments

Number of modes	Fiber type	Mode MUX	MIMO signal processing	Number of wavelengths	Modulation format	$\eta_{spectral}$ (b/s/Hz)	Capacity (Tb/s)	Distance (km)
3	FMF	Phase plate	6 × 6 TDE 401 taps	96	PDM-16QAM	4	57.6	119
3	GI FMF	Phase plate	6 × 6 TDE 801 taps	146	PDM-QPSK	2.53	27.7	500
				16			3	1000
3	GI FMF	Photonic lantern	6 × 6 FDE 1000 taps	60	PDM-QPSK	3	18	1050
6	GI FMF	Photonic lantern	12 × 12 TDE 800 taps	32	PDM-16QAM	5.33	24.6	177
6	FMF	Photonic lantern	12 × 12	255	PDM-QPSK	1.36	34.7	74.17
10	GI FMF	Photonic lantern	20 × 20 FDE 1000 taps	120	PDM-QPSK	2.9	115.2	87
15	GI FMF	Photonic lantern	30 × 30 FDE 1800 taps	12	PDM-QPSK	2.9	17.2	22.8
3	MSI MC-FMF	Silica PLC	6 × 6 parallel TDE, 61 taps	20	PDM-32QAM	6.88	61.97	40.4
3	GI MC-FMF	Silica PLC	6 × 6 parallel FDE, 128 taps	20	PDM-QPSK	2.62	23.6	527
6	GI MC-FMF	Reflective phase plate	12 × 12 TDE 1000 taps	360	PDM-QPSK	4	2050	9.8

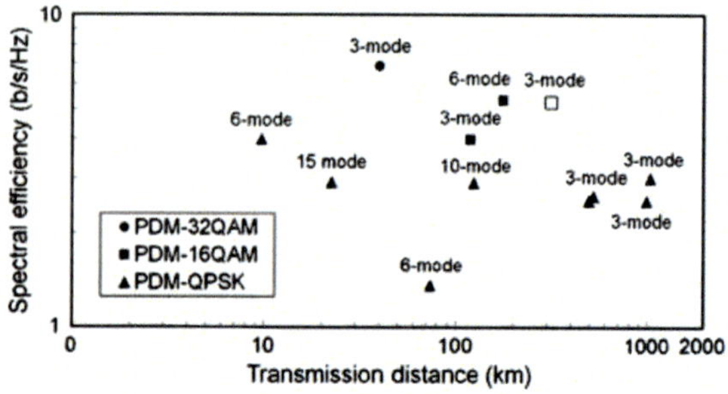

Fig. 5.49 Spectral efficiency versus transmission distance of recent multi-mode transmission experiments

5.4.3 Multi-core Multi-mode Transmission

5.4.3.1 Research on Multi-core Multi-mode Transmission

Research on space division multiplexed (SDM) transmission over a multi-core fiber or a multi-mode fiber began at around 2009 for increasing transmission capacity per fiber of long-distance core networks [111]. As described in earlier chapters, a capacity of Pbit/s, which is an order of magnitude higher than the single-mode fiber (SMF) capacity limit, was demonstrated with multi-core transmission technology. To further increase capacity to reach the ultimate capacity of a hundred times that of a conventional SMF, increasing spatial multiplicity was required. As the significant step toward ultrahigh capacity, dense space division multiplexing (DSDM) with a spatial multiplicity of over 30 per fiber was proposed and demonstrated combining the multi-core and multi-mode transmission [142].

Transmission over a multi-core few-mode fiber (MC-FMF) was first reported in 2014, with a spatial multiplicity of 36 (12 core × 3 mode) and an aggregate spectral efficiency ($\eta_{spectral}$) of 247.9 bit/s/Hz over 40.4 km [131]. A follow-up multi-core multi-mode transmission experiment was reported with a spatial multiplicity of 21 over a 1-km fiber [143]. More recently, a long-distance 12 core × 3 mode DSDM transmission over 527 km employing a low-DMD graded-index MC-FMF and low complexity parallel MIMO frequency-domain equalization [144], transmission of 114 spatial multiplicity (19 core × 6 mode) [145] with 2.05-Pbit/s capacity over a 9.8-km fiber [112], and 108 spatial multiplicity (36 core × 3 mode) using a 5.5-km fiber [146] was reported. Brief description of the three typical multi-core multi-mode transmission experiments will be described in Sect. 5.4.3.2.

Performances of early multi-core multi-mode experiments are summarized in Table 5.2, including spatial multiplicity, fiber type, modulation format, cladding diameter, spatial efficiency ($\eta_{spatial}$), spectral efficiency ($\eta_{spectral}$), aggregate $\eta_{spectral}$, capacity, and transmission distance.

Figure 5.50 shows the spatial multiplicity versus transmission distance of SDM transmission experiments. The multi-core multi-mode transmission scheme made it possible to increase the spatial multiplicity to more than 30 and realize the first DSDM transmission. DSDM transmission was also demonstrated with a multi-core

Table 5.2 Summary of early multi-core multi-mode experiments

Spatial multiplicity	Fiber type	Modulation format	DIA (μm)	$\eta_{spatial}$ (mm^2)	$\eta_{spectral}$ (b/s/Hz)	Aggregate $\eta_{spectral}$ (b/s/Hz)	Capacity (Tb/s)	Distance (km)
36 (12 core × 3 mode)	MSI, heterogeneous	PDM-32QAM	229	874.1	6.88	247.9	61.97	40.4
21 (7 core × 3 mode)	SI	PDM-32QAM	192	725.3	3.8	80	200	1
36 (12 core × 3 mode)	GI, heterogeneous	PDM-QPSK	230	866.5	2.82	94.3	23.6	527
114 (19 core × 6 mode)	GI	PDM-QPSK	318	1435.4	3.03 4	345 456	30.3 2050	9.8
108 (36 core × 3 mode)	SI, heterogeneous	PDM-QPSK	306	1468.6	–	–	–	5.5

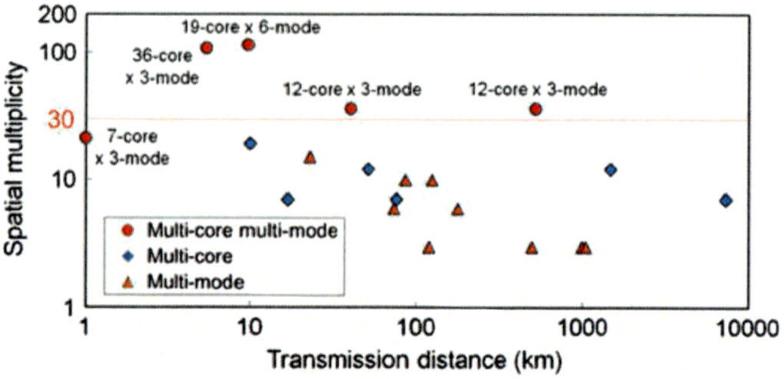

Fig. 5.50 Spatial multiplicity versus distance of recent SDM transmission experiments

fiber transmission line [147]. Furthermore, over a hundred spatial multiplicity and long-distance DSDM transmission were demonstrated.

Figure 5.51 shows the spectral and spatial efficiencies of recent multi-core multi-mode transmission experiments. The spatial efficiency (η_{spatial}) defined in [133] is equal to the spatial multiplicity divided by the fiber cross-sectional area. Among the various SDM transmission categories described in [128], the multi-core multi-mode transmission scheme offers the highest η_{spectral}. Increasing spatial efficiency η_{spatial} by DSDM and spectral efficiency η_{spectral} by dense wavelength division multiplexing (DWDM) will lead to higher transmission capacity, since the capacity is the product of η_{spectral}, η_{spatial}, bandwidth, and fiber area. Challenges remain for multi-core and multi-mode transmission studies to increase transmission capacity and distance for application to future long-distance optical transport systems.

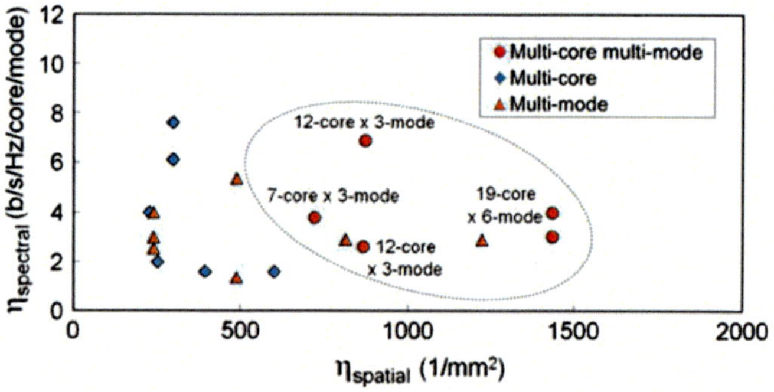

Fig. 5.51 Spectral and spatial efficiencies of recent SDM transmission experiments

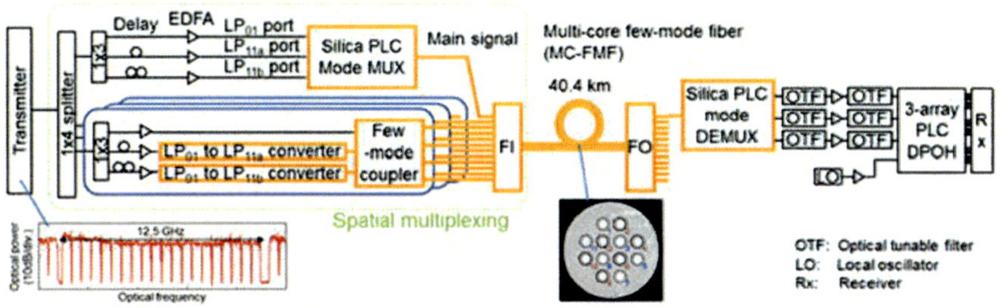

Fig. 5.52 Setup of the first 12-core × 3-mode DSDM transmission experiment

5.4.3.2 Multi-core Multi-mode Transmission Experiments

In this section, typical transmission experiments over a multi-core few-mode fiber are described.

12-Core × 3-Mode Transmission Experiment

In this section, the first DSDM transmission experiment [131] using a 12-core × 3-mode MC-FMF and the first long-distance 12-core × 3-mode DSDM transmission over 527 km with parallel multi-mode recirculating loop system [132] are introduced.

To realize the first multi-core multi-mode DSDM transmission, various technologies were developed including low-DMD low-crosstalk low-loss 12-core × 3-mode transmission fiber [134], physical contact-type fan-in/fan-out (FI/FO) devices, silica planar light wave circuit (PLC) mode multiplexer/demultiplexer (MUX/DEMUX), parallel MIMO time-domain equalization (TDE) digital signal processing (DSP) technique, and silica PLC dual-polarization optical hybrid (DPOH) for 6 × 6 MIMO DSP [131].

Figure 5.52 shows the experimental setup of the world first DSDM transmission experiment using a multi-core multi-mode fiber. At the transmitter, 20 wavelength channels were generated by laser sources and multiplexed into 12.5 GHz-spaced odd and even channels. PDM-32QAM low-baud rate 0.525-GHz × 20 subcarrier signals were digitally generated. The signals were then polarization-division multiplexed (PDM) and split by a 1 × 4 splitter. One of the output ports was used as the main signal. The signal was split into three, relatively delayed, amplified by single-mode EDFAs, and input into three input ports of the silica PLC mode MUX. Eleven additional three-mode-multiplexed signals were prepared by using mode MUX based on LP11 converters and 3 × 4 few-mode combiners. The 12 FMFs were input into the 12-core × 3-mode fiber through the FI device and transmitted over 40.4 km. After transmission, one of the ports of the FO device was selected, and mode demultiplexed by the silica PLC mode DEMUX. Three signals were wavelength filtered and received together by the 3-array silica PLC DPOH module.

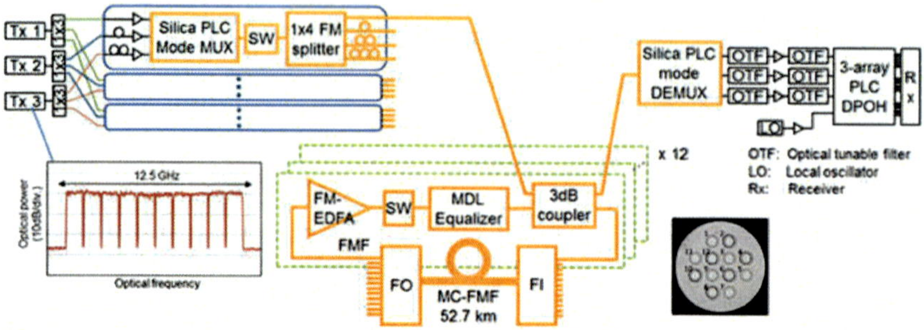

Fig. 5.53 Setup of the 12-core × 3-mode recirculating loop experiment

The acquired data were processed offline by 6 × 6 parallel MIMO TDE with 61 taps per subcarrier. The Q-factors for all 720 tributaries (36 DSDM × 20 DWDM channels) were above the forward error correction limit of 5.7 dB assuming 20% overhead, which confirmed the successful DSDM transmission for the first time using multi-core and multi-mode. The net data rate was 86 Gbit/s, the transmission capacity was 61 Tbit/s, the transmission distance was 40.4 km, the spectral efficiency was 6.88 bit/s/Hz/core/mode, and the aggregate spectral efficiency was 247.9 bit/s/Hz [131].

To realize long-distance multi-core multi-mode transmission, additional technologies were developed including low-DMD graded-index 12-core × 3-mode transmission fiber [148], parallel MIMO frequency-domain equalization (FDE) digital signal processing technique [144], differential-modal-gain-improved few-mode EDFA [135], and mode-dependent loss (MDL) equalizer [133] to reduce mode-dependent gain/loss even after several 100-km transmissions.

Figure 5.53 shows the experimental setup of the long-distance DSDM transmission experiment with multi-mode recirculating loops. At the transmitter, 20 wavelength channels were generated by laser sources and multiplexed into 12.5 GHz-spaced odd and even channels. Three independent sets of PDM-QPSK low-baud rate 1-GHz × 10 subcarrier signals were digitally generated. They were relatively delayed, amplified by single-mode EDFAs, and input into silica PLC mode MUXs as shown in the figure. The 3-mode-multiplexed signals were further split by few-mode splitters, delayed by FMFs, and input into 12 parallel multi-mode recirculating loops. The recirculating loop consisted of few-mode 3-dB coupler, few-mode EDFA, few-mode switch, MDL equalizer, and one of the cores of the 52.7-km spool of the 12-core × 3-mode fiber connected by FI/FO devices. After 10 loops, the signal under measurement was selected, and mode demultiplexed by the silica PLC mode DEMUX. Three signals were wavelength filtered and received together by the 3-array silica PLC DPOH module.

The acquired data were processed offline by 6 × 6 parallel MIMO FDE with 128 taps per subcarrier. The Q-factors for all 720 tributaries (36 DSDM × 20 DWDM channels) were above the forward error correction limit of 5.7 dB assuming 20% overhead, which confirmed the successful DSDM recirculating loop transmission

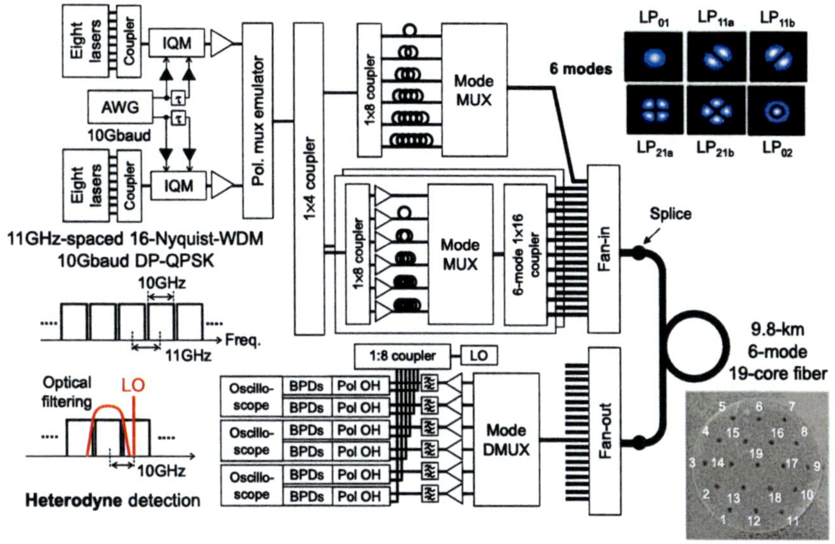

Fig. 5.54 Experimental setup (Reprinted with permission from [145]. © 2016 The Optical Society)

experiment for the first time using MC-FMF. The net data rate was 32.79 Gbit/s, the transmission capacity was 23.6 Tbit/s, the transmission distance was 527 km, the spectral efficiency was 2.62 bit/s/Hz/core/mode, and the aggregate spectral efficiency was 94.3 bit/s/Hz [132].

In the latest studies, a novel CMP scheme and intermodal interference cancelation techniques were proposed to increase transmission distance. A long-distance transmission of 2500 km over the 12-core × 3-mode fiber was demonstrated by employing the CMP and USIC techniques [139]. Further, the transmission distance was extended to over 3000 km by using the CMP and unreplicated parallel interference canceling (UPIC) techniques [149].

19-Core × 6-Mode Transmission Experiment

In this section, we introduce the SDM transmission experiment using a 9.8-km-long six-mode nineteen-core fiber (6M-19CF), resulting in the space division multiplicity of 114 [112, 145].

Figure 5.54 shows experimental setup. In the transmitter, CW lights generated from eight lasers were combined with the frequency spacing of 22 GHz, and then, they were independently modulated by optical IQ modulators (IQMs) for even and odd channels. The IQMs were driven by 10-Gbit/s electrical Nyquist-pulse-shaped binary signals, which are based on pseudo-random binary sequence (PRBS) with the length of $2^{15} - 1$. Combining the even and odd channels with polarization multiplexing, we obtained 16-WDM 10-Gbaud PDM-QPSK signals with the frequency spacing of 11 GHz. The WDM channels were ranged from 193.353 to 193.518 THz.

The WDM signals were split into measured signals and two sets of dummy signals. The measured signals were split into six paths with a relative delay of 50 ns between subsequent paths. The delayed signals were mode-multiplexed by a six-mode multiplexer based on multi-plane light conversion (MPLC) [150]. The sustained modes were four types of LP modes with the degenerated modes, namely LP_{01}, LP_{11a}, LP_{11b}, LP_{21a}, LP_{21b}, and LP_{02}. After that, the mode-multiplexed signal was launched into the measured core of a 9.8-km 6M-19CF through a fan-in (FI) device. The measured tributaries were switched to the measured core in each bit error rate (BER) measurement. For the remaining 18 cores of 6M-19CF, dummy SDM tributaries were generated by other two sets of a six-mode multiplexer with a six-mode 1:16 coupler, which was composed of a tree structure of five 1:4 couplers based on free-space beam splitters. The input powers of all SDM tributaries launched into 19 cores were adjusted to be 11 dBm/wavelength/mode.

The transmission line was composed of the spool of 9.8-km 6M-19CF. The core diameter, core pitch, and cladding diameter of 6M-19CF were 17 μm, 62 μm, and 318 μm, respectively. They were designed in order to suppress the core-to-core crosstalk of less than 30 dB/50 km for all cores. The 6M-19CF has the graded-index profile with $\alpha \sim 2$, and Δ was designed to be over 1.1%. According to the designed index profile, the difference of the effective refraction between LP_{01}/LP_{11} and LP_{21}/LP_{02} is maintained to be over 10^{-3}, whereas that between LP_{21} and LP_{02} is relatively close. As a result, the coupling between the mode groups (LP_{01}/LP_{11} and LP_{21}/LP_{02}) can be suppressed. The transmission loss for LP_{01} input was less than 0.5 dB/km for all cores.

For input and output of optical signals, we used the FI and fan-out (FO) devices fusion-spliced to the input and output edges of 6M-19CF. Based on free-space optics, 19 × six-mode-multiplexed signals were launched into 19 cores of 6M-19CF. The total loss of 6M-19CF including FI/FO devices was maintained to be less than 6 dB for all cores, when LP_{01} was input.

In the receiver, the measured core was selected, and then, the SDM tributaries were spatial-demultiplexed by the FO device and the six-mode demultiplexer. After six demultiplexed signals passed through optical band-pass filters, they were simultaneously detected by six coherent receivers based on the heterodyne reception. A free-running laser with linewidth of less than 100 kHz was used as a local oscillator (LO). The offset frequency between LO and the received signal was set to be 10 GHz.

The six received signals were digitized at 50 Gsample/s using three real-time oscilloscopes. The stored data were processed offline by digital signal processing (DSP) as follows. After frequency shift and rectangle-shaped Nyquist filtering, the set of stored samples for LP_{01}, LP_{11a}, and LP_{11b} and another set of samples for LP_{21a}, LP_{21b}, and LP_{02} were independently processed by 6 × 6 MIMO-based equalization. For three sets of the stored samples for LP_{01}, LP_{11a}, and LP_{11b}, the polarization and mode demultiplexing and signal equalization were performed by a half-symbol-spaced adaptive 6 × 6 MIMO equalizer with 500 taps. The tap coefficients were

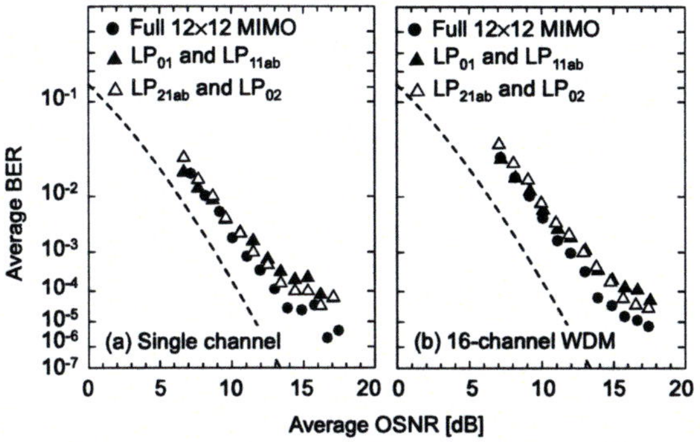

Fig. 5.55 Measured BERs as a function of OSNR in **a** single channel and **b** WDM cases. Dots: all modes in the case of full 12 × 12 MIMO, closed triangle: LP_{01} and LP_{11ab} in the case of 6 × 6 MIMO, open triangle: LP_{21ab} and LP_{02} in the 6 × 6 MIMO, and dashed line: calculated result determined from OSNR (Reprinted with permission from [145]. © 2016 The Optical Society)

adaptively controlled based on the decision-directed least mean square (DD-LMS) algorithm [151]. After the symbols were decoded, bit error was counted. The other three modes such as LP_{21a}, LP_{21b}, and LP_{02} were processed in the same manner.

We evaluated the BER performance of PDM-QPSK signals as a function of optical signal-to-noise ratio (OSNR). The results of the single-channel signal are shown in Fig. 5.55a. Dots indicate the average BER of six modes for the use of full 12 × 12 MIMO with 40 taps. For the use of the parallel 6 × 6 MIMO with 40 taps, closed triangles show the average BER of three modes of LP_{01}, LP_{11a}, and LP_{11b}, and open triangles indicate that of LP_{21a}, LP_{21b}, and LP_{02}. Dashed line indicates the calculated curve determined from OSNR. Compared with the results of full 12 × 12 MIMO, the penalty due to dividing into two sets of 6 × 6 MIMO is less than 1 dB. By dividing into partial MIMO, reduction of the calculation complexity is expected for demodulation of mode-multiplexed signals. Figure 5.55b indicates the measured BERs of the center channel of WDM channels. The penalty due to the use of the partial 6 × 6 MIMO does not increase from the single-channel results, although BER performance was slightly degraded due to residual ASE noises in the image band of the heterodyne detection.

We measured BERs of all SDM and WDM tributaries. The results are shown in Fig. 5.56. In the horizontal axis, the number of SDM tributary is defined by order of core number and mode. For instance, LP_{01}, LP_{11a}, LP_{11b}, LP_{21a}, LP_{21b}, and LP_{02} in the core 1 are SDM tributaries from 1 to 6, respectively. After transmission, OSNR was maintained to be over 25 dB. In this case, the expected BER of the 1 M demodulated bits for each SDM tributary would be less than 1×10^{-6}. For around the half of all SDM tributaries, the error-free operation corresponding to less than 1×10^{-6} was achieved in all WDM channels. Although some bit errors due to coupling between mode groups at splice points between 6M-19CF and FI/FO devices were measured for other tributaries, their BERs did not exceed the BER threshold of 2.7×10^{2} of the LDPC convolutional codes using layered decoding algorithm with 20% overhead

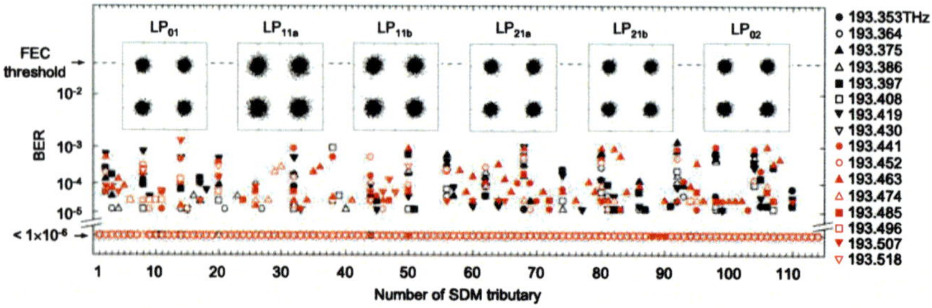

Fig. 5.56 Measured BERs of all SDM tributaries (Reprinted with permission from [145]. © 2016 The Optical Society)

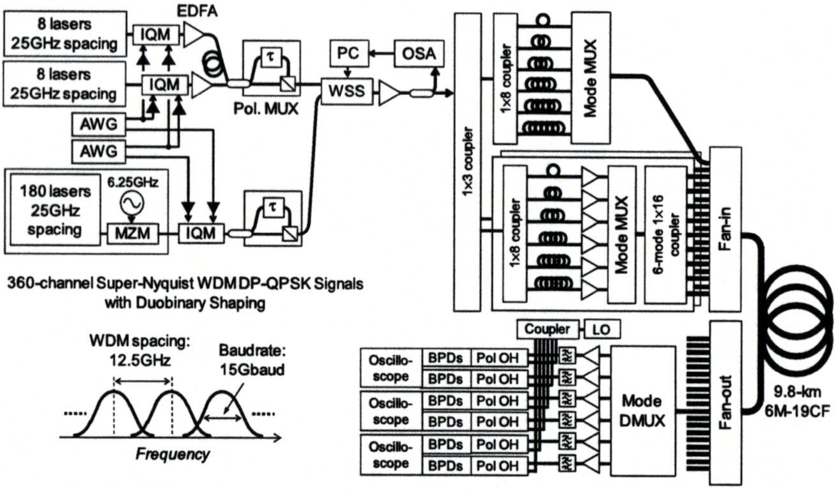

Fig. 5.57 Experimental setup for 2.05-Pbit/s transmission (Reprinted with permission from [112]. © 2015 IEEE)

[95]. The aggregate spectral efficiency was achieved to be over 345 bit/s/Hz (6 modes × 19 cores × 3.03 bit/s/Hz) even in the use of QPSK format.

As mentioned in the previous section, the spatial multiplicity of 114 has been achieved by using 6M-19CF. With the 6M-19CF, the 2-Pbit/s transmission was demonstrated. Figure 5.57 shows the experimental setup. In the transmitter, the even and odd measured channels with eight frequencies each were independently modulated by IQMs, and then, the even and odd channels were combined and polarization-multiplexed. The IQMs were driven by 15-Gbaud duobinary-shaped electrical signals, which were generated by two AWGs for I and Q components.

The third rail was used to load 360 WDM channels for maintaining not only an OSNR but also nonlinear effects. 180 lasers were combined with 25-GHz

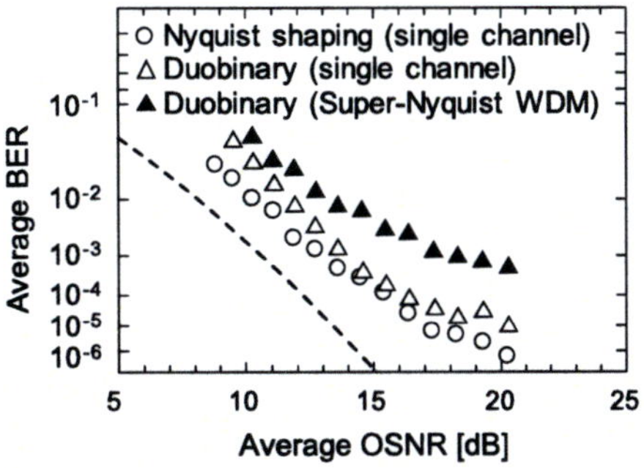

Fig. 5.58 BERs of six-mode-multiplexed super-Nyquist-WDM PDM-QPSK signals (Reprinted with permission from [112]. © 2015 IEEE)

frequency spacing. By modulating them at frequency of 6.25 GHz with the carrier-suppressed condition, we obtained 360 tones with 12.5-GHz frequency spacing, ranging from 191.69375 to 196.18125 THz. All the tones were signal-modulated and polarization-multiplexed in the same manner as the measured channels. After three rails were combined, we consequently obtained 360-channel super-Nyquist-WDM PDM-QPSK signals with a bitrate of 60 Gbit/s including 20% FEC overhead [95].

The WDM signals were split into three branches for measured and dummy SDM channels. The measured signals were divided into six branches with the relative delay of 50 ns between subsequent paths. The delayed signals were six-mode-multiplexed with a mode multiplexer in the same manner as the experiments mentioned in the previous section. After that, the mode-multiplexed signals were launched into the measured core of 6M-19CF. For the remaining 18 cores, dummy SDM channels were generated with two sets of the mode multiplexers and 1:16 six-mode couplers. The input power of the FI device was adjusted to be −24 dBm/wavelength/mode.

The transmission line was composed of 9.8-km 6M-19CF same as that of the previous 114 × SDM experiment. The transmitted SDM channel was demultiplexed with the FO device and the mode demultiplexer. After that, the six signals after the mode demultiplexer were detected by the six coherent receivers based on the heterodyne detection. In this experiment, the frequency offset between LO and the signal was set to be 9.5 GHz. The received electrical signals were digitized at 50 Gsample/s using three synchronized real-time oscilloscopes.

The stored samples were processed offline as follows: The samples were down-converted to base band. After the rectangular-shaped Nyquist shaping, the samples for all modes were simultaneously processed by a half-symbol-spaced 12 × 12 MIMO equalizer. The tap size was set to 1000, and the tap coefficients were updated based on a LMS algorithm. After MLSE was performed for I and Q components of the equalized samples, bit errors were counted.

First, we evaluated the performance of the six-mode-multiplexed PDM-QPSK signals in the back-to-back configuration without any multi-core components. Figure 5.58 shows the measured BERs averaged for the six spatial modes at the center WDM channel. The results of duobinary-shaped PDM-QPSK signals with and without adjacent channels are plotted by closed and open triangles, respectively.

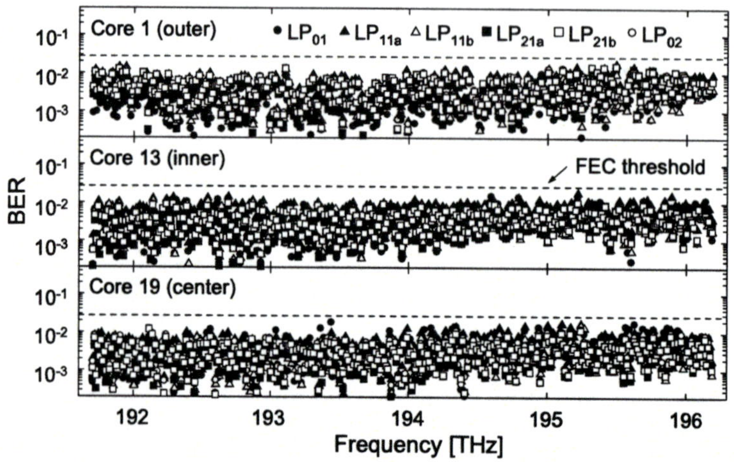

Fig. 5.59 BERs of SDM/WDM channels of outer, inner, and center cores (Reprinted with permission from [112]. © 2015 IEEE)

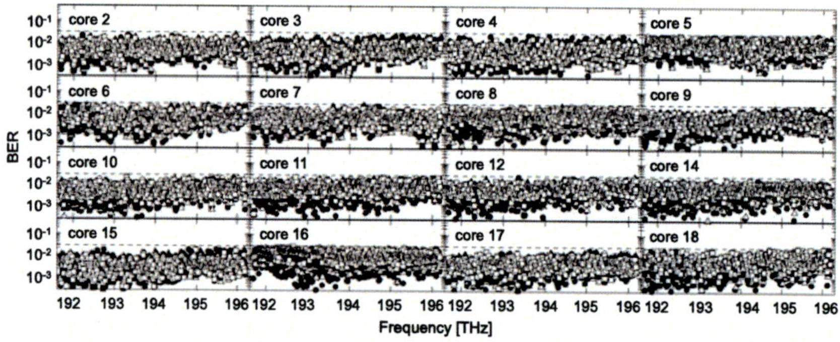

Fig. 5.60 BERs of SDM/WDM channels (Reprinted with permission from [112]. © 2015 IEEE)

For comparison, those of the single-channel Nyquist-shaped signals are plotted by open circles. Compared with the single-channel Nyquist-shaped signal, the OSNR penalty in the super-Nyquist-WDM case was suppressed to be less than 3 dB at BER of 1×10^{-2} even with 12.5-GHz channel spacing for the 15-Gbaud signal.

Finally, we measured BERs of all 41,040 WDM/SDM channels. Figure 5.59 shows the BERs of six-mode-multiplexed WDM signals in the center core (core number 19), the inner core (core number 13), and the outer core (core number 1), respectively. Figure 5.60 shows the results of the remaining 16 cores. The worst value in the measured BERs was 2.3×10^{-2}, which did not exceed the BER threshold of 2.7×10^{-2} of the low-density parity check convolutional codes with 20% overhead. In this experiment, we achieved a transmission capacity of 2.05 Pbit/s (360 WDM $\times$ 114 SDM $\times$ 50 Gbit/s) in the full C band.

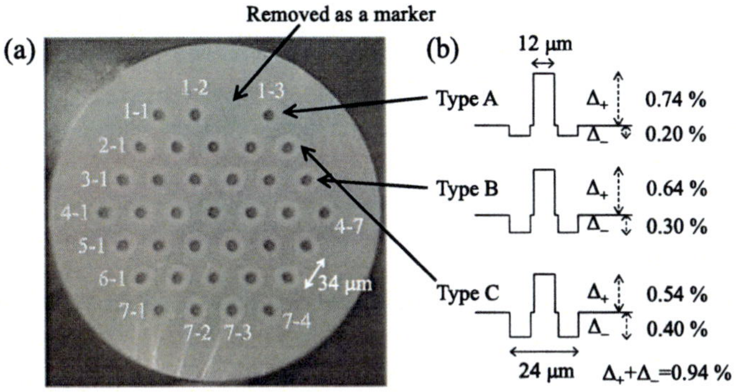

Fig. 5.61 Structure of the 36-core 3-mode fiber. **a** Facet view and core identification numbers. **b** Refractive index profiles of heterogeneous cores (Reprinted with permission from [146]. © 2015 The Optical Society)

After this experiment, the fiber capacity was further increased to 10.16 Pbit/s [152], where the signal bandwidth was expanded from only C band to C + L band and the higher-order modulation formats, 16QAM or 64QAM, and the rate-adaptive FEC with three types of overheads were introduced. In order to maximize the attainable fiber capacity, the optimum modulation format and FEC overhead were selected according to the OSNR of the received signals.

36-Core × 3-Mode Transmission Experiment

Fibers with over 100 spatial channels have been realized with two approaches. One is to increase the propagation modes per core, as demonstrated in [112, 145], and the other is to increase the number of uncoupled cores. In order to minimize the core pitches while suppressing the inter-core crosstalk, heterogeneous core structures as described in former sections have been employed. For realizing a 36-core fiber with 3 modes per core, three different types of cores were used in a hexagonal arrangement [146].

The facet view of the 36-core few-mode multi-core fiber (FM-MCF) is shown in Fig. 5.61a. The 36 cores were hexagonally arranged with a 34-μm core pitch and within a 306-μm-diameter cladding area. Three types of refractive index profiles shown in Fig. 5.61b were employed. Relative refractive indices of difference of cores (Δ_+) ranged from 0.74 to 0.54%, while diameters of cores (12 μm) and depressed claddings (24 μm) were common to all. Also, the relative refractive index differences of the depressed claddings (Δ_-) were designed so that $\Delta_+ + \Delta_- = 0.94\%$, which resulted in similar effective area among all core types. The LP_{21} cutoff wavelengths were primarily subject to Δ_+ and also core positions. The center core with the type A profile was expected to have the longest cutoff wavelength, which was reported to be 1526 nm. The effective index difference between core types should be >0.001

for all combinations of LP_{01} and LP_{11} modes. With bending radius larger than a few centimeters, and assuming the correlation length of 5 cm, the inter-core crosstalk was calculated to be at most −38 dB/km at 1550 nm. The reported transmission properties of the fiber, either measured or calculated, are summarized in Table 5.3. Core identification (ID) numbers were defined as shown in Fig. 5.61a. Also in Fig. 5.62, the measured and calculated differential mode delay (DMD) values are shown. Fluctuation of DMD values was attributed to fabrication error of each core, and still, DMD values were approximately predictable from measured refractive index profile of the fabricated preform.

Feasibility of 108 spatial channel transmission with this FM-MCF was investigated using a coherent WDM transmission setup and spatial/mode multiplexers (MUXs) as shown in Fig. 5.63. 100 GHz-spaced 40 wavelength channel (1532.6–1563.8 nm) 25-GBaud dual-polarization quaternary phase-shift keying (PDM-QPSK) signals were split into three test channels and 105 dummy channels, which were connected with 108 input ports of the 36-core 3-mode MUX based on free-space optics. The three test channels were launched to each mode of the core under test with appropriate time decorrelations as shown in the setup. The 36-core 3-mode MUX, shown in the upper inset of Fig. 5.63, consisted of a common 3-mode MUX configuration shared by all the core channels. The losses for LP_{01} inputs were less than 7 dB, while those for $LP_{11a/b}$ inputs were around 7.5–9.4 dB. A major issue of such free-space spatial/mode MUXs was that long pedestals of the LP_{11} beam profiles after mode conversion in the phase plates couple to un-intended cores and result in inter-core crosstalk. Although the phase plate angles were adjusted with a view to mitigate the inter-core crosstalk [153], significantly large crosstalk was observed for LP_{11} inputs.

The signals of test and dummy channels were transmitted over 5.5 km of the 36-core 3-mode fiber, and then, the signals of the test channels were separated from dummy signals by a free-space spatial demultiplexer, shown in lower inset of Fig. 5.63. Each mode of the test channels was further separated by a standard free-space 3-mode DEMUX and received with three coherent receivers after wavelength demultiplexing. A 6×6 multiple-input and multiple-output (MIMO) equalization was employed offline to resolve the mode mixing at the input/output of the FM-MCF, presumably resulted from the multiplexers and FM-MCF splicing.

The inverse impulse response for each core consisted of two peaks temporally separated by the amount of accumulated DMD as shown in Fig. 5.64a. Although equalization of such widespread response with full-length taps was challenging, a sparse MIMO equalizer [154] with two sets of 201 taps, which covers gray-shaded area of Fig. 5.64a, could sufficiently resolve the mode mixing for all the cores. The bit error rates (BERs) for each core, averaged over three modes, at nine representative wavelength channels (test WDM channels) are shown in Fig. 5.64b. Core-to-core fluctuation of BER was presumably due to variation in the transmission losses and inter-core crosstalk. The BER for the worst core (6–4) was still below the threshold level (3.5×10^{-3}) of present 7% overhead forward error correction (FEC).

Table 5.3 Optical properties of the 36-core 3-mode fiber at 1550 nm

Measured properties	Mode	Type A			Type B			Type C		
	Core	4-1	4-4	5-2	4-2	5-3	6-1	4-3	5-1	6-2
Transmission loss (dB/km)	$LP_{01} + LP_{11}$	0.242	0.266	0.255	0.283	0.292	0.281	0.262	0.308	0.284
Cable cutoff wavelength (nm)	LP_{21}	1458	1526	1478	1351	1357	1346	1199	1235	1245
Calculation from measured refractive index profiles		Mode			Type A (4-1)		Type B (4-2)		Type C (4-3)	
Effective area (μm^2)		LP_{01}/LP_{11}			77/105		74/102		76/110	
					A–B		B–C		C–A	
Core-to-core crosstalk (dB/km)		LP_{01}-LP_{01}/LP_{01}-LP_{11}			−105/−68		−93/−39		−95/−55	
		LP_{11}-LP_{01}/LP_{11}-LP_{11}			−84/−61		−67/−38		−40/−39	

Parentheses indicate core ID numbers. Reprinted with permission from [146], © 2015 Authors

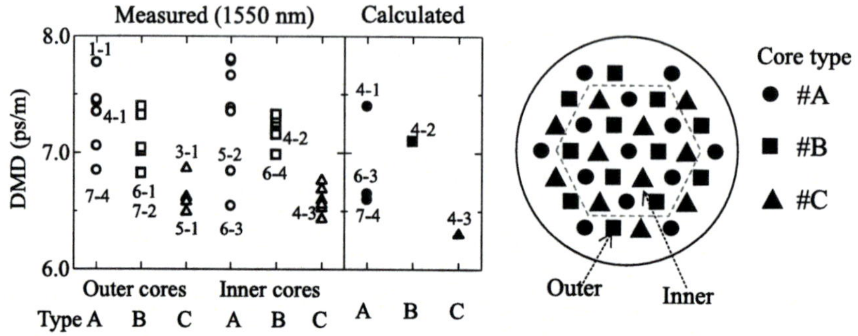

Fig. 5.62 DMD of the 36-core 3-mode fiber at around 1550 nm (Reprinted with permission from [146]. © 2015 The Optical Society)

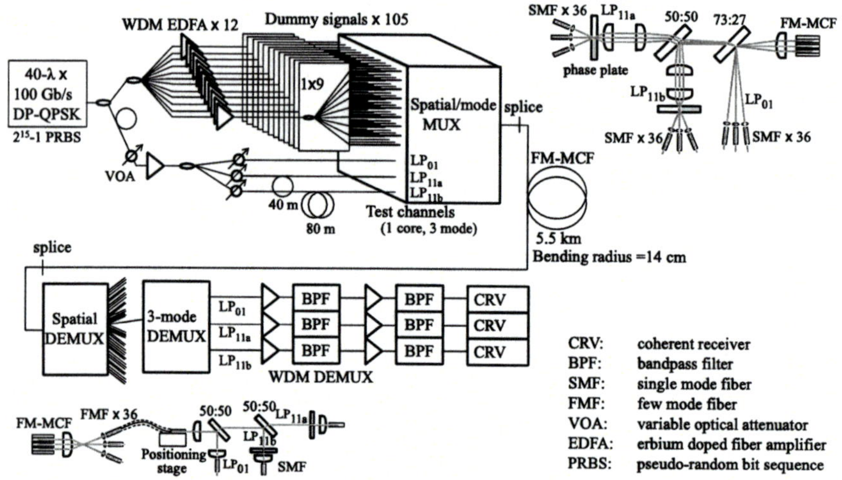

Fig. 5.63 Setup of the 108 spatial channel transmission. Upper and lower insets show configurations of spatial/mode multiplexing and demultiplexing using free-space optics (Reprinted with permission from [146]. © 2015 The Optical Society)

5.5 Signal Processing Technologies

5.5.1 *State-of-the-Art Optical Signal Processing Technologies*

Demands on the speed for the signal processing are rapidly increasing as the data rates in the optical fiber communications are growing. All-optical signal processing using optical nonlinearities meets these demands and has been exploited for last several decades. However, speed of the electrical signal processing is steadily increasing, and LSI chips for 100-Gbit/s signals are available nowadays. Nonetheless, optical

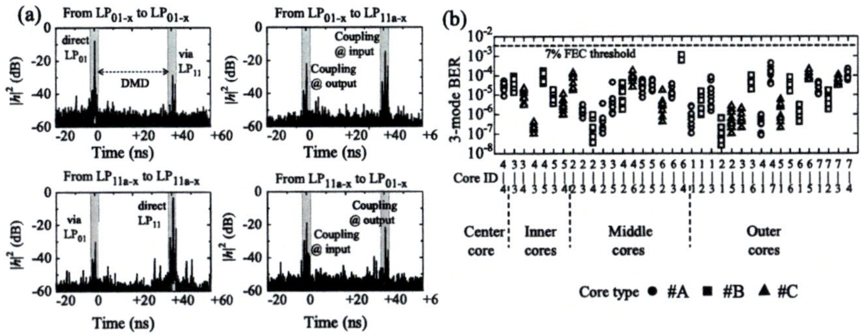

Fig. 5.64 Transmission results of 36-core 3-mode fiber. **a** Examples of inverse impulse responses for core 4–1 at around 1550 nm. **b** BER results for each core and test WDM channels (Reprinted with permission from [146]. © 2015 The Optical Society)

signal processing offers many roles and functions for ultrafast, Ebit/s systems, in conjugation with high-speed electronics [155, 156]. These "optically assisted" signal processing can reduce the power consumption of the network in comparison with all-electronic network. It is also important to develop ultrafast pulse sources as the signal source for data and the pump source for the optical signal processing.

It is the advance in nonlinear materials and devices that makes the practical utilization of optical signal processing in the future. Silica glass fibers or semiconductor waveguides have been used as a fast nonlinear media for all-optical signal processing. Nonlinearities in glass fibers are classified into the nonlinear refractive index change (nonlinear Kerr effect) and the nonlinear scatterings [157, 158]. Nonlinear response in the glass is instantaneous (~1 fs), which is good for ultrafast signal processing; however, the fiber length tends to be long (a few meter to kilometer) due to small nonlinear coefficient in the glass. Semiconductor optical amplifiers (SOAs) have large nonlinearities such as nonlinear gain or refractive index changes, whereas their response is slower than glass, typically a few tens of ps. Silicon waveguides, which are promising for the platform for photonic integrated circuits (PICs), can also be a platform for nonlinear signal processing, but the problem is the additional loss induced by the two-photon absorption (TPA). Thus, there are yet rooms for further development of nonlinear materials and devices for all-optical processing, and new nonlinear materials, such as nonlinear polymers, carbon nanotubes, and graphene.

Mode-locked lasers are used to generate ultrafast pulses whose duration is less than a few ps. Especially, mode-locked fiber lasers can generate high-quality, high-intensity clean (chirp-free or linearly chirped) pulses. They are classified into actively mode-locked fiber lasers using an intracavity modulator and passively mode-locked fiber lasers with an intracavity fast saturable absorbers (SAs). In actively mode-locked fiber lasers, pulse repetition rate can be raised to the integer multiple of the cavity free spectral range (FSR), sometimes a few 10 GHz, by using so-called harmonic mode locking. However, the width of the pulses from actively mode-locked lasers is normally limited to a few ps. For nonlinear signal processing, the pulse width should be as short as possible, preferably less than 1 ps, in order to enhance the peak intensity. In passively mode-locked fiber lasers, the pulse width can be much shorter, and a few tens of fs are feasible in fiber lasers, whereas the pulse repetition rate is not so high, typically a few tens of MHz, since the mode locking is only stable in the fundamental mode-locking state.

In this section, we review the ultrafast signal generation from mode-locked fiber lasers and the optical signal processing in optical fibers and waveguides. We also present novel nonlinear materials for fiber/waveguide-type nonlinear signal processor.

5.5.1.1 Ultrafast Signal Generation

There are two types of mode-locked fiber lasers, actively mode-locked fiber lasers and passively mode-locked fiber lasers [158]. In actively mode-locked fiber lasers (Fig. 5.65a), the cavity loss is modulated by an intracavity modulator with the modulation frequency equal to the roundtrip time $T = nL/c$ (L: cavity length) of the cavity. Then, a single short pulse is formed and circulates in the cavity at the timing so that the pulse can go through the modulator, that is, when the modulator is "on." Thus, a pulse train is obtained through an output coupler at the repetition rate $1/T = c/nL$, which is equal to the free spectral range (FSR) of the cavity. In the frequency domain, this is equivalent to locking the relative phases between adjacent longitudinal modes whose spacing is equal to FSR; therefore, this is called the active mode locking. In contrast, in passively mode-locked fiber lasers, saturable absorbers (SAs) are used instead of the modulators in the active one, as shown in Fig. 5.65b. Saturable absorption is the phenomena where high-intensity light "bleaches" the material and reduces the absorption. In a linear regime, where the incident optical intensity is weak, the SA absorbs the light, resulting in attenuation. When the incident optical intensity is high, saturation of absorption occurs, resulting in decreased attenuation. When a SA is placed in the laser cavity, it will favor pulse generation and suppress CW lasing since the pulse has higher peak intensity. In a sense, the SA is a self-intensity

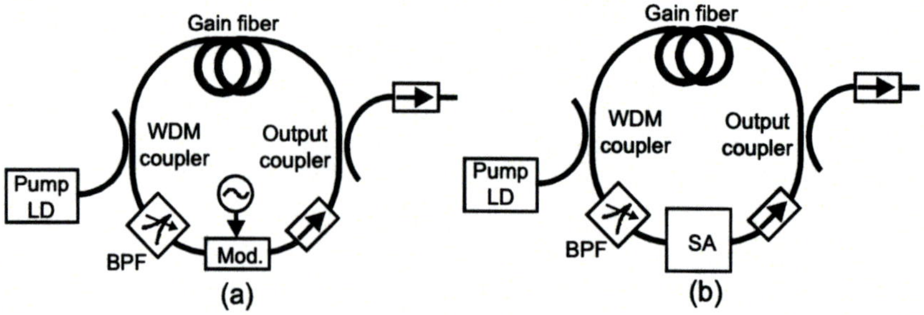

Fig. 5.65 Mode-locked fiber lasers. **a** Active mode locking using a modulator. **b** Passive mode locking using a fast saturable absorber (SA)

modulator. This is called passive mode locking. For short pulse generation, SA has to have the fast response time (or recovery time).

In the mode-locked lasers, a single short pulse circulates in the cavity, and the repetition rate is equal to the FSR of the cavity. This is called fundamental mode locking. Besides, it is also possible to have the two or more pulses circulate in the cavity, which is called the harmonic mode locking. In the harmonic mode locking, the pulse repetition rate is raised to be N times of the FSR ($= Nc/nL$), where N is the number of the pulses in the cavity. It is difficult to realize stable harmonic mode locking in the passively mode-locked lasers, but easy in the actively passively mode-locked lasers simply applying higher modulation frequency ($= N \cdot \text{FSR}$). An issue of the harmonically mode-locked lasers is that they tend to be unstable as the order of harmonics (N) increases. There exist N sets of combinations of longitudinal modes contributing to the harmonic mode locking, and they compete to each other. This is called the super-mode noise. In the fiber lasers, it has been demonstrated to suppress the super-mode noise by inserting the nonlinear media, such as highly nonlinear fibers. Actively and harmonically mode-locked fiber lasers have been demonstrated to be able to generate transform-limited and stable pulse trains with the pulse width of a few ps and the repetition rate as high as 40 GHz [159].

Actively and harmonically mode-locked fiber lasers have been applied to the RZ or soliton signal sources for optical fiber communications. They can also be seeds for the ultra-wideband supercontinuum (SC) generation in highly nonlinear fibers. SC processing can offer even shorter pulses. Pulse generation as short as 170 fs has been reported so far [160], and a transmission experiment using it as a RZ signal source.

For the passive mode locking, fast SAs are crucial. SAs for the passively mode-locked fiber lasers are classified into three types. The first one is the semiconductor-based reflection-type SA, ordinary called semiconductor saturable absorbing mirror (SESAM) [161]. It is a semiconductor multiple quantum well (MQW) Bragg reflector device utilizing the SA property of semiconductor material. It requires a complex and costly fabrication process, and also needs extra optics to couple the light into the optical fiber. The second one is the fiber-based SA, such as nonlinear polarization rotation (NPR), nonlinear optical loop mirror (NOLM), nonlinear amplifying loop mirror (NALM) [158]. NPR is sometimes termed as nonlinear polarization evolution (NPE). They are not a natural but artificial SA, in which nonlinear phase shift by self-phase modulation (SPM) and interference of two modes (two polarization modes in the case of NPR and two opposite direction modes in the case of NOLM and NALM) makes the loss to be a periodic function of input light intensity. It is a very fast SA, and modulation depth can be high. Also, it is very much suitable for fiber lasers since they are all-fiber devices composed of a length of nonlinear fiber and an inline polarizer (NPR), or a fiber coupler (NOLM and NALM), which can be made in an ordinary optics laboratory. However, they are polarization-dependent, and the device length tends to be long, typically longer than a few meters, which sometimes causes instability due to polarization fluctuation. Also, lack of slower (> a few ps) recovery from saturation that exists in natural SAs can make the self-starting of mode-locked operation difficult in many cases. Since it is difficult to apply harmonic mode locking in the passively mode-locked fiber lasers, the high-repetition rate exceeding GHz can only be achieved by shortening the cavity length < a few cm. A GHz-range passively mode-locked glass laser has been reported using highly Er-doped glass and SESAM. It is also necessary to feedback-control the cavity length to match the repetition rate with the external timing, e.g., data rate [162].

As the third type of SA, nano-carbon-based SAs, such as carbon nanotube (CNT)-based SA (CNT-SA) and graphene-based SA (G-SA), have attracted considerable attentions. CNT-SA uses the saturation in resonant absorption between energy levels that appears in semiconducting CNT. The recovery time from the saturation state is inherently short, ~1 ps; therefore it is good for laser mode locking. Absorption wavelength ($\propto$1/energy gap) is controllable by choosing CNT diameter, and typical CNT has absorption in the wavelengths from 1 to 2 μm, which is good for fiber lasers. Graphene has the wavelength-independent absorption as a result of linear dispersion relations of graphene (Dirac cone). G-SA uses the saturation of the wavelength-independent absorption, whose recovery time is even faster than CNT-SA, ~100 ps. They both can be easily incorporated in fiber devices, by simply sandwiching them in between fiber connectors or using evanescent field coupling in side-polished or tapered fibers. These devices offer several key advantages such as: ultrafast plus slow recovery from saturation, small size, low background loss, polarization independence, ability to operate both in transmission, reflection and bidirectional modes, and compatibility to optical fibers. Table 5.4 summarizes a comparison of the three SAs.

Historically, CNTs were first demonstrated to have very fast saturable absorption and be applicable to short pulse generation in fiber lasers in 2004 [163]. Graphene was found to have similar saturable absorption properties to the CNT as recently as 2009, when it was first applied to the mode locking of fiber lasers [164]. Since then, many types of passively mode-locked fiber lasers at various wavelength regions (1 μm, 1.5 μm, 2 μm, etc.) and at various pulse formation mechanisms (soliton, stretched pulse soliton, dissipative soliton, etc.) have been reported by many research institutes [165,

Table 5.4 Comparison of saturable absorbers

	Semiconductor-based devices	Fiber/waveguide	Nano-carbon (CNT, graphene)
Response time	5 ~ 20 ps	~10 fs	1 ~ 0.1 ps
Optical damage threshold	Low	High	Medium
Nonlinear coefficient (W^{-1} km^{-1})	γ_{eff} ~ 10^8	γ ~ 1–10^3	γ_{theory} ~ 10^8
Fabrication	Complex	Simple	Simple
Fiber compatibility	X	o	o
Beam transmission	X	X	o
Integration	o	Δ	o

166]. Here, we introduce our work on a high-repetition-rate mode-locked fiber laser using the CNT-SA, taking advantage of its small size and fiber compatibility.

Typical mode-locked fiber lasers work in the fundamental mode at the repetition rate of a few MHz to a few tens of MHz. This is determined by the cavity length (a few tens to a few hundreds of meters). High-repetition-rate mode-locked lasers at the repetition rate of a few tens of GHz are useful as a signal source for optical communications and nonlinear signal processing. To realize the fundamental mode locking in fiber ring lasers, for example, at 10 GHz, the fiber ring laser cavity has to be reduced to ~2 cm, which is not practical. Fiber Fabry–Perot (FP) laser makes such high-repetition rate possible in conjugation with a highly doped gain fiber and a small-size CNT- or G-SA. We reported a 2-cm-long mode-locked linear fiber laser at 5 GHz repetition rate by using a high-gain Er:Yb-doped fiber together with high finesse fiber cavity and CNT-SA as early as 2005 [167]. It was further developed recently to generate up to 20 GHz repetition rate pulse with 5-mm-long cavity composed of higher-gain Er:Yb-doped fiber [168]. Figure 5.66 shows the optical spectrum and autocorrelation trace of the 10-GHz mode-locked pulse with the 1-cm-long FP cavity. CNT thin film is formed onto the fiber ferrule highly reflective (HR) mirror ($R \sim 99\%$) by the spray method and butt-coupled to the gain fiber in a ferrule to be a sandwiched CNT-SA mirror, as shown in the inset of Fig. 5.66a. The gain fiber is pumped through another fiber ferrule HR mirror. We could obtain nearly transform-limited, very stable mode-locked pulse whose pulse width is ~1 ps. We could generate the SC spanning from 1.4 to >1.7 μm by amplifying and putting the pulse into the highly nonlinear dispersion flattened fiber. We could also realize the 10 GHz mode-locked 1-cm-long FP laser using G-SA [169], but the performances, such as mode-locking threshold

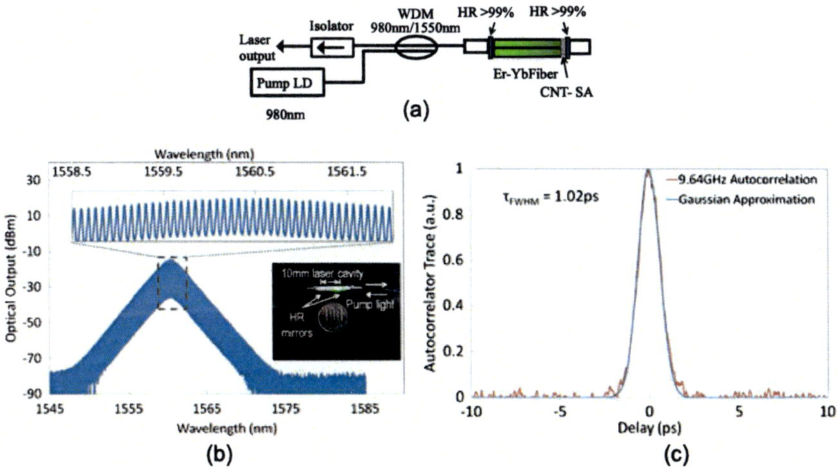

Fig. 5.66 10 GHz, 1-cm-long fiber laser mode-locked by CNT. **a** Laser configuration, **b** lasing spectra, **c** autocorrelation trace. Reprinted with permission from [168]. © 2011 The Optical Society (Reprinted with permission from [166]. 2014 Elsevier)

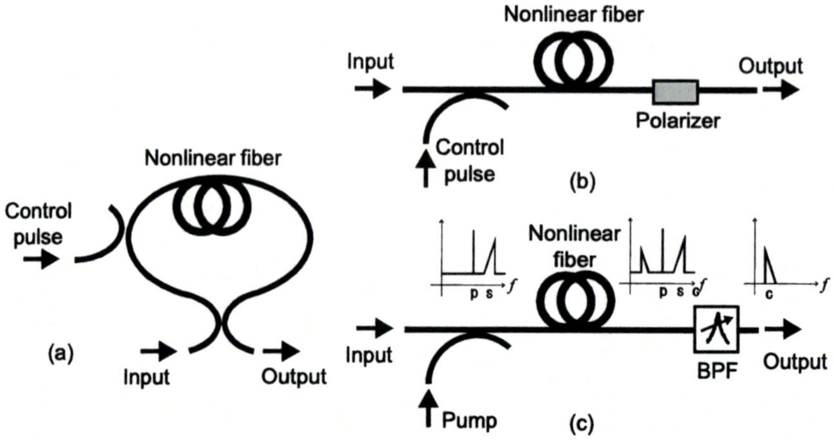

Fig. 5.67 Nonlinear fiber devices. **a** Switching/demultiplexing using fiber nonlinear loop mirror (NOLM). **b** Switching using fiber Kerr shutter. **c** Wavelength conversion and phase conjugation using fiber four-wave mixing (FWM)

and the stability, were not as good as the one with CNT-SA, possibly because of the higher saturation intensity of the G-SA at the wavelength of 1.55 μm.

5.5.1.2 Ultrafast Signal Processing

Nonlinear effects in optical fibers are classified into the nonlinear refractive index change (nonlinear Kerr effect) and the nonlinear scattering. The nonlinear refractive index change is the effect in which the effective refractive index of the fiber is changed according to the light intensity, due to the nonlinear polarization in the glass. The effective refractive index n is given by

$$n = n_0 + n_2 I, \tag{5.4.1}$$

where I is the intensity of the light, n_0 is the linear refractive index, and n_2 is called the nonlinear refractive index [157]. The nonlinear refractive index change causes several phenomena in optical fibers, such as the self-phase modulation (SPM), the cross-phase modulation (XPM), and the four-wave mixing (FWM). When the pump light becomes more intense in the FWM process, the signal light is amplified along with the wavelength-converted idler light, which is called parametric amplification (PA). The phase matching condition has to be satisfied in the FWM process; thus, the pump wavelength has to be set near-zero dispersion wavelength in case of the long optical fibers. The nonlinear scattering phenomena are the stimulated Brillouin scattering (SBS) due to the acoustic phonons, and the stimulated Raman scattering (SRS) due to the optical phonons. It is different from the linear scattering, the Rayleigh scattering,

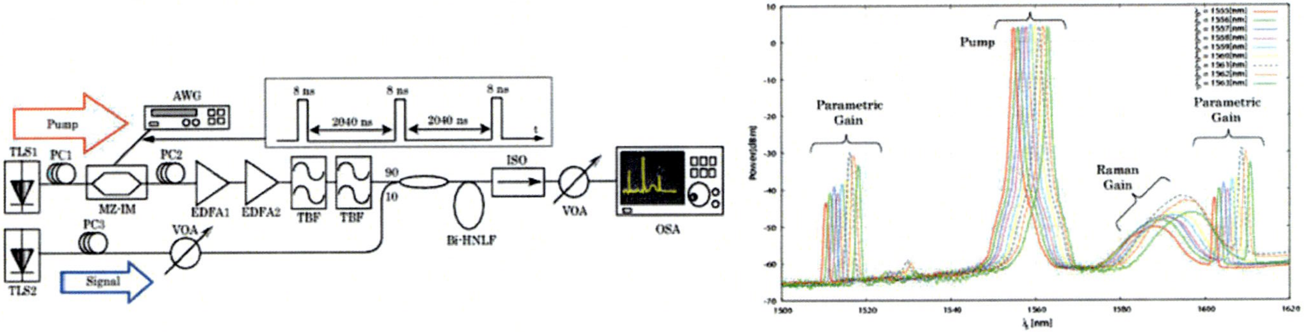

Fig. 5.68 Narrowband parametric amplification (PA) in bismuth-based highly nonlinear fiber (HNLF) (Reprinted with permission from [176]. © 2008 The Optical Society)

in that the scattered lights are shifted in wavelength (~0.1 nm in SBS and ~100 nm in SRS), The SBS limits the incident CW power into the optical fibers.

The nonlinear refractive index n_2 depends on the material. The nonlinearity in optical fibers is also dependent on the field confinement in the fiber, which is defined as the nonlinear coefficient,

$$\gamma = \frac{2\pi}{\lambda} \cdot \frac{n_2}{A_{\text{eff}}}, \tag{5.4.2}$$

where A_{eff} is the effective core area of the fibers. In the standard single-mode fibers (SMFs) or the dispersion shifter fibers (DSFs), they have $\gamma = 1\text{–}2\ \text{W}^{-1}\text{km}^{-1}$. There are several kinds of highly nonlinear fibers (HNLFs), in which n_2 is enhanced and/or A_{eff} is reduced. Conventional fiber-based one using silica glass is called highly nonlinear DSF (HNL-DSF), whose γ is enhanced to be ~20 $\text{W}^{-1}\text{km}^{-1}$ by reducing A_{eff} [170]. Photonic crystal fiber (PCF)-based one can reduce the A_{eff} further and can have higher γ [171], although the loss is higher than the HNL-DSF. The highest $\gamma = 1000\ \text{W}^{-1}\text{km}^{-1}$ has been reported in the fibers using highly nonlinear glass materials, such as chalcogenide or bismuth glasses [172, 173]. Such nonlinear glasses are used to make nonlinear waveguides. However, such nonlinear glasses have high material dispersion, which has to be mitigated for FWM or PA generation by using short length or by canceling with structural dispersion. Table 5.5 summarizes a comparison of various HNLFs.

Figure 5.67 shows the typical constructions of nonlinear fiber devices. For ultrafast signal processing, the nonlinear refractive index change is mainly utilized. Figure 5.67a shows the nonlinear optical loop mirror (NOLM) that is used for ultrafast switching. It is composed of a 50:50 fiber coupler and a length of the nonlinear fiber connected to the same side of the coupler. It is also called as the fiber loop mirror or the fiber Sagnac interferometer. Additional fiber coupler is used to input the control pulses. Without the control pulses, the input pulses, after split into the clockwise and anticlockwise directions and traveling in the fiber, are totally reflected back to the input direction as a result of interference between clockwise and anticlockwise pulses. When the control pulses are input to the clockwise direction, a

Table 5.5 Comparison of nonlinearity, dispersion, and loss in fibers at the wavelength of 1.55 μm

	Nonlinear coefficient γ ($\text{W}^{-1}\ \text{km}^{-1}$)	Dispersion (ps/km/nm)	Loss (dB/km)
Silica SMF	~1	17	0.2
Silica DSF	~2	~0	0.25
Silica HNL-DSF	~20	~0	0.5
Silica HNF-PCF	~10	~0	10
Bismuth HNLF	~1000	−260	800
Chalcogenide HNLF	~1000	−560	1000
Bismuth HNF-PCF	580	−10	1900

phase shift is induced onto the clockwise pulse through the XPM process. When the phase shift is π, the input pulses totally go through to the output directions. Thus, a XPM-based all-optical switching of short pulses can be realized with the control pulses. NOLM has been used to demultiplex the ultrafast time-domain multiplexed (TDM) pulses. Similar functions have also been realized with the fiber Kerr shutter (Fig. 5.67b) using XPM or PA in the nonlinear fibers. Ultrafast TDM signals (160 Gbit/s) have been demultiplexed to 10 Gbit/s at the whole C band [174].

Wavelength conversion of the signal light is possible with FWM in the nonlinear fibers. Figure 5.67c shows the typical construction of the fiber wavelength converter. FWM is the process that, when the pump light at the frequency of f_p and the signal light at the frequency of f_s, a new light is generated at the frequency of $2f_p - f_s$, mirrored position of the signal frequency with respect to the pump frequency. FWM can also be used for demultiplexing, by using the pump light as the control pulses. FWM not only converts the wavelength of the signal, but also inverts the original spectrum, which is called phase conjugation. It has been used for compensating the chromatic dispersion and the nonlinearities in the transmission fiber by inverting the signal spectrum at the midpoint of the transmission [175]. This is called midpoint spectral inversion (MSSI). Since the FWM-based wavelength conversion can be very broadband (up to the dispersion characteristics of the nonlinear fiber), it has been used to convert the wavelengths and invert the spectra of the group of WDM signals. PA, which is basically the same phenomena with FWM, can also be very broadband amplification > a few 100 nm by using small and flat dispersion fiber. Inversely, we have demonstrated very narrowband PA in the short (~1 m) bismuth glass HNLF, as shown in Fig. 5.68 [176].

All-optical 2R (reshaping and regeneration) or 3R (reshaping, retiming, and regeneration) is important for long-distance transmission of RZ pulses. 2R pulse regeneration has been realized using the SPM whose spectral broadening is proportional to the pulse peak power. By spectral filtering after SPM spectral broadening, 2R regenerated pulse is obtained, which is called the Mamyshev regenerator [177]. It is also suggested that higher-order FWM components can be used for 2R regeneration [178].

SRS has been used for broadband amplification, fiber Raman amplifiers (FRAs). Differently from rare-earth-doped fiber amplifiers, FRA can work at any wavelength with proper choice of the pump wavelength. In contrast, SBS is very narrowband (~100 MHz), which is not good as a signal amplifier. But recently, SBS-based slow light phenomenon is attracting many researchers. It is based on the fact that the spectral change in gain accompanies group delay change, i.e., the Kramers–Kronig relation [179]. 150 ns variable delay has been achieved by the SBS in the fiber [179]. SBS bandwidth can be enhanced by broadening the pump spectrum for fast signals, whereas there exists a trade-off between the achievable delay and bandwidth. On-chip SBS delay light has been realized in the nonlinear waveguides [180].

There have been efforts to find new high nonlinear materials for all-optical signal processing. One of them is the nano-carbon material, such as CNT and graphene. We have realized the FWM-based wavelength conversion of 10-Gbit/s signal using the nonlinear refractive index change in CNT (Fig. 5.69) [181, 182] and 2R regeneration

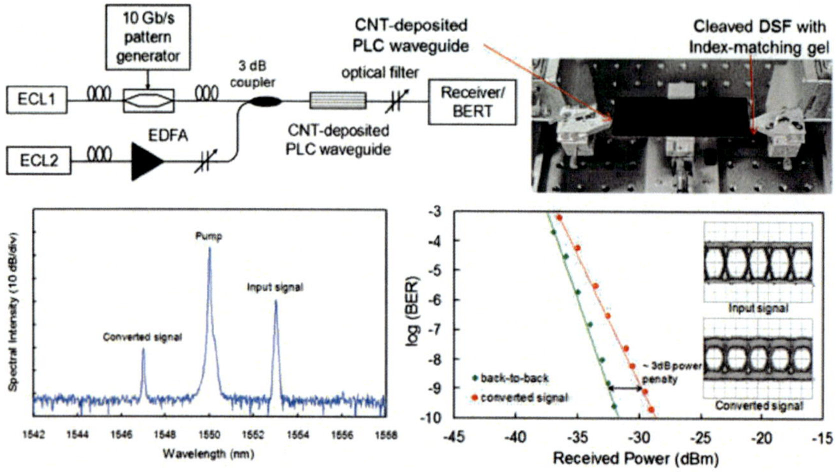

Fig. 5.69 FWM-based wavelength conversion of 10-Gbit/s signal using the nonlinear refractive index change in CNT-deposited PLC waveguide (Reprinted with permission from [182]. © 2010 The Optical Society)

of 10-Gbit/s RZ signal in the CNT-based Mamyshev regenerator [183]. We have also demonstrated the FWM-based wavelength conversion using graphene [184]. Another example is the organic nonlinear materials. By inserting the organic nonlinear materials into the silicon-based slot waveguide, FWM-based demultiplexing of 130-Gbit/s to 10-Gbit/s signals has been demonstrated [185]. Slot waveguide utilizes the light guiding in the very narrow (= a few 100 nm) low-index part in the high-index contrast material, such as silicon. Since the light is confined in the very small region, γ can be very high. $\gamma = 10^5\ \mathrm{W^{-1}km^{-1}}$ has been reported.

5.5.1.3 Summary and Future Prospects

This section described the ultrafast signal generation from mode-locked fiber lasers and the optical signal processing in optical fibers and waveguides. We also presented several efforts searching for novel nonlinear materials. All-optical signal processing is a well-established technology, downsizing and lower cost and power consumption are desired. In that context, small-size high-repetition-rate passively mode-locked lasers and small-size highly nonlinear materials, such as highly nonlinear glasses, organic materials, and nano-carbon materials, are very attractive. With these developments, we expect the deployment of all-optical signal processing in the commercial optical networks in conjugation with high-speed electronics, leading to the realization of the energy-saving "green" photonic networks.

5.5.2 SDM Processing Technologies

In addition to encouraging device integration, SDM technology also has the potential to exploit parallel transmission channels to make optical systems which more efficiently share both physical and signal processing resources between channels. In this section, we describe a range of mutually beneficial technologies including pilot tone transmission, spatial super-channels, and multi-dimensional modulation formats designed to both improve overall performance and improve efficiency for SDM applications over short- and long-haul transmission. In addition to describing the performance and applications of these technologies, we discuss the need for coherence between SDM channels and compare which forms of SDM are suited to each technology.

5.5.2.1 Spatial Super-Channels (SSCs) for Shared Resources

Spatial super-channels were proposed shortly after the first transmission demonstrations, with the aim of sharing digital signal processing (DSP) resources that are common to all cores [85]. The spatial super-channel is comprised of signals of the same wavelength in multiple cores, as shown in Fig. 5.70a [186]. Hardware savings are possible by sharing a single laser between multiple cores, and they have also been used to simplify switching in reconfigurable optical add and drop multiplexers (ROADMs) using MCF links [187, 188]. In addition, SSCs open the prospect of combining with other transmission technologies discussed in this section that may further increase efficiency in SDM systems by using one SDM channel for pilot tone transmission for self-homodyne detection [189] or shared carrier reception [190], Fig. 5.70b, and increasing the number of dimensions over which modulation is applied [191].

In principle, SSCs are compatible with other forms of SDM, but in practice skew variation between channels and inter-channel crosstalk may impact performance [84]. While fixed skew between SDM channels may be compensated with fixed optical delays, dynamic skew may require buffering of individual channels to enable shared

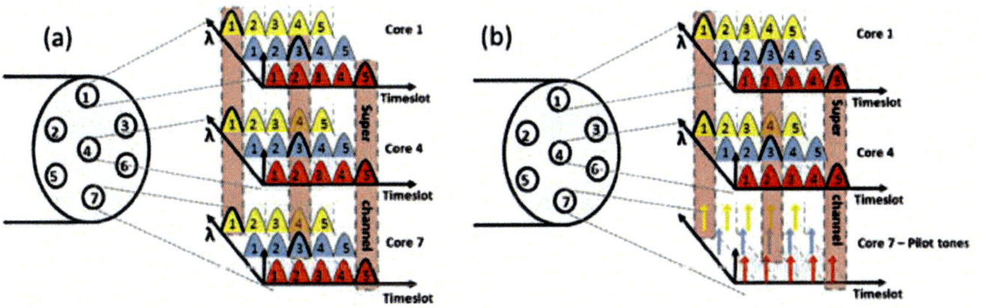

Fig. 5.70 **a** 3-core spatial super-channels in 3 different wavelengths and **b** 2-core spatial super-channels in 3 different wavelengths (Reprinted with permission from [186]. © 2017 IEEE)

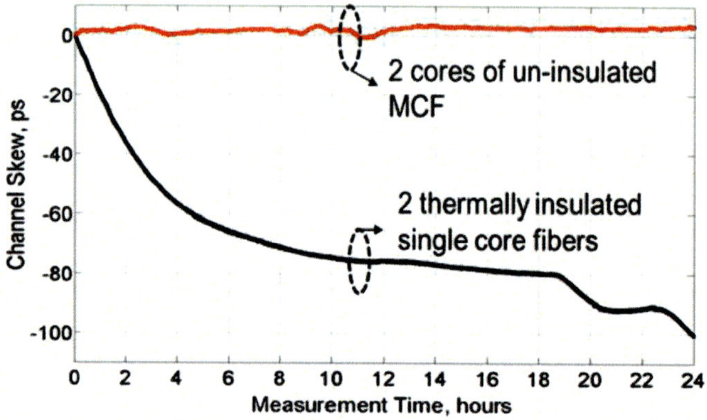

Fig. 5.71 Comparison of dynamic skew between 2 cores of an MCF and 2 insulated single-mode fibers over a 24-h period (Reprinted with permission from [83]. © 2015 The Optical Society)

receiver processing. For this reason, temporal coherence and low dynamic skew are advantageous for technologies based on SSC transmission making it a topic for study. In [85], the phase drift between 2 cores of a MCF was observed to vary gradually over 400 μs period with absolute drift and order of magnitude smaller than the same measurements with 2 single-mode fibers. Similarly, as shown in Fig. 5.71 from [83] the dynamic inter-core skew in a 28.3-km MCF span was observed to vary by only 7 ps in a 24-h period compared to over 100 ps in 2 insulated single-mode fibers. While further study is needed, these studies suggest that MCF, where cores sharing the same cladding experience the same environmental fluctuations, provides advantages for transmission schemes seeking to gain advantages from parallel transmission channels beyond simply a multiplication of capacity. Dynamic skew measurements of multi-element fiber (MEF) and multi-mode-fiber (MMF) have yet to be reported.

5.5.2.2 Self-homodyne Detection in Multi-core Fibers

Self-homodyne detection (SHD) [192, 193] has recently been proposed as a compatible technology for reducing costs in such space division multiplexing (SDM) systems [194, 195]. In SHD, a transmitted pilot tone (PT) originating from the transmitter laser is space- or polarization-multiplexed with the data signal and used as local oscillator (LO) for coherent reception. Phase or near-phase coherency between the data signal and the PT yields phase noise cancelation (PNC) after coherent detection, which can be exploited to reduce the impact of laser phase noise and subsequently enable the use of spectrally efficient high-order quadrature amplitude modulation (QAM) formats. Furthermore, as has also been demonstrated with shared carrier reception schemes in multi-core fiber (MCF) [190], both signal cores and a PT are received with an intradyne detection (ID) receiver. In addition to eliminating frequency offset compensation, PNC in SHD can relax the required rate carrier phase recovery by several orders of magnitude [196, 197]. SHD with a PT transmitted on an orthogonal polarization to the data was demonstrated for a range of multi-level modulation formats [198–200], but suffers from a reduction of spectral efficiency (SE) of up to 50% compared to polarization-division multiplexed (PDM) transmission. Although the SE may be improved by spectrally interleaving the PT with the signal [201], recent work on high channel-count SDM systems [145, 146] has opened the possibility of employing SHD with the PT transmitted through 1 SDM channel and the remainder used for data signals. The SE reduction due to PT transmission compared to an equivalent intradyne scheme becomes inversely proportional to the number of SDM channels. An additional impairment in all SHD systems is the accumulation of noise along the PT transmission, which translates into an optical signal-to-noise ratio (OSNR) penalty with respect to equivalent intradyne detection. For typical system configurations, this penalty ranges from 1 to 3 dB and is largely determined by the bandwidth of the filter required to isolate the PT in a WDM configuration, the ratio of signal OSNR and PT OSNR, and the modulation format [202, 203].

SHD Transmission in MCF

The prospect of employing SHD in multi-mode fiber systems was investigated numerically [204]. However, to date, because of the relative ease of maintaining path length alignment between cores, all experimental demonstrations have used MCF. Initially, the feasibility was investigated experimentally with an external cavity tunable laser (ECTL) for linewidth-sensitive 5 GBaud quadrature phase-shift keyed (QPSK) signals [195], before high-capacity transmission across a span of 10.1-km 19 core fiber was demonstrated using low-cost DFB lasers with MHz linewidth. Firstly, 105-Tbit/s transmission was demonstrated with at 25-GBaud QPSK signals in 125 wavelengths with 50-GHz channel spacing [205] before increasing to 210-Tbit/s WDM-SDM-PDM transmission with the use of PDM-QPSK signals. This transmission rate, assuming a 7% overhead for forward error correction (FEC), gives a SE of 33.4 bit/s/Hz [189]. More recently, long-distance self-homodyne transmission was investigated using a setup with parallel recirculating transmission loops [206], where self-homodyne transmission of 6800 km was observed for BER of 1.5×10^{-2}. This distance was within 1 dB of the distance covered by signals received with an intradyne receiver which was consistent with the measured back-to-back penalty.

Practical Implementation of SHD Transmission System

Many of the benefits of SHD are derived from the phase noise cancelation property of SHD systems. However, optimum performance requires some degree of alignment of the signal and PT path lengths from the point of separation at the transmitter side, up to the balanced photodetectors at the receiver side. The impact of skew between signal and PT has not been widely studied and no universal model yet exists, however [207] presented an analytical model of signal–PT skew for SDM SHD systems that showed

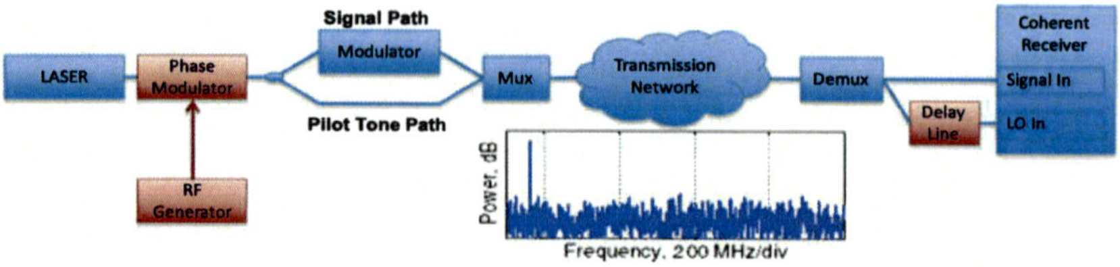

Fig. 5.72 Path length alignment monitoring subsystem (Reprinted with permission from [209]. © 2013 IEICE)

good agreement with experiment for PDM-QPSK transmission. Here, the impact of varying skew and the block length of the feed-forward carrier phase recovery (CPR) module was investigated, showing that while increased skew up to 312 symbols increased the variance of phase noise in the receiver, it could be partially compensated by reducing block length of the CPR module. The impact on required OSNR was investigated in [208], showing that for PDM-QPSK, 100 MHz linewidth and skew of 60 ns (750 symbols) had little impact. However, for 16 QAM signals, the same conditions required an additional 1 dB OSNR compared to low skew and linewidth product, suggesting that the lower phase noise tolerance of higher-order modulation formats also reduces their tolerance to signal–PT skew.

Since inter-channel skew and dynamic skew may also impact performance and degree of complexity required in receiver technology for all SSC transmission technologies, active path length monitoring and alignment may be considered for systems with extreme-phase noise sensitivity. Such a system was implemented in [209] and, as shown in Fig. 5.72, uses variable optical delays to adjust the relative path between signal and PT path based on the feedback signal from a low-frequency phase modulation applied to the optical signal before splitting signal and PT paths. Upon recombination at the receiver side, nonzero path length misalignment will yield a measureable interference signal, as shown in the inset of Fig. 5.72. This signal can be used as feedback to control a variable optical delay line. When perfect path length alignment is achieved, the interference signal is canceled and has no impact on the system performance.

Another implementation obstacle for SHD is ensuring the correct polarization alignment between the PT and the polarization beam splitter (PBS) in the receiver after transmission. Active polarization tracking and control systems have been investigated for more than 25 years [210], and with update rates over 50 krad/s possible [211, 212], it is envisaged that they could be applied to SHD systems. An alternative is to modify the design of the coherent receiver to digitize additional signals obtained from the combined signals of the polarization components after the PBS to reconstruct transmitted signal in DSP regardless of the incoming PT polarization state. Such a receiver was experimentally demonstrated for a polarization-multiplexed PT SHD system [208], and although not yet demonstrated, it envisaged a similar approach applied to SDM SHD systems.

SHD in Networks

In addition to a point-to-point transmission demonstration, the first experiments exploring the feasibility of SHD from a networking perspective have also been demonstrated. Architecture-on-demand (AoD) nodes have been widely proposed and also demonstrated to incorporate SDM links [213, 214], and reconfigurable optical add and drop multiplexer (ROADM) designs for the switching of spatial super-channels [85, 187] in MCFs have also been proposed. In [215], a reconfigurable SDM-WDM hybrid add–drop node capable of routing both signals and PTs was demonstrated for 100-Gbit/s PDM-QPSK signals. SHD has also been used in a larger network demonstration utilizing AoD nodes and software-defined networking (SDN), described more fully in Chap. 6. In this demonstration, both seven-core and 19-core MCFs were used in a network demonstration using spatial super-channels and a variety of modulation formats with both signal and PTs dynamically assigned by an SDN controller over an open flow interface [216, 217]. This demonstration showed that where capacity allows, the advantages of SHD may be exploited even in dynamic networking scenarios by integrating signal and PT management into control plane architectures.

5.5.2.3 Modulation Formats for Multi-core Fiber Transmission

SDM technology in conjunction with SSCs also enables greater design flexibility and greater efficiency by increasing the number of dimensions over which optical modulation is applied. Multi-dimensional modulation formats [87, 218] have previously been studied using the 4-dimensional signal space of I and Q quadratures in 2 polarizations with additional orthogonal dimensions being added by timeslots [219], wavelengths [218], or polarization states [220]. Many such formats exist and have been proposed, typically based on geometric design methodologies such as of spherical cutting of optimal N-dimensional lattices or by coding of a base constellation [87, 221]. Such formats offer improved spectral efficiency or sensitivity which in some cases may be flexibly implemented with the same optical hardware to respond to changing conditions.

In this section, we describe how some families of multi-dimensional modulation format may be applied to SDM and experimental results to date. Potential formats first are compared with 2 primary metrics, the theoretical sensitivity, as denoted by the asymptotic power efficiency (APE) at high optical signal-to-noise ratio (OSNR), and the spectral efficiency (SE), which indicates the achievable bitrate for a given available bandwidth [221, 222]. The APE, subsequently denoted as γ, is defined for an additive white Gaussian noise channel as $\gamma = {d_{\min}}^2 \log 2M/(4E_S)$, where M is the number of constellation points, E_S is the symbol energy, $d_{\min}$ is the minimum Euclidian distance between 2 symbols, and the factor of ¼ normalizes the expression to be 1 for binary phase shift keying (BPSK) and quaternary phase-shift keying (QPSK). The SE is defined per dimension pair as $\mathrm{SE} = \log_2 M/(N_D/2)$, where N_D is the total number of dimensions including quadratures and polarizations. In general, it is desirable to maximize both SE and APE in an attractive format but we note that in practice the sensitivity and SE are often traded to some extent.

Space-Position Modulation

In single-core transmission, the most common technique for increasing the dimensionality of the 4-dimensional optical signal space of in-phase and quadrature dimensions of orthogonal polarizations is in the time domain. Originally named, pulse-position modulation (PPM), it is a subset of a more general modulation scheme known as multi-dimensional position modulation [223], where each dimension may be a timeslot, wavelength or SDM channel and information coded in the presence of light in each dimension. PPM was first investigated with intensity modulated data with higher dimensionality achieved by splitting each symbol period into a frame of K pulse slots. K is typically a power of two to encode an integer number of bits per frame, where the presence or absence of light in one [224, 225] or more [219] slots is used to encode data which may also be used in combination with existing coding techniques [226]. Subsequently, the same principle was combined with coherent quadrature amplitude modulation (QAM) [227] to further improve both γ and SE.

However, although combining QAM with a single slot per frame enables record sensitivity to be achieved [228, 229], it comes at the cost of reduced SE, which may be increased by combining QAM symbols with multi-slot PPM to encode more bits in each symbol slot. Each variant of this family of modulation formats uses a different, fixed number of L pulses per frame of length K to transmit QAM symbols and in combination with QPSK or polarization-switched (PS)-QPSK has been shown capable of simultaneously increasing both the SE and γ compared to the case of transmitting QPSK or PS-QPSK in each symbol slot [223, 230]. Such formats are referred to here as K over L multi-dimensional position modulation (MDPM) since there are $\begin{pmatrix} K \\ L \end{pmatrix}$ combinations of transmitting the pulses and the total number of bits per frame is $\log_2\begin{pmatrix} K \\ L \end{pmatrix}$ BQAM L, where BQAM is the number of bits in each QAM symbol. The particular case where $L = K - 1$ is known as inverse PPM (iPPM), inverse MDPM (iMDMP), or pulse erasure modulation since the position data is modulated in the absence of a QAM symbol in 1 slot. These formats are of similar complexity as conventional PPM and carry an integer number of bits per frame when K is a power of two. Combined with QPSK and $K = 4$, it offers a 1.25 dB increased γ over QPSK with maintained SE and for $K = 8$ both the SE and γ may be increased over QPSK [218].

Since these formats may be generalized as types of MDPM, they may be simply transposed to the spatial domain as space-position modulation (SPM) as fiber modes or cores. Furthermore, each polarization may be considered individually to give space-polarization-position modulation (SPPM) such that each core has two possible slots and it has been shown that individual position modulation over polarizations

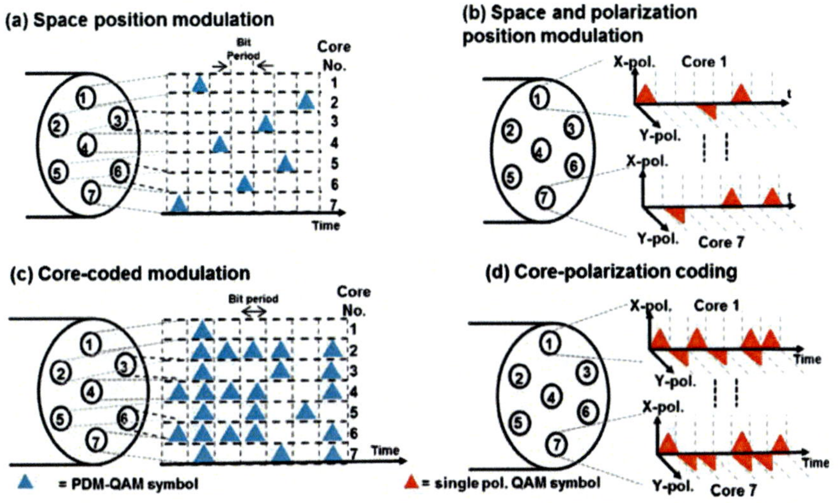

Fig. 5.73 Space-position modulation formats in a 7-core fiber (Reprinted with permission from [191]. © 2014 The Optical Society)

has better performance in terms of both γ and SE compared to position modulation over both orthogonal polarization states [223, 227].

A recent variation of MDPM experimentally investigated with MCF is known as core-coding (CC) or core-polarization coding (CPC) QAM, depending on whether each core with a PDM QAM symbol or each polarization with single polarization QAM symbol is used. These formats combine QAM of dimensionality D, with additional dimensions, K, of an MCF to make a "multi-core" modulation format with total dimensionality $N_D = K \times D$, where K is equivalent to the number of MCF cores for core-coding or twice the number of cores for core-polarization coding. Each format encodes K bits in each symbol timeslot as a binary word of length K, where each bit is determined by the presence or absence of light on a core or core-polarization and we refer to these as intensity modulated (IM) bits. In addition, each illuminated core or polarization carries a QAM symbol containing B_{QAM} bits to code an average of $B_{QAM} \times K/2$ additional bits. Hence, an average number of bits $B_{Tot} = K(1 + B_{QAM}/2)$ are encoded per MC symbol and $B_{Core} = (1 + B_{QAM}/2)$ bits encoded per core. This scheme may be considered as a variant of K over L MDPM, but with a variable number of lit cores/polarizations, as shown schematically in Fig. 5.73 [191]. Figure 5.73 also shows that compared to transmitting the same QAM format in each core or each polarization. The CC and CPC formats transmit symbols only half the time on average, meaning that the average optical power is reduced by 50%. This feature may have consequences in terms of energy consumption with energy savings at the transmitter and optical amplifiers [231], but also potentially reduces the impact of fiber nonlinearities in long, dispersive fiber links. This follows from the well-known Gaussian noise model, according to which the nonlinear distortion is proportional to the cube of the average power [232].

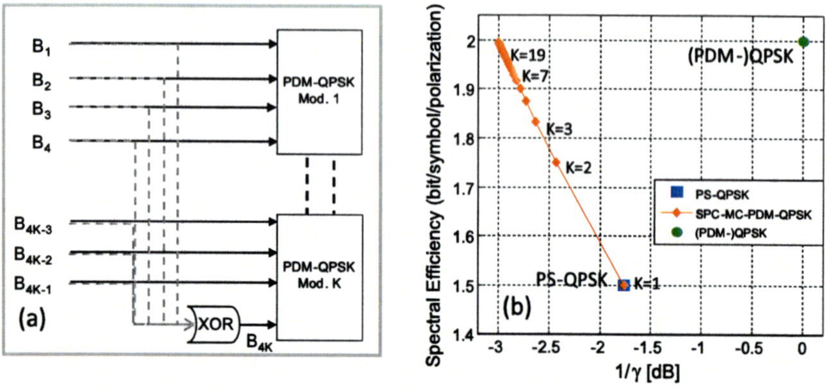

Fig. 5.74 **a** Logical transmitter configuration for SPC-MC-PDM-QPSK based on PDM-QPSK modulators where B1 to B4K denote binary driving bits and **b** SE as a function of asymptotic power penalty ($1/\gamma$) (Reprinted with permission from [191]. © 2014 The Optical Society)

Set-Partitioned Modulation

The simplest kind of coded modulation is set-partitioning (SP) where a subset of symbols is selected from a base constellation. In single-core transmission, SP has been shown to be a simple method for improving γ at the cost of some spectral efficiency for QAM formats [233–236] by using 1 information bit as a single-parity check on all the symbol bits to reduce their error probability. For example, in [236] set-partitioning was applied to PDM-16QAM, transforming gray-coded 256 symbol set of 8 bits into a 128 symbol set carrying 7 bits with the final bit maintaining parity and increasing $d_{\min}$ by a factor of $\sqrt{2}$ compared to PDM-16QAM. This format was labeled 128-SP-QAM [235], but to be consistent with naming strategy used subsequently may also be referred to as single-parity check (SPC)-PDM-16QAM.

Another example of set-partitioning is the four-dimensional modulation format PS-QPSK [222], which is the basis for extending these formats into the spatial domain. PS-QPSK may be described as transmitting independent BPSK symbols in three dimensions, while the fourth is dependent on the other three as an SPC bit [218] that reduces the SE of the format by 25%. This concept was applied to MCF transmission [191, 237], but considering modulation in all cores of a spatial super-channel to be part of the same multi-core symbol. Again using the example of BPSK modulation in each available dimension, this format, referred to as SPC-MC-PDM-QPSK, can be considered as transmitting independent BPSK symbols on each orthogonal polarization state up to $4K-1$ dimensions of an MCF with K cores with the last reserved for the SPC bit, formed as $B_{4K} = \mathrm{XOR}(B_1, B_2, \ldots, B_{(4K-1)})$ where B_n is the binary or BPSK symbol in the nth dimension. This is also illustrated in Fig. 5.74a, which shows the transmitter implementation based on standard PDM-QPSK modulators each with 4 binary driving bits. Again, $d_{\min}$ is increased by a factor of $\sqrt{2}$ giving an advantage in γ but reducing the total number of dimensions from $N_D = 4K$ to $N_D = 4K - 1$. Hence, both the SE and γ depend on the number of available cores according to SE = 2 − (1/2K) and $\gamma = 2 - (1/2K)$ although we note that the identical expression for both does not hold for higher-order SPC QAM formats. The SE as a function of $(1/\gamma)$ for QPSK, PDM-QPSK, PS-QPSK, and SPC-MC-PDM-QPSK with $K = 1, 2, \ldots, 100$ is shown in Fig. 5.74b.

These formats were experimentally investigated as function of the number of cores up to $K = 7$ using noise loading to compare the required OSNR at different BER thresholds. Although the increasing γ for higher core numbers suggest higher sensitivity at high OSNRs, in the noise-limited transmission regime, sharing the parity bit over more bits increases the number of nearest neighbor symbols and increases the probability of multiple errors which may remain uncorrected by the SPC bit. Hence, an additional OSNR penalty of 0.8 dB required OSNR was observed at a BER = 1×10^{-3} when applying the parity check over 7 cores compared to 3. However, when the baud rate of the more spectrally efficient 7-core format was adjusted to provide equal data rate signals, this penalty was reduced to 0.2 dB. In the same study, the 3-core format was shown to increase the transmission distance by 20% for the same data rate of 112.5 Gbit/s. SPC modulation over PDM-16QAM was also investigated for WDM transmission with a fixed channel spacing and baud rate [238], where the same trade-off between spectral efficiency and resilience to noise was observed as the number of cores increases between 1 and 7. For 1, 3, and 7 SPC-MC-PDM-16QAM formats, the transmission distance was increased by 123%, 82%, and 44% compared to PDM-QPSK for a SE reduction of 0.91, 0.3, and 0.13 bit/s/Hz/core, respectively.

5.5.2.4 Coding and Multi-dimension Modulation Formats

The number of spatial channels used in research demonstrations has been steadily increasing in recent years and in 2015 exceeded 100 [145, 146]. This number of channels in combination with higher-order modulation formats means that SSCs with sufficient number of bits for more complex coding blurs the lines between optical modulation format and forward error correction implemented in the optical domain. Furthermore, the drive to gain most efficiency from SDM technology and the need to compensate for imperfect properties of the optical channels have led to various approaches for incorporating elements of coding into the SDM transmission system.

Coded optical modulation using multiple parity checks in combination with single-mode higher-order QAM formats has been widely studied and used to reduce the required OSNR in several high-capacity transmission demonstrations with overheads of up to 25% [239, 240]. Multi-dimensional modulation formats based on short linear block codes in up to 24 dimensions were studied in [87, 241], and as discussed in [238] for application to SSCs, such formats may be flexibly used as a kind of optical inner FEC code to be in used in conjunction with a traditional electronic outer FEC code after detection. The performance of 24D extended Golay coded format was also investigated for 6-mode MMF system by simulation, showing a 4 dB improvement in span-loss budget compared to BPSK [242].

Coded modulation of SSCs has been demonstrated in 2 multi-core fiber experiments. A 16-dimensional format based on the extended Nordstrom–Robinson [243] code applied to PDM-QPSK modulation in a 4-core SSC, to increase the transmission distance by 55% compared to transmitting BPSK in each core. This format constellation was the same in each core with a null field average enabling the use of automatic biasing of optical modulator and use standard DSP algorithms used for with only a modified decoded from standard PDM-QPSK receiver [244]. In [86], PDM-64QAM, 228-bit SSCs in a 19-core fiber were coded with shortened Bose–Chaudhuri–Hocquenghem (BCH) codes [245], showing up to 122% increase in transmission distance compared to uncoded format. More interestingly, such codes reduced the required OSNR to achieve a BER threshold suitable for hard-decision FEC after detection showing that in some cases, combining such codes with HD-FEC can achieve comparable performance as serial post-detection SD-FEC codes for a similar total overhead but with reduced DSP complexity, latency, and required ADC resolution.

The mode coupling and mode-dependent loss in multi-mode fiber systems has led to various forms of space–time coding to exploit the huge bandwidth potential of multi-mode transmission. Space–time coding, where delayed copies of the same bit stream are transmitted on more than one spatial channel to improve detection reliability, was investigated by simulation with and without optical scrambling [246, 247]. Space–time coding was experimentally investigated for few-mode fiber systems in combination with various QAM formats [248] showing both reduced OSNR requirement and SE advantages over equivalent single-core transmission. A similar approach of optical diversity transmission was also proposed to mitigate nonlinear impairments in early MCF research [249]. This approach can give SDM transmission advantages over single-mode transmission, but significantly reduce the achievable SE compared to transmitting independent on each spatial channel. Using spin and orbital angular momentum modes in MMF transmission for multi-dimensional modulation has also been proposed and investigated numerically in combination with LDPC coding [250, 251].

5.5.3 Orbital Angular Momentum

As one of the candidates of SDM technologies, Laguerre–Gaussian (LG) modes carry the orbital angular momentum (OAM) that arises from the rotational Poynting vector on the wave front [252]. (Namely, it is also called as OAM mode.) LG modes are the eigenmodes derived from paraxial wave equation given by $LG_\ell{}^p$, where ℓ is the azimuthal index which represents the winding number of the phase. Usually in communication applications, we neglect the radial index p and hereafter describe LG mode as $LG_\ell{}^0$. The ℓth LG mode has a gradationally changing phase that rotates around a propagation axis over an angle proportional to $\ell \cdot 2\pi$ [253]. The mode is equivalent to the normal Gaussian when $\ell = 0$.

Unlike Hermite–Gaussian modes, HG_{nm} or TEM_{nm}, which are widely used in optical communications, LG modes are characterized by the helical wave front. The projection of the propagation axis on the LG beam indicates a phase singularity, which results in a null intensity in its cross section as shown in Fig. 5.75, which is in contrast to the linearly aligned modes of the TEM beam. The angular momentum carried by TEM has only two-dimensional spin angular momentum associated with right and left circularly polarized states. In contrast, the OAM in LG_ℓ has infinite dimensions in principle since ℓ can take any integer value; hence, it can reach higher dimensions and thereby sustain the multiple orthogonal channels. The information carried by multiple LG modes is thus decomposed into multi-dimensional states of their OAM, providing additional degrees of freedom which can be utilized for an increase of the density of the transferred information.

Several demonstrations of OAM-MUX/DEMUX in free-space propagating beam have been performed [254–256]. On the other hand, the purpose of this chapter is mainly introduction of fiber-coupled OAM.

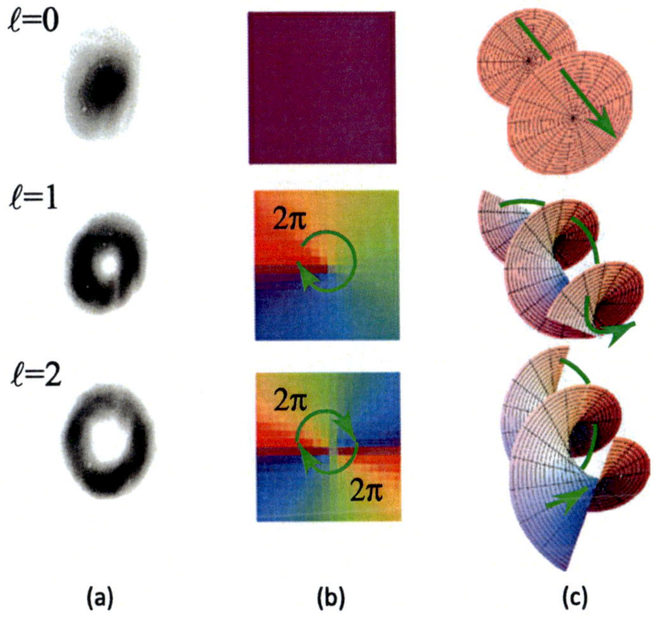

Fig. 5.75 Optical vortices and orbital angular momentum. **a** Beam profiles, **b** phase of cross section, **c** wave front. Courtesy of Prof. Yasunori Toda. © 2021 Yasunori Toda

5.5.3.1 Multiplexing and Fiber Transmission of OAM

Since OAM is carried by wave front of LG mode beam, it suffers from turbulence or distortion in free space. In an analogous way of thinking, the preservation of OAM in fiber propagating mode is an important issue. Before recent SDM era, preliminary trial of LG mode launching in multi-mode fiber and MDM is already presented [258]. A 10-m-long, 50-μm-diameter GI fiber was used with 633-nm He–Ne laser. DEMUX was performed topologically by using spatial mode filter, and no data modulation was tested. The modal crosstalk was fairly large, more than −10 dB in average.

The first demonstration of preservation of OAM and data transmission in telecom wavelength (1.5 μm) was demonstrated by using 7-core MCF [257] as shown in Fig. 5.76a, b. Even each core of MCF was single-mode, and topological layout and relative phase of output of cores could recompose similar phase relation with helical wave front of LG beam at the end of fiber. The cladding diameter was 150 μm which accommodated seven cores. The inter-core crosstalk was effectively suppressed to the level of −70 dB/100 km by adopting trench-assisted index profile. LG mode was generated by spatial light modulator (SLM) from fundamental Gaussian beam and focused on the cleaved facet of MCF. In this experiment, $\ell = 0$ and 1 corresponding to Gaussian mode and doughnut beam which have different data sequences. This doughnut beam was loosely focused as the doughnut beam waist was overlapped six cores of outer ring, while Gaussian beam was focused to match with mode field of center core. Hence, the coupling efficiency to outer cores was not optimized, but the phase relation of each oscillated cores reflects the wave front of LG mode beam. After fiber propagation, the output of fiber was injected into another SLM to discriminate OAM in combination with SMF aperture as spatial filter. SLM can convert the six beam groups to a coupled mode of SMF, and any higher-order OAM was repelled on the facet of SMF. By this manner, multiplexed OAM was successfully distinguished. Figure 5.77 shows measured bit error rate of each data sequence.

In this first demonstration, the fiber length was only 80 cm, and upgraded experiment in 500-m-long MCF was performed afterward [259]. It was found that the polarization of each core was rotating randomly. Because commercially available SLM has polarization dependency, additional polarizer was required before DEMUX SLM to eliminate undesired polarization. Even with such fluctuation, OAM was successfully recovered from MCF output. It means that OAM was preserved by using multiple optical paths accommodated in same cladding regardless of another fluctuation such as polarization rotation.

Other candidate of propagation media is vortex fiber which has circularly symmetric special index profile as shown in Fig. 5.78 [260]. Vortex fiber has strongly separated effective indices of the antisymmetric vector modes which avoids the creation of the conventional LP11 states and can effectively suppress intermodal crosstalk resulting to the preservation of OAM. The prototype of vortex fiber can support first-order antisymmetric modes corresponding to $\ell = \pm 1$. Four OAM modes including two LP01 modes of orthogonal polarization were multiplexed and propagated in vortex fiber. The fiber length was 1.1 km, and it was found that degradation of OAM is asymptotic and occurred within first 100 m. In the demonstration of

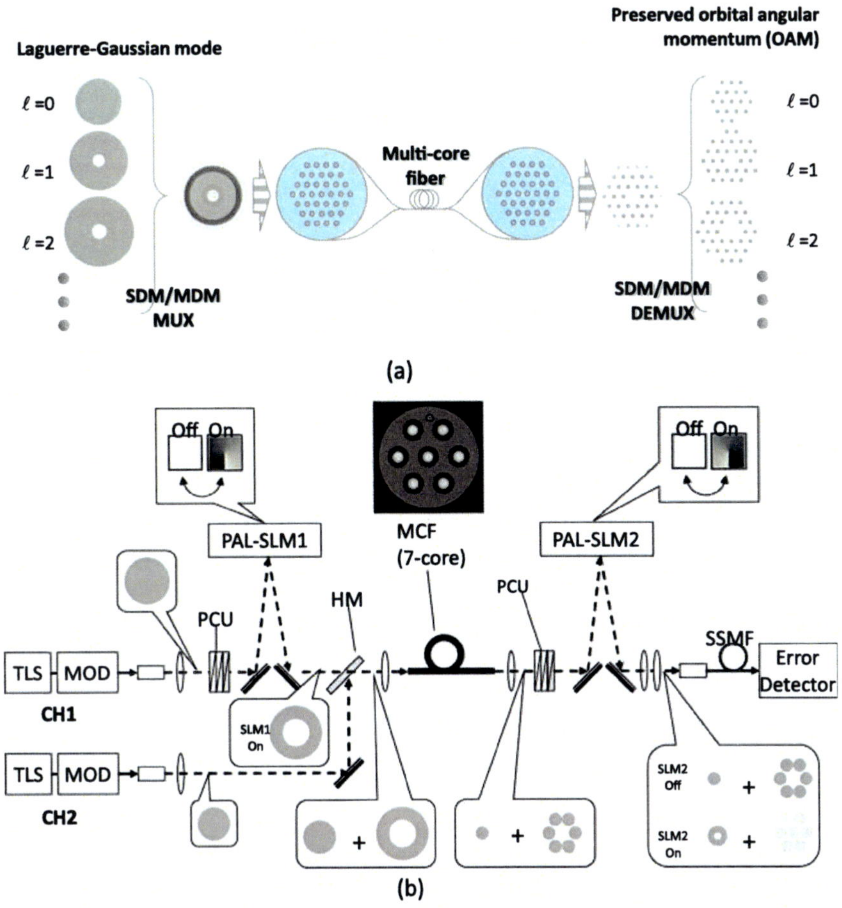

Fig. 5.76 OAM preservation in MCF propagation. **a** Principle of OAM preservation in MCF propagation and **b** experimental setup. TLS: tunable light source, MOD: intensity modulator which provides 10 Gbit/s, PN: 23, OOK data, PCU: polarization control unit, HM: half mirror, SLM: spatial light modulator, SSMF: standard single-mode fiber (Reprinted with permission from [257]. © 2011 Authors)

multiplexed data transmission, two-stage multiplexer was utilized, which was OAM conversion and precise polarization controlling. Four modes were converted from linearly into circularly polarized states in order to avoid creation spurious TE01 and TM01 modes. Even the vortex fiber spool was thermally isolated in a box, and it was observed that some of propagation mode has time-dependent fluctuation. The average intermodal crosstalk was −14.8 ~ −16.1 dB, and multipath interference (MPI) was −19.7 ~ −35.3 dB, respectively. The demultiplexer was also composed from two functions: One is spin sorter based on waveplates and OAM sorter by SLM. There are several results by these approaches which are maximum four-mode

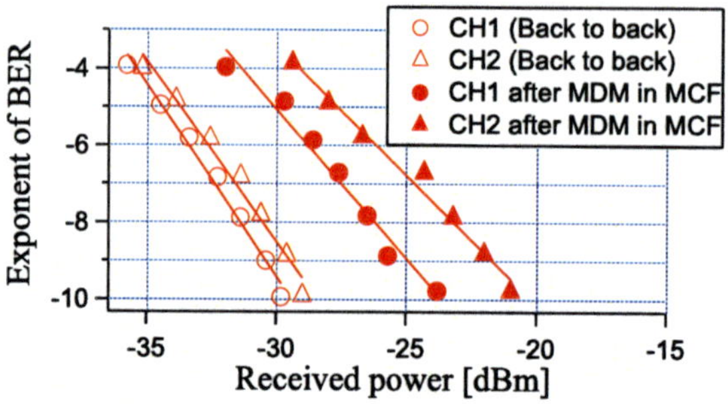

Fig. 5.77 Measured bit error rate (Reprinted with permission from [257]. © 2011 Authors)

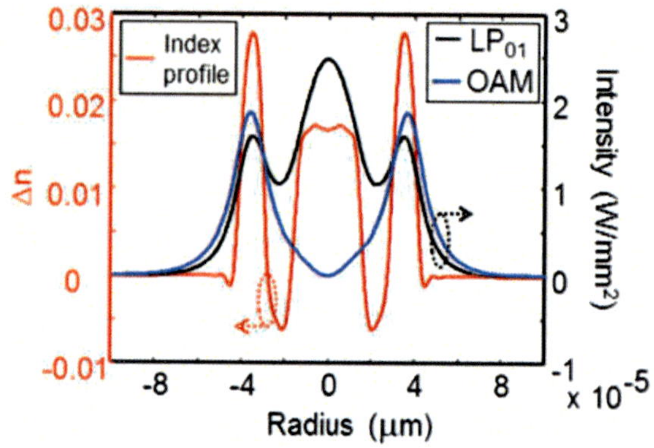

Fig. 5.78 Index profile of vortex fiber (Reprinted with permission from [260]. © 2015 The Optical Society)

MUX/DEMUX and 1.6-Tbit/s transmission by using two OAM modes combined with QAM/WDM.

5.5.3.2 Processing for OAM

Traditional approaches of OAM technology utilize full information of LG mode beam. On the other hand, several demonstrations tried to use spatially partial information of LG mode such as composition of OAM and partial detection. Actually, OAM propagation through MCF described in the previous section can be considered as a variation of such approaches. In the case of 7-core MCF propagation, the output beam from MCF has center beam and six beams on the outer ring. These six beams carry OAM, while they have ±6 satellite components around original mode on OAM spectra. This is topologically corresponding that six segmented doughnut beams have OAM of $\ell = \pm 6$.

In order to generate LG mode beam, SLM, hologram, or phase plate was utilized to convert Gaussian mode to LG mode, conventionally. Recently, other approaches were reported to compose OAM by illuminating several spots of doughnut beam profile. 3D

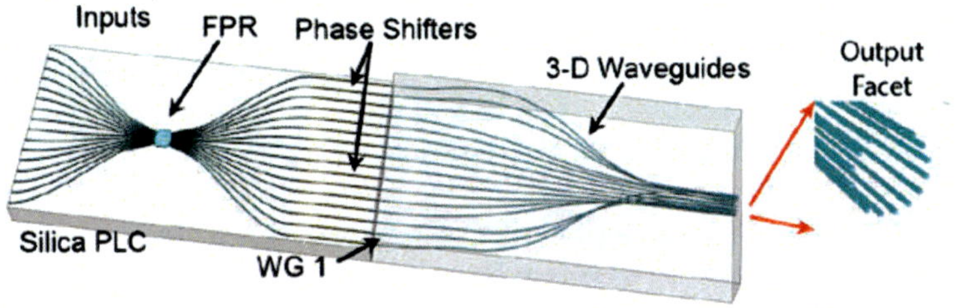

Fig. 5.79 3D waveguide OAM generator (Reprinted with permission from [261]. © 2014 The Optical Society)

waveguide type [261] (Fig. 5.79) needs precise controlling of path length of each spot. On the other hand, silicon circuit can be fabricated quite strict, and it can support much number of illuminated spots [262, 263]. By means of such illumination technique, fine quality of LG mode beam in far field can be achieved. This demonstration indicates the possibility to handle OAM discretely. Almost vulnerability of OAM technology comes from phase fluctuation or spectral spreading on OAM spectra caused by distortion of wave front. Toward more stable systems, digitized approach is a candidate.

Discrete sampling and detection of phase relation in MCF transmission is one of the digitized approaches which can discriminate the limited groups of OAM. As shown in Fig. 5.80 [264], the phase of helical wave front of LG beam rotates $\ell \times 2\pi$ during propagation for the distance of a wavelength and six outer cores of 7-MCF were oscillated according to the phase relation topologically. Therefore, the relative phase difference between core #1 and #4 is $\ell \times \pi$. For example, when two OAM modes were multiplexed (CH1: $\ell = +2$ and CH2: $\ell = -3$), the phase comparison of core #1 and #4 with appropriate delay adjustment can distinguish CH1

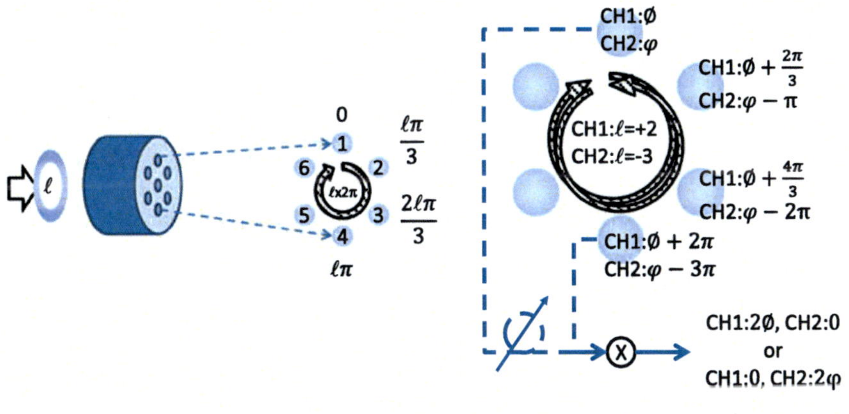

Fig. 5.80 Discrete sampling and detection of OAM propagating MCF (Reprinted with permission from [264]. © 2014 Engineers Australia)

and CH2 because phase differences are π and 2π, respectively. It is also confirmed with the comparison of constructive and destructive interference between cores. Other combination of cores such as #1 and #2, or #1 and #3 will provide more information about relative phase. Achievement of the summarized information of relative phase is equivalent to an OAM detection scheme by using spatial modulation and spatial filter. Moreover, it can be implemented by balanced detection like as digital coherent receiver and is expected to be more robust. By means of such discrete sampling technique, transmission distance expanded from 80 cm to 25 km [265].

References

1. H. Onaka, H. Miyata, G. Ishikawa, K. Otsuka, H. Ooi, Y. Kai, S. Kinoshita, M. Seino, H. Nishimoto, T. Chikama, 1.1 Tb/s WDM transmission over a 150 km 1.3 μm zero-dispersion single-mode fiber, in *Proceedings of Optical Fiber Communication Conference*, paper PD19 (1996)
2. T. Morioka, H. Takara, S. Kawanshi, O. Kamatani, K. Takiguchi, K. Uchiyama, M. Saruwatari, H. Takahashi, M. Yamada, T. Kanamori, H. Ono, 100 Gbit/s x 10 channel OTDM/WDM transmission using a single supercontinuum WDM source, in *Proceedings of Optical Fiber Communication Conference*, paper PD21 (1996)
3. A.H. Gnauck, A.R. Chraplyvy, R.W. Thach, J.L. Zyskind, J.W. Sulhoff, A.J. Lucero, Y. Sun, R.M. Jopson, F. Forghieri, R.M. Derosier, C. Wolf, A.R. McCormick, One terabit/s transmission experiment, in *Proceedings of Optical Fiber Communication Conference*, paper PD20 (1996)
4. S. Bigo, Y. Frignac, G. Charlet, W. Idler, S. Borne, H. Gross, R. Dischler, W. Poehlmann, P. Tran, C. Simonneau, D. Bayart, G. Veith, A. Jourdan, J.P. Hamaide, 10.2 Tbit/s (256 x 42.7 Gbit/s PDM/WDM) transmission over 100 km teralight fiber with 1.28 bit/s/Hz spectral efficiency, in *Proceedings of Optical Fiber Communication Conference*, paper PD25 (2001)
5. K. Fukuchi, T. Kasamatsu, M. Morie, R. Ohhira, T. Ito, K. Sekiya, D. Ogasahara, T. Ono, 10.92-Tb/s (273 x 40-Gb/s) triple-band/ultra-dense WDM optical-repeatered transmission experiment, in *Proceedings of Optical Fiber Communication Conference*, paper PD24 (2001)
6. A. Sano, H. Masuda, T. Kobayashi, M. Fujiwara, K. Horikoshi, E. Yoshida, Y. Miyamoto, M. Matsui, M. Mizoguchi, H. Yamazaki, Y. Sakamaki, H. Ishii, 69.1-Tb/s (432 x 171-Gb/s) C- and extended L-band transmission over 240 km using PDM-16-QAM modulation and digital coherent detection, in *Proceedings of Optical Fiber Communication Conference*, paper PDPB7 (2010)
7. D. Qian, M.-F. Huang, E. Ip, Y.-K. Huang,Y. Shao, J. Hu, T. Wang, 101.7-Tb/s (370 × 294-Gb/s) PDM-128QAM-OFDM transmission over 3 × 55-km SSMF using pilot-based phase noise mitigation, in *Proceedings of Optical Fiber Communication Conference*, paper PDPB5 (2011)
8. M. Nakazawa, M. Yoshida, K. Kasai, J. Hongou, 20 Msymbol/s, 64 and 128 QAM coherent optical transmission over 525 km using heterodyne detection with frequency-stabilised laser. Electron. Lett. **42**(12), 710–712 (2006)
9. J. Hongo, K. Kasai, M. Yoshida, M. Nakazawa, 1-Gsymbol/s 64-QAM coherent optical transmission over 150 km. IEEE Photon. Technol. Lett. **19**(9), 638–640 (2007)
10. M. Nakazawa, Challenges to FDM-QAM coherent transmission with ultrahigh spectral efficiency, in *European Conference on Optical Communication (ECOC 2008)*, Tu.1.E.1 (2008)
11. M. Nakazawa, S. Okamoto, T. Omiya, K. Kasai, M. Yoshida, 256-QAM (64 Gb/s) coherent optical transmission over 160 km with an optical bandwidth of 5.4 GHz. IEEE Photon. Technol. Lett. **22**(3), 185–187 (2010)

12. T. Omiya, M. Yoshida, M. Nakazawa, 1 Tbit/s 256 QAM-OFDM transmission over 560 km with 14.3 bit/s/Hz spectral efficiency, in *OptoElectronics and Communications Conference (OECC 2013)*, MR1-2 (2013)
13. S. Okamoto, K. Toyoda, T. Omiya, K. Kasai, M. Yoshida, M. Nakazawa, 512 QAM (54 Gbit/s) coherent optical transmission over 150 km with an optical bandwidth of 4.1 GHz, in *European Conference on Optical Communication (ECOC 2011)*, PD 2.3 (2011)
14. R. Schmogrow, D. Hillerkuss, S. Wolf, B. Bäuerle, M. Winter, P. Kleinow, B. Nebendahl, T. Dippon, P. C. Schindler, C. Koos, W. Freude, J. Leuthold, 512QAM Nyquist sinc-pulse transmission at 54 Gbit/s in an optical bandwidth of 3 GHz. Opt. Express **20**(6), 6439–6447 (2012)
15. Y. Koizumi, K. Toyoda, M. Yoshida, M. Nakazawa, 1024 QAM (60 Gbit/s) single-carrier coherent optical transmission over 150 km. Opt. Express **20**(11), 12508–12514 (2012)
16. M.-F. Huang, D. Qian, E. Ip, 50.53-Gb/s PDM-1024QAM-OFDM transmission using pilot-based phase noise mitigation, in *OECC 2011*, PDP1.6 (2011)
17. S. Beppu, K. Kasai, M. Yoshida, M. Nakazawa, 2048 QAM (66 Gbit/s) single-carrier coherent optical transmission over 150 km with a potential SE of 15.3 bit/s/Hz. Opt. Express **23**(4), 4960–4969 (2015)
18. M. Terayama, S. Okamoto, K. Kasai, M. Yoshida, M. Nakazawa, 4096 QAM (72 Gbit/s) single-carrier coherent optical transmission with a potential SE of 15.8 bit/s/Hz in all-Raman amplified 160 km fiber link, in *OFC 2018*, Th1F.2 (2018)
19. A. Sano, T. Kobayashi, S. Yamanaka, A. Matsuura, H. Kawakami, Y. Miyamoto, K. Ishihara, H. Masuda, 102.3-Tb/s (224 x 548-Gb/s) C- and extended L-band all-Raman transmission over 240 km using PDM-64QAM single carrier FDM with digital pilot tone, in *Optical Fiber Communication Conference (OFC 2012)*, PDP5C.3 (2012)
20. P.J. Winzer, A.H. Gnauck, S. Chandrasekhar, S. Draving, J. Evangelista, B. Zhu, Generation and 1,200-km transmission of 448-Gb/s ETDM 56-Gbaud PDM 16-QAM using a single I/Q modulator, in *European Conference on Optical Communication (ECOC 2010)*, PD2.2 (2010)
21. G. Raybon, A. Adamiecki, P.J. Winzer, M. Montoliu, S. Randel, A. Umbach, M. Margraf, J. Stephan, S. Draving, M. Grove, K. Rush, All-ETDM 107-Gbaud PDM-16QAM (856-Gb/s) transmitter and coherent receiver, in *ECOC 2013*, PD.2.D.3 (2013)
22. S. Okamoto, T. Omiya, K. Kasai, M. Yoshida, M. Nakazawa, 140 Gbit/s coherent optical transmission over 150 km with a 10 Gsymbol/s polarization-multiplexed 128 QAM signal, in *Optical Fiber Communication Conference (OFC 2010)*, OThD5 (2010)
23. A.H. Gnauck, P. Winzer, A. Konczykowska, F. Jorge, J. Dupuy, M. Riet, G. Charlet, B. Zhu, D.W. Peckham, Generation and transmission of 21.4 Gbaud PDM 64 QAM using a high power DAC driving a single I/Q modulator, in *OFC 2011*, PDPB2 (2011)
24. T. Kobayashi, A. Sano, A. Matsuura, Y. Miyamoto, K. Ishihara, Nonlinear tolerant long-haul WDM transmission over 1200 km using 538 Gb/s/ch PDM-64QAM SC-FDM signals with pilot tone, in *OFC 2012*, OM2A.5 (2012)
25. W.-R. Peng, H. Takahashi, T. Tsuritani, I. Morita, 50-GHz-spaced, 8 x 499-Gb/s WDM transmission over 720-km SSMF using per-channel 41.6-GBd PDM-64QAM, in *ACP 2012*, AF4C1 (2012)
26. S. Randel, D. Pilori, S. Corteselli, G. Raybon, A. Adamiecki, A. Gnauck, S. Chandrasekhar, P. Winzer, L. Altenhain, A. Bielik, R. Schmid, All-electronic flexibly programmable 864-Gb/s single-carrier PDM-64-QAM, in *OFC 2014*, Th5C.8 (2014)
27. J. G. Proakis, *Digital Communications*, 4th edn. (McGraw Hill, 2000)
28. C.E. Shannon, A mathematical theory of communication. Bell Syst. Tech. J. **27**, 379–423, 623–656 (1948)
29. T. Mizuochi, Recent progress in forward error correction for optical communication systems. IEICE Trans. Commun. **E88-B**, 1934–1946 (2005)
30. K. Kasai, A. Suzuki, M. Yoshida, M. Nakazawa, Performance improvement of an acetylene (C_2H_2) frequency-stabilized fiber laser. IEICE Electron. Express **3**, 487–492 (2006)
31. K. Sato, N. Kobayashi, M. Namiwaka, K. Yamamoto, T. Kita, H. Yamada, H. Yamazaki, High output power and narrow linewidth silicon photonic hybrid ring-filter external cavity wavelength tunable lasers, in *ECOC 2014*, PD.2.3 (2014)

32. L. Stolpner, S. Lee, S. Li, A. Mehnert, P. Mols, S. Siala, J. Bush, Low noise planar external cavity laser for interferometric fiber optic sensors. Proc. SPIE **7004**, 700457 (2008)
33. H.T. Quynhanh, A. Suzuki, M. Yoshida, T. Hirooka, M. Nakazawa, A λ/4-shifted distributed-feedback laser diode with a fiber ring cavity configuration having an OSNR of 85 dB and a linewidth of 7 kHz. IEEE Photon. Technol. Lett. **20**(18), 1578–1580 (2008)
34. W. Liang, V.S. Ilchenko, A.A. Savchenkov, A.B. Matsko, D. Seidel, L. Maleki, Whispering-gallery-mode-resonator-based ultranarrow linewidth external-cavity semiconductor laser. Opt. Lett. **35**(16), 2822–2824 (2010)
35. S. Shimotsu, S. Oikawa, T. Saitou, N. Mitsugi, K. Kubodera, T. Kawanishi, M. Izutsu, Single side-band modulation performance of a $LiNbO_3$ integrated modulator consisting of four-phase modulator waveguides. IEEE Photon. Technol. Lett. **13**, 364–366 (2001)
36. N. Kikuchi, Y. Shibata, K. Tsuzuki, H. Sanjoh, T. Sato, E. Yamada, T. Ishibashi, H. Yasaka, 80-Gb/s low-driving-voltage InP DQPSK modulator with an n-p-i-n structure. IEEE Photon. Technol. Lett. **21**, 787–789 (2009)
37. H. Yamazaki, T. Yamada, T. Goh, Y. Sakamaki, A. Kaneko, 64QAM modulator with a hybrid configuration of silica PLCs and $LiNbO_3$ phase modulators. IEEE Photon. Technol. Lett. **22**, 344–346 (2010)
38. D.-S. Ly-Gagnon, S. Tsukamoto, K. Katoh, K. Kikuchi, Coherent detection of optical quadrature phase-shift keying signals with carrier phase estimation. J. Lightwave Technol. **24**, 12–21 (2006)
39. K. Kikuchi, T. Okoshi, M. Nagamatsu, N. Henmi, Degradation of bit-error rate in coherent optical communications due to spectral spread of the transmitter and the local oscillator. J. Lightwave Technol. **LT-2**, 1024–1033 (1984)
40. K. Kasai, J. Hongo, M. Yoshida, M. Nakazawa, Optical phase-locked loop for coherent transmission over 500 km using heterodyne detection with fiber lasers. IEICE Electron. Express **4**, 77–81 (2007)
41. C.R. Doerr, L. Zhang, P.J. Winzer, N. Weimann, V. Houtsma, T.-C. Hu, N.J. Sauer, L.L. Buhl, D.T. Neilson, S. Chandrasekhar, Y.K. Chen, Monolithic InP dual-polarization and dual-quadrature coherent receiver. IEEE Photon. Technol. Lett. **23**, 694–696 (2011)
42. A. Antoniou, *Digital Signal Processing: Signals, Systems, and Filters* (McGraw-Hill Professional, 2005)
43. D. Falconer, S.L. Ariyavisitakul, A. Benyamin-Seeyar, B. Eidson, Frequency domain equalisation for single-carrier broadband wireless systems. IEEE Commun. Mag. **40**(4), 58–66 (2002)
44. M.V. Clark, Adaptive frequency-domain equalisation and diversity combining for broadband wireless communications. IEEE J. Sel. Areas Commun. **16**(8), 1385–1395 (1998)
45. K. Ishihara, T. Kobayashi, R. Kudo, Y. Takatori, A. Sano, E. Yamada, H. Masuda, Y. Miyamoto, Frequency-domain equalisation for optical transmission systems. Electron. Lett. **44**, 870–871 (2008)
46. J.P. Gordon, L.F. Mollenauer, Phase noise in photonic communication systems using linear amplifiers. Opt. Lett. **15**, 1351–1353 (1990)
47. A.D. Ellis, J. Zhao, D. Cotter, Approaching the non-linear Shannon limit. J. Lightwave Technol. **28**(4), 423–433 (2010)
48. R.-J. Essiambre, G. Kramer, P.J. Winzer, G.J. Foschini, B. Goebel, Capacity limits of optical fiber networks. J. Lightwave Technol. **28**(4), 662–701 (2010)
49. C. Paré, A. Villeneuve, P.-A. Bélanger, N.J. Doran, Compensating for dispersion and the nonlinear Kerr effect without phase conjugation. Opt. Lett. **21**(7), 459–461 (1996)
50. X. Li, X. Chen, G. Goldfarb, E. Mateo, I. Kim, F. Yaman, G. Li, Electronic post-compensation of WDM transmission impairments using coherent detection and digital signal processing. Opt. Express **16**, 880–888 (2008)
51. P.K.A. Wai, C.R. Menyuk, H.H. Chen, Stability of solitons in randomly varying birefringent fibers. Opt. Lett. **16**(16), 1231–1233 (1991)
52. K. Kasai, J. Hongo, H. Goto, M. Yoshida, M. Nakazawa, The use of a Nyquist filter for reducing an optical signal bandwidth in a coherent QAM optical transmission. IEICE Electron. Express **5**(1), 6–10 (2008)

53. H. Nyquist, Certain topics in telegraph transmission theory. AIEE Trans. **47**, 617–644 (1928)
54. S. Hara, R. Prasad, *Multicarrier Techniques for 4G Mobile Communications* (Artech House, Boston, 2003)
55. Y. Ma, Q. Yang, Y. Tang, S. Chen, W. Shieh, 1-Tb/s per channel coherent optical OFDM transmission with subwavelength bandwidth access, in *Optical Fiber Communication Conference (OFC 2009)*, PDPC1 (2009)
56. S. Chandrasekhar, X. Liu, B. Zhu, D. Peckham, Transmission of a 1.2-Tb/s 24-carrier no-guard-interval coherent OFDM superchannel over 7200-km of ultra-large-area fiber, in *European Conference on Optical Communication (ECOC 2009)*, PD2.6 (2009)
57. D. Hillerkuss, R. Schmogrow, T. Schellinger, M. Jordan, M. Winter, G. Huber, T. Vallaitis, R. Bonk, P. Kleinow, F. Frey, M. Roeger, S. Koenig, A. Ludwig, A. Marculescu, J. Li, M. Hoh, M. Dreschmann, J. Meyer, S. Ben Ezra, N. Narkiss, B. Nebendahl, F. Parmigiani, P. Petropoulos, B. Resan, A. Oehler, K. Weingarten, T. Ellermeyer, J. Lutz, M. Moeller, M. Huebner, J. Becker, C. Koos, W. Freude, J. Leuthold, 26 Tbits-1 line-rate super-channel transmission utilizing all-optical fast Fourier transform processing. Nat. Photonics **5**, 364–371 (2011)
58. H. Takahashi, K. Takeshima, I. Morita, H. Tanaka, 400-Gbit/s optical OFDM transmission over 80 km in 50-GHz frequency grid, in *European Conference on Optical Communication (ECOC 2010)*, Tu.3.C.1 (2010)
59. X. Liu, S. Chandrasekhar, T. H. Lotz, P. J. Winzer, H. Haunstein, S. Randel, S. Corteselli, B. Zhu, and D. Peckham, Generation and FEC-decoding of a 231.5-Gb/s PDM-OFDM signal with 256-iterative-polar-modulation achieving 11.15-b/s/Hz intrachannel spectral efficiency and 800-km reach, in *Optical Fiber Communication Conference (OFC 2012)*, PDP5B.3 (2012)
60. C. Zhang, Y. Mori, M. Usui, K. Igarashi, K. Katoh, and K. Kikuchi, Straight-line 1,073-km transmission of 640-Gbit/s dual-polarization QPSK signals on a single carrier, in *Proceedings of European Conference on Optical Communication (ECOC)*, Vienna, Austria, 2009, Postdeadline paper PD2.8 (2009)
61. T. Richter, E. Palushani, C. Schmidt-Langhorst, M. Nölle, R. Ludwig, J.K. Fischer, C. Schubert, Single wavelength channel 10.2 Tb/s TDM-data capacity using 16-QAM and coherent detection, in *Optical Fiber Communication Conference (OFC 2011)*, PDPA9 (2011)
62. K. Kasai, T. Omiya, P. Guan, M. Yoshida, T. Hirooka, M. Nakazawa, Single-channel 400-Gb/s OTDM-32 RZ/QAM coherent transmission over 225 km using an optical phase-locked loop technique. IEEE Photon. Technol. Lett. **22**(8), 562–564 (2010)
63. M. Nakazawa, T. Hirooka, P. Ruan, P. Guan, Ultrahigh-speed orthogonal TDM transmission with an optical Nyquist pulse train. Opt. Express **20**(2), 1129–1140 (2012)
64. K. Harako, D. Seya, T. Hirooka, M. Nakazawa, 640 Gbaud (1.28 Tbit/s/ch) optical Nyquist pulse transmission over 525 km with substantial PMD tolerance. Opt. Express **21**(18), 21063–21076 (2013)
65. H. Hu, D. Kong, E. Palushani, J.D. Andersen, A. Rasmussen, B.M. Sørensen, M. Galili, H.C.H. Mulvad, K.J. Larsen, S. Forchhammer, P. Jeppesen, L.K. Oxenløwe, 1.28 Tbaud Nyquist signal transmission using time-domain optical fourier transformation based receiver, in *CLEO 2013*, CTh5D.5 (2013)
66. H.N. Tan, K. Tanizawa, T. Inoue, T. Kurosu, S. Namiki, Seamless spectral defragmentation of Nyquist OTDM-WDM signals in add-drop node for all-optical elastic network, in *ECOC 2013*, We.1.C.5 (2013)
67. K. Kimura, J. Nitta, M. Yoshida, K. Kasai, T. Hirooka, M. Nakazawa, Single-channel 7.68 Tbit/s, 64 QAM coherent Nyquist pulse transmission over 150 km with a spectral efficiency of 9.7 bit/s/Hz. Opt. Express **26**(13), 17418–17428 (2018)
68. K. Harako, D.O. Otuya, K. Kasai, T. Hirooka, M. Nakazawa, High-performance TDM demultiplexing of coherent Nyquist pulses using time-domain orthogonality. Opt. Express **22**(24), 29456–29464 (2014)
69. T. Hirooka, R. Hirata, J. Wang, M. Yoshida, M. Nakazawa, Single-channel 10.2 Tbit/s (2.56 Tbaud) optical Nyquist pulse transmission over 300 km. Opt. Express **26**(21), 27221–27236 (2018)

70. E. Yamazaki, S. Yamanaka, Y. Kisaka, T. Nakagawa, K. Murata, E. Yoshida, T. Sakano, M. Tomizawa, Y. Miyamoto, S. Matsuoka, J. Matsui, A. Shibayama, J. Abe, Y. Nakamura, H. Noguchi, K. Fukuchi, H. Onaka, K. Fukumitsu, K. Komaki, O. Takeuchi, Y. Sakamoto, H. Nakashima, T. Mizuochi, K. Kubo, Y. Miyata, H. Nishimoto, S. Hirano, K. Onohara, Fast optical channel recovery in field demonstration of 100-Gbit/s ethernet over OTN using real-time DSP. Opt. Express **19**(14), 13179–13184 (2011)
71. M. Tomizawa, A. Hirano, Y. Miyamoto, Safety issues in high-power optical fiber communication systems, including distributed Raman amplification systems, in *International Laser Safety Conference ILSC2003* (2003), pp. 291–297
72. T. Morioka, Y. Awaji, R. Ryf, P. Winzer, D. Richardson, F. Poletti, Enhancing optical communications with brand new fibers. IEEE Commun. Mag. **50**(2), s31–s42 (2012)
73. J. Sakaguchi, Y. Awaji, N. Wada, A. Kanno, T. Kawanishi, T. Hayashi, T. Taru, T. Kobayashi, M. Watanabe, 109-Tb/s (7×97×172-Gb/s SDM/WDM/PDM) QPSK transmission through 16.8-km homogeneous multi-core fiber, in *Optical Fiber Communication Conference* (2011), p. PDPB
74. B. Zhu, X. Liu, S. Chandrasekhar, T.F. Taunay, M. Fishteyn, M.F. Yan, J.M. Fini, E.M. Monberg, F.V. Dimarcello, 112-Tb/s (7×160×107Gb/s) space-division multiplexed DWDM transmission over a 76.8-km multicore fiber, in *European Conference and Exhibition on Optical Communication* (2011), p. Tu.5.B.5
75. J. Sakaguchi, B.J. Puttnam, W. Klaus, Y. Awaji, N. Wada, A. Kanno, T. Kawanishi, K. Imamura, H. Inaba, K. Mukasa, R. Sugizaki, T. Kobayashi, M. Watanabe, 19-core fiber transmission of 19 x 100 x 172-Gb/s SDM-WDM-PDM-QPSK signals at 305Tb/s, in *Optical Fiber Communication Conference* (2012), p. PDP5C.1
76. H. Takara, A. Sano, T. Kobayashi, H. Kubota, H. Kawakami, A. Matsuura, Y. Miyamoto, Y. Abe, H. Ono, K. Shikama, Y. Goto, K. Tsujikawa, Y. Sasaki, I. Ishida, K. Takenaga, S. Matsuo, K. Saitoh, M. Koshiba, T. Morioka, 1.01-Pb/s (12 SDM/222 WDM/456 Gb/s) crosstalk-managed transmission with 91.4-b/s/Hz aggregate spectral efficiency, in *European Conference and Exhibition on Optical Communication* (2012), p. Th.3.C.1
77. B.J. Puttnam, R.S. Luís, W. Klaus, J. Sakaguchi, J.-M. Delgado Mendinueta, Y. Awaji, N. Wada, Y. Tamura, T. Hayashi, M. Hirano, J. Marciante, 2.15 Pb/s transmission using a 22 core homogeneous singlemode multi-core fiber and wideband optical comb, in *European Conference and Exhibition on Optical Communication* (2015), p. PDP.3.1
78. T. Kobayashi, H. Takara, A. Sano, T. Mizuno, H. Kawakami, Y. Miyamoto, K. Hiraga, Y. Abe, H. Ono, M. Wada, Y. Sasaki, I. Ishida, K. Takenaga, S. Matsuo, K. Saitoh, M. Yamada, H. Masuda, T. Morioka, 2 x 344 Tb/s propagation-direction interleaved transmission over 1500-km MCF enhanced by multicarrier full electric-field digital back-propagation, in *European Conference and Exhibition on Optical Communication* (2013), p. PD3.E.4
79. K. Igarashi, T. Tsuritani, I. Morita, M. Suzuki, Ultra-long-haul high-capacity super-Nyquist-WDM transmission experiment using multi-core fibers. IEEE J. Lightwave Technol. **33**(5), 1027–1036 (2015)
80. P. Winzer, A. Gnauck, A. Konczykowska, F. Jorge, J.-Y. Dupuy, Penalties from in-band crosstalk for advanced optical modulation formats, in *37th European Conference and Exposition on Optical Communications* (2011), p. Tu.5.B.7
81. F. Ye, J. Tu, K. Saitoh, H. Takara, T. Morioka, Wavelength-dependent crosstalk in trench-assisted multi-core fibers, in *OECC/ACOFT 2014* (2014), pp. 308–309
82. R. Ryf, R. Essiambre, A.H. Gnauck, S. Randel, M.A. Mestre, C. Schmidt, P.J. Winzer, R. Delbue, P. Pupalaikis, A. Sureka, T. Hayashi, T. Taru, T. Sasaki, Space-division multiplexed transmission over 4200-km 3-core microstructured fiber, in *Optical Fiber Communication Conference, 2012* (2012), p. PDP5C.2
83. R.S. Luís, B.J. Puttnam, J.M. Delgado Mendinueta, W. Klaus, Y. Awaji, N. Wada, Comparing inter-core skew fluctuations in multi-core and single-core fibers, in *Proceedings of CLEO*, SM3L.5, San Jose (2015)
84. R.S. Luís, B.J. Puttnam, A.V.T. Cartaxo, W. Klaus, J.M.D. Mendinueta, Y. Awaji, N. Wada, T. Nakanishi, T. Hayashi, T. Sasaki, Time and modulation frequency dependence of crosstalk in homogeneous multi-core fibers. IEEE J. Lightwave Technol. **34**(2), 441–447 (2016)

85. M.D. Feuer, L.E. Nelson, X. Zhou, S.L. Woodward, R. Isaac, B. Zhu, T.F. Taunay, M. Fishteyn, J.M. Fini, M.F. Yan, Joint digital signal processing receivers for spatial superchannels. IEEE Photon. Technol. Lett. **24**, 1957–1960 (2012)
86. B.J. Puttnam, R.S. Luís, J.M. Delgado Mendinueta, Y. Awaji, N. Wada, E. Agrell, Linear block-coding across >5 Tb/s PDM-64QAM spatial-super-channels in a 19-core Fiber, in *Proceedings of ECOC*, paper P.5.6 (2015)
87. D.S. Millar, T. Koike-Akino, S.Ö. Arık, K. Kojima, K. Parsons, T. Yoshida, T. Sugihara, High-dimensional modulation for coherent optical communications systems. Opt. Express **22**(7), 8798–8812 (2014)
88. A. Sano, H. Takara, T. Kobayashi, Y. Miyamoto, Crosstalk-managed high capacity long haul multicore fiber transmission with propagation-direction interleaving. J. Lightwave Technol. **32**(16), 2771–2779 (2014)
89. K. Takenaga, Multicore fiber with dual-ring structure, in *OECC/ACOFT 2014* (2014), pp. 51–53
90. J.G. Proakis, M. Salehi, *Digital Communications*, 5th edn. (McGraw-Hill, 2008)
91. G. Bosco, A. Carena, P. Poggiolini, F. Forghieri, Performance limits of Nyquist-WDM and CO-OFDM in high-speed PM-QPSK systems. IEEE Photon. Technol. Lett. **22**(15), 1129–1131 (2010)
92. J. Li, E. Tipsuwannakul, T. Eriksson, M. Karlsson, P.A. Andrekson, Approaching Nyquist limit in WDM systems by low-complexity receiver-side duobinary shaping. J. Lightwave Technol. **30**(11), 1664–1676 (2012)
93. J. Zhang, J. Yu, Z. Dong, Z. Jia, H.C. Chien, Y. Cai, C. Ge, S. Shi, Y. Chen, H. Wang, Y. Xia, Transmission of 20×440-Gb/s super-Nyquist-filtered signals over 3600 km based on single-carrier 110-GBaud PDM QPSK with 100-GHz grid, in *OFC 2014*, Los Angeles, CA, USA, Th5B.3 (2014)
94. G.D. Forney Jr., Maximum-likelihood sequence estimation of digital sequences and the presence of intersymbol interference. IEEE Trans. Inform. Theory **IT-15**(3), 363–378 (1972)
95. D. Chang, F. Yu, Z. Xiao, N. Stojanovic, F.N. Hauske, Y. Cai, C. Xie, L. Li, X. Xu, Q. Xiong, LDPC convolutional codes using layered decoding algorithm for high speed coherent optical transmission, in *OFC 2012*, Los Angeles, CA, USA, OW1H.4 (2012)
96. S. Chandrasekhar, A.H. Gnauck, X. Liu, P.J. Winzer, Y. Pan, E.C. Burrows, B. Zhu, T.F. Taunay, M. Fishteyn, M.F. Yan, J.M. Fini, E.M. Monberg, F.V. Dimarcello, WDM/SDM transmission of 10 × 128-Gb/s PDM-QPSK over 2688-km 7-core fiber with a per-fiber net aggregate spectral-efficiency distance product of 40,320 km·bit/s/Hz, in *ECOC 2011*, Geneva, Switzerland, Th.3.C4 (2011)
97. Y. Mori, Z. Chao, K. Kikuchi, Novel FIR-filter configuration tolerant to fast phase fluctuations in digital coherent receivers for higher-order QAM signals, in *OFC 2012*, Los Angeles, CA, USA, OTh4C.4 (2012)
98. P.J. Winzer, Spatial multiplexing: the next frontier in network capacity scaling, in *Proceedings of European Conference on Optical Communication*, Paper We.1.D.1 (2013)
99. P.M. Krummrich, Optical amplification and optical filter based signal processing for cost and energy efficient spatial multiplexing. Opt. Express **19**(17), 16636–16652 (2011)
100. K.S. Abedin, J.M. Fini, T.F. Thierry, V.R. Supradeepa, B. Zhu, M.F. Yan, L. Bansal, E.M. Monberg, D.J. DiGiovanni, Multicore erbium doped fiber amplifiers for space division multiplexing systems. J. Lightwave Technol. **32**(16), 2800–2808 (2014)
101. Y. Tsuchida, K. Maeda, R. Sugizaki, Multicore EDFA for space division multiplexing, in *Proceedings of 18th OptoElectronics and Communications Conference*, Paper TuS1-1 (2013)
102. K. Takeshima, T. Tsuritani, Y. Tsuchida, K. Maeda, T. Saito, K. Watanabe, T. Sasa, K. Imamura, R. Sugizaki, K. Igarashi, I. Morita, WDM/SDM transmission of 76 x 128-Gbit/s Nyquist-pulse-shaped DP-QPSK signals over 4,200 km using cladding pumped 7-core EDFA, in *Proceedings of 20th OptoElectronics and Communications Conference*, Paper JThA.13 (2015)
103. K.S. Abedin, T.F. Taunay, M. Fishteyn, D.J. DiGiovanni, V.R. Supradeepa, J.M. Fini, M.F. Yan, B. Zhu, E.M. Monberg, F.V. Dimarcello, Cladding-pumped erbium-doped multicore fiber amplifier. Opt. Express **20**, 20191–20200 (2012)

104. Y. Tsuchida, K. Maeda, K. Watanabe, T. Saito, S. Takasaka, M. Tadakuma, R. Sugizaki, H. Ogoshi, K. Doi, H. Matsuura, K. Takeshima, T. Tsuritani, I. Morita, Cladding-pumped L-band multicore EDFA with reduced power consumption, in *IEEE Summer Topicals*, ME2.2 (2014)
105. S. Takasaka, H. Matsuura, W. Kumagai, M. Tadakuma, Y. Mimura, Y. Tsuchida, K. Maeda, R. Miyabe, K. Aiso, K. Doi, R. Sugizaki, Cladding-pumped seven-core EDFA using a multimode pump light coupler, in *Proceedings of European Conference on Optical Communication*, Paper We.4.A.5 (2013)
106. K. Igarashi, T. Tsuritani, I. Morita, Y. Tsuchida, K. Maeda, M. Tadakuma, T. Saito, K. Watanabe, R. Sugizaki, M. Suzuki, 1.03-exabit/s·km super-Nyquist-WDM transmission over 7,326-km seven-core fiber, in *Proceedings of European Conference on Optical Communication*, Paper PDP3.E.3 (2013)
107. K. Igarashi, K. Takeshima, T. Tsuritani, H. Takahashi, S. Sumita, I. Morita, Y. Tsuchida, M. Tadakuma, K. Maeda, T. Saito, K. Watanabe, K. Imamura, R. Sugizaki, M. Suzuki, 110.9-Tbit/s SDM transmission over 6,370 km using a full C-band seven-core EDFA. Opt. Express **21**(15), 18053–18060 (2013)
108. Y. Kawaguchi, T. Tsuritani, Ultra-long-haul multicore fiber transmission over 5,000 km using cladding pumped seven-core EDFA, in *OECC 2017*, 3-1K-3 (2017)
109. S. Todoroki, *Fiber Fuse* (Springer, 2014)
110. D.G. Foursa, H.G. Batshon, H. Zhang, M. Mazurczyk, J.-X. Cai, O. Sinkin, A. Pilipetskii, G. Mohs, N.S. Bergano, 44.1 Tb/s transmission over 9,100 km using coded modulation based on 16QAM signals at 4.9 bits/s/Hz spectral efficiency, in *European Conference on Optical Communication (ECOC)* (2013), p. PD3.E.1
111. T. Morioka, New generation optical infrastructure technologies: EXAT initiative towards 2020 and beyond, in *OECC2009*, FT4 (2009)
112. D. Soma, K. Igarashi, Y. Wakayama, K. Takeshima, Y. Kawaguchi, N. Yoshikane, T. Tsuritani, I. Morita, M. Suzuki, 2.05 Peta-bit/s super-Nyquist-WDM SDM transmission using 9.8-km 6-mode 19-core fiber in full C band, in *Proceedings of ECOC 2015*, Valencia, PD Papers, PDP3.2 (2015)
113. L. Dou, Z. Tao, L. Li, W. Yan, T. Tanimura, T. Hoshida, J.C. Rasmussen, A low complexity pre-distortion method for intra-channel nonlinearity, in *OFC 2011*, OThF5 (2011)
114. K. Kikuchi, Electric post-compensation for nonlinear phase fluctuations in a 1000-km 20-Gbit/s optical quadrature phase-shift keying transmission system using the digital coherent reciever. Opt. Express **16**(2), 889–896 (2008)
115. A. Yariv, D. Fekete, D.M. Pepper, Compensation for channel dispersion by nonlinear optical phase conjugation. Opt. Lett. **4**, 52–54 (1979)
116. S. Watanabe, T. Chikama, G. Ishikawa, T. Terahara, H. Kuwahara, Compensation of pulse shape distortion due to chromatic dispersion and Kerr effect by optical phase conjugation. IEEE photon. Technol. Lett. **5**, 1241–1243 (1993)
117. T. Umeki, T. Kazama, H. Ono, Y. Miyamoto, H. Takenouchi, Spectrally efficient optical phase conjugation based on complementary spectral inversion for nonlinearity mitigation, in *ECOC 2015*, Valencia, Spain, We.2.6.2 (2015)
118. A.D. Ellis, I.D. Phillips, M. Tan, M.F.C. Stephens, M.E. McCarthy, M.A.Z. Al Kahteeb, M.A. Iqbal, A. Perentos, S. Fabbri, V. Gordienko, D. Lavery, G. Liga, G. Saavedra, M.R. Maher, S. Sygletos, P. Harper, N.J. Doran, P. Bayvel, S.K. Turitsyn, Enhanced superchannel transmission using phase conjugation, in *ECOC 2015*, Valencia, Spain, We.2.6.4 (2015)
119. M. Koga, A. Mizutori, T. Ohata, H. Takara, Optical diversity transmission with signal and its phaseconjugate lights through multi-core fiber, in *OFC 2015*, Th1D.4 (2015)
120. M. Koga, M. Moroi, H. Takara, Optical diversity transmission with WDM signal and phase-conjugate lights through multi-core fiber. Opt. Express **24**(9), 9340–9352 (2016)
121. T. Iida, A. Mizutori, M. Koga, Optical diversity transmission and maximum-ratio combine in multi-core fiber to mitigate fiber non-linear distortion, in *OECC 2012*, Busan, Korea, July 2012, 6B2-5 (2012), pp. 757–758
122. S. Stein, Maximum-ratio combining, in *Communication Systems and Techniques* (IEEE Press, NJ, 1966), pp. 440–441

123. V.A.J.M. Sleiffer, Y. Jung, V. Veljanovski, R.G.H. van Uden, M. Kuschnerov, Q. Kang, L. Grüner-Nielsen, Y. Sun, D.J. Richardson, S. Alam, F. Poletti, J.K. Sahu, A. Dhar, H. Chen, B. Inan, A.M.J. Koonen, B. Corbett, R. Winfield, A.D. Ellis, H. de Waardt, 73.7 Tb/s (96 x 3 x 256-Gb/s) mode-division-multiplexed DP-16QAM transmission with inline MM-EDFA, in *Proceedings of ECOC*, Amsterdam, PD Papers, Th.3.C.4 (2012)
124. E. Ip, M.-J. Li, K. Bennett, Y.-K. Huang, A. Tanaka, A. Korolev, K. Koreshkov, W. Wood, E. Mateo, J. Hu, Y. Yano, 146λ×6×19-Gbaud wavelength- and mode-division multiplexed transmission over 10×50-km spans of few-mode fiber with a gain-equalized few-mode EDFA, in *Proceedings of OFC/NFOEC*, Anaheim, PD Papers, PDP5A.2 (2013)
125. M. Esmaeelpour, R. Ryf, N.K. Fontaine, H. Chen, A.H. Gnauck, R.-J. Essiambre, J. Toulouse, Y. Sun, R. Lingle Jr., Transmission over 1050-km few-mode fiber based on bidirectional distributed Raman amplification. IEEE J. Lightwave Technol. **34**(8), 1864–1871 (2016)
126. R. Ryf, S. Randel, N.K. Fontaine, M. Montoliu, E. Burrows, S. Corteselli, S. Chandrasekhar, A.H. Gnauck, C. Xie, R.-J. Essiambre, P.J. Winzer, R. Delbue, P. Pupalaikis, A. Sureka, Y. Sun, L. Grüner-Nielsen, R.V. Jensen, R. Lingle Jr., 32-bit/s/Hz spectral efficiency WDM transmission over 177-km few-mode fiber, in *Proceedings of OFC*, Anaheim, PD Papers, PDP5A.1 (2013)
127. Y. Chen, A. Lobato, Y. Jung, H. Chen, R.V. Jensen, Y. Sun, L. Grüner-Nielsen, D.J. Richardson, V.A.J.M. Sleiffer, M. Kuschnerov, N.K. Fontaine, R. Ryf, I.P. Giles, R. Chen, V. Carcia-Munoz, A.M.J. Koonen, B. Lankl, nd N. Hanik, 41.6 Tb/s C-band SDM OFDM transmission through 12 spatial and polarization modes over 74.17 km few mode fiber, in *Proceedings of ECOC*, Cannes, Mo.3.3.3 (2014)
128. T. Mizuno, H. Takara, A. Sano, Y. Miyamoto, Dense space division multiplexed transmission systems using multi-core and multi-mode fiber. IEEE J. Lightwave Technol. **34**(2), 582–592 (2016)
129. P. Sillard, M.B.-Astruc, D. Molin, Few-mode fibers for mode-division-multiplexed systems. IEEE J. Lightwave Technol. **32**(16), 2824–2829 (2014)
130. T. Sakamoto, T. Mori, T. Yamamoto, S. Tomita, Differential mode delay managed transmission line for WDM-MIMO system using multi-step index fiber. IEEE J. Lightwave Technol. **30**(17), 2783–2787 (2012)
131. T. Mizuno, T. Kobayashi, H. Takara, A. Sano, H. Kawakami, T. Nakagawa, Y. Miyamoto, Y. Abe, T. Goh, M. Oguma, T. Sakamoto, Y. Sasaki, I. Ishida, K. Takenaga, S. Matsuo, K. Saitoh, T. Morioka, 12-core × 3-mode dense space division multiplexed transmission over 40 km employing multi-carrier signals with parallel MIMO equalization, in *Proceedings of OFC*, San Francisco, PD Papers, Th5B.2 (2014)
132. K. Shibahara, D. Lee, T. Kobayashi, T. Mizuno, H. Takara, A. Sano, H. Kawakami, Y. Miyamoto, H. Ono, M. Oguma, Y. Abe, T. Matsui, R. Fukumoto, Y. Amma, T. Hosokawa, S. Matsuo, K. Saitoh, M. Yamada, T. Morioka, Dense SDM (12-core × 3-mode) transmission over 527 km with 33.2-ns mode-dispersion employing low-complexity parallel MIMO frequency-domain equalization. IEEE J. Lightwave Technol. **34**(1), 196–204 (2016)
133. T. Mizuno, H. Takara, K. Shibahara, A. Sano, Y. Miyamoto, Dense space division multiplexed transmission over multicore and multimode fiber for long-haul transport systems, IEEE J. Lightwave Technol. **34**(6), 1484–1493 (2016)
134. Y. Sasaki, Y. Amma, K. Takenaga, S. Matsuo, K. Saitoh, M. Koshiba, Few-mode multi-core fiber with 36 spatial modes (three modes (LP01, LP11a, LP11b) × 12 cores). IEEE J. Lightwave Technol. **33**(5), 964–970 (2015)
135. H. Ono, T. Hosokawa, K. Ichii, S. Matsuo, M. Yamada, Improvement of differential modal gain in few-mode fibre amplifier by employing ring-core erbium-doped fibre. Electron. Lett. **51**(2), 172–173 (2015)
136. A. Lobato, F. Ferreira, M. Kuschnerov, D. Borne, S.L. Jansen, A. Napoli, B. Spinnler, B. Lankl, Impact of mode coupling on the mode dependent loss tolerance in few-mode fiber transmission. Opt. Express **20**(28), 29776–29783 (2012)

137. R. Ryf, H. Chen, N.K. Fontaine, A.M. Velazquez-Benitez, J. Antonio-Lopez, C. Jin, B. Huang, M. Bigot-Astruc, D. Molin, F. Achten, P. Sillard, R. Amezcua-Correa, 10-mode mode-multiplexed transmission over 125-km single-span multimode fiber, in *Proceedings of ECOC*, Valencia, PDP.3.3 (2015)
138. N.K. Fontaine, R. Ryf, H. Chen, A.V. Benitez, J.E.A. Lopez, R.A. Correa, B. Guan, B. Ercan, R.P. Scott, S.J. Ben Yoo, L. Grüner-Nielsen, Y. Sun, R.J. Lingle, 30 x 30 MIMO transmission over 15 spatial modes, in *Proceedings of OFC*, Los Angeles, PD Papers, Th5C.1 (2015)
139. K. Shibahara, T. Mizuno, D. Lee, Y. Miyamoto, H. Ono, K. Nakajima, Y. Amma, K. Takenaga, K. Saitoh, DMD-unmanaged long-haul SDM transmission over 2500-km 12-core × 3-mode MC-FMF and 6300-km 3-mode FMF employing intermodal interference canceling technique. IEEE J. Lightwave Technol. **37**(1), 138–147 (2019)
140. T. Mizuno, K. Shibahara, H. Ono, Y. Miyamoto, Long-distance PDM-32QAM 3-mode fibre transmission over 1000 km using hybrid multicore EDFA/Raman repeated amplification with cyclic mode permutation, in *Proceedings of ECOC*, Sept 2018, Mo3G.6
141. K. Shibahara, T. Mizuno, H. Kawakami, T. Kobayashi, M. Nakamura, K. Shikama, K. Nakajima, Y. Miyamoto, Full C-band 3060-km DMD-unmanaged 3-mode transmission with 40.2-Tb/s capacity using cyclic mode permutation, in *Proceedings of OFC*, San Diego, Mar 2019, W3F.2
142. T. Mizuno, Y. Miyamoto, High-capacity dense space division multiplexing transmission. Opt. Fiber Technol. **35**, 108–117 (2017)
143. R.G.H. van Uden, R. Amezcua Correa, E. Antonio-Lopez, F.M. Huijskens, G. Li, A. Schulzgen, H. de Waardt, A.M.J. Koonen, C.M. Okonkwo, 1 km hole-assisted few-mode multi-core fiber 32QAM WDM transmission, in *Proceedings of ECOC*, Cannes, Mo.3.3.4 (2014)
144. K. Shibahara, T. Mizuno, D. Lee, Y. Miyamoto, Advanced MIMO signal processing techniques enabling long-haul dense SDM transmissions. IEEE J. Lightwave Technol. **36**(2), 336–348 (2018)
145. K. Igarashi, D. Souma, Y. Wakayama, K. Takeshima, Y. Kawaguchi, N. Yoshikane, T. Tsuritani, I. Morita, M. Suzuki, Ultra-dense spatial-division-multiplexed transmission over 6-mode 19-core fibers. Opt. Express **24**(10), 10213–10231 (2016)
146. J. Sakaguchi, W. Klaus, J.-M.D. Mendinueta, B.J. Puttnam, R.S. Luis, Y. Awaji, N. Wada, T. Hayashi, T. Nakanishi, T. Watanabe, Y. Kokubun, T. Takahata, T. Kobayashi, Realizing a 36-core, 3-mode fiber with 108 spatial channels, in *Proceedings of OFC*, Los Angeles, PD Papers, Th5C.2 (2015)
147. T. Mizuno, K. Shibahara, F. Ye, Y. Sasaki, Y. Amma, K. Takenaga, Y. Jung, K. Pulverer, H. Ono, Y. Abe, M. Yamada, K. Saitoh, S. Matsuo, K. Aikawa, M. Bohn, D.J. Richardson, Y. Miyamoto, T. Morioka, Long-haul dense space-division multiplexed transmission over low-crosstalk heterogeneous 32-core transmission line using a partial recirculating loop system. IEEE J. Lightwave Technol. **35**(3), 488–498 (2017)
148. S. Matsuo, K. Takenaga, Y. Sasaki, Y. Amma, S. Saito, K. Saitoh, T. Matsui, K. Nakajima, T. Mizuno, H. Takara, Y. Miyamoto, T. Morioka, High-spatial-multiplicity multicore fibers for future dense space-division-multiplexing systems. IEEE J. Lightwave Technol. **34**(6), 1464–1475 (2016)
149. K. Shibahara, T. Mizuno, D. Lee, Y. Miyamoto, H. Ono, K. Nakajima, Y. Amma, K. Takenaga, K. Saitoh, Iterative unreplicated parallel interference canceler for MDL-tolerant dense SDM (12-core × 3-mode) transmission over 3000 km. IEEE J. Lightwave Technol. **37**(6), 1560–1569 (2019)
150. G. Labroille, B. Denolle, P. Jian, P. Genevaux, N. Treps, J.-F. Morizur, Efficient and mode selective spatial mode multiplexer based on multi-plane light conversion. Opt. Express **22**(13), 15599–15607 (2014)
151. Y. Mori, Z. Chao, K. Kikuchi, Novel FIR-filter configuration tolerant to fast phase fluctuations in digital coherent receivers for higher-order QAM sig-nals, in **OFC 2012**, Los Angeles, CA, USA, OTh4C.4 (2012)

152. D. Soma, Y. Wakayama, S. Beppu, S. Sumita, T. Tsuritani, T. Hayashi, T. Nagashima, M. Suzuki, H. Takahashi, K. Igarashi, I. Morita, M. Suzuki, 10.16 Peta-bit/s dense SDM/WDM transmission over low-DMD 6-mode 19-core fibre across C+L band, in *ECOC 2017*, Th.PDP.A.1 (2017)
153. W. Klaus, J. Sakaguchi, B.J. Puttnam, Y. Awaji, N. Wada, Free-space coupling conditions for multi-core few-mode fibers, in *IEEE Photonics Society Summer Topical Meeting 2014* (2014), pp. 182–183
154. T. Sakamoto, T. Mori, T. Yamamoto, N. Hanzawa, S. Tomita, F. Yamamoto, K. Saitoh, M. Koshiba, Mode-division multiplexing transmission system with DMD-independent low complexity MIMO processing. IEEE J. Lightwave Technol. **31**(13), 2192–2199 (2013)
155. Special issue on nonlinear-optical signal processing. J. Sel. Top. Quantum Electron. **14**(3) (2008)
156. A.E. Willner, S. Khaleghi, M.R. Chitgarha, O.F. Yilmaz, All-optical signal processing. J. Lightwave Technol. **32**(4), 660–680 (2014)
157. G.P. Agrawal, *Nonlinear Fiber Optics*, 3rd edn. (Academic Press, 2001)
158. G.P. Agrawal, *Applications of Nonlinear Fiber Optics* (Academic Press, 2001)
159. M. Yoshida, K. Kasai, M. Nakazawa, Mode-hop-free, optical frequency tunable 40-GHz mode-locked fiber laser. IEEE J. Quantum Electron. **43**(8), 704–708 (2007)
160. T. Morioka, S. Kawanishi, K. Mori, M. Saruwatari, Transform-limited, femtosecond WDM pulse generation by spectral filtering of gigahertz supercontinuum. Electron. Lett. **30**(14), 1166–1168 (1994)
161. I.N. Duling (ed.), *Compact Sources for Ultrashort Pulses* (Cambridge University Press, 1995)
162. L. Krainer, D. Nodop, G.J. Spühler, S. Lecomte, M. Golling, R. Paschotta, D. Ebling, T. Ohgoh, T. Hayakawa, K.J. Weingarten, U. Keller, Compact 10-GHz Nd:GdVO4 laser with 0.5-W average out-put power and low timing jitter. Opt. Lett. **29**(22), 2629–2631 (2004)
163. S.Y. Set, H. Yaguchi, Y. Tanaka, M. Jablonski, Laser mode locking using a saturable absorber incorporating carbon nanotubes. J. Lightwave Technol. **22**(1), 51–56 (2004)
164. Q. Bao, H. Zhang, Y. Wang, Z. Ni, Y. Yan, Z.X. Shen, K.P. Loh, D.Y. Tang, Atomic-layer graphene as a saturable absorber for ul-trafast pulsed lasers. Adv. Funct. Mater. **19**(19), 3077–3083 (2009)
165. S. Yamashita, A tutorial on nonlinear photonic applications of carbon nano-tube and graphene (invited tutorial). J. Lightwave Technol. **30**(4), 427–447 (2012)
166. S. Yamashita, A. Martinez, B. Xu, Short pulse fiber lasers mode-locked by carbon nanotube and graphene (invited). Opt. Fiber Technol. **20**(6), 702–713 (2014)
167. S. Yamashita, Y. Inoue, K. Hsu, T. Kotake, H. Yaguchi, D. Tanaka, M. Jablonski, S.Y. Set, 5 GHz pulsed fiber Fabry-Perot laser mode-locked using carbon nanotubes. IEEE Photon. Technol. Lett. **17**(4), 750–752 (2005)
168. A. Martinez, S. Yamashita, Multi-gigahertz repetition rate passively modelocked fiber lasers using carbon nanotubes. Opt. Express **19**(7), 6156–6163 (2011)
169. A. Martinez, S. Yamashita, 10 GHz fundamental mode fiber laser using a graphene saturable absorber. Appl. Phys. Lett. **101**, 041118 (2012)
170. M. Onishi, T. Okuno, T. Kashiwada, S. Ishikawa, N. Akasaka, M. Nishimura, Highly nonlinear dispersion-shifted fibers and their application to broadband wavelength converter. Opt. Fiber Technol. **4**(2), 204–214 (1998)
171. J.H. Lee, Novel dispersion properties of photonic crystal fiber. Jpn. J. Appl. Phys. **46**(8B), 5408–5413 (2007)
172. N. Sugimoto, Erbium doped fiber and highly non-linear fiber based on bismuth oxide glasses. J. Non-Cryst. Solids **354**, 1205–1210 (2008)
173. M.D. Pelusi, V.G. Ta'eed, L. Fu, E. Magi, M.R.E. Lamont, S. Madden, D.Y. Choi, D.A.P. Bulla, B. Luther-Davies, B.J. Eggleton, Applications of highly-nonlinear chalcogenide glass devices tailored for high-speed all-optical signal processing. J. Sel. Top. Quantum Electron. **14**(3), 529–539 (2008)
174. S. Watanabe, F. Futami, R. Okabe, R. Ludwig, C. Schmidt-Langhorst, B. Huettl, C. Schubert, H.G. Weber, An optical parametric amplified fiber switch for optical signal processing and regeneration. J. Sel. Top. Quantum Electron. **14**(3), 674–680 (2008)

175. M. Tani, S. Yamashita, Dispersion compensation with an SBS-suppressed fiber phase conjugator using synchronized phase modulation. Electron. Lett. **39**(19), 1375–1377 (2003)
176. K. Seki, S. Yamashita, Narrowband and tunable optical parametric ampli-fication in bismuth-oxide-based highly nonlinear fiber. Opt. Express **16**(18), 13871–13877 (2008)
177. P.V. Mamyshev, All-optical data regeneration based on self-phase modulation effect, in *ECOC98*, Sept. 1998, pp. 475–476
178. S. Yamashita, M. Shahed, Optical 2R regeneration using cascaded fiber four wave mixing with suppressed spectral spread. IEEE Photon. Technol. Lett. **18**(9), 1064–1066 (2006)
179. K.Y. Song, M.G. Herráez, L. Thévenaz, Long optically controlled delays in optical fibers. Opt. Lett. **30**(14), 1782–1784 (2005)
180. B.J. Eggleton, Inducing and harnessing stimulated Brillouin scattering in pho-tonic integrated circuits. Adv. Opt. Photon. **5**, 536–587 (2013)
181. K.K. Chow, S. Yamashita, Four-wave mixing in a single-walled carbon-nanotube-deposited D-shaped fiber and its application in tunable wavelength conversion. Opt. Express **17**(18), 15608–15613 (2009)
182. K.K. Chow, S. Yamashita, S.Y. Set, Four-wave mixing based wavelength conversion using a single-walled carbon-nanotube-deposited planar lightwave circuit waveguide. Opt. Lett. **35**(12), 2070–2072 (2010)
183. K.K. Chow, S. Yamashita, Self-phase modulation based optical regeneration of 10 Gb/s, 1.8ps RZ signal using carbon nanotube device, in *Conference on Lasers and Electro Optics (CLEO2010)*, San Jose, USA, May 2010, no. CWI6
184. B. Xu, A. Martinez, S. Yamashita, Mechanically exfoliated graphene for four wave mixing based wavelength conversion. IEEE Photon. Technol. Lett. **24**(20), 1792–1794 (2012)
185. C. Koos, P. Vorreau, T. Vallaitis, P. Dumon, W. Bogaerts, R. Baets, B. Esembeson, I. Biaggio, T. Michinobu, F. Diederich, W. Freude, J. Leuthold, All-optical high-speed signal processing with silicon–organic hybrid slot waveguides. Nat. Photonics **3**, 216–219 (2009)
186. B.J. Puttnam, R.S. Luís, E. Agrell, G. Rademacher, J. Sakaguchi, W. Klaus, G.M. Saridis, Y. Awaji, N. Wada, High capacity transmission systems using homogeneous multi-core fibers. J. Lightwave Technol. **35**, 1157–1167 (2017)
187. M.D. Feuer, L.E. Nelson, K. Abedin, X. Zhou, T.F. Taunay, J.F. Fini, B. Zhu, R. Isaac, R. Harel, G. Cohen, D.M. Marom, ROADM system for space division multiplexing with spatial superchannels, in *Optical Fiber Communication Conference (Optical Society of America, Technical Digest 2013)*, PDP5B.8 (2013)
188. L.E. Nelson, M.D. Feuer, K. Abedin, X. Zhou, T.F. Taunay, J.M. Fini, B. Zhu, R. Isaac, R. Harel, G. Cohen, D.M. Marom, Spatial superchannel routing in a two-span ROADM system for space division multiplexing. J. Lightwave Technol. **32**(4), 783–789 (2014)
189. B.J. Puttnam, R.S. Luís, J.M.D. Mendinueta, J. Sakaguchi, W. Klaus, Y. Kamio, M. Nakamura, N. Wada, Y. Awaji, A. Kanno, T. Kawanishi, T. Miyazaki, Self-homodyne detection in optical communication systems. Photonics **1**(2), 110 (2014)
190. E. Le Taillandier de Gabory, M. Arikawa, Y. Hashimoto, T. Ito, K. Fukuchi, A shared carrier reception and processing scheme for compensating frequency offset and phase noise of space-division multiplexed signals over multicore fibers, in *Optical Fiber Communication Conference (Optical Society of America, Technical Digest 2013)*, paper OM2C.2 (2013)
191. B.J. Puttnam, T.A. Eriksson, J.-M. Delgado Mendinueta, R.S. Luís, W. Klaus, J. Sakaguchi, Y. Awaji, N. Wada, M. Karlsson, E. Agrell, Modulation formats for multi-core fiber transmission. Opt. Express **22**(26), 32457–32469 (2014)
192. T. Miyazaki, F. Kubota, PSK self-homodyne detection using a pilot carrier for multibit/symbol transmission with inverse-RZ signal. IEEE Photon. Technol. Lett. **17**(6), 1334–1336 (2005)
193. T. Miyazaki, Linewidth-tolerant QPSK homodyne transmission using a polarization-multiplexed pilot carrier. IEEE Photon. Technol. Lett. **18**(2), 388–390 (2006)
194. B.J. Puttnam, J. Sakaguchi, W. Klaus, Y. Awaji, J.-M. Delgado Mendinueta, N. Wada, A. Kanno, T. Kawanishi, Investigating self-homodyne coherent detection in a 19-core spatial-division-multiplexed transmission link, in *Procedings of the European Conference and Exhibition on Optical Communications (ECOC) 2012*, paper Tu.3.C.3 (2012)

195. B.J. Puttnam, J. Sakaguchi, J.-M. Delgado Mendinueta, W. Klaus, Y. Awaji, N. Wada, A. Kanno, T. Kawanishi, Investigating self-homodyne coherent detection in a 19 channel space-division-multiplexed transmission link. Opt. Express **21**(2), 1561–1566 (2013)
196. J.M. Delgado Mendinueta, B.J. Puttnam, J. Sakaguchi, R.S. Luis, W. Klaus, Y. Awaji, A. Kanno, T. Kawanishi, Investigation of receiver DSP carrier phase estimation rate for self-homodyne space-division multiplexing communication systems, in *Optical Fiber Communication Conference (Optical Society of America, Technical Digest 2013)*, paper JTh2A.48 (2013)
197. J.M. Delgado Mendinueta, B.J. Puttnam, J. Sakaguchi, R.S. Luis, W. Klaus, Y. Awaji, A. Kanno, T. Kawanishi, Energy efficient carrier phase recovery for self-homodyne polarization-multiplexed QPSK, in *Proceedings of OptoElectronics and Communications Conference, 2013*, paper ThR3-5 (2013)
198. G.-W. Lu, M. Nakamura, Y. Kamio, T. Miyazaki, 40-Gb/s QPSK and 20-Gb/s PSK with inserted pilot symbols using self-homodyne detection. Opt. Express **15**(12), 7660–7666 (2007)
199. M. Nakamura, Y. Kamio, G.-W. Lu, T. Miyazaki, Ultimate linewidth-tolerant 20-Gbps QPSK-homodyne transmission using a spectrum-sliced ASE light source, in *Optical Fiber Communication Conference (Optical Society of America, Technical Digest 2013)*, paper OThD4 (2013)
200. Y. Kamio, M. Nakamura, T. Miyazaki, 80-Gb/s 256-QAM signals using phase noise and DGD-tolerant pilot-carrier-aided homodyne detection, in *Proceedings of the European Conference and Exhibition on Optical Communications (ECOC) 2007*, paper P089 (2007)
201. M. Sjödin, E. Agrell, P. Johannisson, G.-W. Lu, P.A. Andrekson, M. Karlsson, Filter optimization for self-homodyne coherent WDM systems using interleaved polarization division. J. Lightwave Technol. **29**(9), 1219–1226 (2011)
202. P. Johannisson, M. Sjödin, M. Karlsson, E. Tipsuwannakul, P. Andrekson, Cancellation of nonlinear phase distortion in self-homodyne coherent systems. IEEE Photon. Technol. Lett. **22**(11), 802–824 (2010)
203. R. Luis, B.J. Puttnam, J.-M. Delgado Mendinueta, W. Klaus, J. Sakaguchi, Y. Awaji, A. Kanno, T. Kawanishi, N. Wada, OSNR penalty of self-homodyne coherent detection in spatial-division-multiplexing systems. IEEE Photon. Technol. Lett. **26**(5), 477–479 (2014)
204. Z. Qu, S. Fu, M. Zhang, M. Tang, P. Shum, D. Liu, Analytical investigation on self-homodyne coherent system based on few-mode fiber. IEEE Photon. Technol. Lett. **26**(1), 74–77 (2014)
205. B.J. Puttnam, J.-M. Delgado Mendinueta, J. Sakaguchi, R. Luis, W. Klaus, Y. Awaji, N. Wada, A. Kanno, T. Kawanishi, 105Tb/s transmission system using low-cost, MHz linewidth DFB lasers enabled by self-homodyne coherent detection and a 19-core fiber, in *Optical Fiber Communication Conference (Optical Society of America, Technical Digest 2013)*, paper OW1I.1 (2013)
206. B.J. Puttnam, S.L. Ruben, J.M.D. Mendinueta, J. Sakaguchi, W. Klaus, Y. Awaji, N. Wada, A. Kanno, T. Kawanishi, Long distance transmission in a multi-core fiber with self-homodyne detection, in *Optical Fiber Communication Conference*, Los Angeles, Mar 2015
207. R. Luis, B.J. Puttnam, J.-M. Delgado Mendinueta, Y. Awaji, N. Wada, Photonics in switching 'impact of spatial channel skew on the performance of spatial-division multiplexed self-homodyne transmission systems', in *Proceedings of Photonics in Switching* (2015)
208. R. Luis, B.J. Puttnam, J.-M. Delgado Mendinueta, Y. Awaji, N. Wada, Experimental demonstration of a polarization-insensitive self-homodyne detection receiver for optical access, in *Proceedings of ECOC 2015*, paper Tu.3.4.7 (2015)
209. R. Luis, B.J. Puttnam, J.-M. Delgado Mendinueta, J. Sakaguchi, W. Klaus, Y. Awaji, N. Wada, A. Kanno, T. Kawanishi, In-service method of path length alignment in SDM systems with self-homodyne detection, in *Proceedings of OptoElectronics and Communications Conference 2013*, paper ThR3-3 (2013)
210. R. Noé, H. Heidrich, D. Hoffmann, Endless polarization control systems for coherent optics. IEEE J. Lightwave Technol. **1988**(6), 1199–1207 (1988)
211. B. Koch, R. Noé, V. Mirvoda, H. Griesser, S. Bayer, H. Wernz, Record 59-krad/s polarization tracking in 112-Gb/s640-km PDM-RZ-DQPSK transmission. IEEE Photon. Technol. Lett. **2010**(22), 1407–1409 (2010)

212. B. Koch, R. Noé, V. Mirvoda, D. Sandel, O. Jan, K. Puntsri, 20-Gb/s PDM-RZ-DPSK transmission with 40 krad/s endless optical polarization tracking. IEEE Photon. Technol. Lett. **2013**(25), 798–801 (2013)
213. N. Amaya, M. Irfan, G. Zervas, R. Nejabati, D. Simeonidou, V. Rancano, F. Parmigiani, P. Petropoulos, D. Richardson, J. Sakaguchi, W. Klaus, B. Puttnam, T. Miyazawa, Y. Awaji, N. Wada, I. Henning, First fully-elastic multi-granular network with space/frequency/time switching using multi-core fibres and programmable optical nodes, in *Proceedings of the European Conference and Exhibition on Optical Communication*, London, UK, 16–20 Sept 2012
214. N. Amaya, M. Irfan, G. Zervas, R. Nejabati, D. Simeonidou, J. Sakaguchi, W. Klaus, B.J. Puttnam, T. Miyazawa, Y. Awaji, N. Wada, I. Henning, Fully-elastic multi-granular network with space/frequency/time switching using multi-core fibres and programmable optical nodes. Opt. Express **21**, 8658–8872 (2013)
215. J. Sakaguchi, W. Klaus, B.J. Puttnam, T. Miyazawa, Y. Awaji, J.M.D. Mendinueta, R.S. Luis, N. Wada, SDM-WDM hybrid reconfigurable add-drop nodes for self-homodyne photonic networks, in *Proceedings of the IEEE Photonics Society Summer Topical Meeting Series*, Waikoloa, HI, USA, 8–10 July 2013
216. N. Amaya, S. Yan, M. Channegowda, B.R. Rofoee, Y. Shu, M. Rashidi, Y. Ou, E. Hugues-Salas, G. Zervas, R. Nejabati, D. Simeonidou, B.J. Puttnam, W. Klaus, J. Sakaguchi, T. Miyazawa, Y. Awaji, H. Harai, N. Wada, First demonstration of software defined networking (SDN) over space division multiplexing (SDM) optical networks, in *Proceedings of the 39th European Conference and Exhibition on Optical Communication (ECOC 2013)*, London, UK, 22–26 Sept 2013
217. N. Amaya, S. Yan, M. Channegowda, B.R. Rofoee, Y. Shu, M. Rashidi, Y. Ou, E. Hugues-Salas, G. Zervas, R. Nejabati, D. Simeonidou, B.J. Puttnam, W. Klaus, J. Sakaguchi, T. Miyazawa, Y. Awaji, H. Harai, N. Wada, Software defined networking (SDN) over space division multiplexing (SDM) optical networks: features, benefits and experimental demonstration. Opt. Express **22**, 3638–3647 (2014)
218. T.A. Eriksson, P. Johannisson, E. Agrell, P.A. Andrekson, M. Karlsson, Frequency and polarization switched QPSK, in *Proceedings of the European Conference on Optical Communications (ECOC 2013)*, paper Th.2.D.4 (2013)
219. H. Sugiyama, K. Nosu, MPPM: a method for improving the band-utilization efficiency in optical PPM. IEEE J. Lightwave Technol. **7**(3), 465–472 (1989)
220. H. Bülow, Polarization QAM modulation (POL-QAM) for coherent detection schemes, in *Proceedings of OFC/NFOEC 2009*, San Diego, CA, 22–26 Mar 2009, OWG2 (2009)
221. E. Agrell, M. Karlsson, Power-efficient modulation formats in coherent transmission systems. IEEE J. Lightwave Technol. **27**(22), 5515–5526 (2009)
222. M. Karlsson, E. Agrell, Which is the most power-efficient modulation format in optical links? Opt. Express **17**(13), 10814–10819 (2009)
223. T.A. Eriksson, P. Johannisson, B.J. Puttnam, E. Agrell, P.A. Andrekson, M. Karlsson, K-over-L multidimensional position modulation. IEEE J. Lightwave Technol. **32**(12), 2254–2262 (2014)
224. J. Pierce, Optical channels: practical limits with photon counting. IEEE Trans. Commun. **26**, 1819–1821 (1978)
225. A.J. Philips, R.A. Cryan, J.M. Senior, Performance evaluation of optically pre-amplified PPM Systems. IEEE Photon. Technol. Lett. **6**(5), 651–653 (1994)
226. M. Noshad, M. Brandt-Pearce, Expurgated PPM using symmetric balanced incomplete block designs. IEEE Commun. Lett. **16**(7), 968–971 (2012)
227. M. Karlsson, E. Agrell, Generalized pulse-position modulation for optical power-efficient communication, in *37th European Conference and Exposition on Optical Communications, OSA Technical Digest (Optical Society of America 2011)*, paper Tu.6.B.6 (2011)
228. X. Liu, T. Wood, R. Tkach, S. Chandrasekhar, Demonstration of record sensitivity in an optically pre-amplified receiver by combining PDM-QPSK and 16-PPM with pilot-assisted digital coherent detection, in *Optical Fiber Communication Conference (Optical Society of America, Technical Digest 2011)*, PDPB.1 (2011)

229. A. Ludwig, M. Schulz, P. Schindler, K. Kuder, S. Wolf, C. Koos, W. Freude, J. Leuthold, Stacking modulation formats for highest-sensitivity, in *Advanced Photonics for Communications, OSA Technical Digest* (online), paper JM2B.2 (Optical Society of America, 2014) (2014)
230. J.H.B. Nijhof, Generalized L-out-of-K pulse position modulation for improved power efficiency and spectral efficiency, in *Optical Fiber Communication Conference (Optical Society of America, Technical Digest 2012)*, paper OW3H.7 (2012)
231. R.S. Tucker, Green optical communications—part I: energy limitations in transport. IEEE J. Lightwave Technol. **17**(2), 245–260 (2011)
232. P. Poggiolini, A. Carena, V. Curri, G. Bosco, F. Forghieri, Experimental demonstration of a highly reliable multicore-fiber-based optical network. IEEE Photon. Technol. Lett. **23**(11), 742–744 (2011)
233. G. Ungerboeck, Channel coding with multilevel/phase signals. IEEE Trans. Inform. Theory **28**, 55–67 (1982)
234. L.D. Coelho, N. Hanik, Global optimization of fiber-optic communication systems using four-dimensional modulation formats, in *Proceedings of the European Conference on Optical Communications (ECOC 2011)*, paper Mo.2.B.4 (2011)
235. T.A. Eriksson, M. Sjödin, P. Johannisson, P.A. Andrekson, M. Karlsson, Comparison of 128-SP-QAM and PM-16QAM in long-haul WDM transmission. Opt. Express **21**(16), 19269–19279 (2013)
236. M. Sjödin, P. Johannisson, J. Li, E. Agrell, P.A. Andrekson, M. Karlsson, Comparison of 128-SP-QAM with PM-16-QAM. Opt. Express **20**(8), 8356–8366 (2012)
237. B.J. Puttnam, J.-M. Delgado Mendinueta, R.S. Luís, T.A. Eriksson, Y. Awaji, N. Wada, E. Agrell, Single parity check multi-core modulation for power efficient spatial super-channels, in *Proceedings of the European Conference on Optical Communications (ECOC 2014)*, paper MO.3.3.5 (2014)
238. T.A. Eriksson, R.S. Luís, B.J. Puttnam, J.M. Delgado-Mendinueta, P.A. Andrekson, M. Karlsson, Y. Awaji, N. Wada, E. Agrell, Single parity check-coded 16QAM over spatial superchannels in multicore fiber transmission. Opt. Express **23**(11), 14569–14582 (2015)
239. J.-X. Cai, H.G. Batshon, M. Mazurczyk, H. Zhang, Y. Sun, O.V. Sinkin, D. Foursa, A.N. Pilipetskii, 64QAM based coded modulation transmission over transoceanic distance with > 60 Tb/s capacity, in *Optical Fiber Communication Conference Post Deadline Papers OSA Technical Digest* (online), paper Th5C.8 (Optical Society of America, 2015)
240. H. Zhang, C.R. Davidson, H.G. Batshon, A. Pilipetskii, Coded modulation PDM-QPSK transmission over 6,370 km dispersion managed NZDSF spans with 1.5 to 4 dB margin, in *ECOC 2014*, Cannes, France, Tu.1.5.2 (2014)
241. D.S. Millar, T. Koike-Akino, R. Maher, D. Lavery, M. Paskov, K. Kojima, K. Parsons, B.C. Thomsen, S.J. Savory, P. Bayvel, Experimental demonstration of 24-dimensional extended golay coded modulation with LDPC, in *Optical Fiber Communication Conference Post Deadline Papers OSA Technical Digest, 2014*, paper M3A.5 (2014)
242. S.Ö. Arik, D.S. Millar, T. Koike-Akino, K. Kojima, High-dimensional modulation for mode-division multiplexing, in *Proceedings of OFC*, San Francisco, W4J.1 (2014)
243. A.W. Nordstrom, J.P. Robinson, An optimum nonlinear code. Inf. Control **11**, 613–616 (1968)
244. G. Rademacher, B.J. Puttnam, R.S. Luís, J.-M. Delgado Mendinueta, T.A. Eriksson, Y. Awaji, N. Wada, E. Agrell, Experimental investigation of a 16-dimensional modulation format for long-haul multi-core fiber transmission, in *Proceedings of the European Conference on Optical Communications (ECOC 2015)*, paper P.5.10 (2015)
245. G.C. Clark Jr., J.B. Cain, *Error-Correction Coding for Digital Communications* (Springer, 1981)
246. E. Awwad, G.R.-B. Othman, Y. Jaouën, Y. Frignac, Space-time codes for mode-multiplexed optical fiber transmission systems, in *Photonics for Communications OSA Technical Digest* (online), paper SM2D.4 (Optical Society of America, 2014)
247. E. Awwad, G.R.-B. Othman, Y. Jaouën, Space-time coding and optical scrambling for mode multiplexed optical fiber systems, in *International Communications Conference*, London, ONS-05 (2015), pp. 6846–6852

248. C. Okonkwo, R. van Uden, H. Chen, H. de Waardt, T. Koonen, Advanced coding techniques for few mode transmission systems. Opt. Express **23**(2), 1411–1420 (2015)
249. M. Koga, A. Mizutori, T. Iida, Optical diversity transmission and maximum-ratio combined receiver in multi-core fiber to mitigate fiber non-linear phenomenon. IEICE Commun. Express **2**(2), 67–73 (2013)
250. I.B. Djordjevic, M. Arabaci, L. Xu, T. Wang, Spatial-domain-based multidimensional modulation for multi-Tb/s serial optical transmission. Opt. Express **19**(7), 6845–6857 (2011)
251. I.B. Djordjevic, Energy-efficient spatial-domain-based hybrid multidimensional coded-modulations enabling multi-Tb/s optical transport. Opt. Express **19**(17), 16708–16714 (2011)
252. S. Franke-Arnold, L. Allen, M. Padgett, Laser Photon. Rev. **2**, 299–313 (2008)
253. L. Allen, M.W. Beijersbergen, R.J.C. Spreeuw, J.P. Woerdman, Phys. Rev. A **45**, 8185–8189 (1992)
254. G. Gibson, J. Courtial, M. Padgett, M. Vasnetsov, V. Pas'ko, S. Barnett, S. Franke-Arnold, Free-space information transfer using light beams carrying orbital angular momentum. Opt. Express **12**, 5448–5456 (2004)
255. Y. Awaji, N. Wada, Y. Toda, Demonstration of spatial mode division multiplexing using Laguerre-Gaussian mode beam in telecom-wavelength, in *23rd Annual Meeting of the IEEE Photonics Society* (2010), pp. 551–552
256. J. Wang, J.-Y. Yang, I.M. Fazal, N. Ahmed, Y. Yan, H. Huang, Y. Ren, Y. Yue, S. Dolinar, M. Tur, A.E. Willner, Terabit free-space data transmission employing orbital angular momentum multiplexing. Nat. Photonics **6**, 488–496 (2012)
257. Y. Awaji, N. Wada, Y. Toda, T. Hayashi, World first mode/spatial division multiplexing in multi-core fiber using Laguerre-Gaussian mode, in *ECOC2011*, We.10.P1.55 (2011)
258. S. Berdagué, P. Facq, Mode division multiplexing in optical fibers. Appl. Opt. **21**, 1950–1955 (1982)
259. Y. Awaji, N. Wada, Y. Toda, Observation of orbital angular momentum spectrum in propagating mode through seven-core fibers, in *CLEO2012*, JTu2K.3 (2012)
260. S. Ramachandran, P. Gregg, P. Kristensen, S.E. Golowich, On the scalability of ring fiber designs for OAM multiplexing. Opt. Express **23**, 3721–3730 (2015)
261. B. Guan, R.P. Scott, C. Qin, N.K. Fontaine, T. Su, C. Ferrari, M. Cappuzzo, F. Klemens, B. Keller, M. Earnshaw, S.J.B. Yoo, Free-space coherent optical communication with orbital angular, momentum multiplexing/demultiplexing using a hybrid 3D photonic integrated circuit. Opt. Express **22**, 145–156 (2014)
262. J. Sun, M. Moresco, G. Leake, D. Coolbaugh, M.R. Watts, Generating and identifying optical orbital angular momentum with silicon photonic circuits, in *CLEO2014*, JTh5B.4 (2014)
263. J. Sun, A. Yaacobi, M. Moresco, D.D. Coolbaugh, M.R. Watts, Integrated continuously tunable optical orbital angular momentum generator, in *CLEO2015*, JTh5A.5 (2015)
264. Y. Awaji, N. Wada, Y. Toda, Discrete sampling and detection of multiplexed OAM modes propagating through multi-core fiber, in *OECC2014*, WE9B-4 (2014)
265. Y. Awaji, N. Wada, Y. Toda, Partial phase detection of optical vortices for discrimination of orbital angular momentum. IEICE OCS Tech. Rep. **115**(93), 1–4, OCS2015-10 (2015)

Chapter 6
Network Technologies for SDM

Werner Klaus, Jun Sakaguchi, Benjamin J. Puttnam, Nobuhiko Kikuchi, Yong Lee, and Kenichi Tanaka

Abstract Traffic demands on the Internet are becoming increasingly dynamic and are driving the need for greater routing flexibility in the underlying optical network. Space-division multiplexing (SDM) is mainly seen as a means to increase data throughput and handle exponential traffic growth. But, its role is certainly more diverse. Research on SDM encourages device integration, brings new functionality to network elements and helps optical networks to evolve. As a result, the number of individual components in the future networks will decrease, which in turn will improve overall reliability as well as lower operational expenditure and power consumption. From an application point of view, the prevalence of various band-consuming applications such as video streaming, cloud computing, big data, Internet of Things (IoT), social networking service (SNS) will produce a strong traffic demand especially on short-reach fiber transmission links. Among them, inter- and intra-data center (DC) traffics are expected to grow very rapidly in the coming years, and the development of ultra-high-capacity data interconnection and network technologies is urgently needed. Such short-reach links are promising candidates of the first-stage introduction of SDM technologies not only because of the highest traffic demands but also due to relaxed fiber specifications that come hand in hand with the shorter transmission distance. The first part of this chapter introduces basic technologies devised for network nodes and reviews recent research which has extended the SDM concept to optical switching as well as has utilized SDM-specific features to facilitate network control. The second part then focuses on short-reach systems and discusses an application example of SDM for data center signal protection.

Werner Klaus is a chapter editor.

W. Klaus (✉) · J. Sakaguchi · B. J. Puttnam
National Institute of Information and Communications Technology, Koganei, Tokyo, Japan
e-mail: klaus@nict.go.jp

N. Kikuchi · Y. Lee · K. Tanaka
Research & Development Group, Hitachi Ltd., Kokubunji, Tokyo, Japan

M. Nakazawa et al. (eds.), *Space-Division Multiplexing in Optical Communication Systems*, Springer Series in Optical Sciences 236,
https://doi.org/10.1007/978-3-030-87619-7_6

6.1 Basic Technologies for Network Nodes and Network Control

Introduction

Space-division multiplexing (SDM) is considered the final frontier of multiplexing schemes to deal with the ever-increasing demand for transmission capacity in optical fiber networks. Many impressive experiments based on few-mode fiber (FMF) [1], multi-core fiber (MCF) [2] or few-mode multi-core fiber (FM-MCF) [3] have already clearly demonstrated the potential of SDM. Yet, just having large capacity is not enough for these SDM technologies to be commercially successful or to be accepted as a viable alternative to the already well-established transmission technology based on single-mode fibers (SMFs). As with any new technology, to make SDM attractive to network operators, it has to provide significant benefits in terms of cost and energy per bit. In addition, it must also offer a scenario that allows for a smooth transition from today's constantly growing SMF networks where fiber cables, once installed in the ground or under sea, are the most difficult and expensive part to replace [4]. From a commercial point of view, one may expect, therefore, that the SDM revolution will start on a small scale inside isolated network elements such as transponders, optical amplifiers and optical switches where the cost benefit of integration is the highest. As such, the development of SDM has to closely follow the evolution of today's optical networks and must be able to support photonic switching and routing with reconfigurable optical add/drop capability. In order to succeed, SDM will have to provide network elements with comparable performance but of lower cost than SMF-based devices when processing similar bitrates. Such a cost advantage can only arise if spatial multiplexing inside those elements can sufficiently leverage component sharing and integration.

To date, most research in the field of SDM is still concentrating on the improvement of transmission quality and the capacity-distance product. In contrast, this section will review work, not yet as extensive but still as important, which has been done in bringing SDM to the network nodes, as well as incorporate its concept to manage and optimize data flow in networks. The first part mainly focuses on the wavelength selective switch (WSS), a key enabler of transparent wavelength routing in today's optical networks. In the second part, we then discuss ideas and proposals on how to adapt network control and management to the SDM era as well as on how to utilize the spatial dimension for network optimization.

6.1.1 SDM Switching and Node Elements

6.1.1.1 Basic Components of a Network Node

Next-generation optical networks will most likely have a mesh topology with reconfigurable optical add-drop multiplexers (ROADMs) installed at its nodes. The

ROADM is a sophisticated *optical switch* that allows individual wavelength bands to be dropped from or added to the stream of wavelength-division-multiplexed (WDM) signals passing through the network node [5]. An example of a multi-degree ROADM architecture is shown in Fig. 6.1. The degree of a ROADM refers to how many fiber routes (or fiber pairs as most links are bidirectional) of the network are connected to the ROADM. A wavelength channel can be switched through (express path), removed from the WDM stream by forwarding it to a transponder (drop path) or inserted into the WDM stream from a transponder (add path). The switching performance of a ROADM is measured by functionalities such as colorless, directionless and contentionless (CDC). *Colorless* stands for the ability of the add/drop ports to access any wavelength of the incoming WDM stream, *directionless* stands for the ability to access wavelength channels coming in from any direction, and *contentionless* refers to the ability of not blocking wavelengths that are forwarded to the same drop port (but are processed by different transponders behind that port). The higher the functionality, the more complex and expensive the ROADM becomes. A CDC-ROADM is currently the most flexible but also the most expensive ROADM on the market.

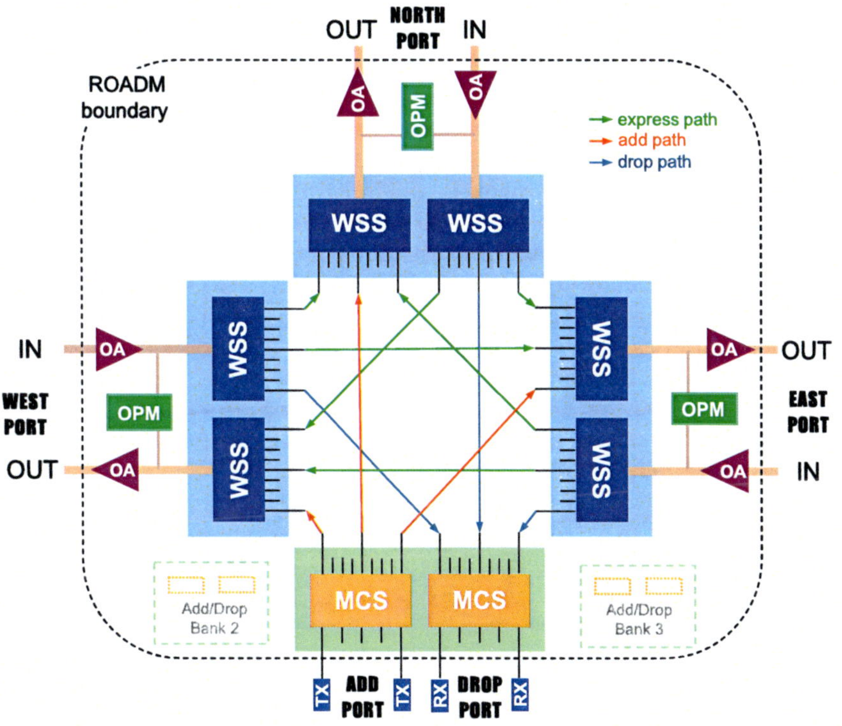

Fig. 6.1 Example of a degree-three CDC-ROADM (WSS: wavelength selective switch, OA: optical amplifier, OPM: optical performance monitor, MCS: multicast switch, TX/RX: transponders with coherent transmitters and receivers)

In addition, simulations of dynamic traffic in realistic networks have shown that in conjunction with a contention-aware routing and wavelength assignment scheme or algorithm, similar low blocking rates could be achieved also without the more expensive contentionless feature [5]. Nevertheless, the additional cost may be justified eventually with the provision of an extremely flexible hardware-based connectivity mechanism in the optical layer. Therefore, the CDC-ROADM is expected to serve best the needs of future networks which will require a high degree of network protection (against interruptions and component failures) as well as mesh restoration, optimization such as wavelength grooming or network defragmentation and most importantly quick adaption to unpredictable traffic patterns including requests from large-bandwidth applications. On top of this, these functions need to be software controllable and provide performance monitoring from a remote location.

On the hardware side, the ROADM typically comprises several wavelength selective switches (WSSs) [6, 7], optical amplifiers, performance monitors, power splitters and spatial switches or a combination thereof, e.g., in form of a multicast switch (MCS). The WSS is the ROADM's key element. Its main job is to optically route wavelength channels between a single port on one side and N ports on the other side (commonly referred to as "1xN"). Since the WSS is optically bidirectional, the flow of wavelength channels can occur either from the single port to any of the N ports, or from any of the N ports to the single port (by means of selective channel blocking). Another job of the WSS is to attenuate each wavelength channel independently for the purpose of amplitude equalizing and power balancing in the network.

Figure 6.2 illustrates the operating principle of a 1×2 WSS. The light of the input fiber carrying an WDM signal is collimated and passes through polarization dispersive and beam shaping optics before it reflects off a diffractive grating that disperses (e.g., horizontally) the WDM signal into many beams with different propagation angles according to their wavelengths. Diffraction efficiency of the grating is increased by aligning the polarization to the grating grooves and illuminating as many grooves as possible with elliptically shaped beams. The beams are then converged by a lens (i.e., optically Fourier transformed) to a steering element which consists either of an array of micro-electro mechanical system (MEMS) mirrors or reflective liquid crystal on silicon (LCoS) pixels. By electronically controlling the tilt of a MEMS mirror or the amount of phase retardation in the LCoS pixels, the wavelength separated beams corresponding to different wavelength channels of the WDM signal are then deflected vertically toward different output ports. The wavelength (frequency) band of the light that is sent to a specific output port is determined by the total width of elements that steer the beam. This width is commonly set in accordance to ITU recommendations [8], e.g., frequency slot widths ranging from 50 GHz down to 12.5 GHz. A steering element may consist of a single micro-mirror in case of MEMS or many pixels in case of LCoS. The higher spatial resolution and flexibility in modulating a beam's amplitude and phase make the LCoS-based WSS currently the preferred choice for applications using a flexible wavelength grid.

The introduction of the WSS in the network nodes marked a major milestone in the network evolution from opaque networks, i.e., networks that perform the switching

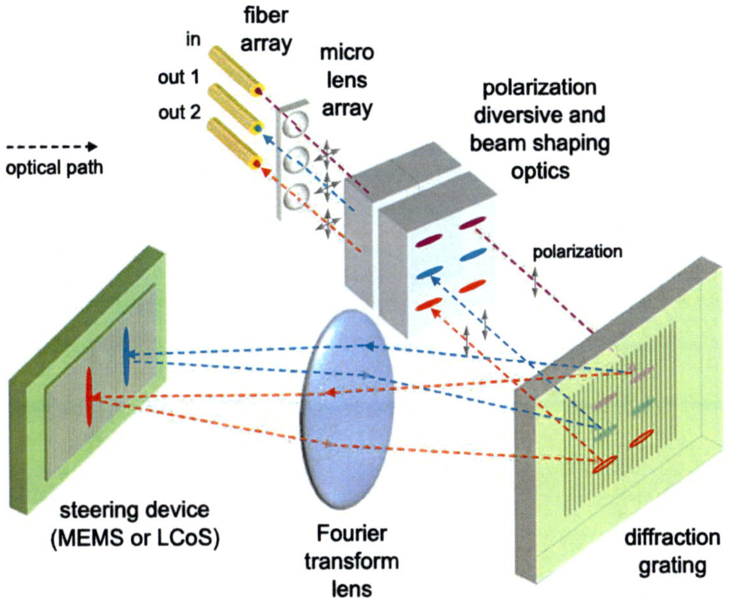

Fig. 6.2 Operation principle of a WSS (some optical paths have been omitted for clarity)

operations in the electrical domain by relying on power-consuming optical-electrical-optical (OEO) conversion to transparent networks, i.e., networks that switch and route WDM signals solely in the optical domain. With such a critical network element at hand, it is clear that incorporating spatial multiplexing into the WSS concept without increasing much the component count and footprint has to be also a primary concern of SDM research.

6.1.1.2 ROADM and WSS for SDM Systems

The first SDM ROADM in conjunction with a simple MCF network was reported by Feuer et al. [9]. It comprised two 1 × 2 WSSs for seven-core MCFs that were made out of one commercial (SMF-based) 1 × 23 WSS by adjusting the output port assignment and MEMS mirror tilt angles. The operating principle of the resulting 7 × (1 × 2) WSS is shown in Fig. 6.3.

Spatial demultiplexers or fan-out elements were used to connect the cores of the MCF to the input ports of the WSS. Inside the WSS, all light beams associated with the same MCF are treated as a single entity and switched together by one steering mirror to one of the two output ports. The spatial extension of the grating (omitted in Fig. 6.3a) allows efficient separation of the wavelengths in parallel, but switching the beams as a group necessitated a larger steering angle, although it was also suggested that interleaving the input and output beams can reduce the steering angle [10].

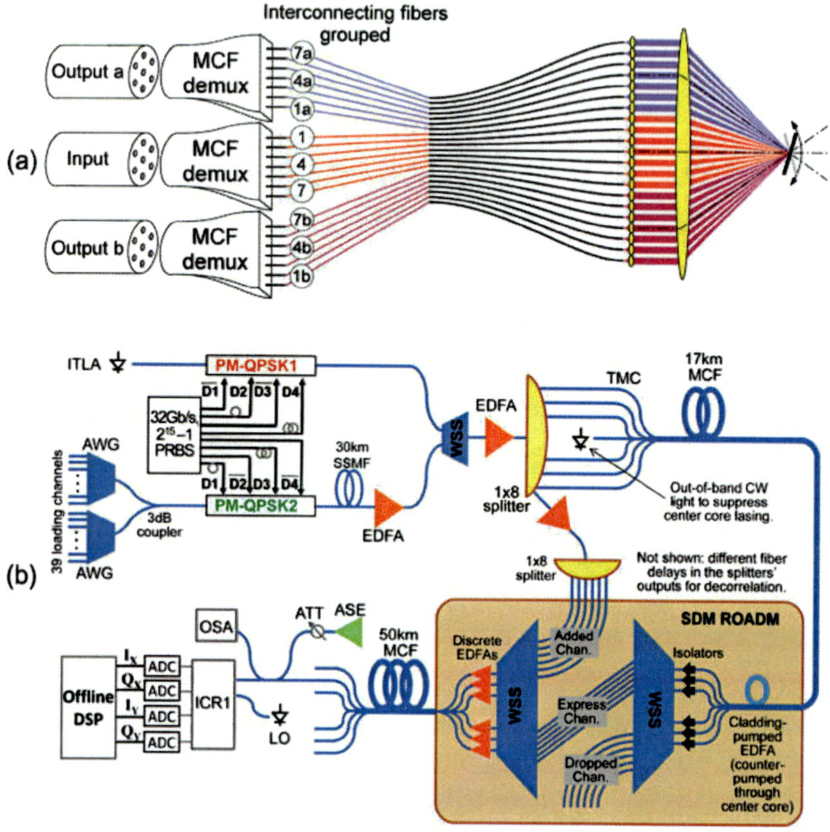

Fig. 6.3 **a** Operation principle of a 7 × (1 × 2) WSS. **b** Experimental setup for MCF transmission experiment with SDM ROADM (PM-QPSK: polarization-multiplexed quaternary phase-shift keying, PRBS: pseudorandom bit sequence, EDFA: erbium-doped fiber amplifier, TMC: tapered multi-core coupler, OSA: optical spectrum analyzer, ASE: amplified spontaneous emission, ATT: attenuator, ADC: analog-to-digital converter, DSP: digital signal processor) (Reprinted with permission from [9]. ©2013 Authors)

As shown in Fig. 6.3b, in the actual experimental setup, the two 1 × 2 WSSs were connected in a back-to-back fashion to provide the express, add and drop paths. Only the six outer cores of the incoming MCF were used as signal channels, while the MCF center core was reserved for pumping backward a cladding-pumped MCF erbium-doped fiber amplifier (EDFA) used as the ROADM's input amplifier. Amplification at the ROADM's output port was carried out by six discrete SMF EDFAs before inserting the light back into the next MCF span by means of a fan-in element. The loss of each WSS was kept below 5 dB at all test wavelengths. However, it was pointed out that the SDM WSS in this particular configuration is not able to correct (or balance) gain differences across the spectra of different cores that typically arise due to non-uniformities in the MCF-EDFA's gain and fan-in/fan-out elements' loss characteristics.

Next, Ryf et al. [11] measured the performance of a WSS when using FMFs with three spatial modes at the input and output ports. The operating principle is similar to the SMF-based device with an LCoS steering device shown in Fig. 6.2, except that the collimator lens array at the input has to collect light from the FMFs. Similar to the previous approach, the spatial modes are treated as a single unit and switched together in the wavelength domain. Since the collimator lens array at the input was not designed for the FMF's smaller numerical aperture (NA), the spectrally separated beams did not properly focus in the LCoS plane causing excess loss and reduced spectral resolution. The measured insertion loss (IL) and mode dependent loss (MDL) ranged from 5.6 to 9 dB and 2.5 to 7.1 dB, respectively. Although the increased MDL restricted the transmission bandwidth, transmission above a Q-factor of 7.7 dB (equivalent to a BER of less than 7.6×10^{-3}) with 32 100 GHz-spaced 30-Gbaud QPSK WDM channels was achieved for up to six consecutive passes through a fiber loop incorporating the FMF-WSS and a 30 km FMF. However, even with FMF-optimized optics, each mode will still have a different spatial extension at the steering device precluding the possibility of obtaining the same spectral resolution for all modes. Different transmission bands and MDL are thus unavoidable and bring about the need for larger guard bands between the channels, as compared to SMF-based devices. The problem escalates with the number of modes the WSS has to deal with.

This problem can be solved by converting the higher-order modes to fundamental modes, e.g., with a photonic lantern (PL) [12, 13]. The PL has a similar job as the fan-in and fan-out elements discussed earlier with respect to MCFs, but instead of breaking out a number of closely spaced cores, it transforms a multi-mode input signal into several single-mode output signals. If the transition from the multi-mode to the single-mode section is lossless (adiabatic), as can be achieved by slowly varying the dimensions of a waveguide in the form of a taper along the propagation direction, then the mode conversion itself becomes also lossless, provided the number of single-mode signals at the output equals the number of the guided modes at the input. Such an approach was adopted by Carpenter et al. [14] who used a commercial twin 1×24 WSS to build a 1×11 FMF-WSS supporting three spatial modes. After mode conversion, two output ports of each PL were connected to one 1×24 input array of the WSS, whereas each PL's third output port was connected to the second 1×24 array. Figure 6.4 shows the MDL and IL characteristics for the spectral passbands ranging from 10 to 100 GHz. In this case, the transmission bands were not MDL limited anymore clearly demonstrating the benefits of manipulating light inside the WSS from only the fundamental mode.

This approach was further extended to six spatial modes by Fontaine et al. [15]. Instead of the typical 1D fiber array at the WSS input, a 2D array was introduced to process simultaneously a larger number of spatial channels. As shown in Fig. 6.5, the output ports of a six-mode PL were connected to six SMF input ports of the WSS arranged in a matrix fashion of two rows and three columns. All six beams associated with one input FMF are then directed via a 2D collimator lens array to the diffraction

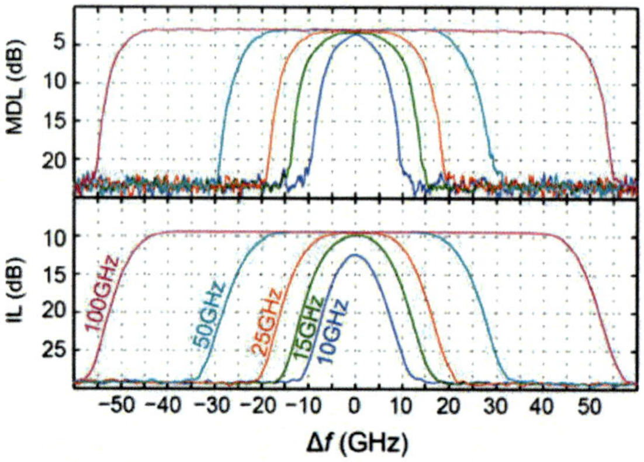

Fig. 6.4 Insertion loss (IL) and mode dependent loss (MDL) characteristics (Reprinted with permission from [14]. ©2014 The Optical Society)

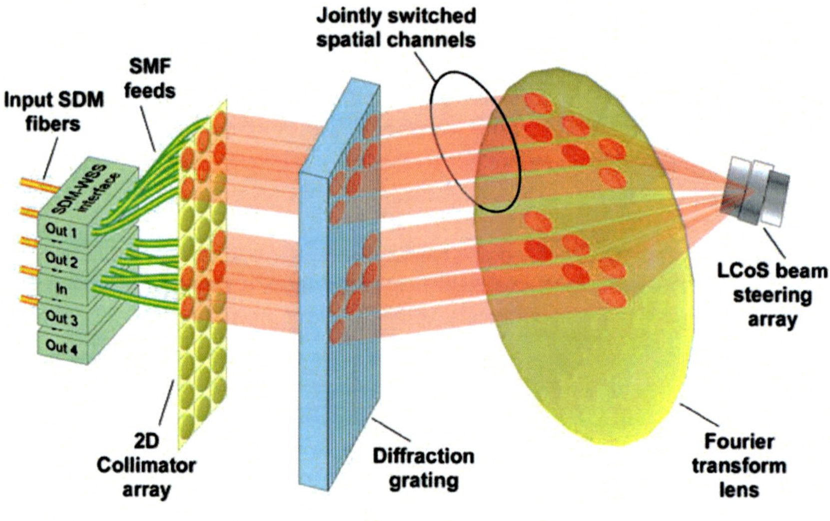

Fig. 6.5 FMF-WSS for six spatial modes (Reprinted with permission from [15].© 2015 Authors)

grating and focused onto the same LCoS steering element that deflects all six beams by the same amount. Since the number of modes to be switched simultaneously is twice the number of columns, the steering action has to skip every other row when deflecting the light to a 2 × 3 array of output ports. The WSS exhibited 5–8 dB insertion loss and a 0.5 dB spectral resolution of 78 GHz for 100-GHz-spaced channels.

When designing a WSS for SDM networks, the additional dimension of space can be utilized in two ways, to increase either the switch's throughput or granularity. Most of the experimental demonstrations are thus far placed priority on increasing throughput by means of "joint switching" of all available spatial channels simultaneously. For an SDM system based on FMF, such a choice is a natural consequence of the unavoidable mode coupling occurring along the fiber and the required MIMO DSP at the receiver. It would simply make no sense to route individual modes in the nodes when they need to be sent through the network as a unit anyway. On the other hand, if the SDM system is based on low-crosstalk MCFs, no constraints are imposed from the side of the transmission medium, and as we will see in the following section, there may be instances in a network where an increased granularity or the availability of an additional switching dimension can be more beneficial than an increased throughput.

Finally, we note that the generalization of the 1xN WSS would be a NxM wavelength selective optical cross-connect (WS-OXC). The WS-OXC is the ultimate flexible switch in a network node that can route any wavelength from any of N input ports to any of M output ports. Such OXCs are very complex (e.g., when realized with cascaded WSSs) and have found thus far only limited application in SMF networks due to cost restrictions. However, as we have seen from the previously discussed examples, the tight transmission band and insertion loss tolerances of WSSs will most likely restrict optical processing inside the WSS to beams with minimum spatial extension, i.e., the fundamental waveguide mode or Gaussian beam. The design principles of OXCs developed today for SMF networks may thus equally be applied to SDM networks in the future. In this respect, a clever design for an all-optical WS-OXC was recently proposed by Fontaine et al. [16]. This optical switch featured a full cross-connect with flexible passbands and had a similar amount of component building blocks and footprint as a single $1 \times N$ WSS. The optical layout and working principle is shown in Fig. 6.6. Although this device was not specifically developed with SDM in mind, it represents a good example of how the need for parallelization and integration stimulates advanced designs that utilize "space" and some yet unexploited parallel processing capabilities of optics more efficiently in network elements.

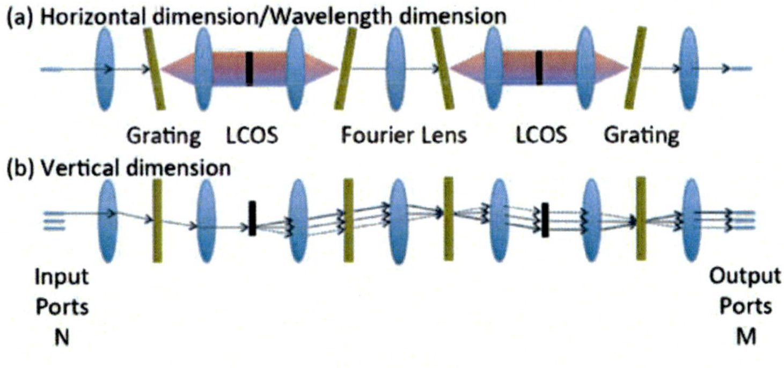

Fig. 6.6 Operation principle of wavelength selective OXC (Reprinted with permission from [16]. ©2013 IEICE)

6.1.2 SDM Network Control and Node Configuration

6.1.2.1 Architecture of Optical Networks

The technological achievements in the field of optical communications over the past few decades have enabled and supported a dramatic growth in traffic on the Internet. As the human mind is intrigued by the idea of accessing any kind of information from anywhere at any time with the highest possible quality, this trend is likely to continue for the coming decades. Mobile broadband and multimedia applications as well as machine-to-machine communication and cloud computing are fueling the need of using the network's optical and electronic resources as efficiently as possible in terms of capacity, cost per bit and power consumption.

One way to increase efficiency in a network is to make it flexible and elastic, i.e., provide some means that let the network dynamically adapt to unpredictable traffic patterns and rapidly establish or tear down connections through software. The goal of the flexible network is to utilize the available bandwidth more efficiently by introducing a frequency grid for the optical carrier frequencies (wavelengths) with a granularity four times finer than the 50 GHz spacing used in traditional networks. An architecture based on such a flexible grid will allow the network operator to match better heterogeneous traffic requests with different bandwidths. The LCoS technology of WSSs is well-suited to deal with such requirements. Note that, in the context of elastic networks, the WSS designed for flexible grids is sometimes referred to as spectrum selective switch (SSS). Elasticity is introduced into a network by employing configurable optical hardware, e.g., the ROADM, and by making the provisioning process of the end-to-end connections software controllable. As data center access and cloud computing increasingly dominate the traffic on the Internet, dynamic configurability and controllability are considered indispensable for provisioning the required quality and performance of future data links.

The network architecture is mainly determined by the hardware installed in a network node and the software controlling it. On the hardware side, various types of advanced optical network elements have been developed over the years of which the ROADM is a prominent example. The introduction of the ROADM has significantly reduced the amount of expensive and power-consuming electronic equipment in the node. It has changed also the notion of a network node of being a point of regeneration to being a point of routing. *Routing* stands for the process of selecting a path through the network based on a certain traffic demand. Another important aspect of network management is *wavelength/spectrum assignment*. The term *spectrum assignment* is commonly used in the context of elastic/flexible optical networks. In combination with the task of routing, these processes are referred to as the routing and wavelength assignment (RWA) problem or the routing and spectrum assignment (RSA) problem. A number of algorithms have been developed to solve these problems, and cost-effective and efficient network design today relies heavily on such algorithms. Another major contribution from the software side toward dynamic networking has been the introduction of software-defined networking (SDN) [17]. SDN acts as a network-wide operating system that allows unified and dynamic control of the data flow through the network devices in a flexible manner. Most importantly, routing tasks and traffic engineering can be performed with software developed independently from the node hardware.

Although the functionalities of ROADMs are being constantly improved, their architecture based on hard-wired device connections may eventually limit them in terms of scalability to deal efficiently with heterogeneous traffic requests in a dynamic network. A possible solution for scalable optical switches is offered by a multi-granular architecture [18]. Increased flexibility and scalability are realized by splitting up the switching process into several layers of switching granularities such as on a per-fiber basis, a per-waveband basis or a per-wavelength basis. A waveband represents here a set of contiguous wavelength channels that is treated as a single unit with respect to add/drop or node bypass. The layers can be implemented either with separate OXCs or collapsed into a single switching fabric. The latter has led to the important concept of Architecture-on-Demand (AoD) [19]. As shown in Fig. 6.7, an AoD node consists of a an optical backplane, e.g., implemented by a large port-count 3D-MEMS switch, to which several pluggable modules are connected providing a set of signal processing functions, such as optical amplification, spectrum/wavelength/subwavelength switching, wavelength conversion, and signal regeneration. Subwavelength switching refers here to the process of subdividing a single wavelength channel further into timeslots by means of a fast time-multiplex scheme.

Based on traffic requirements, an AoD synthesis algorithm calculates a synthetic node design that provides the required functionality. AoD is flexible and modular, facilitates provisioning, simplifies the introduction of new functionalities and supports arbitrary granularities. On the other hand, the development of an AoD algorithm that can calculate the optimal node architecture while taking into account cost and availability of architecture-building modules as well as traffic dynamics may be very complex. Furthermore, the benefit of increased flexibility needs also to be balanced with a larger footprint and power consumption of the node due to the additional MEMS switch [20].

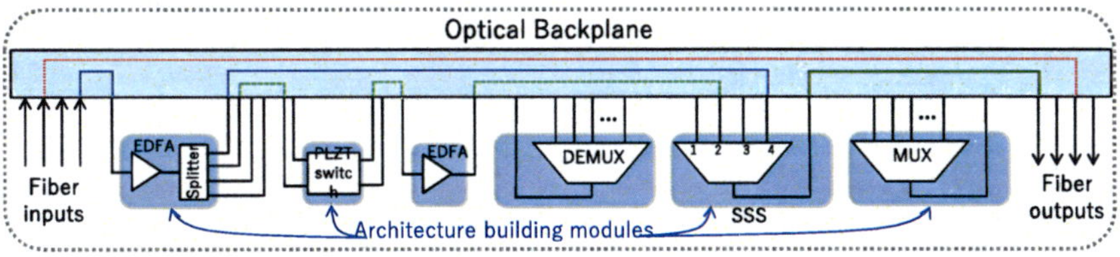

Fig. 6.7 Example of an AoD node configuration

6.1.2.2 Control and Management of SDM Networks

The introduction of SDM into the network will increase capacity and switching flexibility, but it also adds new challenges to an already complex network control. Amaya et al. [21, 22] demonstrated in two experiments how the AoD architecture could be beneficial for SDM networks. In the first experiment, shown in Fig. 6.8, a network was set up comprising four programmable AoD nodes connected via two seven-core MCFs and several SMFs [21]. The optical backplane consisted of a 160 × 160 3D-MEMS optical switch interconnecting the fibers with several wavelength selective switches, optical amplifiers, power splitters and a fast lead lanthanium zirconate titanate (PLZT) switch for subwavelength switching. Switching across the space, frequency and time dimensions with a granularity range of over 6000 was demonstrated in Node 2. The maximum amount of traffic was achieved with core-to-core switching, where 939 Gb/s were switched with a single cross-connection. The minimum amount of traffic allocated was one timeslot in a 64-timeslot frame of a 10G subwavelength channel with 156 Mb/s.

The second experiment demonstrated the use of SDN to control a SDM network consisting of three AoD nodes, linked by several SMFs and two MCFs with 19 and 7 cores, respectively [22]. SDN provides four key features that make it suitable for such heterogeneous networks: Firstly, the data and control plane are separated which allows an independent evolution of the network and control software as well as network virtualization where the SDM infrastructure can be partitioned (sliced) into multiple logically isolated units. Secondly, the network control is centralized which leads to a better overview and stability of the network. Thirdly, open (i.e., vendor independent) interfaces are used between the software/hardware in the control and data planes which achieve plug-and-play operation of the network. And fourthly, programmability of the network by external applications due to the abstraction of the

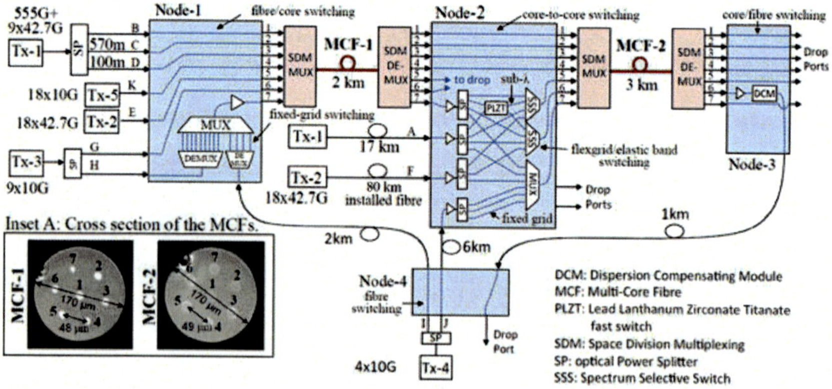

Fig. 6.8 Experimental setup of the multi-granular SDM network (Reprinted with permission from [21]. ©2013 The Optical Society)

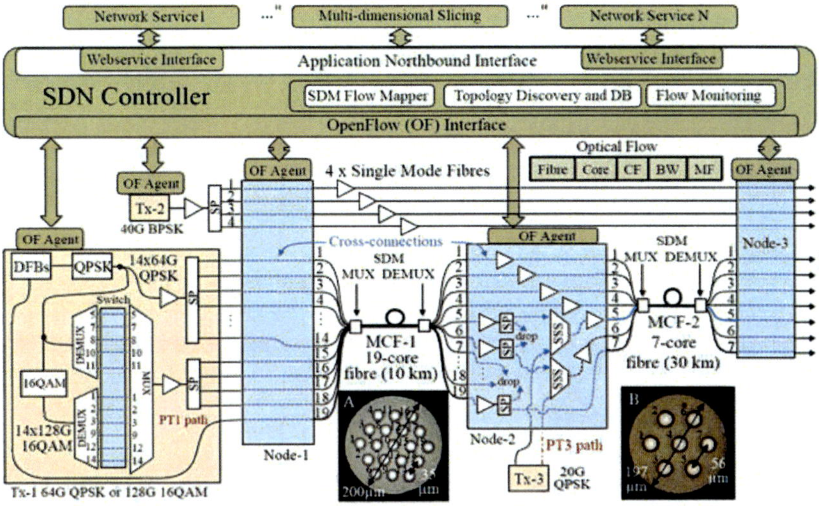

Fig. 6.9 Experimental setup of the SDN-controlled SDM network (Reprinted with permission from [22]. ©2014 The Optical Society)

underlying network infrastructure by the SDN controller enables fast development of new services and a rapid evolution of the network.

As shown in Fig. 6.9, the control plane includes an SDN controller and a multi-dimensional network slicing service (i.e., external applications). Communication between the control and data plane is carried out with an OpenFlow interface at the control plane side and OpenFlow agents at the data plane side. OpenFlow [23] is a standard protocol that acts as the control data plane interface to enable manipulation and centralized control of network devices by the SDN controller.

MCFs are abstracted as a single entity with multiple spatial channels whose information is used by the slicing service for routing spatial superchannels. A spatial superchannel is generated by grouping together signals on a number of spatial channels (i.e., cores) that operate at the same wavelength and thus provision more instantaneous bandwidth to a single request. For example, a 19-core MCF could be abstracted as a superchannel with 1.9 Tb/s per wavelength, while only 100 Gb/s (per wavelength) are carried by each core. On the other hand, SMFs are abstracted as multiple entities each with a single independent spatial channel and, therefore, are not suitable for transmitting spatial superchannels. An example of data flow using these different types of spatial channels is illustrated in Fig. 6.10. The SDN controller translates connectivity requirements such as source, destination and quality of transport (QoT) into physical layer requirements such as number of cores, wavelength, data rate and the modulation format. Each data signal is then assigned to either a single SMF core or a group of MCF cores representing a superchannel slice. In this way, flexible bandwidth provisioning from 40 to 512 Gbps in conjunction with different QoT requirements could be demonstrated.

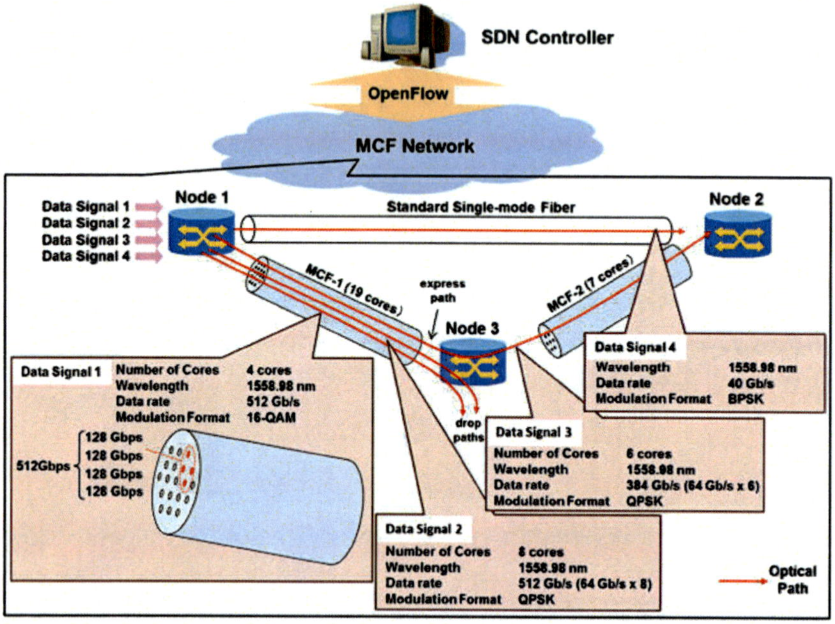

Fig. 6.10 Example of the heterogeneous data flow in an SDN-controlled SDM network with sliced spatial superchannels

The elastic node architecture in combination with a flexible resource allocation gives the network designer more freedom to optimize the traffic routed through the SDM network. Fujii et al. [24], for example, demonstrated how a smart spectrum and core allocation method could reduce spectrum fragmentation, blocking probability and node cost in a seven-core MCF network taking into account realistic topologies. Furthermore, they also showed that by including inter-core crosstalk between adjacent cores and information on the core position in the RSA problem, the impact of crosstalk from the perspective of the network can be significantly mitigated even under high traffic load [25].

6.1.3 *Network Enhancements Due to MCF*

Although the integration of parallel transmission channels almost always entails a certain amount of channel crosstalk, the physics of inter-core crosstalk in MCFs has been well-studied, and an adequate fiber design is likely to be able to mitigate it to negligible levels (see also Sect. 2.2). In the following, we summarize three features that are unique to low-crosstalk MCF transmission and thus may help enhance the performance of future networks.

(1) *Wavelength/spectrum contention management in the optical domain*

In an SMF network, different traffic demands with the same wavelength cannot be routed through the same link due to wavelength contention. To solve this problem without having to resort to OEO conversion, much hope has been placed in research on all-optical wavelength converters based on semiconductor or periodically poled lithium niobate devices as well as highly nonlinear fibers. However, the trend toward using digital coherent transmission techniques in the future optical networks may substantially limit the range of potential solutions. This in turn will make it also increasingly challenging to convert those research outcomes into efficient and practical products due to the intrinsic properties of the optical processes involved [26]. Therefore, it is likely that SMF networks will continue to rely on RWA/RSA algorithms and OEO regeneration to minimize the probability of blocking due to wavelength contention.

In an SDM network based on FMF, traffic requests with the same wavelength (but assigned to a different spatial mode) can be routed through the same link, but the add and drop points of the requests need to be the same due to mode mixing. Therefore, from the viewpoint of wavelength contention management, FMF networks offer no advantage over SMF networks.

In an MCF network, on the other hand, the wavelength contention problem could be further mitigated solely in the optical domain since core (spatial channel) switching can be easily incorporated in the node. A simplified example is shown in Fig. 6.11. When there is an add request with high QoT requirement against less-impaired but already-occupied channel (Core 2), the express traffic with less stringent QoT requirements can be temporarily switched to another (by the same wavelength) unoccupied spatial channel (Core 1) with less favorable transmission characteristics. The risk of having to block a new request can be effectively averted in the optical domain.

(2) *Sliceable spatial superchannels*

A spatial superchannel does not need to occupy all cores of an MCF, but it could also be formed out of one or more subsets of core groups referred to as superchannel slices thereby increasing switching granularity. The size (bandwidth) of these slices can be flexibly adjusted according to traffic demands as was already exemplified in Fig. 6.11. Note that, the concept of the spatial superchannel is based on the fact that

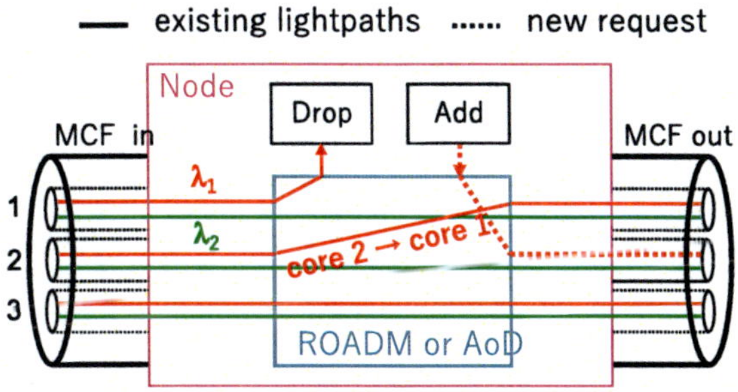

Fig. 6.11 Example of wavelength contention management with MCF

the phase shifts of signals due to environmental changes in MCF cores surrounded by the same cladding are much stronger correlated than in parallel strands of SMFs [27]. These correlations could be utilized to save in DSP complexity by sharing that phase information between DSPs in a master–slave configuration to jointly process the signals on the subchannels of a superchannel [28

(3) *Bidirectional transmission in a single fiber*

Although it is possible to carry bidirectional traffic on a single SMF, it is not common practice to do so in today's networks. Telecom carriers typically light up an additional pair of fibers when they need to add capacity to a link. On the other hand, MCFs are well-suited to handle bidirectional transmission in a single strand of fiber with the side benefit of significantly reducing the effect of inter-core crosstalk when assigning each direction to a core in an interleaved fashion [29, 30].

Conclusions

As network traffic continues to grow, the required physical layer capacity will most likely be provisioned in the near term by deploying more SMFs per link and deriving more flexibility from the underlying network technology. However, as network evolution strongly depends on a reduction of the cost per bit and to an increasing extent the energy per bit, it is also clear that this trend cannot go on forever and it may not take too long before a tipping point is reached where SMF networks with multiple fibers and separate network elements are no longer economically viable. The development of SDM technology and the associated trend toward greater device integration are a consequence of the continuing quest for more efficient optical networks. Besides new fiber types and amplification techniques, advances in node design and network management strategies are crucial to make efficient use of the last unexploited physical dimension in optical communications.

6.2 Application of SDM Technologies in Short-reach Systems

Introduction

The demand for large-capacity data transmission is rapidly growing driven by the expansion of broadband application services [31] in large-capacity short-reach transmission systems due to the ongoing increase in data traffic. Especially, network reliability has become a serious issue [32, 33] in large-capacity transmission systems. Some researchers have recently made avid appeals for the need to prepare for disabilities in communication networks [34, 35]. Space-division multiplexing (SDM) schemes utilizing multi-core fibers (MCFs) [36], on the other hand, have attracted a great deal of attention as promising technologies for overcoming the limitations in not only transmission capacity imposed by conventional single-core single-mode fibers (SMFs) [37] but also cable size at data center (DC) networks and passive optical networks such as fiber to the home (FTTH) [38] in access areas, which have necessitated the use of high fiber counts and high-density cables. MCFs can provide protection lines for redundancy as well as primary high-capacity signal lines in addition to these capabilities and can thus create networks that are extremely reliable.

One such approach can be the network system with optical protection switches and MCFs for high capacity and high reliability: We recently propose its basic concept [39, 40] and have fabricated prototype protection optical switch units (POSUs) with MCFs, as well as compact fan-in/-out (FI/FO) modules and optical connectors for the MCFs. Automatic impairment-aware optical-path switching was experimentally demonstrated by configuring an MCF network with these units and components [41]. In this prototype, we focus on access networks for business [42] residential areas [43] in this study as targets that required reliability on the basis of business continuity plans (BCPs) [44, 45], since access networks have explosively expanded due to carriers and have been alternatively developed and improved. For instance, a DC in a business area has one base node as a rule of aggregation and several accompanying branch nodes [46, 47]. A core switch (SW) is prepared at the base node as an input/output (I/O) interface for access to an external system. The end-of-rack SWs/top-of-rack SWs are prepared at the branch nodes to access servers and storage. The network topology in such a DC is useful for operation under no centralized traffic by means of increasing numbers of aggregation nodes that prevent collisions between data and shorten the access time to servers [48, 49]. Moreover, such networks need to have highly reliable characteristics to prepare for unpredictable problems such as failures or sudden excessive usage.

After an introduction of some types of short-reach systems in Sect. 6.2.1, we discuss in Sect. 6.2.2 a novel MCF link based on a multi-ring structure to achieve robust access networks such as those in DC systems in business areas and FTTH systems in residential areas. We also present a switching control architecture that could suitably manage this network and was effective in recovering from failures that occurred in the network. An MCF connector and an MCF FI/FO module were developed to enable low-loss connection between units. We also experimentally investigated the fundamental operation of automatic impairment-aware optical-path switching when failures occurred in the proposed link.

6.2.1 Types of Short-reach Systems

6.2.1.1 High-capacity Data Interconnection

As the speed of short-reach optical links is getting higher and higher, the use of parallel transmission becomes more and more popular: Table 6.1 shows the examples of high-speed Ethernet standards over a last decade: To achieve higher total capacity, most of standards utilize the plural lanes using parallel optical fibers or a single

Table 6.1 Examples of high-speed Ethernet standards

Standard	Baud rate/format	Type	Reach	Number of lanes[a]
40GBASE-SR4	10.3125 Gbaud/NRZ	4MMF	100/150 m	4
40GBASE-LR4	10.3125 Gbaud/NRZ	SMF/4WDM	10 km	4
100GBASE-SR10	10.3125 Gbaud/NRZ	10MMF	100 m	10
100GBASE-SR4	25.78125 Gbaud/NRZ	4MMF	100 m	4
100GBASE-LR4	25.78125 Gbaud/NRZ	SMF/4WDM	10 km	4
100GBASE-DR[b]	53.125 Gbaud/PAM4	SMF	500 m	1
400GBASE-SR16	26.5625 Gbaud/NRZ	16MMF	100 m	16
400GBASE-DR4	53.125 Gbaud/PAM4	4SMF	500 m	4
400GBASE-FR8/LR8	26.5625 Gbaud/PAM4	SMF/8WDM	2 km/10 km	8

[a]For unidirectional connection
[b]Standardization not completed

optical fiber with wavelength-division multiplexing (WDM) technique. Since the increase of number of lanes (transceivers) leads to significant increase of cost, power and transceiver size, it is kept to less than four in many standard categories.

With the 10-times capacity increase from 40 to 400 GbE, the maximum channel rate is also increased by 10 times from binary 10 Gbit/s (10 Gbaud NRZ) to 100 Gbit/s (50 Gbaud PAM4) with the use of high-speed multilevel modulation [50, 51], which keeps the number of lanes to four. But, the use of 50 Gbaud PAM4 signal is currently limited to less than 500-m SMF categories (100GBASE-DR and 400GBASE-DR4 in Table 6.1), mainly due to the reduction of power budget of high-speed PAM signals. In the case of 400GBASEFR8/LR8 with longer reach using WDM, the number lanes (wavelength) must be increased to eight. Further increase of channel speed to 100 GBaud or beyond is investigated; however, it will face issues of device bandwidth limitation and less transmission reach. For the case of less-expensive shorter-reach interconnection over multi-mode fiber (MMF) commonly used in intra-building or rack-to-rack cabling, the number of signal lanes required for bidirectional connection is increased from 8 to 32 in the case of 400GBASE-SR16 [52].

In the future higher-speed interconnection like 1 Tbit/s or beyond, the number of signal lanes will increase more and more, due to the limits of both physical lane speed and number of wavelength: For example, to realize a 1.6-Tbit/s interconnection, the number of signal lanes (unidirectional) is expected to increase up to 64 for 10-Gbaud binary signaling, 32 for 25-Gbaud PAM4 signaling, 16 and 8 for even 50-Gbaud and 100-Gbaud PAM4 signaling.

Today's parallel interconnection typically uses ribbon fibers with MTP™/MPO-type parallel connectors, which causes significant wiring problems even in today's DCs, network nodes or supercomputers: Jacketed ribbon fiber cables are thick, heavy, costly and difficult to handle. Their connectors are also expensive, bulky and consume precious front or back panel area. Also in large DC networks, the use of parallel multi-mode ribbon fibers causes differential modal skew problem which makes timing adjustment of high-speed signals difficult [53]. Further increase of number of fibers in a single ribbon fiber or connector seems very difficult to achieve.

Future introduction of newly developed MCF-based data interconnection (possibly with even MCF-based ribbon fiber) is expected to relax these wiring problems considerably, reducing the number of fiber cables by one order of magnitude and will give future DCs precious margin for further expansion. At the same time, multi-mode MCF is expected to have much less differential modal skew compared with ribbon MMFs.

Table 6.2 shows recently proposed MCF links for short-reach data interconnections. Requirement on MCF for short-reach interconnections is considerably different from those of core networks: One major requirement is the choice of signal wavelength, and typically 0.85 or 1.3 μm is used in very short reach (up to 100 ~ 300 m) and short reach (up to 40 km) applications. For the very short-reach applications, MCF with multi-mode fiber core [54–56] is preferable to relax the coupling accuracy between optical devices and the fiber core and to reduce assembly cost. Single-mode MCFs are also required for short-reach applications, since transmission reach of high-speed signals over multi-mode core is seriously limited by its modal skew. Clad diameters of MCF should be 125 μm or similar, in terms of fiber handling, bending loss, fiber cable density and backward compatibility. On the other hand, core-to-core crosstalk specifications required for short-reach interconnection can be relaxed drastically since the amount of total crosstalk is proportional to fiber length.

Other considerations are the number of cores and the core arrangement: In terms of efficient use of cores, the core number for short-reach parallel interconnections should be compatible with the number of their signal lanes (e.g., Table 6.1). The core arrangement may be more important in terms of the coupling to optical devices; since fan-in/fan-out devices between MCF and single-core fibers are costly and bulky for low-cost short-reach interconnections, direct coupling between MCF to integrated lasers and photodetectors seems to be very favorable in the future. In some cases, square core arrangement [58, 59] may be favorable, since it is compatible to the arrangement of integrated optical devices.

Table 6.2 Proposed MCF links for short-reach data interconnection

Refs.	No. of core	Core arrangement	Core diameter or MFD/mode	Clad diameter (μm)	Wavelength
[57]	2	Dual (2)	10 μm/single	125	1.3/1.5 μm
[58]	4	Square (2 × 2)	10 μm/multi-	125	850 nm
[54, 55]	7	Hexagonal (6 + 1)	26 μm/multi- (GI)	125	850 nm
[59]	8	Square (2 × 4)	4.9 μm/single	200[a]	1.5 μm
[60]	8	Circular (8)	8.4 μm/single	125	1.3 μm

[a] Estimated from photograph

6.2.1.2 PON and Data Center Networks

Various studies were reported on new applications of MCF/few-mode fiber (FMF)-based SDM technologies for short-reach systems. One of such application areas is access network (e.g., [61, 62]). Zhu et al. [38] proposed the first application of MCF to increase the capacity of PON and successfully demonstrate parallel transmission of seven PON signals over 11.3-km seven-core MCF with 2.5-Gbit/s upward/downward signals and 1:64 splitting.

Further capacity increase was demonstrated by Li et al. [63], with an aggregating downstream capacity of 300 Gbit/s using six MCF cores with 10-channel WDM. Hu et al. [64] demonstrated mode-division multiplexing (MDM) PON with double capacity over using LP00 and LP01 modes of low modal coupling FMF and passive mode multiplexers/demultiplexers.

Another application of SDM technique in PON is the low-loss combining of multiple upstream signals: Instead of using a conventional power combiner, Nakanishi et al. [65] and Cheng [66] give the basic idea of using a mode combiner and a multi-mode (mode coupling) receiver for the lossless combining of PON upstream signals. Fujiwara et al. [67] propose a PLC-based hybrid upstream SM-to-MMF mode combiner/downstream conventional splitter (Fig. 6.12). The 1 × 16 prototype splitter shows 7.7-dB loss reduction from the ideal combining loss (12 dB), although the MMF length is limited to short distances.

Xia [68] further proposes the use of several-km long FMF and a mode combiner replacing conventional single-mode feeder fiber and a power combiner and demonstrated 20-km 3-mode transmission of commercial G-PON signal. Effenberger [62] further introduces the concept of PON upstream upgrade of lossless coupling configuration to mode group multiplexing by adding a mode splitter at the OLT side.

One of the issue of MMF feeder technique is a limited number of independent spatial mode unlike the case of long-distance transmission where MIMO technique is typically used to separate multiple spatial mode, since the use of computational-heavy MIMO calculation is not suitable to access networks. One approach to solve

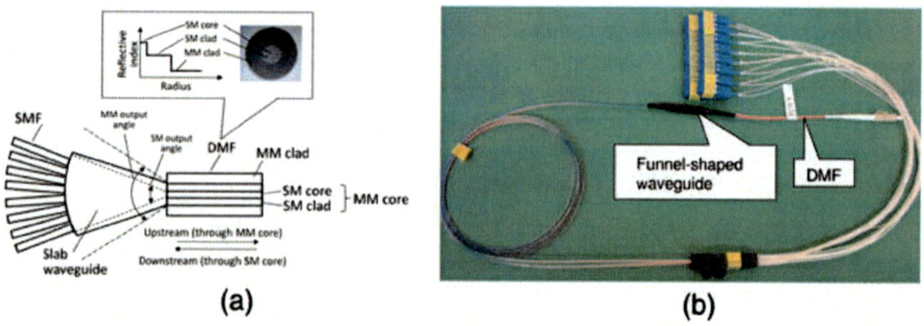

Fig. 6.12 Proposed optical splitter. **a** Configuration. **b** Photograph of 1 × 16 prototype (Reprinted with permission from [67]. ©2014 Authors)

this problem is the use of OAM multiplexing, and Fang [69] proposes its use with free-space optics (FSO) or vortex fiber and successfully demonstrated 20-mode multiplexing of 10-Gbit/s on–off keying (OOK) signals using a 0.4 m FSO setup. On the other hand, Effenberger [62] proposes that the computational load of the MIMO DSP of the FMF mode demultiplexing can be relaxed in the PON application due to a small differential modal group delay and the burst TDM signaling, which tend to enhance modal orthogonality.

The use of optical interconnection and optical SDM switches in the future data center is proposed and discussed in many papers (e.g., [53, 70–72]), and the introduction of MCF/FMF-based SDM technologies is expected in the future. Yan [73] proposes the intra- and inter-data center networking architecture utilizing WDM/TDM/SDM technologies and experimentally demonstrates their feasibility with the prototype utilizing three-element multi-element fiber (MEF) to handle top-of-rack SDM links and SDM-to-WDM conversion techniques based on fiber four-wave mixing (FWM) [74].

6.2.2 Application Example of SDM Technology in Short-reach Systems

A failure recovery system utilizing a multi-core fiber (MCF) link with protection optical switch units was developed to achieve high-capacity and high-reliability optical networks in short-range links. We describe the novel MCF link based on a multi-ring structure and a protection scheme to prevent link failures. Fan-in/fan-out modules and connectors were also developed in order to demonstrate the development status of the MCF connection technology for the link. We demonstrated path recovery by switching operation within a sufficiently short time (< 50 ms), which is required by ITU-T. The selection of a protecting path as a failed working path was also optimized as the minimum passage of units for low-loss transmission. The results we obtained indicate that our proposed link has potential for the network design of highly reliable network topologies in access areas such as data centers, systems in business areas and fiber to the home systems in residential areas.

The remaining part of this chapter is organized as follows: Sect. 6.2.2.1 explains the basic concept of the proposed MCF link, and Sect. 6.2.2.2 describes the POSUs, MCF FI/FO modules and MCF connectors used for the MCF link. Section 6.2.2.3 then discusses the MCF link based on a multiple ring structure, the switching control architecture, failure recovery scheme and switching control architecture. Section 6.2.2.4 details an experimental demonstration for recovery from link failures in the ring structure. Finally, in conclusion, we summarize our findings.

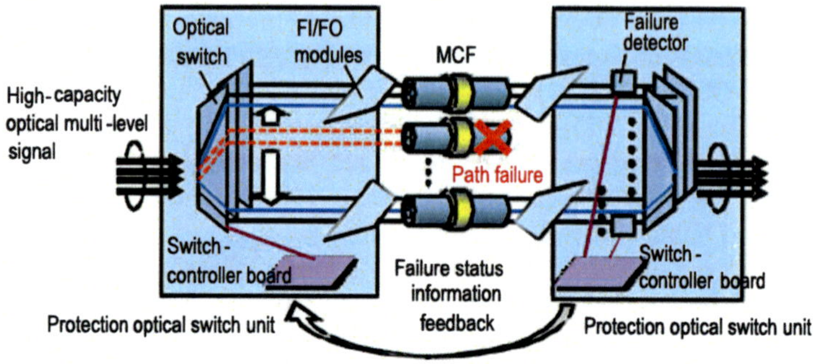

Fig. 6.13 Configuration of the proposed MCF system

6.2.2.1 Concept of Proposed MCF Link

The configuration of the proposed system [40], consisting of POSUs and MCFs, is shown in Fig. 6.13. High-capacity signals for an inter-/intra-data center network or PON network are gathered by the FI/FO modules and input into the MCFs. The protection cores for optical signals in each MCF are reserved for redundancy operation. Switching from a signal line to a protection line occurs immediately after a fiber link failure is detected. With this configuration, the proposed system supports large-capacity, high-density and high-reliability transmission simultaneously by using MCFs and POSUs.

6.2.2.2 Components of the Proposed MCF Systems

(1) *Protection optical switch units*

The fabricated POSU for the MCF link is shown in Fig. 6.14 [40]. It was designed for a link with three seven-core MCFs. The center core carries a control signal; the

Fig. 6.14 Fabricated POSU

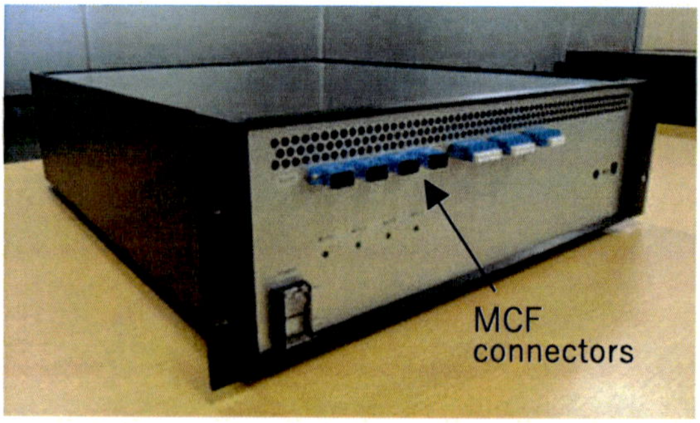

other (outer) six cores are used as main and standby signal lines. The unit size is 430 × 550 × 132.5 mm (corresponding to 3U of a 19-inch rack).

The POSU consists of optical switches, a switch-controller board, a field programmable gate array (FPGA), tap photodiode (PD) arrays and optical couplers. A thermo-optic (TO) switch was chosen as the optical switch because it provides high-speed operation as well as low insertion loss operation.

The tap PD monitors optical power of the main signal lines. In the case that unacceptable loss is detected in the main signal line, the FPGA sends a control signal to the switch-controller board, which then automatically switches to the standby signal line for restoration of communication.

(2) *Fan-in/fan-out modules*

We used a seven-core MCF as shown in Fig. 6.15 with a cut-off wavelength below 1.3 μm designed for applications in DCs [53]. The mode field diameters (MFDs) of the MCFs were 8.6 to 8.8 μm at a wavelength of 1.55 μm. The average core pitch of the MCFs was 40 μm.

An FI/FO module based on free-space optics [75] was fabricated to connect MCFs with the SMFs. The free-space optics-based FI/FO module could be applied to a wide range of wavelengths, any number of cores and any core arrangements; i.e., it had design flexibility.

It was necessary to expand the diameters of the light beams from 8.8 to 10.5 μm and their spaces from 40 μm to more than 0.9 mm to couple each core between the MCF and the SMFs. Figure 6.16 schematically shows the free-space-based FI/FO. Both ends of the MCF and the SMFs were polished flat at an angle. Aspherical lenses

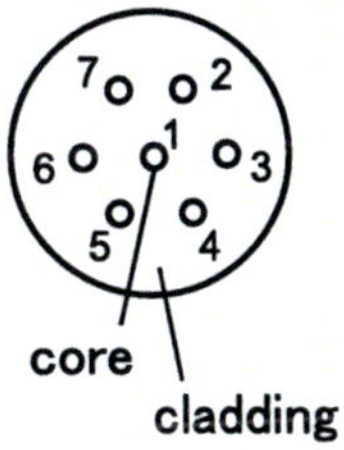

Fig. 6.15 Schematic of cross-section of a seven-core MCF

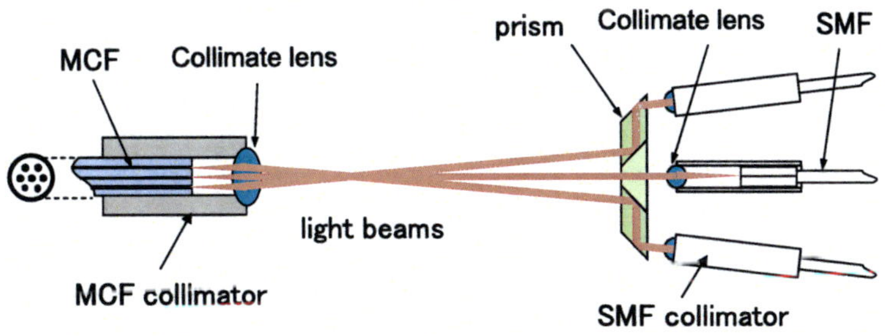

Fig. 6.16 Schematic of free-space-based FI/FO

were located on the ends of the MCF and the SMFs in a confocal arrangement. The diameter of the light beams emitted from the MCF was magnified by 1.2 times at the end of the SMF by these lenses. Six prisms were inserted between six pairs of the lenses to expand the spacing between the light beams. Each facet of the fibers, lenses and prism was coated with an anti-reflective film to obtain transmittance of more than 99% and an optical loss of less than 0.044 dB. The estimated losses between the center core of the MCF and the SMFs were less than 0.26 dB because the light beams passed through six facets of the fibers and lenses in the FI/FO. The estimated losses between the outer cores of the MCFs and SMFs were less than 0.35 dB because the losses of the two facets of the prism in the FI/FO were added to the above.

A SC/PC connector was mounted on the other ends of the MCF, and LC/PC connectors were mounted on the other ends the SMFs. The SC/PC MCF connector is described in Sect. 6.2.2.3–(3) in detail. Figure 6.17 shows a photograph of the fabricated FI/FO. The characteristics of the fabricated FI/FOs are shown in Fig. 6.18. The measured insertion losses of the FI/FOs were less than 0.7 dB (typically 0.3 dB) including the insertion losses of the MCF connectors and return losses that were more than 40 dB.

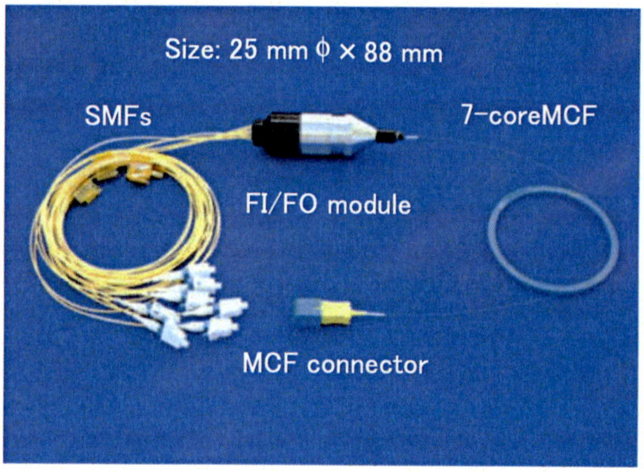

Fig. 6.17 Photograph of free-space-based FI/FO with fibers

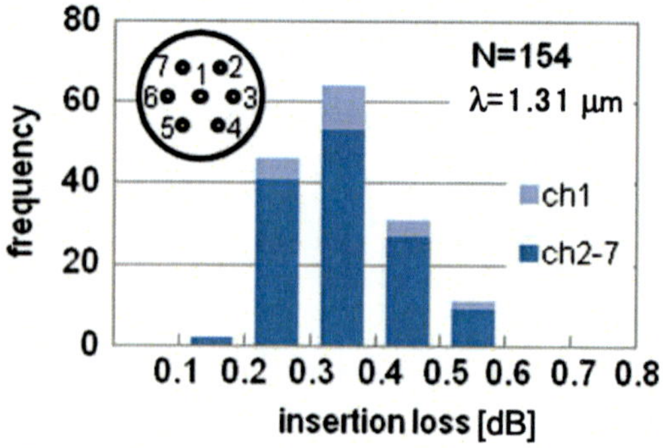

Fig. 6.18 Insertion loss of fabricated FI/FOs measured at wavelength of 1.31 μm

(3) *MCF connectors*

The fabricated MCF connectors were the widely used SC type [76] with a PC structure [77]. Although the conventional SC type is common, a ferrule freely turns in the connector because of the clearance between the connector housing and the ferrule. When the housing is applied to the connector for MCFs, coupling losses in the outer cores of the MCFs are increased by angular misalignment between the two MCFs.

The coupling loss of the fiber pair can be calculated by using the following Gaussian field approximation [78]:

$$\text{Loss} = -10\log_{10}(\eta) \tag{6.1}$$

$$\eta = \left(\frac{2\omega_1\omega_2}{\omega_1^2 + \omega_2^2}\right)^2 \exp\left(-\frac{2d^2}{\omega_1^2 + \omega_2^2}\right), \tag{6.2}$$

where d is the core offset and ω_1 and ω_2 are the mode field radii of the pair of cores. Since the range of core pitch errors is 1 μm, the angular misalignment must be reduced to within 0.7 degrees to satisfy a coupling loss of less than 0.5 dB [79] from (6.1) and (6.2). The ferrule is held with a pressurization spring attached to a plug frame in a connector housing to prevent the ferrule from turning [80]. The pressurization spring is a plate spring formed by using sheet metal processing. It has four claws. As we can see from Fig. 6.19, it holds the flange of the ferrule from four directions without slippage in the plug frame. The plug frame is made of plastic molding and has a space into which the pressurization spring is inserted. The ferrule is thus "floating" in the plug frame being held by the four claws of the pressurization spring, and it can only move in the x- and y-directions. A photograph and characteristics of the fabricated connectors are shown in Fig. 6.20 [80]. The measured insertion losses of the connectors are less than 0.5 dB (typically 0.2 dB), and the return losses are more than 40 dB.

Furthermore, we fabricated LC-type MCF connector by using the same technique as used to fabricate SC-type connector (see Fig. 6.21). The LC-type connector is four

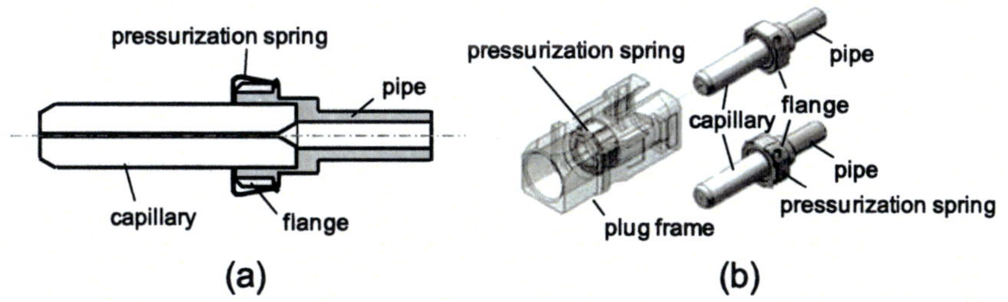

Fig. 6.19 Schematic structure of ferrule fixed in plug frame with pressurization spring: **a** Cross-section of ferrule and **b** Bird's eye view

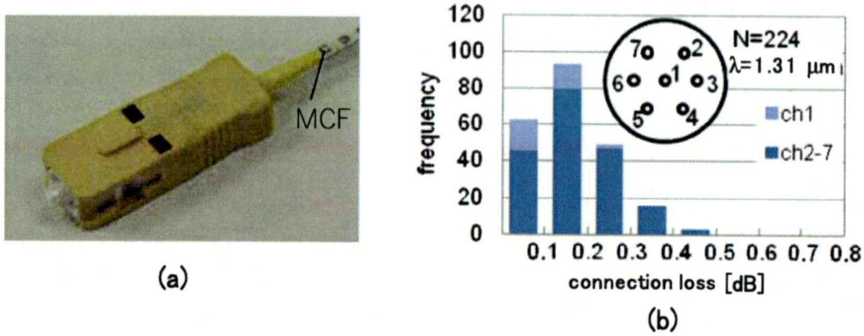

Fig. 6.20 **a** Fabricated SC-type MCF connector **b** Connection loss of the SC-type MCF connectors measured at 1.31 μm wavelength

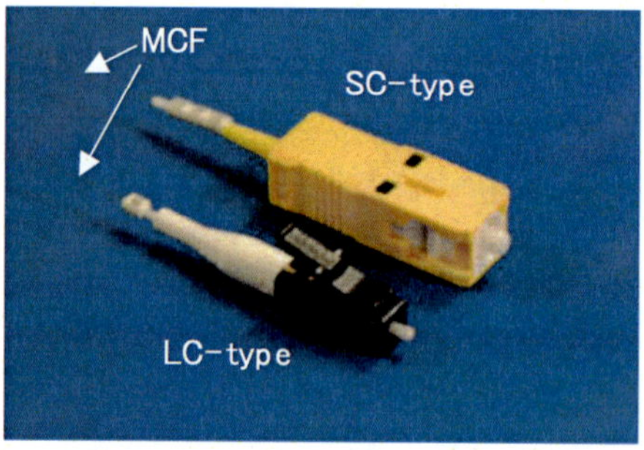

Fig. 6.21 Photographs of a fabricated LC-type MCF connector together with the SC-type connector for comparison

times smaller than SC type and widely used in data centers (see DC standard TIA-942). Insertion loss, PDL, connection loss after mating cycles are almost equivalent to those of the SC-type connector as shown in Table 6.2.

6.2.2.3 MCF Link Based on Multi-ring Structure

Protecting the paths substituted for working paths needs to be adequately arranged in links between the centralized node as the I/O interface and its branch nodes to construct a robust network with high levels of reliability in access areas such as DCs and FTTH systems. We considered a ring network topology and MCF in the present study, which were able to cope with ultra-large data [81, 82]. The ring network topology provided promise in terms of helping to increase performance, e.g., to ensure highly reliable characteristics while maintaining large-capacity transmission in such networks. We, therefore, constructed a link based on a multi-ring structure where an MCF link [82] was used to connect nodes. Such MCF links can make connections between centralized nodes and their branch nodes through various routes.

Figure 6.22 is a schematic of the proposed ring network topology. One POSU with an optical SW is located at a centralized node, and POSUs with optical SWs are

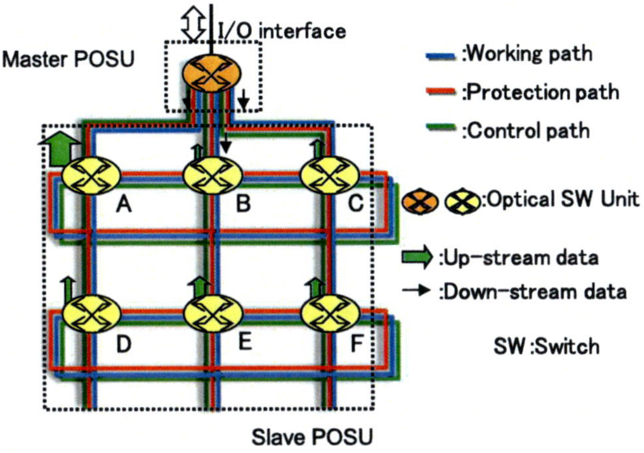

Fig. 6.22 MCF link for highly reliable network

arranged at branch nodes (A, B, C, D, E and F). The link between units is connected by just one MCF constituting a one-ring structure via Nodes A, B and C. An additional unit at Node D can be linked by a cascade connection with the unit at Node A by just one MCF. Similarly, units at Nodes E and F can be linked with units B and C by just one MCF. These links also constitute a one-ring structure by using Nodes D, E and F. The unit at centralized Node Z plays a master role, and the units at its branch nodes of A, B, C, D, E and F play slave roles.

Since this topology can easily be used to set up branch nodes due to cascade connections, it has an advantage in terms of scalability. Moreover, this MCF link based on a multi-ring structure undeniably leads to increased network performance, viz., various kinds of paths (e.g., working, protection and control) can be flexibly laid on the links because of the use of multiple cores in just one MCF. This enables us to secure a protected path substituting for failures in working paths even if network failure occurs. As was previously stated, we could construct a network with strong capacity, scalability, flexibility and reliability characteristics by using an MCF link based on a multi-ring structure.

Furthermore, the MCF link communicates with data between the master unit and the slave units by means of bidirectional transmissions through working/protection paths, as outlined in the upstream/downstream data in Fig. 6.22. The MCF link can assign multiple paths in addition to the working path, thus increasing the transmission capacity from the node when upstream data from a node increases in the ordinary operating mode, as indicated in Fig. 6.22. The units at the nodes can share network information indicating failure alarms/notifications, restoration requests/replies and switch configurations by using the control paths.

However, a control architecture is needed to switch from a working path to a protective path in the event of link or node failure to manage the MCF link based on the multiple ring structure. We proposed a failure recovery scheme to manage the link by means of monitoring signals (FRS-MS) [83] to cope with multiple failures in not only the working paths, but also the protective paths of the link. Figure 6.23 shows a schematic of the link, which is structured between two nodes with switch (SW) units. The link between units is constructed by assigning paths a, b and c as working paths and paths d, e and f as protective paths (Fig. 6.23a). MCFs are all that are needed because they have multiple cores, and we can freely assign working and protective paths to the cores in the link connected by MCFs when designing a network to fit the specific needs of applications, such as high capacity through the use of *switch database management* (SDM) technology [36] and high reliability through using restoration schemes by using protective paths. Moreover, the MCF connection has advantages in terms of handling and the space it occupies in comparison with connections using various kinds of fibers such as ribbon and bundle fibers.

The POSU in Fig. 6.23 has an optical SW for transferring paths, a controller for controlling other units and itself and multiple signal detectors for detecting path

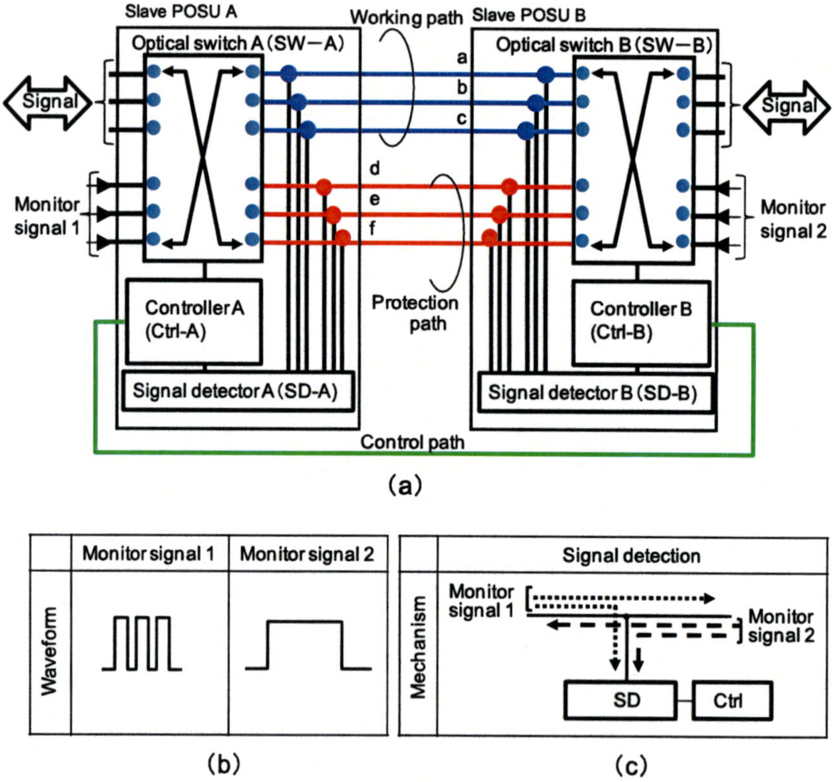

Fig. 6.23 Schematic of failure recovery scheme that sends monitoring signals through protection paths: **a** Switch units consisting of controllers and signal detectors have multiple ports for in/outputting signals and multiple ports for inputting monitoring signals, **b** Waveform of monitoring signals and **c** Mechanism for detecting monitoring signals

failures. The controller (Ctrl-A) in one unit is connected to Ctrl-B in the other by a control path, as seen in Fig. 6.23a. In addition to the input and output ports for the signals, the unit has input ports for the monitoring signals that are needed in FRS-MS. Since monitoring signals such as laser light from the input ports can always be transmitted through protection paths and be observed by the signal detectors (SD-A/SD-B) in this link, we can check the propriety of path usage before switching operations are carried out. In addition, a monitoring signal helps to secure a protective path for the working path because we can use it to find the most available path in the protective paths. FRS-MS can work by doing just one switching procedure due to its control of all-optical POSUs at each node in the MCF link based on the multiple ring structure, which shortens the recovery time.

Moreover, a link based on FRS-MS can distinguish link failure from component failure. For instance, the waveform of Monitoring Signal 1 incoming to POSU A is different from that of Monitoring Signal 2 incoming to POSU B in Fig. 6.23b. The signal detector can, hence, distinguish a monitoring signal from the right from one from the left in Fig. 6.23c. Hence, when SD-A in POSU A detects Monitoring Signal 2 instead of Monitoring Signal 1 and SD-B in POSU B detects Monitoring Signal 2 instead of Monitoring Signal 1, we can determine that there has been a failure in optical switch A (SW-A) as opposed to failure in one of the working paths without failure in optical switch B (SW-B). However, when SD-A in POSU A detects Monitoring Signal 1 instead of Monitoring Signal 2 and SD-B in POSU B detects Monitoring Signal 2 instead of Monitoring Signal 1, we can determine that there has been a failure in one of the working paths as opposed to failure in optical SW-A without failure in SW-B. This function helps to inspect the paths between units, diagnose the optical switches in units and exchange the failure components and fibers, which thereby easily helps to maintain the network. Hence, we could suitably manage the network and consequently increase its reliability by implementing such FRS-MS on FPGA in units at nodes in the MCF link based on the multiple ring structure.

6.2.2.4 Recovery from Failure in an MCF Link

This section describes the experimental setup for testing recovery from a link failure. Next, we will discuss the experimental evaluation in terms of the operation of this MCF link in the event of a link failure.

(1) *Setup for evolution of recovery*

We experimentally investigated fundamental operations with relation to the reliability of the proposed MCF link. Figure 6.24 shows the MCF link based on a one-ring structure, which consists of Unit Z as a master and Units A, B and C as slaves. These latter three units are all located on one ring. Each unit has a 20×20 thermo-optical SW for transferring paths, tap PDs for detecting link failure, lasers with a wavelength of 1.5 μm for monitoring the protecting paths, two kinds of field programmable gate arrays (FPGAs) (FPGA-A and FPGA-B) for controlling the

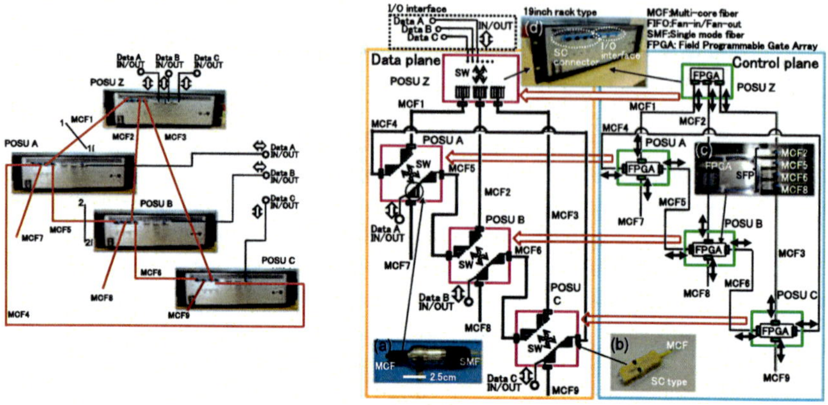

Fig. 6.24 MCF link based on one-ring structure: **a** MCF FI/FO module, **b** MCF connector, **c** FPGA-based control board in unit where outgoing data from control board transmit and receive via optical signals through small form-factor pluggable (SFP) modules and **d** Front panel of unit settled in 19-in. rack. Black arrows indicate bidirectional communication between FPGAs in units, and red-lined arrows indicate control of optical switches by FPGAs in units. Cross-sectional images of 1–1′ in MCF1 and of 2–2′ in MCF5 are depicted in Table 6.3

MCF link (Fig. 6.24c), four fan-in/fan-out modules for connecting the MCF to each port of an SW (Fig. 6.24a) and four MCF connectors for connecting the MCF to a unit (Fig. 6.24b). The FI/FO modules and MCF connectors are explained in detail in the following Subsections (2) and (3). All components in each unit are accommodated in a 19-inch rack, as shown in Fig. 6.24d. Figure 6.24 also shows that this optical SW allows data incoming to a unit to transfer to two units, although the unit is related to four units arranged around itself. For instance, incoming data to Unit A through MCF1 can exit through MCF5 or MCF7 and are prevented from exiting through MCF4, while the FPGA in the unit can communicate with those in all four units located around itself by means of bidirectional transmission using the controlling path. As an example, the FPGA in Unit A can communicate with all other FPGAs in units that are connected by MCF1, MCF4, MCF7 and MCF5. As a result, all FPGAs around the unit can share network information with one another, even if failure occurs anywhere in the network.

Our MCF link was organized with a data plane for data transmission and with a control plane for FPGA-based management, as seen in Fig. 6.24. Incoming Data A,

Table 6.3 Summary of characteristic of LC- and SC-type MCF connectors

	SC type	LC type
Size of cross-section (mm^2)	9 × 7.5	4 × 5
Insertion loss (dB)	< 0.5	< 0.4
PDL (dB)	< 0.03	< 0.01
Connection loss after ten mating cycles (dB)	< 0.4	< 0.5

B and C to Unit Z (downstream) in the data plane can be output from Units A, B and C by working paths that lead to these units by way of MCFs 1, 2 and 3 after Unit Z in the ordinary operating mode. When we refer to the number of unit passages in a path as the "hop count," each path, therefore, corresponds to a hop count of two in the ordinary operating mode. Similarly, Incoming Data A, B and C to Units A, B and C (upstream) can be output from Unit Z together by using the same paths downstream. In contrast, FPGA-A detects network failures in the control plane and determines the protecting path as a switching destination from the failure working path on the basis of the switching control architecture mentioned in the previous section. FPGA-B controls the SW in each unit accordance with a command from FPGA-A, as indicated by the red-lined arrows in Fig. 6.24.

The link between units seen in the data and control planes is connected by just one MCF with seven cores with a length of 100 m. This connection by MCF actually enables us to provide protection and control paths as well as a working path by using just one fiber. Table 6.3 summarizes the assignment of working, protection and control paths for seven cores in the MCFs in the search for designing a network from the viewpoint of reliability. Core 2, Cores 3–7 and Core 1 in MCF1, MCF2 and MCF3 play the roles of working, protection and control paths, respectively, while Cores 2–7 and Core 1 in MCF4, MCF5 and MCF6 play the roles of protection and control paths, respectively. We assigned a control path to Core 1, which was somewhat influenced by crosstalk from the other cores (inset: Table 6.3) [84, 85], because a signal in the control path could be enhanced in terms of the signal-to-noise ratio. Thus, this assignment in the MCF link effectively contributed to securing a protection path when failure occurred in the network.

We used the laser lights of the wavelengths in the C-band in these experiments as the incoming data to the I/O interface in Unit Z. The laser lights of the outgoing data from Units A, B and C were measured with an oscilloscope (Agilent: MSO6104A) through an optical/electronic (O/E) converter and with an optical spectrum analyzer (Anritsu: MS6710A). In addition, all outgoing electronic signals (P1, P4 and P8) from the control board in each unit indicated the detection of network failure by FPGAs, the operation of SW driven by FPGAs and the detection of data signals by FPGA, respectively, and were observed with the same oscilloscope. We could accurately estimate the recovery time by using P1, P4 and P8 from that in Unit Z and trace the protecting path as the switching destination with P4 in all units (Table 6.4).

(2) *Automatic switching for recovery from fiber failure*

Figure 6.25 shows the signals for the outgoing data from Units A, B and C and that of the outgoing electronic signals from the control board in all units when failure occurred in MCF1 between Unit Z and Unit A in Fig. 6.24. Automatic impairment-aware optical-path switching was done in the MCF link based on a one-ring structure in the event of MCF1 failure. As a result, only the signal of Data A with laser light that had a 1.54 μm wavelength, which was transmitted to Unit A from Unit Z, was intercepted and subsequently recovered, as shown in Figs. 6.25a, c, d. We estimated the recovery time to be $\Delta T = 7.4$ ms by using P1, P4 and P8 from the control board in Unit Z, as shown in Fig. 6.25b. This time was sufficiently less than that required by the

Table 6.4 Assignment of working/protection/control paths for seven cores in MCFs

Path name	MCF1/MCF2/MCF3		MCF4/MCF5/MCF6	
	Arrangement	Cross-section	Arrangement	Cross-section
Working	Core2		–	
Protecting	Core3,4,5,6,7		Core2,3,4,5,6,7	
Control	Core1		Core1	

Telecommunication Standardization Sector of the International Telecommunication Union (ITU-T) [86]. This is because the protection path substituted for the failure working path was immediately secured when failure occurred in MCF1.

Figure 6.25 presents signals of Data A and B with laser light that had a wavelength of 1.55 μm, which were transmitted to Unit B from Unit Z, and Data C with a wavelength of 1.56 μm, which were transmitted to Unit C from Unit Z, in the ordinary operating mode (c) and after switching operation (d), when failure occurred in MCF1. The working path in the ordinary operating mode for Data A, which led to Unit A through the working path in MCF1 after unit Z, corresponds to a hop count of two. Data A in the MCF link is recovered with an optical loss of $\Delta L = 7.4$ dB after switching operation, as shown in Figs. 6.25 (c) and (d), which is equivalent to a transmission loss for one unit. We estimated one unit to have an 8.6 dB transmission loss as the sum of the component insertion loss of 0.6 dB at two MCF FI/FO modules, 0.4 dB at two MCF connectors, 4.0 dB at optical SW, 1.2 dB at six fiber junctions and 2.4 dB at four points inserted by tap PD. This means that, a loss for a unit increases in the protection path for Data A. The path, hence, must have a hop count of three, viz., the protection path leads to Unit A through a protection path in MCF4 by way of Unit C through a protection path in MCF3 after Unit Z. Transmission through this path after failure can be explicitly confirmed by the dependence on time with the pulse shape of outgoing electronic signals P4 from Units B, C and Z with no change in dependence on time from that of Unit A, as shown in Fig. 6.25 (e). Therefore, we can draw working paths in the ordinary operating mode and working/protection paths after switching operations in Figs. 6.26a, b. An incoming signal to a unit is only allowed to exit using permitted paths through an optical SW in the MCF link, as shown in Fig. 6.24. However, despite this restriction, we can select a path with the lowest loss from several paths from Units Z to A, except for a path in the ordinary operating mode, due to the switching control architecture proposed in this paper. As a result, the MCF link enables us to suppress any increase in the hop count causing low-loss transmission after switching operation.

Figures 6.25a, c, d indicate that the MCF link based on a one-ring structure kept the outgoing signals of Data B and C transmitting, regardless of failure in MCF1. This meant that, the process of switching to a protection path from a working path for Data A did not interrupt the transmission of Data B or C. This is because MCF3 had

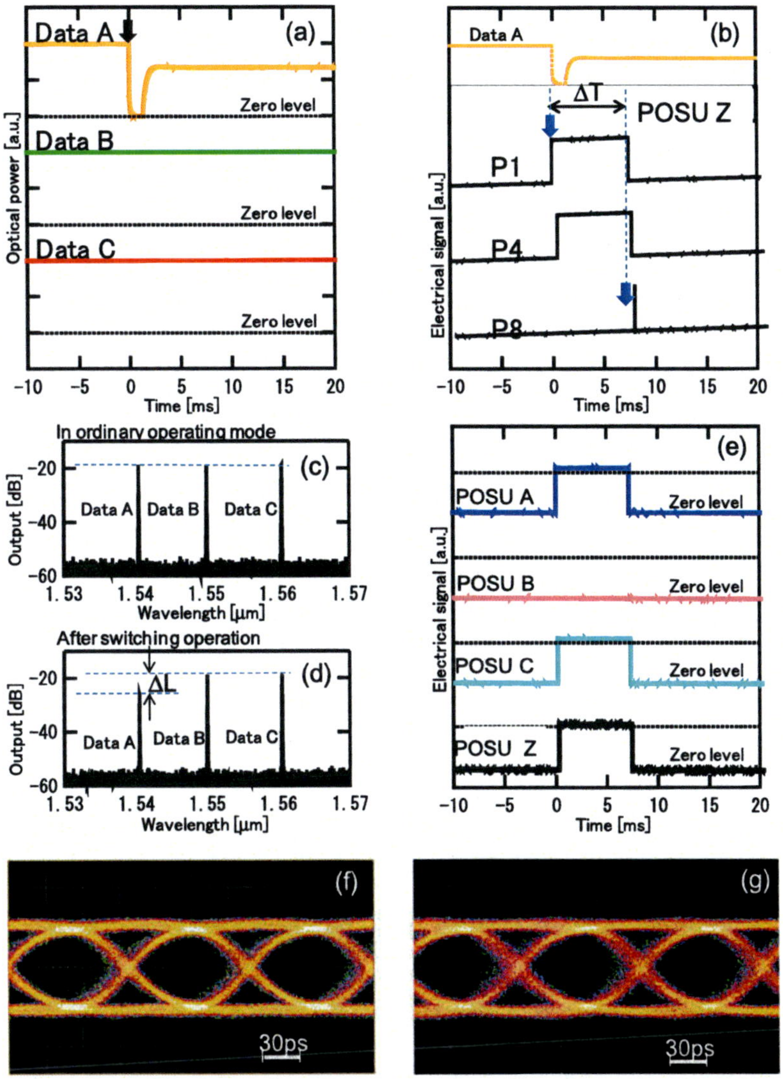

Fig. 6.25 Signals for data transmission in MCF link and that of outgoing electronic signals from control board in units when failure occurred in MCF1 in Fig. 6.24: **a** Dependence of outgoing signals of Data A, B and C on time from Units A, B and C, **b** Dependence of outgoing electronic signals P1, P4 and P8 on time from that in Unit Z, **c** Outgoing signals of Data A, B and C from Units A, B and C in ordinary operating mode, **d** Outgoing signals of Data A, B and C from Units A, B and C after switching operation for recovery and **e** Dependence of outgoing electronic signals P4 from FPGAs on time in Units A, B, C and Z. Electrical eye diagrams of outgoing data from Unit C, **f** In ordinary operating mode and **g** After switching operation for recovery

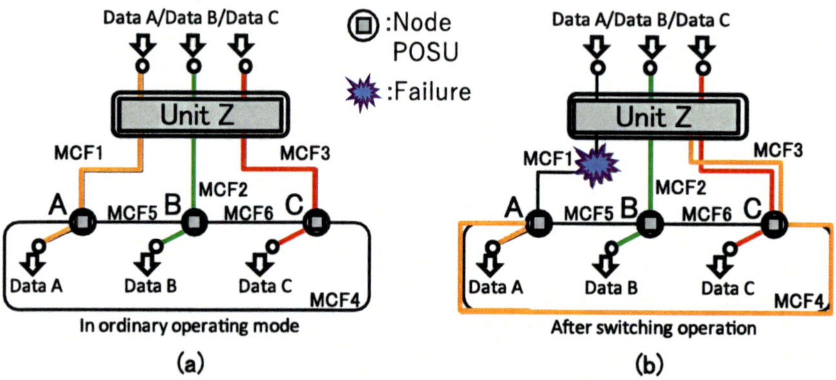

Fig. 6.26 Paths for data transmission in MCF link when failure occurred in MCF1 in Fig. 6.24: **a** Working paths in ordinary operating mode and **b** Working and protection paths after switching operation for recovery, when failure occurred in MCF1

both a working path for accessing Unit C and a protection path for accessing Unit A, as listed in Table 6.3. Figure 6.25e also shows that the optical switch modules in Units A, C and Z operate simultaneously in laying the protection path. We can secure a protection path to substitute for a failed working path for Data A in just one switching process. The relatively short recovery time, as was previously mentioned, can be attributed to this switching process.

We next examined the high bit-rate performance of data transmission between units in the MCF link. Direct digital modulation transfer between units was carried out using a conventional 10 Gbit/s form-factor pluggable (XFP) module modulated by a non-return-to-zero (NRZ) data format with a $2^{31}-1$ pseudorandom binary sequence (PRBS) at the wavelength of 1.55 μm to evaluate data transmission. Figure 6.25 has the measured electrical eye diagrams of outgoing data from Unit A for incoming data to Unit Z. Clear 10-Gb/s eye openings were obtained in ordinary operating mode and after switching operation (see Figs. 6.25f, g). Thus, this indicates that the MCF link could effectively continue data transmission between units even in the event of link failure due to the sophisticated recovery scheme of FRS-MS by using FPGA-based switching units.

Moreover, such a principle of operation in an MCF link based on a one-ring structure can equivalently be applied to that based on a multi-ring structure. Thus, automatic impairment-aware optical-path switching works in the event of multi-failures in the MCF link based on a multi-ring structure.

Conclusions

We presented a novel MCF link based on a multi-ring structure with FPGA-based optical switching units. A switching algorithm implemented in these units was also established in accordance with processes able to cope with link failures in a network, thereby managing this MCF link suitably. Moreover, this link had a connector with a typical connection loss of 0.2 dB, and the FI/FO module had a typical connection loss of 0.3 dB to connect units with the MCF link. We experimentally demonstrated automatic impairment-aware optical-path switching with FPGA-based control in units to verify the fundamental operation of the proposed link when failure occurred in the MCF link based on a one-ring structure. We succeeded in path recovery by switching operation within a relatively short time of 7.4 ms, which was sufficiently less than that required by the ITU-T (i.e., < 50 ms). The selection of a protecting path to substitute for a failure working path was also optimized as the minimum passage of units for low-loss transmission. These results indicate that our proposed link has the potential for network design with a highly reliable network topology in access areas such as those for DCs and FTTH systems.

Acknowledgements We appreciate Dr. T. Kamiya and Prof. T. Morioka for their support in researching short-reach MCF applications. This work was also partly supported by "The research and development project for the ultra-high speed and green photonic networks" of the Ministry of Internal Affairs and Communications, Japan.

References

1. R. Ryf, S. Randel, A.H. Gnauck, C. Bolle, A. Sierra, S. Mumtaz, M. Esmaeelpour, E.C. Burrows, R.-J. Essiambre, P.J. Winzer, D.W. Peckham, A.H. McCurdy, R. Lingle Jr., Mode-division multiplexing over 96 km of few-mode fiber using coherent 6 × 6 MIMO processing. J. Lightwave Technol. **30**(4), 521–531 (2012)
2. H. Takara, A. Sano, T. Kobayashi, H. Kubota, H. Kawakami, A. Matsuura, Y. Miyamoto, Y. Abe, H. Ono, K. Shikama, Y. Goto, K. Tsujikawa, Y. Sasaki, I. Ishida, K. Takenaga, S. Matsuo, K. Saitoh, M. Koshiba, T. Morioka, 1.01-Pb/s (12 SDM/222 WDM/456 Gb/s) crosstalk-managed transmission with 91.4-b/s/Hz aggregate spectral efficiency, in *38th European Conference and Exhibition on Optical Communication (ECOC), paper Th.3.C.1* (2012)
3. T. Mizuno, T. Kobayashi, H. Takara, A. Sano, H. Kawakami, T. Nakagawa, Y. Miyamoto, Y. Abe, T. Goh, M. Oguma, T. Sakamoto, Y. Sasaki, I. Ishida, K. Takenaga, S. Matsuo, K. Saitoh, T. Morioka, 12-core × 3-mode dense space division multiplexed transmission over 40 km employing multi-carrier signals with parallel MIMO equalization, *in Optical Fiber Communication Conference and Exposition (OFC/NFOEC), paper Th5B.2* (2014)
4. P.J. Winzer, Making spatial multiplexing a reality: the future of high-capacity optical networks. Nat. Photon. **8**(5), 345–348 (2014)
5. S.L. Woodward, M.D. Feuer, P. Palarchala, ROADM-node architectures for reconfigurable photonic networks, in *Optical Fiber Telecommunications VIB*, eds. I.P. Kaminow, T. Li, A. Willner (Academic Press, 2013)
6. D.M. Marom, D.T. Neilson, D.S. Greywall, C.-S. Pai, N.R. Basavanhally, V.A. Aksyuk, D.O. López, F. Pardo, M.E. Simon, Y. Low, P. Kolodner, C.A. Bolle, Wavelength-selective 1 × K switches using free-space optics and MEMS micromirrors: theory, design, and implementation. J. Lightwave Technol. **23**(4), 1620–1630 (2005)
7. G. Baxter, S. Frisken, D. Abakoumov, H. Zhou, I. Clarke, A. Bartos, S. Poole, Highly programmable wavelength selective switch based on liquid crystal on silicon switching elements, in *Optical Fiber Communication/National Fiber Optic Engineers Conference (OFC/NFOEC), paper OTuF2* (2006)
8. International Telecommunication Union, Spectral Grids for WDM Applications: DWDM Frequency Grid. ITU-T Rec. G.694.1 (2012)

9. M.D. Feuer, L.E. Nelson, K. Abedin, X. Zhou, T.F. Taunay, J.F. Fini, B. Zhu, R. Isaac, R. Harel, G. Cohen, D.M. Marom, ROADM system for space division multiplexing with spatial superchannels, in *Optical Fiber Communication/National Fiber Optic Engineers Conference (OFC/NFOEC), paper PDP5B.8* (2013)
10. L.E. Nelson, M.D. Feuer, K. Abedin, X. Zhou, T.F. Taunay, J.M. Fini, B. Zhu, R. Isaac, R. Harel, G. Cohen, D.M. Marom, Spatial superchannel routing in a two-span ROADM system for space division multiplexing. J. Lightwave Technol. **32**(4), 783–789 (2014)
11. R. Ryf, N.K. Fontaine, J. Dunayevsky, D. Sinefeld, M. Blau, M. Montoliu, S. Randel, C. Liu, B. Ercan, M. Esmaeelpour, S. Chandrasekhar, A.H. Gnauck, S.G. Leon-Saval, J. Bland-Hawthorn, J.R. Salazar-Gil, Y. Sun, L. Gruner-Nielsen, R. Lingle, D.M. Marom, Wavelength-selective switch for few-mode fiber transmission, in *European Conference and Exposition on Optical Communications (ECOC), paper PD1C4* (2013)
12. D. Noordegraaf, P.M.W. Skovgaard, M.D. Nielsen, J. Bland-Hawthorn, Efficient multi-mode to singlemode coupling in a photonic lantern. Opt. Express **17**(3), 1988–1994 (2009)
13. S.G. Leon-Saval, A. Argyros, J. Bland-Hawthorn, Photonic lanterns: a study of light propagation in multimode to single-mode converters. Opt. Express **18**(8), 8430–8439 (2010)
14. J. Carpenter, S.G. Leon-Saval, J.R. Salazar-Gil, J. Bland-Hawthorn, G. Baxter, L. Stewart, S. Frisken, M.A.F. Roelens, B.J. Eggleton, J. Schröder, 1×11 few-mode fiber wavelength selective switch using photonic lanterns. Opt. Express **22**(3), 2216–2221 (2014)
15. N.K. Fontaine, T. Haramaty, R. Ryf, H. Chen, L. Miron, L. Pascar, M. Blau, B. Frenkel, L. Wang, Y. Messaddeq, S. LaRochelle, R.J. Essiambre, Y. Jung, Q. Kang, J.K. Sahu, S.U. Alam, D.J. Richardson, D.M. Marom, Heterogeneous space-division multiplexing and joint wavelength switching demonstration, in *Optical Fiber Communication Conference and Exposition (OFC/NFOEC), paper Th5C.5* (2015)
16. N.K. Fontaine, R. Ryf, D.T. Neilson, Wavelength selective crossconnects, in *Opto Electronics and Communications Conference (OECC), 2013, paper ThT1.4* (2015)
17. Open Networking Foundation, Software-Defined Networking: The New Norm for Networks. ONF White Paper (2012)
18. X. Cao, V. Anand, C. Qiao, Multi-layer versus single-layer optical cross-connect architectures for waveband switching. IEEE INFOCOM **2004**, 1830–1840 (2004)
19. N. Amaya, G.S. Zervas, D. Simeonidou, Architecture on demand for transparent optical networks, in *International Conference of Transparent Networks (ICTON), paper Th.A1.5* (2011)
20. M. Garrich, N. Amaya, G.S. Zervas, P. Giaccone, D. Simeonidou, Power consumption analysis of architecture on demand, in *European Conference and Exhibition on Optical Communication (ECOC), paper P5.06* (2012)
21. N. Amaya, M. Irfan, G. Zervas, R. Nejabati, D. Simeonidou, J. Sakaguchi, W. Klaus, B.J. Puttnam, T. Miyazawa, Y. Awaji, N. Wada, I. Henning, Fully-elastic multi-granular network with space/frequency/time switching using multi-core fibres and programmable optical nodes. Opt. Express **21**(7), 8865–8872 (2013)
22. N. Amaya, S. Yan, M. Channegowda, B.R. Rofoee, Y. Shu, M. Rashidi, Y. Ou, E. Hugues-Salas, G. Zervas, R. Nejabati, D. Simeonidou, B.J. Puttnam, W. Klaus, J. Sakaguchi, T. Miyazawa, Y. Awaji, H. Harai, N. Wada, Software defined networking (SDN) over space division multiplexing (SDM) optical networks: features, benefits and experimental demonstration. Opt. Express **22**(3), 3638–3647 (2014)
23. N. McKeown, T. Anderson, H. Balakrishnan, G. Parulkar, L. Peterson, J. Rexford, S. Schenker, J. Turner, OpenFlow: enabling innovation in campus networks. ACM SIGCOMM Comp. Comm. Rev. **38**(2), 69–74 (2008)
24. S. Fujii, Y. Hirota, T. Watanabe, H. Tode, Dynamic spectrum and core allocation with spectrum region reducing costs of building modules in AoD nodes. Telecommun. Netw. Strategy Plann. Symp. (Netw.) **2014**, 1–6 (2014)
25. S. Fujii, Y. Hirota, H. Tode, K. Murakami, On-Demand spectrum and core allocation for reducing crosstalk in multicore fibers in elastic optical networks. J. Opt. Commun. Netw. **6**(12), 1059–1071 (2014)

26. E. Ciaramella, Wavelength conversion and all-optical regeneration: achievements and open issues. J. Lightwave Technol. **30**(4), 572–582 (2012)
27. R.S. Luís, B.J. Puttnam, J.-M.D. Mendinueta, W. Klaus, Y. Awaji, N. Wada, Comparing inter-core skew fluctuations in multi-core and single-core fibers, in *Conference on Lasers and Electro-Optics (CLEO), paper SM2L.5* (2015)
28. M.D. Feuer, L.E. Nelson, X. Zhou, S.L. Woodward, R. Isaac, B. Zhu, T.F. Taunay, M. Fishteyn, J.M. Fini, M.F. Yan, Joint digital signal processing receivers for spatial superchannels. Phot. Techn. Lett. **24**(21), 1957–1960 (2012)
29. T. Ito, E.L.T. de Gabory, M. Arikawa, Y. Hashimoto, K. Fukuchi, Reduction of Influence of inter-core cross-talk in MCF with bidirectional assignment between neighboring cores, in *Optical Fiber Communication Conference and Exposition (OFC/NFOEC), paper OTh3K.2* (2013)
30. F. Ye, T. Morioka, Interleaved core assignment for bidirectional transmission in multi-core fibers, in *European Conference and Exhibition on Optical Communication (ECOC), paper We.2.D.5* (2013)
31. S.K. Korotky, Semi-empirical description and projections of internet traffic trends using a hyperbolic compound annual growth rate. Bell Labs Techn. J. **18**(3), 5–21 (2013)
32. Network Disaster Recovery. IEEE Commun. Mag. **49**(1) (2011)
33. S. Secci, S. Murugesan, Cloud networks: enhancing performance and resiliency. IEEE Comput. 82–85 (2014)
34. R.E. Krock, Lack of emergency recovery planning is a disaster waiting to happen. IEEE Commun. Mag. 48–51 (2011)
35. K.T. Morrison, Rapidly recovering from the catastrophic loss of a major telecommunications office. IEEE Commun. Mag. 28–35 (2011)
36. J. Sakaguchi, B.J. Puttnam, W. Klaus, Y. Awaji, N. Wada, A. Kanno, T. Kawanishi, K. Imamura, H. Inaba, K. Mukasa, R. Sugizaki, T. Kobayashi, M. Watanabe, 305 Tb/s space division multiplexed transmission using homogeneous 19-core fiber. IEEE J. Lightwave Technol. **31**(4), 554–562 (2013)
37. H. Takara, H. Takara, A. Sano, T. Kobayashi, H. Kubota, H. Kawakami, A. Matsuura, Y. Miyamoto, Y. Abe, H. Ono, K. Shikama, Y. Goto, K. Tsujikawa, Y. Sasaki, I. Ishida, K. Takenaga, S. Matsuo, K. Saitoh, M. Koshiba, T. Morioka, 1.01-Pb/s (12 SDM/222 WDM/456 Gb/s) crosstalk-managed transmission with 91.4-b/s/Hz aggregate spectral efficiency, in *Proceedings of European Conference of Optical Communication (ECOC), PDP Th.3.C.1*, Amsterdam (2012)
38. B. Zhu, T.F. Taunay, M.F. Yan, J.M. Fini, M. Fishteyn, E.M. Monberg, F.V. Dimarcello, Seven-core multicore fiber transmissions for passive optical network. Opt. Express **18**(11), 11117–11122 (2010)
39. K. Hiruma, T. Sugawara, K. Tanaka, E. Nomoto, Y. Lee, Proposal of high-capacity and high-reliability optical switch equipment with multi-core fibers, in *Proceedings of the 18th Opto Electronics and Communications Conference (OECC), paper ThT1-2*, Kyoto, July (2013)
40. Y. Lee, K. Tanaka, K. Hiruma, E. Nomoto, T. Sugawara, H. Arimoto, Experimental demonstration of a highly reliable multicore-fiber-based optical network. IEEE Photon Technol. Lett. **26**(6), 538–540 (2014)
41. Y. Lee, K. Tanaka, K. Hiruma, E. Nomoto, T. Sugawara, H. Arimoto, Multi-core fiber technology for optical-access and short-range links, in *Proceedings of the 12th International Conference on Optical Internet (COIN), TB2-4* (2014)
42. C.J.S. DeCusatis, A. Carranza, C.M. DeCusatis, Communication within clouds: open standards and proprietary protocols for data center networking. IEEE Commun. Mag. 26–33 (2012)
43. T. Mitsui, T. Sakamoto, K. Hara, N. Yoshimoto, "lexible and scalable PON protection architecture using N:M redundancy toward next generation access network, in *Proceedings of The 17th Asia-Pacific Conference on Communications (APCC)*, pp. 224–229, October (2011)
44. T. Costello, *Business Continuity: Beyond Disaster Recovery*, pp. 62–64, IT Pro September/October (2012)

45. S. Senda, K. Nguyen, S. Yamada, Requirements for resilient information and communication technology, in *Proceedings of 2013 Seventh International Conference on Complex, Intelligent, and Software Intensive Systems (CISIS)*, Taichung, Taiwan, pp. 418–423 (2013)
46. C. DeCusatis, Optical interconnect networks for data communications. IEEE J. Lightwave Technol. **32**(4), 544–552 (2014)
47. S. Azodolmolky, P. Wieder, R. Yahyapour, Cloud computing networking: challenges and opportunities for innovations. IEEE Commun. Mag. 54–62 (2013)
48. A. Vahdat, The impact of mega-scale data centers on internet architecture, in *Proceedings of Optical Fiber Communication Conference (OFC), MD3.1*, San Diego, CA, pp. 131–132 (2009)
49. A. Vahdat, M. Al-Fares, N. Farrington, R.N. Mysore, G. Porter, S. Radhakrishnan, Scale-out networking in the data center. IEEE Micro, 29–41 (2010)
50. B. Welch, G. Nicholl, K. Conroy, J. Maki, D. Lewis, 400G-PSM4: A proposal for the 500 m objective using 100 Gb/s per lane signaling, in Contribution to IEEE 802.3bs 400 Gb/s Ethernet Task Force, Interim Meeting, Pittsburgh, PA, USA, May (2015)
51. R. Hirai, H. Toyoda, N. Kikuchi, Feasibility study of 100 G/lambda Nyquist-PAM4 with commercially available 1.3 μm/1.5 μm EML, in Contribution to IEEE 802.3bs 400 Gb/s Ethernet Task Force, Plenary Meeting, Ottawa, Ontario, CANADA, September (2014)
52. IEEE P802.3bs MMF Ad Hoc, 400 Gb/s 100 m MMF reach objective draft baseline proposal, in Contribution to IEEE 802.3bs 400 Gb/s Ethernet Task Force, Plenary Meeting, San Antonio, TX, USA, November (2014)
53. H. Liu, C. F. Lam, C. Johnson, Scaling optical interconnects in datacenter networks opportunities and challenges for WDM, in *Proceedings of 18th IEEE Symposium on High Performance Interconnects*, pp. 113–116 (2010)
54. B. Zhu, T.F. Taunay, M.F. Yan, M. Fishteyn, G. Oulundsen, D. Vaidya, 70-Gb/s multicore multimode fiber transmissions for optical data links. IEEE Photon. Technol. Lett. **22**(22), 1647–1649 (2010)
55. B.G. Lee, D.M. Kuchta, F.E. Doany, C.L. Schow, P. Pepeljugoski, C. Baks, T.F. Taunay, B. Zhu, M.F. Yan, G.E. Oulundsen, D.S. Vaidya, W. Luo, N. Li, End-to-end multicore multimode fiber optic link operating up to 120 Gb/s. IEEE J. Lightwave Technol. **30**(6), 886–892 (2012)
56. Y. Sun, R. Shubochkin, B. Zhu, Space division multiplexing in access networks, in *Proceedings of SPIE*, 9387, San Francisco, USA (2015)
57. Y. Geng, S. Li, M.-J. Li, C.G. Sutton, R.L. McCollum, R.L. McClure, A.V. Koklyushkin, K.I. Matthews, J.P. Luther, D.L. Butler, High-speed, bi-directional dual-core fiber transmission system for high-density, short-reach optical interconnects, in *Proceedings of SPIE*, 9390, San Francisco, USA (2015)
58. B. Rosinski, J.D. Chi, P. Grosso, J.L. Bihan, Multichannel transmission of a multicore fiber coupled with VCSEL. IEEE J. Lightwave Technol. **17**(5), 807–810 (1999)
59. M-J Li, B. Hoover, V.N. Nazarov, D.L. Butle, Multicore fiber for optical interconnect applications, in *Proceedings 17th Opto-Electronics and Communications Conference, paper 5E4-2* (OECC 2012), Busan, Korea, July 2012
60. T. Hayashi, T. Nakanishi, K. Hirashima, O. Shimakawa, F. Sato, K. Koyama, A. Furuya, Y. Murakami, T. Sasaki, 125-μm-cladding 8-core multi-core fiber realizing ultra-high-density cable suitable for O-band short-reach optical interconnects, in *Proceedings of Optical Fiber Communication Conference (OFC)*, Los Angeles, CA, March, 2015, post-deadline paper Th5C.6 (2015)
61. T. Kuri, H. Harai, N. Wada, T. Kawanishi, M. Hosokawa, Adaptable access system: pursuit of ideal future access system architecture. IEEE Netw. Mag. **26**(2), 42–48 (2012)
62. F.J. Effenberger, Space division multiplexing in access networks, in *Proceedings of SPIE*, 9387, San Francisco, USA (2015)
63. B. Li, Z. Feng, M. Tang, Z. Xu, S. Fu, Q. Wu, L. Deng, W. Tong, S. Liu, P. Shum, Experimental demonstration of large capacity WSDM optical access network with multicore fibers and advanced modulation formats. Opt. Express **23**(9), 10977–11006 (2015)

64. T. Hu, J. Li, P. Zhu, Q. Mo, Y. Ke, C. Du, Z. Liu, Y. He, Z. Li, Z. Chen, Experimental demonstration of passive optical network based on mode-division-multiplexing, in *Proceedings of Optical Fiber Communication Conference (OFC), paper Th2A.63*, Los Angeles, CA, March, 2015
65. K. Nakanishi, S. Yoshida, S. Aoyagi, A low loss multiplexing scheme for PDS system, in *IEICE General Conference, B-10-112*, p. 621 (1997) (in Japanese)
66. N. Cheng, Z. Liao, F.J. Effenberger, Large splitting and long reach passive optical networks with mode coupling receivers, in *Proceedings of 36th European Conference on Optical Communication (ECOC 2010), paper Tu.5.B.3*, Torino, Italy, September, 2010 (2015)
67. M. Fujiwara, K. Suzuki, N. Yoshimoto, M. Oguma, S. Soma, Increasing splitting ratio of 10 Gb/s-class PONs by using FW-DMF, in *Optical Fiber Communication Conference (OFC), paper Tu.2.C.5*, San Francisco, CA, March, 2014 (2014)
68. C. Xia, N. Chand, A.M. Velázquez-Benítez, Z. Yang, X. Liu, J.E. Antonio-Lopez, H. Wen, B. Zhu, N. Zhao, F.J. Effenberger, R. Amezcua-Correa, G. Li, Time-division-multiplexed few-mode passive optical network. Opt. Express **23**(2), 1151–1158 (2015)
69. Y. Fang, J. Yu, N. Chi, J. Zhang, J. Xiao, A novel PON architecture based on OAM multiplexing for efficient bandwidth utilization. IEEE Photon. J. **7**(1), 1–6 (2015)
70. C. Raffaelli, R. Veisllari, Scheduling and performance of hybrid traffic in a data center optical core switch, in *Proceedings of 16th International Conference on Transparent Optical Networks (ICTON), paper B1.1*, Graz, Austria, July, 2014
71. S. Zhong, Z. Zhu, Optical virtual switching (OvS): a distributed optical switching fabric for intra-data center networking, in *Proceedings of 23rd Wireless and Optical Communication Conference (WOCC), paper O2.3*, Newark, NJ, U.S.A., May 2014
72. M. Fiorani, M. Casoni, S. Aleksic, Large data center interconnects employing hybrid optical switching, in *Proceedings of 18th European Conference on Network and Optical Communications (NOC) and 8th Conference on Optical Cabling and Infrastructure (OC&i)*, July, 2013, pp. 61–68
73. S. Yan, E. Hugues-Salas, V.J.F. Rancaño, Y. Shu, G.M. Saridis, B.R. Rofoee, Y. Yan, A. Peters, S. Jain, T. May-Smith, P. Petropoulos, D.J. Richardson, G. Zervas, D. Simeonidou, Archon: a function programmable optical interconnect architecture for transparent intra and inter data center SDM/TDM/WDM networking. IEEE J. Lightwave Technol. **33**(8), 1586–1595 (2015)
74. V.J.F. Rancaño, S. Jain, T.C. May-Smith, E. Hugues-Salas, S. Yan, G. Zervas, D. Simeonidou, P. Petropoulos, D.J. Richardson, Demonstration of space-to-wavelength conversion in SDM networks. IEEE Photon. Technol. Lett. **27**(8), 828–83 (12015)
75. Y. Tottori, T. Kobayashi, M. Watanabe, Low Loss optical connection module for seven-core multicore fiber and seven single-mode fibers. IEEE Photon. Technol. Lett. **24**(21), 1926–1928 (2012)
76. E. Sugita, R. Nagase, K. Kanayama, T. Shintaku, SC-type single-mode optical fiber connectors. IEEE J. Lightwave Technol. **7**, 1689–1696 (1989)
77. K. Shikama, Y. Abe, S. Yanagi, T. Takahashi, Physical-contact conditions for multicore fiber optional connectors, in *Proceedings of Optical Fiber Communication Conference (OFC), paper OM3I.1*, Anaheim, CA, March (2013)
78. D. Marcuse, Loss analysis of single-mode fiber splices. Bell Syst. Techn. J. **56**, 703–718 (1977)
79. International Standard, IEC 60874-14-5 First edition 1997–06. Connectors for optical fibres and cables. Part 14-5: Detail specification for fibre optical connector type SC-PC untuned terminated to single-mode fibre type B1.
80. E. Nomoto, K. Hiruma, T. Sugawara, K. Tanaka, and "connectors for optical fibres and cables", SC-type multi-core optical-fiber connectors using a pressurization spring. Opt. Rev. (2015). https://doi.org/10.1007/s10043-015-0.127-0
81. N. Farrington, A. Andreyev, Facebook's Data Center Network Architecture, in *Proceedings of IEEE Optical Interconnects Conference*, Santa Fe, New Mexico, May (2013)
82. B. Zhu, T.F. Taunay, M.F. Yan, M. Fishteyn, G. Oulundsen, D. Vaidya, 70-Gb/s multicore multimode fiber transmissions for optical data links. IEEE J. Lightwave Technol. **22**(22), 1647–1649 (2010)

83. K. Tanaka, Y. Lee, E. Nomoto, T. Sugawara, H. Arimoto, Experimental evaluation of recovery from multiple failures in multi-core fiber links using FPGA-based optical switch units. IEEE J. Lightwave Technol. **33**(1), 201–211 (2015)
84. J. Sakaguchi, Y. Awaji, N. Wada, Fundamental study on new characterization method for crosstalk property of multi-core fibers using long wavelength probe signals, in *Proceedings of Optical Fiber Communication Conference (OFC), OW1K.1*, Anaheim, CA, March (2013)
85. T. Hayashi, T. Nakanishi, T. Sasaki, K. Saitoh, M. Koshiba, Dependence of crosstalk increase due to tight bend on core layout of multi-core fiber, in *Proceedings of Optical Fiber Communication Conference (OFC), W4D.4*, San Francisco, CA, March (2014)
86. Telecommunication Standardization Sector of International Telecommunication Union (ITU-T). Transmission System and Media. Digital System and Networks G.841

Chapter 7
High-Power Issues

Toshio Morioka, Kazi S. Abedin, Nobutomo Hanzawa, Kenji Kurokawa, Kazunori Mukasa, Ryo Nagase, Hidehiko Takara, Shin-ichi Todoroki, Makoto Yamada, and Shuichi Yanagi

Abstract Fiber fuse, a thermal destruction phenomenon, which propagates toward optical sources and permanently damages optical fiber cores, connection points (connectors and splicing) as well as optical components sets the ultimate input optical power limits into optical fibers, thus determining the ultimate data capacity in optical fiber communication systems. This chapter describes basic properties of fiber fuse, followed by that of optical communication fibers, its detection and halting (blocking) methods, fiber fuse-based incidence as well as fiber fuse-tolerant fibers. Lastly, safety

---

Toshio Morioka is a chapter editor.

T. Morioka (✉)
Technical University of Denmark, Kongens Lyngby, Denmark
e-mail: tomo@fotonik.dtu.dk

K. S. Abedin
LGS Innovation—CACI, Reston, VA, USA

N. Hanzawa
NTT Access Network Service Systems Laboratories, NTT Corporation, Tsukuba, Ibaraki, Japan

K. Kurokawa
Kitami Institute of Technology, Kitami, Hokkaido, Japan

K. Mukasa
Furukawa Electric Co., Ltd., Kameyama, Mie, Japan

R. Nagase
Chiba Institute of Technology, Narashino, Chiba, Japan

H. Takara
National Institute of Technology, Okinawa College, Nago, Okinawa, Japan

S. Todoroki
National Institute for Materials Science, Tsukuba, Ibaraki, Japan

M. Yamada
Osaka Metropolitan University, Sakai, Osaka, Japan

S. Yanagi
NTT Advanced Technology Corporation, Musashino, Tokyo, Japan

M. Nakazawa et al. (eds.), *Space-Division Multiplexing in Optical Communication Systems*, Springer Series in Optical Sciences 236,
https://doi.org/10.1007/978-3-030-87619-7_7

issues of optical communication systems from the viewpoint of IEC laser safety standardization are described.

7.1 Fiber Fuse

7.1.1 Basic Properties

Although nearly 180 papers have been published on the catastrophic fiber fuse phenomenon since its discovery in 1988 [1, 2], we still cannot prevent it completely except by limiting the light power to below the threshold for fiber fuse propagation, which is about 1.4 W for a standard single-mode fiber (see Fig. 7.1). This is because we still have little knowledge about the factors controlling fiber fuse initiation and the true nature of the reaction region, which moves at about 1 m/s with strong light emission. This section summarizes recent advances on these issues in relation to single-mode fibers including work on (1) a quantitative evaluation of fiber fuse initiation [3], (2) a dissipative soliton model, and (3) void morphology related to the energy flow in the reaction region. Some ways are also suggested for investigating this phenomenon in advanced optical fibers. A more detailed description is available in a recently published monograph [4].

7.1.1.1 Initiation and Its Probability

A fiber fuse is initiated in a heated segment of an optical fiber where the propagating light is strong enough to allow a fiber fuse to propagate. Silica glass optical fibers are known to become absorptive when they are heated above 1000 °C [9–11] (see Fig. 7.2). This induced absorbance triggers "an avalanche of phonons," i.e., an accelerated heating, that generates a small high-temperature region surrounded by

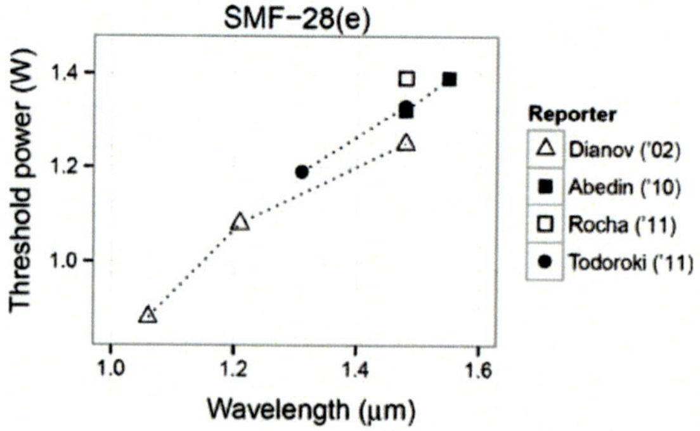

Fig. 7.1 Wavelength dependence of threshold power for fiber fuse propagation through SMF-28 and SMF-28e (Corning). *Source* [5–8] (Reprinted with permission from [4]. ©2014 NIMS)

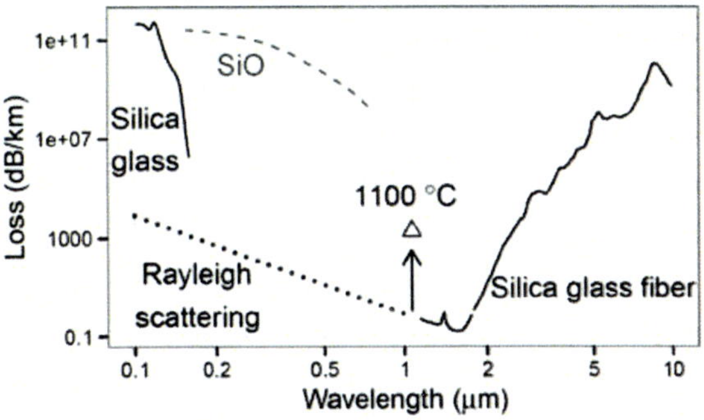

Fig. 7.2 Loss spectrum of silica glass [12], SiO [13], and silica glass optical fiber [14, 15]. A heat-induced loss increase is shown as a pair consisting of an arrow and an open triangle [9] (Reprinted with permission from [4]. ©2014 NIMS)

molten glass. This region sometimes begins to move toward the light source, and this phenomenon is recognized as a fiber fuse.

Its physical scheme resembles a grass fire (see Fig. 7.3a, b), namely a solitary fire wave or a bright spot persists as an irreversible reaction region that is fueled from the front, emitting light and heat all around, and leaving cinders or damage behind it. Both systems can be described as a "dissipative soliton" [16] (see Fig. 7.3c). The temperature of the hottest area in the silica fiber is estimated to be several thousand °C or more [17]. The fiber can retain this hot spot without deformation because it moves faster than the heat conduction to the fiber surface (see Fig. 7.3d).

Accordingly, fiber fuse initiation is a process in which an optical fiber is heated to form a moving reaction region that balances incoming light and energy dissipation (heat and light). Its transition is expected to be manifold and very often fails because

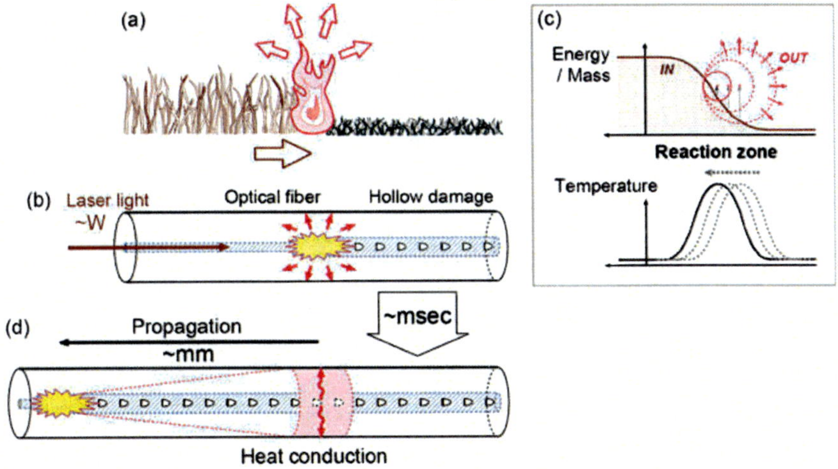

Fig. 7.3 **a, b** Comparison of a grass fire and a fiber fuse with arrows showing the energy flows into/from the propagating reaction zone. **c** Concept of a dissipative soliton showing energy/mass flow and temperature distribution. **d** Geometric relation between fiber fuse propagation and heat conduction (Reprinted with permission from [4]. ©2014 NIMS)

continuous local heating induces the unmanageable growth of the reaction region and deformation of the optical waveguide. Thus, this process should be analyzed statistically under fixed experimental conditions that include fiber geometry, irradiation power, and time.

The most sensitive of these conditions is heating. Although the simplest heating approach involves bringing the end of the fiber into contact with a metallic object, any small gap between the two surfaces prevents the formation of a closed reaction region. This situation is improved when we use highly Co-doped borosilicate glass instead of metal because the beam exiting the fiber (wavelength: 1.4–1.8 μm) softens the glass surface before contact and the consequent melt covers the fiber endface [3]. This gap-free contact reduces the uncontrollability and enhances stable heating.

Figure 7.4a–d shows a typical example of this initiation process. The absorptive glass approaches the fiber end from the right while being irradiated with the outgoing beam (a), the heated melt expands to cover the fiber end (b–c), and a fiber fuse is initiated (d). Their relative positions are precisely controlled with an automatic XY stage as shown in Fig. 7.4e.

Figure 7.5 shows some damaged fibers that were irradiated for a maximum of 5 s. Among the initiated samples, the irradiation time, *t*, is generally related to the void volume inside the fiber (a, b). On the other hand, the failed samples exhibit damage

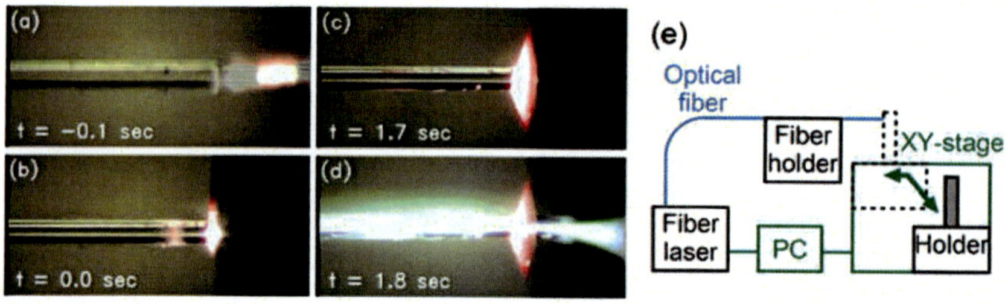

Fig. 7.4 **a–d** Snapshots of reaction area; **a** just before contact, **b** in contact, **c** just before initiation, **d** at initiation, and **e** experimental setup of automatic feeder. A video clip is available at http://doi.org/10.11503/nims.1001

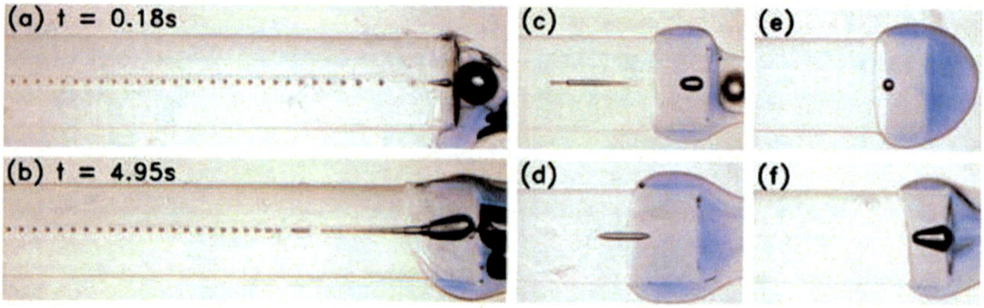

Fig. 7.5 **a, b** Various voids left after fiber fuse initiation and (c–f) five seconds laser irradiation without initiation. *t*: irradiation time required for fiber fuse initiation. Fiber: LEAF, laser: 1480 nm, 5.8 W

with various shapes despite experiencing the same irradiation time (c–f). Each sample has a stable structure, which is unchanged by any further irradiation. Consequently, the fiber fuse appears without being trapped by these deadlock conditions.

Then, 333 trials were performed using 5-s-long irradiation with various fibers, laser powers, and fiber end temperatures. The fiber tip was heated with a micro-ceramic heater (5 × 5 × 1.7 mm) as shown in Fig. 7.12a. The results are plotted in Fig. 7.6 with closed circles and crosses for initiated and failed samples, respectively. Each initiation probability corresponds to the length of a closed red circle segment along the y-axis. Figure 7.6a, b, e shows the irradiation power dependence of three kinds of fibers at room temperature and (c, d) for SMF-28e+ at two elevated temperatures.

It is reasonable that the minimum laser power for initiation decreases with the mode field diameter (MFD), namely in the order (e), (b), and (a). However, the increase in fiber temperature did not enhance the initiation as shown in (e), (d), and (c). This is probably because external heating does not assist any thermal confinement

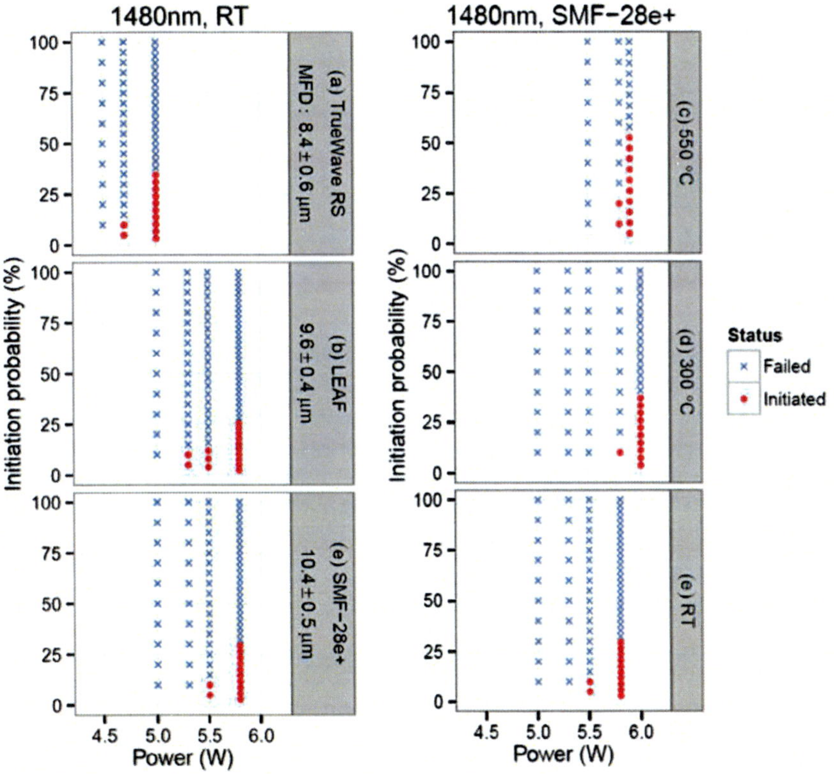

Fig. 7.6 Fiber fuse initiation probabilities, shown as the length of the closed red circle segment, for three kinds of fibers at RT (room temperature; **a**, **b**, **e**) and for SMF-28e+ at two elevated temperatures (**c**, **d**)

at the core but may trap the reaction area in one of various deadlock configurations such as those in Fig. 7.5c–f. In other words, a steep temperature gradient around the fiber core is necessary for fiber fuse initiation.

We should note that a fiber fuse can be initiated with lower laser power if we are free from the experimental conditions applied above. The purpose of this experiment is to achieve a relative comparison under different conditions.

Consequently, a highly Co-doped borosilicate glass was found to be a useful heat generator for evaluating the fiber fuse initiation tendency. It is reasonable to expect that this method will be applicable to state-of-the-art fibers including holey fibers, multi-mode fibers, and multi-core fibers.

In addition, another quantitative evaluation method is proposed recently where the sample is exposed with 1-s-long arc discharge generated with a fusion splicer [18, 19].

7.1.1.2 Propagation and Its Energy Flow

Compared with the fiber fuse initiation process, the propagation behavior appears to be simpler to investigate. Many researchers have determined the minimum power for stable propagation, *P*th, as seen in Fig. 7.1, which is known to increase with MFD [6, 20–22]. When the pump power is reduced below this value, the fiber fuse loses its energy flow balance and disappears. Namely the receiving light energy becomes smaller than the sum of the work required for moving and the energy dissipation.

However, this simple scheme is secured by the following two conditions: (1) The fiber is single mode, and (2) the pump power is constant along the fiber length. To deepen our understanding of fiber fuse propagation, we have to observe its response to a modulated energy flow. This knowledge will be useful when it comes to handling advanced fibers and designing fiber fuse termination devices.

Here we focus on the void train in damaged fibers because it contains a historical record of the energy flow in the reaction region. In other words, the void forming action of the surrounding glass melt is modified by the energy flow modulation. Therefore, we can obtain fundamental knowledge about fiber fuse propagation by analyzing the void pattern in typical single-mode fibers.

First, let us review the fiber fuse propagation modes under equilibrium conditions [23]. The propagation state is classified into three modes depending on the shape of the reaction region confined in the core: cylindrical, unimodal, and unstable. As the pump power increases, the volume of the reaction region increases, and its shape is expanded along the fiber. This behavior is recorded in the top part of the damage train left in the core region (see Fig. 7.7).

In the cylindrical mode, the top void contains a cylindrical part as shown in Fig. 7.7e, f. In addition, the following voids look like bullets and appear periodically. This interval increases with the pump power. Figure 7.7c, d is classified as the unimodal mode. The diameter of the top void is equivalent to that in the cylindrical mode. As the pump power approaches *P*th, the top void shrinks as shown in Fig. 7.7a,

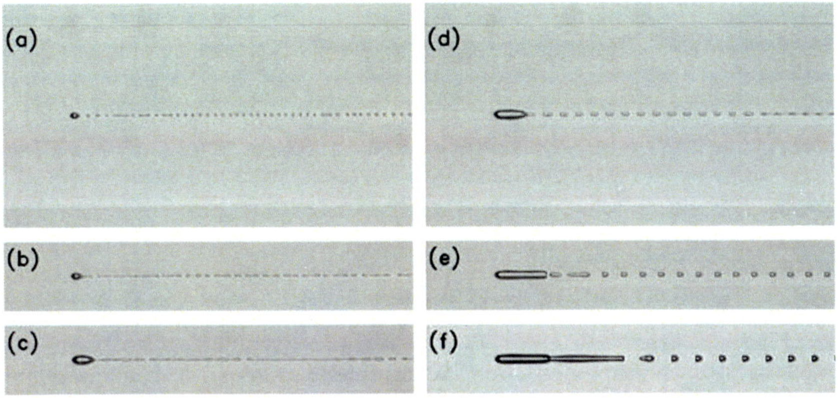

Fig. 7.7 Optical micrographs showing the front part of the fiber fuse damage generated in single-mode silica glass fibers (LEAF, corning). The pump laser powers are **a** 1.26 W (propagation threshold), **b** 1.3 W, **c** 1.5 W, **d** 2.0 W, **e** 3.2 W, and **f** 5.3 W. Wavelength: 1480 nm (from a Raman fiber laser, PYL-10-1480, IPG Laser). The two thin lines at the top and bottom of (**a**) and (**d**) are the edges of the fiber, whose diameter is 125 μm

b. Sometimes the void train contains a periodic segment and/or a pair consisting of a long void and a void-free segment [4, 24].

These void trains are known to be modified by a sudden pump power change ($\Delta t = 0.2$ ms) [23]. Typical examples are shown in Fig. 7.8. This is caused by the delayed response of the surrounding glass melt. When the pump power is reduced, the reaction region shrinks at once, but the viscous melt cannot follow it immediately and the tail of the hollow space is quenched as shown on the left in Fig. 7.9. By contrast, the surrounding melt slowly inflates after the power increments and the periodic void formation is suspended until the inflation is complete (on the right in Fig. 7.9).

These phenomena appear in few-mode fiber driven by the interference between the LP_{01} and LP_{02} modes [25]. The resulting void train is a repetition of a 340-μm-long sequence consisting of a void-free segment, a long void, and a few periodic voids. The void sequence becomes more complicated in multi-mode fibers [26] and double-clad fiber lasers [27].

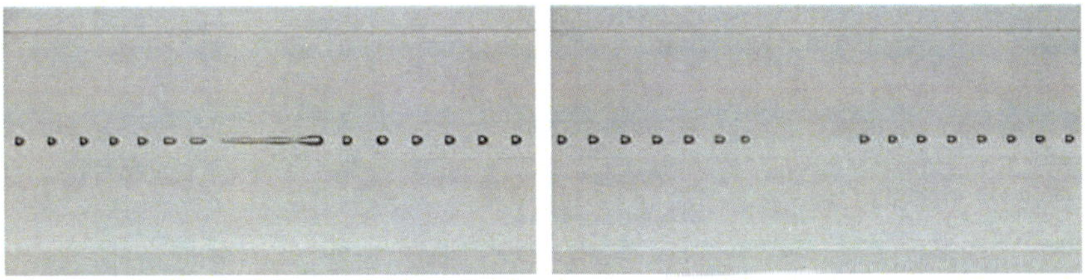

Fig. 7.8 Photographs of irregular segments left at the decrement (left) and increment (right) points of the pump power. Fiber: SMF-28e, laser: 1480 nm, 6.5 → 5.6 W (left) and 4.9 → 5.5 W (right) (Reprinted with permission from [4]. © 2014 NIMS)

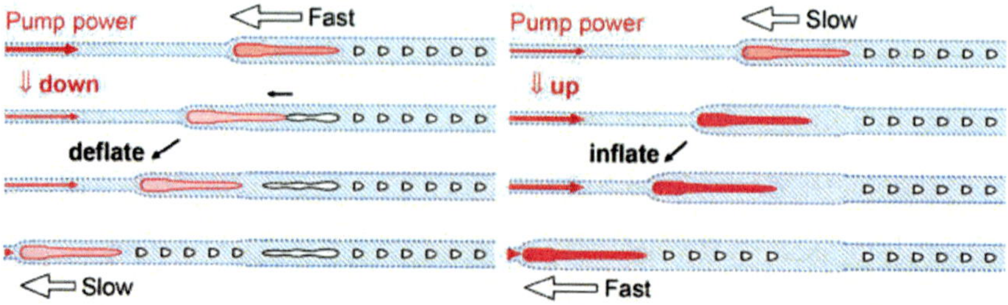

Fig. 7.9. Formation mechanisms of irregular void patterns shown in Fig. 7.8

A fiber fuse gains extra energy from its emission if the fiber geometry allows return paths [4]. The simplest case is seen in a bare fiber, where the emission from the top reaches the tail of the reaction region through the cladding modes (see Fig. 7.10a). This self-pumping results in an elongated tail or a volume expansion of the reaction region as shown in Fig. 7.11. This elongation occurs when the length of the reaction region exceeds $2r\ \tan\theta_c$, where r is the fiber radius and $\theta_c = \arcsin(n_{air}/n_{silica\ glass})$ ~ 44°, or the pump power exceeds 7 W (wavelength: 1480 nm) for SMF-28e+ fiber [4]. This critical length decreases in holey fibers as shown in Fig. 7.10b, where r is the distance between the core and the air/glass interface.

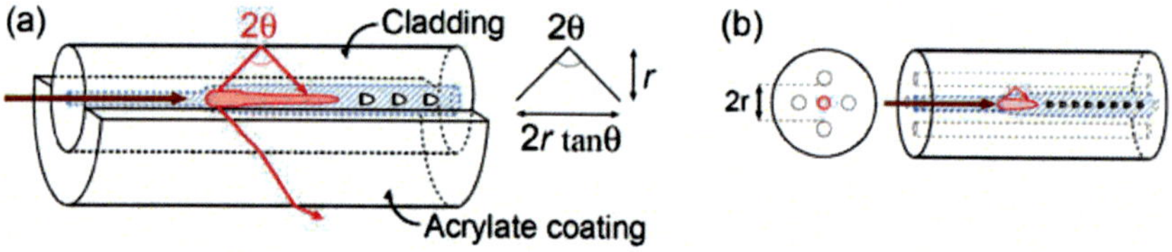

Fig. 7.10 **a** Geometry of cladding mode propagation from a fiber fuse (top) and its suppression by acrylate coating (bottom). **b** Self-pumping scheme in a holey fiber (Reprinted with permission from [4]. ©2014 NIMS)

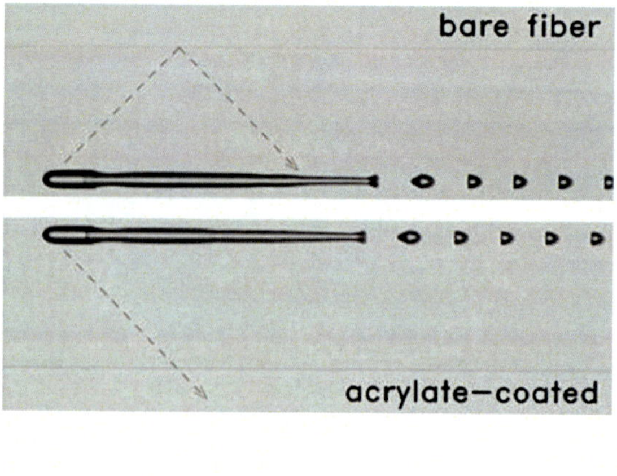

Fig. 7.11 Differences between the damage sites left by a fiber fuse in a bare fiber (top) and in an acrylate-coated fiber (bottom). The pump light is 8.9 W at 1480 nm. Fiber: SMF-28e (corning) A movie version of this photograph is available at http://doi.org/10.11503/nims.1002 (Reprinted with permission from [4]. ©2014 NIMS)

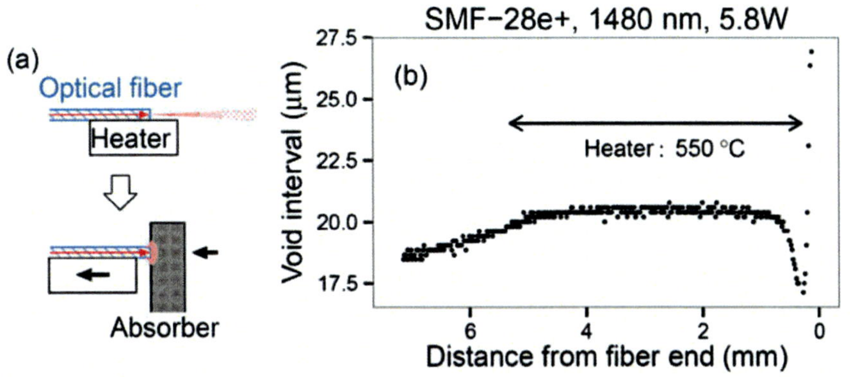

Fig. 7.12 **a** Extension for heating a fiber tip installed in an experimental setup shown in Fig. 7.4. The width of the heater was 5 mm. **b** Positional dependence of void interval. A fiber fuse was initiated at 0 mm after 1.7 s of laser irradiation. The large variation near the fiber end was caused by the transition from the non-equilibrium state of initiation

The reaction region also gains extra energy from external heating. This phenomenon is recorded in the void train as void interval modulations. Figure 7.12b shows the positional dependence of the void interval left near the initiation point at a fiber end. A five-mm-wide heater was located under the fiber tip as shown in Fig. 7.12a and shifted aside before the absorber touched the fiber end. The void interval clearly increased in a segment in contact with the heater.

Consequently, a fiber fuse in single-mode fibers leaves its energy flow information in a void train. Further case studies have been described elsewhere [4, 19]. The above is expected to be useful when analyzing the action of termination devices presented in the following section.

7.1.2 *Fiber Fuse of Optical Communication Fibers*

7.1.2.1 Introduction—Fiber Fuse Problem of Current Telecom Fibers

In this section, we will report current problems caused by high optical power inputs, such as fiber fuse in telecom fibers and damages to the fiber coating caused by bending the telecom fibers.

Recently, the transmission capacity has been increased by using high-density wavelength division multiplexing (WDM) transmission. Naturally, the optical power inputted to the fiber core has been largely increased. For example, the power of semiconductor lasers in 1500 nm band reaches ~ 400 mW. Furthermore, distributed Raman amplification, which realizes a wide-band flat gain by using the transmission fiber as an amplification medium, sometimes requires the excitation power exceeding 1 W [28]. Under these circumstances, new problems such as damages to optical fibers and devices as well as human safety issues become more important. In particular, the phenomenon called fiber fuse, which melts parts of the optical fiber core, propagates toward the light source emitting visible light potentially causes catastrophic damages not only to the optical fiber itself but also to the light source. Therefore, the fiber fuse potentially will become a serious problem in the future [29, 30]. Other serious problems would be a temperature increase or, in the worst case, fire of the fiber coating caused by the bending of the fiber under the high-power transmitting condition. The bending can be accidentally added to the fibers during installing them, for example.

Currently, conventional single-mode fibers (ITU-T G.652: SMF) and dispersion-shifted fibers (ITU-T G.653: DSF) are widely used. However, we have confirmed even the SMF, which has a larger mode field diameter (MFD), has a possibility to cause the fiber fuse phenomena. In this section, we will also describe the recent experimental results on fiber fuse and damages of the coating by bending optical fibers.

7.1.2.2 Fiber Fuse of Telecom Fibers

Fiber fuse is a phenomenon which propagates along the fiber by making many holes (voids) in the fiber core as shown in Fig. 7.13. Though the shapes of holes change depending on the condition, they typically are bullet shapes. Fiber fuse continues until we shut down the light source or decrease the power lower than the propagation threshold power and, eventually, it can reach to the light source and damage it. Therefore, it is extremely important to understand the threshold power against fiber fuse before designing the real systems. Taking these issues into account, we performed experiments which intentionally start fiber fuse using two light sources at 1064nm and 1467 nm in the system shown in Fig. 7.14. It should be noted that the fiber fuse does not always occur under the high-power circumstances. According to a report [31], the resistance of the silica glass against the high power is more than 10 GW/cm^2, and conventional SMFs should have enough resistance against the power more than 1W. However, if there is an incident in the fiber under the high-power condition, such as power absorption by substances or optical power concentration by the multiple refractions due to the fiber break, fiber fuse occurs at the point and

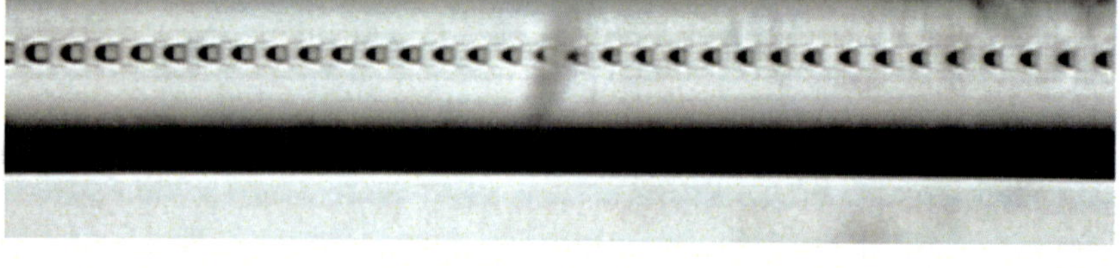

Fig. 7.13 Optical fiber after fiber fuse (Reprinted with permission from [32]. ©2003 Furukawa)

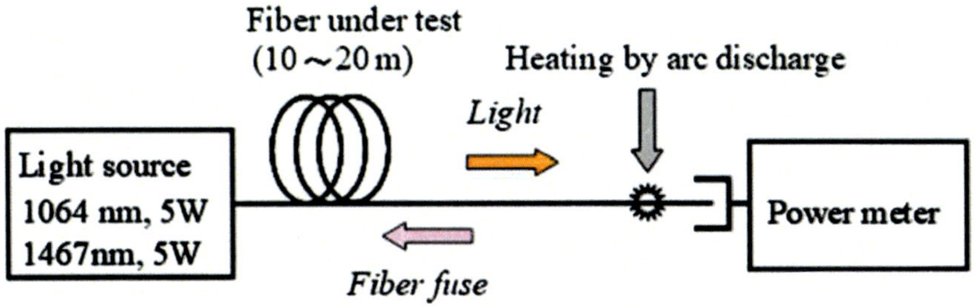

Fig. 7.14 Measurement setup for fiber fuse threshold power (Reprinted with permission from [32]. ©2003 Furukawa)

Fig. 7.15 Fiber fuse threshold (wavelength dependence) (Reprinted with permission from [32]. ©2003) Furukawa)

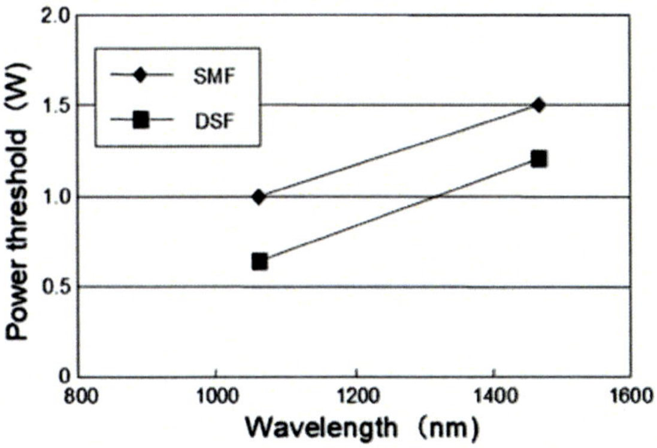

propagates toward the light source. In our experiments, a part of the fiber transmitting high powers (around 5 W) was heated to more than 1000 °C by the arc discharge in order to start the fiber fuse, and the power was gradually reduced until it stops. The power where the fiber fuse stops was defined as the propagation threshold power.

Figure 7.15 shows the measurements results of the fiber fuse threshold power of SMF and DSF measured at two wavelengths. The reason why the DSF has a smaller threshold than that of the SMF is the smaller MFD. The same reason can be applied to why the threshold power at 1064 nm is lower than that at 1467 nm. We also measured the thresholds at 1467 nm for various types of fibers which have different MFDs, and the measured results are shown in Fig. 7.16. The fibers used for the experiments are dispersion-compensating fiber (DCF) for SMFs, DSF, and SMF with enlarged MFDs (20 and 30 μm). The results are a little counterintuitive; the threshold is proportional to the MFD, not effective core area (A_{eff}). If it is proportional to A_{eff}, the measured data can be fitted with a quadratic curve in blue, not linear one in red. Therefore, it is clear that enlarging MFDs (and A_{eff}) is an effective way to increase the fiber fuse threshold power.

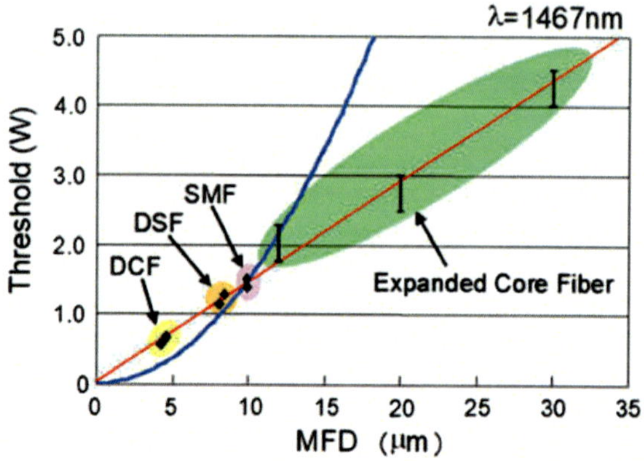

Fig. 7.16 Fiber fuse threshold power (MFD dependence) (Reprinted with permission from [32]. ©2003 Furukawa)

7.1.2.3 Damage to the Coating of Telecom Fibers by Bending

Beside the fiber fuse, there is another big issue, namely damage to the fiber coating caused by bending the fibers under the high-power condition. Basically, the permitted bending diameter in the case of SMFs is determined to be more than 60 mm ϕ from the aspect of fiber reliability. However, there is a potential risk that the fiber would be accidentally bent to a smaller diameter, for example, during the construction. If the fiber under high-power transmission is bent to a small diameter, leakage power from the core will be absorbed by the coating, and the temperature of the coating will increase. If the power is high, the temperature might reach the temperature level which damages the coating. Therefore, we performed qualitative analyses on the damages to the fiber coating when short-term bending was added to the fibers under high-power transmission.

Figure 7.17 shows the experimental setup to study the coating damages. The center wavelength was 1480, nm and the maximum power was 3 W. The sample fiber coatings are 0.25 mm ϕUV-cured coatings (transparent, white and green) and 0.9 mm ϕ nylon coating (white). The fibers under the test were wound at one turn

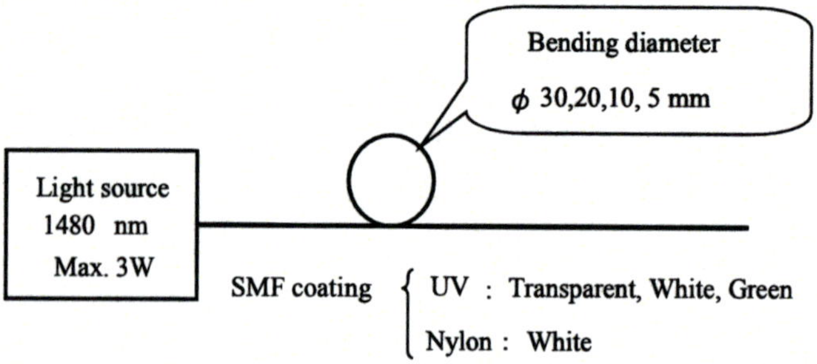

Fig. 7.17 Experimental setup to evaluate coating damage (Reprinted with permission from [32]. ©2003 Furukawa)

for diameters of 30, 20, 10, and 5 mm ϕ. Before the experiments, we measured the temperature changes of the fibers which were bent to 3.0 mm ϕ with 3 W input. Figure 7.18 showed the measured temperature against time for UV white coating and UV transparent coating. It was indicated that the temperature increased rapidly and reached at the equilibrium within 5 min. Based on the result, the high power was input into the test samples for 5 min.

The experimental results are shown in Table 7.1, which qualitatively evaluate the damage of the coating by four categories. "Bending" indicates the bending of the coating was kept by the coating stiffening. "Colored" indicates the coating became brown color. "Melt" indicates the coating melted, and the fiber came out to the outside. Finally, "Burned" literally means the coating burned with fire. In the case of nylon coating with a bending diameter less than 5 mm ϕ, it melted with 1 W power and burned with 3 W. In the case of UV cured coating with a bending diameter less than 5 mm ϕ, all of the fibers were colored with 3 W power, and all the fibers, except for the transparent one, were bent or colored with 1 W power. The results indicate the use with diameters less than 5 mm ϕ potentially is very dangerous. Even in the case

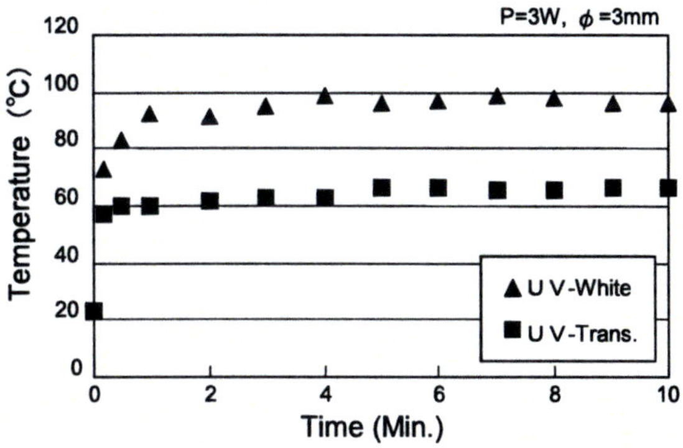

Fig. 7.18 Temperature change of coatings (Reprinted with permission from [32]. ©2003 Furukawa)

Table 7.1 Experimental results of coating damage. UV TP[1]: UV Transparent (Reprinted with permission from [32]. ©2003 Furukawa)

Input 1480 nm	Bending ϕ (mm)	Kinds of coating			
		UV TP[1]	UV white	UV green	Nylon white
1 W	30	No change	No change	No change	No change
	20	No change	No change	No change	No change
	10	No change	Bending	Bending	Bending
	5	No change	Bending	Colored	Melt
3 W	30	No change	No change	No change	No change
	20	No change	No change	Bending	No change
	10	No change	Bending	Bending	Bending
	5	Colored	Colored	Colored	Burned

of bending diameter of 10 mm ϕ, all the fibers, except for the transparent UV-cured coating, were bent by the coating stiffening. As a result, to maintain the reliability of the transmission lines, we need to maintain the bending diameters to be > 20 mm ϕ in the case of 1 W input and > 30 mm ϕ in the case of 3 W input. It should be added that transparent UV-cured coating has a higher resistance against high input power than colored UV-cured coating.

The temperature distribution of the fiber coating measured by the thermo-viewer is shown in Fig. 7.19. The highest temperature among them is plotted in Fig 7.20. In particular, in the case of white nylon coating with 5 mm ϕ bending diameter and 3 W input, we observed the temperature is higher than 160 °C. This temperature is higher than a melting point of the nylon but lower than the burning temperature. Therefore, it is assumed that the melting of the nylon coating caused a fiber break and the following power radiation caused the burning.

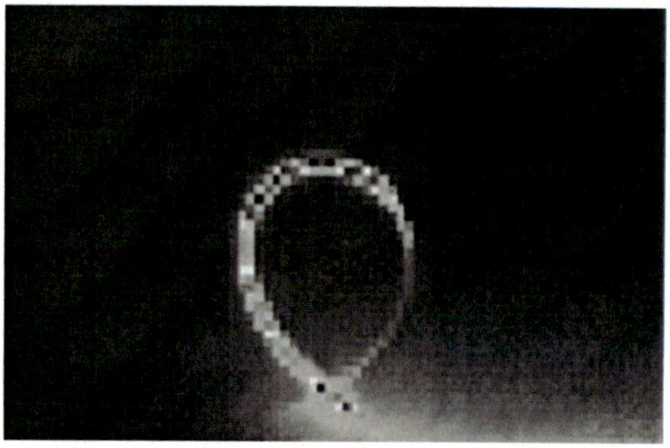

Fig. 7.19 Observed view by a thermo-viewer (Reprinted with permission from [32]. ©2003 Furukawa)

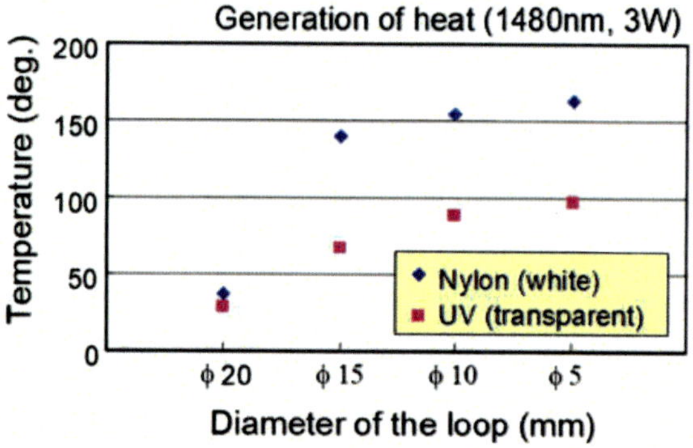

Fig. 7.20 Coating temperature against bending diameters (Reprinted with permission from [32]. ©2003 Furukawa)

7.1.2.4 Summary and Two Potential Approaches to Solve the Fiber Fuse Issues

From the results mentioned in the previous section, we figured out that the threshold power of SMFs at 1467 nm is about 1.5 W. In other words, we need to suppress the input power less than that value, ~ 1.5 W, to avoid the fiber fuse. However, it is expected that the input power will get higher in the future, and we need to consider the technologies to avoid fiber fuse as well as the further studies of its mechanism.

As for the stopping technology when the fiber fuse accidentally happens, a system, which monitors the output power and immediately stops the input when the output power becomes zero or lower than the threshold value, has been suggested. Figure 7.21 shows output power measured by power meter in the system, and this result indicates that the system stops immediately after the occurrence of fiber fuse. By using this system, we can minimize the damage because the speed of fiber fuse is around ~ m/s. In particular, the automatic power reductionAPR) system, which has been standardized as ITU-T G.664 and will be described in detail in 7.1.4 as a new stopping method of light source in the case of fiber fuse, is very effective for not only human safety aspect but also saving the system. In addition, we figured out from our experiments that it is effective to have a large MFD portion within the connector. For example, as shown in Fig. 7.22, we fabricated an SC connector which contains the MFD expansion parts with MFD of ~ 30 μm, and we confirmed the fiber fuse, which occurred at 4 W input power at 1467 nm, has successfully stopped at the fabricated

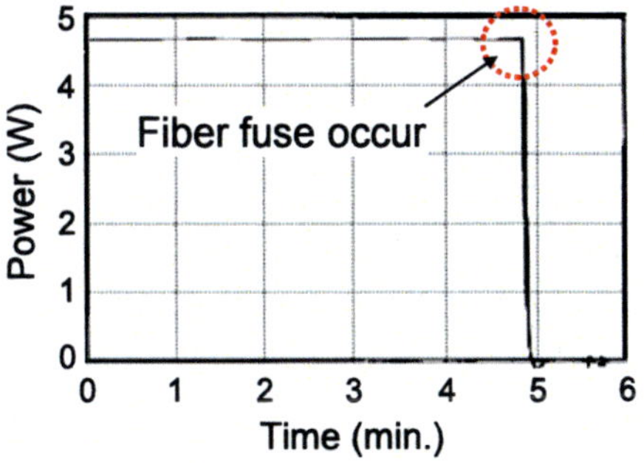

Fig. 7.21 Monitoring results of the output power (Reprinted with permission from [32]. ©2003 Furukawa)

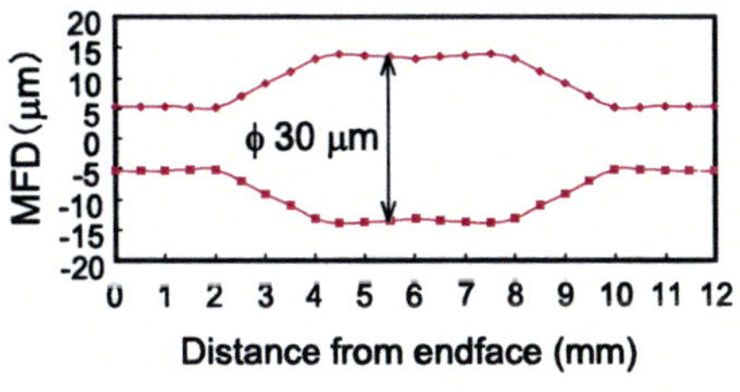

Fig. 7.22 MFD profile of the connector (Reprinted with permission from [32]. ©2003 Furukawa)

connector. These approaches, which prevent damages of the system by the fiber fuse, will be important in terms of avoiding catastrophic damages to the system.

On the other hand, increasing resistance of optical fibers against high-power input by improvements of optical properties is another effective way to avoid not only fiber fuse but also nonlinear signal distortions. However, enlargement of A_{eff} or MFD, which is effective to suppress the fiber fuse and nonlinear effects, generally accompanies the bending loss increase [33]. If the bending loss increases, the escaped power damages the coating as described in the previous section. Therefore, it is important to expand A_{eff} keeping low bending loss property. The optimizations of the refractive index profiles of SMFs and holey fibers have been investigated to solve this trade-off. Photonic bandgap fibers (PBGFs) have also been studied for future transmissions since they show unique properties due to the air-core photonic bandgap guidance.

7.1.3 Detection Methods of Fiber Fuse

When a fiber fuse starts, the optical discharge propagates unimpeded back along the fiber toward the light source, leaving the fiber core permanently damaged and unable to transmit light. The velocity at which fuse propagates is known to depend on the power of the incident laser radiation as well as the type of fiber, specifically the fiber core size and the doping concentrations. The fuse propagation can be halted by reducing the power below a threshold power that depends on the fiber type. In order to minimize the damage of the transmission fiber due to fiber fuse, a number of methods have been developed recently that quickly detect a fuse propagating in remote location and hence allow terminating the optical source.

In this section, three novel detection schemes by monitoring the electrical spectrum of back-reflected light using an optical time domain reflectometer (OTDR) and monitoring the peculiar luminescence observed during the fiber fuse propagation are described.

7.1.3.1 Fuse Detection Through Monitoring the Electrical Spectrum of Back-Reflected Light

When the fiber fuse starts propagating, it leaves behind periodic micro-voids. The periodic voids, shown in Fig 7.23, are reportedly created as a consequence of the Rayleigh instability that is due to capillary effects in the molten silica that surrounds the vaporized fiber core [34]. The periodic change in refractive index along the fiber due to the presence of the voids results in distributed reflection from the air–glass interface. When new voids are formed from the leading bubble surrounded by the molten silica, the shape of the void changes with time that resulted in a periodic modulation of the Fresnel reflection at the air–glass interface and thus an intensity modulation of the back-reflected laser radiation. By observing the characteristics of

Fig. 7.23 Microscope image of the damage caused by the fiber fuse effect in a single-mode optical fiber

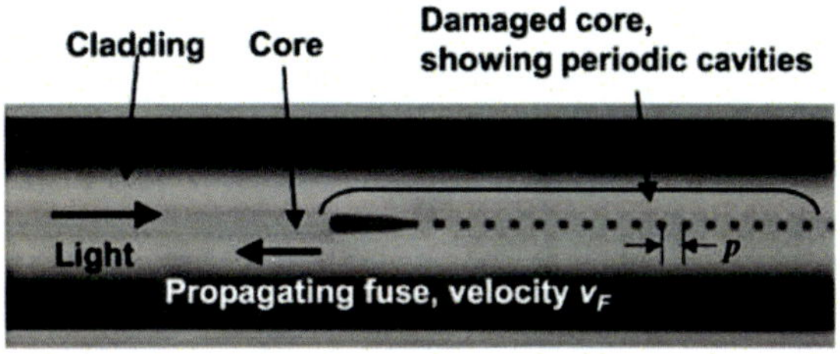

the electrical spectrum of the back-reflected light, it is thus possible to identify the presence of a propagating fuse [35–37].

Figure 7.24 shows a schematic diagram of fiber fuse detection system that is based on the above principle. Light that is back-reflected toward the source during fiber fuse propagation is detected using a photodetector, and its output is analyzed using an electrical spectrum analyzer in the frequency range of 20 KHz–1 MHz.

Before initiation of the fiber fuse, the RF spectrum of the back-reflected light is free from any frequency components, as shown by the dashed curve in Fig. 7.25. However,

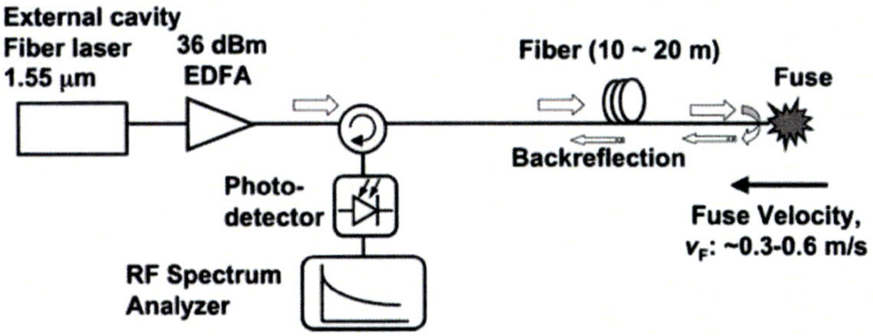

Fig. 7.24 Schematic of fiber fuse detection system based on monitoring the electrical spectrum of the back-reflected light

Fig. 7.25 Electrical spectrum of the back-reflected light from a propagating fiber fuse is shown in solid line. The spectra before the fuse are shown by dotted lines

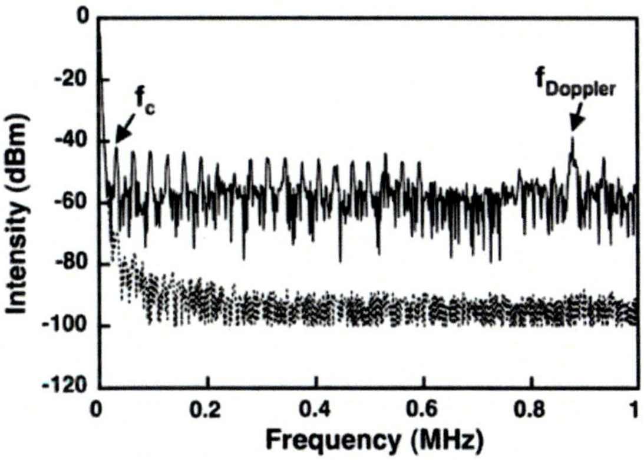

as soon as the fiber fuse is initiated, the intensity of the electrical signal in the RF spectrum increases drastically. One could clearly see the fundamental modulation frequency component, f_c, as well as many of its harmonics in the spectrum. The intensity is ~ 40–50 dB greater than the noise floor. The fundamental frequency f_c ($= v/p$), 31.06 kHz, yields a pitch p of 14.5 μm, which matches perfectly with the pitch observed in the damaged fiber. In addition, the electrical spectrum also shows a frequency component near 876.6 KHz, corresponding to the Doppler frequency shift, $f_{\text{Doppler}} = 2nv/\lambda$. Using, $n = 1.5$ and $\lambda = 1.55$ μm, the velocity of fuse propagation, $v = 0.45$ m/s, can be readily determined.

7.1.3.2 Real-Time Monitoring of Fiber Fuse Using an Optical Time Domain Reflectometer (OTDR)

When a fuse occurs, the Fresnel reflection comes mostly from the leading edge of a propagating fiber fuse, allowing us to identify such event by using an OTDR [38]. As the fuse occurs, a sharp reflection point in the OTDR trace that corresponds to the leading edge of the fuse can be observed. Recently, an OTDR-based monitoring scheme has been developed to detect fuse occurring at remote location (over a km). The system is effective even in the presence of strong spurious backscattering, such as spontaneous Raman scattering in a transmission system employing distributed Raman amplification.

To demonstrate, an experimental setup as shown in Fig. 7.26 is used [38]. A high-power cascaded Raman laser operating at a wavelength of 1480 nm is employed to initiate a fuse and an OTDR for monitoring and detection of fiber fuse. The interrogating pulse emitting the OTDR is 10-ns-long and had a spectral width of about 15 nm and a peak wavelength of ~1540 nm. The laser and the OTDR are connected through a WDM coupler and a bandpass filter placed after the OTDR preventing the broad Raman ASE induced by the pump from entering the OTDR. In

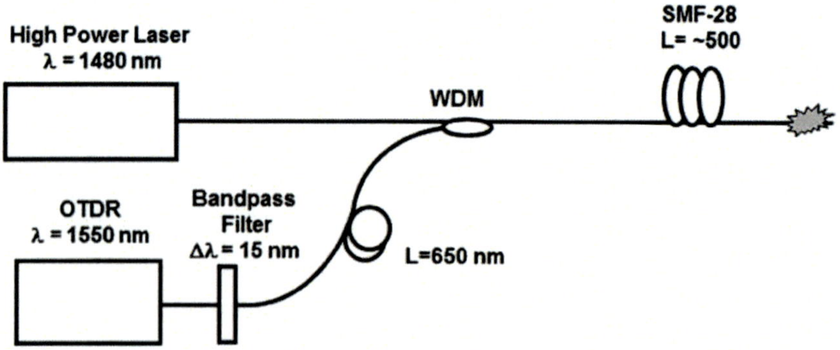

Fig. 7.26 Fiber fuse detection using OTDR (Reprinted with permission from [38]. ©2010 The Optical Society)

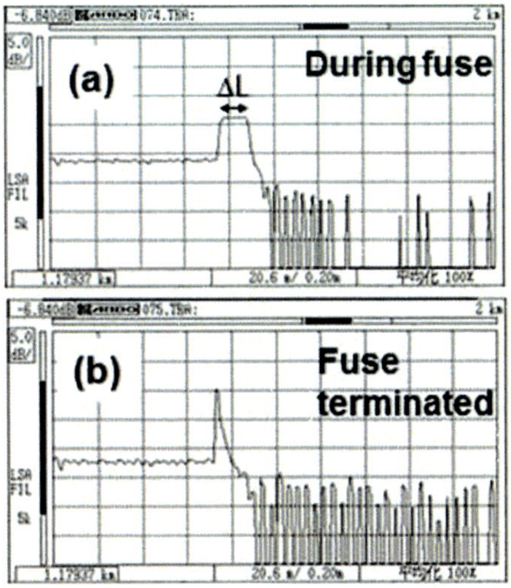

Fig 7.27 OTDR traces taken during fuse propagation (**a**) and after terminating the fuse (**b**). Integration time was 30 s, and each division in the traces corresponds to 5.1 m

addition, a long piece of SMF (~ 650 m) is also installed in the path of the OTDR to remain outside the dead zone of OTDR measurement.

Figure 7.27a shows the OTDR trace measured over an integration time of 30 s. As the fuse propagated, and OTDR measurements are integrated over a fixed time interval, an OTDR trace with a distinct a flat region can be observed. As the leading edge of the fuse where reflection occurs advances with time, the width ΔL of this flat plateau becomes proportional to the integration time (Δt). From ΔL and Δt, the speed of fuse propagation v can be readily measured from $v = \Delta L/\Delta t$. An OTDR trace taken after the fuse is terminated (Fig. 7.27b) on the other hand, shows a narrow peak due to reflection from the leading edge of the damaged fiber.

7.1.3.3 Detection by Monitoring the Peculiar Luminescence Observed During the Fiber Fuse Propagation

One of simple systems that accurately detects and certainly stops a propagating fiber fuse is to monitor the peculiar luminescence observed with the fiber fuse phenomenon, and the high-power light source output that causes the fiber fuse is discontinued based on the monitoring result [39]. Figure 7.28 shows the configuration of the detection and termination system and the electrical schematic, respectively. A Si photodiode (PD) is located on the side of an optical fiber for monitoring the peculiar luminescence (radiated visible light) which is propagating toward a fiber laser [39]. The output current of the PD is supplied into a distinction circuit with a latching function through current/voltage conversion circuit (TIA). The output of the decision circuit is connected to the relay circuit which controls the interlock of the light source. The decision circuit is composed of a comparator and a flip-flop circuit. If the Q output becomes high level, the output of the fiber laser will be discontinued

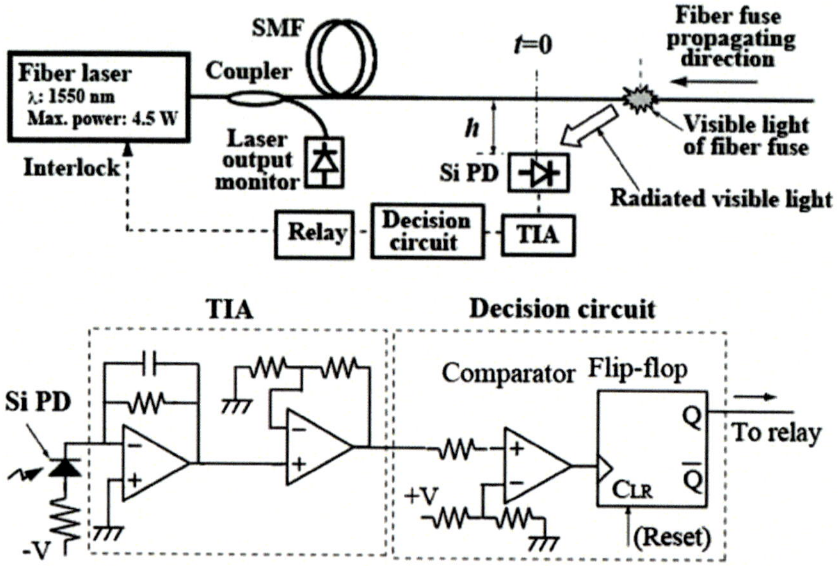

Fig 7.28 Configuration and electrical schematic of detection and termination system (Reprinted with permission from [39]. ©2013 IEICE)

by changing interlock into an open state through the relay which is connected to the Q output. Figure 7.29 shows the signals which observed the output from the TIA, the Q output of the decision circuit, and the fiber laser output monitored with the oscilloscope, when radiated light by fiber fuse propagates the front of the PD. Fiber fuse formation power and the propagation velocity of fiber fuse were 3 W and 0.49 m/s, respectively. As shown in this figure, it was confirmed that the Q output becomes

Fig 7.29 Signals which observed the output from the TIA, the Q output of the decision circuit, and the fiber laser output monitored, when radiated light by fiber fuse propagates the front of the PD (Reprinted with permission from [39]. ©2013 IEICE)

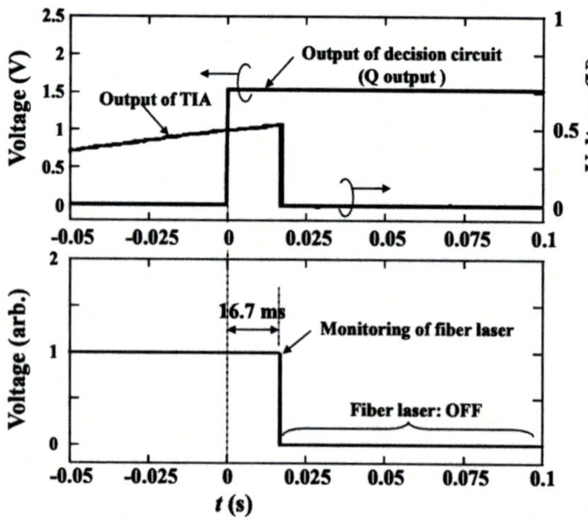

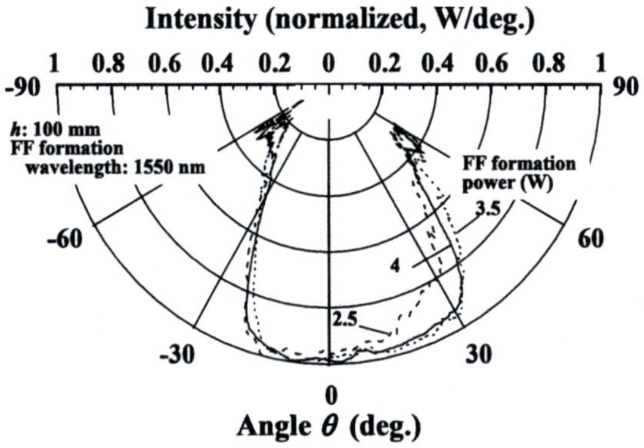

Fig. 7.30 Estimated radiation patterns of peculiar luminescence observed during the fiber fuse propagation (Reprinted with permission from [40]. ©2015 Authors)

high level, after the PD receives the radiated light generated by fiber fuse, and the output from fiber laser could stop certainly after 16.7 ms. It is considered that the delay time after the Q output becomes high level until the output of the fiber laser stops is the operating time of the fiber laser itself, since the response time of our detection and termination system is in the order of a microsecond.

In addition, the radiation pattern of the peculiar luminescence was estimated from the photocurrent waveforms which was obtained by the measurement setup as shown in Fig. 7.28, because it is important when designing the system. It is found that the maximum luminescence power was observed around the perpendicular direction to the propagation direction of the optical fiber as shown in Fig. 7.30 [40].

7.1.4 Halting (Blocking) Methods of Fiber Fuse

Several approaches to preventing the catastrophic damage caused by a fiber fuse have been proposed. These techniques include the use of passive fiber fuse terminators and the application of a device that can rapidly detect a fiber fuse by monitoring back-reflected light. Once back-reflected light from a fiber fuse is detected, this device terminates the fuse by deactivating the light source. In this section, we describe techniques for the passive termination of fiber fuses.

7.1.4.1 Introduction

One technique for passive termination of fiber fuses is to decrease the optical intensity by expanding the mode field diameter (MFD). A down-tapered fiber [41, 42] and a thermally diffused expanded core (TEC) fiber [43] are used for the MFD expansion. Another technique is the use of optical fibers with air holes. Passive fiber fuse terminators with hole-assisted fibers (HAFs) [44] and hollow optical fibers [45] have been reported. Hollow optical fibers consist of a central air hole surrounded by a high-index ring core and silica cladding. Another approach involves the use of optical fibers with decreased cladding thickness which is achieved via etching. The deformation or even destruction of the cladding under the high temperature and pressure of the optical discharge can halt the fiber fuse [46].

In Sect. 7.1.4.2, we describe a fiber fuse terminator with a TEC fiber, and a passive fiber fuse terminator with HAF is described in Sect. 7.1.4.3. Further, in Sect. 7.1.4.4, we discuss the importance of a method for rapidly deactivating the light source when the fiber fuse is initiated, which should be used in addition to a passive fiber fuse terminator.

7.1.4.2 Fiber Fuse Terminator with a TEC Fiber

A fiber fuse may occur when watt-level optical signals are transmitted through optical fiber cable. When a fiber fuse occurs, it does not stop but proceeds toward the light source with increasing optical power, and when it reaches the light source, the device may break. Therefore, a fiber fuse termination function is essential for securing the safe operation of high-power optical network systems. The fiber fuse can be stopped by reducing the optical power density in the core region. We can reduce the optical power density in the core region by expanding the core diameter. A thermally diffused expanded core (TEC) fiber is one of the means to reduce the optical power density in the core region. However, when the TEC is formed by parching the fiber with a burner, its mechanical strength is deteriorated. The MU-type fiber fuse terminator (FFT) was proposed [43] by installing the TEC fiber in the ferrule part as shown in Fig. 7.31.

Using the FFT with 20-μm MFD TEC fiber, the fiber fuse propagation was stopped when the 1.48 μm Raman laser input power fell to 1.1 W for a DSF fiber with a mode field diameter (MFD) of 8 μm. The fiber fuse propagating speed depends on the incident light power. Because the MFD of the TEC region is 2.5 times larger than that of usual DSFs, the TEC is considered to be effective for fiber fuse termination with an incident power of 2 W. As the length of TEC region was 11 mm and the ferrule was 16 mm length, we could build the entire TEC region into the ferrule. With both end faces of the ferrule in the same way as with usual fibers, low connection

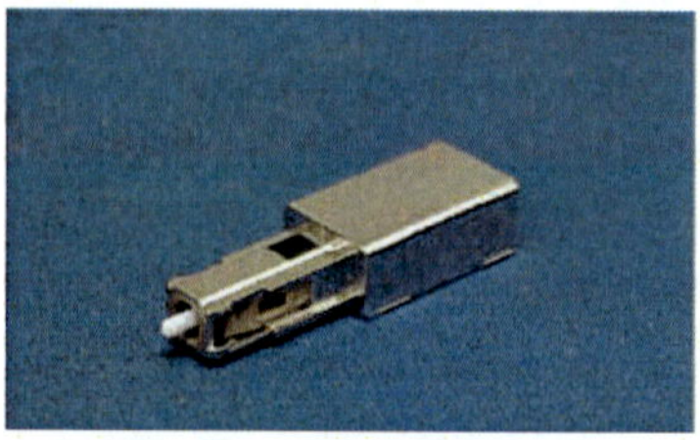

(a) Fiber fuse terminator (FFT)

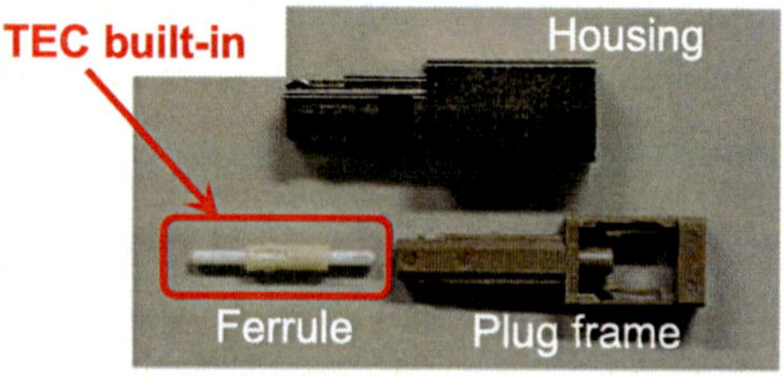

(b) FFT component parts

Fig. 7.31 Fiber fuse terminator structure and experimental setup

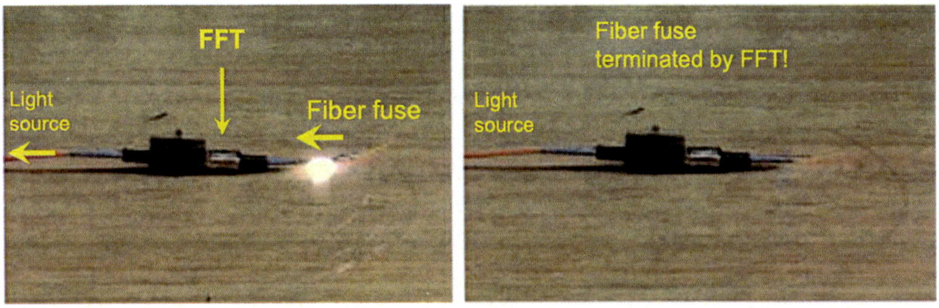

(a) fiber fuse is propagating (b) fiber fuse is terminated by the FFT

Fig. 7.32 Fiber fuse terminator structure and experimental setup

loss in the optical function part of less than 0.05 dB is expected. Figure 7.32 shows that (a) the fiber fuse is propagating, and (b) the fiber fuse is terminated by the FFT. The FFT with a 20-micron mode field diameter TEC can terminate a fiber fuse below a 2 W optical input, and the terminator with a 30 μ mode field diameter TEC can be used for a 3 W optical input.

7.1.4.3 Fiber Fuse Terminator with HAF

Recently, we reported the development of a compact fiber fuse terminator consisting of a short length of HAF, which can be used for a significantly higher input power than fiber fuse terminators with the MFD expansion [44]. Figure 7.33 shows the structure

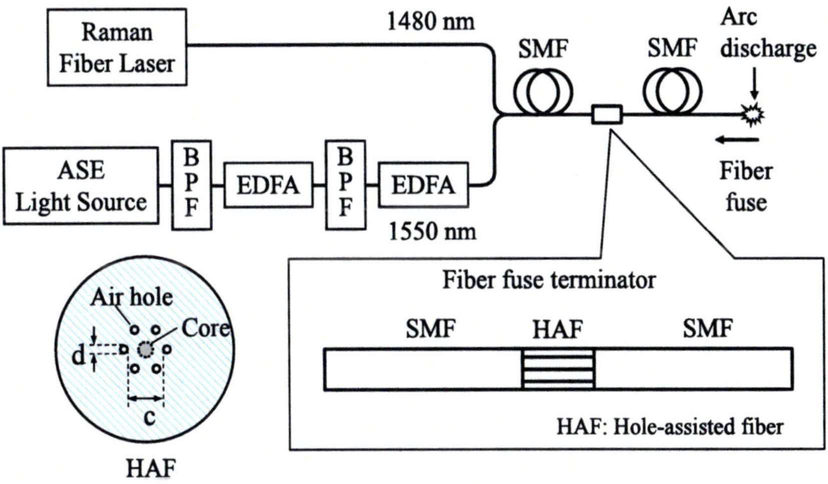

Fig. 7.33 Fiber fuse terminator structure and experimental setup (Reprinted with permission from [44]. ©2013 IEICE)

of our fiber fuse terminator and the experimental setup. The fiber fuse terminator consists of a short length of HAF spliced with conventional SMFs. The total splice loss between the HAF and SMFs was 0.1 dB. We used 1.4- and 0.41-mm-long HAFs with six air holes in our experiment, and the HAF core diameter was 9 μm. The diameter of the inscribed circle linking the air holes (c) and the HAF hole diameter (d) were 17.1 and 4.6 μm, respectively. As can be seen in Fig. 7.33, our fiber fuse terminator is compact and has a simple structure with low loss. It is easy to handle and can be installed in an optical connector in a similar manner to a fiber fuse terminator with a TEC fiber [43].

The light source used for the initiation and propagation of the fiber fuse consisted of a Raman fiber laser and an amplified ASE operating at wavelengths of 1480 and 1550 nm, respectively. Two continuous wave (CW) lights were simultaneously multiplexed with a WDM coupler and then guided into the fiber fuse terminator. We initiated a fiber fuse with an arc discharge by heating a SMF that was spliced to the fiber fuse terminator and observed the dynamics of the fiber fuse propagation and its termination in the HAF of the fiber fuse terminator using a high-speed camera.

Figure 7.34 shows the dynamics of the fiber fuse termination near the splice point between the HAF and SMF. The HAF length was 1.4 mm, and the total power input into the HAF was 22 W. The input powers at 1480 and 1550 nm were 7.9 and 14.1 W, respectively. These images were obtained at intervals of 0.1 ms using a high-speed camera operating at 10^5 fps. As shown in Fig. 7.34c, d, the size of the optical discharge decreased after it penetrated the HAF. Since an optical discharge is considered to be a high-temperature fluid at high pressure [47], its front shape shown in Fig. 7.34c indicates a jet of high-temperature fluid entering the air holes. Then,

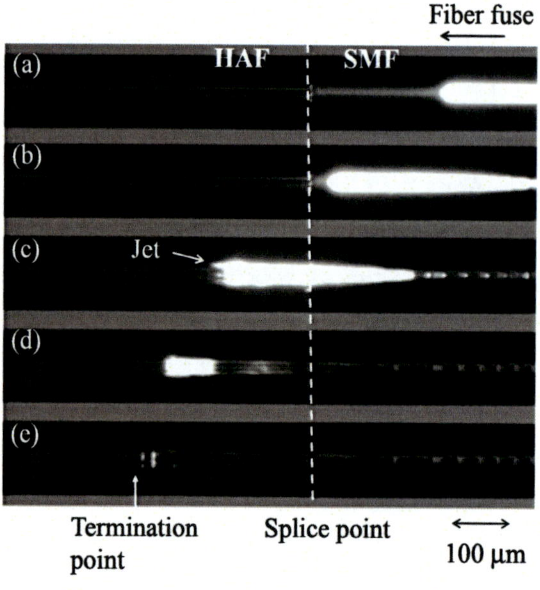

Fig. 7.34 Dynamics of fiber fuse termination near splice point between HAF and SMF. The HAF length is 1.4 mm, and the input power is 22 W. The images were obtained at 0.1 ms intervals (Reprinted with permission from [44]. ©2013 IEICE)

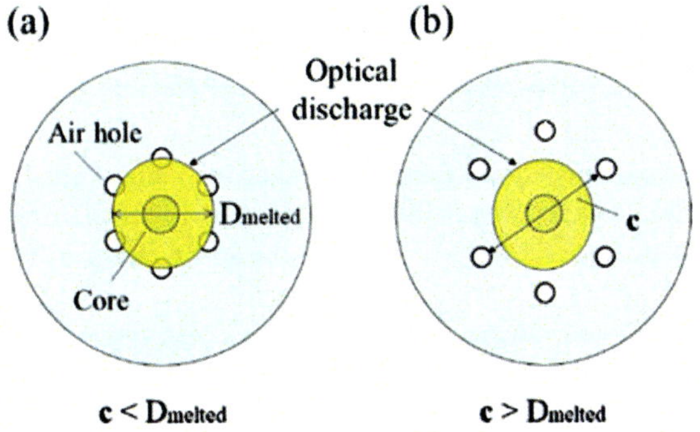

Fig. 7.35 Relationship between **c** and D_{melted} in HAF

the optical discharge was terminated, as shown in Fig. 7.34e. The optical discharge penetrated 0.3 mm into the HAF. Therefore, we found that we could terminate the fiber fuse propagation by using a fiber fuse terminator consisting of a HAF of only 1.4 mm in length, when the input power was 22 W. The mechanism of fiber fuse termination in HAF can be explained as follows [48]. Because the diameter of the optical discharge (D_{melted}) at an input power of 22 W exceeds 23 μm [49], which is significantly larger than the **c** in the HAF, the high-temperature fluid reaches the air holes, as shown in Fig. 7.35a. At the same time, the jet penetrates the air holes. This results in a reduction in the pressure of the high-temperature fluid. Simultaneously, the fluid temperature decreases, and the fiber fuse is terminated. In fact, when D_{melted} is significantly smaller than **c** in HAF, as shown in Fig. 7.35b, the fiber fuse propagates in the same manner as in SMF [49, 50].

The above experimental result indicates that our fiber fuse terminator composed of HAF is attractive as a passive fuse terminator.

7.1.4.4 Passive Fiber Fuse Terminators and Automatic Power Reduction (APR)

The passive fiber fuse terminators described in Sects. 7.1.4.2 and 7.1.4.3 are attractive for halting fiber fuses. However, in addition to the application of a passive fiber fuse terminator, it is important to note that a method for rapidly deactivating the light source when the fiber fuse is initiated would be useful. This is because a passive fiber fuse terminator stops the fuse propagation only, but has no function as regards reducing the optical power at the fuse termination point. In fact, it has been reported that an optical fiber was heated and burned by the light scattered from a bubble train at the fuse termination point at an input power of over 3 W [51].

ITU-T recommendation G.664 provides optical safety procedures and requirements for optical transport systems [52]. It also provides automatic power reduction (APR) procedures for systems employing discrete optical amplification and for systems based on distributed Raman amplification, which apply when optical power loss occurs within one section of the main optical path. When the optical powers of the current optical transport systems are increased significantly, which should be achieved in the near future, a corresponding APR procedure will be developed. We believe that the APR functions of such optical transport systems are useful for terminating fiber fuses, because the optical signal is lost when a fiber fuse occurs. ITU-T recommendation G.664 states that the sum of the power from all optical outputs should be reduced in the impacted optical transmission section within a certain period (a maximum of 3 s) following the interruption of the optical transmission section continuity. The specific shutdown time is dependent on the actual operational optical power and can be calculated from the maximum permissible exposure specification. For example, the calculated shutdown time is approximately 1 s for a maximum permissible optical power of 4 W, provided that the nominal ocular hazard distance (NOHD) is 25 cm [53]. Since the fiber fuse velocity in an SMF is approximately 0.6 m/s for a 4 W input, the fiber fuse propagates 0.6 m during a shutdown time of 1 s. Therefore, in addition to the APR function of optical transport systems, we believe that the passive fiber fuse terminator is important as regards the prevention of the equipment damage caused by a fiber fuse that occurs very near to the equipment output end.

7.1.4.5 Conclusion

We have described techniques for the passive termination of a fiber fuse. Further, we have shown that fiber fuse terminators consisting of a TEC fiber and/or a short length of HAF are attractive as passive fuse terminators. We have also suggested that the APR function of optical transport systems is useful for fiber fuse termination when applied to systems with high input powers.

7.1.5 Fiber Fuse-Based Incidence

Once the fiber fuse is initiated, a bubble train forms in the fiber core after the fiber fuse, and it propagates toward the high-power optical source until the optical power in the core falls below the threshold fiber fuse power. The fiber in which the fiber fuse was formed cannot be used for optical transmission, because optical signals are scattered by the bubble train which is formed by the fiber fuse phenomena. Furthermore, recently heating and burning of an optical fiber have been reported when the bubble train is formed in the middle of a fiber as a result of high-power light being launched again into the fiber [54–56].

In this section, scattering, heating, and burning characteristics of the bubble train that is formed in the middle of a fiber as a result of the fiber fuse phenomenon are described. In addition, the safety of the scattered light based on laser product safety standards is discussed, because these characteristics are also important in terms of safety.

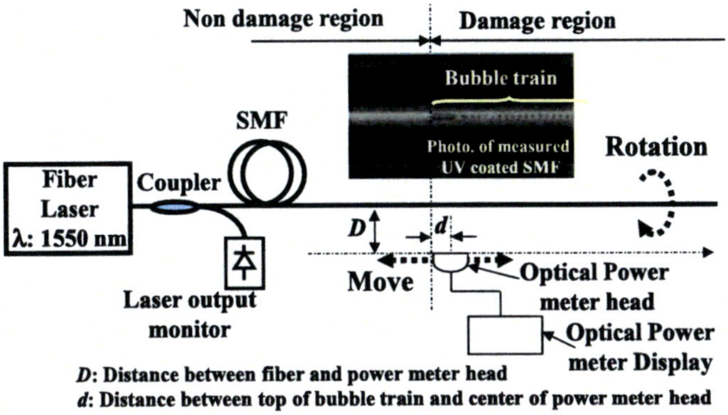

Fig. 7.36 Setup for measuring scattering characteristics of light from bubble train (Reprinted with permission from [54]. ©2011 Authors)

7.1.5.1 Scattering Characteristics

Figure 7.36 shows the measurement setup of the scattering characteristics of the light from a bubble train [54, 55]. The scattered optical light was measured using an optical power meter. The measured fiber section in which the bubble train was formed in the middle of a fiber was straight, and the optical power meter head was moved along the measured fiber section. The fiber section was also rotated while the rotation angle dependence of the scattered light was being measured. The relationship between *D* (the distance between the fiber and the power meter head) and the scattered power for various *d* values (the distance between the top of the bubble train and the center of the power meter head) and the relationship between the rotation angle and the scattered power for various *d* and *D* values are shown in Fig. 7.37a, b, respectively. [55] The fiber fuse formation power was 4 W. The launched optical power was 0.42 W. It is found that the light was scattered cylindrically, and we confirmed experimentally that almost all the light was scattered at the top of the bubble train, i.e., the boundary between the undamaged and damaged regions of the fiber.

7.1.5.2 Heating and Burning Characteristics

The heating characteristics of a fiber or a cable in the middle of which a bubble train had formed when a high-power light was launched into it were measured using infrared thermography. The measured fibers and cables were various commercially available UV-coated fibers, nylon-jacketed fiber, flame-retardant polyethylene (FRPE) and polyvinyl chloride (PVC) jacketed fibers, and FRPE and PVC fiber cables. Figure 7.38 shows example thermographs of the UV-coated fiber and FRPE-jacketed fiber at a launched power of 3 W [55]. It is confirmed from these thermographs that there is a noticeable temperature increase at the boundary between the

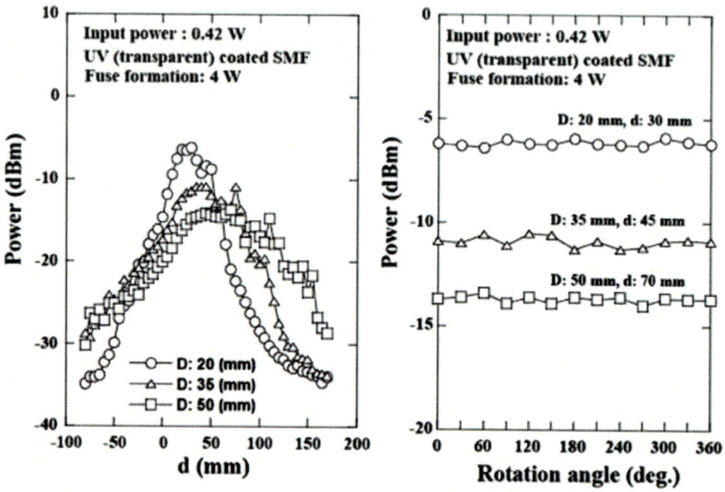

Fig. 7.37 Typical scattering properties (Reprinted with permission from [55]. ©2012 IEICE)

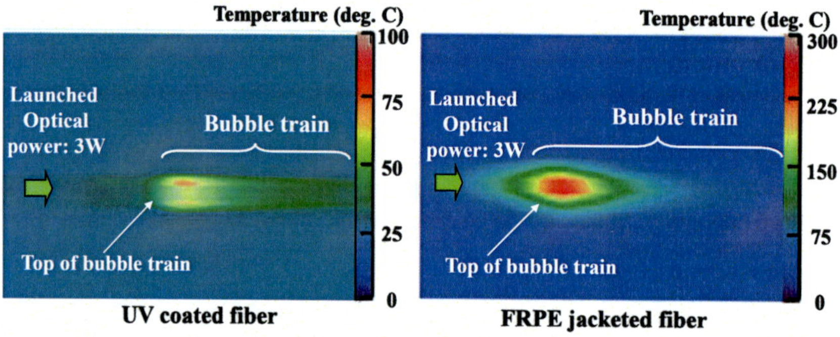

Fig. 7.38 Example thermographs for UV-coated fiber and FRPE-jacketed fiber (Reprinted with permission from [55]. ©2012 IEICE)

undamaged and damaged regions. Figure 7.39 shows the maximum fiber temperature as a function of launched optical power for various measured fibers [55]. The fiber fuse formation power was 3 W. The temperature was measured within less than a minute after a high-power light had been launched into the fibers and before burning and destruction. The maximum fiber temperature increases as the launched optical power is increased for all of the fibers and cables. The maximum temperature of the nylon, FRPE, and PVC jacketed fibers was over 200 °C and that of the FRPE and PVC fiber cables was over 600 °C, whereas the temperature of the UV-coated fiber was around 100 °C. Figure 7.40 shows the maximum fiber temperature as a function of total time after light was launched into various fibers and cables [55]. The fiber fuse formation power was 3 W. All of the fibers and cables we measured were

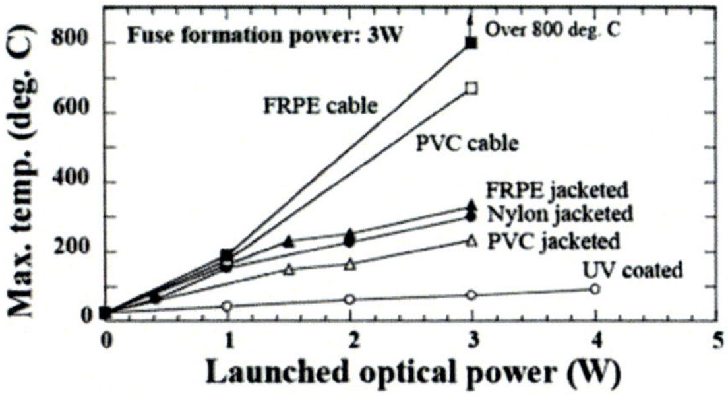

Fig. 7.39 Maximum fiber temperature as function of launched optical power (Reprinted with permission from [55]. ©2012 IEICE)

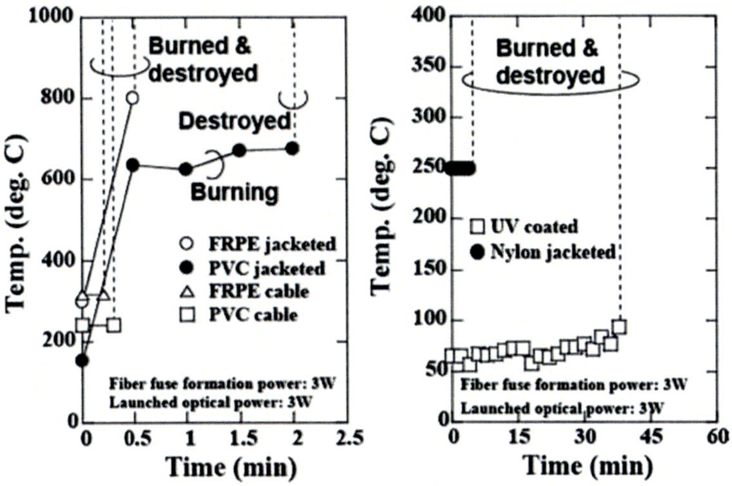

Fig. 7.40 Maximum fiber temperature as function of total time after launching light into various fibers (Reprinted with permission from [55]. ©2012 IEICE)

burned and destroyed when the launched optical power was 3 W. The photographs taken after the fibers were burned and destroyed as shown in Fig. 7.41 [55]. Since the burning spread along the UV-coated and nylon-jacketed fibers, the photographs of these two fibers were taken after the fire had been extinguished, and we found that these fibers posed a problem as regards flammability. On the other hand, although PVC and FRPE-jacketed fibers and PVC and FRPE fiber cables burned and broke, it is confirmed that the fire did not spread.

7.1.5.3 Safety of the Scattered Light Based on Laser Product Safety Standards

Figure 7.42 shows the setup for measuring the light scattering characteristics from a bubble train formed by the optical fiber fuse phenomenon [56]. Since almost all

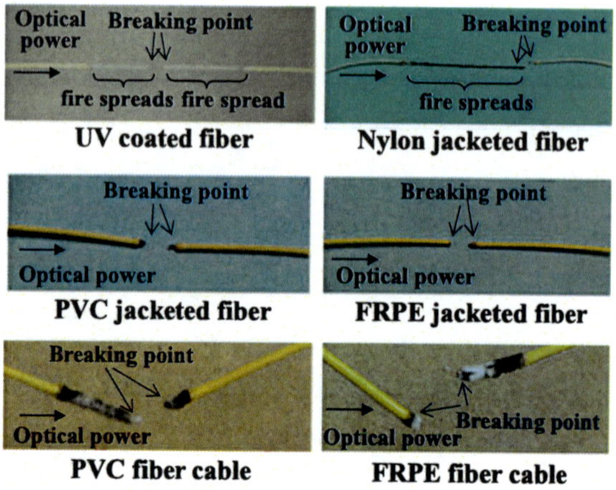

Fig. 7.41 Typical photographs after fibers and cables were burned and destroyed (Reprinted with permission from [55]. ©2012 IEICE)

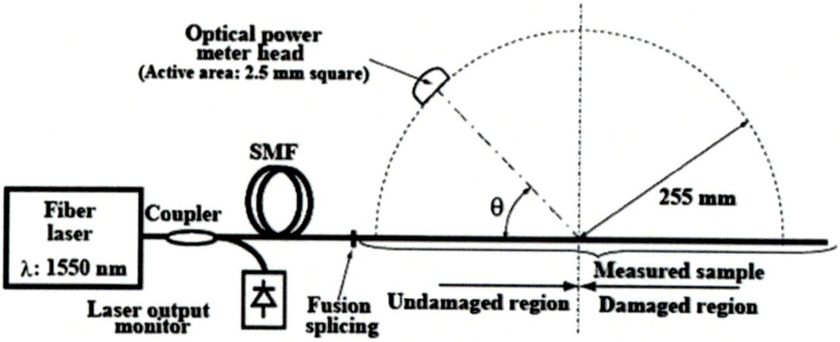

Fig. 7.42 Light scattering characteristic measurement setup (Reprinted with permission from [56]. ©2012 IET)

the optical light is scattered at the top of the bubble train when high-power light is launched again into the fiber [55], the scattered light power was measured using an optical power meter by moving it around a circle centered on the top of the bubble train. The distance between the top of the bubble train and the power meter was 255 mm. The measured samples were UV-coated SMFs (SMF-28, UV coating color: transparent) in which a bubble train was formed in the middle of the fiber with fiber fuse formation powers of 4, 3, and 2 W. The light scattering characteristics were measured for samples with and without the UV coating. The optical power launched into the fiber was up to 4 W for the fiber without UV coating and 0.5 W for the fiber with UV coating to prevent the fiber from burning, which occurs when the power exceeds a few watts.

Figure 7.43a shows the light scattering characteristics when optical powers of 0.5, and 1, 2 and 4 W were launched into the fiber [56]. The fiber fuse formation power was 3 W. It is found that the scattering properties were almost the same whatever

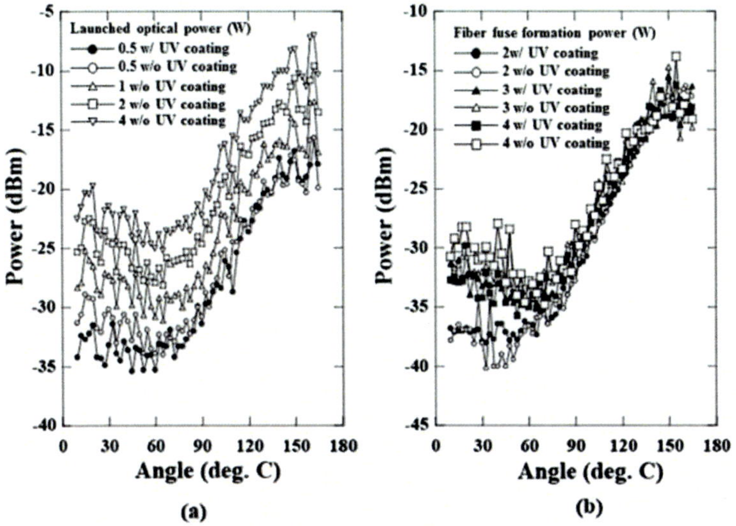

Fig. 7.43 3 Scattering properties **a** dependency of optical power launched into the sample for a fiber with a fiber fuse formation power of 3 W, **b** dependency of fiber fuse formation power at launched optical power of 0.5 W (Reprinted with permission from [56]. ©2012 IET)

the optical power launched into the fiber and with or without the UV coating. The scattering properties of the samples where the fiber fuse formation power differs are shown in Fig. 7.43b [56]. The launched optical power was 0.5 W. We found that the maximum scattering angle of the light scattered at the top of the bubble train was around 160° with or without a UV coating and regardless of the fiber fuse formation power.

Figure 7.44 shows the relationship between the maximum scattered optical power and the optical power launched into the fiber with or without the UV coating [56]. The maximum scattered optical power was measured for a sample in which a bubble train had formed with a fiber fuse formation power of 4 W, and the maximum scattered optical power was obtained at a scattering angle of 165°. The maximum scattered optical power for both the fiber with and without UV coating increases in proportion to the launched optical power, and the value is higher for the fiber without the UV coating. This figure also shows the accessible emission limits (AELs) of the class 1, 1M, and 3R products specified by the laser product safety standard IEC 60825-1, which we converted to match our measurement conditions. It is clear that the extrapolation line for the fiber with the UV coating is sufficiently lower than the AEL for class 1M products, and the maximum light power scattered from the UV-coated optical fiber meant that it could be considered as a class 1M product when the incidence light power launched into the optical fiber is several watts.

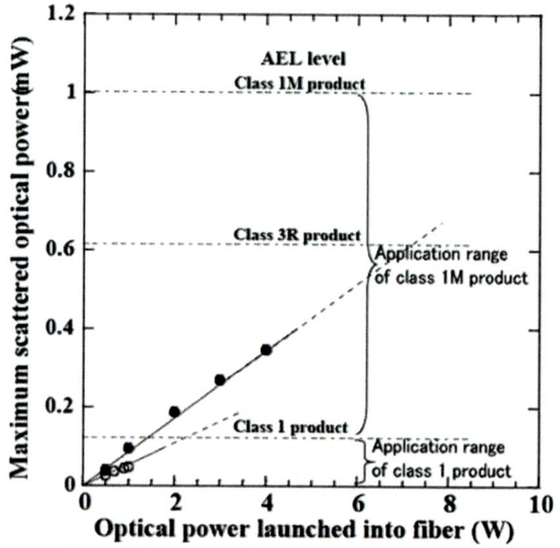

Fig. 7.44 Relationship between the maximum measured scattered optical power and the optical power launched into a fiber with or without UV coating (Reprinted with permission from [56]. ©2012 IET)

7.1.6 Fiber Fuse Tolerant Fibers

Enlarging mode field diameters (MFDs) and adapting structures in cross section with air holes can be considered as a way of increasing the propagation threshold of fiber fuse. Although the propagation threshold of fiber fuse is proportional to the MFD [20, 21], improving propagation threshold was difficult by enlarging the MFD. On the other hand, it is reported that the propagation threshold powers for hole-assisted fibers (HAFs) and photonic crystal fibers (PCFs) were greatly increased [49, 57, 58]. Since the propagation threshold can be also increased in a hollow optical fiber [59], the cross section of fiber with air holes was considered to be effective as a fiber fuse-tolerant fiber. In this section, we describe the fiber fuse tolerance in HAFs and a PCFs.

Figure 7.45 shows cross sections of an HAF and a PCF. Table 7.2 shows the structural parameters of the HAFs and PCFs in this study. We used six kinds of HAFs and a PCF with air holes at different positions. Figure 7.46 shows the experimental setup. The light sources used for the initiation and propagation of the fiber fuse were a Raman fiber laser and an amplified ASE source operating at wavelengths of 1480 nm and 1550 nm, respectively. The amplified ASE was obtained using an ASE light source, bandpass filters (BPFs), and EDFAs. Two CW lights were multiplexed with a WDM coupler to obtain a high input power of more than 10 W and then guided into a test fiber after passing through Fiber 1. We initiated a fiber fuse by heating Fiber 2 with an arc discharge, and we examined its propagation in the test fiber. We also observed the dynamics of fiber fuse termination at a splice point between Fiber 2 and a test fiber using a high-speed camera.

Table 7.3 shows our experimental results for fiber fuse propagation. The fiber fuse

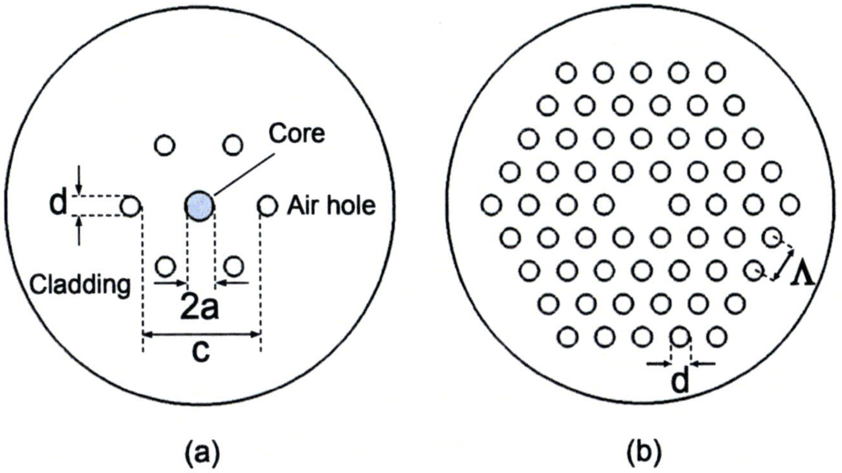

Fig. 7.45 Cross sections of **a** a hole-assisted fiber (HAF) and **b** a photonic crystal fiber (PCF)

Table 7.2 Fiber parameters of HAFs and PCFs

	c/2a	$d/2a$	d/Λ	Λ (μm)	Number of holes	$2a$ (μm)
HAF-A	3.4	0.5	–	–	6	9.0
HAF-B	1.9	0.5	–	–	6	9.0
HAF-C	1.1	0.5	–	–	6	9.0
HAF-D	2.3	0.4	–	–	10	9.0
HAF-E	2.3	0.3	–	–	10	9.0
HAF-F	2.5	1.0	–	–	6	8.0
PCF	–	–	0.7	7.6	60	–

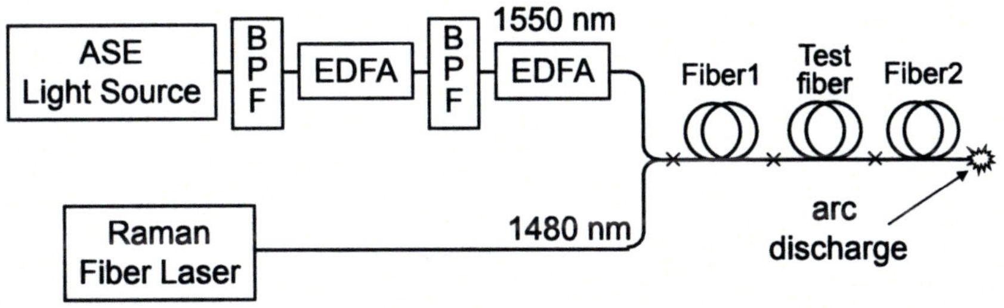

Fig. 7.46 Experimental setup

Table 7.3 Results of fiber fuse propagation

	Fiber 1	Fiber 2	Fiber fuse propagation
HAF-A	SMF	SMF	Propagation
HAF-B	SMF	SMF	No propagation
HAF-C	DSF	DSF	No propagation
HAF-D	SMF	SMF	No propagation
HAF-E	SMF	SMF	Propagation (when nearly *Pth*)
HAF-F	SMF	SMF	No propagation
PCF	DSF	DSF	No propagation

propagation was observed in HAF-A and HAF-E. This fiber fuse propagation in the HAFs and the PCF can be explained by the relationship between the parameter c in Fig. 7.45a and the diameter of the melted area (D_{melted}) [49]. The melted area is caused by the fiber fuse propagation, and D_{melted} is considered to be almost the same size as the plasma. Figure 7.47a, b shows the picture of D_{melted} and the D_{melted} versus the input power at a wavelength of 1550 nm in a conventional SMF, respectively. The D_{melted} has an input power dependency and its size decreased rapidly at an input power of near the propagation threshold as shown in Fig. 7.47b. The D_{melted} also has an MFD dependency, and the $\mathrm{D_{melted}}$ at a wavelength of 1310 nm in a conventional SMF in [4] was smaller than that of Fig. 7.47b when the input power was 5 W.

We consider that the suppression effect of fiber fuse propagation in a HAF when c is less than D_{melted} because the D_{meteld} is almost the same size as the plasma during the fiber fuse propagation.

Figure 7.48a–c consists of seven successive pictures taken at intervals of 50 μs that show the fiber fuse termination at a splice point between Fiber 2 and HAF-B, HAF-C, PCF at an input power of 12.0 W, respectively. These pictures were obtained with a high-speed camera operating at 10^5 frames per second (fps). As shown in Fig. 7.48a,

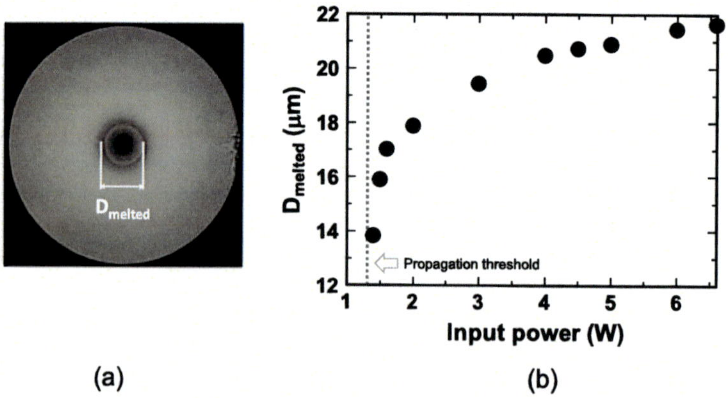

Fig. 7.47 Input power dependence of D_{melted} in an SMF at 1550 nm

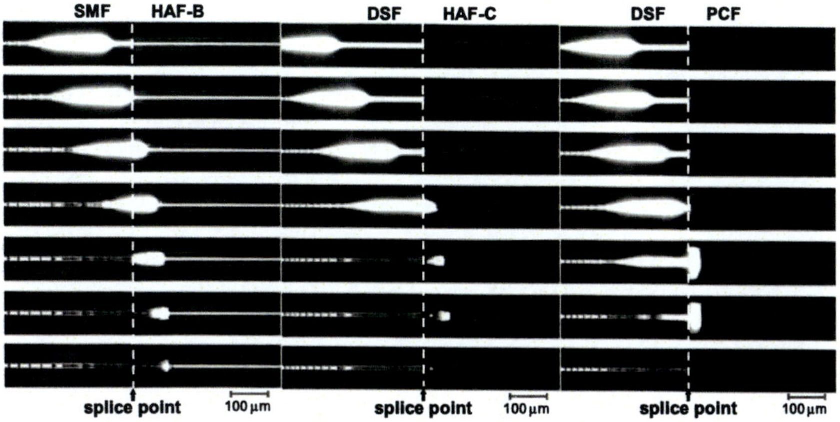

Fig. 7.48 Fiber fuse termination at a splice point between Fiber 2 and HAF-B, HAF-C, PCF. **a** HAF-B. **b** HAF-C, **c** PCF

the plasma was smaller after it had passed through the splice point and the fiber fuse terminated around 110 μm from the splice point at an input power of 12.0 W. We also observed that the air holes were completely destroyed by the fiber fuse from the splice point to near the termination point.

Figure 7.49 shows the variation in penetration length with input power in HAF-B, HAF-C, and PCF. Since the penetration length is the smallest in PCF, we consider that this result is caused by the values of c and d. The c values of HAF-B, HAF-C, PCF were 17.1 μm, 9.5 μm, and 9.9 μm, respectively. Since the parameters of c in HAF-C and PCF have smaller than that of HAF-B, the penetration length is small. Although the c values of HAF-C and PCF were almost the same, d of 5.3 μm in PCF was larger than that of HAF-C with d of 4.5 μm. Therefore, the penetration length of PCF was smaller than that of HAF-C. Since the HAFs with the same c value and different d values were observed to have different fiber fuse penetration lengths [48],

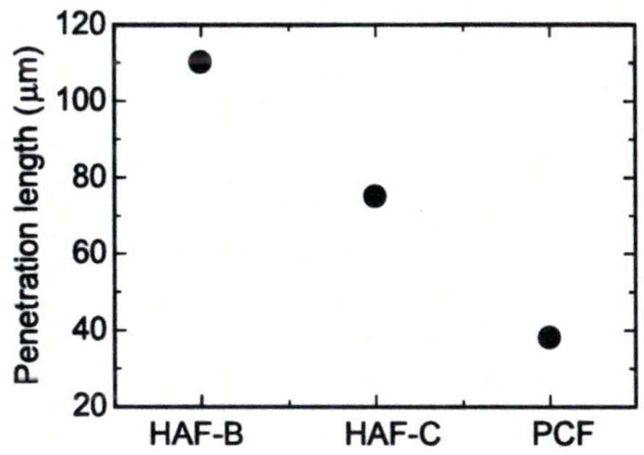

Fig. 7.49 Fiber fuse penetration lengths at a splice point between Fiber 2 and HAF-B, HAF-C, PCF

the suppression of fiber fuse propagation in HAFs requires enough air hole diameters in order to reduce the pressure of plasma.

We consider that fibers with air holes in cross section are effective as a fiber fuse-tolerant fiber for future systems where the input power tends to increase with an increase in the multiplexing number per fiber when SDM technology is used.

7.2 Safety of Optical Communication Systems (From the Viewpoint of IEC Laser Safety Standardization)

7.2.1 Standardization of Laser Safety

The importance of optical safety standardization has been increasing with the recent trend toward higher power for higher speed and capacity and longer transmission distance in optical fiber communication systems. Standardization related to optical communication system safety is being discussed in the international standardization organizations IEC and ITU–T (Fig. 7.50). That process is being led by IEC TC 76 (optical radiation safety and laser equipment), and IEC 60825-1 (safety of laser products—Part 1: equipment classification and requirements) specifies safety standards for laser products [60]. Safety standards for optical fiber communication systems are also specified in IEC 60825–2 (safety of laser products—Part 2: safety of optical fiber communication systems) [61].

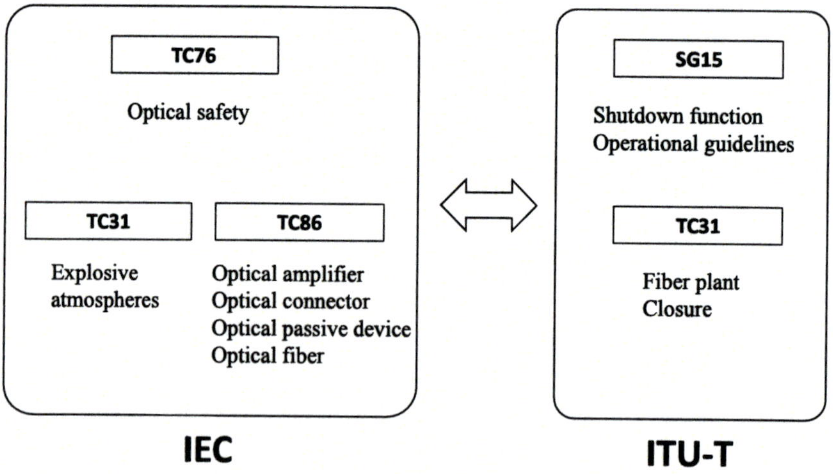

Fig. 7.50 Organizations involved in international standards concerning laser safety

7.2.2 IEC Standards

IEC60825-1 specifies the basic items listed below for safety standards related to laser equipment.

(1) Maximum permissible exposure of light radiation (MPE)
(2) Extended source viewing and intrabeam viewing
(3) Classification and AEL
(4) Time base
(5) Handling repeatability pulsed and modulated lasers
(6) Handling multi-wavelength light sources
(7) Measurement conditions for class 1M and class 2M.

In particular, the new classes 1M and 2M were added in 2001. These new classes are safe for inspection with the naked eye, although damage may result from viewing through a lens, and the permissible power restrictions for laser optics that have a wide emission angle (fiber, laser diodes, etc.) are relaxed.

In 1993, IEC established the IEC60825-2 specifications for the safety of optical fiber communication systems. Those specification documents introduced the concepts of location classes and hazard levels and relaxed the in-fiber power restrictions with an automatic power reduction mechanism.

7.2.3 Safety Specifications for Optical Fiber Communication Systems

Light is not emitted externally in the normal operation of optical fiber communication systems, so that they are recognized as Class 1, and there is no danger. However, risk may occur during maintenance and inspection or when fibers are cut. The concept of hazard level (HL) has been introduced to safety specifications for optical fiber communication systems to indicate the potential latent risk involved in maintenance, inspection, and fiber cutting. The seven HL classes correspond to the AEL values of each class of the new specifications (1, 1M, 2, 2M, 3R, 3B, and 4). Examples of the power limits for various hazard levels are listed in Table 7.4. In a communication system in which a 1550 nm optical signal is transmitted in a standard single-mode fiber (SMF), for example, the power limit for hazard level 1M is 136 mW. However, it must be noted that the permissible HL when an optical connector is disconnected is lower than that when a fiber is broken.

Considering that the knowledge and awareness of safe laser handling vary among the persons that access these systems, three types of locations are established (unrestricted location, restricted location, and controlled location). An unrestricted location is a location that is open to the public. A restricted location is a facility in a factory or commercial location that is not open to the public. A controlled location is a location where an engineering or administrative control is present to make it

Table 7.4 Examples of optical power limit for various hazard levels

Wavelength	Hazard level			
	1	1M	3R	3B
1420 nm	10 mW (+ 10 dBm)	115 mW (+ 20.6 dBm)	([a])	500 mW (+ 27 dBm)
1550 nm	10 mW (+ 10 dBm)	136 mW (+ 21.3 dBm)	([a])	500 mW (+ 27 dBm)

[a]If the applicable limit of hazard level 1M is larger than the limit of 3R and less than the limit of 3B, hazard level 1M is allocated

inaccessible, except to authorized personnel with appropriate laser safety training. The permissible HL for optical fiber communication systems depends on the type of location.

The condition for setting the HL is not the power within the fiber, but whether the time of exposure after a fiber is broken or a connector is disconnected is one second or 3 seconds or more (the time is different for different locations). Accordingly, the operating power within the fiber can be increased by using a mechanism that automatically reduces the optical power on the transmitter (APR) when a break in the optical signal is detected. If the power within the fiber can be reduced by APR to the class 1 AEL or less within one or three seconds, the hazard level will be 1 regardless of the optical power within the fiber. However, the power within the optical fiber is not unlimited, because the cumulative exposure up to completion of the APR operation must be no more than the MPE.

The power within the fiber P (W) for which the cumulative exposure during operation of the APR is the maximum permissible exposure [MPE (J m^{-2})] is given by the following equation.

$$P = \frac{\pi d^2 \mathrm{MPE}}{4t} \frac{1}{1 - \exp\left[-0.125\left(\frac{\pi \omega_0 d}{\lambda\,\mathrm{NOHD}}\right)^2\right]} \tag{7.1}$$

In the above equation, ω_0 is the mode field diameter (m), d is the limiting aperture diameter (m), t is the power shutdown time (s) for the end of APR operation, $NOHD$ is the hazard distance (m), and λ is the wavelength (m). Examples of the optical power within the fiber that produce the MPE are listed in Table 7.5, where the values are for SMF. Each location has different values for the maximum permissible time of power reduction and hazard distance (NOHD: Nominal Ocular Hazard Distance). For example, in a controlled location, an optical power within the fiber of up to 2.6 W is permitted when a 1550 nm optical signal is completely attenuated within 3 seconds by APR.

The values in Tables 7.4 and 7.5 are from IEC 60825–2 edition 3.2. IEC/TC76/WG5 has been periodically discussing revisions to IEC 60825–2, so the power limits for each hazard level may change. Refer to the latest specifications for the current values.

Table 7.5 Examples of the maximum output power during the shutdown time for optical fiber communication systems having APR

	Location type		
Wavelength (nm)	Unrestricted shutdown time: 1 s NOHD 10 cm (mW)	Restricted shutdown time: 3 s NOHD 10 cm (mW)	Controlled shutdown time: 3 s NOHD 25 cm (mW)
1480	650	288	1774
1550	1273	428	2640

7.2.4 Measures for Implementing Higher Optical Power

Expansion of the emission angle from the fiber is one way to maintain safety in systems that use higher optical power. Expansion of the emission angle decreases the optical power which enters the eye, so the restriction of optical power can be relaxed for classes 1M and 2M (Fig. 7.51). Accordingly, the use of optical fiber that has a wide emission angle is effective for safety in a high-power optical fiber communication system. However, while there is safety for the naked eye, there is a possibility for damage when the light is viewed through a lens. Also, there is no less risk of damage to skin.

Recently, an optical connector that has a lower-power density achieved by broadening the beam diameter to prevent damage to the connector end surface by high-power light from the fiber has been developed. However, the emission angle from the fiber with wide beam diameter is generally small, so the optical power conditions for classes 1M and 2M are severe and safety may be a problem. Higher optical power in future systems requires research and development on fibers and connectors that satisfy both safety for the human body and prevention of damage to the optical fibers.

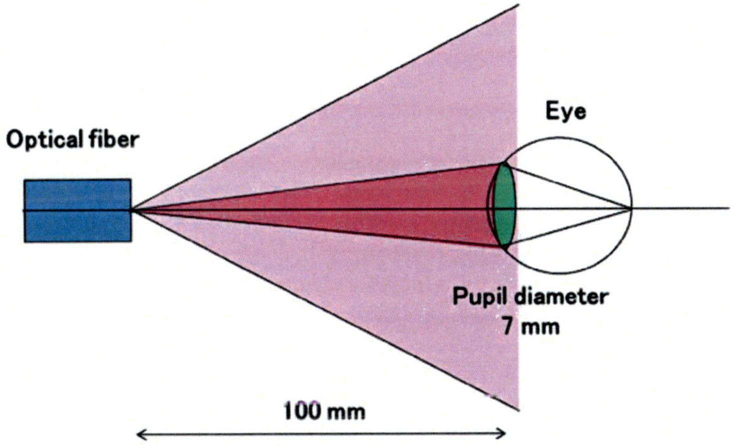

Fig. 7.51 Measurement setup in class 1M, 2M

Another measure that is being considered for higher power in optical communication systems is shortening of the shutdown time of the automatic power reduction mechanism (APR) to increase the permissible fiber power. The maximum optical power from the fiber for the specified times of exposure ($t = 1$ s or 3 s) for evaluating exposure level when a fiber is broken or a connector is disconnected is shown in Table 7.5 for each location type, but as we see from (7.1), the input power can be increased by shortening the power shutdown time. If a power shutdown time of 1 s or less is possible, the permissible power within the fiber is 1 W or more. However, the prevention of fire hazard from damage or ignition of optical fibers or optical components is important as well as restrictions on direct exposure of the human body for higher power systems with watt-class optical powers.

7.2.5 Conclusion

The recent trend toward higher power in optical fiber communication systems to achieve higher speed and capacity and longer transmission distance has been accompanied by increasing importance of standardization related to laser safety, and specifications for the safety of laser equipment and optical fiber communication systems are being set and revised by IEC. Those specifications add the new classes 1M and 2M, introduce the concepts of location types and hazard levels, and include the relaxation of optical power restrictions by using an automatic power reduction mechanism to lower safety risk. For the future, measures such as expansion of the emission angle and reduction of the automatic power shutdown time are being considered to allow relaxation of restrictions on the optical power within the fiber to allow further increases in optical power.

For future high-power optical fiber communication systems, it is important to fully maintain safety by investigating materials, devices, equipment, and systems in order to realize both protecting people from optical radiation and preventing damage to fibers and optical components.

References

1. R. Kashyap, K.J. Blow, Observation of catastrophic self-propelled self-focusing in optical fibres. Electron. Lett. **24**(1), 47–49 (1988)
2. D.P. Hand, P.S.J. Russell, Solitary thermal shock waves and optical damage in optical fibers, the fiber fuse. Opt. Lett. **13**(9), 767–769 (1988)
3. S. Todoroki, Quantitative evaluation of fiber fuse initiation probability in typical single-mode fibers, in *Optical Fiber Communication Conference*. Optical Society of America, W2A.33 (2015)
4. S. Todoroki, *Fiber fuse—light-induced continuous breakdown of silica glass optical fiber* (NIMS Monographs, Springer Japan, Tokyo, 2014)
5. K.S. Abedin, M. Nakazawa, T. Miyazaki, Backreflected radiation due to a propagating fiber fuse. Opt. Express **17**(8), 6525–6531 (2009)

6. E.M. Dianov, I.A. Bufetov, A.A. Frolov, V.G. Plotnichenko, V.M. Mashinskii, M.F. Churbanov, G.E. Snopatin, Catastrophic destruction of optical fibres of various composition caused by laser radiation. Quantum Electron. **32**(6), 476–478 (2002)
7. A.M. Rocha, F. Domingues, M. Facão, P.S. André, Threshold power of fiber fuse effect for different types of optical fiber, in *The 13th International Conference on Transparent Optical Networks (ICTON 2011)*, Stockholm, Sweden, pp. 1457–1549, Tu.P.13 (2011)
8. S. Todoroki, Threshold power reduction of fiber fuse propagation through a white tight buffered single-mode optical fiber. IEICE Electron. Express **8**(23), 1978–1982 (2011)
9. R. Kashyap, Self-propelled self-focusing damage in optical fibres, Lasers'87, in *Proceedings of the 10th International Conference on Lasers and Applications*, ed. by F.J. Duarte (STS Press, McLean, VA, 1988; Lake Tahoe, Nevada, USA, Dec 7–11, 1987), pp. 859–866
10. D.P. Hand, J.E. Townsend, P.S.J. Russell, Optical damage in fibres, the fibre fuse, in *Digest of Conference on Lasers and Electro-Optics*. Anaheim, US, Paper WJ1 (1988)
11. D.A. Dvoretskiy, V.F. Hopin, A.N. Gur'yanov, L.K. Denisov, L.D. Ishakova, I.A. Bufetov, Optical losses in silica based fibers within the temperature range from 300 to 1500 K, science and education. Electron. Sci.-Tech. J. **5** (2013) (in Russian)
12. H.R. Philipp, Silicon dioxide (SiO_2) (glass), in *Handbook of Optical Constants of Solids*, ed. by E.D. Palik (Academic Press, New York, 1985), pp. 749–763
13. H.R. Philipp, Optical properties of non-crystalline Si, SiO, SiO_x and SiO_2. J. Phys. Chem. Solids **32**(8), 1935–1945 (1971)
14. T. Izawa, S. Sudo, *Optical Fibers: Materials and Fabrication* (KTK Scientific Publishers, Tokyo, 1987)
15. H. Kanamori, H. Yokota, G. Tanaka, M. Watanabe, Y. Ishiguro, I. Yoshida, T. Kakii, S. Itoh, Y. Asano, S. Tanaka, Transmission characteristics and reliability of pure-silica-core single-mode fibers. J. Lightwave Technol. **4**(8), 1144–1150 (1986)
16. N. Akhmediev, P.S.J. Russell, M. Taki, J.M. Soto-Crespo, Heat dissipative solitons in optical fibers. Phys. Lett. A **372**(9), 1531–1534 (2008)
17. E.M. Dianov, V.E. Fortov, I.A. Bufetov, V.P. Efremov, A.E. Rakitin, M.A. Melkumov, M.I. Kulish, A.A. Frolov, High-speed photography, spectra, and temperature of optical discharge in silica-based fibers. IEEE Photon. Technol. Lett. **18**(6), 752–754 (2006)
18. S. Todoroki, Quantitative evaluation of fiber fuse initiation with exposure to arc discharge provided by a fusion splicer. Sci. Rep. **6**, 25366 (2016)
19. S. Todoroki, Fiber fuse phenomenon. J. Plasma Fusion Res. **10**, 505–508 (2018). ((in Japanese))
20. N. Nishimura, K. Seo, M. Shiino, R. Yuguchi, Study of high-power endurance characteristics in optical fiber link, in *Technical Digest of Optical Amplifiers and Their Applications* (TuC4), pp. 193–195 (2003)
21. K. Takenaga, S. Omori, R. Goto, S. Tanigawa, S. Matsuo, K. Himeno, Evaluation of high-power endurance of bend-insensitive fibers, in *Proceedings of Optical Fiber Communication/National Fiber Optic Engineers Conference (JWA11)* (2008)
22. H. Takara, H. Masuda, H. Kanbara, Y. Abe, Y. Miyamoto, R. Nagase, T. Morioka, S. Matsuoka, M. Shimizu, K. Hagimoto, Evaluation of fiber fuse characteristics of hole-assisted fiber for high power optical transmission systems, in *Proceedings of the 35th European Conference on Optical Communication (P1.12)* (2009)
23. S. Todoroki, Fiber fuse propagation modes in typical single-mode fibers, in *Optical Fiber Communication Conference* (OSA Technical Digest Optical Society of America, 2013). Paper JW2A.11
24. S. Todoroki, Origin of periodic void formation during fiber fuse. Opt. Express **13**(17), 6381–6389 (2005)
25. I.A. Bufetov, A.A. Frolov, A.V. Shubin, M.E. Likhachev, S.V. Lavrishchev, E.M. Dianov, Propagation of an optical discharge through optical fibres upon interference of modes. Quantum Electron. **38**(5), 441–444 (2008)
26. D.D. Davis, S.C. Mettler, D.J. DiGiovani, Experimental data on the fiber fuse, 27th annual boulder damage symposium: laser-induced damage in optical materials, in *SPIE Proceedings*, eds. by H.E. Bennett, A.H. Guenther, M.R. Kozlowski, B.E. Newnam, M.J. Soileau, vol. 2714 (Boulder, CO, USA, 30 Oct. 1995), pp. 202–210

27. H. Zhang, P. Zhou, X. Wang, H. Xiao, X. Xu, Fiber fuse effect in high-power double clad fiber laser, in *Conference on Lasers and Electro-Optics Pacific Rim (CLEO-PR)* (Paper WPD-4) (2013)
28. Y. Emori et al., Less than 4.7 dB noise figure broadband in-line EDFA with a Raman amplifier-1300 ps/nm DCF pumped by multi-channel WDM laser diodes, in Technical Digest OAA 98, Vail CO, July, 1998, paper PD3
29. R. Kashyap, Self-propelled self-focusing damage in optical fibers. Electron. Lett. **24**(1) (1988)
30. D.P. Hand, P.S.J. Russell, Solitary thermal shockwaves and optical damage in optical fibers: the fiber fuse. Opt. Lett. **13**, 767 (1988)
31. R.H. Stolen, *Optical Fibre Communications*, in eds. by S.E. Miller, A.G. Chynoweth (Academic Press, 1979)
32. K. Seo, N. Nishimura, M. Shiino, R. Yuguchi, H. Sasaki, Evaluation of high-power endurance in optical fiber links. Furukawa Rev. **24**, 17–22 (2003)
33. K. Mukasa et al., New type of dispersion management transmission line for long-haul high-capacity transmission SubOptic'01 proceeding, T.4.2.4 (2001)
34. R.M. Atkins, P.G. Simpkins, A.D. Yablon, Track of a fiber fuse: a Rayleigh instability in optical waveguides. Opt. Lett. **28**, 974–976 (2003)
35. K.S. Abedin, T. Morioka, Remote detection of fiber fuse propagation in optical fibers, in *Optical Fiber Communication Conference, OThD5* (2009)
36. K.S. Abedin, M. Nakazawa, T. Miyazaki, Backreflected radiation due to a propagating fiber fuse. Opt. Exp. **17**, 6525–6531 (2009)
37. K.S. Abedin, T. Miyazaki, M. Nakazawa, Measurement of spectral broadening and Doppler shift on backreflections from a fiber fuse using heterodyne detection. Opt. Lett. **34**, 3157–3159 (2009)
38. K.S. Abedin, M. Nakazawa, Real time monitoring of a fiber fuse using an optical time-domain reflectometer. Opt. Express **18**, 21315–21321 (2010); T. Kinoshita, N. Sato, M. Yamada, Detection and termination system for optical fiber fuse, in *OptoElectronics and Communications Conference, WS4-6* (2013)
39. T. Kinoshita, N. Sato, M. Yamada, Detection and termination system for optical fiber fuse, in *OptoElectronics and Communications Conference, WS4-6* (2013)
40. M. Yamada, O. Koyama, Radiation characteristics of peculiar luminescence which observed with optical fiber fuse phenomenon, in *The 7th International Symposium on Ultrafast Photonic Technologies (ISUPT2015) and International Symposium on Extremely Advanced Transmission Technology (EXAT 2015)*, p. 7 (2015)
41. D.P. Hand, T.A. Birks, Single-mode tapers as 'fiber fuse' damage circuit-breakers. Electron. Lett. **25**(1), 33–34 (1989)
42. A.M. Rocha, G. Fernandes, F. Domingues, M. Niehus, A.N. Pinto, M. Facao, P.S. Andre, Halting the fuse discharge propagation using optical fiber microwires. Opt. Express **20**(19), 21083–21088 (2012)
43. S. Yanagi, S. Asakawa, M. Kobayashi, Y. Shuto, R. Nagase, Fiber fuse terminator, in *CLEO-PR 2003*, vol. 1, p. 386 (2003)
44. K. Kurokawa, N. Hanzawa, Suppression of fiber fuse propagation and its break in compact fiber fuse terminator, OECC/PS 2013, WS4-5 (2013)
45. W. Ha, Y. Jeong, K. Oh, Fiber fuse effect in hollow optical fibers. Opt. Lett. **36**(9), 1536–1538 (2011)
46. E.M. Dianov, I.A. Bufetov, A.A. Frolov, Destruction of silica cladding by the fuse effect. Opt. Lett. **29**(16), 1852–1854 (2004)
47. D.P. Hand, P.S.J. Russell, Solitary thermal shock waves and optical damage in optical fibers: the fiber fuse. Opt. Lett. **13**(9), 767–769 (1988)
48. K. Kurokawa, N. Hanzawa, K. Tsujikawa, S. Tomita, Hole-size dependence of fiber fuse propagation in hole-assisted fiber (HAF), in *Proceedings of the 17th Mirooptics Conference (H-30)* (2011)
49. N. Hanzawa, K. Kurokawa, K. Tsujikawa, T. Matsui, K. Nakajima, S. Tomita, M. Tsubokawa, Suppression of fiber fuse propagation in hole assisted fiber and photonic crystal fiber. J. Lightwave Technol. **28**(15), 2115–2120 (2010)

50. K. Takenaga, S. Tanigawa, S. Matsuo, M. Fujimaki, H. Tsuchiya, Fiber fuse phenomenon in hole-assisted fibers, in ECOC 2008, P.1.14 (2008)
51. M. Yamada, O. Koyama, Y. Katsuyama, T. Shibuya, Heating and burning of optical fiber by light scattered from bubble train formed by optical fiber fuse, in OFC 2011, JThA1 (2011)
52. ITU-T Recommendation G.664, Edition 4.0 (2012)
53. IEC Technical Report IEC 61292-4, Edition 2.0 (2010)
54. M. Yamada, O. Koyama Y. Katuyama, T. Shibuya, Heating and burning of optical fiber by light scattered from bubble train formed by optical fiber fuse, in *Proceedings of OFC/NFOFC2011, JThA* (2011)
55. M. Yamada, A. Tomoe, T. Kinoshita, O. Koyama, Y. Katuyama, T. Shibuya, Heating and burning of optical fibers and cables by light scattered from bubble train formed by optical fiber fuse. IEICE Trans. Commun. **E98-B**(9), 2638–2641 (2012)
56. M. Yamada, A. Tomoe, H. Takara, Light scattering characteristics of a hole formed by a fiber fuse. Electron. Lett. **48**(9), 519–520 (2012)
57. E.M. Dianov, I.A. Bufetov, A.A. Frolov, Y.K. Chamorovsky, G.A. Ivanov, I.L. Vorobjev, Fiber fuse effect in microstructured fibers. IEEE Photon. Technol. Lett. **16**(1), 180–181 (2004)
58. K. Takenaga, S. Tanigawa, S. Matsuo, M. Fujimaki, H. Tsuchiya, Fiber fuse phenomenon in hole-assisted fibers, in *Proceedings of the 34th European Conference on Optical Communication (P.1.14)*, pp. 27–28 (2008)
59. W. Ha, Y. Jeong, K. Oh, Fiber fuse in hollow optical fibers. Opt. Lett. **36**(9), 1536–1538 (2011)
60. IEC60825-1 Safety of laser products—part 1: equipment classification and requirements
61. IEC60825-2 safety of laser products—part 2: safety of optical fibre communication systems

Chapter 8
Japanese National Projects on SDM Technologies

Yoshinari Awaji, Yutaka Miyamoto, Tetsuya Miyazaki, Toshio Morioka, Itsuro Morita, Kazuhide Nakajima, Hidehiko Takara, and Takehiro Tsuritani

Abstract This chapter briefly describes pioneering Japanese national projects on SDM technologies, namely i-FREE (Innovative Optical Fiber Technologies: 2010–2012), i-ACTION (Innovative Optical Communication Infrastructure: 2011–2015), and i-FREE^2 (Innovative Optical Fiber and Communication Technology for Exa-bit Era with SDM: 2013–2018), which led to setting a series of transmission capacity records per fiber of 1 Pbit/s (2012), 2 Pbit/s (2015), and 10 Pbit/s (2017) as well as capacity-distance product record of 1 Ebit/s·km (2013).

8.1 Introduction

One of the important achievements by the EXAT initiative is that it has led to the creation of a series of Japanese national SDM projects as early as 2010 to further develop the ideas proposed by EXAT. Three major projects were created, which are

Yoshinari Awaji is a chapter editor.

Y. Awaji (✉) · T. Miyazaki
National Institute of Information and Communications Technology, Koganei, Tokyo, Japan
e-mail: yossy@nict.go.jp

Y. Miyamoto
NTT Network Innovation Laboratories, NTT Corporation, Yokosuka, Kanagawa, Japan

T. Morioka
Technical University of Denmark, Kongens Lyngby, Denmark

I. Morita · T. Tsuritani
KDDI Research Inc., Fujimino, Saitama, Japan

I. Morita
Waseda University, Shinjuku, Tokyo, Japan

K. Nakajima
NTT Access Network Service Systems Laboratories, NTT Corporation, Tsukuba, Ibaraki, Japan

H. Takara
National Institute of Technology, Okinawa College, Nago, Okinawa, Japan

M. Nakazawa et al. (eds.), *Space-Division Multiplexing in Optical Communication Systems*, Springer Series in Optical Sciences 236,
https://doi.org/10.1007/978-3-030-87619-7_8

i-FREE (Innovative Optical Fiber Technologies: 2010–2012), i-ACTION (Innovative Optical Communication Infrastructure: 2011–2015), and i-FREE^2 (Innovative Optical Fiber and Communication Technology for Exa-bit Era with SDM: 2013–2018). In 2016, another new project "R&D of Space-Division Multiplexing Photonic Node" (2016–2020) started. Furthermore, an EU-Japan coordinated R&D project on "Scalable And Flexible optical Architecture for Reconfigurable Infrastructure (SAFARI)" (2013–2017) has been created linking the Japanese EXAT community and European partners, commissioned by the Ministry of Internal Affairs and Communications (MIC) of Japan and EC Horizon 2020. The Japanese national projects have been leading the most advanced research field on SDM technologies, setting various world records such as 1-Pbit/s transmission (2012) [24], 1-Ebit/s·km (2013) [25, 28, 30], 2-Pbit/s (2015) [49, 50], and 10-Pbit/s transmission (2017) [51].

8.2 i-FREE

During the first national project "i-FREE" (Innovative Optical Fiber Research for Exa-bit Era), we mainly focused on the research on multi-core fibers (MCFs). In this sub-chapter, we would like to describe major results with respect to design, fabrication, and characterization of MCFs and their interoperability trial.

8.2.1 *Various MCFs*

Japanese five fiber manufacturers, Fujikura, Furukawa Electric, Hitachi Metals (former Hitachi Cable), Mitsubishi Cable Industries, and Sumitomo Electric Industries, participated in the "i-FREE" project and developed various MCFs. Figure 8.1 shows example MCFs fabricated in the project period. Cross-sectional photos are

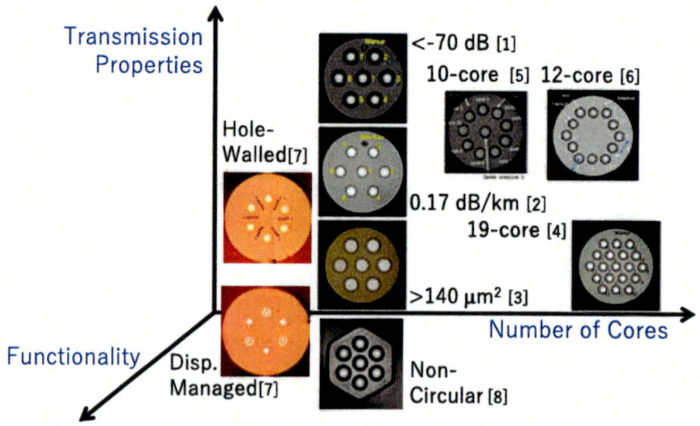

Fig. 8.1 Various MCFs developed under the "i-FREE" project. Cross-sectional photos are mapped as a function of number of cores, transmission properties, and their functionality

mapped as a function of the number of cores, transmission properties, and functionality. In the earlier stage of "i-FREE," MCFs with hexagonal core arrangement were mainly discussed with viewpoints of core number and transmission properties. A seven-core MCF is the most basic structure, and Hayashi et al. reported an MCF with extremely low cross talk of less than -70 dB [1] and with sufficiently low attenuation of 0.17 dB/km [2]. Imamura et al. realized an MCF with an enlarged effective area A_{eff} of more than 140 μm^2 using a hexagonal seven-core arrangement [3]. Sakaguchi et al. realized a 19-core MCF, which was the largest core number obtained in the "i-FREE" project and reported its applicability to SDM-WDM-PDM (space/wavelength/polarization division multiplexing) transmission [4]. Matsuo et al. proposed MCFs with a circular core arrangement in order to optimize the spatial density (number of cores) and cross talk between neighboring cores (transmission properties) simultaneously [5, 6]. In the middle and later periods of "i-FREE," MCFs with additional functions were investigated. Yao et al. reported a hole-walled MCF and clarified its impact on the cross talk reduction [7]. He also proposed that MCF composed of two different cores and revealed that it can be used for realizing a dispersion managed transmission line by concatenating it so that the different cores are spliced [7]. Tanaka et al. realized an MCF with non-circular cladding in order to support angular alignment in splicing process [8].

We have made much progress in developing design principles of MCFs. Koshiba et al. clarified the applicability of both coupled-mode and coupled-power theorems to cross talk in MCFs [9] and succeeded in deriving its analytical expression [10]. It should be noted that the development of various MCFs under the "i-FREE" project was effectively supported by these progresses in the design principles.

8.2.2 *Characterization of MCF*

Fiber characterization and input/output connection technologies are the most important subjects for developing MCFs. Figure 8.2 shows schematics of the representative characterization techniques developed in the "i-FREE" project. Nakazawa et al. realized the cross talk measurement by using a multi-channel optical time domain reflectometry (OTDR) technique and revealed that the proposed technique can be used to evaluate the longitudinal distribution of coupling coefficients of MCFs [11]. Ohashi et al. revealed that the longest cutoff wavelength of MCFs can be evaluated by utilizing a simultaneous launching and detection scheme [12].

Kokubun et al. investigated the applicability of integrated waveguides as a fan-in/fan-out (FI/FO) device for MCFs. They designed and fabricated a laminated polymer waveguide fan-in/fan-out for a 19-core MCF as shown in Fig. 8.3 [13]. The result showed the feasibility of the proposed technique and applicability to the MCF with various core arrangements.

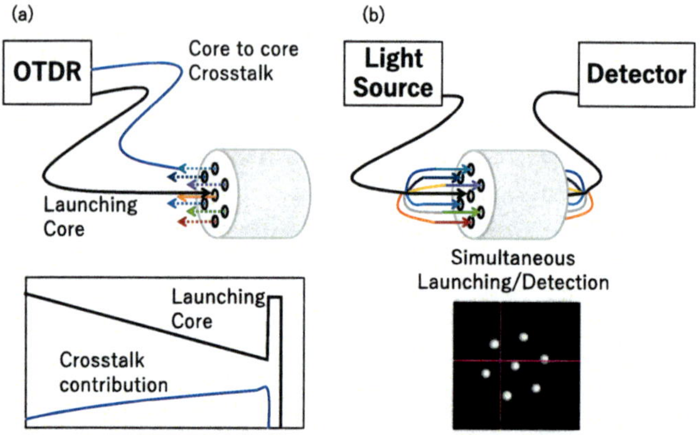

Fig. 8.2 Schematics of the characterization techniques for MCFs developed under the "i-FREE" project. **a** multi-channel optical time domain reflectometry (OTDR) techniques for measuring the longitudinal cross talk characteristics. **b** Cutoff wavelength measurement technique using a simultaneous launching/detection scheme

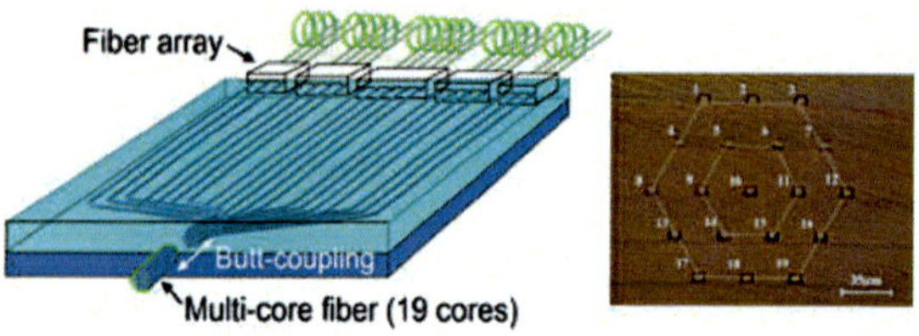

Fig. 8.3 Schematic view of a laminated polymer waveguide fan-in/fan-out device designed for a 19-core MCF. The photo on the right shows the cross section of the waveguide (Reprinted with permission from [13]. ©2015 IET)

It can be said that collaborative research on design, fabrication, and characterization techniques effectively accelerated the development of MCF technologies during the "i-FREE" period.

8.2.3 Interoperability Trial

The first trial of multi-vendor interoperability using MCF technologies was demonstrated at the exhibition of CLEO-PR and OECC/PS 2013 held in Kyoto, Japan, as the collaborative work of "i-FREE" and "i-ACTION" (See clause 8.2 for more details in the section of "i-ACTION"). Figure 8.4 shows a schematic diagram of the multi-vendor interoperability trial. The inset photo shows the overview of the exhibition. In this trial, ten MCFs were fabricated by Fujikura, Furukawa Electric, Hitachi Metals, Mitsubishi Cable Industries, and Sumitomo Electric. The MCFs

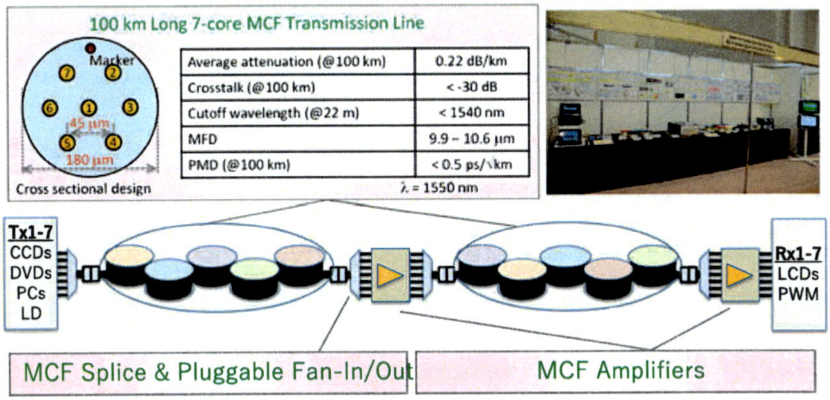

Fig. 8.4 Schematic diagram of multi-vendor interoperability trial using MCF technologies developed under the "i-FREE" and "i-ACTION" projects. An inset photo shows the overview of the exhibition

were designed to have hexagonally close-packed seven cores, and we set a target core pitch and cladding diameter values of 45 μm and 180 μm, respectively. The MCF samples were fusion spliced, and two sets of 50-km-long MCF transmission lines were constructed. The properties of the transmission lines are summarized in the inset table. The average attenuation at 1550 nm was 0.22 dB/km including fusion splice losses. The cross talk at 1550 nm and 100 km was controlled to less than -30 dB. Then, a 100-km-long transmission system was composed of using the MCF transmission lines, fan-in/fan-out devices (provided by Furukawa Electric, Mitsubishi Cable Industries, Optoquest, and Sumitomo Electric), and core-pumped MCF amplifiers (provided by Furukawa Electric, and NTT). In this trial, each core of the MCFs was simply used as SDM resource, and different contents, such as charge-coupled device (CCD), digital versatile disc (DVD), laptop screen, and so on, were transmitted simultaneously. Our first trial proved that the MCF-based SDM transmission can be a promising candidate for future ultra-large capacity transmission systems.

8.3 i-Action

The second national project "i-Action" (Research and development of innovative infrastructure for optical communications) is a five-year project from FY2011 to FY2015. In this project, eighteen research organizations both from industries and universities in Japan have been involved, and technologies on optical amplifier and connection with MCFs have been developed. System evaluations by combining them have also been conducted as shown in Fig. 8.5.

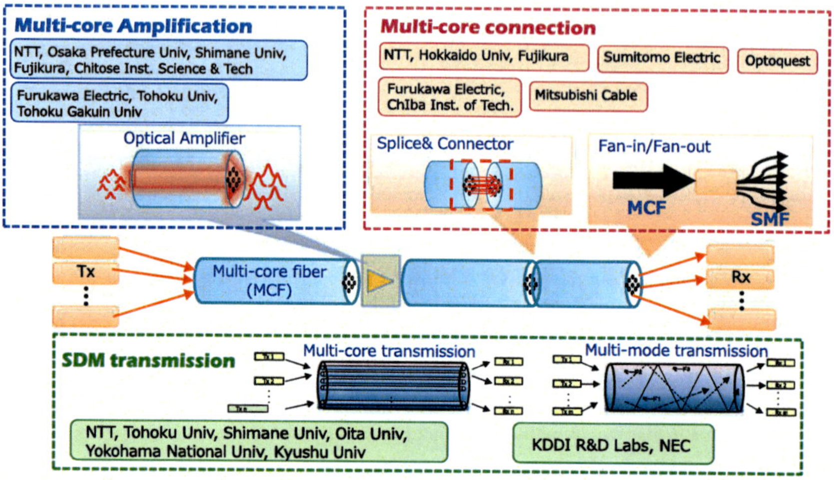

Fig. 8.5 i-Action project

8.3.1 Research on Optical Amplifier

Optical amplifiers are indispensable for long-distance transmission systems. For SDM transmission with MCFs, EDFAs consisting of Er-doped MCFs (MC-EDFAs) have been fabricated [14–16] in "i-Action". There are two schemes for realizing MC-EDFAs as shown in Fig. 8.6. One is a core-pumped scheme in which single-mode pump laser diodes (LDs) are used to pump each of the cores [14]. In this scheme, the required number of pump LDs linearly increases with the number of core. To simplify the configuration of MC-EDFAs, a cladding pumping scheme which uses a single high-power LD as a pump source and does not require any fan-in and fan-out devices between MCFs and MC-EDFs has also been proposed [15, 16]. With the cladding pump scheme using a single multi-mode LD for pumping, a reduction of power consumption is also expected.

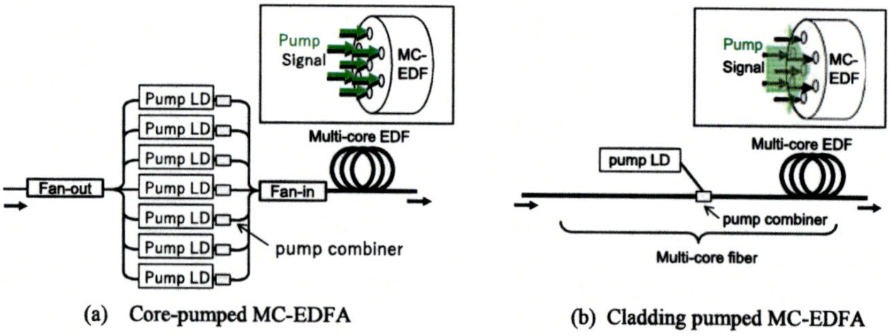

Fig. 8.6 Configurations of MC-EDFAs

Distributed Raman amplifiers (DRAs) with MCFs have also been studied [17], since DRA has an advantage in terms of its low noise characteristics. In addition, remote optically pumped amplifier (ROPA) for unrepeatered transmission systems with MCFs has been developed, and 120.7 Tb/s (7 cores × 180 × 95.8 Gb/s) unrepeatered transmission over 204 km was demonstrated [18].

8.3.2 Research on Connection

In MCF systems, simple coupling devices between MCFs and fan-in and fan-out (FI/FO) devices between MCFs and single-mode fibers (SMFs) shown in Fig. 8.7 are indispensable. A fusion splicing technology is also important. For MCF connection, both alignment and axial rotation adjustment are required. In the "i-Action" project, various types of connection devices have been studied to reduce the connection loss. For FI/FO, a fiber bundle-type [19, 20] and lens coupling-type [21, 22] devices have been developed as shown in Fig. 8.8. In both types of FIFO devices, the connection loss of less than 0.5 dB was achieved for an MCF with seven cores.

MCFs with a non-circular-shaped cladding have also been studied for easier axial rotation adjustment as shown in Fig. 8.9. An MCF connector with a hexagonal-shaped cladding was assembled, and a connection loss around 0.5 dB was achieved [23].

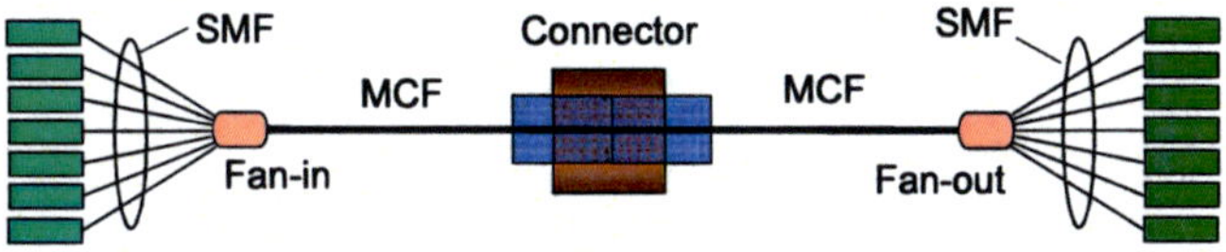

Fig. 8.7 MCF connection

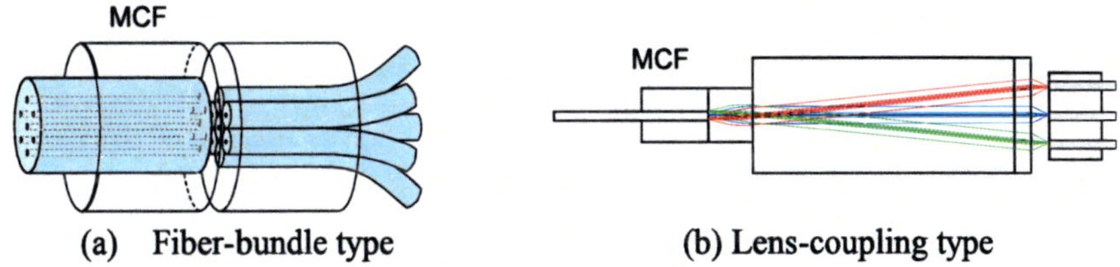

Fig. 8.8 Fan-in/fan-out devices for MCF

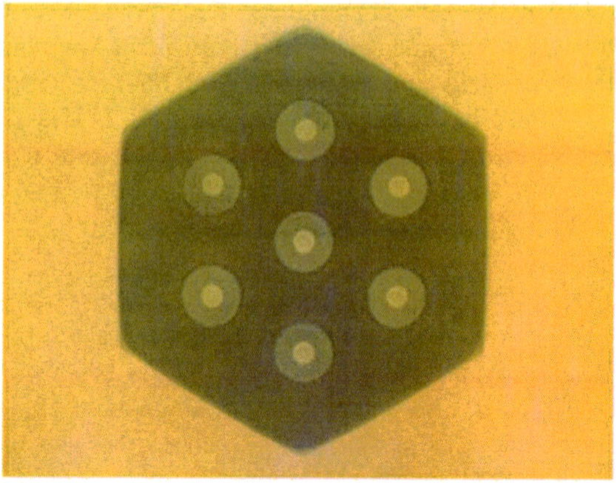

Fig. 8.9 MCF with a hexagonal-shaped cladding

8.3.3 Transmission System Technology

By using the abovementioned key components, high-capacity transmission experiments with MCFs were conducted as shown in Fig. 8.10. In this figure, the results achieved in "i-Action" are shown with filled symbols.

The capacity of 1 Pbit/s was achieved in 12-core MCF transmission over 52 km with an aggregated spectral efficiency of 91.4 bit/s/Hz per fiber [24]. In long-haul transmission experiments, a high-capacity–distance product (CDP) was achieved with uncoupled MCFs. With a 12-core MCF, 688 Tb/s (12 cores × 748 λ × 77 Gb/s) transmission over 1500 km was achieved with a propagation-direction interleaved configuration to suppress the cross talk between adjacent cores [25]. In this transmission experiment, the CDP of 1.03 Ebit/s·km was obtained. Transoceanic class transmission was also achieved by employing a seven-core MCF with suppressed

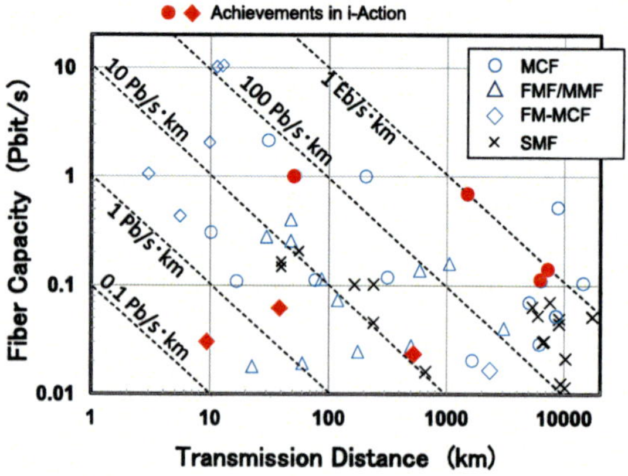

Fig. 8.10 Major high-capacity transmission experiments

inter-core cross talk [26–28]. The first demonstration of transoceanic class transmission with MCFs was 28.8-Tbit/s (7 cores × 40 λ × 103 Gb/s) transmission over 6160 km [26]. Then the fiber capacity and transmission distance were improved by introducing a more spectral efficient modulation format and expanding the gain bandwidth of MC-EDFA in the seven cores × 201 λ × 100 Gb/s transmission experiment over 7326 km [28]. The obtained CDP in this experiment has also exceeded 1 Ebit/s·km.

8.4 i-FREE²

The latest national project of "i-FREE²" (Innovative optical fiber and communication technology for exa-bit era with SDM) was launched for developing the results obtained with "i-FREE." In this project, two groups have been conducting intensive research to realize a 100 Pbit/s·km class transmission using the SDM technologies. In this section, we would like to introduce the latest research on MCF and FM-MCF technologies.

8.4.1 MCFs for 100 Pbit/s·km Transmission

Figure 8.11a shows the transmission capacity per fiber as a function of transmission distance in recent transmission experiments using the SDM technologies. Red triangles, blue circles, red dots, and blue diamonds indicate experimental results for MCFs, few-mode fibers (FMFs), few-mode multi-core fiber (FM-MCFs), and coupled-core

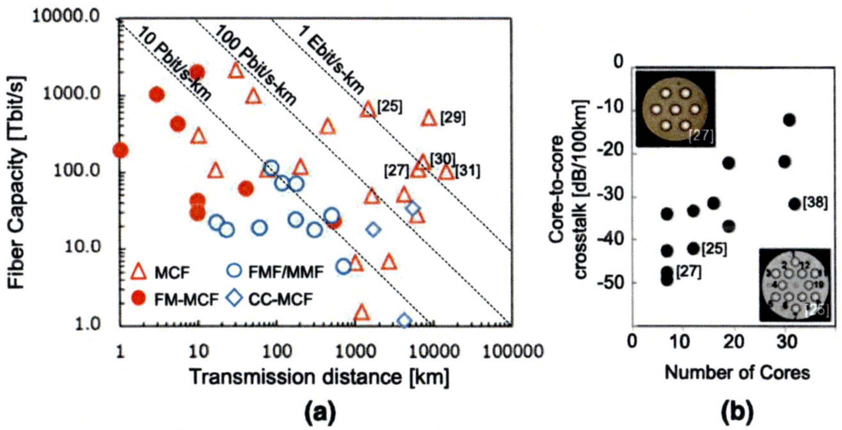

Fig. 8.11 **a** Transmission capacity per fiber using SDM technologies in recent experiments as a function of transmission distance. **b** Core-to-core cross talk as a function of number of cores in fabricated MCFs. Inset: cross sections of trench-assisted seven-core fiber [27] and double-ring structure 12-core fiber [25]

multi-core fibers (CC-MCFs), respectively. From this figure, SDM technology for potentially achieving 100 Pbit/s·km class transmissions with over 100Tbit/s was only MCF transmission. In particular, only seven-core [30] and 12-core [25, 29, 31] MCFs were possible to achieve 1-Ebit/s·km transmission. Figure 8.11b shows core-to-core cross talk as a function of number of cores in fabricated MCFs [24, 25, 32–38]. This is because the core-to-core cross talk of less than − 40 dB/100 km is required to achieve 1Ebit/s·km transmission (ex. PDM-QPSK, 7300 km transmission [30] and PDM-16QAM, 1500 km transmission [25]). In light of this, trench-assisted core profile [39] and heterogeneous core arrangement [40] have been studied in our projects so that high-density MCFs [38] with low core-to-core cross talk could be achieved in a cladding diameter of less than 250 μm. In addition, it would be important to mitigate connection losses (splicing and connectors) between MCFs for long-haul transmission. We also attempted to reduce the fluctuation of core pitches along the length of MCF in the fabrication process so that precise core arrangement would be achieved.

Progress in the fabrication technique is one of the key issues for realizing a 100-Pbit/s·km transmission using MCFs. Table 8.1 summarizes the features of four different fabrication methods of MCF preforms. Currently, the method (A) "stack and jacketing" and method (B) "drilling" have been used commonly. In the method (A), multiple core rods are stacked to form a desired core arrangement. Then, the stacked rod is jacketed by the silica glass to form the outer cladding. This technique is suitable when we fabricate MCFs with hexagonal core arrangement by stacking the rod with a uniform diameter. However, it becomes difficult when rods with different diameters are needed and/or the number of rod becomes large. Thus, the technique has the disadvantage regarding the size extension and core arrangement. In the method (B), several holes are directly created in a uniform cladding glass in order to insert the core rods. The drilling technique provides the big advantage on the core arrangement since it can directly control the core position. However, it is still difficult to fabricate large preforms because the drilling accuracy tends to degrade by increasing the preform length. Therefore, it is important for both methods (A) and (B) to improve their fabrication processes in order to realize the long length of MCFs while keeping the sufficient uniformity.

Table 8.1 Comparison of various fabrication techniques for forming MCF preforms

Preform fabrication process	(A) Stack and jacketing	(B) Drilling	(C) One step cladding formation	(D) One step core and cladding formation
Size extension	−	−	+	++
Loss reduction	−/+	−/+	++	++
Core arrangement	−	++	++	+
Difficulty	+	+	++	+++

As described above, both methods (A) and (B) require an additional process, jacketing or drilling, compared with that in the fabrication of conventional single-core fiber. The other method (C) "one step cladding formation" may reduce the complexity of the fabrication process. If we can construct the outer cladding around the multiple core rods similar to the conventional vapor deposition technique [41], it is expected to improve the jacketing process and reduce the difficulty of core arrangement in the current method (A). Ultimately, the method (D) "one step core and cladding formation" may solve the size limitation and complexity of the current fabrication process simultaneously, although it is difficult to control the multiple burners precisely in the vapor deposition technique to form the multi-core arrangement while keeping the sufficient longitudinal uniformity. However, it is worthwhile to consider the possibility of these novel fabrication techniques (C) and (D) in order to make progress in real implementation of MCF technology. Thus, we recognized the improvement and/or refinement of the MCF fabrication technique as one of the most important challenges in the "i-FREE^2" project.

8.4.2 *FM-MCF Technology*

The number of cores in the MCF would be limited to ~ 32 in an appropriate cladding diameter due to mechanical reliability. Therefore, mode multiplexing combined with MCFs is an effective way to further increase the SDM multiplicity. So far, few-mode MCFs (FM-MCFs) with the SDM multiplicity of over 100 have been reported [42–44] as shown in Table 8.2. In [44], an ultra-high-density SDM fiber (6-mode 19-core fiber)

Table 8.2 FM-MCFs with over 100 SDM multiplicities

	Cross section of fiber	Spatial multiplicity	Transmission length (km)	Cladding diameter (μm)
[42]	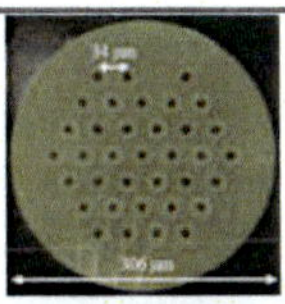	36 × 3 modes 108	5.5	306
[43]	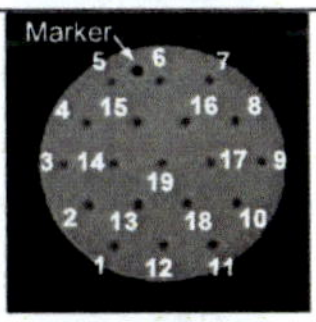	19 × 6 modes 114	9.8	318
[44]		19 × 6 modes 114	8.5	246

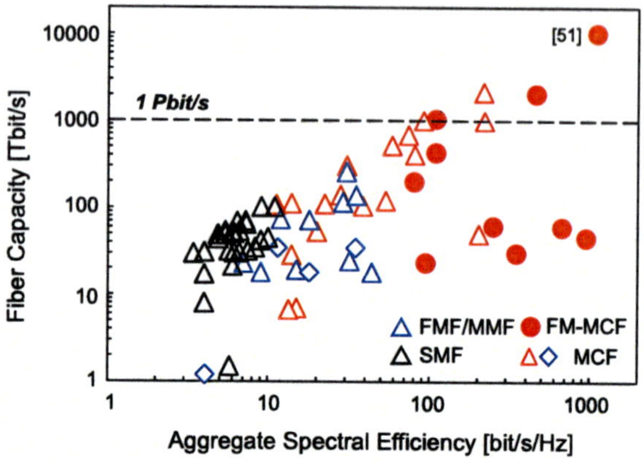

Fig. 8.12 Transmission capacity per fiber using SDM technologies in recent experiments as a function of aggregated spectral efficiency

was fabricated in the cladding diameter of less than 250 μm. Figure 8.12 shows the transmission capacity per fiber as a function of aggregated spectral efficiency in recent transmission experiments using the SDM technologies. Highest aggregated spectral efficiency of around 1 kbit/s/Hz and world-record-class transmission capacity of 2 Pbit/s have been achieved in the FM-MCF transmission [45, 46, 50] followed by a 10-Pbit/s experiment [51].

Characterization of FM-MCFs is mandatory to make progress in the convergence of core and mode multiplexing, and it was one of the important objectives of the ongoing project. Table 8.3 summarizes the key parameters FM-MCFs which should be evaluated as well as corresponding test methods. The mode exciting ratio is the basic information for FMFs. It can be directly observed by measuring the impulse response, or alternatively the difference in bending loss sensitivity among the transmission modes can be used for evaluation. Cross talk is the most important property of the SDM fibers, and the traditional power transmission method and multi-channel OTDR technique [11] can be used. Here, it should be noted that FM-MCFs contain both intra- and inter-core modal cross talk, and it is very important to evaluate the influence of each contribution on the transmission properties. Both

Table 8.3 Key parameters of FM-MCFs and corresponding test methods

Parameters	Corresponding test methods
Mode exciting ratio	• Impulse response • Bending
Intra/inter-core modal crosstalk	• Power transmission • OTDR
Cutoff wavelength	• Multimode launching
Mode field diameter, effective area	• Near field scanning • Far field scanning • Variable aperture

Selective mode excitation

• Spatial type
• Fiber type
• Waveguide type

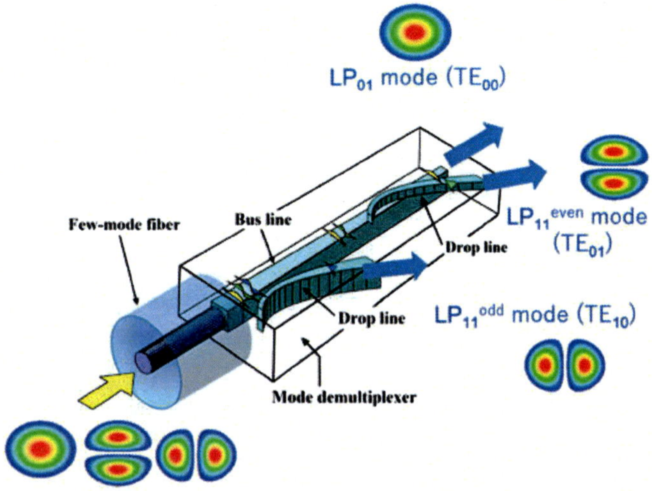

Fig. 8.13 Schematic image of mode multiplexing/demultiplexer using the stacked waveguide structure (Reprinted with permission from [48]. ©2014 IEEE)

cutoff wavelength and mode-field diameter (effective area) are also important properties of optical fibers. Traditional measurement techniques [47], such as multi-mode launching, near-field scanning, far-field scanning, and variable aperture, can be used for FM-MCFs. However, the cutoff wavelength measurement tends to be difficult as increasing the transmission mode. The definition and analyzing procedure of the field pattern will also be important when we consider the measurement accuracy of MFD/A_{eff} in each transmission mode.

Here, the development of selective mode excitation techniques greatly supports the characterization of individual transmission mode. Kokubun et al. investigated a mode multiplexing/demultiplexing technique by designing a stacked waveguide as shown in Fig. 8.13, and he revealed the applicability of the proposed technique to degenerated modes [48]. It is expected that these collaborative researches on FM-MCFs will open up the possibility toward exa-bit class ultra-large capacity transport.

References

1. T. Hayashi, T. Taru, O. Shimakawa, T. Sasaki, E. Sasaoka, Low-crosstalk and low-loss multi-core fiber utilizing fiber bend, in *Proceedings of OFC, OWJ3* (2011)
2. T. Hayashi, T. Taru, O. Shimakawa, T. Sasaki, E. Sasaoka, Uncoupled multi-core fiber enhancing signal-to-noise ratio. Opt. Express **20**(26), B92–B103 (2012)
3. K. Imamura, H. Inaba, K. Mukasa, R. Sugizaki, Multi core fiber with large A_{eff} of 140 μm^2 and low crosstalk, in *Proceedings of ECOC, Mo.1.F.2* (2012)
4. J. Sakaguchi, B. Puttnam, W. Klaus, Y. Awaji, N. Wada, A. Kanno, T. Kawanishi, K. Imamura, H. Inaba, K. Mukasa, R. Sugizaki, T. Kobayashi, M. Watanabe, 19-core fiber transmission of

19 × 100 × 172-Gb/s SDM-WDM-PDM-QPSK signals at 305 Tb/s, in *Proceedings of OFC, PDP5C.1* (2012)
5. S. Matsuo, K. Takenaga, Y. Arakawa, Y. Sasaki, S. Tanigawa, K. Saitoh, M. Koshiba, Large-effective-area ten-core fiber with cladding diameter of about 200 μm. Opt. Lett. **36**(23), 4626–4628 (2011)
6. S. Matsuo, Y. Sasaki, T. Akamatsu, I. Ishida, K. Takenaga, K. Okuyama, K. Saitoh, M. Koshiba, 12-core fiber with one ring structure for extremely large capacity transmission. Opt. Express **20**(27), 28398–28408 (2012)
7. B. Yao, K. Ohsono, N. Shiina, K. Fukuzato, A. Hongo, E. Sekiya, K. Saito, Reduction of crosstalk by hole-walled multi-core fibers, in *Proceedings of OFC, OM2D.5* (2012)
8. M. Tanaka, M. Hachiwaka, Y. Fujimaki, H. Taniguchi, Butt joint of hexagonal cladding multi-core fiber (in Japanese), in IEICE Technical Report, OFT2012-68, pp. 77–82 (2013)
9. M. Koshiba, K. Saitoh, K. Takenaga, S. Matsuo, Multi-core fiber design and analysis: coupled-mode theory and coupled-power theory. Opt. Express **19**(26), B102–B111 (2011)
10. M. Koshiba, K. Saitoh, K. Takenaga, S. Matsuo, Analytical expression of average power-coupling coefficients for estimation intercore crosstalk in multicore fibers. IEEE Photon. J. **4**(5), 1987–1995 (2012)
11. M. Nakazawa, M. Yoshida, T. Hirooka, Nondestructive measurement of mode couplings along a multi-core fiber using a synchronous multi-channel OTDR. Opt. Express **20**(11), 12530–12540 (2012)
12. R. Okuno, R. Fukami, M. Ohashi, Y. Miyoshi, Cutoff wavelength measurement of two core multi-core fiber, in *Proceedings of OECC, MS1-3* (2013)
13. T. Watanabe, Y. Kokubun, Stacked polymer waveguide type fan-in/fan-out device for dense multi-core fibre. IET Optoelectron. **9**(4), 158–162 (2015). https://doi.org/10.1049/iet-opt.2014.0137
14. Y. Tsuchida, K. Maeda, K. Watanabe, T. Saito, S. Matsumoto, K. Aiso, Y. Mimura, R. Sugizaki, Simultaneous 7-core pumped amplification in multicore EDF through fibre based fan-in/out, in *Proceedings of ECOC2012, Tu.4.F.2* (2012)
15. Y. Tsuchida, K. Maeda, K. Watanabe, T. Saito, S. Takasaka, M. Tadakuma, R. Sugizaki, H. Ogoshi, K. Doi, M. Matsuura, K. Takeshima, T. Tsuritani, I. Morita, Cladding-pumped L-band multicore EDFA with reduced power consumption, in *Proceeding of IEEE Summer Topicals*, Paper ME2.2 (2014)
16. H. Ono, K. Takenaga, K. Ichii, S. Matsuo, T. Takahashi, H. Masuda, M. Yamada, 12-core double-clad Er/Yb-doped fiber amplifier employing free-space coupling pump/signal combiner module, in *Proceedings of ECOC2013, We.4.A.4* (2013)
17. K. Suzuki, H. Ono, T. Mizuno, Y. Hashizume, Y. Abe, T. Takahashi, K. Takenaga, S. Matsuo, H. Takara, Pump light source for distributed Raman amplification in MCFs with LD sharing circuit. IEEE Photon. Tech. Lett. **24**(11), 1937–1940 (2012)
18. H. Takara, T. Mizuno, H. Kawakami, Y. Miyamoto, H. Masuda, H. Ono, S. Asakawa, Y. Amma, K. Hirakawa, S. Matsuo, K. Tsujikawa, M. Yamada, 120.7-Tb/s (7 SDM/180 WDM/95.8 Gb/s) MCF-ROPA unrepeatered transmission of PDM-32QAM channels over 204 km, in *Proceedings of ECOC2014, PD3.1* (2014)
19. K. Watanabe, T. Saito, K. Imamura, M. Shiino, Development of fiber bundle type fan-out for multicore fiber, OECC2012, 5C1-2 (2012)
20. Y. Abe, K. Shikama, S. Yanagi, T. Takahashi, Physical-contact-type fan-out device for multicore fibre. Electron. Lett. **49**(11), 711–712 (2013)
21. Y. Tottori, H. Tsuboya, T. Kobayashi, Multi functionality demonstration for multi core fiber fan-in/fan-out devices using free space optics, OFC2014, Th2A.44 (2014)
22. O. Shimakawa, H. Arao, M. Harumoto, T. Sano, A. Inoue, Compact multi-core fiber fan-out with GRIN-lens and micro-lens array, OFC2014, M3K.1 (2014)
23. M. Hachiwaka, M. Tanaka, Y. Fujimaki, H. Taniguchi, FC connector of hexagonal cladding multi-core fiber. Laser Eng. **41**(6), 442–446 (2013) (in Japanese)
24. H. Takara, A. Sano, T. Kobayashi, H. Kubota, H. Kawakami, A. Matsuura, Y. Miyamoto, Y. Abe, H. Ono, K. Shikama, Y. Goto, K. Tsujikawa, Y. Sasaki, I. Ishida, K. Takenaga, S. Matsuo, K.

Saitoh, M. Koshiba, T. Morioka, 1.01-Pb/s (12 SDM/222 WDM/456 Gb/s) crosstalk-managed transmission with 91.4-b/s/Hz aggregate spectral efficiency, ECOC2012, Th.3.C.1 (2012)
25. T. Kobayashi, H. Takara, A. Sano, T. Mizuno, H. Kawakami, Y. Miyamoto, K. Hiraga, Y. Abe, H. Ono, M. Wada, Y. Sasaki, I. Ishida, K. Takenaga, S. Matsuo, K. Saitoh, M. Yamada, H. Masuda, T. Morioka, 2 x 344 Tb/s propagation-direction interleaved transmission over 1500-km MCF enhanced by multicarrier full electric-field digital back-propagation, ECOC2013, PD3.E.4 (2013)
26. H. Takahashi, T. Tsuritani, E. L. T. de Gabory, T. Ito, W. R. Peng, K. Igarashi, K. Takeshima, Y. Kawaguchi, I. Morita, Y. Tsuchida, Y.Mimura, K. Maeda, T. Saito, K. Watanabe, K. Imamura, R. Sugizaki, M. Suzuki, First demonstration of MC-EDFA-repeatered SDM transmission of 40 × 128-Gbit/s PDM-QPSK signals per core over 6160-km 7-core MCF, ECOC2012, Th.3.C.3 (2012)
27. K. Igarashi, K. Takeshima, T. Tsuritani, H. Takahashi, S. Sumita, I. Morita, Y. Tsuchida, M. Tadakuma, K. Maeda, T. Saito, K. Watanabe, K. Imamura, R. Sugizaki, M. Suzuki, 110.9-Tbit/s SDM transmission over 6370 km using a full C-band seven-core EDFA. Opt. Express **21**(15), 18053–18060 (2013)
28. K. Igarashi, T. Tsuritani, I. Morita, Y. Tsuchida, K. Maeda, M. Tadakuma, T. Saito, K. Watanabe, K. Imamura, R. Sugizaki, M. Suzuki, 1.03-Exabit/s·km super-nyquist-WDM transmission over 7326-km seven-core fiber, ECOC2013, PD3.E.3 (2013)
29. A. Turukhin, H.G. Batshon, M. Mazurczyk, Y. Sun, C.R. Davidson, J.-X. Cai, O.V. Sinkin, W. Patterson, G. Wolter, M.A. Bolshtyansky, D.G. Foursa, A. Pilipetskii, Demonstration of 0.52 Pb/s potential transmission capacity over 8830 km using multicore fiber, in *Proceedings of ECOC2016* (2016)
30. K. Igarashi, T. Tsuritani, I. Morita, Y. Tsuchida, K. Maeda, M. Tadakuma, T. Saito, K. Watanabe, K. Imamura, R. Sugizaki, M. Suzuki, Super-Nyquist-WDM transmission over 7326-km seven-core fiber with capacity-distance product of 1.03 Exabit/s·km. Opt. Express **22**(2), 1220–1228 (2014)
31. A. Turukhin, O.V. Sinkin, H. Batshon, H. Zhang, Y. Sun, M. Mazurczyk, C. Davidson, J-X Cai, M. A. Bolshtyansky, D. Foursa, A. Pilipetskii, 105.1 Tb/s power-efficient transmission over 14,350 km using a 12-core fiber, in *Proceedings of OFC2016, Th4C.1* (2016)
32. J. Sakaguchi, Y. Awaji, N. Wada, A. Kanno, T. Kawanishi, T. Hayashi, T. Taru, T. Kobayashi, M. Watanabe, 109-Tb/s (7 × 97 × 172-Gb/s SDM/WDM/PDM) QPSK transmission through 16.8-km homogeneous multi-core fiber, in *Proceedings of OFC/NFOEC2011, PDPB6* (2011)
33. B. Zhu, T.F. Taunay, M. Fishteyn, X. Liu, S. Chandrasekhar, M.F. Yan, J.M. Fini, E.M. Monberg, F.V. Dimarcello, 112-Tb/s space-division multiplexed DWDM transmission with 14-b/s/Hz aggregate spectral efficiency over a 76.8-km seven-core fiber. Opt. Express **19**(17), 16665–16671 (2011)
34. H. Takara, H. Ono, Y. Abe, H. Masuda, K. Takenaga, S. Matsuo, H. Kubota, K. Shibahara, T. Kobayashi, Y. Miaymoto, 1000-km 7-core fiber transmission of 10 × 96-Gb/s PDM-16QAM using Raman amplification with 6.5 W per fiber. Opt. Express. **20**(9), 10100–10105 (2012)
35. A. Sano, H. Takara, T. Kobayashi, H. Kawakami, H. Kishikawa, T. Nakagawa, Y. Miyamoto, Y. Abe, H. Ono, K. Shikama, M. Nagatani, T. Mori, Y. Sasaki, I. Ishida, K. Takenaga, S. Matsuo, K. Saitoh, M. Koshiba, M. Yamada, H. Masuda, T. Morioka, 409-Tb/s + 409-Tb/s crosstalk suppressed bidirectional MCF transmission over 450 km using propagation-direction interleaving. Opt. Express **21**(14), 16777–16783 (2013)
36. J. Sakaguchi, B.J. Puttnam, W. Klaus, Y. Awaji, N. Wada, A. Kanno, T. Kawanishi, K. Imamura, H. Inaba, K. Mukasa, R. Sugizaki, T. Kobayashi, M. Watanabe, 19-core fiber transmission of 19 × 100 × 172-Gb/s SDM-WDM-PDM-QPSK signals at 305 Tb/s, in *Proceedings of OFC2012, PDP5C.1* (2012)
37. J. Sakaguchi, W. Klaus, B.J. Puttnam, J.M.D. Mendinueta, Y. Awaji, N. Wada, Y. Tsuchida, K. Maeda, M. Tadakuma, K. Imamura, R. Sugizaki, T. Kobayashi, Y. Tottori, M. Watanabe, R.V. Jensen, 19-core MCF transmission system using EDFA with shared core pumping coupled via free-space optics. Opt. Express **22**(1), 90–95 (2014)

38. T. Mizuno, K. Shibahara, F. Ye, Y. Sasaki, Y. Amma, K. Takenaga, Y. Jung, K. Pulverer, H. Ono, Y. Abe, M. Yamada, K. Saitoh, S. Matsuo, K. Aikawa, M. Bohn, D.J. Richardson, Y. Miyamoto, T. Morioka, Long-Haul dense space division multiplexed transmission over low-crosstalk heterogeneous 32-core transmission line using partial recirculating loop system. J. Lightwave Technol. **35**, 488–498 (2017)
39. K. Takenaga, Y. Arakawa, S. Tanigawa, N. Guan, S. Matsuo, K. Saitoh, sM. Koshiba, Reduction of crosstalk by trench-assisted multi-core fiber, in *Proceedings of OFC/NFOEC2011, OWJ4* (2011)
40. Y. Amma, Y. Sasaki, K. Takenaga, S. Matsuo, J. Tu, K. Saitoh, M. Koshiba, T. Morioka, Y. Miyamoto, High-density multicore fiber with heterogeneous core arrangement, in *Proceedings of OFC 2015, Th4C.4* (2015)
41. T. Izawa, S. Sudo, F. Hanawa, Continuous fabrication process for high silica fiber preforms. IECE Japan **E-62**(11), 779–785 (1979)
42. J. Sakaguchi, W. Klaus, J.M.D. Mendinueta, B.J. Puttnam, R.S. Luis, Y. Awaji, N. Wada, T. Hayashi, T. Nakanishi, T. Watanabe, Y. Kokubun, T. Takahata, T. Kobayashi, Realizing a 36-core, 3-mode fiber with 108 spatial channels, in *Proceedings of OFC2015, Th5C.2* (2015)
43. K. Igarashi, D. Souma, Y. Wakayama, K. Takeshima, Y. Kawaguchi, T. Tsuritani, I. Morita, M. Suzuki, 114 space-division-multiplexed transmission over 9.8-km weakly-coupled-6-mode uncoupled-19-core fibers, in *Proceedings of OFC2015, Th5C.4* (2015)
44. T. Sakamoto, T. Matsui, K. Saitoh, S. Saitoh, K. Takenaga, T. Mizuno, Y. Abe, K. Shibahara, Y. Tobita, S. Matsuo, K. Aikawa, S. Aozasa, K. Nakajima, Y. Miyamoto, Low-loss and low-DMD few-mode multi-core fiber with highest core multiplicity factor, in *Proceedings of OFC2016, Th5A.2* (2016)
45. K. Igarashi, D. Soma, Y. Wakayama, K. Takeshima, Y. Kawaguchi, N. Yoshikane, T. Tsuritani, I. Morita, M. Suzuki, Ultra-dense spatial-division-multiplexed optical fiber transmission over 6-mode 19-core fibers. Opt. Express **24**(10), 10213–10231 (2016)
46. D. Soma, Y. Wakayama, S. Beppu, K. Igarashi, T. Tsuritani, H. Taga, I. Morita, M. Suzuki, 665 and 947 b/s/Hz ultra-highly aggregate-spectral-efficient SDM/WDM transmission over 6-mode 19-core fibre using DP-16QAM/64QAM signals, in *Proceedings of ECOC2016, Th.3.2.C* (2016)
47. ITU-T Recommendation G.650.1, Definitions and test methods for linear, deterministic attributes of single-mode fibre and cable (2010)
48. T. Watanabe, Y. Kokubun, Stacked waveguide type mode-evolutional multi.demultiplexer for LP_{01} $LP_{11}{}^{a}$ and $LP_{11}{}^{b}$, in *Proceedings of ECOC, We.1.1.7* (2014)
49. B.J. Puttnam, R.S. Luís, W. Klaus, J. Sakaguchi, J.-M.D. Mendinueta, Y. Awaji, N. Wada, Y. Tamura, T. Hayashi, M. Hirano, J.R. Marciante, 2.15 Pb/s transmission using a 22 core homogeneous single-mode multi-core fiber and wideband optical comb, ECOC 2015, PDP.3.1 (2015)
50. D. Soma, K. Igarashi, Y. Wakayama, K. Takeshima, Y. Kawaguchi, N. Yoshikane, T. Tsuritani, I. Morita, M. Suzuki., 2.05 Peta-bit/s super-nyquist-WDM SDM transmission using 9.8-km 6-mode 19-core fiber in full C band, ECOC 2015, PDP.3.2 (2015)
51. D. Soma, Y. Wakayama, S. Beppu, S. Sumita, T. Tsuritani, T. Hayashi, T. Nagashima, M. Suzuki, H. Takahashi, K. Igarashi, I. Morita, M. Suzuki, 10.16 Peta-bit/s dense SDM/WDM transmission over low-DMD 6-mode 19-core fibre across C+L band, in *ECOC 2017 Th.PDP.A.1* (2017)

Chapter 9
Concluding Remarks

Yuichi Matsushima and Takeshi Kamiya

Abstract Over the past decade, intensive studies of space-division multiplexing (SDM) technologies based on multi-core fibers or few-mode fibers have been done in the world. In the next decade, SDM technologies will open the way to real deployment in many application fields. It should be necessary for the deployment to continue many studies and further efforts.

9.1 Introduction Historical

Worldwide the amount of Internet traffic is increasing at an annual growth rate of more than 30%. Optical fiber transmission will undoubtedly continue demanding ever-increasing capacity for the rapidly growing amount of global traffic generated by big data applications. Current long-haul and large capacity optical fiber transmission systems are based on wavelength-division multiplexing (WDM) systems with conventional single-mode fibers, but the capacity of these transmission systems is limited. Over the past decade, this has given rise to intensive studies of space-division multiplexing (SDM) technologies utilizing multi-core fibers or few-mode fibers. The main motivation driving the search for a fundamentally new modulation format came from the recognition that a shortage in transmission capacity (the so-called capacity crunch) exists, because the global transformation into an IT-based society that was initiated at around the year 2000 requires the capacity for limitless amounts of traffic. In 2006, Desurvire wrote an article entitled "Capacity Demand and Technology Challenges for Lightwave Systems in the Next Two Decades" [1]. It was also recognized that the upper limit of input optical power per fiber may define the transmission capacity limit (Chap. 7). In Japan, the

Yuichi Matsushima and Takeshi Kamiya are chapter editors.

Y. Matsushima (✉)
Waseda University, Shinjuku, Tokyo, Japan
e-mail: matsushima@waseda.jp

T. Kamiya
The University of Tokyo, Bunkyo, Tokyo, Japan

M. Nakazawa et al. (eds.), *Space-Division Multiplexing in Optical Communication Systems*, Springer Series in Optical Sciences 236,
https://doi.org/10.1007/978-3-030-87619-7_9

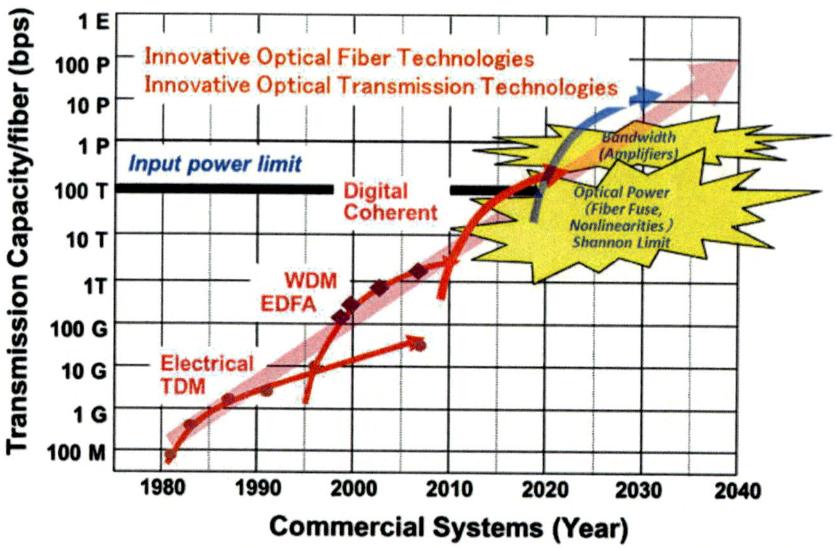

Fig. 9.1 Schematic view of roadmap for optical fiber transmission systems

National Institute of Information and Communications Technology (NICT) started the **EX**tremely **A**dvanced **T**ransmission (EXAT) initiative to study potential candidates for "extremely advanced transmission technologies" in 2008 [2], after which a variety of proposal-based research programs followed. Especially, three research targets were highlighted: multi-core fiber, multi-mode control, and multi-level modulation (3 M) (Chaps. 1 and 8).

Figure 9.1 shows the schematic history and future trend of commercial optical fiber transmission systems including the so-called capacity crunch point at 100 Tbit/s of single-mode optical fiber transmission. After current digital coherent systems, SDM technologies are expected to play an important role in such huge capacity systems.

The EXAT initiative in Japan has triggered R&D that is intensively focused on multi-core fiber and many efforts to realize SDM are in progress. Increasing the capacity by more than 1000 times would necessitate the use of an optimized combination of the abovementioned 3 M approach.

The multi-level modulation of optical signals has the advantage of increasing spectral efficiency. The principle underlying digital coherent detection was proposed by Kikuchi in 2005 [3], and the real-time receiver DSP, which was implemented as a result of collaboration between NTT, Fujitsu, NEC, and Mitsubishi Electric, has led to the global penetration of 100 G transceiver sets since 2012 [4]. Efforts are continuing to implement 400 G and 1 Tbps transceivers and to jump up to the 1024 QAM scheme developed by Nakazawa (Sect. 5.2) [5]. The technical challenges associated with digital coherent technologies were extensively described in the book "High Spectral Density Optical Communication Technologies," which was edited by Nakazawa et al. [6].

To date, many kinds of SDM technologies have been developed very rapidly, and many successful transmission experiments have also been demonstrated as described in previous chapters and a current review report [7]. Transmission capacities exceeding 1 Pbit/s per fiber utilizing multi-core/few-mode fiber and multi-level modulation have already been demonstrated [8–10] and seem to be ready for real services in the near future. Next, we have to consider schemes that would enable the deployment of SDM technologies into real future systems.

9.2 Challenges of SDM Deployment

Even though the excellent capabilities of SDM technologies have been demonstrated at major international conferences of optical communications, their deployment in real-world systems requires additional effort with many key points that need to be considered. These points may be expressed by the following set of questions:

(a) Which is the best application field?
(b) What scheme would allow migration from existing systems?
(c) How would the cost be evaluated?
(d) What would the standardization scheme be?
(e) When would the new system be started?
(f) Who would be the major players?

These questions are closely related to each other. The component cost could be substantially reduced by increasing the market size; hence, worldwide standardization would be beneficial. However, a detailed discussion of standardization may be time consuming, thereby hindering early deployment. A business exploiting specific transmission routes carrying heavy traffic such as ultra-realistic video services (Sect. 1.3.1) or datacenter-based services might become a front-runner. Both careful studies and courageous decisions are required.

9.3 Major Application Fields

SDM technologies have been developed to extend the transmission capacity especially in high-capacity and long-distance systems such as core network systems and submarine cable systems. Core network systems consist of many existing optical networks based on traditional single-mode fibers. Therefore, worldwide deployment would require standardization. Future large-scale replacement or newly constructed optical networks would present opportunities for the deployment of SDM; thus, it may take time for this technology to come to fruition (Sect. 1.4). Submarine optical cable systems seem to constitute "green field" development, which means the first stage of the system is not standardized. The space factor is a very important issue in submarine systems because of the limited space which is to locate undersea optical cables

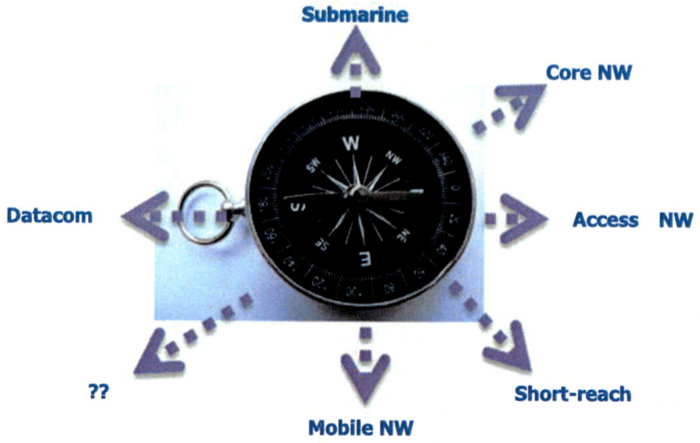

Fig. 9.2 Which approach is suitable for the first deployment of SDM?

and repeaters. We should recall the historical reasons as to why WDM systems were firstly deployed as submarine cable systems rather than terrestrial systems (Sect. 1.5). SDM essentially offers high space efficiency as well as WDM. The next jump in the transmission capacity of submarine systems may come from SDM when the electric power consumption of the optical amplifier in the repeater nearly reaches the same level as that of current commercial systems.

Candidate application fields of SDM technologies are suggested in Fig. 9.2. In addition to core network and submarine systems, access networks/mobile networks and short-reach communication systems seem to be suitable application fields for which a rapid increase in the capacity of traffic can be foreseen and more flexible networks as well as WDM-PON (passive optical network) systems would be preferable. However, as the cost balance will be a key issue in these fields, very low construction costs are recommended (Sects. 1.3.2 and 6.2).

The datacom field, which includes intra-datacenter signal transmission involving signaling between large numbers of racks/boards, is a promising area for SDM application. Datacenters are expected to require very large capacity signal transmission in combination with simple connections between optical fiber cables, because the space factor is indispensable. Even now, many fiber cables are connected in a complicated way. SDM offers a solution for this complexity. The integration of the optical components of SDM would also substantially reduce the number of components in a datacenter. The use of SDM fiber transmission over short distances would obviate the need for optical amplifiers. Furthermore, datacom systems seem to be "green field" implementations, where the preferred network should only be determined by an operator of the particular datacenter. However, datacom demands real-time communication and low electric power consumption, which requires the MIMO chip in mode-division multiplexing (MDM) to be carefully designed for actual applications.

9.4 Additional Research Efforts Needed

Thus far, studies of SDM technologies have mainly focused on point-to-point transmission to achieve a very large transmission capacity exceeding 1 Pbit/s/fiber. However, the networking design and architecture of real systems are important and need to be given the necessary consideration. The elastic node architecture of SDM will open the way to additional benefits such as high-contention resolution and much higher throughput by utilizing new dimensions of SDM. Fundamentally, new frontiers of SDM applications may be considered in the future. Exploration of the potential of the new field calls for the contribution of many ideas from various research areas other than optical communication and for researchers from these areas to become involved in the research field of SDM technologies.

9.5 Summary

In summary, current demands for large amounts of communication traffic are expected to lead to a capacity crunch of optical fiber network systems. The most powerful candidate for pursuing a solution to this quandary is SDM. In Japan, the EXAT initiative, which was started in 2008, has promoted intensive R&D mainly focused on multi-core and multi-mode fiber in addition to multi-level modulation. Many efforts to realize SDM systems are currently underway. Attaining a thousand-fold capacity increase would necessitate an optimized combination of *multi-level* modulation, together with the use of *multi-core* fiber and *multi-mode* control (3 M). The deployment of SDM demands a new infrastructure which would take time. Could an earlier market possibly arise in the field of datacom transmission or optical submarine cable systems? The next key issues in SDM are the integration of optical components, efficient optical amplification, switching and networking as well as standardization and international collaborations.

References

1. E.B. Desurvire, Capacity demand and technology challenges for lightwave systems in the next two decades. J. Lightwave Technol. **24**(12), 4697–4710 (2006)
2. T. Morioka, New generation optical infrastructure technologies: "EXAT Initiative" towards 2020 and beyond, in *Optoelectronics and Communications Conference (OECC), FT4*, Hong Kong (2009)
3. S. Tsukamoto, D.-S. Ly-Gagnon, K. Katoh, K. Kikuchi, Coherent demodulation of 40 Gbit/s polarization-multiplexed QPSK signals with 16 GHz spacing after 200 km transmission, in *Optical Fiber Communication Conference (OFC), PDP29*, Anaheim (2005); K. Kikuchi, Phase diversity homodyne detection of Multi-level optical modulation with digital carrier phase estimation. IEEE J. Sel. Topics Quantum Electron. **12**(4), 563–570 (2006)

4. E. Yamazaki, S. Yamanaka, Y. Kisaka, T. Nakagawa, K. Murata, E. Yoshida, T. Sakano, M. Tomizawa, Y. Miyamoto, S. Matsuoka, J. Matsui, A. Shibayama, J. Abe, Y. Nakamura, H. Noguchi, K. Fukuchi, H. Onaka, K. Fukumitsu, K. Komaki, O. Takeuchi, Y. Sakamoto, H. Nakashima, T. Mizuochi, K. Kubo, Y. Miyata, H. Nishimoto, S. Hirano, K. Onohara, Fast optical channel recovery in field demonstration of 100-Gbit/s Ethernet over OTN using real-time DSP. Opt. Express **19**(14), 13179–13184 (2011)
5. M. Yoshida, S. Beppu, K. Kasai, T. Hirooka, M. Nakazawa, 1024 QAM, 7-core (60 Gbit/s $\times$ 7) fiber transmission over 55 km with an aggregate potential spectral efficiency of 109 bit/s/Hz. Opt. Express **23**(16), 20760–20766 (2015)
6. M. Nakazawa, K. Kikuchi, T. Miyazaki (eds.), High Spectral Density Optical Communication Technologies
7. Research highlight on spatial division multiplexing SDM. IEEE Photon. Society News. **29**(5), 4–20 (2015)
8. H. Takara, A. Sano, T. Kobayashi, H. Kubota, H. Kawakami, A. Matsuura, Y. Miyamoto, Y. Abe, H. Ono, K. Shikama, Y. Goto, K. Tsujikawa, Y. Sasaki, I. Ishiba, K. Takenaga, S. Matsuo, K. Saitoh, M. Koshiba, T. Morioka, 1.01-Pb/s (12 SDM/222 WDM/456 Gb/s) crosstalk-managed transmission with 91.4-b/s/Hz aggregate spectral efficiency, in *European Conference on Optical Communication (ECOC), Th3 C.1*, Amsterdam (2012)
9. B.J. Puttnam, R.S. Luís, W. Klaus, J. Sakaguchi, J.-M. Delgado Mendinueta, Y. Awaji, N. Wada, Y. Tamura, T. Hayashi, M. Hirano, J. Marciante, 2.15 Pb/s transmission using a 22 core homogeneous single-mode multi-core fiber and wideband optical comb, in *European Conference on Optical Communication (ECOC), PDP3.1*, Valencia (2015)
10. D. Soma, K. Igarashi, Y. Wakayama, K. Takeshima, Y. Kawaguchi, N. Yoshikane, T. Tsuritani, I. Morita, M. Suzuki, 2.05 Peta-bit/s super-nyquist-WDM SDM transmission using 9.8-km 6-mode 19-core fiber in full c band, in *European Conference on Optical Communication (ECOC), PDP3.2*, Valencia (2015)

Index

M. Nakazawa et al. (eds.), *Space-Division Multiplexing in Optical Communication Systems*, Springer Series in Optical Sciences 236,
https://doi.org/10.1007/978-3-030-87619-7